Mexico
BIOGRAPHY OF POWER

Mexico

BIOGRAPHY OF POWER

A History of Modern Mexico, 1810–1996

ENRIQUE KRAUZE

Translated by Hank Heifetz

HarperCollins*Publishers*

 For information address HarperCollins Publishers, Inc., 10 East 53rd Street, New York, NY 10022.

HarperCollins books may be purchased for educational, business, or sales promotional use. For information please write: Special Markets Department, HarperCollins Publishers, Inc., 10 East 53rd Street, New York, NY 10022.

FIRST EDITION

Designed by Alma Hochhauser Orenstein

Library of Congress Cataloging-in-Publication Data

Krauze, Enrique.
Mexico : biography of power : a history of modern Mexico, 1810–1996 / by Enrique Krauze ; translated by Hank Heifetz.
p. cm.
Includes bibliographical references and index.
ISBN 0-06-016325-9
1. Mexico—History—1810– I. Title.
F1231.5.K72 1997
972'.04—dc20 96-33046

---

97 98 99 00 01 ❖/RRD 10 9 8 7 6 5 4 3

For Isabel, León, and Daniel

History endures in Mexico. No one has died here, despite the killings and the executions. They are alive—Cuauhtémoc, Cortés, Maximilian, Don Porfirio, and all the conquerors and all the conquered. That is Mexico's special quality. The whole past is a pulsing present. It has not gone by, it has stopped in its tracks.

—José Moreno Villa

Providence willed my history to be the history of Mexico since 1821.

—Antonio López de Santa Anna

CONTENTS

III The Revolution

IV The Modern State

V Past and Future: The Decline of the System

ILLUSTRATIONS

Maps

PREFACE

"Jerusalem needs a psychoanalyst," the Israeli poet Yehuda Amichai said to me one autumn evening in 1989, as we looked out a window of his home at one of the walls of that city. His words made me think of another ancient city, as scored and burdened by the past as his Jerusalem. Immersed in a present of tensions and conflicts that sometimes appear insoluble, both cities have been the site of mythical pilgrimages, theological revelations, and historical conquests. Both still show signs of the different cultures that have inhabited them. And both cities once saw themselves as the center of the world. "Mexico," I thought, "The city and the nation could also use a psychoanalyst."

This book is not a psychoanalysis of Mexico or a psychohistory of its "representative men." But it is a history of Mexico in which the principal protagonist is history itself—the many forms in which the past has not disappeared but has persisted within the memories and day-to-day reality of the Mexican people. The weight of the past has sometimes been more present than the present itself. And a repetition of the past has sometimes seemed to be the only foreseeable future. In certain areas of Mexican life, the past has survived as a legacy of stability and cohesion; at other levels it exists in the form of unresolved, partially repressed conflicts, always ready to burst through the surface of the present. And in Mexico, as in all countries with ancient cultures, our view of the past that was actually experienced is influenced by the past as it came to be remembered, reconstructed, and sometimes, for ideological purposes, invented. One of the duties of a historian is to separate the past as it was from all the superimpositions of imagination.

The formative period of Mexican history was the century and a half (1600–1750) known as the Era of the Baroque. The initial cycle of material and spiritual conquest, which had begun in 1521 with the arrival of Hernán Cortés, was more or less over. The country was in the process of creating the fundamental characteristics of its culture, a complete body of values (in life, art, ethics, and in intellectual, political, and religious forms) stemming from the mixture of Indian and Spanish elements. New Spain—thanks to mostly Indian hands and mostly Spanish concepts—gradually developed into a multicolored Mexican mosaic of customs and traditions. This process of *mestizaje* ("racial mixing") is absolutely central to the history of Mexico. No other country in the Americas experienced so inclusive a process. In Mexico, *mestizaje* was not only ethnic but intensely cultural. It permeated every area of life and became the framework and substance of a society that is in many ways both an innovation and an experiment.

During this Baroque Era, New Spain remained a society immersed in its own creation. It would prove resistant to the political and intellectual currents of the European Enlightenment that reached Mexico at the end of the eighteenth century. New Spain would be suspicious of the outside world and poorly prepared to embrace the new values proposed by western Europe. "Mexico was born with its back turned," the poet Octavio Paz has written. Its back was turned to the modern world, toward which the country born in 1821 could only aspire, while steadily and nostalgically looking toward a past that reluctantly it would have to renounce (at least in part). Beginning with the War of Independence (1810–1821), Mexico—or more precisely its elite leadership—essentially became a being with two faces, one assuming that a return to the past was possible, the other yearning to wipe it all out and begin anew.

Mexico's problems in establishing a firm connection with the modern world are rooted in certain traditions, especially in what might be termed its theological-political framework. The concentration of power in the hands of one person has been a phenomenon all too common in Latin American societies (and elsewhere), but in its specifically Mexican manifestation both the Indian and Spanish traditions have come into play. Mingling characteristics from both sources, a peculiar type of leadership has developed in Mexico, less connected to the person of a particular leader than to the institution of personal power. Combining two traditions of absolute power—one emanating from the gods and the other from God—this political *mestizaje* conferred a unique connection with the sacred on Mexico's succession of rulers. The European revolutions of the nineteenth and twentieth centuries were carried out in the name of some idea, some ideology, some "ism." In Mexico the suffix *ismo* finds its most important political usage linked to a man of power: Juarismo, Porfirismo, Zapatismo, Cardenismo, Salinismo.

This book threads the lives of the most important leaders during the last two centuries into a single biography of power, but I am in no way subscribing to an outmoded (and unacceptable) great-man theory of history. What I hope to convey is that in Mexico the lives of these men do more than represent the complexities and contradictions of the country they came to govern or in which they took center stage for a time at the head of armies fighting for change or for a return to the past (or for both). The accidents of their individual lives also had an enormous effect on the directions taken by the nation as a whole. Personal characteristics and events that in a moderately democratic country might be mere anecdotes—interesting, amusing, or trivial—can in Mexico acquire unsuspected dimensions and significance. An early psychological frustration, a physical defect, a family drama, a confused prejudice, a tilt one way or the other in a man's religious feelings or his passions, even a local tradition automatically accepted could literally alter the fate of Mexico, for better or for worse. Biography is a genre that has not been much cultivated in Spanish and Spanish-American culture. Mexico is no exception, and this seems to me a strange thing, given the personal impact of men like these leaders on all aspects of national life. It is this long-lasting historical reality on which I have based my approach in a book that is essentially a political history of the last two centuries in Mexico.

Each section has a different theme. Each is written in a somewhat different (and I hope appropriate) style. The introduction centers on a single event, the lavish celebration of the centenary of Mexican independence in 1910 by the government of the aged dictator Porfirio Díaz, and it offers a rapid cross section of Mexico in that year, at the very threshold of the great Revolution. Part 1, "The Weight of the Past," is descriptive and analytical, presenting the five major leitmotifs of Mexican history, which spiral and return throughout the book. From then on narrative takes over. In part 2, "Century of Caudillos," history unfolds at a classical pace, with the kind of elaboration characteristic of the nineteenth century itself. In "The Revolution" (part 3) the rhythm of events and the writing changes, racing along rather like the Revolutionary caudillos themselves on horseback at the head of their troops. With "The Modern State" (part 4), Mexican history moves toward a stasis and I try to combine narration with an explanation of the unique formation, structure, and rise of the modern Mexican system. The final part, "Past and Future: The Decline of the System," is more of an essay and sketch dealing with the most recent decades of Mexican history and including a strong element of personal interpretation and observation by someone who has been both an observer and participant during these turbulent decades. In the view of that observer (myself), the period is marked by the steady—and wildly punctuated—collapse of the system, but a true history of it cannot yet be written. It does seem clear that the Mexican institution of the imperial presidency is under siege, that the call for democracy is growing stronger and stronger, but no one can say how it will all come out, because history is affected by frameworks and patterns and human intentions but also by luck and chance.

SOURCES AND ACKNOWLEDGMENTS

In most cases, I have used texts in Spanish (which have then been directly translated into English) even when translations exist in English or—in the case of some scholarly works—when the books were originally written in English. The notes indicate where the original English has been used (as in most of the quotations from John Reed).

For part 1 I have used a number of historians and chroniclers, old and new. My greatest debts to modern historians are to Luis González y González on cultural *mestizaje*, Richard M. Morse and José Miranda on the political characteristics of New Spain, and David Brading on the whole nature of the viceroyalty, especially the Church. (And my thanks to Hugh Thomas—in his *Conquest of Mexico*—for some fresh insights on the personality of Cortés.)

Part 2 uses both primary sources (the most important being various collections of documents on the War of Independence, especially the recent compilations by Carlos Herrejón), as well as a number of secondary sources. I've favored authors close to the time of the events (Lucas Alamán, José María Luis Mora, Lorenzo de Zavala, Francisco Bulnes, Justo Sierra, Emilio Rabasa) but also important historians who have worked in this century: notably Luis Villoro, Ernesto Lemoine, Luis González, Egon Cesar Conte Corti, Ralph Roeder, Carleton Beals. *Historia Moderna de México*, under the supervision of Daniel Cosío Villegas (who wrote the sections on political history), was essential to me for the period from the late nineteenth century to the Revolution. Justo Sierra (1848–1912)—though I do not always agree with him—has been an important inspiration. An accomplished biographer, he was also one of the few historians who attempted a general overview of the Mexican past. Transferring his religious feelings to history, he made the past of Mexico almost a sacred text but never without applying the criticism and perceptive understanding needed to balance the record of so complex a country as Mexico.

For part 3, I have used many newspapers and journals of the period and researched personal archives (especially for Madero and Calles) as well as books and monographs written during and after the wars. Among recent historians, the most important to me have been Stanley Ross, Charles Cumberland, José Valadés, John Womack Jr., Jesús Sotelo Inclán, Berta Ulloa, Ramón Puente, Jean Meyer, and Luis González. From Richard M. Morse I drew my inspiration for one of the central structuring ideas of the book—the survival within the modern Mexican state of the political "architecture" of the colonial past.

For part 4 I am indebted to the essays of Daniel Cosío Villegas and Octavio Paz. There are relatively few secondary sources. Among them, the *Historia de la Revolución Mexicana* (1910–1964) published by the Colegio de México was very useful. But oral history and extensive interviews were my primary raw material for this section, and I spoke with a very wide range of participants and witnesses. Presiding over this section, as teacher and inspiration, is the historian and essayist Daniel Cosío Villegas, like Sierra a priest of history but more incisive and critical and—in confrontation with the powers of his time—a much braver man.

For the final section, the work of Gabriel Zaid has been important to me, but the general interpretation of the period comes from my own observations, as a committed and independent observer—since 1968—of my country's political and social reality.

A complete list of the people who have contributed in one form or another to the creation of this book would be immense, and I must limit myself to those who have directly helped me. I am especially grateful to my editor, Cass Canfield, Jr., who participated in articulating its form and was understanding enough to wait a long time for a long book; to my friend Hank Heifetz, who not only translated and edited the book but truly helped to create it. I am grateful to those who aided me with my research: Margarita de Orellana, Cayetano Reyes, Javier García-Diego, Aurelio de los Reyes, Alejandro Rosas, José Manuel Villalpando, Fernán González, Ana María Cortés, Ricardo Pérez Montfort, Javier Bañuelos, Ana María Serna, Álvaro Vázquez, Leonardo Martinez Carrizales, Jeannette Porras, Greco Sotelo, Marco Antonio Maldonado. And to those who did editorial work: Fernando García Ramírez, Pedro Molinero, and Xavier Guzmán; and to the translators Iris Mielonen,

Charlotte Broad, and Anthony Stanton, who worked on early versions of some of this material; to Moisés González Navarro, Solange Alberro, Nigel Davies, and Josué Sáenz, who read and commented on various portions in manuscript; to my informants, especially Reneé González Salas vda. de Valdés, Rafael Carranza, Hortensia Calles vda. de Torreblanca, Antonio Mena Brito, Antonio Ortiz Mena, and Gustavo Díaz Ordaz Jr. (who was kind enough to give me access to the unpublished memoirs of his father); and to the friends who have also been teachers: Octavio Paz, Luis González, Gabriel Zaid, Alejandro Rossi, Jean Meyer, and Fausto Zerón-Medina. And my wife and sons—to whom this book is dedicated.

HISTORICAL CHRONOLOGY

1325	Founding of the city of Mexico-Tenochtitlan by the Mexicas, later known as the Aztecs.
1440–1487	The Aztecs greatly expand their power and empire under Emperor Moctezuma I.
1502	Moctezuma II becomes emperor of Tenochtitlan.
1519	Hernán Cortés lands his forces on the shores of Mexico.
1520	Death of Moctezuma II. He is replaced by Cuitláhuac, who reigns for only eighty days and dies of smallpox (a disease brought by the Spaniards). Cuauhtémoc, the last Aztec emperor, continues to resist the Spaniards.
1521	Tenochtitlan falls to the Spaniards and their Indian allies.
1524	The conversion of the Indians to Christianity begins.
1528	Arrival of the first bishop of Mexico, Fray Juan de Zumárraga.
1533	Inspired by the *Utopia* of Thomas More, Vasco de Quiroga founds the first hospital-town in Michoacán.
1535	Antonio de Mendoza is named the first viceroy of New Spain.
1542	The Spanish Crown issues *Las Nuevas Leyes* (The New Laws) to protect the Indians.
1551	The University of Mexico is founded.
1571	The Inquisition is established in Mexico City.
1700	Philip V becomes king of Spain, and the Bourbon dynasty replaces the dynasty of the Hapsburgs.
1767	Expulsion of the Jesuits.
1803	The traveler and scientist Alexander von Humboldt visits Mexico.
1810	The priest Miguel Hidalgo y Costilla begins the War of Independence against Spain.
1811	Hidalgo is defeated and executed. José María Morelos y Pavón takes command of the insurrection.
1813	Morelos convokes the first Mexican Congress, which formally declares Mexican Independence.
1814	The Congress issues the first Constitution of Mexico.
1815	Morelos is defeated and executed.
1821	Agustín de Iturbide unites various forces and gains the Independence of Mexico.
1822	Iturbide is proclaimed emperor.
1823	A rebellion led by Antonio López de Santa Anna forces the abdication of Iturbide.
1824	The first Constitution of independent Mexico formally establishes a federal republic.
1833	Santa Anna becomes President for the first of eleven times.
1836	The State of Texas declares its independence from Mexico and begins a war against the central government. Santa Anna is defeated by the Texans.
1838	French forces attempt to occupy Veracruz and are defeated by Santa Anna.
1845	Texas becomes part of the United States of America.
1846–1848	War between Mexico and the United States, ending with the defeat of Mexico and a peace treaty that cedes more than half the territory of Mexico to the United States.
1853–1855	Final presidency of Santa Anna.
1855	Forces under the leadership of Juan Álvarez and Ignacio Comonfort overthrow Santa Anna. The call is issued for a Constitutional Convention to create a new Constitution.
1857	A new and liberal Constitution is approved, preceded by a series of laws directly opposing the interests of the Church and Mexican conservatives.
1858–1861	The War of the Reform between Liberals and Conservatives.
1861	The Conservatives are defeated. President Benito Juárez suspends

payment of the foreign debt for two years. France, England, and Spain sign an agreement intended to compel Mexican payment of the debt.

1862 The French Army, supported by Mexican Conservatives, invades Mexico. The War of the French Intervention begins.

1864 The French Army and Mexican Conservatives establish the Second Mexican Empire, crowning the Austrian archduke Maximilian von Hapsburg emperor of Mexico.

1867 The Liberal armies defeat the Empire. Maximilian is executed. Juárez reestablishes the Republic.

1872 Death of Juárez. Sebastián Lerdo de Tejada becomes President of Mexico.

1876 Porfirio Díaz overthrows Lerdo de Tejada and becomes President. He will reelect himself seven times, and his dictatorship, the "Porfiriato," will last thirty-four years.

1910–1911 A revolution led by Francisco I. Madero overthows the Díaz regime.

1913 A military coup led by Victoriano Huerta overthrows Madero, who is later murdered. Venustiano Carranza leads a rebellion against Huerta. After the victory, the Revolutionaries fight among themselves. The forces led by Carranza defeat Francisco (Pancho) Villa and Emiliano Zapata. Carranza becomes President and convokes a new Constitutional Convention.

1917 A new Constitution is issued. Carranza becomes Constitutional President.

1920 Carranza is overthrown and dies in an ambush. New elections lead to the presidency of Álvaro Obregón.

1924 Plutarco Elías Calles becomes President.

1926–1929 Conflicts between the government and the hierarchy of the Catholic Church lead to the Cristiada, a widespread revolt in central and western Mexico.

1928 Obregón is elected President again and assassinated a few months later. Emilio Portes Gil becomes provisional President.

1929 Plutarco Elías Calles forms the Partido Nacional Revolucionario (National Revolutionary Party, or PNR).

1934 Lázaro Cárdenas becomes President of the Republic.

1938 Cárdenas nationalizes the oil industry. The PNR changes its name to the Partido de la Revolución Mexicana (PRM).

1940–1946 Presidency of Manuel Ávila Camacho.

1946–1952 Presidency of Miguel Alemán Valdés.

1946 The PRM is restructured for the last time and renamed the Partido Revolucionario Institucional (PRI).

1952–1958 Presidency of Adolfo Ruiz Cortines.

1956–1958 Labor unrest with a new teacher's union at the forefront. The movement is defeated by government repression.

1958–1964 Presidency of Adolfo López Mateos.

1958–1959 Labor unrest by the Railroad Workers Union. The movement is repressed and its leaders jailed.

1964–1970 Presidency of Gustavo Díaz Ordaz.

1968 A large and important Student Movement ends with police and army firing on students at the Plaza of Tlatelolco.

1970–1976 Presidency of Luis Echeverría.

1976–1982 Presidency of José López Portillo. His administration bases the national economy on large, newly discovered oil reserves. A drop in the international price of oil precipitates one of Mexico's worst economic crises.

1982–1988 Presidency of Miguel de la Madrid.

1988–1994 Through elections widely regarded as fraudulent, Carlos Salinas de Gortari becomes President. He initiates important economic changes, privatizes many state enterprises, and signs the North American Free Trade Agreement with the United States. An economic recovery begins.

1994 A rebellion breaks out in the southern state of Chiapas, led by the Zapatista Army of National Liberation and commanded by a charismatic leader known as Subcomandante Marcos. The official presidential candidate of the PRI, Luis Donaldo Colosio, is assassinated during his campaign. Ernesto Zedillo, who replaces him, is elected President. The flight of foreign investment capital provokes the worst economic and financial crisis in modern Mexican history.

Mexico Today

INTRODUCTION: PAST, PRESENT, AND FUTURE

The Centenary Fiestas

Mexico, September 1910, and a double cause for celebration: the centenary of the War of Independence and the eightieth birthday of President Porfirio Díaz. In the capital and other major cities the days were filled with banquets, patriotic ceremonies, garden parties, outdoor festivals, and parades of bands and wagons featuring allegorical representations—live or in effigy—of the great events of Mexican history. In the evenings, sumptuous balls and receptions and theatrical performances drew elegantly dressed crowds to the handsome colonial buildings with patriotic slogans and images outlined in patterns of light. It was the Mexican belle époque at the point of its greatest splendor.

Nearly every country in the world that maintained diplomatic relations with Mexico sent special envoys to that lavish celebration of the Fiestas del Centenario. Day after day, new public buildings and services were inaugurated, testifying to the progress that at last, after long centuries, had become a standard feature of Mexican life. Less than fifty years earlier, anyone who traveled the ancient royal highways and horse trails dating from the days of the viceroyalty did so at the risk of their lives or at least their possessions.

In 1910 things were very different. When the envoys arrived, either by boat at Veracruz or by train on one of the two modern railroad routes from the northern border to Mexico City, they would have seen evidence of the solid infrastructure that "Don Porfirio"—as almost all Mexicans respectfully called him—had constructed across the country

Centenary Independence Parade, Mexico City, 1910
(Collection Jesús Montes de Oca)

since his ascent to power more than three decades before, in 1876. The ports had grown, the railways were excellent, and telephone, telegraph, and postal services were in full operation. Once in Mexico City, the foreign delegates were invited to attend scientific congresses and expositions or the inauguration of schools that included the reopening of the National University. (Its predecessor, the Royal and Pontifical University of Mexico, founded in 1551, had been closed midway through the nineteenth century.) They could attend ceremonies for new hospitals, an asylum, hospices, a workers' park, and a penitentiary, all equipped with the most up-to-date facilities. Outstanding technical achievements such as the seismology station and the drainage system were completed and inaugurated during the celebrations. The former would allow better detection and study of earthquakes, almost as frequent and deadly as the political tremors of the nineteenth century had been before Don Porfirio's seven terms in office. The latter—a costly system of tunnels, sewers, and channels—was designed to solve the severe problem of flooding in Mexico City, a burden for each and every viceroy, emperor, and President of the country since Hernán Cortés made the decision to construct his new capital on the ruins of Tenochtitlan, the Aztec city of lakes.[1]

Peace was the first and most sacred word of the age, already christened the Era Porfiriana. The regime took great pride in the fact that it had maintained peace for thirty-four years. Don Porfirio could not claim all the credit. Historical forces had prepared not only Mexico but the rest of the Western world for this long period of calm. The liberal, nationalistic, and socialist revolutions of nineteenth-century Europe had come to a halt in the seventies. From then until the outbreak of the First World War, Europeans were absorbed in assimilating their colonial wealth and enjoying unprecedented material expansion. It was a period of long-reigning monarchs—Queen Victoria, the Prussian Kaiser Wilhelm II, the Austrian Emperor Franz Joseph—and of intellectual and artistic pomp and pageantry. With the Civil War behind it, the United States would, in a few years, complete its march to the west and was also advancing on the frontier of industrial and technological progress with remarkable speed. The Americans had not forgotten the doctrine of manifest destiny, which encouraged them to dominate the whole continent. But their drive toward annexation at the expense of Mexico was nearly spent. Though Central America and the Caribbean were still felt to be under constant threat from the United States, Mexico—during the final decades of the nineteenth century—had become a target more for economic than for territorial penetration. With foreign powers now posing no immediate military threat, Mexico was in a position to benefit from the winds of material progress blowing from the United States and Europe.

The country was in grave need of that change of air. Ever since Mexico had gained its independence in 1821 after a long decade of war, successive governments—during almost every one of the fifty-five years up to 1876—had confronted civil struggles and foreign invasions. They had also, over a period of decades, been violently dismantling the hierarchy of privileges inherited from colonial times (especially those of the army and the Church). In the midst of all this, a new nation had to be firmly established—a secure and clearly defined territory, a legitimate government, a secular state, a constitutional structure.

The Centenary Fiestas provided a perfect, symbolic opportunity for Mexico to affirm its commitment to an all-inclusive peace. The country was willing to heal old wounds, among them its differences with Spain, the imperial power that had ruled the vast territory formerly known as New Spain for three hundred years (1521–1821), when the colony had included not only the Mexico of 1910 but part of Central America and the four North American states of New Mexico, Arizona, California, and Texas. Throughout the nineteenth century, Spain had remained distant and sometimes openly hostile toward its former colony, even taking military action against Mexico in 1829 and 1861. By 1910, times had changed, and as a sign of this the Spaniards returned the standard, uniform, and pectoral cross of the second-greatest hero of the War of Independence—José María Morelos, whom they had executed in 1815. "Long live your great President," the Spanish ambassador exclaimed at the ceremony, to which Díaz responded with emotion: "Long live Spain, our great mother."[2]

France—whose clothing fashions, styles in art, literary tastes, and political ideas had been so admired by the Mexican elite since the mid-eighteenth century—finally admitted that it too had treated Mexico unfairly during the nineteenth century. Its first war against Mexico in 1838 was little more than a skirmish, but the second had been a full-scale intervention. While the United States had been embroiled in its Civil War, Napoleon III, partly inspired by his Spanish wife Eugenia de Montijo's dreams of reconquest, had ordered the military invasion of Mexico—a war that lasted from 1862 to 1867 and was an attempt to establish a permanent French presence on the continent. The luckless Maximilian von Hapsburg was installed as emperor of Mexico in 1864, only to be executed in 1867 before a firing squad. Porfirio Díaz had been one of the leading republican *caudillos* (military leaders) in the war against the French, but that was all history now. During Díaz's long years of power, Paris had once again captured Mexico's cultural imagination, and relations between the two countries amounted to little less than a love affair, at least on the Mexican side. Well-to-do Mexicans dreamed of Paris, traveled to Paris, built their homes in the styles of Paris. Academic textbooks, ideas, and curricula were a faithful copy of France's, while the French—as their contribution—invested in Mexican banks and mines. (And businessmen of French descent owned the most important stores of the capital.) Both countries, in that September of 1910, made symbolic gestures to confirm this reconciliatory mood. Mexico laid the foundation stone of a monument to Louis Pasteur, and France handed back the ceremonial keys to Mexico City, seized on June 10, 1863.[3]

Although the delegation from the United States did not return anything their country had taken in the way of tokens, keys, or territory, they did make an effort to show, during the Centenary Fiestas, that they were ready to be friends with their apprehensive neighbors. There were reasons for Mexico's continuing distrust. Before turning to France for inspiration, the founders of the Republic of Mexico had greatly admired the United States. The first federalist Constitution of Mexico (issued in 1824, three years after independence was declared) had been based partly on the American Constitution, which *El Sol* (*The Sun*), a newspaper of the time, had considered "one of the most perfect creations of the spirit . . . the foundation for the most uncomplicated, most liberal, most fortunate government in history."[4]

Unfortunately—in contrast to its domestic affairs—the United States managed foreign policy by a much different standard: the doctrine and practice of manifest destiny. The first military applications of this doctrine were support for the secession of Texas in 1836, its subsequent annexation in 1845, and the war against Mexico in the following year. On September 16, 1847—the anniversary of Mexico's independence—the troops of General Winfield Scott had raised the Stars and Stripes over the National Palace. Nothing could erase this insult from the collective Mexican memory. A year later, when the fever for gold flared up in California, Mexico had to surrender the richest, though almost uninhabited half of that territory to the United States. Nor would the discord end there. Within a short time, Mexico lost another stretch of land along its northern border, and in 1859, at the height of the Guerra de Reforma ("War of the Reform") between Liberals and Conservatives, Juárez's Liberal government was on the point of accepting the proposals of President Buchanan, under which Mexico would virtually become a protectorate of the United States, in exchange for economic and military aid against the Conservatives. This project for American expansion—actually articulated in a formal treaty—fell through because of several unexpected happenings, among them a veto by the United States Senate and the outbreak of the American Civil War.

The defeat of the Confederacy changed relations between the two countries, favoring economic over territorial expansion. During a visit to Mexico in 1880, the former American president Ulysses S. Grant described Mexico as a "magnificent mine" that lay waiting south of the Rio Grande for North American interests. His words symbolized the beginning of the new era of "peaceful penetration."[5]

Porfirio Díaz had forgotten none of this, but from the moment he took power he had carefully and efficiently managed his diplomatic relations with Mexico's intimidating northern neighbor. President Sebastián Lerdo de Tejada, whom Porfirio Díaz had deposed, used to say: "Between weakness and power—the desert."[6] Díaz had been quick to learn that peace between the two unequal neighbors depended on changing one of those words, and he did it—from *desert* to *railroad*. Throughout the years of Porfirismo, investments replaced invasions. The Americans were freely operating in every area of the Mexican economy: mining, railroads, banks, oil, industry, and agriculture. Their only limit came from the competition of European investors (especially British and Germans) to whom Porfirio Díaz's government, in an effort to maintain a balance, shrewdly offered some degree of preference. More than $2 billion of foreign money—two-thirds of the total capital investment—had flowed into Mexico since President Díaz had given the nation structure and stability. Capital for railways, mines, factories, and plantations arrived at the rate of $200 million a year. By 1910, the U.S. stakes in Mexico amounted to 38 percent of the total foreign investment, surpassing that of England, its traditional competitor. Though the British maintained an overall predominance throughout Latin America (55 percent of all foreign investment), their share in Mexico was now only 29 percent, spread across mines, banks, and oil wells. Two-thirds of the railways in Mexico were owned by Americans.[7]

Díaz had noted with concern a growing crescendo in Theodore Roosevelt's policy of "the big stick." The occupation of Cuba and the Philippines and the increasing presence of the United States in Central America were lighting sparks of zealous nationalism among many Mexicans, reminding them of old grievances. William Howard Taft had become president of the United States in 1908, and the relations between the two countries had recently cooled because of a series of minor problems that irritated the North Americans: Díaz's diplomatic flirtation with the Japanese empire, his denial of access to a bay on the Pacific coast traditionally used by the U.S. Navy for training exercises, and his support for an opposition government in Nicaragua. These moves had sometimes exasperated President Taft as much as Taft's tolerance of certain "rebellious" Mexicans on the border had irritated Díaz.[8]

Despite all this and the growing sense of nationalism among Mexico's working and middle classes (who resented the presence and influence of foreigners on their life and economy), the Centenary Fiestas encouraged both governments to bury their recent differences for a while. It was in this spirit that Díaz laid the foundation stone of a monument to George Washington and the United States delegation paid homage to the *Niños Héroes* ("The Child Heroes"), six young cadets who, according to patriotic legend, had chosen to die in September 1847 rather than surrender the last military bastion in Mexico City—the Castle of Chapultepec—to the invading U.S. Army. The shining white monument erected in their memory was decorated once a year by American veterans, but in 1910 the ceremony included phrases of mutual esteem and admiration. "You have set us an example on many occasions," Porfirio Díaz acknowledged, "particularly at that transcendent and historic moment of our Independence." Mr. J. J. Slayden, the delegate from Washington, had even stronger words of praise for his host:

> As Rome had its Augustus, England its Elizabeth and its Victoria, so Mexico has its Porfirio Díaz. Everything is going well in Mexico. Under Porfirio Díaz a nation has been created.[9]

The second sacred word of the Porfirian trinity was *order,* which Díaz had succeeded in imposing primarily by force. There had been no mercy for those who dared to interfere with the stringing of thousands of miles of telegraph wire along the bandit-ridden roads and the construction of thousands of post offices throughout the country. In a 1907 interview with American journalist James Creelman, President Díaz had said:

> We began by making robbery punishable by death and requiring the execution of offenders within a few hours after they were caught and condemned. We ordered that wherever telegraph wires were cut and the chief officer of the district did not catch the criminal, he himself should be brought to justice; and in case the wires were cut on a hacienda, the owner who had failed to prevent it should be hanged from the nearest telegraph pole. Notice that these were military orders. We were relentless. Sometimes we were relentless to the point of cruelty. But it was all necessary then to the life and progress of the nation. If we committed acts of cruelty, the end has justified the means. . . . It was better that a little blood should be spilled to avoid more being spilled later. The blood that was spilled was bad blood; what was saved was good blood.[10]

With the same heavy hand he applied to bandits, kidnappers, and criminal gangs, Don Porfirio moved against the Yaquis in the northwest and the Mayas of Yucatán, the only Indians who had remained in rebellion through much of the nineteenth and into the twentieth century. To break their resistance, he had no qualms about burning villages, transporting Mayas to an inferno in the state of Oaxaca euphemistically called the National Valley, or sending the Yaquis to a remote jungle on the edge of the Yucatán peninsula.

Much more was implied by the notion of order. Besides suppressing rebellious Indians, annihilating bandits, and (at the beginning of his long reign) crushing pockets of military resistance, Díaz decided, in his own words, to "bridle the horses" by introducing an all-encompassing program of political control and centralization. According to the Liberal Constitution of 1857 in force at the time, Mexico was a "representative, democratic and federal Republic." But this did not describe the present political reality.

The division of government into three branches had been effectively annulled by the concentration of power in the executive. Díaz had weakened and corrupted the Legislature by making it a mere adjunct of the Presidential Chair. The bothersome business of electing candidates was conveniently overcome by appointing them. Don Porfirio knew the background of each nominee and supervised the drawing up of the list. Among the 227 representatives nominated in 1886, 62 came from Oaxaca, the President's home state. The Legislature was to become a gentlemen's club for the friends of Don Porfirio and his wife. Díaz would pay no attention to geographical location or place of birth. A progovernment writer from Veracruz might wake up one morning to find himself representing a district in the depths of Sonora. One timid critic dared to say that the Senate had become a "political mausoleum." No presidential initiative was ever questioned, and nothing moved in the Legislature without the consent of "the Great Elector." A similar process of servitude neutralized the judiciary as Don Porfirio freely appointed and removed judges.[11]

The idea of "democratic representation" prescribed in the Constitution had also been emptied of meaning. Elections at every level were only a formality. Don Porfirio reelected himself every four years from 1876 onward, except for one four-year term from 1880 to 1884 when he hand-picked his compadre General Manuel González and placed him in that sacred locus of Mexican politics, the Presidential Chair. (To become a compadre in Mexico—the "godfather" of another man's child—is to create a strong and solemn personal bond. Díaz was, for these few years, governing through a man even closer than a friend.) One of the paradoxes of history is that Porfirio Díaz had chosen the revolutionary slogan of "Valid Voting, No Reelection" in 1871 when he led a rebellion against President Benito Juárez, who had reelected himself three times. Now, in 1910, Díaz was beginning his eighth term.

Federalism was another fiction. The state governors were all "mini-Porfirios," fiercely loyal to their master, who kept a very close eye on them. Bernardo Reyes, the governor of the state of Nuevo León and a true proconsul for Don Porfirio in the northeast, received daily instructions, reports, and suggestions from the President concerning issues as varied as elections for the Legislature and the judiciary; pardons for condemned prisoners; tactics for subjugating the old caciques (regional autocrats), recommendations concerning the property of local aristocratic families; lists of bandits for transportation to Yucatán; internal problems between the states of Nuevo León and Coahuila; the case of a priest who had been robbed in Monclova; the suppression of rebels; how to deal with Yankees who were attempting to take over the railroad from Monterrey to the Gulf of Mexico ("they can't be allowed to make a joke of Mexicans"); and hundreds more.[12]

The press was the last stronghold of original, classic political liberalism, which had originated during the War of Independence and briefly flowered in the period known as the *República Restaurada* (Restored Republic) under Presidents Juárez and Lerdo de Tejada (1867–1876). During Díaz's reign, the press was hobbled by the *Ley Mordaza* (Gag Law). A journalist could be imprisoned for a "psychological crime" or even through a report to the police of his "intentions."[13] Time and again prison would silence the few voices who dared to criticize the dictator by publishing statements like this one (from an editorial—on October 7, 1886—in *El Monitor Republicano*):

> The practice of debasing a country in order to make it rich and happy is too high a price to pay. Democracy may be a fiction and freedom a hoax, but so is the prosperity of the nation without democracy and freedom.[14]

The Mexican political press—all 130 publications—initially resisted being muzzled. But the indifference of the Mexican middle-class public (who were the readers of the newspapers), living in their more or less illusory world of material well-being, was more effective than the arrests, suppression of news, closures, and occasional assassinations. The press lost its resolve. By the end of the century, political freedom and democracy did, indeed, appear to be a hoax. Significantly, from 1896 on, *El Imparcial*, subsidized by and representing the government, became the most widely read newspaper in Mexico.[15]

The intellectuals did not really give Don Porfirio any trouble either. He had a phrase from his childhood that he applied to them—"This rooster wants corn!"—and he would throw them some grains. Díaz bought several generations of intellectuals and critics. He would hire them to draft statutes and speeches, but he actually distrusted and despised them as "experts in profundity." He employed them in insignificant posts with even more insignificant stipends. They had to be broken and kept "on a short rein."

Andrés Molina Enríquez, the eminent sociologist who in 1909 published the book that, in a sense, predicted the Mexican Revolution, *Los grandes problemas nacionales*, asserted that Díaz ran an "integral" or "total" (*integral*) government. He was correct. Porfirian *order* was achieved by *integrating* into the person of the President the real powers—national and local politico-military leaders (caudillos and caciques) and army generals—as well as the formal powers (governors, senators, representatives, and judges). And by neutralizing dissident voices.[16]

Through the principle of "*pan y palo*" ("bread and the bludgeon"), a careful balance of persuasion and violence, Porfirio Díaz "bridled" the Mexican political "horses" and had remained for decades in the Presidential Chair—master, as the old saying went, of the "lives and haciendas" of his country.

Once on the path of *progress*, the third key word of the age, there had been no stopping Don Porfirio. The first and most decisive step he took was to construct a network of railways across the nation. "The railway has played a great part in bringing peace to Mexico," Díaz used to say. The 638 kilometers of track in existence before 1876 linked the capital with Veracruz and with Querétaro. By 1910, 18,642 new kilometers had been laid, crisscrossing the whole country. During the same period, agriculture had increased by 4 percent, mining by almost 8 percent. And the growth in general was both extensive and diversified, swelling and broadening Mexico's export trade. Mexican silver circulated everywhere, but the country now produced industrial metals as well. In 1903, the first steelworks in Latin America, Fundidora de Fierro y Acero de Monterrey, was founded. In 1893, Díaz's brilliant finance minister, José Ives Limantour (the true mastermind behind Porfirian progress), had achieved the unimaginable: the first budget surplus in Mexico's history. Up until 1894, the governments of independent Mexico had run on a chronic internal deficit. From then on, they managed to maintain a surplus for many years. Ever since the 1880s Porfirio Díaz had been skillfully renegotiating Mexico's external debt, developing terms that were advantageous to Mexico. The government had always been burdened not only with its internal deficit but with huge obligations to foreign creditors. In 1861 Juárez even had to suspend payment of the external debt, precipitating the French invasion. But now, by the turn of the century, Mexico's credit was good around the world. The Mexican silver peso had become an internationally valid means of exchange, recognized as far away as China. In 1910 the government had an accumulated reserve of 75 million pesos.[17]

Porfirio Díaz firmly believed that he himself was the embodiment of his country's good health. Those who met him would immediately notice "a certain athletic bearing," which was never to leave him. Federico Gamboa, a famous writer of the period, said: "His physique promises an incalculable life span; almost treelike, it might have been hewn with an axe from oak . . . an upright, well-built, broadshouldered and barrel-chested body, with the quick glances of a big cat."[18]

This was precisely the image Don Porfirio wanted to project. He looked tall without being so and always rode tall horses. Every morning he arose to the trumpet notes of reveille, did his exercises, and took a cold bath. The story goes that one day some doubtful foreign investors who were visiting the military academy with Limantour were surprised to see an old man lifting weights in the superbly equipped gymnasium. "Ah, yes," replied Limantour, "it is President Díaz. He comes here every morning."[19] The sight was enough to convince the investors of his "incalculable life span." "At my age," Díaz once said, "I am happy to be in robust health . . . I would not exchange it for all the millions of the American oil barons."[20] Andrew Carnegie may very well have agreed. He called Díaz "the Moses and Joshua of Mexico."[21]

The *Grito*

On the night of September 15, the special envoys stood on the illuminated balconies of the National Palace and watched the fiesta of all fiestas on the Mexican civil calendar: the *grito de independencia*, the "cry of independence." A hundred years earlier (less a few hours) at dawn on Sunday, September 16, 1810—while Napoleon's troops were occupying Spain and King Ferdinand VII was still in captivity—Miguel Hidalgo y Costilla, a fifty-seven-year-old priest from an old family of *criollos* (Mexican-born Spaniards) had suddenly begun to harangue his parishioners in the small town of Dolores in the state of Guanajuato, "seducing them" (according to a chronicle of the time) to rise up in arms—even with stones, slings, sticks, or spears—in order to defend their religion against the "French heretics" who had occupied Spain since 1808 and now threatened to come over to the Americas. What Hidalgo intended—and accomplished—was to launch his flock against the hated *gachupines* (Spaniards born in Spain and living in Mexico) "who had been exploiting the wealth of the Mexican people with the greatest injustice for three hundred years." Within a month, he had been joined by more than fifty thousand men, mainly Indians from the poorest levels of society. Attracted by his religious magnetism and by other, less noble motives, this multitude devastated the cities of San Miguel, Celaya, and Guanajuato and were on the point of entering Mexico City when Hidalgo ordered them to retreat.

A few months later, in July of 1811, he was tried by the Inquisition, condemned by the civil authorities, and executed. But by then the seed had begun to sprout. It took the form of a long and violent social earthquake, almost without precedent in New Spain or the Americas: the Mexican War of Independence—a truly popular movement led by four hundred armed parish priests—only to be compared in its fury with the uprising of black slaves in Santo Domingo in 1801 and the Indian rebellion of Tupac Amaru (1781) in Peru.

Not many remembered the revolutionary aspect of the War of Independence on that night of nights in 1910. As in every other year, what really mattered was going to the *Zócalo* (central plaza) to participate in the ritual of the *grito*. According to witnesses present at the original event, Hidalgo and then his followers had shouted, "*¡Mueran los gachupines! Viva la Virgen de Guadalupe!*" ("Death to the Spaniards! Long live the Virgin of Guadalupe!"), but after one hundred years, time, good manners, and secularization had transformed the ritual from the call for a holy war to a peaceful, patriotic affirmation. At 11:00 P.M. on that September 15, 1910, President Porfirio Díaz stood on the main balcony of the National Palace and once again rang the same bell Hidalgo had rung in Dolores. He shouted several *vivas*: "Long Live the Heroes of the Nation!" "Long live the Republic!" Below him, in the majestic *zócalo* that from the days of the Aztecs had been the ceremonial heart of the Mexican nation, a hundred thousand voices shouted in reply, "*¡Viva!*" But why had the President delivered this *grito* on the night of September 15 rather than at dawn on September 16, when it all really began? A minor historical license: September 15 was the Day of Saint Porfirio (a Greek saint of the fourth century) and the birthday of President Porfirio Díaz.[22]

> Death to the Spaniards! Is there any Mexican who has never in his life shouted these sacrosanct words?[23]

This statement, by the famous radical Ignacio Ramírez, is a succinct expression of the Liberal interpretation of Mexican history. According to this viewpoint, Mexico had been created in only a century. The construction of a sovereign and independent nation had begun in 1810 and been carried on with infinite difficulty. This process of nation-building completely eclipsed the three-hundred-year history of Spanish rule (1521–1821), seen at best as a time of "laborious and deficient gestation" or, at worst, as a long historical night that had fortunately ended in the birth cry (*grito*) of a nation. Did it make any sense, argued the Liberals, to copy forms and manners from those "dismal" and "superstitious" colonial times of stagnation or, even worse (and this was the only point of partial contact with the Conservatives), from the "revolting human slaughter of the indigenous theocracies?" Whatever remained of those periods was a residue of the past. One had to escape, to free oneself from it. Escape to the future was the foundation of the Liberal attitude.[24]

Liberal is a key word in the modern history of Spain and her old domains, including Mexico. In the Golden Age of Spanish literature (in *Don Quixote*, for example), the adjective *liberal* simply meant "generous." As a result of the Napoleonic Wars in Spain (1808–1814), generating that vacuum of power within which the independence of Spain's American colonies was forged, the word took on a new meaning. *Liberal* in the Spanish world became a noun and the name of a political faction. Under the influence of the French Charter of 1793, the first constitution of Spain was drawn up in 1812 in the city of Cadiz by the "liberals." The authors viewed it as a "sacred code," and two years later it was to be a model for the first Mexican constitution. It aimed at reversing two centuries of decadence through radical reform, turning away from the absolutist tradition of "one sword, one monarch, one faith," which the liberals saw as a dangerous obstacle to any attempts at navigating the seas of the modern world.

In Mexico it was some time before the liberals formally adopted the name: up to the middle of the century they still called themselves "the Party of Progress" in opposition to the conservative party, which they called "the Party of Regression." Nevertheless, the new spirit and ideology were already evident in several of the leaders of the War of Independence. After the achievement of independence in 1821, the first formally liberal generation appeared. They read the works of classic French and English liberalism, and their major voice was the solitary figure of José María Luis Mora (1794–1850), a former priest who had been converted to the ideas of Montesquieu, Benjamin Constant, and Jeremy Bentham. In the middle of the nineteenth century, a second generation began to form around the traumatic experience of the war against the United States. It was the generation of romantic Liberals, made up of readers and translators of Lamartine, Michelet, Byron, Victor Hugo, and Alexandre Dumas. Its outstanding representatives were Ignacio Ramírez (1818–1879), Guillermo Prieto (1818–1897), and Ignacio Manuel Altamirano (1834–1893). In the War of the Reform against the Conservatives and in the War of the Intervention against the French army, these Liberals gradually adopted openly Jacobin attitudes, inspired by their passionate readings of the literature of the French Revolution. With the triumph of the Republic in 1867 and then when Díaz finally took power nine years later, a third generation of Liberals, at least in name, took the stage—the "positivist" or "evolutionist" Liberals, who began to intellectually justify the "lenient" and "benevolent" dictatorship of Porfirio Díaz through an eclectic mixture of the theories of Comte, Darwin, and Spencer. Although the generation included several thinkers of interest, the supreme high priest of evolutionist Liberalism in Mexico was the historian and educator Justo Sierra (1848–1912).

In his *Evolución política del pueblo mexicano* (first published in 1901), a rich and sympathetically inclusive overview of the centuries since the coming of the Spaniards, Justo Sierra restated the view of Mexican history that had by then become standard for the Liberal interpretation. There were clear-cut periods moving in a single, steady, ascending direction. After the "atrophied" and "inferior" world of the aboriginal civilizations, the first historical step was achieved: the "enforced revolution" of the Conquest. The conquistadors and their descendants had established not a modern colony with a capitalist orientation (as the English or Dutch would do a century later) but an almost feudal domain. The fortress of New Spain was enclosed within another fortress, that of Spain itself, and it reproduced the isolation and stagnation of the mother country. The material life of countless "centers of exploitation" was ruled by a "rusting but dreaded" sword, while moral and social life was controlled by "the cross on that sword." The fact that two very different dynasties had reigned during the viceregal period—the Hapsburgs (1521–1700) and the Bourbons (1700–1821)—was a minor detail in Sierra's history. To him, nothing had really changed because of the "fictitious move" from one to the other. At the end of the colonial period, Mexico was not a *patria* ("fatherland"—a key concept of the Liberal ideology) but merely an embryonic *nacionalidad* (nationality.) Only with independence had the national character ("that living organism formed by factors of race, environment, religion, language and custom") begun to assume an autonomous form, transforming nationality into "a legal entity," into a nation.[25]

According to Sierra, the new nation had paid a high price for that "first violent acceleration" of its historical evolution. After the achievement of independence in 1821, a period of anarchy, poverty, dishonor, and international isolation had begun, culminating in the amputation of more than half the national territory in 1848. Soon after, and as a direct result of that trauma, came the wars of the mid-nineteenth century: "the Second Revolution" (which Sierra believed would be the last). Through these Wars of the Reform and the French Intervention, the history of Mexico continued its ascent, this time toward emancipation not from Spain but from the residual patterns and customs of the colonial regime. The Liberal party had become not merely the only party but, according to Sierra, the true embodiment of the *patria*.[26]

In 1910 only a few pure classical liberals and a handful of anarchists would refuse to join the universal praise for the "unquestionable achievement of the present administration." Sierra wrote that everything appeared to be "a dream": the country at peace, the national economy flourishing, the railroads uniting Mexico itself, and the ports joining the country to the world—"now everything is real, everything is on the move and on course."[27]

Yet Justo Sierra, who was still after all a Liberal, had no illusions about the political reality. "The country as a whole" had "created Porfirio Díaz's power through a series of transfers, of illegal abdications." And this was dangerous, "extremely dangerous for the future, since it conditions habits that prevent men from governing themselves, without which there may be great individual men but no great nations."

Mexico was living through a period of "spontaneous Caesarism," a "social dictatorship" that had to be transitory "because of the necessary process of evolution." The surprising conclusion of Sierra's history was also a prediction: "The political evolution of Mexico has been sacrificed to the other phases of its social evolution. . . . The whole social evolution of Mexico will have been abortive and in vain, if it does not achieve the full goal of freedom."[28]

> "Long live the Virgin of Guadalupe and death to the Spaniards!" . . . A cry of death and desolation that I heard a thousand times in my early youth and after so many years it still resounds in my ears with a terrifying echo.[29]

Although the author of these lines, the historian Lucas Alamán, founder of the Mexican Conservative party, had died in 1853, his interpretation of history continued to attract followers in 1910. According to the Conservative version of Mexican history, Mexico had been born not in 1810 but in 1521, with the Conquest. Rather than escape from the colonial past, Mexicans should remain faithful to it, treasure it, and in a sense recover it. Alamán's purpose in writing his *Theses [Disertaciones] on the History of the Mexican Republic from the Conquest to Independence* (1844) had been "to change completely the [Liberal] conception . . . of the Conquest, the Spanish dominion and the way Independence was achieved."[30] He would later expand his original intentions, because he considered it necessary to study not only New Spain but also Spain itself, as the source of Mexico's language, religion, and forms of government. The keys for "understanding our own history" lay in the "splendor and decadence" of Spain, in being able to learn from her wisdom and her errors.

According to the Conservative version, the history of Mexico after the *grito* of 1810 had all gone downhill—it had been a *caída* (fall). "September 16 is a dark date for the Mexican nation. . . . On that day in Dolores, Hidalgo raised the standard of revolution which, spreading rapidly, resulted in the devastation of the country." Mexico was living an error in the midst of a century of error: "In this 19th century that terms itself philosophical and in which all notion of honor and loyalty has been destroyed, what remains is only the physical and material. We have sacrificed those principles which were formerly the foundation of society and which have gradually been reduced to empty and insignificant terms."[31]

Even though the source of his "notion of honor and loyalty" had its roots in his own biography—he came from a rich and pious creole family of mining entrepreneurs—his first intellectual inspiration had come from reading Edmund Burke's *Reflections on the Revolution in France* (1792). Fundamental to Alamán's thinking was Burke's key objection to the French Revolution as a destructive leap in history and social organization, contrary to what Burke considered the normal flow of nature—an "earthquake" rather than a legitimate process of growth.

Alamán would decide that Mexico had *forced* her historical nature. The Revolution of Independence had been a sudden and unnecessary earthquake: It could have been achieved by peaceful means and, more important, without breaking ties to the Spanish and Catholic tradition. Unlike the United States, which (in Alamán's view) had opted for tradition by accommodating the forms and customs of the British colonies, Mexico "had destroyed all that formerly existed."[32]

The Paseo and the City

The War of the Reform was the final means of settling the conflict between these two versions of history: one yearning to escape from the past and the other to preserve it. Under the leadership of Benito Juárez, the Liberals had won that war and its aftermath, the War of the French Intervention, a victory that they considered a "Second Independence." Their triumph officially banished the Conservative version of history. From then on, schoolchildren learned *historia patria* (the history of the fatherland), described by Justo Sierra as "a patriotic religion that unites and unifies us" through "holy love" and "deep devotion" for the (Liberal) heroes.[33]

The envoys to the Centenary Fiestas could get an idea of the official version of Mexico's history by simply strolling along the main avenue of the city and looking at the statues and monuments on the sidewalks and on the traffic circles filled with plants and flowers. It was the famous Paseo de la Reforma, which connected the legendary Castle of Chapultepec—the former palace of the Spanish viceroys constructed in the favorite forest of the Aztec emperors—with the park of the Alameda, whose paths, fountains, and pavilions had been a place for the citizens of the capital to wander and relax since the seventeenth century. The Paseo had been built by the emperor Maximilian for his wife, Carlota, after the model of the Champs-Elysées. That "Causeway of the Empress" had been renamed after the fall of his dreamlike empire. Over the years it had come to represent a public lesson in *historia patria,* the same script that was studied in state schools like a "patriotic catechism."

Across from the handsome mansions belonging to the rich families of the capital who had accepted them as honored guests, the envoys could *read* this Liberal version of history, along the almost three-kilometer length of the Paseo de la Reforma. In the traffic circles of the Paseo (modeled after the *étoiles* of Paris) stood statues heavy with significance. Nearest the Alameda, dating from the last years of the viceroyalty, was the equestrian image of Emperor Charles IV. Despite their rejection of the colonial past, the Liberal governments had allowed the figure to remain in place, where it had been since 1852, as a decoration and also because the citizens of the capital had forgotten who it represented. They called it "The Little Trojan Horse." In the next circle, on the way to Chapultepec, stood a monument dedicated to the discovery of America. It had been placed there in 1877 under the patronage of a wealthy businessman and represented Christopher Columbus with four of the principal missionary friars who had brought Christianity to the new Spanish possessions. In that same year—the first of his many in power—Porfirio Díaz intervened in the process of embellishing the Paseo, to ensure that the remaining circles would reflect Liberal and not Conservative mythology. He issued a decree that sealed the future of the Paseo de la Reforma as the ultimate Liberal avenue:

> Wishing to beautify the Paseo de la Reforma with monuments worthy of the culture of this city, the sight of which may recall for posterity the heroism with which the nation struggled against the Conquest of the Sixteenth Century and then for Independence and in the present for the Reform, the honorable President of the Republic has ordered that a monument dedicated to Guatimotzin [Cuauhtémoc] and the other leaders who distinguished themselves in the defense of the fatherland be erected in the circle situated to the west of the one occupied by the statue of Columbus; in the following circle another dedicated to Hidalgo and the other heroes of the War of Independence; and in the next [and final circle] a monument to Juárez and the other leaders of the War of the Reform and the Second Independence.[34]

The monument to Cuauhtémoc, last emperor of the Aztecs, had been installed in 1887. The special envoys to the Centenary Fiestas were themselves present at the inauguration of the Column of Independence on September 16, 1910, and the Hemicycle dedicated to Benito Juárez two days later. Though the monument to Juárez was not placed in a circle of the Paseo but within the Alameda, the government of Porfirio Díaz, beginning at the end of the century, had arranged an additional support for the *historia patria* along the famous street. On both sides of the paseo, near the tall ash and eucalyptus trees, each state of the Republic would raise statues of its two principal heroes, from the War of Independence or the War of the Reform. And the Paseo de la Reforma gradually filled with sentinels of bronze.

For the Liberals, the history of the country before 1810 only made sense insofar as it converged with or prefigured the Liberal victory over the Conservatives who supported the ephemeral empire of Maximilian. From the viewpoint of the *historia patria,* the great caudillos Cuauhtémoc, Hidalgo, and Juárez shared a common quality: All of them had fought for the fatherland against an enemy from abroad. Cuauhtémoc foreshadowed Hidalgo, Hidalgo followed Cuauhtémoc and led directly to Juárez, the triumphant avenger who fulfilled the dreams of both his predecessors. All the heroes shared the same nature: tragic, stoic, backs to the wall, and nearly always defeated. Except for Juárez, a victor who died in his bed, the other heroes of the pantheon (Aztecs, Insurgents, "Child Heroes," and many of the Liberal leaders) had "fallen for the fatherland."

If the Paseo de la Reforma was the deliberate inscription, the sculptural libretto of the Liberals, the ancient center of Mexico City spoke of the other history, which, according to the Conservative version, had created Mexico.

If the envoys looked carefully, they could discover the traces of a fascinating archaeological history. Fires, floods, frequent earthquakes, the hand of man and time had almost completely worn away the city built by Cortés "with the stones of the idols and heathen temples"—that model Renaissance city of the first viceroys, which several visitors had compared to Venice. But the period still had left its mark in the layout of the streets (on a straight, orderly grid) and in a few churches and buildings. Much more present, though also worn down by the years, was the Baroque city of the seventeenth century. To it belonged the monasteries and colleges of the various orders of friars (Franciscans, Dominicans, Augustinians, Jesuits), the twenty convents that had existed since the viceroyalty, the hospitals of the religious orders, and a number of civil buildings. The neoclassical city of the late eighteenth century had contributed some impressive structures to the urban landscape but not without destroying—through a change in aesthetic taste—many of the facades and retables of the Baroque. A century later, the Jacobin impetus during the War of the Reform quickened this process, wiping out much of the architectural heritage of the colony.

Nonetheless, in 1910 all the styles of the viceroyalty cohabited in the noble religious and civil architecture of the old city, which a famous traveler of the nineteenth century had called "The City of Palaces." The cathedral itself was the emblem of this shared presence. Begun in 1573 and completed in 1813, it included elements of all styles and periods and had something to say to all sensibilities. It was the solid existence of the Catholic and viceregal past in a Mexico that was trying to establish its modern, secular identity.[35]

If the envoys sharpened their eyes, they could discover, under the colonial city, a level still more ancient, disdained by both Liberals and Conservatives: the city of the Indians. In 1910, Indians still traveled the canals that reached as far as the center of the city, rowing their supply boats, the *trajineras*. They came from the outskirts, bringing vegetables and flowers to the central markets. These canals were the last traces of the city on the lake that the Spaniards had so admired.

The three great avenues that led to the central plaza followed the course of the great Aztec avenues, and like many other streets, villages, and locations within the country, they had kept their original names in Náhuatl. The plaza itself (known as the Plaza Mayor or more commonly as the Zócalo), an immense quadrangle flanked by the cathedral, by the old viceregal palace now converted into the Palacio Nacional, by the city hall and other impressive buildings and gates, echoed the words of Bernal Díaz del Castillo in his *True History of the Conquest of New Spain*:

> And after we had closely observed and considered all that we had seen, we looked again at the great plaza that held a multitude of people, some buying and others selling, so that the noise and hum of the sounds and the words resounded from there for a league, and among us there were soldiers who had been in many parts of the world, even in Constantinople, and in all of Italy and in Rome, and they said that never before had they seen a plaza so harmoniously arranged, of such size and filled with so many people.[36]

The noise and the hum of the Indian voices could still be heard, mingled with the voices of other eras. In 1910, an attentive ear could catch them all: the bargaining over a sale in Náhuatl, a mass in Latin, a patriotic discourse in Spanish, a pedantic young gentleman dropping French phrases.

In which mirror was Mexico most faithfully reflected? In the Paseo de la Reforma, pride of the Liberals, or in the old center of the city, the sanctuary of the Conservatives?

The Liberals had some truths on their side. They were right to celebrate the creation of an autonomous state, secular and formally republican, together with a modest internal market and a society of individuals who were at least in theory (and to a degree in practice when compared with colonial times) free and equal before the law. All of this had been accomplished against the difficult heritage of a colonial state ruled by Church and Crown, a closed and captive economy, and a society that was passive and hierarchic. But the Conservatives were also right to emphasize the many, varied values (ethical, aesthetic, religious) that stemmed from New Spain. These had formed the Mexican identity and were not to be dismissed, in the words of Justo Sierra, as a "laborious and deficient gestation."

On the other hand, both versions of history were also wrong. Though neither would ever admit it, each side idealized its favored events and heroes. Each side had its own misreading of reality. Throughout the nineteenth century the Liberals believed that Mexico could move into a modern future (republican, capitalist, federal, and democratic) merely by establishing goals and most especially through drafting laws. The Porfirian Liberals exaggerated the degree of modernity they had achieved and the speed of present and future progress. The Conservatives, in turn, idealized many ideas and institutions of the enclosed and self-absorbed colonial past, the restoration of which, in purely practical terms, would have been impossible in the modern world of the nineteenth century.

But from the heights of "Peace, Order and Progress" that the country believed it had reached during the Porfirian period, the disputes over Mexico's history seemed to have become more and more academic. Nobody was unaware that it was precisely these differing views of the past that had split the Mexican elite of the nineteenth century into two camps so ferociously opposed that they had to go to war. But it did seem as if the argument had been settled at last. The Conservatives had been politically defeated and their heroes had been expelled from the history textbooks and from the Paseo de la Reforma, but they would continue to prosper economically and bring up their children in the Roman Catholic faith, practicing it with no interference. And more—they could hold the highest political posts as long as they professed to be moderately liberal in thought and did not flaunt their religious beliefs.

What was even more surprising was the fact that many values of Conservative Mexico had been tacitly appropriated by the Mexico of the Liberals. Porfirio Díaz himself was the best example of this odd mixture. Officially, he was President of a representative, democratic, and federal republic; in practice, he was the paternal dictator for life of an absolute and centralized monarchy: just the kind of viceregal leader for whom many Conservatives had fought their wars. Unsurprisingly, at the turn of the century, several influential figures in the Porfirian regime considered themselves "Liberal conservatives."

Revolution

But history had an immense revelation in store for Porfirio Díaz. The regime had intended 1910 to be a patriotic "holy year" in honor of the *patria*—the apotheosis of the past hundred years. In reality, it was the tumultuous beginning of a new century, one that would move in directions never dreamed of within the Liberal script. "A holy year" it would be, but with a sanctity quite different from what the Porfiristas had imagined.

The new revolution began on November 20 of that same year. According to the official version of history, the upheaval of 1810 should never have happened again. The primary purpose of the War of Independence had been to free Mexico from Spain and to begin the process that would culminate in the Liberal *patria*. There was no need to study or comprehend the nature of the struggle of 1810 beyond its significance for the achievement of independence. None of its revolutionary episodes, ideas, or characters seemed to offer any lessons for the present.

But in reality they offered a great lesson. The Porfiristas could not *see* either the former or the coming revolution because they were trapped within their linear conception of history. To begin to comprehend the new revolution, it was not necessary to choose the opposite conception (the cyclical worldview of the Aztecs) or the nostalgic Conservative notion, which denied validity to everything created in the nineteenth century. What was needed was an understanding of the true complexity of Mexican history. In 1810, the vast colonial order that had endured for three centuries had exploded into a revolution whose declared objective—though of course not the only one—was to escape from that past in order to construct a future. Since that moment, Mexico had been not merely the site of an *ascent* or a *fall*, but the historical arena of various open and unresolved *tensions* between two historical orientations: the weight of the past and the call of the future.

Throughout the nineteenth century the country had turned its back on the past and primarily looked toward the future—but always with enormous cost and trouble and in modes that were ambiguous, uneven, and often contradictory. The long Porfirian dictatorship was the living proof of this unresolved tension. In the economic sphere it seemed the very image of progress, but in the area of politics it had, in some measure, restored the old colonial order with certain pre-Hispanic features. In 1910 the internal tensions of the time—based not in some idealized world but in the daily life of the Mexican people—reached a new point of explosion. But this time it was for the opposite reasons compared with 1810. It was not the future calling for escape from the past. It was the past demanding adjustments.

Porfirian Mexico read its past badly and therefore also its present historical position and its future dangers. In Mexico, as opposed to other countries with ancient histories (Greece or Persia or Egypt), the past—both Indian and viceregal—was still alive and functioning, still influential, still charged with unresolved possibilities. The reasons for this survival had of themselves a long history.

I

THE WEIGHT OF THE PAST

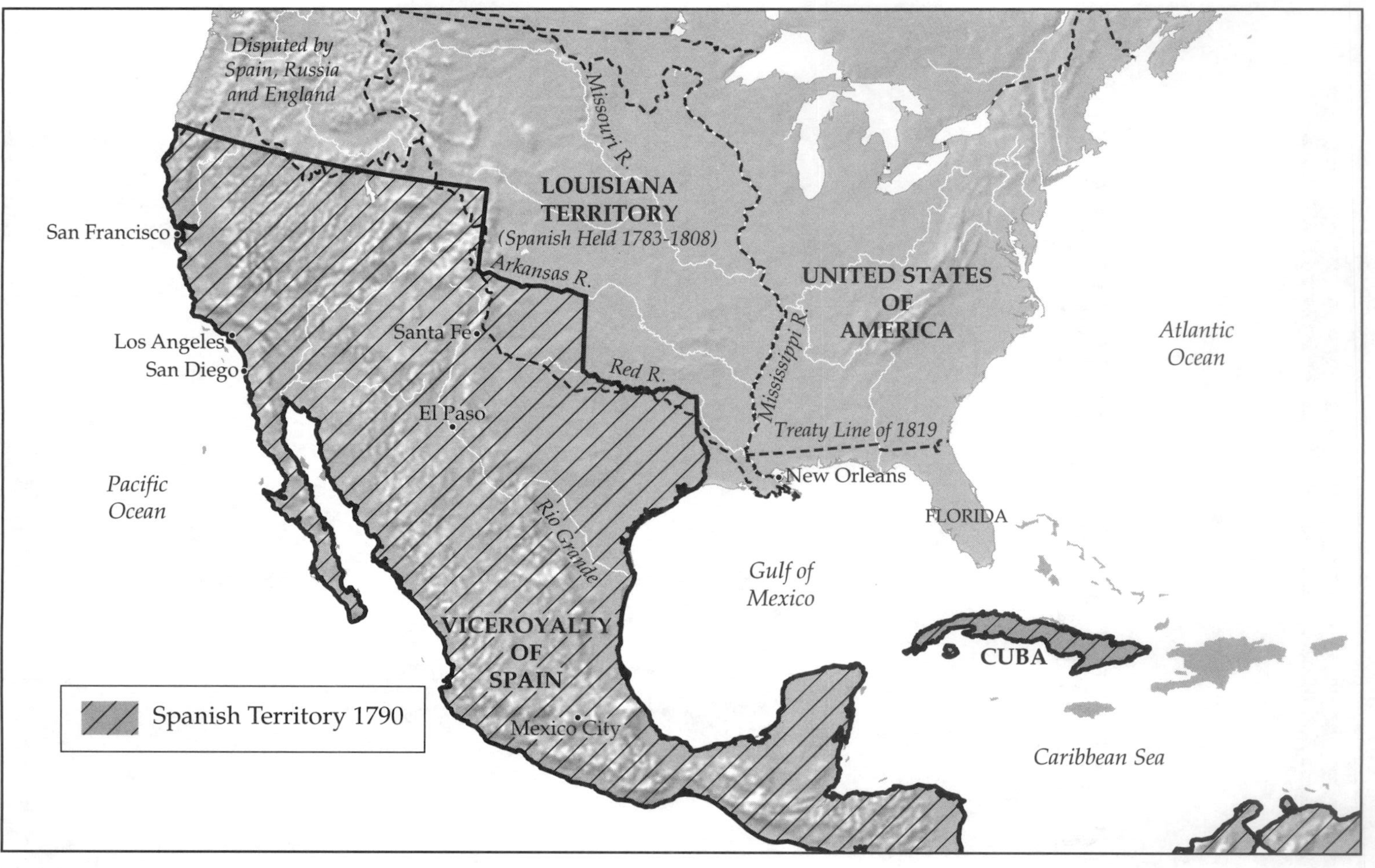

Spain in North America, circa 1790

1

THE CHILDREN OF CUAUHTÉMOC

The pyramidlike monument to Cuauhtémoc on the Paseo de la Reforma is decorated with symbols of various pre Hispanic cities: the friezes of Mitla; the columns of Tula; the cornices of Uxmal; the shields, war dresses, and weapons of Tenochtitlan. Porfirio Díaz had unveiled the monument at a ceremony held in August of 1887. Seated on a chair that recalled the throne of the Aztec monarchs, he listened to laudatory speeches in Spanish and poems in Náhuatl that acclaimed "the feats of Cuauhtémoc, the last of the Aztec emperors, and the other caudillos who distinguished themselves in defense of their Nation." Cuauhtémoc (whose name means "Falling Eagle") had led the Aztecs in their final, furious resistance to the conquistadors of Cortés after the death of Emperor Moctezuma, who had welcomed the Spaniards. At either side of the monumental complex, impressive bas-reliefs show climactic episodes in the life of Cuauhtémoc. The events chosen for portrayal are his imprisonment and his torture. In the scene of the imprisonment—based on the testimony of Cortés himself—Cuauhtémoc puts his hand on Cortés's dagger and asks to be killed. The other relief depicts his further sufferings. After the fall of Tenochtitlan in August 1521, despite white-hot irons pressed to the soles of his feet, Cuauhtémoc refuses to reveal the location of Moctezuma's treasure.

Raised high on the pedestal of the monument, the statue of

Cuauhtémoc's martyrdom
(Photo by Victor Gayol, Editorial Clío)

Cuauhtémoc, with its haughty expression, exemplified the neo-Aztec fashion in academic art, a style that had been growing steadily more popular since the victory of the Liberals. Paintings and sculptures would portray Indians with Apache faces and Apollonian bodies draped in Roman togas, participating in idealized scenes of "the ancient world." The Aztec past had such ornamental appeal that the first Mexican brewery, which opened in the industrial city of Monterrey in 1890, was named Cervecería Cuauhtémoc. It used a drawing of the statue as its trademark.[1]

Ideological manipulation of the ancient past was an old Mexican custom. When they consolidated their empire, the Aztec emperors took the codices recording their true nomadic origins and had them burned. Then they had history rewritten to link them directly with the Toltecs—the great classical civilization of Mesoamerica—and the mythical founder of that culture, Quetzalcóatl. They were familiar with the abandoned cities of Teotihuacan and Tula, "coming from a time that no one now can describe."[2] They called the Toltec culture "our possession, that which we must preserve." The Porfiristas, conscious of the power of history as a source of legitimacy, had taken the same route as the Aztecs. Crossing out their Catholic and Spanish origins, they were stretching a bridge toward the Aztec world.

The rewriting of history was not the only way to maintain an Aztec presence, in support of the present order. Even the practice of erecting statues to honor great leaders could be traced to the Aztec past. Following a Toltec custom attributed to Quetzalcóatl, who, "when he went away, left his appearance sculptured on sticks and stones," the fourteenth-century *tlatoani* (emperor) Moctezuma I had ordered his image and that of the powerful Tlacaélel, his brother and counselor, to be carved on the slope of the hill of Chapultepec, "in perpetual memory, as reward for our work, so that, on seeing the faces, our sons and grandsons may remember our deeds and strive to imitate them."[3] Scattered remains of those immense stone sculptures could still be seen in 1910, at the foot of Chapultepec Castle, near the monument to the Niños Héroes.

Well into colonial times, and thanks to the copious and meticulous works of ethnography produced by the Spanish missionaries with the help of Indian informants, the creoles of New Spain also resorted to an ideological manipulation of the past. What arguments could they muster against their ethnic brothers, the Spaniards from the Iberian peninsula, who were given preference in almost all areas of the life of New Spain while the *criollos* were ranked as second-class Spaniards? They had only one strong claim—they had been "born in these lands," in "our Mexican home."[4] To enrich this single assertion, the mere geographical fact of birth soon came to be nourished with every conceivable item of cultural separation. The creoles, through the force of history, would try to build a case for themselves as legitimate heirs to the land of New Spain. "In the Mexican Emperors may be found," the prolific creole writer and scientist Don Carlos de Sigüenza y Góngora wrote in 1680, "what others had to beg for in fables."[5] Sigüenza wrote the biography of the nine Aztec emperors, favorably comparing the political virtues of each with analogous Greek or Roman rulers. In his *Ancient History of Mexico* (1780), the Jesuit humanist Francisco Javier Clavijero ascribed to the civilization of the Mexicas a classical rank equal to that of Greece and Rome: "The state of culture of the Mexicans when the Spaniards discovered them greatly surpasses that of the Spaniards themselves when they came to be known by the Greeks, the Romans, the Gauls, the Germans, and the Bretons." To prove it, he examined all the areas of ancient Aztec life, weighing their characteristics with an impressive combination of comprehensive sympathy and Olympian objectivity, qualities required more than anything for his analysis of the human sacrifices:

> I admit that the religion of the Mexicans was very bloody and that their sacrifices were extremely cruel . . . but there is no nation in the world that has not sometimes sacrificed victims to the God they adored. . . . In respect to cannibalism, their religion was without doubt more barbarous than those of the Romans, Egyptians and other civilized nations but . . . it was less superstitious, less ridiculous and less indecent.[6]

In 1794 the priest Servando Teresa de Mier publicly asserted the identity of Quetzalcóatl and the apostle Thomas. Not only did he agree with Sigüenza and Clavijero in their view of the indigenous world as a classical past, but he was willing to transfer Christian legitimacy from the three colonial centuries to the Toltec culture. These historical assertions of the Mexican creoles—and idealized identification with the world of the Indians—were bound to play an important role in the War of Independence. The chronicler Carlos María de Bustamante, another prominent creole, invoked the "Spirits of Moctezuma and Cuauhtémoc" at the height of the war, in order to proclaim "the renewal of the Mexican empire."[7]

Needless to say, when independence became fact in 1821, the new nation did not reinstate the Mexican Empire. But the country took its name from the original tribal name of the Aztecs, and for the emblem on its flag it used the mythical symbol of the foundation of the city of Mexico-Tenochtitlan: an eagle perched on a nopal cactus holding a writhing serpent in its beak.

The mestizo Liberals too put the past to political use. The strange thing was that their opinion of the Indian past was essentially as negative as the view of the creole José María Luis Mora, father of Mexican liberalism, who ridiculed "the myths of grandeur, prosperity and enlightenment surrounding the ancient Mexicans." Guillermo Prieto, Juárez's friend and poet, felt "the greatest repugnance for the idolatrous religion, the grimly renowned sacrifices, the accursed offerings to the gods [in] the somber and melancholy world of the indigenous race." The radical Juarista, Ignacio Ramírez, said that "it was a time when terror made the whole social body tremble and the people consisted of subjects and slaves."[8]

The later, Porfirian generation softened this vision but did not completely abandon it. In his books and compendia, even in his *Catecismo de historia patria*, Justo Sierra referred to the "more or less useless civilizations" of the Mexican past, though he tried to emphasize the positive aspects of both heritages—the Indian and the Spanish—and to play down areas of conflict. He pointed to the "extraordinary vitality" of the ancient Mayas and the magnificent mosaic of cultures (Olmec, Tarascan, Mixtec, Zapotec, Maya, Aztec, and dozens more scattered throughout the land) that had constituted Mexico for almost thirty centuries. But he was convinced that the Conquest had been a dramatic step forward on the path of human evolution. Mexican indigenous history presented a "picture of ravenous gods and terrified multitudes, surrounded by sacrificers, everything smeared in black. . . . This delirium of blood had to end."[9]

And yet, the Juarista and Porfirista Liberals identified with the Aztec past through Cuauhtémoc, the young nobleman who had decided to die fighting along with his people rather than surrender to the invader (though his death would actually come later, as a captive of Cortés). But it was a political, not a cultural identification. This patriotic cult of Cuauhtémoc was relatively new. It was only during the nineteenth century, when Mexico for the first time became involved in wars against foreign invaders, that the ruling elites really began to see themselves represented in the last emperor of the Aztecs. The veneration appeared in literature before it entered politics. On Independence Day, September 16, 1839, shortly after the French attempt to seize the port of Veracruz, Ignacio Rodríguez Galván, a young poet, wrote his *Profecía de Guatimoc* (*Prophecy of Guatimoc*). It presented the ghost of the unhappy emperor mourning for the blood spilled by the "European barbarian" who had made

The innocent welcoming bed
A home of lust and of horrors.

But the poem also announced that vengeance was on its way:

Tremble! Feel yourselves quiver! O kings of Europe!
He who deals death, death will be his in return![10]

A generation later, after their defeat of the French interventionist troops and the execution of Maximilian (descendant of the same Hapsburgs whose conquistadors had defeated the Aztecs), it was possible for the Liberal republicans—led by their Indian President, Benito Juárez—to feel that Galván's prophecy had been realized and that Cuauhtémoc had finally been avenged. In 1869, two years after his triumph, Juárez himself unveiled the first monument to Cuauhtémoc, a small bust on a boulevard in the outskirts of the city.

The regime of Porfirio Díaz not only seized on the historical parallel between the Aztecs and the republicans; it converted indigenism, and particularly the mystique of the Aztecs, into an effective state ideology, with great collateral benefits to culture and knowledge. Never before had the government given so much support to Mexican archaeology and the study of what Justo Sierra called "the aboriginal cultures." Sierra wrote that, before the Porfirian era, "our ruins were dying. They were the ruins of ruins." Under Díaz, the State again became trustee of the Indian ruins, responsible for saving them, for protecting, conserving, and restoring them. Don Porfirio ordered the reconstruction of well-known sites, among them Mitla, in his native Oaxaca, where he had played as a child, and financed explorations to search out and excavate other areas. He sent delegations to display pre-Columbian antiquities at international exhibitions, founded the Trust for the Inspection and Conservation of Archaeological Monuments, placed a ban on the export of artifacts and monoliths and provided for their proper exhibition by renovating the National Museum. And he encouraged research, the teaching of ethnology and archaeology, the diffusion of information, all of which was reflected in a magazine of high scholarly quality, the *Anales del Museo*, which was published throughout the Porfirian era. It was due to the sponsorship of Porfirio Díaz that the Seventeenth International Conference of Americanists was held in Mexico City during the Centenary Fiestas.[11]

The speech Sierra delivered at the inauguration of that event reflects the real stance of his time toward the indigenous past. Years before, at a similar meeting, he had said, "I come from a land that has here and abroad been called the Mexican Egypt; an immense necropolis where various civilizations rest in their successive levels."[12] In 1910 he mentioned with pride the "centuries turned to stone by the crude and interesting remains" and praised the recent reconstruction of ancient sites under the auspices of President Díaz, that "great venerator of our history." He then invited the Americanists, and particularly foreign archaeologists, historians, and ethnologists, to investigate ancient Mexican civilizations. But Mexico, Sierra implied, had more urgent duties. In 1910, "this country has not lost an ounce of its religious devotion to its history . . . [but] it lives as if possessed by a fever for the future."[13]

During the era of Don Porfirio, "the religious devotion to history" bore fruit in academic circles where the indigenous past of Mexico was examined with the same committed interest that scholars had devoted to the ruins of Egypt. The enthusiasm would broaden knowledge of the past, but in the political arena it was placed at the service of "the fever for the future," officially encouraged by the State. Ignoring the cultural plurality of pre-Hispanic history, the Porfiristas completed the identification of their State with the powerful and centralist Aztec Empire and with its clearest symbol: Cuauhtémoc, the last emperor. And he became, in official legend, the first and archetypal Mexican caudillo, the founder and symbol of Mexican nationalism, the remote ancestor of the country's present *tlatoani*, Porfirio Díaz.

The principal pre-Hispanic cultures were dead (though not without leaving their traces) in such essential areas as religion and political organization, but other aspects were very much alive, modified by contact with the Western world yet also capable of altering whatever they came to touch. During almost four centuries, Mexico had become a fascinating and dynamic laboratory for the ethnic and cultural convergence known as *mestizaje* (mixing), even if not all the Indians wanted or were able to become part of this experience.

In 1910, one-third of the total population of 15 million was still of pure Indian stock. They were members of different "nations," like the Tarahumaras in the mountains of Chihuahua, the Mayos and Yaquis in the fertile valleys of Sonora and Sinaloa, the Coras and Huicholes in the mountains of Nayarit, the Tarascos in the hills and along the lakes of Michoacán, the Mazahuas, Nahuas, and Otomís scattered throughout the mesa of central Mexico, the Tzeltales, Tojobales, Chontales, and Tzotziles in Chiapas, the Zapotecs, Mijes, Zoques, Huaves, and Mixtecs in Oaxaca, the Huastecos and Totonacos in Veracruz, the Mayas in the Yucatán peninsula. In the center and south of the country, Indians made up between 50 and 75 percent of the population, and some regions were almost entirely Indian. More than a hundred indigenous languages and dialects were still spoken.[14]

The fact that the Indians had survived was equally apparent in many provincial towns and even in the capital itself, but the Porfiristas—dedicated admirers of the dead Indian—cared little about studying or learning from the present, living Indian and even less about identifying with him. The Indian was the burden of the Mexican nation.

Several centuries earlier, the opposite had seemed true. The Indian then had been seen as the hope of Mexico. In each age, the dominant value system would dictate the way in which the indigenous races were perceived and interpreted. During the sixteenth century, it was the great Spanish theologians who became involved in the debate over the Indians. Never before in the universal history of conquest had men argued so heatedly about the moral and theological implications of their enterprise.[15]

The imperialist interpretation of the Conquest (stridently represented by Juan Ginés de Sepúlveda) justified the war against the Indians on the grounds of their allegedly natural vices and defects: They were subhumans, sodomites, barbarians, cannibals, cowards, idolaters, liars, and depraved idlers. Their backwardness prevented them from freely submitting to the law; they were "slaves by nature."[16]

Totally opposed to this position was the Dominican Bartolomé de las Casas. (He was its greatest opponent, but he was by no means alone. Among others, his fellow Dominican Francisco de Vitoria—one of the most distinguished theologians and jurists of sixteenth-century Spain—had written, "It chills the blood to hear the assertion that Indians are not men but apes").[17] The tireless Las Casas crossed the ocean eight times in his efforts to document the worthiness of the pre-Columbian civilizations and the evil of their destruction by the conquistadors. In his many writings, he examined various features of pre-Columbian life, including the worship of idols and human sacrifice, to argue for their value on not only a natural but a religious level.

Las Casas had irritated the landowners in his own diocese of Chiapas by denying them the sacraments if they refused to liberate their Indians. And he had collected many testimonies to Spanish mistreatment of the Indians. With his *Apologética historia sumaria*, he produced an original study in comparative ethnography, based on the writings of the first Franciscans and very much including his own experiences and the oppression he had witnessed on the islands of the Caribbean.[18]

His goal was to show that the pre-Columbian cultures were not only comparable but in many ways superior to other ancient pagan cultures. The stoic instructions delivered by Aztec fathers to their sons; the beauty and sophistication of their arts; the precision of their astronomical observations, and their use of a calendar system to record the passage of time; the elaborate organization of their government, education, army, religion, trade, and legislation—all more than met the great Aristotle's own prescriptions for a civilized society. Las Casas even saw the ritual of human sacrifice as a twisted, satanic but ingenuous exaggeration of a fervent desire to serve God. And he argued that their worship of idols was a favorable preparation for a later devotion to Christianity. After presenting all his research and reasoning, Las Casas proposed that the Indians be given back their lands, arguing that if his advice went unheeded, they would eventually rebel against the tyranny of Spain.[19]

The Franciscan friars in New Spain did not go to such extremes. They saw the Conquest as the design of Divine Providence. But like nearly all the other friars and priests, they agreed that the Indians were a perfect flock for the labor of conversion.

The Franciscans together with members of other mendicant and preaching orders (Augustinians, Dominicans, Jesuits, Carmelites) and representatives of the regular (or "secular") clergy carried out the "spiritual conquest" of Mexico during the three centuries of colonial rule but especially in the decades immediately following the Conquest. Many of them shared their lives with the Indians in every corner of the country from Yucatán to remote California. The *padrecitos* (beloved fathers), as the Indians soon came to call them, were at the heart of one of the most extraordinary chapters in the religious history of the West: the conversion of millions of Indians to Christianity. In part a response to the Indians' sense of spiritual deprivation (they had been "orphaned" by the collapse of their divinely sanctioned governments, the overthrow of their powerful gods), the conversion was no rational decision, though the Indians learned the new prayers and dogmas. And even though the missionaries, especially in the first decades, not only offered spiritual support and genuine love but also zealously persecuted Indians who continued to worship "idols," the Indians never entirely gave up their old beliefs. Catholicism (unlike North American Protestantism) could eventually accommodate elements of the older religions, accepted tacitly and often in the dress of Christian saints.[20]

The conversion of the younger generations was especially important to the missionary fathers, and as a result, they put great energy and fundamental, constructive effort into basic education. Men like Fray Pedro de Gante ("the first who taught them how to read and write") learned the languages of the Indians and devoted their lives to instructing the young.

A mixture of benevolence, paternalism, social concern, and intolerance shows up in the vivid report dispatched (on August 27, 1529) to the king of Spain by Franciscan and Erasmist Fray Juan de Zumárraga, the first bishop of Mexico:

> We are very busy with our continuous and great work in the conversion of the infidels of whom . . . over a million people have been baptized, five hundred temples of idols have been razed to the ground and over twenty thousand images of devils that they adored have been broken to pieces and burned. . . . And . . . the infidels of this city of Mexico, who in former times had the custom of sacrificing each year over twenty thousand human hearts to their idols, now make their offerings to God instead of to the devils. . . . Many of these children, and others who are older, know how to read, write, sing and sound the proper pitches for singing. . . . They watch with extreme care to see where their parents hide their idols, and then they steal them and faithfully bring them to our friars. For doing this, some have been cruelly slain by their own parents, but they live crowned in glory with Christ. . . . Each one of our monasteries has next to it a house in which children are taught and where there is a school, a dormitory, a dining hall and a chapel for devotion. . . . Blessed be the Lord for everything. . . . [21]

The spiritual conquest used many methods, but the fathers nearly always relied on the senses and on the appeal of the arts, almost never on arguments directed at the intellect. They allowed the Indians to sing and dance before the Christian images as they had before their gods. They gave them sacramental theater, music and choirs, mural paintings, elaborately impressive church facades and altarpieces. Rather quickly, the Indians' initial distrust and the missionaries' initial intolerance would both soften. A true encounter would become possible: the embrace that established Mexican spirituality.

The encounter went both ways. The missionaries educated the Indians, but they also began to learn about them and value their culture. The most outstanding study was *The General History of the Things of New Spain*, a complete encyclopedia in twelve volumes on every aspect of pre-Columbian Aztec life, originally compiled in Náhuatl over a period of more than thirty years by Fray Bernardino de Sahagún (1499?–1590).

He used detailed questionnaires, patiently collected oral and pictographic testimonies from elderly "informants" in various parts of the country, critically compared their information, and then ordered it to be copied down and translated into Spanish. He was assisted by a number of young Indians who in time would become important translators from their own language into Spanish and Latin. Modeled on the *Natural History* of the Roman historian Pliny, the work moved from the "supernatural" to the "natural" level. In the first section (a detailed description of the religious ideas of the Aztecs, their ceremonies, their common practice of human sacrifice and cannibalism) Sahagún demonstrated, to his own satisfaction, that the Indians were possessed by Satan. The "Mexica" were a people damned "because God hates idolaters more than any other kind of sinner." But Sahagún's attitude changed drastically when he turned to the "natural" sphere, where almost everything seemed to him worthy of admiration: the rigorous and ascetic education transmitted by the elders to the young at home and in schools with "arguments of excellent moral doctrine," the beautiful architecture and political organization of Tenochtitlan, "which was another Venice," their "very pleasing and even very mystical" style of song, their poetic "excellences," their "marvelous language and very refined metaphors and admirable admonitions."[22] Indeed, "The Indians were so trampled and destroyed, they and all their things, that no sign remained of what they were before. And so they were considered barbarians and people of low degree. . . . When truly . . . they are ahead of many other nations that are arrogant about their degree of refinement."[23]

Vasco de Quiroga—a secular priest who adapted the ideas of Thomas More's *Utopia* with considerable success to the Indians of the ancient Tarascan Empire in Michoacán—said that the indigenous people were like "soft wax" without a sign of "arrogance, ambition or covetousness." All they lacked was faith "to become perfect and true Christians." Quiroga—though unusual for the unique quality and endurance of his achievement—is perhaps the most outstanding example of the missionary love for the Indians, in his case free of intolerance, strongly constructive, but still protectively paternal.[24] In Spain he had been a judge in a *morisco* area (inhabited by Moors forced to convert to Catholicism—or leave the country—when Ferdinand and Isabel completed the Christian Reconquest of Spain). He had shown respect and tolerance toward the *moriscos,* and he brought the same qualities to Mexico.

Quiroga began to create the "hospital-towns" (*pueblos-hospitales*) in 1532. At Santa Fe, near Mexico City, he acquired land and installed a group of Indian families in ten houses that shared a common patio, a church, and cells for the friars. Santa Fe rapidly expanded: seventy houses, several orchards, fields of maize, wheat, barley, and flax, pens for animals and poultry. Years later Quiroga extended his experiment to the lake towns in the old Tarascan region of Michoacán. There he founded Santa Fe de la Laguna, where his utopia put down strong roots and began to multiply across what is now Michoacán.[25]

Each town would center around a "hospital" in the medieval sense of the word, an institution that would welcome not only the sick but also the poor, the hungry, and even travelers in need of a place to stay. Within the hospital-towns, the main features of life were the practice of religion, work in the fields, and the learning of skills. Children were required to go to mass and catechism, taught to read and write in Spanish, and trained in a specific craft: bricklaying, tanning, carpentry, metalworking, ceramics, or textile production. Three kinds of authority were recognized: natural authority (the Indian patriarchs), the *principales* elected by vote of the heads of each household, and the *rector* (community priest). The use of domestic servants was not allowed, and collective cooperation was encouraged in the fields and in public projects.[26]

These hospital-towns were an economic and social success. In a short time Quiroga established almost a hundred new communities in different parts of Michoacán. To avoid both internal and intercommunity competition, Quiroga assigned one trade or craft to each town, thus creating a network of exchange all over the Tarascan area.

Not only was Quiroga's basic idea for the hospital-towns drawn from Thomas More's *Utopia*, but so were a number of his specific measures: communal and family-style organization, authority of the elders, community use of the produce of collective work, elimination of private ownership of land, and the idea of trade apprenticeship.

In Pátzcuaro, in 1540, Quiroga, who was now bishop of Michoacán, established the Colegio de San Nicolás, a seminary where creoles and mainland Spaniards studied side by side with Indians and mestizos.[27] His hospital-towns survived throughout the colonial period (despite some attempts to seize their lands). In 1776 they were functioning under the same rules laid down by Quiroga in 1563.

In 1996 the town of Santa Fe de la Laguna still exists as an Indian community, with the same hospital and church built by Quiroga. In the Tarascan area many communities still practice the same crafts assigned to them by the bishop. And in other towns the hospitals still stand, enduring emblems of a man venerated by the descendants of those Indians who called him "Tata Vasco."

Fray Julian Garcés, another famous missionary father, wrote in a letter to Pope Paul III: "[The Indians] are remarkably good-natured . . . modesty and composure is their nature . . . if they are ordered to sit, they sit, and if they are standing, they remain standing, and if they are kneeling, they kneel. . . . Who can doubt that, as time moves on, many of these Indians will become very saintly and will shine with every virtue?" It is believed that Garcés's letter had a direct influence on the papal bull *Unigenitus Deus* (issued in 1537), which stated that the Indians were endowed with reason and souls and could therefore become Christians.[28] But the influence of Bartolomé de las Casas on the New Laws of the Indies for the Good Treatment and Preservation of the Indians (1542) was even more extensive and important. "Humanity is one," Las Casas had said. In the great debate on the human nature of the Indians, the Spanish Crown gave complete and conclusive support to the position of Las Casas against Sepúlveda, declaring that the Indians were "free vassals."

The New Laws became the moral charter for the founding of New Spain and the historical basis for the relationship between the dominant power and its weakest subjects. Without denying the Indians' natural equality, the Crown recognized that the Indians had less power than the Spanish colonists and therefore placed them under a system of permanent protection. A charitable state, ruled by Christian morality, was expected to correct all the injustices the Indians had suffered. The viceroy was given the title of "Protector of the Indians," and they could lodge a complaint with him at any time. Indigenous authorities, mores, and practices were to be respected, provided they did not violate the Catholic religion. The Indians should be encouraged to form separatist communities and "republics," and they ought to have absolute ownership of their communal lands. To forestall any danger of encroachment on these lands, they were forbidden to deal commercially with their immediate neighbors. A special court was set up—the Indian Tribunal—to adjudicate controversies with Spaniards. Those Spaniards who employed Indians were ordered to convert them, treat them well, supply them with food, lodging, clothes, medical attention, and a priest, and pay them in real money. Of course the New Laws, as noble as they sounded, did not prevent the exploitation of Indian labor, but they did put an end to the concept that Indians were property and natural slaves of the Spaniards.[29]

At the beginning of the twentieth century, a politician close to Porfirio Díaz claimed that "one white immigrant was worth more than five Indians."[30] Although other writers of the Porfirian age held similar views, they generally based their negative portrayal of the Indian purely on character and custom (the so-called innate Indian qualities, like obsequiousness, apathy, distrust, hypocrisy, unproductiveness, and timidity).

Except in a very few cases, they did not make direct claims of racial superiority—for which any one of numerous examples in recent Mexican history would have been refutation enough. But Benito Juárez was, of course, the best of all. What other country in the Americas could boast of having had an Indian President?

And yet this isolated case was counterproductive. Here was an Indian who had risen to the height of power entirely through his own efforts. Surely this proved—to the most influential writers of the time—that only the march of progress, the hard, cruel process of natural struggle and selective assimilation, could gradually rescue the indigenous races. Emilio Rabasa recorded years later that "no special effort was taken toward civilizing the Indians, nor did we know what that effort might be, which might lead through direct action to the transformation of the race."[31]

By 1910 the Porfirian intellectuals had in fact come to agree, though they may not have realized it, with Ginés de Sepúlveda and the imperialist school of thought about the Conquest. They condemned precisely those characteristics of the Indians that the missionaries had believed would ensure their redemption. Francisco Bulnes, an acid polemicist and shrewd historian, wrote:

> The Indian is disinterested, stoical and unenlightened; he despises death, life, gold, morals, work, science, pain and hope. He dearly loves four things: the idols of his former religion, the land that feeds him, personal freedom, and alcohol which induces morose and silent deliria. He is a man who ought to dress in a shroud and give away his magnificent teeth, since he does not laugh, talk or sing and almost does not eat. . . . Why work if he cannot own anything? After he had just been robbed by the Conquistador, along came the friar, the cacique, the municipality, the small-time lawyer, anyone at all. The Indian belongs to everyone who wants to dominate him.[32]

Justo Sierra, as a young man, had heard Juárez propose his own radical solution to the problem: "I wish that Protestantism would become Mexicanized and convert the Indians; they need a religion that would make them read instead of wasting their savings on candles for the saints."[33]

Justo Sierra himself was less extreme. His generous imagination sought a way in which to shape another kind of religious influence. He thought the secret might lie in treating the Indians as they had once been treated, in attempting their "social redemption" by applying a modern version of the evangelical work of the Franciscan friars in the sixteenth century. He felt that this civic mission—compulsory education—would work miracles. Rabasa had raised the question, "What good is school? The Indian who may know how to read and write has gained nothing by it. What will some other Indian gain if he learns to read and write?" But Sierra would not agree. With all the strength of his religious soul, he would die believing in the redemptive power of education. New missionaries would be needed though, to spread the gospel of learning; and under Porfirio Díaz, they were nowhere to be found.[34]

For the Porfiristas, the dead Indian was a fossil from a remote and symbolic past, almost totally alien to their everyday experience. The living Indian was his real heir, all that really remained of him, but the memory of the dead Indian served the political purpose of legitimizing the State, while the living Indian was a blemish on the landscape of modern, progressing Mexico. To hide the flaw, Díaz even banned the wearing of traditional white cotton clothes for Indians within the boundaries of Mexico City.

During the Centenary Fiestas, Porfirio Díaz took all the foreign envoys to visit the ruins at Teotihuacan, climbed to the top of the Pyramid of the Sun, dined with them in the "Porfirio Díaz Grotto," and was photographed standing next to the Aztec Calendar Stone, that petrified symbol of the past. But behind the life and times of the "great venerator of the past" was a very different truth: The arrogant Mexico of 1910 had turned away from its deep, indigenous roots.

2

THE LEGACY OF CORTÉS

One of the traffic circles on the Paseo de la Reforma was dedicated to Christopher Columbus. Statues of the "Great Navigator" could also be found in other cities of the Porfirian republic. But if some curious envoy to the Centenary Fiestas had wanted to visit a monument to Hernán Cortés, he would have been surprised to learn that nothing—no public image, no street name, not even a plaque—existed to commemorate him. Only one private institution retained, within its walls, a bust of Cortés. It was the old Hospital de Jesús, founded by the Conquistador himself, which had then been in operation for nearly four centuries.

"In Mexico absolutely everything that exists has its roots in that prodigious Conquest," Lucas Alamán wrote in his *Disertaciones*. "The Conquest was the means through which the civilization and religion of this country were established and Don Hernán Cortés was the extraordinary man sent by Providence to achieve this purpose." But in Mexico, even as Alamán wrote these words, Cortés was considered the greatest villain in the history of the country.

He had not always filled that role, though it was Spaniards—later followed by the English—rather than Indians who had created the "black myth" of Cortés. For "the Apostle of the Indians," Bartolomé de las Casas, Cortés had been a tyrant, a usurper of kingdoms, a criminal who deserved to be beheaded.[1] Even Cortés's own followers wrote with remorse about the horrifying massacre of Indian noblemen in the plaza of Cholula, triggered by Cortés when the Conquistador claimed to have discovered a conspiracy. (He was actually making a calculated move to

Moctezuma
(Biblioteca de Arte Ricardo Pérez Escamilla [BARPE])

Hernán Cortés
(Patronato del Hospital de Jesús)

intimidate and terrify the Cholulans, ordering his musketeers and harquebusiers to massacre hundreds of astonished and unarmed people.) Cortés had been responsible for many other cruelties, and he was especially criticized for the torture and execution of Cuauhtémoc. But with all that and in spite of losing much of the royal favor that had once been his, the Conquistador enjoyed unrivaled renown during his lifetime. Three centuries later, with the primary exception of Alamán and his conservatives, Cortés was almost universally despised.

Curiously enough, José María Luis Mora, Alamán's liberal adversary, showed the same respect as his conservative counterpart toward the Conquistador: "The name of Mexico is closely interwoven with the memory of Hernán Cortés . . . while the former exists, the latter cannot perish."[2]

But the victory of the Liberal republicans over the Conservative imperialists in the Wars of the Reform and the French Intervention strongly nurtured the "black myth" of Cortés. To the romantic Liberals, the Conquest represented not the founding of a nation but its violation. For them, Cortés was "that great outlaw who has only been raised to the level of hero by his own good fortune and the interests of Spain . . . because his life was no more than a web of vile actions and betrayals . . . of perfidy, murder and cruelty."[3]

Well into the epoch of Porfirio, another Liberal, the writer Ignacio Manuel Altamirano, a pure-blooded Indian, founder of the literary review *Renacimiento* (which initiated a distinctly national Mexican literary culture beyond specific ideologies), took for his own the words of Heinrich Heine: "It was no more than a captain of bandits who with his insolent hand wrote in the book of fame his insolent name: Cortés!"[4]

The power of the black myth of Cortés grew to such a degree that not even the Conservatives could deny it: "Our acknowledgment for the hero," wrote the historian Manuel Orozco y Berra, but "never our love for the Conquistador!"[5]

Yet in the mood of reconciliation that characterized the last days of the Porfiriato, Justo Sierra—though he strongly condemned the violence of the Conquest and assigned Cuauhtémoc "a gigantic moral superiority over his conqueror"—would assert that the Conquest had acquired the character of an "incomparable and admirable enterprise" and Cortés the rank of the "founder of [Mexican] nationality."[6]

On the morning of September 15, a day before the centenary of Mexico's independence, Porfirio Díaz invited the visiting envoys to watch an unusual spectacle from the principal balcony of the National Palace. In the Zócalo, some fifty thousand spectators applauded the brief presentation. A hundred amateur actors staged a re-creation of the historic encounter between Moctezuma and Hernán Cortés. Drums, trumpets, and crossbowmen heralded the arrival of Cortés. The Conquistador, mounted on his gray stallion and followed by his Spanish captains, Indian allies (the Tlaxcalans), and the famous Indian princess, "La Malinche," his lover and interpreter, approached from one direction. "Moctezuma's entourage," said the reports, "was even more brilliant. It brought alive, for our imagination, the memory of that court of the Mexican emperors, haughty with all the natural wealth put to use in adornments and also with the fierceness of its warriors." Jaguar knights, eagle knights, priests, archers, women, and lords marched in procession; and behind them, carried on his palanquin of gold, came the emperor Moctezuma:

> Treading on the carpets that his servitors spread before him, the emperor Moctezuma advanced to meet the Spaniard who had descended from his horse. Cortés approached with the intention of embracing the monarch, but when he was told that this could not be done, he hung a string of glass beads around the neck of Moctezuma.[7]

It was a historical pantomime that would not have been performed fifty years earlier. The scene chosen was a symbol of the conciliatory attitude toward the past so characteristic of the Porfirian age. The wound suffered at the time of the Conquest—reopened during the War of Independence—seemed to have finally healed by 1910. Instead of presenting a re-creation of the massacres of the Conquest (and emphasizing the split between the two opposing views of Mexican history), the organizers—Justo Sierra surely among them—wanted to stress the opposite movement: toward *convergence*. The idea was a departure from the standard Liberal version of history, but it had a profound resonance. For Cortés and Moctezuma had moved along their different paths toward a meeting as major protagonists of history who were destined to belong (by nature and fate) to different genres of literature—Cortés to epic and Moctezuma to tragedy.

The melancholy Moctezuma had carried the burden of a fatalistic vision of history that—through a series of dreams, prophecies, and bad omens—had foretold the fall of his empire. The chronicles describe him as a man "of a very grave and calm appearance." After he ascended to the throne in 1503, "he had been the greatest butcher . . . only so as to be feared and reverenced." Before dying, his ally, the king of Texcoco, had said, "Soon Moctezuma will see and experience what must come to him because he has wanted to do more than the god himself who determines these things."[8] From that moment on, a kind of cosmic fear possessed the emperor. He ordered his representation carved on the mountain of Chapultepec, and when he saw it finished, "he could not stop crying." Every week he would spend time contemplating it, and he was heard to say: "If our bodies were as lasting . . . as this painted effigy . . . who would be afraid of death? . . . But I have to die and only this memory can remain of me. . . . How may I avoid the disaster, the evil that I expect?"[9]

Apparently, at least in passing, he considered fleeing his responsibilities and giving up the throne. There is the story of him hiding, overcome with depression, in the cave of a priest on the hill of Chapultepec. The priest, berating him, is supposed to have said: "Mighty lord, what is this? What will they think of Mexico, of the heart of all the land? Return, Lord! to your state and your throne and give up this foolishness! Think of the dishonor you do to us all!"[10]

He sought answers from his astrologers, his magicians, his diviners, and when they could not give him any satisfactory advice, he punished them, threw them into prison, killed their families. Very soon after, when the news reached him of those white, bearded men sailing on "towers or small hills," Moctezuma caught a glimpse of his destiny. And he surrendered to it.

Cortés could be none other than Quetzalcóatl himself, the founder of the great Toltec civilization.[11] "He must have returned to claim what is his," Moctezuma is supposed to have said, "for they are only on loan to me—this throne, this place, this power."

Hernán Cortés was the spiritual antithesis of Moctezuma—the offshoot of an expanding linear faith. While one sat and waited, imprisoned within his circular cosmogony, the other charged on, driven by his upbringing and character and the will to spread the Cross, in his search for "honor and glory" and gold.[12] He was drawn like steel to the magnet of his opponent's symmetrical weakness. And yet a strange attraction developed between them in the long hours they spent together (first as emperor of the Aztecs and envoy of the kingdom of Spain, then as prisoner and warder).[13] It was as if both wished to study and understand the other. This converging encounter between Moctezuma and Cortés would in the end be more representative of the history of Mexico than the war to the death between Cuauhtémoc and Cortés.

And yet, if the Porfiristas in 1910 no longer considered Cortés the "great outlaw," there were few who could understand the dimension and range of his work as a founder of the nation. He himself had written, "[Mexico] is the cloth that I spun and wove."[14]

In his *Disertaciones sobre la historia de la Republica Mejicana*, Alamán gives a detailed argument for the importance of Cortés as a founding father of the Mexican nation. He compares the Conquest of Mexico with the conquests of Rome, which "united all known nations under the same laws, gave them the same language and universal civilization, thus facilitating the process of conversion to Christianity." Alamán's overview depicts Cortés as first of all a master builder. He ordered his men to construct Mexico City from the ruins of Tenochtitlan with the stones of what had been Indian structures ("it will be the most majestic and populous city in the inhabited world") and to lay the foundations for other cities and towns. The expeditions he sent out to the four directions on missions of conquest were all successful; his captains and soldiers conquered the other great Indian empire of the time (the Tarascan, in Michoacán), defeated twelve other nations, and marched into modern-day Guatemala and Honduras. He began to introduce the Spanish system of local government as soon as he arrived on the mainland of Mexico; he summoned the twelve Franciscan friars, the so-called Twelve Apostles, to come and convert the Indians; he made detailed and paternal regulations on the personal labor of the Indians and decreed the death penalty for any Spaniard found guilty of stealing from them (the land as a whole and the gold of Moctezuma presumably not included). And he had taken Moctezuma's advice and respected the boundaries of indigenous communities as well as many of their economic and political practices, while ordering his men to send for their wives and permanently settle in the New World.[15]

Most of all, he wanted to populate the whole of the territory, so vast and diverse, which he had christened New Spain, to convert its inhabitants to Christianity and to protect and instruct them, in order to prevent a repetition of what had happened in the Caribbean where the Spaniards "had exhausted, destroyed and then abandoned" the islands. In 1844, three years before the U.S. invasion of Mexico, Alamán would write: "What he did in our country during the three years immediately following the Conquest by far exceeds what has happened in the United States, and, given the circumstances, it is hard to believe that one man could do so much."[16]

Alamán had personal connections with the achievements of Cortés. At the time he was writing, he was administrator of the estate of the Duke of Monteleone, a descendant of the Conquistador. Alamán not only supervised these haciendas in the old (and immense) "Marquisate of the Valley," which the Crown had granted to Cortés, he was also in charge of the Hospital de la Purísima Concepción in Mexico City (better known as the Hospital de Jesús). Much of Mexican history had passed through the arcades, halls, and patios of the handsome building, without altering the mission Cortés had assigned to it in his will. What greater proof of the ultimate results of the Conquest—Alamán may have felt—than this institution dedicated to charity?

In 1910 the doors of the hospital were still open to the sick, but few remembered its founder and few read the *Disertaciones* of its former administrator. If Cortés's stature as a master builder was not well known, the side of his character that was pure Mexican went virtually unrecognized. Nobody thought of Cortés as a conqueror enchanted by the land, climate, plants, women, and skies of Mexico. In 1910, two black *zapotes*—a Mexican tree—were still growing in the orchard of the house in Spain at Castilleja de la Cuesta where, in 1547, Cortés had died a miserable death, neglected by the king and heavily in debt. "I always wanted to discover as many secrets as possible about that place," he once wrote. In his majestic palace in Cuernavaca he kept a collection of indigenous religious art and idols that many Spaniards considered blasphemous but which Cortés appreciated and cherished, "because he wished that these kinds of idols be remembered." In his will he had asked to be buried in New Spain. His remains would be returned to Mexico in the seventeenth century, but they were not laid to rest within the monastery specified in his will. Instead they were moved from church to church until finally hidden (in 1836) by Alamán in the Hospital de Jesús itself. The first meeting between Cortés and Moctezuma had happened at exactly that spot, on November 8, 1519. "In gratitude," Cortés had ordered the construction of the hospital. His remains were still there in 1910. Nobody in Mexico had discovered them or seemed to be interested in finding them.[17] (These remains were only located in 1946, when a copy of Alamán's account of his actions—sent and kept in secret for over a hundred years—was found in the archives of the Spanish Legation in Mexico City.)

Both myths about Cortés, the "white" and the "black," were relevant. Both encounters—Cortés with Moctezuma, Cortés against Cuauhtémoc—had taken their separate courses and would mark the entire future history of Mexico. Cuauhtémoc had truly been the hero of his people and Cortés truly the cruel conqueror. Moctezuma had yielded to Cortés, been deceived by Cortés, and lost his empire to him, but the two had also come together in their probing relationship—constrained by a tragic flow of events and a mutual confrontation with the unknown—yet nevertheless real and perhaps even vibrant. For all his heroism, Cuauhtémoc had been less the founder of Mexican nationality than the nineteenth-century symbol of a fatherland willing to defend itself to the death against foreign assault. Even given Moctezuma's insecurity and Cortés's arrogance, it was they who had created a new nationality at the instant they met.

Within the Conquest of Mexico, there had been in reality many conquests. Some had been terrible, like the fall of Tenochtitlan, but others had been gentler, closer to the positive meaning of the term in the world of love.

Even though they had their modern problems with the Indians, it was a consolation for some Porfirian thinkers to compare the historical fate of the Mexican Indians with that of their counterparts in South America (especially in Argentina and Uruguay, where they had been almost totally exterminated).

They could have drawn an even more illuminating example from Peru. Though the long, drawn-out conquest of Peru may have cost even more blood than Mexico, it had not had such a profound transforming effect on the indigenous population. In Peru, it had begun with Pizarro's assassination in 1532 of the emperor Atahualpa (who had put up no resistance and had already become a Christian), and after four decades of war among Spaniards and between Spaniards and Indians, it had ended with the traumatic hanging, in a public ceremony, of the last Inca emperor, Tupac Amaru. His execution was witnessed by thousands of grieving Indians. These were wounds that would never heal.

And while the colonization of Mexico remained much closer to the protective guidelines laid down under the New Laws and the humane teachings of Bartolomé de las Casas, the conquerors of Peru opposed that legislation with their swords, supporting naked domination with the ideas of Juan Matienzo, a disciple of Las Casas's great opponent, the imperialist theologian Ginés de Sepúlveda. The Indians were "naturally born or brought up to serve, and they benefit more from serving than ruling, and you should know that they are born to serve, because, as Aristotle said, to these beings Nature gave stronger bodies and less knowledge."[18] In Peru, *mestizaje* (the mixing of races) remained much less common and conversion less widespread and far more strongly resisted.

The Porfiristas did point to the example of the United States, where, after a war of expansion and attrition, the Indians were kept in virtual captivity on their reservations. Emilio Rabasa wrote: "In Mexico, the Indians form part of the nation; as it progresses, so do they. In the United States, each time the nation takes a step forward, it drives the Indians into a new form of exile. If progress in Mexico is slow, one ought to bear in mind that this country has not shrugged off its burden in order to speed up the process."[19]

There had, of course, been cases of resistance in Mexico after the conquest of Tenochtitlan and during the sixteenth century. The most striking example was the Yucatán peninsula. Unlike the Aztecs, the Mayas—divided at that time into a number of small "principalities"—had not deified the Spaniards and had fought against them for many years. The Spanish victory took the form of an alliance between the descendants of the first conquistador of Yucatán (Francisco de Montejo) and the caciques of the most powerful "principalities."[20] Many of those Indian ministates remained independent well into the seventeenth century. The conquistadors had soon realized that this vast expanse of limestone—dry and flat with patches of forest—lacked not only mines and fertile land but also water. The only thing not in short supply was indigenous labor, which led to a much more prolonged and intense exploitation of manpower than in other regions of the country. While the Spanish language spread through almost all of Mexico, in Yucatán Mayan not only survived but eventually became the mother tongue of the white minority. Like the Peruvian Indians, the Mayas surreptitiously retained many of their pagan ceremonies and beliefs, including a book of chronologies that—within its cyclical formulations—predicted the defeat and disappearance of the Spaniards. All these factors, plus the large number of Indians and small number of Spaniards, gradually created that wall of bitterness between the Indian and the white man that would lead to the Caste War of the nineteenth century.

Although Yucatán was always an extreme case, the Mayas were not the only group who opposed the Spaniards. The Aztecs of course fought ferociously to defend their city, and in 1521, when all was lost, many of the defenders left alive—men, women, and children—had thrown themselves into the canals of Tenochtitlan. They would be remembered as a classic example of resistance. The Indians in Chiapas put up a fight; so did the Zapotecs in Oaxaca. The Mixes, of the same region, were never defeated. The Spanish conquest of the other great kingdom of central Mexico—the Tarascans of Michoacán—was marked by a cruelty comparable to Peru, an incomprehensible ferocity given the friendly acceptance these Indians gave the Spaniards. (They were dragged along as beasts of burden by the conquistador Nuño Beltrán de Guzmán on his expeditions farther north, and—apparently out of sheer sadism, after frequent tortures to extract treasure—he burned their king alive).[21]

In the north-central part of the country (San Luis Potosí, Coahuila, Zacatecas, Durango, Nuevo León), it took four decades (1550–1590) to subdue the nomad Chichimecs; and some indigenous groups—like the Huicholes, the Tarahumaras, the Seris, and the Tepehuans—lived in relative isolation throughout the colonial period. Others farther north—in Sonora, New Mexico, and California—were never subdued.

But despite these examples of protracted defiance, the version of Indian response to the Conquest offered by the Liberal historians (centering around the unifying national myth of Cuauhtémoc's resistance) was far from being the whole truth. At the other extreme from the Mayans were those indigenous nations who remembered the Conquest as a kind of golden age. If Cortés had not been aided by the Cempoaltecs, the Huejotzincas, the Texcocans, and the Tlaxcalans,[22] his five hundred men would have been literally devoured by the hundreds of thousands of Aztec warriors. To reward them, the Tlaxcalans were given special status as free allies: They fought with the Spaniards in their war against the Chichimecs, they colonized the north of Mexico and founded several cities (Querétaro, Saltillo, San Luis Potosí) where they introduced customs—like their favorite foods and their multicolored sarapes—still in use during the Porfirian age. The Otomís earned a similar status. They were completely impoverished in 1910, but the heroic feats of their ancestors (in behalf of and in league with the Spaniards) were remembered and treasured. Led by the Otomí conquistador Don Pedro Martín del Toro (who is portrayed in Otomí drawings of the seventeenth century wearing a crown and Spanish clothing), they had fought in the Great Chichimec War and had founded several mining cities in the heart of the country (Guanajuato and Sombrerete).[23] Even the Aztecs recovered from the trauma of conquest and soon joined forces with the Spaniards halfway through the sixteenth century to suppress indigenous rebellions in the west of Mexico (the conflict known as the Guerra del Mixtón). In the case of each of these groups, the conquered themselves became conquerors.

Between the extremes of these "Hispanicized" Indians and those Indians who resisted the Spaniards or remained firmly rooted within their cultures lay the option of coexistence. In contrast to the bitter racial conflicts that marked the history of Peru (and parts of the Mexican south), the Indians and the descendants of the conquistadors throughout most of the country interrelated and softened the edges of racial difference. There were barriers and prejudices and great class distinctions but not true ethnic hatred. Bartolomé de las Casas and other Spanish theologians had fought to create such an atmosphere, fiercely defending the natural freedom of the Indian and arguing for equality between Indian and Spaniard. And even the one dark mark against Bartolomé de las Casas—the proof that saintly men can also be blind—was eventually resolved by the process of *mestizaje*. Las Casas—to free his Indians of the burden of slavery—argued that Africans should be imported to do the work of slaves. Africans were brought in and they were certainly slaves. They were used primarily along the coasts though also in mining work in the Bajío, but the numbers were always comparatively small and intermarriage with Indian women legally freed the children from slavery. When the institution of slavery was abolished—at Independence—there were only about ten thousand black slaves throughout Mexico. Despite their low rank in the colonial hierarchy, they too would join in the process of creating a nationality where the color of one's skin would not fundamentally determine the fate of a human being.[24]

Although serfdom and slavery did not disappear in the Spanish dominions, slaves in New Spain were never, for the most part, treated as they were in Anglo-Saxon possessions. (Yucatán and Chiapas well into the nineteenth century were notable exceptions.) The will to coexist and even to mingle gave rise gradually to a new collective protagonist in the history of Mexico: the mestizo.

Over most of the country, the Indians would try to become part of the new civilization that had conquered them. The creole Spaniards born in Mexico, looking back to a pre-Hispanic past that was not really theirs, would claim an indigenous heritage to distinguish themselves from the privileged mainland Spaniards. So the Indians turning toward the future and the creoles looking toward the past would both move down the road to that racial and cultural fusion called *mestizaje*.

In a famous passage, Justo Sierra summarized the spirit of reconciliation of 1910, within which the idea of the historical convergence of Spaniard and Indian now had a place: "As Mexicans we are the children of two nations, of two races; we were born from the Conquest; our roots are in the land that the indigenous peoples inhabited and also in the soil of Spain. This fact dominates our whole history; to it we owe our soul."[25]

3

THE MESTIZO FAMILY

In the historical parade of September 15, 1910, only one of the major figures was a woman, but she stood out prominently. She had been as critical as Moctezuma to the project of the Conquest of Mexico. Official history remembers her only negatively, the Indian interpreter who aided Cortés from the moment he landed, his concubine, the traitor to her race without whom the Spanish victory would have been impossible, the woman whom Bernal Díaz del Castillo described as "meddlesome and shameless as she was attractive," she who had plotted against Cuauhtémoc and finally persuaded Cortés to hang him from a tree—the Princess Malintzin, who would take the Christian name of Doña Marina and for whom time would reserve a disrespectful appellation that came to stand for everything that does harm to itself or its own: "La Malinche." Given the historical repudiation of her memory, it is strange that the descriptions of the parade report no hissing from the public when she passed. Perhaps there was none. Perhaps the spirit of reconciliation prevailed over prejudices. Had not Justo Sierra written that Mexican nationality was born "from the first kiss of love" between Cortés and Malintzin?[1]

If not from the first kiss, then from the one nine months before their mestizo son was born in 1522. Cortés named him Martín, after his own father, and legitimized him through a papal bull of 1529. As a reward for her services, Doña Marina received a field near Chapultepec, a portion of the orchard of Moctezuma, and a building site where she constructed the house she would live in till her death in the middle of the sixteenth century. In 1524 she was married to the conquistador Juan Jaramillo. For her dowry, Cortés gave her two villages in Veracruz and an estate in Xilotepec.[2]

If Martín Cortés was the first mestizo, he was also the first Mexican, but his condition would not become the norm for many years. In biological terms, the process of *mestizaje* took centuries. Around 1600, there were only a few thousand mestizos in a population of perhaps a few million. In 1803 the German traveler and scientist Alexander von Humboldt—whose *Political Essay on the Kingdom of New Spain* first brought the country to the broader notice of Europe—estimated the population as 41 percent Indian, 39 percent mixed (mestizos and blacks included), and 19 percent white. Not long after, sometime during the nineteenth century, the mestizos became the majority and the proportion of Indians and whites steadily declined. Martín Cortés had won the battle.[3]

The history and character of Spain were partly responsible for the success of the process of *mestizaje*. Though it had expelled the Moors and the Jews, Spain remained (especially in Andalusia and Cortés's native province of Extremadura—racially mixed areas that produced many of the conquistadors and immigrants) a vast ethnic and cultural melting pot. If a Spaniard became a prisoner of the Moors and happened to fall in love, he might very well, as classical Spanish literature recounts, live in delightful captivity. Well aware of the natural desire of many conquistadors to be captivated, the Spanish Crown explicitly advised them to marry indigenous women, especially if they were daughters of the nobility, the caciques. The colonizers, beginning with Cortés, generally preferred to emulate their ancestors in the wars against the Moors by baptizing Indian women and then taking them "as if they were wives," to be their mistresses.

Aside from matters of religion, in which it was fiercely exclusivist, Spanish culture tended toward mixing and variety. The Spaniards had far less color prejudice—and a much longer history of encounters with other races—than the primarily Protestant English settlers of North America. And the Catholic religion (at least among its most truly "catholic" exponents) allowed them to conceive of themselves and the Indians as belonging to the same natural and spiritual realm. Their aim, at least in Mexico, was not to annihilate but to assimilate.

Other, less benevolent factors contributed to the process of *mestizaje*. Despite their mass baptism, the Indians very soon learned that Christian liberty and natural equality meant little when compared to the biological disasters that would now assault them. With the conquistadors came diseases against which the Indians had no immunity. These began to run wild toward the end of the sixteenth century. Smallpox, which the Indians called *Cocolitztli*, was the most lethal. The epidemics claimed far more victims than all those who had ever been sacrificed on the thirsty altar of the god Tezcatlipoca. In 1521 there were probably ten million inhabitants in the country conquered by the Spaniards. A hundred years later there were fewer than a million.[4] In their desperation, some of the Indians, already fleeing from life into the daily drunkenness of pulque, chose to kill their children, leap from peaks to their deaths, or simply lose the will to live, letting themselves starve, "letting themselves die." Diego Garcés wrote in 1579: "When they fall ill they have the most perverse habit of refusing to take food or even look at it, and not their son or wife nor anyone else would ask them to eat, and so they die like brutes."[5]

"In the days when they were heathens," states the *Relación de la ciudad de Pátzcuaro*, in the sixteenth century, "they lived more healthily and longer and multiplied more and never suffered from the pestilence that has fallen upon them in these times, the cause of which may be attributed to the Will of God. The elders say that when they were infidels, they had less vices and did not eat or drink as they do now."[6]

In the popular language a phrase would take hold, to be used for centuries about any deep trouble: *Está del cocol* ("it's part of the smallpox"), as if the people of Mexico, somewhere deep in their collective memory, could not forget this final grim twist to the conquest of their kingdoms and the overthrow of their gods.

The huge decline in the indigenous population by the end of the sixteenth century affected the process of *mestizaje* in two ways: It reduced the vast demographic gap between Indians and Spaniards, and among many of the surviving Indians it triggered a reflex of flight from their tribal areas. Their escape toward the haciendas, the mines, the factories, into the Spanish cities and towns seemed a movement toward freedom and relative well-being. No one knew this better than the Indian women, those descendants of "La Malinche" whose desire to have children with Spaniards could be seen as an instinct for biological survival.[7] Cortés himself fathered children with two of Moctezuma's daughters (Isabel and Marina), and he always watched over them. Two generations later, a mestizo named Cortés-Moctezuma would found the wealthy mining city of Zacatecas.

Toward the end of the eighteenth century, Juan Solórzano Pereyra—a major compiler of information on the life, laws, and institutions of Spanish America—was to write, "Many Indian women leave their Indian husbands, or hate and desert the children they have born to them, considering them as subject to taxations and required personal service. And they desire, love, and cherish more those they bear from Spaniards outside of marriage."[8] A century later, Humboldt wrote, "Indian women who still have some property would rather be joined by marriage to the conquerors than be treated contemptuously as Indians."[9]

Rejected by the Indians as Spanish, by the Spanish as Indian, the offspring of those unions, the mestizo, was throughout the colonial period an insecure, bitter, and displaced person. Generally he was burdened with the sin of being born a bastard. *Mestizaje* was almost always synonymous with illegitimacy. If they were legitimate or legitimized children—like Martín Cortés—mestizos could in theory be treated like Spaniards (though in reality not even the creoles achieved the status of the Spanish-born). They could for example become priests or nuns. Yet it rarely happened. "Ordinarily," noted Solórzano Pereyra, "they are born from adultery or from other illicit and punishable unions, because there are few Spaniards of honor who marry Indian or black women, since they would be despised for such a failing."[10]

Since the real issue was illegitimacy and not the color of their skins, mestizos—even if they looked the part—tended not to accept their condition in public. As a result, the colonial censuses hardly recognized *mestizaje* and usually included mestizos under the rubric of "Castes," on rare occasions as "Spaniards." To make room for themselves in a society that denied them access to public and religious offices or even entrance into prestigious guilds, the mestizos would normally hide their status, which fed strong and repressed feelings of resentment.[11]

"Indian women are weak and crazy about black men," one viceroy complained to Philip II, without realizing that liaisons with blacks were also a form of manumission. The children would be freed from the Indian taxes and labor exactions and also from black slavery, though not from the "infamous mark" of their Negro origin, which involved still more limitations on their possibilities for honor and labor in that hierarchical society. It was natural that the *castas* (an overall term for many different color variations of black descent) would take an active part in the War of Independence. An ethnic undercurrent ran through the whole movement, and so did the demand for full equality regardless of color. But by that time nature had taken great strides toward resolving the problem. The very abundance of Caste designations, localized in a whole string of names (*sambo*, *mulato*, *saltapatrás*, *coyote,* and many others) made the distinctions innumerable and finally of no importance.

After independence, Liberal legislation officially obliterated these differences. While the color of one's skin remained, and would always remain, a major factor in Peruvian discrimination, the purely racial problem in Mexican history was largely resolved through *mestizaje*.

(The only great exception has been the Mayan-speaking zone of Yucatán and Chiapas, which would experience the War of the Castes—fueled by racial hatred between Indians and whites—in the nineteenth century.) In Peru, mestizos have come to be known by the derogatory appellation *cholo*. No such strongly charged term existed in Mexico. Nearly everyone, after all, was a mestizo.

The viceroys had a habit of officially despising the ethnic "stain" spreading throughout New Spain without ever suspecting what a miracle it was. Considering how many Indian peoples and cultures were exterminated across the rest of America, this convergence called *mestizaje*, with its admirable spiritual foundations—and despite the terrible contributing factor of millions of Indian deaths by disease—helped to save Indian cultures and the Indians themselves. It is one of Mexico's most original contributions to the social and moral history of the Western world.

"The Indians have become us," wrote Justo Sierra. By "us," he meant the "mestizo family." But the transformation across the centuries had been not unilateral but reciprocal, Indian to Spaniard and Spaniard to Indian. The decisive *mestizaje* had been neither ethnic nor biological but cultural. The conquest of Mexico can be seen as a long process in which the cultures of the conquerors and the conquered flowed together into a new synthesis. The resultant mixture was infinitely varied though marked always by a gravitation of the Indian toward the Spaniard.

The mix varied from region to region. The largest area of the country but also the least populated was the north, from the desert of Baja California to the fertile plains of the northeast and the dry cattle-herding regions of the northern plateau. In colonial times it had been home to a few indigenous nations who lived either in isolation or "reduced" into Christian missions in the rugged mountains. The north had mainly been a frontier for creole settlers: cattlemen and landowners, small ranchers, miners, merchants, whose descendants fought against the Apaches and the Comanches. It was natural that a self-sufficient, practical, and individualist culture should develop in this area. Nevertheless the language, the food and eating habits, the dances, even the fighting spirit of these people were molded by the Indians among whom they lived.

Between the two great mountain chains that descend from the north along the coasts of the Pacific Ocean and the Gulf of Mexico until they form a knot across the neck of the country, there stands the great mass of the central and western plateau. For some, this was "the basic area of Mexico." Since it had been the central stage for the material and spiritual conquest, others called it "Old Mexico." It had the greatest density of population, was the granary of the country, grew a variety of products including cotton and sugar, nourished a fishing industry and extensive herds, exploited long-standing silver mines, and produced impressive handicrafts. With no wide rivers to navigate but full of volcanoes to contemplate and fear, it had been—along with the eastern coastal strip on the Gulf of Mexico with its many crops and valuable tropical woods—the heartland of the mestizo fusion.

In the south and the southeast—from the tangled hills of Oaxaca to the jungles and high valleys of Chiapas and the limestone plateau of Yucatán, the people and culture were still primarily Indian, existing on a subsistence economy of corn, beans, squash, and the production of handicrafts. In the south there had never been much interaction between creole and Indian communities. The city of Mérida was typical. Surrounded by a resentful constellation of Mayan communities, it was known as the "white city." Its creole minority owned the enormous sisal hemp haciendas, which were still farmed—even at the beginning of the twentieth century—by labor working in conditions very close to slavery. The landowners considered themselves "a divine caste" whose children had to be educated in New Orleans or Europe, certainly not Mexico City. Their luxurious mansions full of marble made Mérida look like a small Parisian suburb—in the midst of the Mayan communities. Similar, though on a smaller scale, was San Cristóbal de las Casas in Chiapas, where a closed group of white creoles (the *coletos*), with the aid of some mestizos (the ladinos), despised and exploited the Indians, who were not even allowed to walk freely in the streets. In the province of Oaxaca the density of the Indian population was similar to Yucatán and Chiapas (more than 75 percent), but there was less resentment. The indigenous nations lived apart from one another while cultural *mestizaje* was further advanced.

A collection of mixed cultures flourished in hundreds of subregions within these three zones. There were large black enclaves on the tropical Pacific and Gulf coasts, Indian colonists from Tlaxcala in the cities of the north, isolated creole groups in the deep south, and many indigenous communities or creole cities and towns in the mestizo heartland: a true Mexican mosaic.

In 1910 most of the country continued to be what it had been since the time of the Conquest: a vast, dynamic, and largely peaceful laboratory for cultural convergence—with only occasional ethnic conflicts. In material culture and the details of daily living, Indian traditions were still strongly present. The plough had replaced the *coa* (a narrow, long-handled digging stick), donkeys did the work of the *tameme* (the Indian porter), and a shirt and pants were worn instead of the traditional loincloth. Corn, beans, squash, chile peppers, cacao, and turkey were still the basic ingredients of the Mexican diet, but the Indians, who had been primarily vegetarians, long ago had begun to eat the meat of animals imported by the Spaniards: cows, chickens, and especially pigs. In pre-Hispanic times, one type of "tamal" (made of crushed maize and wrapped in a maize husk) was filled with human flesh, but after the Conquest pork was used.[12]

The cuisine of mestizo Mexico would ultimately show a dominant Indian character in its standard dishes (the *mole* sauces, *pozole* stews, cornmeal tortillas) and even in its kitchen utensils. The destructive plague of drunkenness that had persisted throughout the colonial period and well into the Porfirian age (in 1910 there were still about a thousand pulque taverns in Mexico City) was, paradoxically, also a product of cultural convergence. In the pre-Hispanic world, it was a serious crime (except for old people) to drink at any time other than during religious rites. The freedom the Spaniards brought loosened this particular restriction. Another clear case of indigenous survival was a belief in magic, animism, and folk medicines. Creoles would visit Indian *curanderos* ("healers") and willingly swallow the herbs they prescribed to cure (or inflict) all kinds of illnesses (including love).

In 1910 a sense of communal existence, of indigenous roots, still marked life in the villages and the relationships between relatives, friends, and neighbors. In many places the usual house continued to be a pre-Hispanic one with walls of adobe and reed grass, roofs of thatch or wooden shingles. But the furnishings were enhanced with the tables and chairs of Europe. To the Indian communal baths (the *temazcales*)—complained a chronicler—"came men and women all naked, of all classes: Indians, mestizos, mulattos and Spaniards." Such baths still existed in 1910. The colorful pre-Hispanic markets (*tianguis*), renowned for their profusion of flowers and medicinal herbs, were a feature of every community throughout the country, with the difference that they now used mainly hard currency (not only cacao or bartering) and ran according to the Christian calendar (market day was usually Sunday, when people went to mass). Some of the popular dances and the use of certain hallucinogenic plants (peyote, mushrooms, cannabis) had been passed down from pre-Hispanic times. And so had the custom of lavish fiestas, on which everyone spent more than they could afford. But they now honored the village saint rather than the ancient gods.[13]

Numerous Indian words were incorporated into Spanish for the names of plants, animals, places, objects, and people. Throughout the country, towns and villages were given dual names (always the name of a saint followed by an Indian name, like San Miguel Chapultepec). Tones of voice, gestures, and idiomatic expressions all revealed the influence of the indigenous "polished, polite and curious mode of speech, with such preamble, tactfulness and rhetorical devices, untaught and natural . . . as though it had been created by a lifetime in court. . . ."[14] (A descriptive phrase based on this traditional Indian politeness would later become current across Latin America: *cortés como un mexicano*, "courteous as a Mexican.") Through the widespread institution of the Indian "nannies" ("Nana Felipa," wrote José Joaquín Fernández de Lizardi, the first Mexican novelist, was "servant, sister, friend, daughter, mother of my mother"), the Indians also left their mark on lullabies, games, stories, superstitions, spells, songs, and musical rhythms. The custom of honoring the dead by eating a meal around their graves in the cemeteries was another clear example of Indian influence. This was not, as such, a pre-Hispanic custom (the indigenous peoples, as a rule, had buried their dead under their houses and not in cemeteries). Nevertheless, the way in which people communicated with the dead, the familiarity with which the living treated death, and the ritualistic offerings they prepared for their dead contained an undeniable indigenous "flavor" that had been fully adopted by the mestizo culture.

In values and customs of an intellectual, ethical, aesthetic, political, and religious nature, the Spanish imprint had grown steadily stronger. The alphabet replaced the hieroglyphs. Spanish gradually supplanted Náhuatl and dozens of other indigenous languages. The doctrine taught on the facades of the churches filled the role of the ancient pre-Hispanic education. The ethic of renunciation preached by the missionaries was easily accepted and absorbed into the stoic Indian morality. While the Indian hands that built the churches and created the sculptures and the paintings gave them all a special quality, the themes were invariably Catholic. The Spanish laws and institutions of government, like the municipal council (*cabildo*), were adopted everywhere, even in the indigenous communities. And though the ritualism, idols, and beliefs of the Indians often continued to exist behind (or absorbed into) the altars, a fervent adoption of Christianity was perhaps the most striking element of Mexican colonial history. Some steps in this steady process happened overnight—human sacrifice ended with the Conquest—and others took time, such as the transformation of the ancient settlements (the *altepeme*) into villages (*pueblos*).

The nineteenth century accelerated the process of incorporating the Indians not only into Spanish and Catholic culture but now also into that of western Europe. Schools, newspapers, modern transportation, new manners and habits, even the Wars of the Reform and the French Intervention little by little promoted the birth of a national and even international consciousness, beyond the local patriotism of the *patrias chicas* ("little fatherlands"). The Liberal legislation that broke down ethnic barriers and legislated individual equality provoked defensive reactions in the indigenous communities and villages accustomed to the relative protection of the viceregal laws, but it also permitted many individual Indians to ascend the social, economic, and political ladder. Juárez in politics and Altamirano in literature were the two paradigms, but they were not in any way unique. In the Conservative and Liberal camps, in the Church, the military, and even in business, individuals of Indian origin were increasingly present, with no one seeing them as such or focusing on their origins. Juárez was Juárez the Señor, Juárez the Lawyer, Juárez the President, not Juárez the Indian. This almost total absence—at a national level—of sentiments of racial discrimination was an inheritance from the theologians and missionaries of the sixteenth century, the spiritual "founding fathers" of Mexico, but it acquired definitive legal status in the nineteenth century.

Culturally, Mexico went on being many Mexicos. The tesserae of the Mexican mosaic were layered together in endless variety. Most Mexicans lived in small villages, of fewer than five hundred inhabitants. (There were more than 100,000 of them in 1910.) Sometimes the differences between villages separated by only a few kilometers could be unfathomably deep. Nor were all the Indians able to transform themselves into the "us" of Justo Sierra. But in 1910 that "us" was already the mainstream of the life of the nation: a family solidified by a high degree of ethnic tolerance. Precisely because cultural *mestizaje* had been (and to a certain degree still was) the central fact of their history, the people of 1910 did not speak of themselves or their fellow men in terms of color, and rather than use the adjective *mestizo*, they would simply say, "Mexican."

4

THE SPANISH CROWN

At another of the traffic circles on the Paseo de la Reforma stood the equestrian statue of Charles IV, originally donated by the Spanish monarch to his loyal subjects in 1803. It was the only official allusion to the three centuries of Spanish rule (1521–1821) and the twelve monarchs who ruled during that period, five from the royal house of the Hapsburgs (1521–1700), six from the house of the Bourbons (1700–1821). It had once stood in the heart of the Zócalo, the Central Plaza of Mexico City. In 1823 Lucas Alamán had to save it (along with the remains of Cortés) from popular rage, when the people were about to destroy it as a symbol of colonial oppression. The statue stood in the courtyard of the Pontifical University until 1852, when it was transported to a location that would eventually become the last traffic circle before the Alameda on the Paseo de la Reforma.[1]

Porfirio Díaz would leave it there. But the major ideologists of Liberal romanticism would have done much more than shift the statue of Charles IV from one location to another: They would have melted it down. Mexico's birth as an independent nation coincided with its rejection of the colonial tradition. All the Liberal thinkers of the nineteenth century shared a common conviction: The new country should make "a fresh start." All previous history had been a nightmare during which the progress of the country had been retarded for at least three centuries.

Not a single statue existed to commemorate any of the sixty-three viceroys of New Spain. Nothing (except the city itself) was there to remind the citizens of the first viceroy, Antonio de Mendoza (1535–1550), who had built roads and schools and created the first publishing house in Latin America. His aesthetic vision had shaped the city of Mexico. It was he who had conceived the urban design of the capital, drawing his inspiration from Alberti, the great Italian Renaissance architect, for the rectangular grid of the streets and the proportions and arrangement of squares and avenues. A mid-sixteenth-century chronicler would describe the city as "very finely constructed with streets that are very long, wide and straight with the most beautiful paving."[2]

Nor was there any statue of the second viceroy, Luis de Velasco, who founded the first university in the Americas in 1551. Nor any image of the greatest theologico-political figure of the Mexican Baroque, the man who embodied both "Majesties" in one: Juan de Palafox y Mendoza (1600–1659), who was briefly viceroy, then bishop of Puebla and also archbishop of Mexico. Or any of the other Hapsburg viceroys who opened factories for the weaving of fabrics and the production of wool; constructed drainage works to fight the constant floods in Mexico City; founded more than 150 cities and hundreds of hospitals, colleges, and seminaries throughout the kingdom; built walls and fortifications to protect the ports of Campeche, Veracruz, and Acapulco from constant raids by pirates; and expanded the frontier to New Mexico and Texas. They did much more as well—bringing water to the cities, a mail service to the country, enforcing the abolition of slavery, while, at the center of the kingdom, as in the time of the Mexicas, they adorned the city of Mexico.[3]

The Bourbon viceroys, imbued with the spirit of "enlightened despotism," were no less active in construction of buildings and public works, in the encouragement of agriculture and the sciences and arts. Only the name of a small street commemorated the second count of Revillagigedo, who during his term of office (1789–1794) built many of the beautiful neoclassical buildings that, according to Humboldt, "would have graced the finest streets of Paris, Berlin and St. Petersburg"[4] and would make the capital famous as "The City of Palaces." All of those buildings were still in use in 1910, exhibited with pride to the envoys during the Centenary Fiestas.

Spain had constructed the human map of New Spain, and New Spain had passed on the heritage to Mexico. As early as 1604, one of the first poets ever to write about Mexico, Bernardo de Balbuena, in his *Grandeza Mexicana* ("The Grandeur of Mexico") described (as a broad flow of action):

> *the location of the famous city of Mexico,*
> *the origin and grandeur of its buildings,*

the horses, the streets, the trade, the courtesy,
the learning, the virtues, the range of occupations,
and its luxuries, its opportunities for pleasure,
its endless spring with all the signs of the season,
its distinguished government, religion and high rank.[5]

Three centuries later, in the age of Don Porfirio, there was no official recognition of the fact that this grandeur, still evident in the urban landscape of Mexico, was a legacy of the viceroyalty.

With all this fever for construction, the most important architectural legacy of the colony was not in stone or city planning but in the political ideas and institutions of men.

The Hapsburg Catholic monarchy (1516–1700) had considered itself the legitimate heir to the two founding pillars of Western civilization: the Roman Empire and Christianity. The Spaniards, new Romans of the sword and the Cross, had been "appointed by Providence" to civilize and convert the New World. In the mid-sixteenth century, with the conquered territories under the control of the conquistadors and the conquistadors under the control of the royal authorities, the Hapsburgs wanted to build an order that would endure. Due to the political and religious conflicts of the sixteenth century, the Spain of Philip II (the austere, absorbed, monastic monarch who ruled his vast empire from the stone fortress of El Escorial) became the stronghold of the Counter-Reformation and the arrangements of the Council of Trent, geared toward a militant reform of the Church and confrontation with its Protestant and Islamic enemies. In the political sphere, it carried the ideas of Saint Thomas Aquinas to their practical and ultimate consequences.[6]

The essential message of neo-Thomism (advocated by the philosophers Francisco de Vitoria and Francisco Suárez, among many others) was an organic conception of power. The Hapsburg state, which adopted and implanted this system, was like an architectural construction within which people were meant to see themselves as parts of an almost mystical body, a single unified whole. The metaphors vary: The structure was a giant made up of several other, smaller bodies with the king for its head; or a house with many rooms and floors rising from the inferior to the supreme; or a family with the monarch for its father. Government was conceived as all-inclusive and static: "an ordered whole in which the collective will and the prince are harmonized in the light of natural law and in the common interest."[7]

Within this political construction, the viceroy "represented and incarnated the Majesty of the King," sharing to that extent in the divine origin of royalty. The viceroy held immense power over the whole territory of New Spain, even though his actions were restricted in various ways: hierarchically (by the king and the Council of the Indies in Spain), politically (by the power of the Church, especially the archbishop), and legally (by the Audiencia, the high court of the colony). In addition to these authorities, there were inspectors or auditors (*visitadores*), periodically sent by the king to supervise and sometimes punish or even remove the viceroy. And there was a time limit on his power. He would be removed, on the average after five or six years, and then undergo a kind of audit called the *Juicio de Residencia*, the "Judgment on His Term of Office."[8]

Like the proconsuls of Rome, the viceroy had to deal with an immense range of issues. He governed and administered (as the governor of the kingdom). He judged (as chief of the high court). He was superintendent of the royal treasury, commander-in-chief of the army, and vice patron of the Church. And the Hapsburg viceroy for New Spain had another, specifically local role. He was "the father to all the people."

Because the Indians were legally protected—as enshrined in the New Laws of the Indies that Bartolomé de las Casas had fought for and won—the patriarchal features of the Spanish political system had become more pronounced in New Spain. This affected not only the Indians but all classes and ethnic groups. The viceroy had to settle disputes between institutions, families, and individuals, and even was called upon to save marriages. One viceroy would say that he simply had "too much to do and not enough time."[9]

In theory at least, all commercial business within the colony had to pass through his hands. As if the viceroyalty were part of his personal patrimony, he could lawfully favor or even conduct business deals, and he could sell favors and public positions. (The sale of public posts was a common practice in the world of the Hapsburgs and a considerable source of income for the Crown. It was one way in which officials could materially—and legally—benefit from their positions of authority.)[10] Due to their physical distance from the mother country, viceroys were instructed by the monarchs in Spain to act "as you consider it to be proper," and this in practice meant expanding their regulatory powers. They were directly responsible for many of the rules ordering social and economic life in New Spain.

The viceroy's right to grant favors (*mercedes*) was especially valued. "Favors" could involve gifts of money, the cancellation of debts, or assignments of various kinds of land and property. Viceroys could change anyone's fortunes "with the stroke of a pen." A man could be rescued from obscurity or poverty, or else he could be ruined by a decree of exile, although there was always the possibility of appeal to the Audiencia, which—throughout the colonial period—provided an effective counterbalance. As limbs are governed by the brain, all lesser authorities (governors, mayors, justices-in-chief) were appointed by the viceroy. The intellectuals (the *letrados*) were for the most part docile court employees, while the universities were nurseries for the civil service and the ecclesiastic bureaucracy. Political opposition, dissidence, or a free press were not only impossible, they were unthinkable. Outside this huge construction there was nothing but emptiness, a space only the mind of a heretic could even imagine.[11]

New Spain, one of the most complex and diversified societies that has ever existed, organized its ethnic policy in a similar, hierarchical order. Each group was treated differently. Mainland Spaniards and creoles enjoyed full rights, but in practice the *criollos* had no access to high posts in politics or in the Church. Indians, who were subject to a vast body of custodial legislation, received protective treatment from the viceroy and had access to a special tribunal, the Juzgado de Indios. As for the other racial groups of "dishonorable origins" (mestizos and Castes), they had no legal prospects of social mobility and could not enter the religious orders. But these lowest groups were not completely excluded from advancement because the society, ethnically, was somewhat more porous than it appeared and there were a limited number of opportunities available to them.[12] The basic racial equality they would gain through *mestizaje* was still in the process of development.

The social architecture followed a similar pattern. New Spain was a stratified society divided into separate units known as corporations (*corporaciones*). They were formed on the basis of common connections, whether by blood or office or activity. They ranged from such important forces as the Church, the army, or the university to merchant and craft guilds, members of town councils, exploiters of some particular resource like the mines, all the way down to people with—for instance—a devotion to the same patron saint. These bodies enjoyed specific and special privileges. The corporations had their own law codes that guaranteed their particular legal rights, and they could be judged in their own courts. They were granted commercial or other monopolies and the sole right to wear the costumes or insignia of their group.

Corporativism was the normal form of life for everyone in this society. At the height of the system, most of the population (except for the Indians, black slaves, and Castes) would belong to some corporation. Each group had its own special place in the architecture. It was not a horizontal society of equal units but a markedly unequal hierarchy. Yet there was one—and very central—equality among them. The corporations did not relate freely with one another. Authority—the Crown in the person of the viceroy—had granted them their privileges and their very existence. The most essential connection was not with other corporations but with the viceroy at the apex of the system.[13]

Clearly the corporate structure limited individual freedom, but the remarkable thing was that these restrictions were imposed without repression. For those inhabitants of New Spain who accepted its legitimacy—and nearly all of them did across the three centuries of its existence—it appeared to be a natural *shared order*. In 1700, at the end of the Hapsburg period, Spain did not have to worry about the internal stability of its sprawling colony, which had a population of three million, a bureaucracy of very moderate dimensions, and an army so small that it was no more than a palace guard. Disturbances or revolts were surprisingly few.

The key to the success of the system lay in a combination of authority and flexibility. Checks and balances were achieved not through a division of powers but through competition and balance between the political bodies of the architectural construction. The discretionary powers of the viceroys were wide-ranging, but the limits remained deliberately undefined and were always subordinate to the head: the supreme authority of the king. The occasional popular rebellions were aimed at officials directly responsible for a grievance, never at the sovereign.[14]

The symmetrical main square of the capital and of every substantial Mexican town symbolized this organization of Hapsburg Mexico: the religious power (the cathedral), the temporal power of the kingdom (the palace of the government), and the city hall (the seat of corporate civil society). It was a political order made not for change and development but for stability and endurance.

In 1700 the dynasty of the Bourbons moved into this edifice of Hapsburg colonial society. They would be the owners, living at the apex, until 1821. The age of the Enlightenment drew them toward an activist "scientific" approach to government—most especially after Charles III, the classic "enlightened despot," came to the throne in 1759. They had no intention of destroying the architectural structure, but they felt a need for improvements. In every possible way, they wanted to encourage the renewal of the Spanish Empire, to rekindle the glories of the sixteenth century. Their "revolution . . . in government" primarily involved an opening on the economic front—the encouragement of free trade—and a tightening of political measures: rigid centralization and a modern bureaucracy, for both the homeland and the colonies.[15]

The medieval structure of the economy of New Spain was gradually loosened. The privileges of merchant and craft guilds were abolished; trading monopolies were broken; manufacturing was encouraged; new fiscal and banking measures were introduced. But always the objective was to increase the revenue of the Bourbon state, from "those territories that we want to squeeze,"[16] as a viceroy of Peru crudely remarked.

In economic terms the new scheme proved successful. New Spain was already the source of more than 20 percent of the empire's global revenue; it was where more than 60 percent of the empire's silver was mined. By the end of the eighteenth century the economy was flourishing, thanks to the reforms implemented by José de Gálvez, the powerful *visitador general* ("inspector general") of New Spain from 1765 to 1772. Gálvez also embodied—on the political front—the new rigor in government and increasing emphasis on military force. He greatly expanded the size of the army and the defenses of New Spain, created the Royal Gunpowder Factory, ferociously repressed indigenous rebellions, and signed dozens of death sentences.[17]

The income from New Spain for the Bourbons' European wars swelled from 5.5 million pesos in 1763 to 10 million in 1798 and 21 million in 1820. The architectural structure of traditional authority was still in place, but the Bourbons had—with apparently excellent results for the Crown—loosened the economic order and (with strongly economic motives) begun to restrict the power of the Church. In 1810 New Spain seemed to be one of the richest countries in the world. But it was also—now seen against the background of the recent American and French Revolutions—even more authoritarian than under the Hapsburgs. When the Bourbons (in 1767) had taken their most severe action against the interests of the clergy and the religious loyalties of the Mexican people by expelling the Jesuits (with later, more severe repercussions never imagined by the Crown), there had been riots and brief uprisings in protest. The Bourbons responded—in part—by posting a proclamation (and assertion) on the walls of Mexico City:

> THE SUBJECTS OF THE GREAT MONARCH WHO OCCUPIES THE THRONE OF SPAIN SHOULD LEARN ONCE AND FOR ALL THAT THEY WERE BORN TO OBEY AND REMAIN SILENT AND NOT TO THINK OR GIVE THEIR OPINIONS ABOUT THE HIGH MATTERS OF GOVERNMENT.[18]

And Alexander von Humboldt—who in so many ways foresaw the future of Mexico—would comment not only on the riches but on the vast inequalities of Bourbon New Spain during his visit of 1803. "Mexico is the country of inequality," he would write and propose a series of reforms designed to further liberalize the economy and especially to improve the condition of the poorest and reduce the great gaps between the classes. The Bourbons ignored him. And in 1910—during the Centenary Fiestas celebrating the imperial presidency of Porfirio Díaz—the memory of Alexander von Humboldt would be honored again not with actions but this time in stone, one more statue, a donation by the German embassy.

5

THE MOTHER CHURCH

At the Centenary Fiestas, Porfirio Díaz, like an old lover, poured out his secret passion for "our great mother" Spain, but he also sometimes said that "bullrings and pawnshops" were the only living heritage of the three centuries of Spanish rule.[1] The comment was a remnant of his Jacobin upbringing, which, though moderate, was no less persistent. Spain had obviously given Mexico much more than "bullrings and pawnshops," much more than the Spanish language spoken by 87 percent of its fifteen million inhabitants. It had also left them a culture closely linked with the Catholicism practiced by almost the entire population.

It was immediately visible in the country's architecture. In 1910, just as in 1810, 1710, or 1610, you only had to glance at the surrounding landscape to observe a presence as typical as the volcanoes: the towers of the churches. The view of the Valley of Mexico from Chapultepec Castle—Porfirio's favorite place—was a maze of cupolas overshadowed by the majestic twin towers of the cathedral.

These churches were silent witnesses to the different periods of Mexican colonial history. The fortress-style monasteries (ramparted memories of the war of the *Reconquista* against the Moors) were an expression of the sixteenth century—the age of the spiritual conquest—as were a multitude of churches in Renaissance style, often provided with outdoor "open chapels" to shelter the devotions of the huge indigenous population. The century of the Enlightenment, which had come late to Mexico and affected only the topmost level of society, was

Our Lady of Guadalupe
(Private Collection)

represented by the neoclassical churches, their pure, cold forms which the people had transformed through the festive colors of their religious devotion.

All the arts—painting, sculpture, poetry, theater, music, and even Mexican cuisine—had flourished in the shade of the Church, above all during the long period of the Baroque (1600–1750). From this era come the four great individual jewels of the Mexican Baroque: the churches of the Sagrario Metropolitano of Mexico City, Santa Clara in Querétaro, San Francisco Javier in Tepotzotlán, and Santa Prisca in Taxco. Sculptures of saints in tragic poses, effigies of the suffering Christ, mournful paintings in chiaroscuro depicting scenes from the Gospels, hermetic symbologies in stone—all were visual messages designed to induce the faithful to live by a strict ethical code that stressed many forms of abstinence and only a few positive virtues. In the cloisters of monasteries, in the church squares, before the stone facades and gilded altar screens carved and painted by the hands of Indians and displaying their lessons of sacred history, Mexican Catholicism had established itself as the primary force for the integration and self-identity of the Mexican people.[2]

Consumed by their "fever for the future," the Porfiristas in power were contemptuous of the living presence of the Catholic past. As professed or sincere nonbelievers—atheists, agnostics, Freemasons, or believers only in the Supreme Architect of the world or the goddess Reason—they saw Catholicism as an inconvenient vestige of the past, an uncivilized symbol of fanaticism in a country that now prided itself on being secular and modern. But popular faith and religiosity were as old as the Mexican people. Much older, in its rituals and attitudes, than even the religion brought by the Spaniards.

The prophecies of Bartolomé de las Casas had been realized. Many of the features of the Indian religions had played a key role in the adoption of Catholicism. Certain Indian ideas—gods sacrificing themselves for men, the need to do penance, the stoic acceptance of suffering, the evanescence and futility of earthly pleasures—resembled those of the new faith, which the Indians embraced with enthusiasm. In their mass conversions, the friars wanted to see the theological reversal of human sacrifices, the convincing light of Christian truth, the image of a new Israel in the Promised Land. But conversion had been more a transference of religious rites than of dogmas and profound arguments. The widespread and rapid conversion that so amazed the friars owed as much to the receptive matrix of indigenous religion as it did to the devoted commitment of the *padrecitos*.

The Indians were *religiosísimos* ("extremely religious"), as the friars discovered, but the fathers did not fully understand the *inclusive* nature of the old religions. For the Indians, new gods presented no problem. They could be worshipped right alongside the older deities. In addition to the great gods concerned with the creation and operation of the world and the nature deities (of rain, fire, water) within the indigenous pantheon, there were gods of every kind: regional, local, and household gods. The calendar was carefully designed so that a festival could be held in honor of each one. To both Catholic and indigenous religious feeling, images were central. The appeal of sight and sound spoke much more directly to the Indians than religion communicated through preaching or the written word. ("They come from the remotest areas to listen to music. This is how we make converts, not through sermons," wrote Bishop Juan de Zumárraga).[3] The need to worship all these gods covered the land with temples. For the Indians, the temple was not only a sacred site but the highest symbol of kingdom and community. When the temples disappeared, each settlement demanded its own Christian church with its own local god: a patron saint.

Jesus Christ was merely another, though prominent, saint. A symbol of social and political unity, the saint was the father of the Indians and the guardian of the land. Every member of the community participated in the fiesta held in his honor. Besides the parish saints, there were household saints, which the Indians would tend and "serve" with candles, incense, and flowers. "Our gods have died and we die with them," the Aztec nobility had sung during the Conquest. In the sixteenth century, most Indians had witnessed, suffered, or at least remembered the death of those gods, but they had not died with them. Being "extremely religious," they adopted the saints, the new gods, the workers of miracles, and once again decorated the landscape of the country with their temples.[4]

Centuries later, in 1910, Mexican Catholicism had lost none of these rituals or any of its devotion. In most towns and villages throughout Mexico, but particularly in the heartland of its pre-Hispanic and colonial past—the central and southern regions—people still lived in the light and shade of the religious ideas, symbols, buildings, liturgies, and holy days of the seventeenth and sometimes of the sixteenth century. The *Santoral* or Calendar of Saints' Days continued to register the days of worship—Holy Week, Christmas, All Saints' Day (November 1), All Souls' Day or the Day of the Dead (November 2). Holy days celebrating figures born in the New World were added to the universal Christian calendar (like San Felipe de Jesús—a Mexican Jesuit martyred in Japan—or Santa Rosa de Lima). An old saying was still current: "To every little chapel, there comes its little fiesta." Each city, town, and parish noisily celebrated the day of its patron saint with bands of music, confetti, fireworks, displays of fruit and flowers, fairs and processions.

The feast day of the Virgin of Guadalupe—December 12—was perhaps the greatest occasion on the Church calendar. Year after year, an enormous assemblage of her faithful followers would pay their homage on that date to the Virgin, staying up all night in the courtyard before her shrine. In 1895, by decree of Pope Leo XIII, the Virgin of Guadalupe was officially crowned "Queen of Mexico." In December 1910, one hundred thousand of her devotees managed to enter the basilica to worship her, though many who had made the journey were left outside.[5]

Even the most casual observer of the social scene in Mexico City would have noticed, in 1910, the long processions of Indian men, women, and children, with their brightly colored blankets and bare or sandaled feet, moving continually from all parts of the valley and down from the mountain passes to kneel before the altar of the sacred image of the "Dark Virgin," "Our Mother of Guadalupe," which they believed had appeared on the cloak of the pious Indian Juan Diego in 1531. Mexicans might argue about the "Fathers of the Patria" but there was no discussion as to who was the Mother of every Mexican, whether Liberal or Conservative. At the height of the War of the French Intervention, Guillermo Prieto wrote a sonnet to the Virgin, asking her support against the bishops and in favor of the Liberal cause:

Pity your children, loving Mother! . . .
Raise up, Oh Mother of God! raise up triumphant
the cause of free men, that we so love![6]

The Virgin of Guadalupe has been Mexico's supreme expression of popular religiosity. Like the Virgin of Loreto and the Virgin of Los Remedios, the white Virgin of Guadalupe had arrived from Spain (specifically from Extremadura, the native province of Cortés and many other conquistadors) in the sixteenth century. The Spaniards had worshipped her in a small hermitage in the Tepeyac hills north of Mexico City. For reasons shrouded in myth and legend, but most probably because she is supposed to have appeared to the Indian Juan Diego at a site where an indigenous goddess had been revered—the missionaries recorded the name of the goddess as Tonantzin (Our Mother)—the Spanish cult image was replaced by that of a dark-skinned Virgin. According to skeptics, the new image had been painted in the middle of the sixteenth century and "placed" in the hermitage by order of the second bishop of New Spain, Alonso de Montúfar, in an effort to strengthen the secular church against the competition of the mendicant orders. And it seems relevant that one of the most illustrious Franciscans, Sahagún himself, criticized the cult of the Virgin of Guadalupe as a remnant of idolatry. But for the multitude of her devotees, historical truth had little importance. The Virgin had *appeared*, like the burning bush to Moses, on the cloak of Juan Diego.[7]

As early as the sixteenth century, the Indians brought her offerings and gifts of food, but it was not until midway through the seventeenth century that various creole priests (the most important was Miguel Sánchez, who wrote a tract about her) began to popularize the Virgin of Guadalupe. Through the Virgin, they were attempting to create a potent symbol that could anchor the hold of Christianity on the people of Mexico. The story of her appearance on the coarse cloak of the Indian stems from this time. The titles of the first publications about the Virgin speak for themselves: *The Happiness of Mexico* or *The Miraculous Origin of the Sanctuary of the Virgin Mary of Guadalupe* or *The Northern Star on the Height of Mount Tepeyac, the Shore of the Tezcucan Sea . . . to Light the Faith of Indians; to Set Spaniards on the Path of Virtue, to Calm the Tempestuous Flooding of the Lake.* Around 1660 a broad new avenue was built from the center of Mexico City to the Sanctuary at Tepeyac. It was flanked by chapels representing the mysteries of the rosary. In 1695 work began on a large and splendid new church, for which a spacious square was designed to accommodate the thousands of Indian pilgrims who arrived daily from every part of the country to dance and sing hymns for the Virgin in their mother tongues.[8]

During the outbreak of a plague in 1737, the Virgin was proclaimed patron of Mexico City; ten years later she would be recognized as the universal patron of the whole country. Long before that, she had become the symbol of the mestizo union between creoles and Indians, a representation of the process that formed the Mexican nation. In her name, her iconography, her formal religious significance, she was a Spanish Virgin. But just as the ancient cultures had escaped from the boundaries of their indigenous origins and, modified by Spanish influence, found their survival within a mestizo identity, so the color of the Indian found a refuge in the dark skin and soft smile of that image, creating a Virgin for themselves, a *Mexican* Virgin. From the seventeenth century on, the Virgin of Guadalupe would become a central figure in all the swings of Mexican history.

In his *Political History*, Justo Sierra wrote that Mexico had been conquered by "the Spanish sword," then controlled in turn by "the cross on that sword."[9] Yet, to the surprise of his generation, the quarrel between those "Two Majesties" was not over. The Porfirian "policy of reconciliation" was only a brief parenthesis of peace in a very long and unresolved history rooted, like so many other things Mexican, in the century of the Conquest.

After the Council of Trent in 1554, the New World became the setting for a vast experiment in religious organization based on the consolidation of the Church's apostolic authority. The Church stood at the very heart of the life of New Spain. In the first place of course, it offered the holy sacraments. It administered, celebrated, and validated the ceremonies that bind and unbind human beings before the eyes of God—births, communions, marriages, and deaths. Throughout the three centuries of Spanish rule, the Church would also continue its mission of conversion and it would monopolize education within the universities and colleges (founded from the sixteenth century on) and in the seminaries created by order of the Council of Trent. Its hospitals and charitable activities offered care and assistance to widows and orphans, to the sick and hungry and poor or to the victims of the frequent plagues, famines, and earthquakes. The complex bureaucracy of the Church gave it direct control over Spanish and Indian confraternities, church-related construction and pious activities, funds from ecclesiastical foundations and endowments and over many other social practices and institutions related to the health of the faithful and the salvation of their souls. And the Church would summon its flock to celebrate the religious festivals that were so important a feature of their faith and their pleasure. Offering sacraments, education, charity, employment, recreation, and overseeing public morality as well, the Church was everywhere. It was an all-encompassing power.[10]

In order to attend to its countless duties toward the *other* world, the Church kept its feet firmly on the ground in *this* world. The varied activities of each diocese were carefully organized. A body of professional bureaucrats, superintending the management and investment of tithes—the major source of income for the secular clergy—served the cathedrals. The mendicant orders, especially the Augustinians and Jesuits (the Franciscans respected their vow of poverty), became the primary landowners and large-scale hacendados of the colony until the middle of the eighteenth century. One of the typical characters of the Mexican countryside was the priest-farmer who owned sugar mills and cattle.[11]

In addition to its own economic privileges, the Church was society's principal banker and—through the *fueros* ("exclusions" or "special privileges")—enjoyed total immunity from the royal courts, maintaining its own special ecclesiastical tribunals. It was hardly surprising that so many young men in New Spain wanted to become priests. By the 1650s there were 4,000 secular priests and 3,000 priests in the mendicant and preaching orders, serving a population of 1,625,000 people. Far too many, even in the opinion of one very pious viceroy.[12]

Justo Sierra felt that the primary economic problem of New Spain had been the fact that its wealth, especially the land, had been frozen in the hands of the Church. "Without the circulation of wealth," wrote Sierra, "social growth is stunted." It was true—he said—that the Church aided the poor, yet by doing so it had encouraged begging. Certainly it had educated the young, both mainland Spaniards and creoles, but in empty scholasticism ("there was no philosophy, no contact with those ideas that illuminated the intellectual heavens of the century of Descartes, Newton and Leibnitz"). It offered spiritual comfort, but its doctrine was "pure superstition in the paralyzed mind of the Indian" and "devout practice with no substance whatsoever" among the other classes. The literature of the Baroque period was "profuse and diffuse," while its greatest writer, Sor Juana de la Cruz, was "clever and sentimental." Nor did his condemnation exclude the treasures of Baroque architecture and sculpture, so important a feature of many Mexican cities and towns. He instinctively rejected that frugal, ascetic, stoic, fanatical, and indolent Baroque world inhabited by verbose sermons and saints in tragic poses: "In the evening shadows the monasteries gave the city a sinister, medieval appearance."[13]

And yet, in many of its spheres of operation, the actions of the Church were in tune with the fundamental values of Christianity taught by the missionaries. This was true above all of its charity. The official philanthropy of the government in 1910 was certainly no greater. But it was in other areas, like education, where the influence of the Church was clearly harmful. It was responsible for the cloistered, scholastic, rhetorical, and, above all, the intolerant strain in Mexican thought, evident in 1910 not only among the few Catholic or Conservative thinkers, but also among the Porfiristas themselves, presumptuous about their Liberal lineage and professing the positivism of Auguste Comte as if it were a new and infallible faith. Colonial education throughout the Baroque period had been governed by the principle that students should be taught not how to think and observe but only to believe and learn by rote. (Fray Francisco Naranjo, a respected scholar, memorized the whole of Aquinas's *Summa Theologica*.) Degrees in higher education—bachelor, licentiate, master, and doctor—were conferred by the old university only on students of the arts, law, canon law, theology, and (beginning in 1640) medicine. Until well into the eighteenth century in New Spain, Aristotle was seen as the sum total of all knowledge. Anyone who deviated from the canon ran the risk of being denounced, investigated, and sometimes punished by the tribunal of the Inquisition, which could put on trial those who owned, were suspected of owning, or even made a favorable reference to a banned book. Except for a few scholars such as Carlos de Sigüenza y Góngora, whose studies wavered between astronomy and astrology, Baroque culture—immersed in theology or at most in the latest trends of neo-Platonism—expressed little interest in the scientific study of reality or technical innovations. Newton's insights took a hundred years to reach New Spain.[14]

In his evaluation of the art and culture of the colony, Justo Sierra was certainly wrong. Across three centuries, there had been, in poetry and theater, for instance, moments and creators of the very first order. The most outstanding literary figure was Sor Juana Inés de la Cruz, perhaps the greatest poet of Mexico. Born from an illegitimate union in 1651, she had shown, from childhood, an amazing aptitude for learning. Like many other women of the time, she entered a convent, not primarily to follow a religious vocation but seeking a protective and, in her case, relatively free environment in which to pursue her studies. Sor Juana, a natural scientist in a country closed to science, had a library of four thousand books and possessed an exceptional knowledge of languages, philosophy, theology, astronomy, painting, and music. She published poems, allegorical and religious plays, comedies, meditations, letters, and many other texts for which she was applauded not only in New Spain—where she was worshipped—but in Spain itself, where she came to be known as the *Décima Musa*, the "Tenth Muse." Her consuming passion was to satisfy her "tremendous" love for scientific, objective truth. Sor Juana once wrote that "angels were only more than man because they understand more."[15] The course of her unique achievements was eventually cut short by the ecclesiastical authorities, whose envy of her intellect and hatred of her as a woman—and a very beautiful one—moved them to take repressive actions. Her confessor constantly pressured her to give up writing and to flog herself in penitence. Although she wrote her defense in a famous autobiographical letter—an ode to learning sung against the descending darkness—she eventually surrendered all her books, stopped her writing, and began to mortify her flesh and expiate her "sins." She was no longer Sor Juana but "Juana, the worst of them all." During the last years of her life, when Mexico City was struck by an Indian rebellion, floods, and epidemics, she spent her days tending the sick and the wounded. Her death would come in 1695—and it would be the death of a "handmaiden of the Lord"—from plague contracted while caring for the sisters of her order.[16]

* * *

The Church faced few problems like Sor Juana. Its all-embracing structure made room for everyone without distinction and without the need for threats. In New Spain, there was no locus, no apparent reason for dissidence. Despite the notorious *autos-da-fé* against crypto-Jews ("judaizers") in the middle of the seventeenth century, not even the work of the Inquisition in New Spain came close to the cruelty of its namesake in the motherland.[17] And yet, during the colonial period, the Church frequently became involved in controversies concerning matters of this world. The problems faced by the entire society were reproduced in a much more articulate and pronounced fashion *within* the Church, presumably because it played a prominent role in the collective consciousness and also because the clergy, on the whole, had the best education.

There were endless battles between creoles and Spaniards, within the orders, and within the secular clergy. Some of them were settled peacefully by agreements to alternate in office, but in other cases the "ethnic" confrontation led to ongoing tension. The most serious controversy of all, however, was not ethnic but political: between the orders—particularly the Jesuits, the richest order—and the secular clergy. It was a many-sided juridical dispute involving questions of jurisdiction (who should run the parishes?), hierarchy (who should assign priests?), and, more than anything else, economic power (who should collect and claim the tithes?).[18]

The orders had come to be a horizontal constellation of utopian enterprises—an autonomous, free, and decentralized universe with friars and Indians working to build a Christian community. The secular clergy was grouped into a rigid vertical and centralized structure, a spiritual bureaucracy acting on behalf of the Spanish Crown. Though the pendulum of power finally swung toward the secular clergy in the mid-eighteenth century, the controversies between the two branches of the Church would continue until the very end of the colonial period.

In the New Spain of the Hapsburgs, conflicts between Church and Crown were few and far between and seldom very serious, but when one did present itself, the Church invariably won. It was inevitable. To the people, their bishop represented the Eternal King, and he was much more real to them than the Spanish monarch. (Not once during the three centuries of the colony did a Spanish monarch brave the Atlantic to visit his dominions overseas.) When a grain shortage caused a popular uprising in Mexico City in 1624, it was the sight of the crucifix in the archbishop's hand that calmed the multitude, just as they were about to storm the viceregal palace with the cry: "Long live the king!

Death to bad government!" Fifty years later, the Marquis of Mancera, another viceroy, pointed out to his successor that there were a "large number of cases" in which the ecclesiastical authorities had infringed upon the civil authorities' power by interfering in "legal, governmental and military matters." In an attempt to avoid these "perilous reefs," he concluded, "on specific occasions I have had to yield."[19] And he had no alternative. Ever since the spiritual conquest, the priests of Mexico held much greater and more direct power than any that might be imposed by Spain or for that matter Rome. They could count on the zealous loyalty and devotion of the people. No figures in society were more venerated than the "honored priests." The Jesuit Francisco Javier Clavijero noted that in Náhuatl the Indians had an endlessly long word with which they addressed them: *Notlazomahuizteopixcatatzin*, "my valued, honored father and venerated priest."[20]

To a large extent, people from all walks of life conducted their lives according to the most stringent demands of the Spanish Counter-Reformation. Women who had sinned or seen visions often repented by entering convents (and went on having visions), and men entered the priesthood for secure material—and possibly eternal—lives. Everyone's day was governed by the peal of the church bell. The people would attend to the announcements of the Holy Inquisition proclaimed by criers in the streets or written out in the edicts attached to the walls; they would listen to the priests' long cautionary sermons; they would—or at least they would try to—comply with the self-denial, austerity, and abstinence from worldly comforts demanded by the voice in the confessional. If they were wealthy, they would compete with one another to see who could build the most lavish church or bequeath the richest hacienda to the Society of Jesus. If they were poor, they could find solace in the splendor of the rites, learn sacred history from the church facades (whose images they reproduced on the modest, illuminated altars in their own homes), and participate—with great enthusiasm, devotion, and a spirit of penitence—in the processions honoring the saints. Whether rich or poor, Spanish, creole, mestizo, Caste, or Indian, they freely submitted to the authority of the Church, which filled their lives with constant religious practices. And pious creoles of the Baroque period forged, with the help of mestizo artists and Indian labor, an artistic landscape of the sacred, formed of richly organized exteriors and constantly surprising interior grottoes of whirling gold, peopled by saints and angels of both Spanish and Indian inspiration—so sensuous and luxurious that it seemed to have been inspired by visions of paradise. Perhaps never before had religion in Mexico been so vital, so varied, and so widely shared as it was during the 150 years of the *barroco*, when Mexico—even under the early Bourbons—was still dominated by the Cross.[21]

"The political structure and the religion of a State have too much influence on the soul of a nation,"[22] Clavijero had written at the end of the eighteenth century. It was a perfect epigraph for the kingdom of New Spain, but during its Hapsburg phase, it was religion that had greater weight than political structure.

The marriage between the "Two Majesties," with a tilt toward the theocratic pole, lasted until the middle of the eighteenth century. As a "Guadalupana" identity (of devotion to the Virgin of Guadalupe) was being forged throughout the country, each diocese seemed to have its miracles or saints or tireless laborers for the glory of God: monks supposed to have walked across rivers on the backs of crocodiles; nuns famous for their sanctity; lay sisters possessed by divine or demonic visions; priests traveling from village to village to preach in the streets; missionaries initiating a second spiritual conquest and preaching the faith to nomadic tribes in the new northern and eastern provinces (California, Texas, Chihuahua, New Mexico); Franciscan friars founding new apostolic colleges; historians recording the achievements of their religious orders; clerics planning new institutions for retreat and devotion. While the Western world was in the midst of the Enlightenment, Mexico was characterized by a flourishing Catholicism, closed off into itself.[23]

Then, suddenly, the "Century of the Enlightenment" came to New Spain. Scholasticism and pious oratory began to fall out of fashion; scientific and technical studies were encouraged; newspapers, magazines, and novels began to appear. Even the secular clergy—by then the majority—picked up axes to destroy Baroque altars and raise, in their place, sober and harmonious neoclassical structures, which, contrary to their expectations, were harshly resisted by the devout. For the first time in the history of New Spain, the religious feeling of the people clashed with modern taste. The Mexico of the mestizos, Indians, and Castes remained loyal to the intensity of the Baroque, loyal to itself.

And the pendulum of power began to shift toward royalism. The first sign was a decree issued by the Bourbon Crown: The parishes still administered by the orders were to be handed over to the secular clergy. The process of centralization that had threatened the semiautonomous life of the friars and the Indians ever since the sixteenth century now took a decisive leap forward. The orders responded with more vehemence than success. Some, like the Dominicans in Oaxaca or the Franciscans in Yucatán, gave sound reasons (such as the influence they had always exerted on the large indigenous population) and succeeded in holding on to part of their parishes. Wealthier orders, like the Augustinians of Michoacán[24] who owned enormous haciendas, engaged in a long and difficult struggle with successive bishops until an agreement was reached to retain some of their landed property in exchange for surrendering schools and parishes they had founded in the sixteenth century.[25]

The secular clergy firmly believed that these measures had strengthened them. But they were mistaken. By collaborating with the royal authorities to disempower the orders who had founded Mexican Catholicism, they were ultimately contributing to their own disempowerment.

The Crown had in fact made use of the secular clergy for its own specific purposes. After the disastrous Seven Years' War (1756–1763), Charles III, the new Bourbon king, realized that any attempt to reform the economy of the state would necessarily affect the Church. Several court authors, José de Campomanes in particular, denounced the Pope's temporal authority and the right of the Church to take any kind of legal or coercive action in civil life, which they argued was exclusively the responsibility of the State. But the monarchy—on the other hand—had the right to place a limit on the number of priests and control the wealth of the Church simply by exercising its power of patronage (*patronato*). It could, for example, expropriate church lands, a process known as disentailment (*desamortización*), or suppress its privileged exemptions (*fueros*) and its special tribunals. Some steps were taken in this direction both in Spain and in the American dominions, but none was so radical—in its scope and effects—as the expulsion (in 1767) of the wealthy and influential order of the Jesuits from all territories ruled by the Spanish Crown.

While five hundred Jesuit priests were traveling across the country to set sail from Veracruz toward their exile in the Papal States, Indian rebels in Michoacán, Guanajuato, and San Luis Potosí opposed the expulsions with fire and sword. The Crown responded ferociously. By command of the *visitador* Gálvez, houses were sacked and burned, Indians were hanged, and the heads of rebels were displayed in the city plazas. The Crown expropriated the vast properties formerly owned by the Society of Jesus and leased them to private entrepreneurs. As a next move—unexpected by the secular clergy and based on new legislation issued in 1786—civil authorities then took over the entire process of collecting and administering the distribution of tithes. The secular clergy, who had wanted to displace the mendicant orders and reign supreme, discovered that they too were being displaced by the Crown, the overwhelming power during the Bourbon era.

* * *

The expulsion of the Jesuits was, at least in one respect, a costly mistake for the Bourbons. It generated a new spirit of Mexican patriotism that fed on the long-standing resentments of the creoles. The exiled Jesuits had been more than farmer-priests. In New Spain, and particularly in Valladolid, the capital of the diocese of Michoacán, the generation of Jesuit humanists born between 1720 and 1730 had used its power and wealth to teach and practice empirical sciences and critical thought. They were the first opponents of the scholastic tradition in New Spain. Father Rafael Campoy proposed "looking for the truth in everything, making a detailed investigation of every single thing, deciphering enigmas, distinguishing the known from the uncertain, rejecting the inveterate prejudices of man and passing from an old to a new form of knowledge."[26]

The creole Jesuits in exile (Francisco Javier Clavijero, Francisco Xavier Alegre, Diego José Abad, José Luis Maneiro, and others) made a fundamental contribution to Mexican thought and culture. They helped to *rediscover* Mexico.[27] Banished from their own land to live in Europe, offended by the current European stereotypes of the "inferior," "degenerate," and "monstrous" nature of the geography and people of the Americas, the Mexican Jesuits compiled the first inventories of the riches of *their* nation. Like Sahagún and Las Casas, they studied the ancient history, archaeology, language, geography, and ethnography of New Spain. Digging into the past, they wrote memorable works (like Clavijero's *Historia antigua de México*) in which they identified with the indigenous races through the bond of their common homeland. Looking into the future, they exuded optimism, based on the natural wealth of their country ("the best of all those circled by the sun"). It was the first time that the creoles, as a small but collective voice, had taken a firm historical stand. Their optimism initiated a new kind of nationalism (fierce, distrustful of foreigners, linked to the land and thus to the Indians), which would dominate the Mexican mind for centuries.[28]

Though these Jesuits would die in exile, their scientific spirit and their urge toward self-knowledge lived on. The physical defeat of the Jesuits was transformed into an intellectual victory. The seed of nationalism bore its first political fruit toward the end of the eighteenth century, when Fray Servando Teresa de Mier produced his theological conjectures geared toward claiming a creole right—based on their native birth—to the political and economic power then monopolized by fifteen thousand Spaniards in a country of six million. Clavijero had compared the Aztec past with the Greek and Roman world. In his famous sermon delivered in the Colegiata of Guadalupe (for which he was expelled from the country), Fray Servando went even further. Considering himself a direct heir of the mendicant friars and inspired particularly by Fray Bartolomé de las Casas, he built a mythological bridge linking the creoles to the Aztecs and denying the divine rights of the Spanish Crown after the Conquest. Fray Servando asserted that the root of the Náhuatl word *mexi* was the same as the Hebrew word for the Messiah. As some sixteenth-century scholars had maintained, the ancient Mexicans might be none other than one of the lost tribes of Israel. And Quetzalcóatl, who had visited and instructed the Indians and then gone away again—Servando asserted—must have been the apostle Thomas, bringing some touch and leaving some trace of Christianity in Mexico long before the Spaniards. From claims like this, it was only a small political distance to the demand for independence.[29]

By the time Napoleon invaded Spain in 1808, the Jesuits and Fray Servando had armed the creoles with an arsenal of arguments. And since there was no legitimate Spanish monarch, Spanish America—originally an independent kingdom—could break its pact with the mother country and *regain* its autonomy. When the Revolution of Independence set New Spain on fire in 1810, the optimistic views of the Jesuits about the future of their country and Fray Servando's far-fetched theories about its foundations (humanism and theology, history and myth united) formed the foundation of Insurgent ideology.

To Mexico, the Enlightenment brought a multiple awakening to empirical science, modern philosophy, ancient Mexican history, the geography and natural resources of the territory, and also to a sense of its needless dependence on a nation that did not provide any advantages but only exploited the colony to finance disastrous European wars. And there was in addition a vague awareness of the ideas underlying the American Revolution in the English colonies and the struggle for freedom and equality of the French Revolution, despite the attempts by the Inquisition to exclude all these ideas from Mexican soil. Every aspect of this awakening shared one common characteristic. Like so many features of New Spain, they had been conceptualized by priests in the bosom of that complex and omnipresent institution: the Catholic Church. And therefore all the historical changes proposed or suggested by this process had to confront one assumption: the need to respect the greatest of the traditions of New Spain—that of the Church.

The Bourbons were not disposed to respect it. They went so far as to move against the sacrosanct *fueros* (ecclesiastical privileges and immunities), which inspired (in 1799) a famous petition (*representación*) from the liberal bishop of Michoacán, Abad y Queipo (himself a Spanish-born *gachupin*), directed to King Charles IV. Alexander von Humboldt had consulted with Abad—who was the most knowledgeable man in New Spain on economic matters—when he visited the country in 1803, at the height of Bourbon prosperity. Humboldt had come away influenced by advice and arguments that led to the proposals he offered—in his *Political Essay*—to an inattentive Bourbon government. Abad y Queipo had inspired Humboldt's suggestions of legal equality for all citizens, the end of taxes and labor exactions imposed on the Indians, complete freedom for them and the Castes, agricultural reforms including the division of large estates and the assignment of uncultivated land to be worked by the landless, as well as greater economic diversification. But now, when it came to the Church, Abad y Queipo insisted that "any decrease in the immunities enjoyed by the Spanish clergy, immunities which are the foundation of the monarchical form of government, may well affect the monarchy." This argument would apply to Spain, but it was even more relevant to New Spain: "The American clergy is the only class that, because of its concern for spiritual and civil well-being, has gained power and esteem in the heart of the people."[30]

The lower classes had become subjects of the Crown because of "the religious faith dwelling in the depth of their hearts, which was the consequence of the preaching and advice given from the pulpit and in the confessional by the ministers of the Church." According to Abad, "religion . . . softened the chains of servitude." The corollary was obvious: By depriving the clergy of any of their *fueros*, the monarchy would be doing serious damage to a group that formed the only link between the people and the Crown. It would, in other words, come close to starting a revolution in New Spain.[31]

The Crown turned a deaf ear. But worse was yet to come. In 1804 the Crown issued the most onerous decree in colonial history. It was part of the process of paying its debts for the war against England, which had ended with defeat at the Battle of Trafalgar. Through this decree, the *Consolidación de Vales Reales* ("Consolidation of Royal Vouchers"), the Crown took over the right to collect, for itself, all debts owed to the Church. The problem—and it immediately became a major one for landowners, mine owners, cattle ranchers, most of the medium or substantial property owners of the colony—was that the Church in New Spain, for lack of other alternatives, had long functioned as a low-pressure banking institution. It was the primary source of loans for property improvement, issued on a long-term repayment basis (sometimes fifty years) and at no more than a very moderate 6 percent per annum interest rate—because anything higher would incur the sin of "usury." The collateral on these safe and easy loans was the property itself, thereby mortgaged to the Church. These were comfortable, mutually beneficial arrangements, with the Church lending out funds that it had received as legacies from pious parishioners and using the interest payments for its pastoral work. Suddenly—through the stroke of a royal pen—the Bourbon Crown created its right to demand the immediate payment of all these debts in full, offering nothing but a voucher (*vale*) in return—a promise of compensation in some uncertain future.[32] There was a storm of protest from all the major institutions of the colony, secular and clerical, but the measures were enforced by a venial viceroy, anxious to please Madrid, where the monarchy was almost bankrupt.

In the grip of this brutal levy, both clergy and landowners saw the accumulated capital of centuries disappearing, going off to feed the insatiable needs of the Bourbon court and its policies in distant Europe. For the Church, this was the last straw in a program of assault on its autonomy and wealth (and it would be in fact be the last major blow inflicted by the Crown). For many debtors, it meant utter financial disaster. Abad y Queipo wrote that the *Consolidación* was no less than "a universal sequestration," which could only swell resentment against the mother country. If debtors could not meet these unexpected demands—and most could not—the Crown would put a lien on the collateral: land, herds, mines. Many creole hacendados had to auction their property to meet the required payments. One of those affected was Abad's close friend and fellow priest Miguel Hidalgo y Costilla. In 1809, the Bourbons would finally respond to the rage and impoverishment of their colony's elite by canceling the writ and returning sequestered property. But by then it was too late for many ruined hacendados—and for the future of Bourbon rule in Spain's oldest American dominion.[33]

Four months before Hidalgo's uprising, in May of 1810, Abad begged the Council of Regency then ruling in Spain to dispatch more troops to Mexico. Ferdinand VII was a captive in France, and Napoleon had invaded Spain. "Any disturbance of public order in this country will necessarily leave the most horrendous anarchy in its wake."[34] Abad was acting as the appointed bishop of Michoacán, but without confirmation from the king or the Pope, perhaps because of his illegitimate birth. Social inequality, which he took such care to describe in his "Representations" to the Crown, was the spark that could begin the "conflagration." He was fearful of an impending disaster. "What would be the logical outcome of a revolution fought by these different classes, by this opposition of conflicting interests and passions? Mass destruction, the downfall and devastation of the country."[35]

Abad y Queipo's prophecy was all too rapidly realized. The four hundred priests who led the Insurgency of 1810 fought not only for the independence of their country but also for the full reinstatement of the ecclesiastical *fueros*, privileges depleted by the Bourbons and now threatened with extinction—according to the Insurgents—by the mainland Spaniards living in New Spain who would "hand over the kingdom" to the "heretical and ungodly French" and to Napoleon, an "instrument of Satan." The fighting slogan of the priest Mariano Matamoros, Father Morelos's closest lieutenant, was "To Die for the *Fueros*."[36]

Morelos would adopt Abad's progressive social, ethnic, agrarian, and economic proposals. In fact, the Insurgents and Abad, who turned out to be their greatest enemy, only disagreed on one point: the call for independence. Mexico would be awakened from its "colonial siesta" by a revolution inspired, led, and also condemned by priests.[37]

II

CENTURY OF CAUDILLOS

For three centuries, the traditional Mexican order of things created in the aftermath of the Conquest had remained in place.

Much had happened culturally. Socially a whole new ethnic world had been created. Even politically there had been a great shift in the tone and nature of power from Hapsburg rule to the height of Bourbon autocracy. But the architectural structure of regal, religious, and corporate authority stood firm, even if there had been renovation in some parts of the building and changes in interior decoration according to the tastes of one period or another. But beginning in 1810, a new kind of earthquake ran great cracks through its foundations.

All across Latin America in the early nineteenth century, the crumbling of the Spanish political order led to the rise of the caudillos. They were the strongmen, the new *condottieri*, the chieftains, the masters of "lives and haciendas," the inheritors of the Spanish and Moorish archetype of the warriors who raised their shining swords. The Mexican caudillos, like their counterparts farther south, fought for independence from Spain. But unlike those others, they were priests. Miguel Hidalgo and José María Morelos (and the hundreds of priests who fought as their lieutenants) had a quality that went beyond mere charisma: a religious aura, a connection with the divine. For some of the many who followed them, they represented Divine Providence. For others, they were to be worshipped in themselves or they were surely destined to become the founders of a theocratic state.

Following the initial emergence of these caudillos in 1810 came an ephemeral, tragic, but meaningful period filled with political and religious tensions, and then an age (1821–1855) dominated by typical Latin American creole caudillos, resembling those to be found elsewhere throughout the former Spanish colonies. One of them, acclaimed as "the Man of Destiny," loomed over his time like a melodramatic colossus: the uncrowned monarch, Antonio López de Santa Anna.

But there was something in the creole mentality—in Santa Anna himself and other caudillos and not only military commanders but also intellectuals like Alamán and Mora—that prevented them from consolidating a nation. Although they had the capacity and intellectual powers to establish a new order—some looking toward the future, eager for a state that would be republican, secular, democratic, and constitutional; others turning toward the past, nostalgic for a Catholic, hierarchical, centralized society—none of them could carry it off. And worse, they presided over a time of anarchy, economic impoverishment, the loss of national territory, and, above all, violence: revolutions, foreign interventions, civil struggles.

Clearly charisma was not enough to reconstruct the lost order of society or to build a new one. Mere, empty charisma was in a sense the primary obstacle to any progress. At this moment Benito Juárez, a pure-blooded Indian, appeared on the scene, to consolidate a new political order. Instinctively, Juárez baptized the new legitimacy with waters drawn from the ancient well of the Aztec *tlatoanis* or, more precisely, from that of his calm, severe, and melancholy Zapotecan ancestors. Under Juárez, Mexico acquired a political structure that was a strange historical compromise between the past and the future: a monarchy in republican clothing, which guaranteed civil rights and freedoms inconceivable during the colonial period. To achieve this form cost years of bloodshed.

Juárez was the central figure in the bitter conflict that tore Mexico apart in the nineteenth century. It was almost a war of religion, without precedent in the history of Latin America. Mexico had gained its independence from Spain but not from the colonial order, because the Roman Catholic Church still maintained its historic role, near the center of power. The nascent Liberal state had no alternative but a fight to the death with this parallel state. It could only be defeated by a caudillo who was prepared to make a stand for the dual cause of the liberal Constitution of 1857 and the Laws of the Reform (both of which modified the historical role played by the Church), while remaining a man inspired by an ancient and uncompromising religious fervor. Juárez was that caudillo-President-*tlatoani.* But he could not have moved forward at all without the ideological support of the Liberal, primarily mestizo intellectuals of his own generation.

Leading the country through the War of the French Intervention (1862–1867), Juárez defeated the man Napoleon III had chosen to be emperor of Mexico—Maximilian of Hapsburg, whose reign was Mexico's second imperial dream. The first emperor had been the creole caudillo Agustín de Iturbide, crowned Agustín I in 1822 after finally completing the process of Mexico's independence from Spain. Both emperors lived out a paradox. They were dedicated to the cause of their century—freedom—but they represented the tradition of other centuries—that of absolute power. Both were monarchs with liberal convictions, who would not survive that inner conflict.

Almost all the Mexican heroes of the nineteenth century, Liberals and Conservatives alike, died as martyrs. The two caudillos from Oaxaca who sacralized the office of the President were exceptions: Juárez and Díaz, the priests of power. In the struggles between Liberals and Conservatives (and within the victorious Liberal camp itself), the "fever for the future" and the longing for the past were the major forces swinging the pendulum of Mexican history.

Nothing can better illustrate the tension between these two nineteenth-century tendencies than the biographies of the caudillos: priests, intellectuals, army officers, civilians, Presidents, and emperors. All of them wanted to guide the country toward prosperity. But each of these men had an individual history, a different notion of the past and of the future. And each caudillo's familial, geographical, ethnic, religious, social, economic, political, professional, even his psychological conditioning was transferred to the history of the country, creating a cumulative biography of power.

6

THE INSURGENT PRIESTS

Father Miguel Hidalgo was part of an old tradition of creole patriotism common to all of Spanish America but dating back, in New Spain, to the sixteenth century and strongly reinforced by the work of the creole Jesuits of the Enlightenment in the mid-eighteenth century. He himself was part of the regular clergy, not a member of any order, but when the Bourbon kings (wary of their prestige and influence and interested in their wealth) had expelled the Jesuits in 1767, Hidalgo was a student at Valladolid, a major center of Jesuit influence, and he would always remember those priests who had placed so much value on the history and natural riches of Mexico. Almost half a century later, Hidalgo declared: "We will freely exploit the incredibly rich products of our country and within a few years its inhabitants will enjoy all the delights of this vast continent."[1]

Hidalgo had other, less idealistic reasons for demanding that the creoles be granted their rights. Through the *Consolidación de Vales Reales* of 1804, the Spanish Crown had put a lien on Hidalgo's haciendas and those of his family, threatening to auction them off unless the sudden, exorbitant demands for payment were met. Hidalgo had been pushed to the edge of bankruptcy, and his younger brother Manuel had been so disturbed that he went mad and would die within a few years, in 1809.

When he launched his revolution in 1810, Miguel Hidalgo had long been recognized as "one of the best theologians in his diocese,"[2] "a man deeply versed in literature and with a very broad range of knowledge in

Miguel Hidalgo y Costilla
(Private Collection)

all areas."[3] The son of the manager of a hacienda, he was born in 1753 on the Hacienda de Corralejo, in the state of Guanajuato. He studied theology in the city of Vallodolid (now called Morelia), where he witnessed the expulsion of the Jesuits. His fellow students called him "the Fox," a nickname that, according to the Conservative Alamán, corresponded "perfectly to his crafty personality."[4] Graduating in 1782, he became a professor (for ten years) and ultimately Rector at the famous seminary of San Nicolás in the same city. Two theological treatises (one in Spanish, one in Latin) that he published in 1784 were praised by the great theologian Joseph Pérez Calama, who described him as "a young man who, like a giant, surpasses many of his elders."[5]

Later, in his own parishes, though he was still a man much admired, somewhat different views were also expressed about him. At the beginning of the century, the Inquisition began to receive accusations against Hidalgo, who was then priest of the rich parish of San Felipe Torres Mochas. The complaints were of two kinds: moral and theological. His accusers did not doubt that he had a "fine brain," a "cheerful" disposition, and was "shrewd in the area of letters," but he also had some extravagant habits. He was a "professional gambler and as such was dissolute," "free in his treatment of women," and given to "continual amusement" to such a degree that "in the house of this Hidalgo there has been enough riotous celebration to turn it into a little France," a place where you find "male and female musicians, games of chance and dancing"; "he keeps an entire orchestra on salary, whose directors dine with him and are treated like family."[6] The problem, nevertheless, was not so much what Hidalgo did as what he thought. To judge by the denunciations, the curate was almost a heretic. Among the many charges raised against him, Hidalgo was said to have denied the existence of hell ("Don't believe that, Manuelita," a close friend—perhaps too close—confessed to have heard him say. "These things are deceptions.") He mocked Saint Teresa ("a deluded woman, who, because she whipped herself, fasted a lot and did not sleep, saw visions"). He preached intellectual licentiousness (the Bible ought to be "studied with the freedom of mind to discuss what we want without fearing the Inquisition"). He read banned books and even went to the extreme of jovially arguing (in the confessional, according to some) that "fornication is not a sin."[7]

The cycle of denunciations ended shortly afterward, and though they began again in 1807, none of them did any practical harm to the priest. Aside from the fact that many of the accusations—particularly the theological ones—were probably exaggerated and distorted, Hidalgo could count on important support: He was highly esteemed by the civil and ecclesiastical authorities of his district and diocese, who for the most part looked kindly on the new air of liberty at the beginning of the new century. Hidalgo's biographers from every age have discounted these indictments, putting them down to envy or associating them with subsequent events in his life (the Inquisition file was in fact reopened during the final trial to which he was subjected). But at least with the moral charges, these accusations show an internal agreement that suggests they were true. Hidalgo was not only a restless priest but an eccentric one, a free and brilliant man, who attracted—and seduced—the most enlightened of his contemporaries but unsettled the more rigid and conservative. They vaguely sensed in him the seed of something new and disconcerting.

Within Hidalgo there were many Hidalgos, and all of them were equally eccentric. As a priest, while ministering to his flock (in the parishes of Colima, San Felipe, and, finally, Dolores), he showed very little interest in administrative work or in celebrating many masses. Instead he enjoyed preaching, where he could apply his knowledge of theology, and he showed great care and concern in taking confessions from the sick and the dying. This paternal attitude of the priest Hidalgo showed above all in his treatment of the Indians. He learned their language and he taught them various crafts and trades. It was as if he were trying to transform theology into charity.

If in his youth he had been primarily a contemplative person, over time he grew more active and became an innovative and hardworking businessman. Besides managing the small family haciendas, he developed an interest in beekeeping, curing skins, manufacturing pottery, cultivating vineyards, and in his last parish, the village of Dolores, he planted an extensive acreage of mulberry trees in order to raise silkworms. The bishop of Michoacán, Abad y Queipo himself, recognized that Hidalgo had introduced "this very important industry" into the region: "The priest Hidalgo has started this in the parish of Dolores with good results though on a small scale."[8]

The reverse side of this display of activity was his spendthrift and disorganized character. Evidence abounds of his procrastination in paying off debts and, later, of his appetite for gambling and extravagance. On the point of departure from Colima for San Felipe, he forgot to pay off money he owed. Once in San Felipe, he incurred another debt with a certain Sr. Ignacio Soto, who pressured the bishop until he finally confiscated a third of Hidalgo's official earnings to settle the grievance. The Colegio de San Nicolás demanded payment of a sum of money, and the tedious inquiries went on for years. In another case of an upaid debt, the vicar-general himself had to reprimand Hidalgo. Overall, this priest's activities were as incessant as they were unreflective. Alamán tells a curious and relevant anecdote about Hidalgo and his silkworms:

> Bishop Abad y Queipo once asked [Hidalgo] what method he was using to chop the leaves and distribute them to the worms according to their age, and to separate the dry leaves while keeping those still on the trees clean, about which the books on the subject go into detailed warnings. [Hidalgo] replied that he did not follow any order in particular, but just threw the leaves as they fell off the trees and the worms could eat them as they wished.[9]

Bishop Manuel Abad y Queipo was one of the most fascinating intellectual personalities of the Enlightenment in New Spain. He kept abreast of the latest developments and political theories in Europe while at the same time composing his famous series of "Representations" in which he outlined the socioeconomic problems of New Spain and warned of the dangers of a social uprising. Hidalgo was one of his friends and a member of his intellectual circle.

Despite their closeness, the secular classics valued by Hidalgo were not those (Adam Smith, Jovellanos, the Physiocrats) esteemed by Abad. Still less was Hidalgo concerned with Voltaire, Rousseau, or the French Encyclopedists, whose writings he probably did not know at all. Hidalgo's century was the seventeenth rather than the eighteenth. He translated Racine and Molière (*Tartuffe* was his favorite work), read La Fontaine, and played Rameau on the violin. The "little France" of his house in Dolores was more an art center than an intellectual salon. He liked to read history—and not only ecclesiastical history—but his favorite historian continued to be Bossuet with his belief in Divine Providence. It was not Montesquieu—frequently quoted by Abad—but the Spanish neoscholastic thinkers of the sixteenth and seventeenth centuries who guided Hidalgo in political matters. Despite recent censorship by the Crown, students of the seminary where Hidalgo had studied were by no means unfamiliar with the assertion of the Jesuit Francisco Suárez (1548–1617) that "sovereignty lies essentially with peoples and not with kings; the kings exercise it by consent of the people and with the understanding and indispensable condition that it should be exercised for the benefit and utility of the people, and in the event of its abuse, kings can be deposed and war may even be declared upon them, because peoples are superior to kings."[10]

These surprising, pre-Rousseauian theories on the tyrannical nature of domination are also found in many other authors, like the Jesuit Mariana, who wrote a justification for the killing of monarchs. They were the theoretical and moral reservoir that the creoles of Mexico City's *ayuntamiento* (municipal council) used to support their claim that the people were now sovereign, after Napoleon's invasion of Spain in September of 1808 and the imprisonment of King Ferdinand VII. Even the viceroy, José de Iturrigaray, agreed to the devolution of legal authority to the *ayuntamiento* (a quasi-democratic body that had represented the *vox populi* in Spain ever since the Middle Ages), but a coup d'état against him led by wealthy Spanish merchants thwarted this first bid for creole autonomy.

Two years after these events, which had become well known throughout New Spain (the viceroy had been deposed, the members of the *ayuntamiento* imprisoned and some of them killed), Hidalgo was doing some highly specialized reading: *A Dictionary of Arts and Sciences* that included an article on artillery tactics and the manufacture of cannons and also the book of Bossuet's *Universal History* that dealt with the conspiracy of Catiline against the Senate of Rome. His purpose was not innocently intellectual. He was himself conspiring with several creole officials against the Spanish government. In September, moved to sudden action through a betrayal of the rebellion he was planning (and the rumor of an invasion by the forces of Napoleonic France), he launched an attempt to claim power for the people. But this time—though Hidalgo used the same theories to justify his actions—power was to be seized not by their creole representatives on the Mexico City council, but by the people themselves, by the masses. He was acting "on the authority of my nation"; he had been chosen "by the Mexican nation to defend its rights"; and he would seek the restitution of "the holy rights granted by God to the Mexicans, usurped by cruel, bastard and unjust conquerors."[11]

On the morning of September 16, he rang the cathedral bell to summon his Indian parishioners into the church square of Dolores and delivered his famous cry (the *grito*): "Death to the Spaniards! Long live the Virgin of Guadalupe!" (though disagreements exist on exactly what he did say). We know he called on the Indians to open the jails of Dolores, free the prisoners, and lock up the Spaniards, and that he authorized the sacking of houses and haciendas belonging to the Spanish-born (*gachupines*) and allowed his followers to kill and satisfy their instincts for revenge. That same day, Hidalgo, possessed by what he would later call his "frenzy," had taken a canvas of the Virgin of Guadalupe from the nearby sanctuary of Atotonilco and attached it to a stick, creating a standard for his struggle. In San Miguel El Grande, it is certain that he did cry out, "Take! My children! Because everything is yours!"[12] Within days, the whole central region of the country, the

Bajío, went up in flames. Following close at the heels of the undermanned and disorganized official regiments commanded by creole officers (Hidalgo never even had a hundred creole followers) came nearly twenty thousand men of the humblest origins, his flock, Hidalgo's army: "an unruly mob of Indians and peasants, with stones, with sticks, with crude lances, without organization of any kind. . . . Mixed with the half-naked, hungry hordes were countless women dressed in rags . . . there were whole families. . . . It was like the ancient Aztec migrations."[13]

With Hidalgo approaching, most of the Spanish population in the city of Guanajuato shut themselves up behind the stone walls of the Alhóndiga de Granaditas, a building used for storing grain. As if history were taking an atrocious revenge for the massacres of Indians by the conquistadors at Cholula and the great temple of Tenochtitlan, Indians and Castes from the city itself joined Hidalgo's Indian brigades in slaughtering all the Spanish men. Captives were beaten to death. Women and children were spared, though one child was hurled from the walls to die on the stones below. Among the survivors were a young creole widow and her son with the surname Alamán, whom Hidalgo protected. (They had not taken refuge in the Alhóndiga.) The son, Lucas, then eighteen years old, witnessed and described the terrible events: Spaniards throwing money from the granary, trying to satisfy the greed of the Indians; Hidalgo's troops searching the corpses for tails, the mark of the Jew according to medieval legend and so a potential proof of heresy and infamy; terrified women fleeing across the terrace roofs toward neighboring buildings. A mob ran wild, looting and wrecking houses and businesses. It was a frenzy fueled by racial and social hatred.

When a fellow priest asked him to explain the nature of his struggle, Hidalgo, already acclaimed as the "Captain-General of America," answered that "he would have found it much easier to explain what he would have liked the revolution to be, but even he did not really understand what it was."[14] His personal motive was something nearly every creole wanted: to gain independence from Spain. What he presumably "would have liked" was to incorporate the creoles into his army, but they would not follow him, because for them the events from the very beginning were an upheaval aimed at eliminating the entire white population of the country, creoles and native Spaniards alike (about a million out of a total population of six million). Hidalgo would later confess that he knew of no way to ignite the war other than the one he put into effect: using the prestige of his priesthood to appeal to the elemental passions of his Indian parishioners, among them plunder and revenge.

Hidalgo had no broader military strategy. Nor did he have a clear idea of the new nation for which he was fighting. "The revolution," said Abad y Queipo, "was like his raising of silkworms."[15] The intensity of Hidalgo's actions steadily mounted—based on a single, fevered purpose. He wanted to destroy the old order, to cure its social and ethnic injustices, to avenge the old grievances of the creoles and to avenge Manuel, the brother who had died. He wanted a universal conflagration.

When Hidalgo occupied Valladolid, and his old friend Abad y Queipo excommunicated him, he responded by issuing a decree to abolish slavery. This must have been one of the earliest of such formal declarations, if not the very first, in the Americas. His conscience was clear: He was the leader of a just war and he had "decided not to make any settlement which did not provide for the liberty of the nation and the rights which the God of nature granted to all men, rights that are truly inalienable, and they should be protected with rivers of blood if it becomes necessary."[16]

Yet both Alamán, the conservative, and the liberal writer José María Luis Mora would accurately point out that the principal driving force of the revolution for the people who followed Hidalgo was something else, though Mora calls it "superstition" and Alamán "religion." They were struggling, said Aldama, one of Hidalgo's lieutenants and few creole followers, "for a sacred liberty, but not for French liberty against religion."[17] Allende, Hidalgo's second-in-command and also a creole, would assert that the Europeans—that is, those born on the Iberian peninsula—"had become Frenchified and corrupted."[18] Through the sermons of their priest, the masses came to share the same belief—they were defending religion against the French heretics and their concrete incarnations, the *gachupines*. It was a war against the devil, in favor of God and with the aid of the Virgin. Alamán, who was an eyewitness, recalled that Hidalgo's soldiers—some of them without weapons—had placed "the image of the Virgin of Guadalupe, which was the insignia of their enterprise, on sticks or on reeds painted different colors," and "they all wore a print of the Virgin on their hats." Among the troops, one would hear that "the Priest is a saint," "the Virgin speaks to him several times a day." And that all they wanted was "to go to Mexico City in order to put His Worship the Priest on the throne."[19]

At the end of October, that caravan of an army halted at Monte de Las Cruces, near Mexico City. Following Hidalgo was his horde of at least eighty thousand Indians (and Castes) armed with lances, stones, and sticks, so thoroughly ready to plunder Mexico City that they had brought sacks with them for carrying off what they would seize. Suddenly, the fire of the royalist artillery spread real fear, for the first time, among Hidalgo's soldiers, who in their desperation tried to charge the big guns and stop up the smoking mouths of the cannon with their huge, straw sombreros. Hidalgo disregarded Allende's advice, as he had done in other situations, and refused to make a direct assault on the capital. Was it his Spanish roots that made him make this decision, to prevent another massacre of Spaniards? Or his discretion, an unwillingness to see the deaths mount up among his Indians? The reasons remain a mystery. But his failure to act was the beginning of his downfall.

Hidalgo withdrew toward the west and set up his headquarters in Guadalajara. While in that city, he issued two important decrees meant to remedy social and agrarian injustices. One abolished taxes ("that disgraceful exaction . . . we have endured for three centuries as a mark of tyranny and servitude");[20] the other ordered the restitution of lands to Indian communities, since "it is my will that they should be enjoyed only by the natives, in their respective villages."[21] And he envisioned summoning a congress, "to be made up of representatives from all cities, towns and places in this kingdom, who—while keeping as their principal goal the maintenance of our religion—would pass mild and benevolent laws suited to the particular circumstances of each *pueblo*: then they will govern with the gentleness of priests, will treat us like their brothers, will banish poverty. . . ."[22]

By then an imperial use of the first person had become habit. Hidalgo was letting himself be treated like a sovereign. He lavishly made official appointments; he lived surrounded by guards; he would walk arm in arm with a lovely young woman and allow himself to be addressed with the title "Most Serene Highness." He attended banquets, dances, ceremonies, plays, parades, gala functions where he accepted the homage of politicians, military men, and priests in the midst of banners, flags, exquisite refreshments, bursts of music, and peals of bells. Guadalajara was now his "little France," and Hidalgo was the Sun King.

That priest-king could be munificent to some and terrible to others. "In Guadalajara," Mora writes, "according to decisions that Hidalgo made in secret . . . a bullfighter named Marroquín would take out groups of Spaniards at night, while the city was quiet, and bring them to the Salto ravine . . . and there he would cut their throats."[23] Through an offering of immunity and protection, Hidalgo had attracted *gachupines* from the surrounding area and concentrated them in the Colegio de San Juan and in the local seminary. Some had brought their families with them. When the real reason behind this assembly was revealed—Marroquín appeared with his knife—Allende, who for some time had disagreed with "that rogue of a priest" as he called him,[24] entertained the idea of poisoning Hidalgo. He did not do it, nor could he prevent the month of massacres that ran from the Feast of the Virgin of Guadalupe on December 12, 1810, to January 13, 1811. According to Mora, the creole Abasolo—another of the great Insurgent caudillos—"saved many of them by helping them to escape, by hiding them or, as he did on two occasions, by wresting them out of Marroquín's hands while he was taking them out to cut their throats."[25] At his trial several months later, Hidalgo confessed,

> those executed at his order . . . [in] Guadalajara . . . added up to some three hundred and fifty, among whom were a Carmelite lay brother and, if his memory served him well, someone from the Order of San Diego, though he could not say whether he was a lay brother or a priest. . . . What is true is that not one of those killed at his order were given a trial nor was there any reason to do so, since he knew perfectly well that they were innocent.[26]

During the colonial period, on the whole, truly indigenous uprisings were fires that quickly went out. In Guadalajara, through exhaustion, satiation, or for motives as mysterious as the sudden disappearance—which Hidalgo could not explain—of the image of the Virgin from their hats (the truth seems to be that some of his soldiers were removing it, as a sign of their displeasure at his failure to attack Mexico City), the Indians began to leave him. For this and other reasons—disorder among the remaining recruits, the vagueness and therefore limited appeal and power of his program, disagreements among his commanders, the emergence of a burdensome and extortionate bureaucracy, and the firm resistance of the royalist troops who defeated his disorganized armies at Puente de Calderon—Hidalgo was forced north toward the interior eastern provinces.

In Saltillo he was welcomed by two groups of Insurgents—under Captains Menchaca and Colorado—whose ranks included contingents of Comanche Indians. On seeing them, Hidalgo was enraptured. "They had their bodies painted with differently colored stripes," a witness noted, "and they were wearing buffalo hides."[27] It would be the last time that the creole priest would be able to preach to an indigenous flock, as different as they were from his parishioners in Dolores. His words touched on the key theme of creole self-identity, the connection of the creoles with the Indians through the abuses they had both suffered at the hands of the Spaniards:

> ... I have come from the south, from making war on the Spaniards in order to tear the country out of their hands, for it does not belong to them and they have held it for a long time with cruelty and tyranny and grave damage to the original inhabitants, the children of the nation; and your ancestors the Indians acted like heroes without knowing it when they saw they had no hope of fighting the conquerors and rather than endure humiliation, they took refuge in the mountains.[28]

After the sermon, he continued moving north, perhaps with the intention of crossing into the United States. But he never reached the border. A royalist officer—learning that Hidalgo was advancing through his desert area with a small group of men—set a trap for him by closing all the water holes except one. With the sun and the sand as his allies (and including a pretended welcome among his weapons, as if he were going over to the Insurgents), the officer—Ignacio Elizondo—lured Hidalgo into an ambush and took him alive, along with Allende and other leaders of the Insurgency.

In the city of Chihuahua, he was tried both by a military court and then by the Inquisition. During the first trial, he recounted the details of his short-lived campaign. He declared himself "sincerely repentant." He took responsibility for the massacres of Spaniards in Valladolid and Guadalajara and claimed that it was because he had wanted to please "the Indians and the lowest rabble . . . the only ones who desired these scenes." As his only justification, he offered the argument, "the need that they had of people for their undertaking and the need to involve the masses did not permit him to be too scrupulous about the means to accomplish this purpose."[29] He described his "unpremeditated" enterprise as "an inconceivable act of indiscretion and a frenzy" but said that he had acted "in good faith." Experience had disillusioned him: "His plan for independence would have undoubtedly ended in total anarchy or total despotism." He wanted his declaration to reach the ears of all Americans, "because it is in line with everything he feels in his heart."[30]

There is no reason to suppose that Hidalgo's remorse was insincere. Nor to doubt that his intention had been—as he himself said—"that of placing the kingdom at the disposition of Don Ferdinand VII."[31] Hidalgo had desired independence as if it were a vague utopia, something that would just happen, the fruit of a miracle as sudden and incomprehensible as the revolution he had unleashed with his preaching. That was why he never dared to declare himself openly in favor of independence. It would have been like substituting himself for

Divine Providence. In his classic creole soul, resentment against the *gachupines* was counterbalanced by a traditional loyalty to the Crown. Hidalgo's perception of the state was in no way that of the later Liberals. He had no clear political alternatives on which to raise up anything new. He was a creole educated within the monarchy and trapped within the monarchy, although he resented the tyranny and despotism of the Spanish government that had "enslaved America for three hundred years . . . and considered Americans unworthy of all distinction and honor."[32] Hidalgo wanted both monarchy and liberty, and when he realized that the two were incompatible, he abandoned himself to "regal pomp and ceremony." He lived out, with "frenzy," an imperial fantasy.

He was, first and foremost, what he had always been: an eccentric theologian, somewhat lost on the practical roads of life. Joseph Pérez de Calama was right when he said that Hidalgo had become "a light set out on the heights, a city on a mountain,"[33] but the light that flooded this theological city was not rational or constructive. It was blinding. It did not illuminate; it inflamed. Hidalgo's special qualities—his Christian charity, capacity for direct contact with the Indians, his "very refined powers of argument" and incessant enterprises, his lack of order, extravagance, festive manner, magnetism—did not disappear during the Insurgency. Beginning with the *grito,* they exploded and they spread like a fire, and like fire they were extinguished. The documents of the period present his enterprise as a sweeping seduction. Possessed by a violent, vengeful, vindictive state of mind and soul—his *frenesí* (frenzy)—Hidalgo seduced the indigenous hosts who followed him because he embodied the etymology of his office (*sacerdote*, from the Latin *sacer dux,* he who leads to the sacred), but—even though he was ambivalent about it—he did choose to guide them in another direction, a historic leap not toward tradition but toward liberty.

In Chihuahua, with death awaiting him, Hidalgo had a chance to respond to the old theological accusations that had accumulated against him since 1780 and were restated by the Inquisition. It is a moment in which we can hear the ordered discourse, firmly based and exceptionally clear, of the man who had been one of the best theologians of his era. He successfully refuted twelve charges brought against him. If he had any weakness in issues of religion, it was, he concluded, on account of "the vicissitudes of my misery . . . and not because I intended to deceive."[34] His error was other: not theological but moral. He had become a leader of the people through his abilities that were the "plague of my seduction"; he had professed ideas that "I abjure, detest and retract. . . . I have already confessed that these run contrary to the morality of Jesus Christ, for which I weep with bitterness and hope that the kindness of our Lord will help me to beg for His compassion."[35]

In the space of a few months, that erudite theologian had moved between two unfathomable mysteries, that of evil, "carried out by his order" against the innocent, and that of mercy, to which in his final hour, humbly, he entrusted himself.

After Hidalgo had been ritually defrocked and stripped of the robes of his office, he was searched. The judge wrote: "They found on his body the sovereign image of Our Lady of Guadalupe, dripping with sweat . . . embroidered with silk on parchment."

On July 30, 1811, he was executed (as were his closest aides—Allende, Aldama, and Jiménez). Three lines, one after another, of nervous soldiers put bullets into Hidalgo's arm, belly, and spine but failed to kill him. Pedro Armendáriz, the commander of the firing squad, ordered two men to follow him as he approached the priest lying in agony. They were ordered to press their rifle muzzles to Hidalgo's heart. Armendáriz would remember that, as he gave the order to fire, Hidalgo "stared straight at us with those beautiful eyes."[36]

The heads were displayed in cages on the four walls of the Alhóndiga de Granaditas, where the Spaniards of Guanajuato had been massacred. There they remained for ten years until Mexico won its independence in 1821.

For the War of Independence had not ended with the death of its first leader. It would go on, under the command of parish priests who were, unlike Hidalgo, mostly mestizos.

Hidalgo had died blindly, without knowing that his "unpremeditated and frenzied enterprise" was successfully spreading across almost half the viceroyalty of New Spain and that the caudillo he had anointed in the south, a man who had been "ignored and slighted at first" (as Alamán wrote), was "growing in power and importance, rising on the horizon like those storm clouds that come up in a part of the southern sky and soon cover an immense area of the country, announcing their arrival with all the fanfare of a terrible storm."[37] He would become known as "the Servant of the Nation," a term he himself coined from his reading of the Gospel of Saint Mark.

José María Morelos y Pavón (born in 1765) was, like Hidalgo, from Michoacán and a native of its capital, Valladolid. But he was a mestizo, of much more humble origins. As a young man, he worked for ten years on a ranch rented by his uncle, where he constructed buildings and watched over animals, did some traveling, and learned to be a cowboy. At night he would study, following an inclination passed down

José María Morelos y Pavón, 1814
(Private Collection)

from his mother's side (she was the daughter of a schoolteacher). Purely on his own, he learned a great deal about grammar, for which he had a passion. And in 1789, specifically to pursue the study of grammar, he enrolled in the Colegio de San Nicolás, whose Rector was Don Miguel Hidalgo. From there he moved on to the Tridentine Seminary and graduated in 1789. Profound religious faith but also economic necessity brought him to the priesthood. His father was an irresponsible man, a carpenter who would frequently drift away from home, leaving his wife in need over long periods of time.

From the very beginning, the priest Morelos accepted both his sacerdotal and economic duties with an equal sense of responsibility. Shortly after his ordination, he began to minister to impoverished villages in the hot country (*tierra caliente*) of Michoacán, where there was "bad weather and little profit."[38] He was one of those "unattached priests," according to Abad y Queipo, "who hold no benefice or receive any money from the government but exist only on the small returns of their office."[39] But Morelos—without reducing his priestly commitments (instead he did even more)—also became a modest but active businessman, organizing a group of mule-drivers to move grain, brandy, and cattle from the village of Nocupétaro to Valladolid, where, thanks to his efforts, his sister and her husband owned a house and a store. But though he put together some savings and material possessions, he was not a man interested in becoming rich. He placed what he had at the service of "the charity which has always motivated me."[40] It was in this same village of Nocupétaro that he—largely with his own hands—built a church of which he proudly said, "Next to the church of Cuzumala, it is the best in the *tierra caliente*."[41] And it was with the people of Nocupétaro, a village of Indians and mestizos, that he compared the Indians of another village, Carácuaro, when he complained about them and they in turn complained about him. Because Morelos was not merely a protective and compassionate *padrecito* but someone who made demands. He wanted a flock not of eternal children but of men able to work in the world. He despaired, in a letter to his bishop, of the "idleness and vices" of the Indians of Carácuaro compared with the dedicated labor of the men of Nocupétaro. And he asked that the men of Carácuaro be compelled to work "because they are very much vagabonds." The Indians responded, "He yells at us and he makes us angry and he treats us badly."[42] But elsewhere in the dusty and unhealthy villages of the "hot country" Morelos was known for his acts of charity toward the sick and dying and toward his fellow priests.

He wrote during those years: "I am a miserable man, more so than most, and my nature is to be of service to an honest man, to lift him up when he has fallen, to pay his bills when he has no money and to offer whatever manner of assistance I can to someone in need, whoever that may be."[43] He was a priest committed to overseeing the spiritual and material well-being of his people. But he was no otherworldly saint. Alamán mentions that "he fathered various children with anonymous women of the people."[44]

In 1808, when Napoleon invaded Spain, Morelos said that he was "totally ready to sacrifice my life for the Catholic religion and the liberty of our sovereign."[45] Two years later there came the alarming rumor of a French invasion of Mexico and the ecclesiastical authorities asked for donations to prepare for the attack, appealing to the patriotism and religious loyalties of the clergy. Morelos donated a whole month's salary. In the autumn of 1810 he learned that the teacher he had admired, Miguel Hidalgo, had taken up arms. The two met in a town near Valladolid, and Morelos accepted the responsibility of raising the revolution in the south. He was forty-five years old.

His unquestionable success for more than three years owed something to the nature of the region where he operated. Wilder and more mountainous, less developed than the Bajío—the mining and agricultural region north of Mexico City where Hidalgo had fought—the south contained many different Indian groups and large Caste settlements but fewer mestizos and creoles. In this physical and social landscape Morelos could practically create his campaign from the ground up.

His performance as a general was, in itself, the best confirmation of Allende's complaints about Hidalgo ("Neither were the men amenable to discipline," Allende had declared at his trial, "nor was Hidalgo interested in regulations.")[46] Morelos was a very different person. He drew inspiration from the *Military Instructions* of Prussia's Frederick the Great and, above all, from his own practical skills. He carefully chose his lieutenants—the Galeana brothers and the Bravo brothers—from among the mestizo *rancheros* of the southern mountain region. And a sense of order governed every action Morelos took.

He converted the loose divisions into regiments and brigades, he managed the treasury with total honesty and attention, he created arms workshops, gunpowder factories, and foundries for lead and copper. Promotion in Morelos's camp was based on the single criterion of merit: "I have not wished to promote the higher-ranking officers on my staff, but intend to reward only those who first set foot within enemy strongholds and not those who have left the plough to be colonels and do not carry out their duties."[47] Among his comrades—as among those Indians in the village of Carácuaro—his discipline was criticized as despotism. Morelos replied to the charge by saying: "To have no captain without a company, no colonel without a regiment, no brigadier without a brigade argues not despotism but good order."[48]

In contrast to the numbers and considerable resources on which Hidalgo had been able to rely, the victorious progress of Morelos demonstrated that the action of thousands of properly trained, uniformed, and organized young men was much more effective than that of an erratic, frenetic, and amorphous mass. "Entire villages keep wanting to follow me, eager to be with me in the fight for Independence," he wrote at the beginning of his campaign in November 1810, "but I prevent them from doing so, telling them that their help would be much more powerful through working the land to supply food for us who launch ourselves into war."[49] Ardor was not enough to make a revolutionary. Sometimes, in Morelos's judgment, it was really a hindrance. In September of 1812 he wrote to the members of an impromptu revolutionary junta formed in Naulingo: "Although your graces hope to canonize this meeting, you are really turbulent and subversive. The junta of Naulingo is in every respect invalid and at fault, because there is a Supreme Junta, the sovereignty of which is legitimate."[50]

Ignacio Rayón, the caudillo of the National Supreme Junta, was Hidalgo's comrade. Rayón was not an Insurgent priest but a civilian who had taken up arms. A businessman involved in mining and a lawyer, a patriot, a man of culture and an admirer of Cromwell, Rayón had been chosen as the commanding general of the Insurgency by the rebel leaders meeting at Zitácuaro after Hidalgo's death. For over two years Morelos continued to acknowledge his leadership and borrowed many ideas and initiatives from him, such as hierarchical structure and insignia within the movement, the minting of coins, the administration of justice, and so on. Most important, Morelos at first accepted almost all the points proposed by Rayón in his *Elementos constitutionales*. In this work, which Morelos called "our constitution," Rayón devised a formula that he felt could resolve the apparent contradiction between independence from Spain and loyalty to the Crown: "Sovereignty arises directly from the people, resides in the person of Ferdinand VII and is exercised by the American National Supreme Congress."[51] In other words, it is born here, resides there, and returns here again because it is exercised here.

But then Rayón and the Junta gradually descended into an excessive fixation on legal details and paperwork without end, while Morelos was engaged in his efficient campaigns, almost always successful. City by city, the south fell into his hands: Acapulco, Tixtla, Izúcar, Taxco. At the beginning of 1812, Morelos's troops fortified themselves within the city of Cuautla. Calleja, the commander-in-chief of the royalist forces, encircled them. Morelos waited in vain for the Junta to come to his aid and deliver the coup de grace to the besiegers, who thought their enemies would surrender in a few days. But Morelos's soldiers held out for sixty-three days. Bewildered, Calleja himself reported to the viceroy:

> If the constancy and activity of the defenders of Cuautla were moral and directed toward a just cause, they would some day deserve a distinguished place in history. Pressed hard by our soldiers and seriously in need, they are cheerful on every occasion. They bury their dead with peals of bells to celebrate the glorious death and with shouting and dances and drunkenness they make an occasion of the returns from their frequent sallies, whatever the outcome may have been, imposing a death sentence on anyone who talks of misfortunes or surrender. This cleric is a second Mohammed, who promises resurrection of the flesh and afterwards paradise with the enjoyment of every passion for his fortunate Muslims.[52]

One resource this Mohammed used to boost the morale of the men during the siege was his sense of humor. "His style tended very much toward the burlesque," Alamán said.[53] On one occasion, for instance, Rayón warned him, in a secret message, that the viceroy had sent a "stout and potbellied" assassin to poison him. Morelos replied in his inimitable style: "There's no potbelly here but me, and since I've been ill, it has trimmed down."[54]

Another great strength was his personal courage, which Alamán described as "calm, without excitement, without hotheadedness."[55] He withstood the cold of the mountains, falls from his horses, leg injuries, wounds, and illnesses that brought him close to death. The handkerchief he always wore on his head was probably meant to reduce the pain of his frequent migraine headaches as well as to protect him from the fierce sun of the south. He would move tranquilly through heavy crossfire. And the courage was passed on to his army who, following their leader's example, stood ready to "die or conquer" the royalists, whom Morelos had christened "the dragon from hell."[56]

Looting had been a constant of Hidalgo's revolution. In his campaigns, Morelos condemned "robberies and extortions that bad Americans inflict on their neighbors and on villages . . . [damaging] the positive opinion they had formed of our behavior."[57]

To one of his chief deputies, Valerio Trujano, he advised:

> Take measures, sir, against anyone who makes the mistake of wronging his neighbor by robbing or plundering, and whoever it

> may be, sir, even if it should be my father, you will order a hood to be put over his head, give him the sacraments and finish him off by shooting him within three hours, if the theft is more [than one peso], and if it is less, refer him to me so that I can send him off to prison . . . if there are many at fault, sir, you will execute every tenth man and the other nine in each group should be sent to me all roped together so that they may be imprisoned as well.[58]

Morelos was as unforgiving with his enemies as he was with looters. But he was not bloodthirsty. "None of us harms families nor are we the wild beasts you make us out to be!" he exclaimed before the terrified priest of Tixtla who expected a repetition of Hidalgo's Guanajuato massacre. Although he did not believe in being magnanimous and complained that "too much clemency has been directed toward the guilty,"[59] Morelos did not wish to annihilate those who opposed him. "What can the nation—or I, for that matter—gain from extermination?" Morelos would write a year after Cuautla to Vélez, the commander he was besieging in Acapulco, "Our system is not bloody, but humane and liberal."[60] According to the creole chronicler Bustamante—who at times would add a touch of fantasy to his memories—Morelos received Vélez's sword with these words: "Long live Spain, but Spain our sister, not the master of America."[61]

He was a leader primarily motivated by political and military aims, not by ethnic hatred or the desire for social revenge. He could "summarize his plan" in "few words": "The nation wants the government to be in creole hands, and since there has been no will to hear this, it has taken up arms to make itself understood and obeyed."[62]

Morelos did not confuse war with theological raging. At times he would free defeated enemy troops or invite them to swell the ranks of the Insurgents. Some he imprisoned. (He had the idea that, after final victory, he would banish dangerous enemies to an island.) Only on rare occasions did he allow vengeful anger to guide his actions. The grimmest of these would take place much later, in reprisal for the trial and execution of his right-hand man, the priest Mariano Matamoros. To the viceregal authorities, Morelos had proposed the exchange of dozens of Spanish prisoners for this single life. Informed of what they had done, he ordered the execution of a hundred men.

According to Alamán, the most objective of his adversaries, Morelos was endowed with insight, clarity, cheerfulness, severity, valor, loyalty, discipline, scrupulousness, originality, honesty, and, as if this were not enough, deep religious faith ("he would always take confession before going into action").[63] One can understand how this new

Mohammed had aroused not only the adherence but the love of his faithful followers. During the siege of Cuautla, the voices of the people sang this song:

For a corporal I'll give two reales
For a sergeant twice the price.
But for my general Morelos
I'll give my whole heart![64]

Morelos finally located a weak point in the royalist troop emplacements and led his forces out by night, at first catching the enemy unaware but then having to launch a running gunfight during which he had a mount shot out from under him and broke a rib in the fall. But the escape was successful and the armies of Morelos passed the rainy season of 1812 training in Tehuacán.

Several months later, at the high point of their success, they captured Oaxaca. While in that city, Morelos issued a number of decrees and regulations concerned with order and good government. He was now the supreme master of much of the center and south of the country—from Colima to Guatemala, including Oaxaca, the south of Puebla, and the southern parts of the provinces of Veracruz and Michoacán. The following year he committed his first serious military mistake, wasting precious time besieging Acapulco. By then his relations with the Supreme Junta had broken down. The troubles between Morelos and Rayón had been triggered by Rayón's excessive zeal for inspections, sending his representatives to look over the troops or make judgments on the authority of the Caudillo of the South. In his irritation, Morelos wrote to the Junta, "Your disagreements have been of service to the enemy."[65] He had no particular interest in leading the movement, but he suspected that for practical reasons it would be better if he did. Insurgents from the various provinces understood that the supreme command legitimately belonged to Morelos. And he accepted it without commotion. In a single move, he stripped Rayón of his positions, abolished the moribund Junta, and in its place convoked the Congress of Anahuac, which would first meet in the town of Chilpancingo in September of 1813. He invited the former members of the Junta to join this new Congress, but in order to maintain his position as supreme executive and military commander, he excluded himself from membership.

Aside from his impressive campaigns and military victories, which ultimately led to nothing, the most important contribution of the struggle headed by Morelos was to introduce a body of highly original ideological arguments that legitimized the revolution. It combined economic, political, and social proposals that were completely modern with ancient messianic themes. Morelos did not see the struggle for independence merely as a matter of arms and politics. For him it was, in the Christian sense of the term, a *mission.*

Perhaps because he came from a lower social position than Hidalgo and also because he was a mestizo, the ideal of social equality within the country took on as much importance for his program as that of national independence from Spain. It is worth noting that he drew more followers from the Castes ("dishonored by law as descendants of black slaves") and those of pure African descent than from the Indians. But part of Morelos's mission was to better the lives of both the Indians and the Castes. On their behalf, beginning in September of 1813, Morelos would expound a series of ideas similar to those "liberal and charitable" proposals that Abad y Queipo, in 1799, had unsuccessfully presented to the Spanish Crown: policies like freedom of contract and movement for the Indians and the Castes, a general abolition of the taxes and labor exactions required of them, the end of their legal inferiority, and permission for them to farm uncultivated lands. From his day-to-day experience as a parish priest in the poor villages where he had served, Morelos had become conscious of the social inequality between the impoverished, dark-skinned Mexico of the Indians and Castes and the affluent, white Mexico of the Spaniards and creoles. Out of these memories he drew his ideology. It was not, as with Hidalgo, a "cry" for a class war—an obsession with the past—but an invitation to reach out and build an agreement, pointing toward the future. Morelos wanted harmony between all the inhabitants of the country, excluding only the mainland Spaniards.

In November 1810, three years before the first meeting of the Congress, Morelos had issued the internal regulations for his army, prefiguring his image of an ideal society:

> If any sign of disturbance is observed between the Indians and the Castes, such that the Indians or the blacks want to move against the whites or the whites against the mulattos, he who raises his voice should be immediately punished. . . . Officers and noncommissioned officers will reach a consensus on the measures to be taken . . . punishing public sins with the greatest Christianity . . . each in agreement and brotherhood with the others.[66]

A year later he explained that "a very serious mistake" was about to throw the inhabitants of the area into "the most horrific anarchy."[67]

Morelos felt the need to declare that the *only* aim of his struggle was to see the government pass from the hands of the Europeans to the creoles. With the triumph of his cause would come a new and harmonious ethnic, economic, and social order. In 1812 he made the following pronouncement from Oaxaca in his peculiar and comic style: "Let that exquisite mouthful of conditions (Indians, mulattos and mestizos) be abolished by calling them one and all Americans."[68]

On the day before the first meeting of the Congress of Chilpancingo, which would formally declare Mexican independence, Morelos verbally sketched his personal utopia in the presence of Andrés Quintana Roo, a lawyer he trusted and who was deeply moved by his words:

> I would like us to make the declaration that there is no nobility but that of virtue, knowledge, patriotism and charity; that we are all equal, since we all come from the same origin; that there are no privileges or ancestral rights; that there is no . . . reason for slavery, since the color of the face does not change the color of the heart or of one's thoughts; that the children of the peasant and the sweeper should have the same education as the children of the rich hacendado; that every just claimant should have access to a court which listens, protects and defends him against the strong and the arbitrary . . . [that the laws should be] such that they . . . narrow the gap between wealth and poverty and increase the daily wages of the poor. . . .[69]

Around the middle of the century, Alamán claimed to see socialist features in the ideology of Morelos. But what Morelos had done was combine modern egalitarian and liberal ideas (drawn from liberal Spanish journals influenced by the French Revolution) with a return to the ancient notion of Christian peace, the reign of equality in which all people are brothers and sisters in *caritas*: "That there will be no distinction of conditions and we will all be called Americans, so that we may look on each other as brothers and live in the holy peace that our Redeemer Lord Jesus Christ left with us when he made his triumphal entrance into heaven. . . ."[70]

This combination of modern ideals and ancient religious paradigms and beliefs also marked his political thought. By adopting the system of representation, separation of the three branches of government, civil rights, and freedom of expression, the new nation would be a republic like its admired northern neighbor. Except in religion. It had to be a Catholic republic. One of Morelos's twenty-three *Sentimientos de la Nación*—a detailed program of republican modernity and religious traditionalism for the new nation—affirmed: "A person may lose the status of citizenship for heresy or apostasy." That the principles of republicanism and democracy could coexist with ancient ideas of religious militancy and total intolerance should come as no surprise, because from the very beginning echoes of the medieval wars of religion had resounded through this revolutionary movement.

The religious call felt by the Insurgent masses (the principal popular incentive in Hidalgo's uprising) was, with Morelos, given a theological formulation. At the same Colegio de San Nicolás where Hidalgo had been Rector, Morelos had come into close contact with the political theories of the Spanish neoscholastic thinkers. Drawing the obvious conclusions in his circumstances, he reasoned that it was *lawful* to wage war against a "heretical," "impious," "idolatrous," and "libertine" government of "Jacobin terrorists . . . in the service of Bonaparte."[71] It was *lawful* as well to wage war against the Spanish king because, ever since they had come to power in the middle of the eighteenth century, the Bourbons had been stripping away the old *fueros*, the Church's exemptions and privileges. This had been happening both in Spain and in the colonies, and now was the time for the Insurgent caudillos of New Spain—most of them priests—to recover those rights. And so the political ideas of the Insurgents involved a profound contradiction: Their fight against the absolute power of the Spanish monarchy was a liberal struggle, but it was also a fight to support the spiritual and temporal power of the Church, an attempt to restore the privileges of the other traditionally absolute power in Mexico. And in this it was antiliberal. But the Insurgents saw no contradiction. They equated civil liberty with the freedom of the Church as opposed to the Crown.

Not even the great South American republican Simón Bolívar, devoted to the teachings of Plutarch and Montesquieu, was alarmed by this strange mixture of politics and religion. He saw it as an effective tactic: "Fortunately," he wrote in 1815, "the leaders of the independence struggle in Mexico have put fanaticism to use by proclaiming the famous Virgin of Guadalupe as the queen of the patriots, praying to her in times of hardship and displaying her on their flags. In this way, political enthusiasm has been mixed with religion to produce a vehement fervor in favor of the sacred cause of liberty. The veneration for this image in Mexico far exceeds the greatest reverence that the shrewdest prophet might inspire."[72]

Though accurate on the effect of this resort to religion, Bolívar misread the motives of the caudillos. If for Hidalgo, who was more a man of the Enlightenment, using the Virgin as his standard involved a measure of opportunism, Morelos saw the Virgin not only as the protector of his cause but, in a certain sense, as its principal protagonist.

Morelos would display his genuine devotion to the Virgin even in minor details. Within the Constitution itself, December 12 was consecrated to the "Patron of Our Liberty." He would attribute his victories to the "Empress of Guadalupe." He asked the members of a newly founded town council to swear to "defend the mystery of the Immaculate Conception of Our Lady,"[73] and he used the emblem of the Virgin of Guadalupe as the seal of the Congress of Chilpancingo, before whom he made the statement: "New Spain puts less faith in its own efforts than in the power of God and the intercession of its Blessed Mother, who appeared within the precincts of Tepeyac as the miraculous image of Guadalupe that had come to comfort us, defend us, visibly be our protection."[74]

One of his officers went so far as to change his name from Félix Fernández to Guadalupe Victoria (combining reverence for the Virgin with a note of military triumph). In 1824 he was to become the first President of Mexico.

Morelos lived at the beginning of the nineteenth century, but his voice, like his favorite reading, went back to the age of the Bible. For him it was natural to compare the destinies of Mexico and the people of Israel: "The cause we defend is just: the Lord of Hosts who protects it is invincible."[75] This atmosphere is clearly reflected in the words of one of his followers, in the *Correo Americano del Sur* of August 12, 1813: "The third year of our glorious insurrection. Our brothers who work day and night like the Israelites in Egypt tending the cane and the land to fatten the fortune of the new Pharoah! Heaven has sent you a Moses and a Joshua to deliver you from this insulting bondage!"[76]

His combination of modern politics and the role of a biblical prophet would lead to a dramatic conclusion. Paradoxically, Morelos was to lose all his military power when he put it at the service of the most modern (and, at that time, most useless) of his creations: the Congress of Chilpancingo.

His luck began to turn near the end of 1813, for a number of reasons. In December of that year, Agustín de Iturbide defeated his troops at Valladolid, the birthplace of both leaders. Shortly afterward, Morelos lost Matamoros and then Galeana. And there was the attitude of Rayón. Hurt and embittered by his fall from power, he refused to keep Morelos informed on the shifting situation in the Bajío, which he knew well and Morelos did not. But perhaps the worst and most fundamental problem was the growing pressure to submit to the will of the Congress. A man so eminently practical as Morelos found it difficult to win battles while protecting a group of lawyers who followed him from village to village composing a constitution under the trees. The Congress would later deprive him of his presidential authority and even discussed the idea of sending him back to his parish in Carácuaro. Once when the Congress criticized his military strategy, he shot back: "When the Lord speaks, the servant ought to keep quiet—that's what my parents and teachers taught me."[77] It was very clear to Morelos that dissension in the ranks of the Insurgency and "diversity of opinions would prevent taking necessary measures," but he also felt that the Congress was immensely important, more important to him (as time would show) than his own life. In Guadalajara, Hidalgo had lived out an imperial fantasy. Morelos would submit himself to another kind of fantasy, this time a republican one.

The Congress finally finished drafting a constitution in October of 1814 in the village of Apatzingán, not far from the poor parishes where Morelos had been a priest. Modeled on the constitutions of the French Revolution, it gave overwhelming power to the legislature as opposed to the executive and judicial branches. We know that Morelos considered some of the provisions "impractical." But despite his reservations, it was still his happiest hour. Now there was a Constitution. "Throwing off his natural restraint, lighthearted and cheerful, he danced and he embraced everyone."[78]

Shortly afterward, still protecting his congressmen, Morelos was forced into an unequal battle, which he could have and would have avoided if he had been willing—impossible for him—to abandon the civilians and lead his troops in a rapid escape. The Insurgents were routed and Morelos himself fled on foot. The usual "pursuit" (*alcance*) of cavalry cutting down fugitives after a battle narrowed into a hunt, for a single man. Insurgents captured alive were interrogated: Had they seen Morelos? One of them gave the troops a direction. Morelos—unarmed, hiding in a thicket of bramble bushes with no chance to escape—saw soldiers approaching, led by a royalist lieutenant whom he knew, a man named Matías Carrasco, once an Insurgent. With the same calm that he would maintain with almost no weakening until the final moments of his life, Morelos straightened up and said: "Sr. Carrasco, it seems that we know each other."

He was taken back to Mexico City in a triumphant military procession. People crowded out their doors to watch him pass, some to admire him in silence, others to openly curse him, but all to see, with their own eyes, the man who had already become a legend. He would have to go through three trials—by the ecclesiastical authorities, the military, and the Tribunal of the Inquisition.

All of New Spain followed the proceedings, with deep though varied feelings. He was charged, among many other things, with heresy. Condemned to be defrocked and executed, he was told that he would be denied the last rites of the Church unless he publicly repented and repudiated his revolutionary cause. And he did so, and denounced by name his comrades-in-arms still in the mountains ("the only acts of weakness," Alamán wrote, "that he showed throughout his trials"). By the next day, when Morelos was defrocked, the moment of weakness had passed. Alamán's description is based on the words of an eyewitness:

> The Bishop of Oaxaca, dressed in the robes of his office, awaited him in the chapel which was at the entrance to the courtroom. Morelos, wearing the ridiculous costume they had dressed him in, carrying a green candle in his hand, accompanied by servants of the Inquisition, had to walk across a wide space. The large crowd, straining to catch a glimpse of him, stood on the benches. . . . Morelos, eyes lowered, but looking composed and walking with a measured pace, approached the altar . . . he put on the vestments of a priest and he dropped to his knees before the bishop.[79]

Morelos heard the terrible words: "We take from you the power to offer the sacrifice of God and to celebrate mass. . . . With this unfrocking we take from you the power you received through the anointing of hands . . . we deprive you . . . we remove you from office, we degrade you, we strip you . . . we cast you out, like an ungrateful son, from the inheritance of the Lord."[80] And then, in the words of Alamán: "Everybody was moved by this imposing ceremony; the bishop broke down and wept. Only Morelos—with a strength so beyond any ordinary measure some termed it insensitivity—remained calm, his expression did not change, and only during the actual act of humiliation did one see him shed a few tears."[81]

The Holy Court of the Inquisition believed that Morelos's trial would set an example, that it would cut off the insurrection at its roots. They could not know that the trial would be the last of its kind in the history of Mexico. Even Alamán, the harshest critic of the Insurgents, condemned those who had condemned Morelos: "The last thing they could charge Morelos with was heresy."[82]

His Calvary was yet to come. On his knees, Morelos listened to his death sentence. The next day, he entered a closed carriage accompanied by a monk. They took the road to the sanctuary of the Virgin of Guadalupe. "Morelos," wrote Alamán, "as they rolled along, recited prayers and the psalms *Miserere* and *De profundis* which he knew by heart. . . . When they arrived at the sanctuary, he managed to drop to his knees, despite his shackles."[83] He would be executed on the patio of an old mansion of the viceroys, in the arid village of San Cristóbal Ecatepec. When he looked out at that landscape, Morelos said, "Where I was born, it was the garden of New Spain." With genuine appetite he drank some chickpea broth, and he smoked his usual cigar. He requested a crucifix, and he said, "Lord, if I have done well, You know it; and if badly, I take refuge in Your infinite mercy." He heard the words of the priest: "You should not forget that our redemption is at hand."[84]

Alamán recounts, "They gave the order to fire and the most impressive man produced by the revolution in New Spain fell pierced by four bullets. But he was still moving and moaning and they shot four more, which extinguished what life there was left in him."

The Congress was dissolved and some Insurgents pleaded for pardon, but the lieutenants closest to Morelos were still in arms when Mexico finally won its independence in 1821.

All the historians of the nineteenth century, both Liberals and Conservatives, respected Morelos. Alamán especially valued his religious faith—his loyalty to the traditions of the past—yet saw him as an enemy. Mora stressed the opposite, his republicanism—his vision of the future—and praised him above all for "his outstanding moral qualities."[85] The Liberal romantics exalted him at times beyond any measure of reality, without understanding him at all. For them he was merely the military hero of the Insurgency and the brave follower of Hidalgo.

Not only the historians but also the caudillos of each camp admired Morelos. Strange as it may seem, Maximilian of Hapsburg was the first to erect a statue to Morelos in Mexico City. During the inauguration, he said:

> We saw the humble man of the people prevail on the battlefield; we saw the modest priest govern the provinces under his command in the difficult moments of their painful regeneration, and we saw him die physically, spilling his blood as a martyr for liberty and Independence; but the moral spirit of this man lives on in our fatherland and the triumph of his principles is the basis of our nationality.[86]

When Juárez restored the Republic, he authorized the creation of a new state south of Mexico City, to be named Morelos. But perhaps it was the last mestizo caudillo of the century who felt closest to the first. For Porfirio Díaz, the affinity with Morelos was not only ethnic and social but also military (he had memorized every detail of Morelos's campaigns) and political. He was deeply moved when the Marquis of Polavieja, during the Centenary Fiestas, returned Morelos's standard, uniform, and pectoral cross; so moved, in fact, that this man who was normally as voluble as the Sphinx went so far as to say:

> It never occurred to me that I would have the good fortune to witness this memorable day, on which these hands of an old soldier are anointed by the uniform that covered the breast of a brave man, felt the heartbeat of a hero and offered intimate protection to that exalted spirit that fought against the Spaniards, not because they were Spaniards but because they opposed his ideals.[87]

Morelos may have been admired by all, but they did not completely understand him. The Liberals forgot what was conservative in his life, the Conservatives what was liberal.

Much more than Hidalgo, the "Servant of the Nation" represented the dilemma of Mexico, a country in permanent tension between deep-seated tradition and inescapable modernity, between the devoted religious project of the sixteenth-century missionaries and the republican and liberal currents of the Western world.

In his *Historia política de México* Justo Sierra asked the generous understanding of his readers for "those early fathers of the Republic [who] clung to their religious beliefs as they would to a life raft. When they said God and Nation, they expressed all the love and faith in their hearts and souls. Children of this century, which is dying a skeptical, disillusioned and cold death, let us learn to respect and admire those who placed their faith and hope in a single religion, even to the steps of the scaffold."[88]

What Sierra could not grasp was that religion for "that early father of the Republic" was not "a life raft" but an essential part of his ongoing life. Yet his will to overcome the burdens and inequalities of the past was even greater. And for that he took up arms.

The nineteenth century may have been dying—skeptical, disillusioned, and cold—but not down in the Mexico of the people. That Mexico went on as it was, identifying its spiritual faith and its worldly expectations with a single religion. It went on as it was, crucified between the two words that formed a living slogan for the man whose prophetic life foreshadowed the future tensions of Mexico: *religión y patria*.

7

THE COLLAPSE OF THE CREOLES

In 1821 Mexico was born to an independent life with immense expectations based on its legendary wealth. It seemed as if the prophecy of the famous German traveler, Alexander von Humboldt, was about to come true: "The vast kingdom of New Spain, well cultivated, could by itself produce everything that commerce goes searching after throughout the rest of the world."[1]

It was a golden year in the history of the nation. In September the country saw the final chapter of an independence movement that was rapid, bloodless, orderly, and above all successful—totally different from Hidalgo's attempt of eleven years before. The unbelievable had happened, in just seven months. Through the bold initiative of the royalist leader Agustín de Iturbide, the country's creole classes—priests, soldiers, businessmen, and professionals—reached out to large sections of the rural and urban populace under the protective mantle of an agreement reached in the small southern town of Iguala. The pact was due to the efforts of two men who had been stubborn enemies since the days of Morelos: Vicente Guerrero, a mestizo, the last leader of the Insurgents, and Iturbide himself, a creole, who had opposed Hidalgo and defeated Morelos. With the passing of time, it would be said that Indians—the Tlaxcalans who supported Cortés—had accomplished the Conquest of Mexico in 1521, and that Spaniards were responsible for

Agustín de Iturbide, 1822

(Museo Nacional de Historia. Instituto Nacional de Antropología e Historia)

the achievement of independence in 1821—the colonial settlers in Mexico who, fearing the imposition of the new liberal Constitution of Cadiz, found their savior in the caudillo Iturbide. There is a good deal of truth to this paradox, but it ignores two elements without which the achievement of independence cannot be understood: the tenacity of the Insurgent guerrillas and Iturbide's biography.

Born like Morelos in Valladolid, Iturbide came from a rich family.[2] When he was only fifteen, in 1798, he was already managing his father's Hacienda de Quirio. At twenty-two, he enlisted as a cadet in the Valladolid Regiment of Provincial Infantry. During the same year he married Ana Huarte, daughter of the most powerful Spaniard in the region. When the first stirrings and conspiracies of the independence movement began in Mexico City and Valladolid in 1808–1809, the social position of the young Iturbide was almost diametrically opposed to that of the creole priest who was eventually to become the leader of the Insurgency. Even as Hidalgo was burying his brother Manuel and finally repossessing the haciendas impounded by the Bourbon government, Iturbide was acquiring a hacienda free and clear—San José de Apeo, not far from Hidalgo's property. It was not surprising that the Iturbide family supported the government against bids for creole autonomy or that Iturbide turned down Hidalgo's offer to join his cause with the rank of lieutenant general. The sacking of his parental hacienda and the flight of the family to their house in Mexico City gave him even more reason for enlisting in the royalist ranks.

From his first encounter with the Insurgent troops, on the Monte de las Cruces, Iturbide was to have a phenomenal military career. "I was always happy during the war," he wrote many years later in his *Memoirs*; "victory was the inseparable companion of the troops I commanded. I did not lose one battle; I defeated as many enemy troops as attacked me or I encountered, even when they outnumbered me, as they often did, by ten to one."[3] Despite the arrogance, this is no exaggeration. His war record—like his own detailed war diary—is a list of how he took fierce caudillos prisoner, captured fortifications difficult to assault, and—most striking—won victories over the most illustrious Insurgent leaders: Liceaga, Rayón, and even Morelos himself at Valladolid.

What his *Memoirs* totally omit is that element in his character described by contemporaries from all factions: his cruelty. As Lorenzo de Zavala would later write: "Over a period of nine years he was known for his brilliant campaigns and for his cruelty toward his fellow countrymen."[4] And Alamán said that "he left a trail of blood in his wake." Though Alamán was no friend of Iturbide (there had been some personal problems over business dealings), he liked the rebels even less.

But he found Iturbide "harsh beyond measure with the Insurgents, [a man who] sullied his victories with a thousand acts of cruelty and the drive to enrich himself by any manner or means."[5]

In his *Historia de México,* Alamán would document Iturbide's excesses with a vengeance. Iturbide would freely execute enemy soldiers and innocent civilians, often without even allowing them the last rites of their religion. He gave orders to burn any and all villages that had been occupied by the Insurgents. He was equally merciless toward his own soldiers whenever he observed the least sign of cowardice, on occasion having soldiers shot by lot for what he saw as group cowardice.

Perhaps the cruelest thing he did was issue a series of edicts, at the end of 1814, directed against female relatives of Insurgent soldiers, or those merely suspected (sometimes wrongly) of being so. Women and their children would be crowded together in miserable conditions. (Within one group, smallpox spread without Iturbide showing any concern.) Some women prisoners, who had been "entombed" by his orders, asked several years later to be "sent to Purgatory, which we judge to be a place less asphyxiating than the one we are in now." In one of his edicts he specified that "one woman in every ten, five or three arrested at any distance from their disloyal fathers, husbands or brothers, etc. should be beheaded without possibility of reprieve whenever the traitors committed specific outrages."[6]

Whatever may have been the psychic sources of Iturbide's cruelty, one of them surely was his deep hatred of the Insurgents. For him, they had done nothing but lay waste to the country, destroy fortunes and resources and "the order of the army." They were responsible for the deaths of thousands and for breeding racial hostility toward the white man. They were "infesting the country."[7] The enmity between Hidalgo and Iturbide, the two *hacendados* from Michoacán, was so intense perhaps because it was a hatred between brothers with the same heritage—the major creole leader of the rebellion and the major creole commander of the royalists. Unfortunately for Mexico, the poisonous weed of this hatred did not die in 1821: It grew with the century, collecting other causes, assuming other names.

For Iturbide, the other side of his cruelty was ambition so great that, as Abad y Queipo asserted: "It would not be strange if, as time passes, he himself should be the one who brings about independence."[8] It was a perceptive statement. Alamán describes him at the beginning of 1815, on the day of the only battle in which luck was not on his side:

> Sitting against a boulder with General Filisola, Iturbide lamented the senseless waste of blood, calling Filisola's attention to the ease

> with which independence could be achieved, simply by negotiating an agreement between the Mexican troops fighting under the royal flags and the Insurgents. But considering the complete disorder of the Insurgents and the atrocious system they had proposed, he concluded by saying that it was necessary to defeat them before thinking of putting into effect any orderly plan of action.[9]

Filisola, Alamán goes on to say, agreed with Iturbide's opinion of the situation, and Iturbide then added: "Perhaps the day will come when I shall remind you of this conversation, and I shall rely on your support."[10] Nevertheless, in the summer of 1814 as the Insurgency was losing ground, Iturbide's ambition was trained on other targets: glory, honors, the cross of the Order of San Fernando, and other awards he believed that he deserved.

Chance and fortune would lead him elsewhere. Beginning in 1816, Iturbide found himself involved in a sensational public scandal concerning his moral conduct during the years of the war. A Dr. Labarrieta—his bitterest detractor—detailed a series of thefts, plunderings, burnings, and illicit commercial transactions for which Iturbide had been responsible. The defenses put forward by the former commander of the royalist troops and by the viceroy, Félix María Calleja, were not enough to clear his name. Iturbide then hired a lawyer to represent him before the Crown, but could not win the complete exoneration he wanted. "He was acquitted," writes Alamán, but "he did not wish to return to his former position of command."[11] He would only be satisfied with a clear-cut, not just general vindication and one thing more: the Cross of Isabel la Catolica, which Calleja had just been awarded for achievements similar to Iturbide's own.

He waited in vain. In 1818 he rented a hacienda close to Mexico City, which he must have mismanaged judging by the loans he began to incur. Alamán says:

> In the prime of life, with an outstanding appearance, pleasing and refined manners, an agreeable and ingratiating way of speaking, well received by society, he devoted himself immoderately to the dissipations of the capital. . . . In such pastimes he rapidly frittered away most of the wealth he had created through his businesses in the Bajío, leaving him in a very poor financial state, when the reestablishment of the Constitution and the resulting opportunities opened a new field to his ambition for glory, honors, and wealth.[12]

It was then that he must have remembered his conversation with Filisola. Now he was experiencing, in his own person, the oppression of his condition as a creole. And suddenly he was able to unite and resolve his own life and that of his country through a single redemptive action. His outraged honor and that of Mexico were worth an embrace with Vicente Guerrero, the last of the Insurgents.

According to the *Plan de Iguala*, the unified army of the two caudillos was to be described as *trigarante* because it would guarantee three fundamental principles: the unity of all ethnic and social groups, the exclusive rights of Roman Catholicism, and absolute independence from Spain. Ties were not to be broken with Spain; they were instead to be loosened. There would not be even a hint of parricide. According to the Treaties of Córdoba signed in August 1821 by Iturbide and the last viceroy of New Spain, the new nation would adopt a constitutional monarchy as its system of government and would invite a Bourbon prince to fill the throne. Some of the phrases of the *Plan de Iguala*, delivered in a public address by Iturbide, convey a sense of the almost messianic happiness of the moment:

> With the growth of towns and learning, since we now know of the many varieties of natural opulence in this soil, the wealth of its minerals, the advantages of its geography, the injuries caused by the distance of the central power from this separate entity, as well as the fact that by now the branch is equal to the trunk—the public and general opinion . . . is for absolute independence from Spain.[13]

On September 27, 1821, Iturbide's thirty-eighth birthday, the ten thousand men of the "Trigarantine Army," royalists and Insurgents united, entered Mexico City. It was the first time that the independence movement had shown its face in the City of Palaces[14]—its first and definitive appearance. The army's tricolored flag symbolizing the gist of the Program of Iguala was so popular that it would come to be adopted, with minor alterations, as the national flag. Against a white background, representing the purity of the Catholic religion, flanked by a green that referred to independence and a deep red that conveyed the memory of Spain, the flag bore the emblem of the mythical foundation of Mexico-Tenochtitlan by the Aztecs: an eagle perched on a nopal cactus and devouring a serpent. "That September twenty-seventh," wrote Lucas Alamán, "was . . . the only day of pure enthusiasm and pleasure, without any admixture of sad memories or foreshadowings of new misfortunes, that Mexicans have ever enjoyed."[15] Mexico was born of multiple reconciliations, of an embrace between royalists and Insurgents, between *gachupines*, creoles, Indians, African-Mexicans, mestizos, and

Castes, between the pre-Hispanic past and the three centuries of colonialism, between the branch and the trunk.

At that moment of expansion and optimism, merely to contemplate the map of imperial Mexico must have been astonishing: The country ran from the Arkansas River and California in the north to Central America in the south, from the long coast of the Pacific to the Gulf of Mexico. It included almost all of the southern part of North America and most of Central America. Iturbide could contemplate the future "intimate union" of the new empire with the island of Cuba and regard the United States with a mixture of admiration and apprehension.[16] The situation in Texas particularly concerned him "because of the former government's neglect of this very interesting part of the Empire."[17]

While he waited for formal Spanish approval of the Treaties of Córdoba—a document that would confirm Mexican independence—Iturbide shared power with a provisional council of regents and with the National Congress, created to draw up a constitution for the new nation.

All the repressed anguish of a decade of war, noble expectations, and ignoble ambitions rose to the surface in a torrent of eulogies, poems, hymns, phrases, or epithets dedicated to the "Father of the Nation," the "Virile Man of God," the "Pillar of the Church," the demigod named Agustín de Iturbide. "It is you who have torn the mane of the Spanish lion to shreds," sang a young poet, while another raised him to the rank of "a new hero without precedent in history."[18] Every adjective of praise in the Spanish language was applied to describe that special envoy of Providence: Iturbide the magnificent, the incomparable, the liberator, the magnanimous, without equal, the immortal, the illustrious.

A few months later, when the fiesta of optimism was still in full swing, Spain made it clear that it would not send a prince to rule the Mexican Empire and, in fact, it did not officially recognize Mexico's independence until fifteen years later. The Vatican took a similar position. In the midst of all the euphoria, a sense of being orphaned suddenly clouded over the historical baptism of the new nation. Unlike Portugal, which had granted Brazil its independence during the same period and had sent a prince to govern the country, Spain denied its paternity and cut the cord of imperial legitimacy with Mexico. It was universally accepted that the only solution to this parental rejection was to create ex nihilo, rather as Napoleon Bonaparte had, an alternative paternity. By "Divine Providence and the will of the Congress of the Nation," which is to say, according to tradition and according to law, Agustín de Iturbide would be both anointed and elected "Constitutional Emperor of Mexico."[19]

The coronation took place on July 21, 1822. Amid the chorus of pomp and circumstance, there were dissonant notes to the ceremony. As Zavala wrote, it was if they were trying "to transplant institutions . . . to America that elsewhere derive their sanctity specifically from a tradition and a history."[20] The participants and the audience seemed to know they were part of a puppet show, a parody where the gestures, the words, the feelings were not quite right. Iturbide broke with custom, for instance, by himself crowning his wife. Congress, in the person of its president, crowned Iturbide. While the crown was being placed, the president of the Congress—a personal friend of the new emperor—joked about not letting the crown fall. But during the ceremony, the crown almost did fall off. Most striking of all was Iturbide's speech after swearing his oath. He declared that *he should not be obeyed* if he did not respect the Constitution, which the Congress would formulate. It was as if he were speaking not as an emperor but as a member of Congress. Words like these were not really self-critical, but because they depreciated his own authority they were damaging to himself and they were clear signs of insecurity. Instead of celebrating his accession to the throne, he spoke of being "bound with golden chains, overwhelmed with immense obligations," and added that "splendor, exaltation and majesty are only vain toys," as if he were already an old and disillusioned king.[21]

Four days of fireworks followed the coronation. Praise and flattery poured in from all sides. But Iturbide, who had not usurped his crown, felt like a tormented usurper. A week after he became emperor, he confided his thoughts in a letter to Simón Bolívar, who of all men in Spanish America could best understand them: "How far I am from considering this a benefit, when it has placed this burden on my shoulders that overwhelms me!" And he added, "I lack the strength needed to hold the scepter; I refused it and finally consented in order to avoid evil consequences for my country, which is on the verge of once again yielding, if not to the ancient slavery, then surely to the ills of anarchy."[22]

During the ten eventful months of Iturbide's reign, the real economic situation of the "opulent empire" became all too clear. Mines had been abandoned during the war, haciendas devastated, and Mexico's incipient industrial development brought to a standstill. There had been an immense flight of capital abroad—more than a hundred million dollars or pesos (the currencies were at parity), ten times Mexico's annual budget. There was only one word for the situation: bankruptcy. No elaborate calculations were needed to prove it. All you had to do was listen to the songs being sung in the streets:

I am a soldier of Iturbide
and have seen the Three Guarantees.
I stand guard in my bare feet
and I go hungry every day.[23]

On the diplomatic front, the picture was no less menacing. Without credit abroad, without recognition from the United States or England, vehemently rejected by Spain, the Vatican, and the member states of the Holy Alliance, Mexico's only apparent hope was to form some connection with the Great Colombia of Bolívar. The empire was isolated.

But the fundamental problem—as had happened with Morelos—was Iturbide's rivalry with Congress. Separated from the trunk and rejected by it, the branch had split in two: emperor and Congress. That collective father had anointed him emperor and felt that it retained rights over him; and Congress tried to exercise them from the very first day. The legislature objected to his veto power, obstructed efficient management of the economy, blocked the emperor's creation of a supreme court; and across the connecting threads of secret Masonic meetings, the legislators wove conspiracies and plans to depose Iturbide. Suddenly, in a strange inversion of historic roles, the emperor would act in a "republican" manner—dividing his power, trying to share it with the legislature—while the Congress would take positions that were imperialist and absolutist. This behavior on the part of Congress was noted by a perceptive English observer who pointed out—though he also did not believe in Iturbide's selflessness—that the deputies "had grafted the boldest theories of the French school onto the rigorous despotism in which they had been raised."[24]

There was bound to be a crisis. In Cromwellian style, Iturbide's next move was to dissolve Congress and put an ineffective and unpopular junta in its place. At this moment a young and imperious creole brigadier made his appearance on the scene. He came from the city of Veracruz, the richest, gayest, most spendthrift and morally easygoing port in the country. Like Iturbide, he had been a royalist, but unlike Iturbide, in his own part of the country he was a king. Among his pursuits were gambling, womanizing, cockfighting, delivering speeches, flattery, and forging commercial documents. His father had been an army officer who was always in debt and his mother a great beauty rumored not to believe in God; he had Gypsy blood in him, and one of his uncles had seen no contradiction in being at once a priest and a bullfighter. This reckless and irresponsible brigadier was the first to "pronounce" against Agustín I; he rose up in arms against "absolutism" and suddenly discovered his republican convictions. He seemed, as Iturbide put it, a "volcanic genius"; and the rumble even ran through his name—Antonio López de Santa Anna.

Although Iturbide could have used the full power of his army to crush the rebellion, he did not. His passivity, impotence, and lack of self-confidence remain a mystery, especially in the light of his outstanding military career against the Insurgents. Perhaps because he had been responsible for so much bloodshed, he decided not to "shed one drop more."[25] He was afraid to act not because he was afraid of his enemies or because he lacked the means or even because he had any doubts about the general support of the military. But he was afraid of anarchy, of public opinion attributing whatever actions he might take to his "private interests," to his "desire to keep wearing the crown which he had accepted only so as to be of service to the nation."[26]

Whatever his reasons may have been, he stoically reconciled himself to the humiliating sequence of events: reinstatement of the old Congress, abdication (on March 19, 1823), ridicule, and exile. The man chosen by fate was fated to become the "unjust man," the "traitor," the "tyrant," and the "new Caligula." Congress condemned him to death, should he ever return to Mexican soil.

He spent time in Italy, where he wrote his memoirs; and toward the end of 1823 he heard news of a possible invasion of Mexico by the Holy Alliance, in support of Spain. The threat would be aborted by English threats of military action and the announcement of the Monroe Doctrine by American president James Monroe. But to Iturbide—at the time and in his circumstances—the threat of reconquest seemed very real. His personal funds were shrinking (as an administrator he had been an honest man), and he was in England, where he received letters imploring him to return. Though advised against it by none other than the great South American caudillo San Martín (who was himself in exile), Iturbide, fearful for Mexico and eager for renewed glory, set sail for home. He went unarmed, accompanied by some of his family.[27]

He was unaware of the sentence of death. When he landed at the port of Soto la Marina on the Gulf of Mexico, he was arrested by one of his former lieutenants, who hesitated between carrying out the death sentence himself or passing the case on to the local state congress of Tamaulipas, meeting in the town of Padilla. Iturbide was sent there, where he asked what crime he had committed to deserve this punishment. Nobody listened to him. The congress of Tamaulipas took on the roles of judge and army. He was condemned to die on July 19, 1824, before a firing squad next to a chapel like so many other Mexican heroes. With his back to the wall, he said, "I die with honor, not as a traitor . . . I am not a traitor . . . I do not say this full of vanity, which I am very far from feeling."[28] Before the firing squad itself, facing his own death, he had excused himself again. Was he perhaps asking pardon for the blood that he himself had spilled, in the days of the Insurgency, unsure that he could ever expiate it even before the rifles of his executioners?[29]

Some years later, Simón Bolívar, the Great Liberator, shared his thoughts about the ephemeral Mexican Empire with his close friend, General Santander:

> That Iturbide had a somewhat meteoric career . . . bright and swift like a brilliant shooting star. If fortune favors daring, I do not know why Iturbide was not favored, because daring was always his guide. . . . Anyhow, this man had a singular destiny; his life served the freedom of Mexico and his death its repose. I frankly confess that I never cease to be amazed that a man as ordinary as Iturbide could do such extraordinary things. Napoleon Bonaparte was called upon to do marvels, but Iturbide was not; and, for that very reason, he did greater things than Bonaparte. God free us from meeting with his luck as he has freed us from following the same course in life, even though we can never free ourselves from the same ingratitude.[30]

Alamán would say that Iturbide's reign, rather than an empire, had been "a theatrical performance or a dream."[31] Now it was the turn of Congress, which had banished Iturbide, to offer the option of a republic. It would prove to be as dreamlike and theatrical an experience as the empire.

In 1824 Mexico adopted its first federal Constitution, elected its first President—Guadalupe Victoria, a veteran of the War of Independence—and chose a representative Congress and a judiciary, all closely modeled after the United States of America. Under the Constitution, each state of Mexico was to be a free and sovereign body, with its own constitution, electing its own governor, legislature, and judiciary.[32] Many individual rights were guaranteed, with the striking exception of religion. Only Catholicism was legally permitted.

Very soon, the ideal Republic had to deal with the real facts of political life in Mexico. Among other things, freemasonry was spreading and the rival Masonic lodges were undermining the representative and democratic republican government. The future directions of the nation were not decided legally in the Congress, supposedly duty-bound to entertain all opinions, or in the press, which fought for the expression of these opinions, but in the shadows, behind closed doors during the lodge meetings of the York Rite Masons (who were anti-Spanish, radical, pro-American, federalists, and embryonic liberals) and of the Scottish Rite Masons (who were pro-English, moderates, centralists, and embryonic conservatives). It was there that the future of the country would be determined, through military conspiracy, bribery of deputies, the rigging of elections, and the use of public money and institutions to back electoral campaigns. By 1828, the underground struggle between the lodges had been settled in favor of the York Lodge, founded by Joel R. Poinsett, the first United States ambassador to Mexico.[33] In that same year, the two factions within this lodge each nominated a presidential candidate. The moderates chose General Manuel Gómez Pedraza, and the radicals selected Vicente Guerrero, hero of the independence movement. Following Gómez Pedraza's victory at the polls, Guerrero organized a rebellion. He was supported by several "democratic" intellectuals and, of course, by the indispensable Santa Anna, who by this time had several riots to his credit. After a mob had sacked and plundered the Spanish market in the center of Mexico City, Guerrero was declared President. Bolívar, who always kept a watchful eye on the misfortunes of Spanish America, wrote:

> The casual right to usurpation and pillage has enthroned itself as King in the capital of Mexico and in the provinces of the Federation. A barbarian from the southern coast [Guerrero] . . . has climbed over two thousand corpses, at the cost of twenty million pesos seized from private ownership, to reach the highest post of government. . . . Unable to become President under the law or at the polls, he has allied himself with General Santa Anna, the most perverse of mortals. First, they destroy the Empire and execute the Emperor, since they could not reach the throne; then they set up the Federation along the guidelines of other demagogues as immoral as themselves, in order to gain possession of the provinces and even the capital. . . . Guerrero . . . Santa Anna . . . What men or what demons they are![34]

Bolívar was mourning the violent end of a republican dream. Military coups led by Santa Anna had brought down both the empire and the Republic, but Bolívar had hoped that the bloody end of the empire—a government that had been a mistake—would bring "repose" to the "opulent nation." The republican coup against republicanism wounded him much more. The events in Mexico nourished the opinion

Bolívar had been forming about the cruel destiny of Spanish America: "There is no good faith in America, not among nations. The treaties are papers; the constitutions are books; the elections are combats; freedom is anarchy; and life a torment."[35]

Several months later, Guerrero, like Iturbide, would be deposed by Congress and declared "mentally incapable of governing."[36] While Iturbide had chosen exile, Guerrero returned to his familiar world of guerrilla warfare. In 1831 the new military government paid a Genovese sailor named Picaluga fifty thousand pesos to give them Guerrero. With flattery and effusive respect, Picaluga lured Guerrero onto his boat at Acapulco, then had him seized and delivered to the Mexican authorities at the port of Huatulco in the state of Oaxaca. On February 13, 1831, Guerrero would face a firing squad in the orchard of an old chapel near the state capital. That volley put an end to Mexico's first republican dream.

The new nation slowly would begin to face reality. The populated regions of the country were not nearly as rich in resources as Humboldt had imagined. The immense deserts of the north were as unwelcoming as the forests on the Gulf Coast. Anyone who wanted to reach the fertile land of the Central Plateau from the Atlantic or the Pacific Ocean had to cross one of the two rugged mountain ranges that, running from the north along either coastline, posed difficulties of passage not only for people and their goods but also for the winds and rain.

In the countryside, the essential unit was the hacienda—self-governing, unproductive, and aristocratic, more an echo of feudal times than of modern capitalist exploitation (there were some six thousand of them in the country). Mexican silver had been a fundamental source of wealth for the Spanish Crown, but eleven years of civil war had closed down many of the mines or at least paralyzed their operation; modern industry was practically nonexistent; and the lack of navigable rivers lent an air of "aristocratic sterility" to the Mexican countryside. Beyond the populated regions, a prodigious wealth of natural resources lay waiting in the vast expanses of the north, in an area touched by only a few colonists, adventurers, and missionaries. But there were virtually no Mexicans who could appreciate or exploit them or even suspect that they existed. At the start of its independent life, the new nation was not even close to having a reliable map of its territories and borders or any inventory of its natural riches.

To control that vast land of more than four million square kilometers, Mexico could count on a population of barely seven million people. Most of them were still Indian, and 90 percent were living in small villages or isolated clusters of huts. But these natural and demographic drawbacks were a minor consideration compared with the political incompetence of Mexico's leaders. The country—centuries behind at birth for its task of creating a regime of civil liberties and economic well-being—would soon lose precious decades in civil discord, finally leading to bankruptcy, internal violence, and—in the forties—war against foreign invaders and the dismemberment of its territory.[37]

Even though Mexico was no longer a colony or had any place in the supranational order of the Spanish Empire, it was not yet a nation. It was an assemblage of villages, settlements, and provinces isolated from one another, without any conception of politics, even less of nationality, and controlled by the strong men of each locality. These figures had sprung up like mushrooms, not only in Mexico but everywhere in Latin America, born and bred—as much as the caudillos—by the collapse of the colonial order. Though the term for them would be different in the Pampas or in Venezuela or in Mexico, they were much alike. Validated by their personal strength, by the power and prestige they had won during the Wars of Independence, by the terror they inspired in their communities, and the benefits that they promised, these men became local monarchs. The name for them in Mexico—*cacique*—was an Indian word for chieftain that stemmed from the Caribbean. Since the earliest period of the colony, it conveyed the idea of absolute—almost theocratic—authority and was clearly rooted in indigenous tradition. But the caciques were local, while the typical Mexican caudillos, those military chieftains who had "risen with and seized the kingdom" during the War of Independence, did not limit their action to provinces but extended their activity to the entire country and sought power over the entire nation. The caciques were generally mestizos (sometimes Indians or Castes), claiming the loyalty of ethnically similar followers. Though Morelos had been an important exception, the caudillos, at least until the middle of the century, were primarily creoles.

The creole elite and their caudillos were given a unique opportunity to exercise unlimited power, but as the years went by, they would prove—to an unbelievable degree—that they could not create an economically solvent and politically stable state. Second-class Spaniards during the colonial period and favored Mexicans after the proclamation of independence, the creole caudillos seemed not to understand the most elementary aspects of economic life and showed almost total inexperience in the arts of government and diplomacy. Spain had deprived them of that training. During three centuries of domination, the Crown had always named the governing authorities. Very rarely were they drawn from the native-born population. And of course Spain had long since ceased to be a model of political and economic efficiency, so that even if the branches showed some similarity to the trunk, Mexico's political future did not look promising.

It would be a long and painful apprenticeship for the creoles. Their lack of experience showed in infinite ways. The disorganized, short-sighted, and unproductive management of the National Treasury and international loans (which came, at first, from England) was just one example: The ongoing expenses, particularly of the army, ate up the budget. Another was the persistence of ethnic and social hatreds originating during the War of Independence: In 1828 the government expelled all Spaniards from Mexico. But perhaps the most characteristic error of the period was the unrealistic idealism of the laws and their resulting devaluation. Many leaders in the public life of the Mexican state had inherited the modes of Spanish political practice, putting their trust in fixed principles, concentrating obsessively on the formal and abstract aspects of constructing a nation and ignoring practical, concrete necessities. Between 1822, the year of Iturbide's enthronement, and 1847, the decisive hour of the U.S. invasion, Mexico lived in a permanent condition of disruption and poverty. The country endured fifty military regimes, swinging from a federalist republic (1824–1836) to a centralist government (1836–1847) and back again to a federalist republic from 1847 on. It suffered losses of territory, including the irrevocable secession of Texas and the secession of Yucatán (which lasted for a decade, to be reversed in 1847). But the governments still found time to convoke five constitutional conventions and promulgate a charter, three constitutions, a reform act, and innumerable state constitutions, each one driven by the notion of final, national redemption.

Closely linked to all this legal labor was its counterpart: the "uprising," the barracks revolt, which would sometimes be preceded by a formal proclamation. Under any circumstances, before each coup, the military officer in command (usually Santa Anna) felt obliged to fire off a volley not of powder but of theatrical rhetoric, the *pronunciamiento*: "Mexicans! . . . Soldiers of freedom!" and so on. In the process, as the programs became more "perfect," the military coups designed to impose them became more frequent and the laws fell into greater disrepute. Mexico came to be known as a country "of revolutions." So relentless were these "pronouncements" from the barracks, and the gun battles immediately afterward, that the citizens began to take them, as a chronicler of the period reported, "with a festive air, dancing and singing, eating and drinking," and "they almost feared the establishment of peace." The people themselves developed a catchphrase to herald the uprisings: "*Ahí viene la bola.*" ("Here comes the mutiny!").[38]

In 1831, ten years after the proclamation of independence, the country was still meandering aimlessly along. This sad state of affairs was true not only of Mexico but also of other countries throughout the former Spanish America. The creole minority was suffering from a severe crisis of political identity and an increasing lack of confidence. Bolívar, who had died a year earlier, had written in 1826, "I feel in my bones that only a competent despotism can succeed in America," but then said in 1827 that "dictatorship is the reef for republics."[39]

None of the Great Liberator's projects for South America were realized. Like his Mexican counterparts, Bolívar produced idealized laws, but however sublime they might be, they had little effect on the real course of events. "We find America ungovernable," he concluded. "He who serves a revolution ploughs the sea."[40]

The implicit notion of power that dominated the creole mentality—traditional, hierarchical, corporative, the patrimony of the Spanish Crown—had been lost with independence. It was very hard for these men to grasp the modern concept of republican and representative power emanating from a constitution. In 1827 the founding father of Mexican liberalism, José María Luis Mora, had written: "The most important thing . . . is to reduce the motives, real or imagined, which could lead to the concentration of a great accumulation of authority and power in the hands of one man. . . . The love of power, innate in man and always on the increase in the process of government, is much more to be feared in republics than in monarchies."[41]

He felt that the great danger for Mexico lay in the appearance of a Mexican Bonaparte, wearing a mask of liberal forms and appearances, who under the guise of representing the country would excite the public mind with promises and praises until the public became his slave and political freedoms would be annulled. Mora felt the best way to prevent this happening was through the precepts of classic, constitutional liberalism.

The founding father of Mexican conservatism, Lucas Alamán, was calling for a strong executive (ideally a European monarch who would arrive without his army) advised by technically competent councilors, and a centralized administration that could show an iron fist together with the will to neutralize the legislative congresses and what he considered their partisan baggage ("the party spirit stains everything that falls within its power and influence").[42] And he wanted a vigorous military but also an independent judiciary.

Unfortunately, neither Alamán nor Mora were given much of an opportunity to experiment with their projects for the nation. Claiming to represent that nation, dressed in liberal (as well as, indiscriminately, conservative) forms and styles, a creole caricature of Bonaparte had, as Mora feared, mounted the national stage, where he would linger until 1855.

"The history of Mexico since 1822," wrote Lucas Alamán, "might accurately be called the history of Santa Anna's revolutions. . . . His name plays the major role in all the political events of the country and its destiny has become intertwined with his."[43]

If Iturbide's empire had been the dramatic enactment of a tragedy, Santa Anna's eleven presidencies might be compared to the less distinguished genre of operetta, in which he played several roles. One of them was "the conspirator." Sometimes, a strident "pronouncement" was enough to topple a government. On other occasions, he would incite the rebels in secret and, when he realized that the government had more soldiers than he did, he would change sides and boldly join the peacemakers. But afterward he would always retire with his profits in money, power, and prestige to Manga de Clavo, his luxurious hacienda in the state of Veracruz. Smothered in the admiration of his public and feigning the most bucolic of lives, "concerned only with his crops and his livestock," Santa Anna would wait for the next call of destiny, which never failed to come and nearly always worked out to his advantage.[44]

Santa Anna played an important role in the uprising of Vicente Guerrero against the elected President Manuel Gómez Pedraza. Audacity was his special military quality—the virtue of a gambler. Besieged in Oaxaca during the uprising, he came up with a whole series of schemes, an entire range of roles. By night he would raid business establishments and religious institutions to hunt for supplies; he dressed up as a woman to carry out his personal espionage on the enemy; he climbed the high walls of a convent and disarmed its military defenders; he disguised himself and some of his followers as friars, then summoned the faithful to mass and, closing the doors when the church was full, exacted heavy loans from the congregation. Another role he was particularly good at playing was "the reckless soldier." He performed it in 1829 when Spain landed an expeditionary force to attempt the reconquest of Mexico. Saber in hand, Santa Anna commanded the battalions that stopped the Spaniards in their tracks at the port of Tampico. His attack on the Spanish headquarters was, in the words of the distinguished general Manuel Mier y Terán (who had been one of Morelos's few creole officers), "a master stroke of boldness."[45] Promoted to major general, Santa Anna was now a national idol. He had become "the hero of Tampico,"[46] "the fearless son of Mars," "the support of a people."[47] There was a momentary rebirth of

Antonio López de Santa Anna, 1853
(Private Collection)

the old creole optimism, barely distinguishable from the breathless enthusiasm of a gambler. Santa Anna was the new man of destiny.

If it was a caudillo's *personal* attributes that were to guarantee the only legitimacy possible at that time, then Mexico's problem was that Santa Anna did not have the diligence, tenacity, patience, or even the will to rule needed by a leader. He had other qualities, which the liberal intellectual Lorenzo de Zavala had detected in 1826. For one, as a soldier: "The soul of the general cannot be contained by his body. He lives in perpetual agitation, he gets carried away by an irresistible desire to acquire glory. . . . Defeat . . . maddens him . . . then he abandons himself to a feeling of weakness though not cowardice. He pays no attention at all to strategy."[48]

And also there was his profile as a politician. The most outstanding feature within that dramatic role was that it showed no principles at all: "He is a man who has within him some force always driving him to take action but since he has no fixed principles nor any organized code of public behavior, through his lack of understanding he always moves to extremes and comes to contradict himself. He does not measure his actions or calculate the results."[49]

In 1833 the "Man of Destiny" was elected President and given his first opportunity to govern the country, but governing was not for him. It annoyed him and bored him, and perhaps it frightened him. Pleading illness, he did not even attend his inauguration but remained in retirement at his hacienda, leaving Dr. Valentín Gómez Farías, the liberal Vice President, in command. During Santa Anna's absence, Gómez Farías, with José Luís Mora as his advisor, enacted an imposing series of reforms directed against the corporate privileges of the Church in economic, judicial, political, and educational affairs. Santa Anna kept himself informed and cleverly kept his distance, attending to which way the wind was blowing. It was not long before General Durán "pronounced" against the government, raising the cry of *Religión y fueros*. General Mariano Arista was ordered to extinguish the rebellion, but the two generals met and decided to proclaim Santa Anna "supreme dictator."[50] According to a later statement by Arista, it was General Santa Anna himself who was conspiring against . . . President Santa Anna. Whether this was true or not, the proposed title angered him. In his "Manifesto to the Nation" of June 1833, Santa Anna gave his support to the reforms.[51] (At another time, of course, he would say that he opposed them.) And then he made a statement that rings false in the context of his capriciousness and cynicism but one that his own actions had just confirmed: "I detest military dictatorship."[52] More precisely, he detested the direct and daily exercise of power.

Santa Anna's contemporaries and later historians have interpreted this and his numerous other retreats to Manga de Clavo (supposedly for health reasons) as merely a manipulative device, an even stronger proof of his will to power. But the truth seems to be different. It is much more likely, in line with the whole political tradition of the Mexican creoles, that Santa Anna in a sense *shunned* power. Manuel Mier y Terán—the most honest, intelligent, and professional soldier among creole officers of the period and one of the few who had fought with the Insurgents—once described his personal attitude toward power, which he shared with Hidalgo, who had refused to take the capital, and with Iturbide and with Santa Anna: "I am not a politician, nor do I like that career, which brings only worries and enmities; I am a soldier by profession. . . . I love this profession because I believe it to be honorable."[53] For those men who had passed their childhood and youth under colonial rule and in the War of Independence, politics was "a profession," strange and disagreeable, a matter of factions and cabinets, pacts and compromises. None of them saw the construction of a nation in practical terms or understood the need to consolidate a state. They were drawn not by a will toward power nor even toward wealth but by a military longing for glory.

Those most aware of their psychological situation mourned their own incompetence and warned that they—the creole soldiers—were losing control of the country. Mier y Terán chose a highly symbolic death. Driven to despair by the people's demand that he become a presidential candidate and by the imminent secession of Texas (over which he had jurisdiction for several years), he would commit suicide, throwing himself on his sword, at the exact spot where Iturbide had been executed. Minutes before, he had asked his secretary to bury him in the same tomb and place him in the same position as the emperor, his body in a posthumous embrace with the liberator.

But Santa Anna showed his fear of power not in a tragic but in an operatic mode. What new part could he play from his remote presidential seat at Manga de Clavo? Not cut out for the role of the administrator involved in the day-to-day running of the country or of the legislator seeking, as Bolívar had, to impose his principles (Santa Anna had none), he chose a parody of Washington ("always my model, the most virtuous of men")[54] in retirement at Mount Vernon—though only for a while of course: "[I have resolved] to interpose the supreme authority—that has been granted to me—between the belligerent parties, to listen to their complaints and to establish myself as the peaceful arbitrator of their disagreements."[55]

These "belligerent parties" had emerged during the former decade but did not formally align themselves until after the invasion by the United States. They were the party of "respectability" led by Alamán (to

Mora the party of "retrogression"), and the party of "progress" led by Mora (the party of "demagogy" for Alamán).[56] Lawyers and professionals from the growing middle class—especially in the provinces—endorsed Mora's party, as did the powerful local caciques, the "Santa Annas" of each region, especially those in the north, hardened by their incessant wars with the nomadic Indians, ready to defend federalism and slowly forming the military nucleus of the future Liberal party. Alamán was supported by most of the landowners and the two corporate bodies eager to retain their privileges and *fueros*: the army and the Church. But during the thirties and forties, the Conservative and Liberal parties were still in the process of formation.

In retirement as usual at his hacienda, enjoying the "love of his countrymen," Santa Anna had sworn never to return to public life unless a "daring enemy" threatened the nation. In 1836 the enemy dared. That year, two territories at either extreme of the country—Yucatán and Texas—rejected the new centralist Constitution and seceded from the nation. With amazing dynamism, Santa Anna raised an army of six thousand men and marched northward to subdue the secessionist settlers. Now playing the role of the "Napoleon of the West," he threatened to advance on Washington if the United States should dare to assist the Texan secessionists. In recruiting and inciting his troops, he was his inimitable self: "I have sworn that my sword will always be the first to deliver a stroke upon the presumptuous neck of the enemy."[57] At the end of March 1836, even the cerebral and pragmatic Lucas Alamán described the situation in the style of a confident idealist: "Señor Santa Anna has so prevailed over the Anglo-American colonists who have rebelled in Texas that we may consider the matter over and done with."[58] Three months later, he had changed his tune:

> In no time at all, the political affairs of this country have taken a terrible turn. . . . [General Santa Anna,] lured on, without doubt, by the idea of finishing the war himself with a decisive blow, and perhaps jealous of the glory his lieutenants had acquired, advanced with a small body of troops, rashly and without taking proper precautions, and was defeated and taken prisoner on the 21st of April. . . . General Santa Anna signed an armistice solely to gain his own freedom and as a result of this agreement he is expected soon in Veracruz.[59]

During his time in captivity, the Texans threatened to lynch him as revenge for the atrocities he and his troops had committed at the

Alamo. He was then delivered to President Andrew Jackson in Washington, and would later sign a treaty in favor of Texan independence. Immediately afterward he disembarked at Veracruz—this time to a cool reception—and retired to Manga de Clavo. There he published a "Manifesto" exonerating not only his campaign—undertaken, according to him, by men who were inexperienced, tired, hungry, moving over difficult terrain, far too dependent on his influence and his person—but also blaming that most Mexican institution of the *siesta*, since he himself, his officers, and many of his men were sleeping when Sam Houston launched his attack at the battle of San Jacinto: "I never thought that a moment of rest . . . could be so disastrous."[60]

As Iturbide and Terán had feared and foreseen, Mexico lost the vast territory of Texas. Santa Anna, for his part, had suffered a loss much more damaging to his self-respect. He had lost the love of his countrymen. To win it back, he needed to lay a new bet, double or nothing. The following year, Providence, in the form of a squadron of French warships that had seized Veracruz, gave him his opportunity.

He threw himself into feverish, energetic action. Fighting like a common soldier and with unquestionable courage, Santa Anna forced the enemy "to meet him with fixed bayonets."[61] He took a wound in the leg. Feeling his death approaching—or pretending that it was—he wrote a pathetic statement that concluded: "Do not deny me the one title I want to leave to my children, that of being a *good* Mexican."[62] Providence took his bet and let him off cheaply; he lost his left leg. The people, moved to compassion, once more adored him. "A heroic attitude, a romantic compliment," Justo Sierra would write during the era of Porfirio Díaz; "and the entire nation leaped to its feet for that Don Juan of the *pronunciamiento*," the "great seducer" for whom "the Republic was a mistress . . . a concubine."[63]

Soon he moved back into his true vocation of conspirator, arousing the emotions of the public with his words ("which," according to Carlos María de Bustamante, "had some quality of inexplicable superiority"),[64] placing and removing army officers from the Presidential Chair and Congresses in and out of the legislative chambers. The revolution of the moment, led by Santa Anna, recalled those of the past and anticipated those to come. As Frances Erskine Inglis—the American wife of the Marquis Angel Calderón de la Barca, Spanish ambassador to Mexico—wrote in her famous book, *Life in Mexico During a Residence of Two Years in That Country* (published in 1843), "It seemed like a game of chess, in which the kings, castles, knights, and bishops are making different moves, while the pawns look on without taking part in the game."[65] And it was all managed so that Santa Anna would be implored to do "whatever he considered best for the happiness of the nation."[66] Then he would return once again to make his triumphal entry "*en roi*, with a cortege of splendid carriages and magnificent horses,"[67] to again become a patron of the arts, lay the foundation stone of a new theater named after him, attend the opera, pose for his statues (on horseback or on foot), discuss his guards' haircuts with his barber, introduce new tariffs and taxes, exact loans and confiscate money at will (even from the untouchable and distrustful clergy), attend parades, fiestas, and Te Deums, tearfully bury his wife, and, forty-one days later, marry a very beautiful fifteen-year-old. And last on this list, but by no means least, one day he would inter, with great pomp and circumstance, with speeches, funeral marches, and honors, what was closest to his heart: the leg amputated after his encounter with the French at Veracruz.

Despite all this activity, Santa Anna would get bored, and he would spend his time in a place that became, during his many terms of office, a branch of the National Palace: the famous cockfighting pit of San Agustín de las Cuevas. There, at the cockfights, "which fascinate him," the young writer Guillermo Prieto saw and described him:

> Santa Anna was the heart and soul of this emporium of confusion and licentiousness. He was something to see at the fights, surrounded by the leading loan sharks of the city, taking the money of others, mingling with employees and even with junior officers. He borrowed money but did not repay it, was praised for contemptible tricks as if they were charming manners, and when it seemed that he was growing tired of the matches, the fair sex would grant him their smiles and join him in his antics. . . . That was where Santa Anna held court.[68]

He never stopped making bets, not only in the pit, on the cocks, but on the political life of the nation. More pronouncements, more conspiracies, more periods in exile. The same people who would acclaim him at the cockfights would then curse him, but in the same festive spirit, whenever they saw him fall from power. "The seething multitude moved," Prieto recalled, "toward the theater and instantly demolished Santa Anna's plaster statue. They rushed furiously to the Pantheon of Santa Paula and with savage ferocity disinterred Santa Anna's leg, playing games with it and making it an object of ridicule."[69]

Although Santa Anna was, without doubt, a primary reason for the restlessness, disorder, vacillation, and disorientation that Mexico endured during the first decades of its independent life, he was also the consequence of those conditions, its personalized expression. All the typically *criollo* defects that Alamán detected in him—his "inner lack of consistency," for instance, or his "enterprising spirit which had neither a set plan nor a fixed purpose"—reflected collective states of mind among many of the creole elite.[70] As one of his numerous biographers wrote, Santa Anna merely "went along with the tumultuous order of the day"; he was the "barometer of national upheavals," the "specter of the society, of its romanticism and megalomania."[71]

That he was, in very concrete ways, and not only in his political life. His passion for gambling was an example. The Marchesa de Calderón described "the habit of gambling, in which . . . [Mexicans] have indulged since childhood, and which has taught them that neither lofty words nor violence will restore a single dollar once fairly lost," enjoyed by the common people "who court fortune in the open air" as well as by the "great gentlemen . . . seated around the green card table, grave as a Council of Ministers."[72] This nearly universal affection for "the mania" of taking risks, as well as the excessive Mexican courtesy that so attracted the attention of the marchesa (the everyday use of such formulas as "at your orders," "I am your servant," "to serve your honor"),[73] even the continuing abundance of European carriages and elaborately costumed servants—what did they mean in a nascent republic if not the tenacious persistence of courtly forms? Mexico seemed to regret that it had not been, and was not, a monarchy.

But the traditional source of legitimate power was gone, and the new legitimacy—to be based on law and authorized by a regime that would be fully republican, representative, and democratic—was not yet in place. In Santa Anna there was a semblance—an often grotesque mixture—of royal and popular legitimacy combined. His "royal" aura sprang partly from his personal magnetism ("Santa Anna's look is everything: he searches and seizes you with it," Prieto would say)[74] but also from the fidelity with which his person—his personality—reflected the attitudes of that society and especially of the creoles: drawn to aristocratic style, to a belief in saviors, to flash and glitter and the charm of the gambler. Alamán tried to explain his power to Santa Anna himself, who surely did not need to understand it because he lived it, made use of it, and sometimes even endured it: "The nation . . . has entrusted you with a power like that on which societies were first formed . . . because it springs from the direct manifestation of the will of the people, which is the presumed origin of all public authority."[75]

But as for ideas, did Santa Anna have any? We do have at least one recorded. To Joel R. Poinsett, who visited him while he was imprisoned in Texas, Santa Anna said that he thought a hundred years would go by before his people would be fit for liberty. He felt that all they could understand was despotism but there was no reason why despotism could not embody wisdom and virtue.

Lucas Alamán had opposed going to war with the United States and thought it preferable to accept the annexation of Texas rather than allow that irretrievable loss to trigger a full-scale war. When the Americans invaded at the beginning of 1847, he desperately looked to Europe as a last resort: "We are lost, with no way out, if Europe does not soon come to our aid." It was "the most unjust war in history, provoked by the ambition not of an absolute monarch but of a republic that claims to be at the forefront of nineteenth-century civilization."[76]

Once again, the man of destiny hopped onto the stage. He had only one leg and was now fifty-two years old, but he was the same as ever. With astonishing speed, he raised and aroused an army of eighteen thousand men in San Luis Potosí, which marched, almost without provisions, 450 kilometers in a rapid twenty-five days over difficult terrain. On February 23, 1847, he fought the first battle, in Angostura in the province of Coahuila, against the American troops under the command of General Zachary Taylor. Following this battle with no clear winner or loser, Santa Anna moved his army, "with incredible speed," wrote Alamán, "to defend the passes through the mountain range in the state of Veracruz and, having been defeated there [by Winfield Scott in Cerro Gordo], he raised yet another army to defend the capital."[77] Meanwhile, at the end of May 1847, Alamán had written to the duke of Monteleone: "It is impossible for a nation to go on for any length of time this way without being annihilated."[78] One month later, he saw that the end was at hand and suggested the reasons:

> It seems reckless of Scott to march with such a small army (12,000 men) against a city of 180,000 inhabitants plus a very considerable garrison—which outnumbers the attacking army—and to have no communication with the coast; despite this, there is no doubt in my mind that he will take the city, because our army consists mainly of recruits under the command of generals who are renowned for how fast they can flee, and nothing will move the masses, who are watching all this as if it were happening in a foreign country. That is how weary they are after so many revolts. This will all be over very soon.[79]

On August 19, through his spyglass from the roof of his house in the capital, Alamán watched the defeat of the Mexican army at the battle of the Padierna Hills. He saw how General Valencia held out as long as he could, and how General Santa Anna failed to come to his aid. (The two generals had quarreled, in anger and envy, on the very day of the battle.) The Americans continued to advance, winning victories at Churubusco, Molina del Rey, and finally the Forest of Chapultepec. In these encounters, Santa Anna led his troops and fought "fully and bravely . . . directly facing enemy fire."[80] But there was no chance of winning. Santa Anna himself said that he and the other generals did not even deserve to be corporals. On September 16, 1847, "the detestable star-spangled banner" was flying from the National Palace.[81]

Within a few days, everything had become strangely normal again. "The troops occupying the city," Alamán wrote, "do not bother anyone. It seems that we are getting used to them."[82] American officers would often visit the Hospital de Jesús, founded by Cortés and now run by Alamán, and they would ask him to show them the portrait of the Conquistador, "at which they would gaze with considerable veneration."[83] On December 2, exactly three centuries after the death of Cortés, Alamán wrote to his employer, the Duke of Monteleone: "In that era, who would ever have imagined that, three centuries after the death of the great Conquistador, the city that he raised from its foundations would be occupied by an army from a nation that had then not even begun to exist?"[84]

The drama appeared to be over. Santa Anna resigned from the presidency "that he had found as wearisome as it was bitter" and went into exile.[85] His only aspiration had been, as he said, to serve "the good of my beloved patria."[86] This time his destination would be a remote one: Colombia, where he bought an extravagant mansion. The American troops left Mexico early the following year. Under the Treaty of Guadalupe Hidalgo, signed in February 1848, Mexico, like Santa Anna, suffered a mutilation—the loss of half its richest territory. "The most unjust war in history" had ended.[87] As a sign of the times, North American prospectors struck gold in California nine days later.

Mora was then the Mexican ambassador to London. He had tried to persuade Lord Palmerston that the British government should intervene in the conflict, but even the offer to sell the two Californias to England would not sway him. Mexico had, in Palmerston's view, shown "poor judgment" in not recognizing Texas, and he refused to involve his country in the war. "The Mexicans," he disdainfully admonished Mora, "should put their hands to work and build a solid and lasting nation."[88] Months earlier, Mora had tried, unsuccesfully, to obtain the aid of France. No one was willing to help Mexico. And with the United States, as he said to a friend, "Every peace treaty signed between Mexico and the United States is nothing more than a truce on the part of the United States, to give them time to prepare the way for a new invasion."[89]

Mora's conversation with Palmerston had an additional theme: the terrible "War of the Castes," which had broken out in Yucatán while the American troops were advancing on the capital. This was a racial, territorial, and, ultimately, historical conflict between the Maya Indians and the sparse white population of the Yucatán. The English in the territory of Belize had been selling weapons to the Mayas, who were devastating the cities where the whites were concentrated. In 1848, two years before his death, while what he saw with horror as a nightmare of social revolution was spreading across Europe, Mora sensed in this Yucatán uprising, which he called "the War of Colors," a new episode of the Insurgency that had led to Independence—but with an additional weight of unbelievable atrocities. His family, like that of Alamán, had been ruined by the ravages of Hidalgo's "hordes," but whatever had happened then could not be compared to the furies unleashed in Yucatán. Whole groups were massacred; women were gang-raped; people were burned and skinned alive. Mora wrote: "The War of Colors is the worst Mexico has ever had to endure, because it is bound to end with the extermination of one of the contending parties and within the natural order of things it is the smaller party that will perish."[90]

Mora, from his European exile, feared that all the creoles in Yucatán would be exterminated. The only means to avoid it was "to expel all the colored elements from the peninsula and to multiply those of the white race."[91] Alamán, within Mexico where he was now excluded from power, was afraid that similar things might happen at the center of the country when the U.S. troops pulled out. And then suddenly, the European hurricane of that year of 1848 reached the center of Mexico in the form of an unexpected but long-brewing insurgency: sporadic warfare between the villages (inhabited by mestizos, but with deep indigenous roots) and the haciendas for possession of the land. The conservative Alamán, who owned haciendas (and administered those of the Duke of Monteleone), showed the same fears for his race and class as the liberal Mora: "The internal conflict will take the form of a caste war . . . and since the whites are the smallest group, they will have to perish and with them all the properties that belong to them."[92]

The "wars of colors"—both in Yucatán and at the center of the country—touched the same painful and sensitive chord in both *criollos*. It took them back to the War of Independence, to the open confrontation between Indians and creoles. The renewal of these clashes in 1848 threw creoles of all political shades into a life-and-death situation, "us"

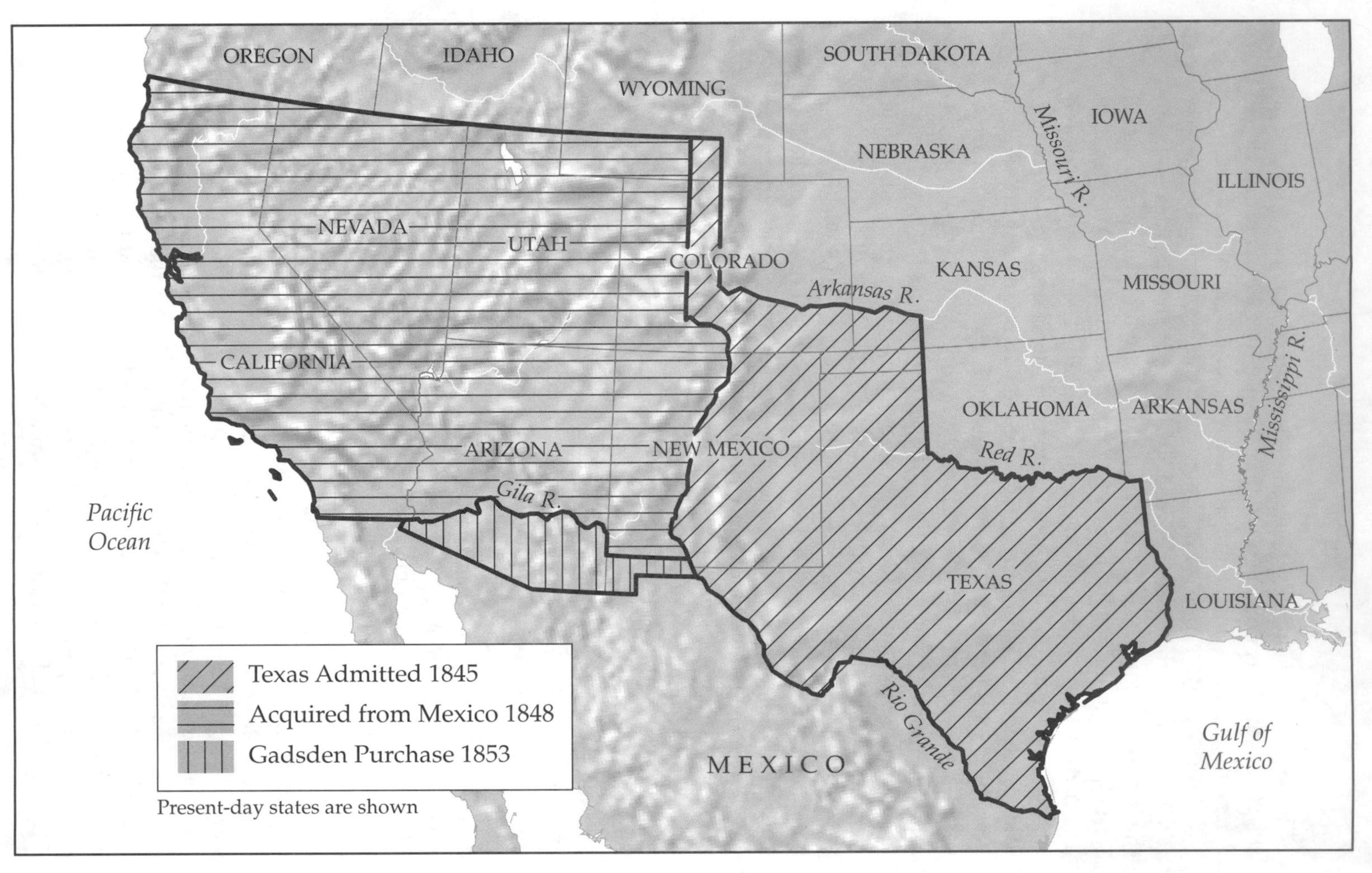

Mexican Border with the United States, circa 1855

versus "them." Confronted by the violent Indian rejection of their people and their social position, those creoles born and raised during the colonial period easily slid back toward the form of power they knew best: that of Spain. The harshness of their judgments on the Indians reflected reality less than it did their own historical despair. The country was slipping out of their hands. They had a vague feeling that it would never belong to them again.

"Death to the *gachupines*!" That "fearsome cry of death and desolation" that Alamán had heard so often in his youth still resounded in his ears forty years later.[93] It was like starting all over again or as if nothing had ever begun. It was like losing everything. The apocalypse seemed to have come to Mexico, but it was really only the apocalypse of the creoles.

In 1850 Mexico found itself at the opposite extreme from its foolish optimism of 1821. There was the war that had been lost against the United States, the continuing War of the Castes in Yucatán, and agrarian revolts in the heartland, a state of permanent war against the nomadic Indians (especially the Apaches and Comanches), who—because of the annexation of Texas, which they used as a safe home base for their raids into Mexico—were devastating all the states of the north (Sonora, Chihuahua, Coahuila, and Tamaulipas) just as in colonial times and even raided mining cities in states far from the border (Durango, San Luis Potosí, and Zacatecas). And as if there were not wars enough across that bellicose geography, Mexicans had to deal with the everyday battles against the gangs of *bandidos* who infested the "royal highways" inherited from the viceroyalty.

The eight million inhabitants were still not enough to populate the two million square kilometers of territory remaining after the war with the United States. Practically no immigrants came to Mexico (except for the families of some French businessmen in the forties), and the infant-mortality rate was very high. Debt servitude was still the normal custom on the basic rural economic unit—the hacienda. Aside from a nucleus of modern textile factories in Puebla and the continuing production from the mines, industrial development was far too slow, and the same was true of modern communications. While the United States had rapidly extended its railroad network to the West and, on the Mississippi and Ohio rivers, the steamboat was in full operation, mules and horses were Mexico's means of transport, at great risk to lives and goods. The coach between Mexico City and Guadalajara—the two largest cities—took seven days. Only in 1857 was the first railway line inaugurated in the capital, but its purpose was not precisely commercial: It linked the city center to the nearby Sanctuary of the Virgin of Guadalupe. To make matters even worse, the successive governments were weighed down by an external debt of 52 million pesos, which they could not shake off, while they were also becoming enslaved to a large and growing group of bureaucrats and soldiers.

The infant country was on the verge of falling apart. In the final volume of his *Historia de México* (published in 1852) Alamán wrote:

> To see in so few years this immense loss of territory; this bankruptcy of the treasury, leaving us with so grievous a debt; this destruction of a courageous and blossoming army, so that we no longer have any means of defense; and above all this complete extinction of public spirit, which has swept away every idea with a national content: not finding any Mexicans in Mexico and contemplating a nation which has moved from infancy to decrepitude without having enjoyed any more than a glimpse of the vigor of youth. . . . [94]

Nevertheless Alamán did not think that the disaster of Mexico was complete or fatal or incapable of being reversed. Nothing thus far had worked. Therefore the system had to be changed. Mexico could not be careless because its northern neighbor was far too powerful. In the event that the southern slave states were to separate from the American Union, which Alamán prophetically argued in 1852 was all too likely, the new nation would annex new land, bring in its slaves, and "subject Indians and Castes of the country it occupied to more or less rigorous servitude."[95] If this should come to be the case, and it could easily happen, Mexico would not be the only loser. Spain would be stripped of Cuba and Puerto Rico, England of Jamaica and the Antilles. The expansionist impulse of the Americans might reach as far as Panama, from there to dominate the trade of the Pacific.

While Alamán proposed his version of the return to tradition—a powerful executive, a centralized admininistration, a greatly weakened Congress—Mora, from France, seeing the same desperate situation, called instead for a federal and representative republic. His countrymen had been exposed far too long to the whims of a single "strong man": Santa Anna, the "Attila of Mexican Civilization."[96]

José María Luís Mora died in Paris on the anniversary of the storming of the Bastille. The final paradox in Lucas Alamán's life was that he died—on June 2, 1853—one month after his appointment to the most important ministerial post in the cabinet of who else but Antonio López de Santa Anna, summoned by his countrymen to assume an eleventh and final presidency. Had Alamán lived, he would have at last had the power to alter the political system of the country.

Without the intellectual and moral presence of Alamán, Santa Anna saw no way to handle this last presidency other than in his usual manner, the only style he knew and enjoyed. He behaved as if he were a king. (Though even the cockfighting pit had begun to offer him less pleasure, because the gamblers there cheated in order to let him win, and he knew it.) He threw himself into every excess of imperial ostentation except for crowning himself emperor. He formed an escort of hussars who wore Swiss-style dress uniforms. He created so close a relationship between the Church and the State that he named the archbishop a state councilor. He ordained by decree that the Jesuits could return to Mexico. In an attempt to raise funds, he introduced outrageous taxes, such as one on the amount of dogs per home. With his enemies he was unyielding. He did not kill them, but he—the conspirator par excellence—enacted a Conspiracy Law and persecuted not only written opinion but rumors and murmurs. Those whom he considered particularly dangerous he banished. Two governors who agreed neither with his goals nor his methods ended up in New Orleans: Melchor Ocampo of Michoacán and Benito Juárez of Oaxaca.

In international affairs, the eleventh term in office of His Most Serene Highness brought the final, humiliating chapter in the war with the United States to a close: Mexico lost another stretch of territory—the Mesilla Valley, now part of southern Arizona. Shortly afterward, Santa Anna received a visit from Antonio García Cubas, a diligent cartographer, who unrolled a carefully drawn map of Mexico before and after the war. The President burst into tears: For the first time, he could measure what the country had lost. His tears were emblematic of the humiliation of the creoles who as a group, in 1821, had experienced the greatest wave of optimism in Mexican history—a rush of belief in their new country as the very center of the world. And now they were left with only a national (and personal) sense of mutilation.

The nation of Mexico did not prosper during the imperial farce of His Most Serene Highness, but the genre of opera, significantly, reached its zenith. Performances at El Gran Teatro de Santa Anna would be announced as:

> Dedicated to His Most Serene Highness, Major-General, President of the Republic, National Hero, Knight Commander of the Great Cross of the Royal and Distinguished Order of Carlos III and Grand Master of the National and Distinguished Order of Guadalupe: Don Antonio López de Santa Anna.[97]

The opera was the perfect genre for that creole minority who only at rare moments had known how to mark their civil action with an epic, dramatic, or tragic meaning. It was the irresponsible fiesta at the edge of the abyss—the easiest way to escape from reality, to compensate for resounding defeats with fanfares and exaggerated displays of feeling. And the same was true of the only work premiered before the court of Santa Anna (among all the operas and operettas and plays) that was destined to be remembered and still is heard and performed: the national anthem of Mexico.

It is the Mexican "Marsellaise," but it was a Marsellaise at that time without any support in history. As a symbolic and operatic triumph presented on the stage, it served to mask the grim reality of political and military defeat. Its theme was war. The actual word appears seven times in the anthem and in many other places through images and synonyms. The poet speaks to warriors and the *patria*. Between each of its ten stanzas, the chorus repeats a verbal cannonade:

Mexicans! When you hear the cry of war!
Ready your swords and your horses!
And let the earth tremble to its core
At the sonorous roar of the cannon![98]

The frustration of that society, after the sad outcome of the last war, was offered a compensation of military sound effects. A volley of *r*s runs through the anthem: *guerra* (war), *guerrero* (warrior), *horrísono* (terrible-sounding), *rugir* (roar), *hórrido* (hideous), *rayo* (thunderbolt), *derrumbar* (demolish), *torrente* (torrent). The finger of God would grant peace to the fatherland. Amid tombs, sepulchres, swords, escutcheons, standards soaked by "waves of blood," the bugles of war, graves, and crosses, Mexicans would swear never to take up arms again "in a struggle between brothers." Direct from a Santa Anna manifesto came the first two lines of one stanza:

But if a foreign enemy should dare
To profane your soil with his steps,
Oh beloved Patria! Think that heaven
Gave you a soldier in every son![99]

The national anthem made its debut on September 15, 1854. Several months later, the Liberal revolution of Ayutla would overthrow Santa Anna, who as usual went into exile. But this time he would stay there for twenty-one years. When he returned, even his name had become an old memory. His wife was quick to perceive this pathetic death in life. She began to approach people in the street and ask them to visit him. They should address him with respect and implore him to recount some of his exploits. The final stanza of the national anthem promised soldiers "a glorious memory" and "a sepulchre of honor."[100] When Santa Anna died in 1876, he was granted neither.

The former omission is understandable, but the latter is harder to explain. If Santa Anna had died when he was victorious (1829 or 1838) or even when he was defeated (1836 or 1847), he would have been granted a place in Mexican history as a hero or perhaps as a martyr. The truth is that up until 1847 everybody thought of him as the savior of the nation. The vices and virtues that so marked him were not his alone but those of the Mexican people who incessantly sought him out and welcomed him, cheered and cursed him. Every Mexican, on more than one occasion, had been a follower of Santa Anna.

Behind the fall of Santa Anna and the passing of Mora and Alemán, the great *criollo* thinkers of the first half of the century, lies a much more essential fact: the defeat of the creoles. In little more than thirty years, they had lost their historical opportunity. The country would now be governed by a group of young mestizos who were closer to Mexican soil, closer to indigenous roots. Born after 1810, their minds and memories were not burdened by systems inherited from the colony or from Spain. They were the first children of Mexican independence. This transference of power from creole to mestizo hands would be accomplished through a personage who would, like Santa Anna but in an utterly different direction, interweave his biography with that of Mexico for fifteen decisive years: a Mexican who ethnically and culturally predated the birth of Mexico and preceded the Spanish Conquest: the Zapotec Indian, Benito Juárez.

8

THE INDIAN SHEPHERD AND THE AUSTRIAN ARCHDUKE

The movement from creole to mestizo power did not take the form of a peaceful transfer of credentials. It required a long and bloody civil war that broke out at the beginning of 1858 and ended three years later. Compared to it, all the Mexican revolutions since the achievement of independence would seem to have been little more than games of toy soldiers. The essential motive for the *pronunciamientos*, insurrections, and barracks uprisings of the Santa Anna era had been the seizure of power by one general (more often than not Santa Anna) through displacing another (frequently Santa Anna). Any ideological impulse was secondary and often only a pretext. But toward the middle of the century, the situation changed. After the defeat by the United States and faced with the dilemma, foreseen by Alamán, of either consolidating the nation or seeing it disappear, the nature of the nation itself had to be defined. But agreement, consensus, peaceful discussion seemed impossible. The definition would come through war, a fierce religious and ideological struggle that would become known as the War of the Reform. In the intellectual realm, it would be foreshadowed by a memorable polemic (between two memorable men) over the practical responsibilities and perquisites of the Catholic Church.

One of them was Melchor Ocampo (1814–1861):

> What use are so many churches in a country which can barely afford to maintain one? Of harmfully multiplying the number of festivals . . . of encouraging idleness, drunkenness and other vices . . . and of giving priests a surplus of income without their having earned it through doing anything really useful? . . . Unhappy Indians whose wealth goes up in the smoke of the candles, the censers and the fireworks![1]

The author of these lines was a rare figure in Mexican history, not because of his birth (he was a mestizo orphan who never knew his parents) or his social situation (he was left at the hacienda gate of a wealthy woman in Michoacán, who brought him up as her own), or even his profession of scientist (a botanist, linguist, astronomer) and world traveler, but because of his religious and political positions. As he said himself, he was a "fiercely independent" man, probably an atheist, and a pure liberal.[2]

In 1851 Ocampo engaged in the most important debate of the century, on a theme that had never been publicly posed before in Mexico. It seemed that he was condemned never to know the identity of key figures in his life, because his opponent was an anonymous "priest" from Maravatío, a prosperous town near the hacienda Ocampo had inherited from the woman who had raised him. The anonymous priest was probably Clemente de Jesús Munguía, the bishop of Morelia, Ocampo's contemporary and fellow student at the Tridentine Seminary of Valladolid (which had been renamed Morelia in honor of the Insurgent hero). Munguía in his youth had been an open-minded cleric but then had become a man of theological learning as vast as it was fervent and intolerant. Oddly enough, he had a special quality that he shared with Ocampo: the same uncertainty of birth. The polemic between these two foundlings lasted from March to November of that year and had national repercussions. The text that provoked the furious response from "a priest" was Ocampo's "Representation" to the Congress of the State of Michoacán, in which he asked that parochial fees be "reformed." All he was really requesting was that the poor should pay less for the sacraments and other services offered by the Church.[3]

For many years Ocampo had been developing his ideas concerning "the spiritual tolls that have been impoverishing the flock without any progress whatsoever," but it was perhaps one incident that moved him to take action.[4] A worker on his hacienda had asked a priest to give his son a free "Christian burial" because he had no money for the fees. The priest had refused, contending that "this was what he lived on."

The poor man had asked, "Sir, what shall I do with my dead son?" and the priest had answered, "Salt him and eat him."[5]

Ocampo's arguments went far beyond the immediate motive for his *representación*. He touched upon all the sore spots in the relationship between Church and State. To begin with, he defended "the natural right of every man to worship God according to the intuitions of his conscience." It was a principle he drew not only from Kant but from his own experience. His travels throughout Europe had inspired him with a spontaneous sense of tolerance. They had shown him that the world was larger and more various than his Hacienda of Pateo or the country around Maravatío (the presumed home of "the priest").

> What should I do when I see dancing and shouting in church; when I see some Protestant closing himself up with his family to read the Bible; when, if I return to Rome and feel like going into a synagogue, I see the Rabbi opening the Sanctum Sanctorum or when I enter a Catholic church and see the Armenians and Copts celebrating their rites; what should I do when I see some devout Muslim performing his ablutions?[6]

His personal answer was simple: the broadest acceptance of the other and of the beliefs of the other—an attitude of tolerance.

"Stop there, Sr. Ocampo," shouted "the priest" in print. The only way to worship God, internally and externally, was as prescribed by the Church. "Luther looked upon his own intuitions with unbridled passion. Oh, how malignant his intuitions turned out to be . . . ! The wisest, the happiest, the most fortunate of all mortals is the one who empties his heart of the scum of his own intuitions."[7] The "pestilential doctrines" that "emanated" from Ocampo's paradoxes could have only one objective, and as he contemplated it, "the priest" shuddered, "Look, Michoacán! toward where we are going to finish though Sr. Ocampo does not realize it—with *freedom of religion and freedom of conscience*. These two concepts, as ungodly as they are fatal, currently serve the interests of European socialism and if God should decide to punish us by spreading them among us, it is certain we would end in universal devastation."[8]

The heart of the controversy was a disagreement about the spiritual services of priests. For Ocampo they were poor, insufficient, and expensive. Priests did not regularly preach, and when they did, they did it in a selective spirit, with disdain for the poor, for servants, for children; they did not tend the sick as the Gospels told them to do ("heal the sick, cleanse the lepers");[9] they did not officiate at the funerals of the poor, who "were buried like animals."

Ultimately Ocampo saw the oppressive clerical fees as the final knot in a vicious circle that he had already studied: debt servitude. If the priest wanted too much, the peon would have to borrow money from his employer to pay, one of the ways in which he continually increased his indebtedness. (The word for the process in Mexico was *endrogarse*, "to become addicted.") The situation created a complicity between the landowner and the priest, at the cost of the peon: "As in the times of Abraham, the peons and workers born on the haciendas belong to them and are bartered or claimed and exchanged and sold and inherited as are herds, tools and land."[10]

"The priest" refused to admit the problem even existed. Like all men, the peons were "free and inalienable."[11] If they remained (even reluctantly) on a hacienda, it was because they owed money there and not because they themselves were part of the estate.

From the beginning, the priest's greatest anxiety was that Ocampo had aired his ideas *in public*. The right way, the Christian way, would have been to present his concerns not before the state congress, which had no right—in the priest's opinion—to deal with the issue, but in private, before the ecclesiastical authorities: "So as not to alarm the faithful by stirring up the bitterest ill-feeling against their priests, [it should be conducted] in secret so that a remedy might be found for this ill without diminishing the respect due to the clergy, even when they may be at fault."[12]

What Ocampo wanted—according to the anonymous priest—was "to stoke up the fire that is drawing us into it," to encourage "a horrible change that will entomb us in the abyss," to unleash "devastation" and "conflagration." His ideas were in line with socialist thought, with the "monstrous heretics," with the revolution in Italy that was threatening the Pope and the Papal States. And also they were fundamentally "Lutheran."[13] The "plague" of the "ruinous French Revolution" had finally reached Mexico, with its train of atheism, its laws banishing priests, its confiscation of the sacred property of the clergy.

In his final reply, the priest suggested the possibility of excommunicating Ocampo and then moving on to other "means of action." Without in any way yielding, Ocampo decided that it was time to stop the polemic.[14]

The argument between these two orphans from Michoacán—one the son of nature and the other of the Church—was the first cloud in a sky that, within a few years, would turn completely dark. It was a clear prediction of the hurricane. With a priest like this, no Liberal could establish a dialogue. The difference of opinion was to have a much more profound effect than that between Alamán and Mora, because it called into question not ideas and projects but the most practical interests. Without knowing it, Ocampo and Munguía had staged in writing the first act of the War of the Reform, which would be fought between two irreconcilable factions. On the one hand would be the Liberals, basing their program on the Enlightenment values of individual freedom and responsibility; on the other, the Conservatives, standing on the traditional values embodied in the Church. The ancient tension had reached the breaking point: a constitutional future or the theocracy of the past?

By the early 1850s, the parties were clearly defined. In large part it was a generational battle between the intellectual disciples of Mora and those of Alamán. The Liberals were civilians and military men under forty, mestizos rather than creoles, and mostly from the center and north of the Republic. On the Conservative side were politicians, clergy, and military men some ten years older than their rivals. They were urban, rich, and, for the most part, creoles. Among the Liberals there were several genuinely federalist governors and quite a few caciques, federalist out of personal interest and hatred for the centralist generals (typical products of the Santa Anna period) rather than through any ideological conviction. Sympathizers also came from the middle class and were found, in steadily increasing numbers, among landowners and entrepreneurs who saw economic opportunities in the imminent sale (or privatization) of Church lands. The Conservatives were supported by the onerous bureaucracy of the capital city, by the "respectable people," and, of course, by the clergy.

There were remarkable men on both sides.[15] Among the Conservatives, no one could could compare with Miguel Miramón, a young general born in 1831. He had been a cadet at the Military College, a "child hero" who had survived the war against the North Americans, a member of Alamán's Conservative party, and an instructor in artillery tactics at the Military College. He was a believer in "fatherland and family," and both his friends and enemies called him "the Maccabeus," because he genuinely seemed to resemble that ancient biblical soldier who had liberated Israel from the Greek invaders and desecrators of the Temple of Solomon.[16] There were other military men of note, though none with Miramón's stature, among the Conservatives: Leonardo Márquez (who would soon draw attention for his bloodthirsty actions), Félix Zuloaga, and two local Indian caciques who showed great military ability: Tomás Mejía, master of the Sierra Gorda of San Luis Potosí, and the mysterious Manuel Lozada, the "Tiger of Álica" whose goal was to reestablish the indigenous kingdom of his homeland, Nayarit. Both Mejía and

Lozada came from a significant historical tradition: the Indian opposition to liberal policies that had resulted in communal Indian lands being put up for sale to private owners.

During the war, the young Maccabeus would be the choice of the Conservatives to become President of Mexico. The total concentration of civil and military power in the hands of a single caudillo was a feature of the epoch that should have drawn more attention at the time. It was the final result of decades of belief, bequeathed by Santa Anna, in a man of destiny, and its focus on new faces was due to the passing away of the older, discredited, and ineffective military class. Another mark of the Conservative leadership was the lack of any real quality among its civilian members. Perhaps, in part at least, Alamán's former dominance over the principles of the party had filled all the intellectual space. There were various old priests and aristocrats whose only desire was to rigidly apply Alamán's program, with the addition of one new clause: the restoration of the Mexican Empire under a European Catholic prince.

The Liberals were totally different men. Their young military leaders acted out of vocation and conviction, not because they were professionals. They had spent their childhood and youth in a country torn by revolutions. Now was the chance to carry out *their* revolution. The case of Porfirio Díaz, the political chieftain of the Isthmus of Tehuantepec, was typical and by no means unique. He had risen against Santa Anna and would never again lay down his arms. In the Liberal armies there were farmers, teachers, lawyers, mining entrepreneurs, bookkeepers, journalists. One of the most important military contingents consisted of the "frontiersmen" (*fronterizos*). These ranchers, traders, smugglers, and caciques from the northern states were very similar to their North American counterparts. Toughened by their battles against Apaches and Comanches, alien to the ideology of conservatism and clericalism, they had been turned into liberals and individualists almost at the moment of their birth into the wars of the harsh north.

The commander of the Liberal armies, Santos Degollado, was a new incarnation of the armed man of religion. During the War of the Reform, "Don Santos," as everyone called him, would accumulate many more defeats than victories, but he remained military leader throughout most of the war because he could unite men behind him and turn their various commitments into a single will. With his sermons on the "holy cause of democracy," Degollado would raise armies, lose them, and then raise them again. He was a visionary and perhaps the finest example of a fusion between inner Catholicism and political Liberalism. Justo Sierra would later write about him:

> He fought for the Reform with all his soul, with a great faith in his ideal and without abandoning a grain of his religious soul. When he tried to strictly limit the power of the Church, it was because the Church had lost its way and denied the teachings of Christ. *He* was the Catholic, the canon lawyer, the theologian; it was the bishops who were impious . . . the only Church that opposed freedom was the Church of oppression and tyranny, of the Inquisition and the sinister monarchs clothed in black of the House of Austria. The impiety lay in wanting to block the advance of new ideas and the ascent of the people within an environment forged by those ideas.[17]

A phrase overheard at night around a Liberal campfire summed up the circle of loyalty between Degollado and his followers: "We are ready to die for Don Santos, only so that Don Santos may know we were ready to die for him."[18]

The civilian wing of the Liberals was even more impressive. There were several first-rate writers and journalists: the poet and journalist Guillermo Prieto; Francisco Zarco, the most accomplished journalist of the nineteenth century; the essayist and editor Ignacio Ramírez, the first confessed atheist in Mexico and the founder of the Institute for Arts and Sciences of Toluca, where he trained Ignacio Manuel Altamirano, a pure-blooded Indian destined to be the most influential literary editor of nineteenth-century Mexico. There were several highly accomplished jurists, at least one socialist landowner (Ponciano Arriaga), and the economist Miguel Lerdo de Tejada. The most prominent intellectual figure was Melchor Ocampo. And there was Benito Juárez.

Throughout the year of 1856, many of these Liberals—with the exception of Juárez—had been members of a Constitutional Convention that had been called to provide the nation with its new Constitution. The importance of those sessions would be later described by the leading Liberal historian of twentieth-century Mexico, Daniel Cosío Villegas: "Mexican history has black and shameful pages that we would love to eliminate; it has heroic pages that we would like to see printed in large letters; but our history has one unique page in which Mexico conveys the sense of being a mature country fully at home within the democracy and liberalism of modern Western Europe. That page is the Constitutional Convention of 1856."[19]

In February 1857, this Liberal generation would approve the first fully liberal Constitution in Mexican history. It made all individuals equal before the law and consecrated wide-ranging liberties (freedom to express ideas in public, freedom of teaching, of the movement of people, freedom of association and conscience); it extended civil protections (abolition of privileges and special courts, abolition of prison for debtors, free defense in every civil or penal trial, and especially the right of habeas corpus in cases of abuse of authority); it gave the legislative power supremacy over the executive power; and it stipulated the free election of judges to the Supreme Court.

According to Cosío Villegas, the 1857 Constitution reflected the spirit of its creators. These men, "fiercely, proudly, arrogantly, absurdly, irrationally independent," had been formed in the chaotic era of the tyrant Santa Anna, who arbitrarily ousted governments and Congresses, whose only conviction was his desire for personal gain and glory.[20] They had driven him from power in 1855, and now it was their duty to vaccinate the country against any new strain of the Santa Anna "disease." It was Ocampo who gave the best definition of the difference in attitudes between Liberals and Conservatives:

> Unfortunately, the Liberal party is essentially anarchic; it will only cease to be so after thousands and thousands of years. Our *criterion for truth lies in the mutual gloss of the senses, or in rigorously logical inductions that agree with experience;* the criterion of our enemies is authority. Thus, when they know that it is the king or the Pope who commands . . . they obey blindly; whereas when *we are commanded and when we are given explanations as to the whys and the wherefores, we murmur and we are slow to move, or we may refuse to obey and rebel.* For each Liberal is a Liberal to the degree in which he knows how to or desires to liberate himself, and our enemies are all equally servile and all almost equally passive pupils, in need of protection. It is hard to be a Liberal in everything, for one has to have the courage to be a man in everything. . . .[21]

Yet the central figure among the Liberal leaders, the eldest and most experienced of them, did not really fit Ocampo's description. In his own stern and idolatrous fashion, he believed religiously in the Constitution, in the sacredness of the presidential position, in the obedience and authority due the Constitution and the President. But he was not a dictator who, in the creole manner, exercised power for the sake of power. Benito Juárez detested what he called "the odious banner of militarism."[22] Even less was he a daring apostle of freedom, a Christian fundamentalist from the catacombs lost in the nineteenth century like Degollado. He would wield power the way his Zapotec ancestors had done: as a responsibility of the utmost seriousness.

Ever since his days as governor of Oaxaca, he had longed for and even partly discerned his role in history. This "child of the people" would continue the endless path of his emancipation: his own and that of his brothers—the Indians (and the mestizos).[23] He foresaw that the flock needed a shepherd. In January 1858 he assumed the role. He would fill it for fifteen years, not missing a single day. Without taking his biography into account, we cannot hope to understand either the triumph of the Liberals in the War of the Reform or the course of Mexican history in the nineteenth century.

Distant from Mexico City and separated from it by almost impassable country, the province of Oaxaca formed a closed world.[24] Its capital was a white and Spanish city, noted for books and culture. Surrounding it was an impenetrable, indigenous mosaic. Twenty different languages, twenty distinct nations, linked only by their peculiar brand of Christianity, shared the mountains and valleys of that Indian universe remote from Hispanic civilization. The Zapotecs were one of these ethnic groups. With their calm, proud, and reserved nature, they had been politicians and traders in the pre-Hispanic world, only to be subdued by the might of their bitter enemies, the Mixtec warriors. With the passing of the centuries, both groups had lost all memory of their heroic period but not of their customs. The extremes of poverty, drunkenness, and unhealthy living conditions were features that alternated with a kind of natural stoicism, with sobriety, industry, and a resistance to suffering, all softened by two healing balms: music and religious faith.

Deep within the heart of this sanctuary, Benito Juárez was born in 1806. Perhaps he was not aware of the advantage he held over the other twenty families in the Zapotec town of Guelatao: He was descended from the old indigenous nobility, those who "gave the orders."[25] According to an ancient ruling by the Crown, only indigenous rulers could own sheep in Oaxaca. After losing his parents at an early age, Juárez was a shepherd for his uncle and spent his childhood fulfilling this lonely responsibility. Shortly before he turned twelve, encouraged perhaps by a sister who was working as a servant in Oaxaca, he felt the need to master grammatical Spanish and to flee to the white city. Between Guelatao, in the cold and rugged mountain range of Ixtlán, and the capital city of Oaxaca, there lay sixty kilometers and something more: centuries of civilization. "Isolated in his Indian environment," Justo Sierra wrote, "unable to speak the language of Castile, confined in the prison house of his own language," he felt the powerful drive to escape: toward the Spanish language, the mingling of races, civilization, the future.[26]

A lay devotee of Saint Francis, Antonio Salanueva, took the young Indian in as an apprentice bookbinder and, two years later, enrolled him in the Council Seminary, the only educational establishment of the capital city. In 1821 Juárez began to study Latin grammar, in 1824 scholastic philosophy, and later moral theology. At the age of twenty-two he was clearly heading straight toward an ecclesiastical career, and then, "out of curiosity or because the study of theology was distasteful to me . . . or because of my natural desire to pursue another profession,"[27] he made the decision to begin the study of law at the new Institute of Arts and Sciences.

Abandoning the Seminary was a risky step in a society that shunned the idea of change. In Oaxaca, as in many provincial cities—even more so because of its isolation—the colonial past was still intact despite the fact that New Spain was now called Mexico. The days were measured by the church bells that summoned people to mass and regularly, from dawn to dusk, punctually marked the times of prayer. The main form of entertainment was attending the processions and public celebrations held in each parish. Indians from the mountains would descend into the city, and once there they would carry the litters that bore the images of the saints and the relics and the lanterns. Accompanying the processions would be the sounds of small drums and high-pitched flutes. Behind the Indians, the friars would march, dressed in the habits of the different orders (Dominicans, Augustinians, Franciscans, Carmelites), then the secular clergy, and then the flagellants whipping their shoulders bloody, the civil authorities, the military, and a musical band usually out of tune. Beggars and men of power, priests and soldiers, whites and Indians—the Church bound them all together in religious communion.

"Oaxaca was a city that lived in the shadow of a monastery," wrote Justo Sierra. "There everyone was a friar or wanted to be one."[28] Juárez had been no exception. Fearful of the wrath of God and the omnipresence of the devil, marked by a deep, inner piety, Juárez participated in the processions and responded to the church bells. Fifty years earlier he would have taken holy orders. But Oaxaca, in spite of its conservatism, was slowly awakening to the new age. Since 1827 the Institute of Arts and Sciences had been competing with the Council Seminary for the vocations of the young. Several more open-minded priests, lawyers, and doctors taught at the Institute. The majority of these teachers were loyal Catholics who felt a need to keep scholastic works on the shelves but also to open windows toward intellectual freedom, new professions, and the sciences. These men were not really liberals, and they were certainly not antireligious: They merely wanted to create a civil branch of the Seminary. Benito Juárez was their disciple. Obscurely, blindly, he experienced the move from the Seminary to the Institute as one more step in his personal liberation.

At the beginning of the 1830s, Juárez was elected to the city government as a councilman.[29] He began to make a reputation for himself as an expert on legal matters, working not only as a lawyer in the court of justice but also writing for the press and eventually filling various positions in the three branches of state government: in the judiciary, as a legislator, and in the important post of Minister of Government—in 1842—when General Antonio de León was governor (during one of Santa Anna's presidencies). Biographies of Juárez often skim over those years, but he too delivered his eulogies and made his professions of loyalty to Santa Anna. Like all the rest of Mexico, there was a time even for Juárez to be a follower of Santa Anna. But nevertheless, in each of these public positions (and as a deputy to the federal Congress as of 1847) he demonstrated a sense of responsibility that was rare, especially in those frivolous times.

Juárez was an Indian who would never again respectfully lower his gaze. The will to emancipate himself was visible in every area of his life. In 1843, at the age of thirty-seven, Juárez advanced one more step on his personal path to transformation. He married a woman of almost totally Spanish descent: Margarita Maza. The children he had with Margarita (he had fathered other children, before his marriage) would no longer be Indians like himself; they would be mestizos: children of an Indian father and a Spanish mother, not the opposite as in the vast majority of cases. In this uncommon instance, a white woman had been conquered by an Indian, not a native woman by a Spaniard.

Juárez was not an intellectual but rather an unusual politician, who expressed himself best not in words but in attitudes and actions. There is nevertheless an early speech, delivered on September 16, 1840, that sheds some light on his ideological formation. It is one of the rare moments when we can almost *hear* the young Juárez openly describe his vision of Mexico and his personal vocation.

He was delicate in his references to Spain. He had no intention of reproaching her or "opening up wounds that should be left to heal." But it seemed clear to him that Spain had left Mexico with a political system ruled by "antisocial maxims." Above all she had "neglected" the education of Mexicans, and had "imposed" doctrines of blind obedience, "created classes with divergent interests . . . isolated, corrupted, intimidated and divided." The result: "Our poverty, our brutalization, our degradation and our enslavement for three hundred years."[30] To whom was Juarez referring with that word *our*? Not to all Mexicans but to the Indians: "There is more: the torpid poverty that submerges our brothers, the Indians. The heavy exactions that still burden them. The pitiful abandonment to which their primary education has been reduced."[31]

These were the "relics of colonial government" that impeded the "consolidation" of independence. In order to overcome these obstacles it was necessary to put into practice the same "social virtues" that drove Hidalgo to lead "a handful of Indians whose only weapons were slings, sickles and sticks." In contrast to the dispirited speeches of the *criollos,* Juárez spoke with his vision fixed on the future. Real independence, the fear and respect of enemies, peace, harmony, the era when Mexico could be the "classic land of honor, moderation and justice" were still to come. If the "holy tree of liberty" were to put down "truly deep roots," someone would have to plant the seed. And later, in the fullness of time, it was he, Juárez, who would do it.[32]

On October 29, 1847, he was named temporary governor of Oaxaca. In 1818, with his infinite patience and stoicism, he had walked to the white city from Guelatao. Walking away from the "torpid poverty" into which he and "his brothers" had been born, Benito Juárez had now become the first Indian governor in the Mexican republic.[33]

The threat of a United States invasion reaching the territory of Oaxaca provided Governor Juárez with a priceless ally: the "venerable clergy . . . of the state." In January 1848, he had asked the Church authorities to use their public sermons to "stir up" the people in defense of the two sacred institutions that were in danger of being lost: religion and the *patria.* The bishop of Oaxaca assured him that he was busy "reviving" public awareness by every means within his power.[34]

The feared invasion did not materialize, but the pact between the civil and ecclesiastical authorities lasted for the whole period of Juárez's governorship. In April of the same year, the state government secured the clergy's collaboration in the building of several schools and roads. The governor returned the favor by decreeing that courses on "ecclesiastical history" would again be taught at the Institute of Arts and Sciences. No one would be able to say that, during his term of office, this institution had been a home to heretics, unbelievers, or corrupters of the young. Shortly afterward, he issued various orders to the chief military commander of the state. They were not military commands but militantly religious. They ordered salvos and an escort to solemnize the celebration of December 12 (the date of the appearance of the Virgin of Guadalupe) and another escort to accompany the procession in honor of Our Lady of Solitude. In addition to these measures, Juárez defended the clergy exactly where Melchor Ocampo had attacked them: on the issue of their material interests. Far from legislating to limit or modify ecclesiastical charges, the civil authorities were to increase their persuasive and coercive intervention to ensure "the maintenance of the ministers of the religion in which we believe."[35]

All documents of the period, and not only those of Oaxaca, were signed with the motto "God and Freedom." But the most remarkable thing about Oaxaca was the close relationship between the two spheres. There came a moment when this confluence of roles reached the level of fusion. Imperceptibly, Juárez was presenting himself to the legislature of Oaxaca as the bearer of an almost messianic mission. If it was not heard or read as such, it was because, unlike Santa Anna, Juárez did not brandish the soul of a savior, nor did he look like one. He resembled a Zapotec idol, an imperturbable god, stonelike, dressed always in a dark frock coat. But Juárez had already totally assumed this destiny for himself: "God and society have placed us in these positions so as to bring happiness to our communities and avoid the evils which could come to them."[36]

He was a savior in the form of a firm, diligent, and severe shepherd. The ancient pre-Columbian chronicles recount that, for their rulers, the Indians chose men who seemed the most serious and melancholy: those able to govern out of silence.

Very soon Providence put him to the test. In the second half of the century, Oaxaca was struck by an epidemic of cholera. In less than two years more than two thousand people would die across the state. Confronted with this disaster, the governor took certain health measures (speeding up the construction of hospitals) and several that were entirely spiritual. On July 3, 1850, *The Chronicle,* the official government newspaper, announced that the state government, invited by the venerable Cathedral Chapter, together with all the authorities, would attend Three Days of Public Prayer to be held for the purpose of "begging our God of Forgiveness to free us from the terrible scourge of cholera morbus."[37] Two days later Juárez would join the procession: "He walks for a while with his arms crossed in imitation of the Divine Savior . . . he murmurs prayers . . . he falls to his knees before the tabernacle and remains there . . . while the priests chant the *Miserere mei Deus.*"[38]

With uncharacteristic objectivity, one of his hagiographers observed that during his period of government "he never ceased to attend the solemn ceremonies of the Church and take his seat under the chancel, on a carpeted platform with a prie-dieu and cushions and a chaplain would say the confession and the credo and then bless him." He was "an old-fashioned Catholic."[39]

Juárez's government proceeded to build a new kind of legitimacy based on observance of the law. The governor gave his public position a religious aura, trying to stay close to the clergy and their legitimizing pres-

Benito Juárez
(Centro de Estudios de Historia de México [CONDUMEX])

ence, invoking God and Providence, and diligently attending church ceremonies. He did all this without hypocrisy but rather with a natural ease, in order to dignify and strengthen the effectiveness of his civil mandate.

His government introduced a style of command that stressed budgetary discipline, regular payment of public employees, arbitration of disputes and interests in a climate of conciliation and understanding, a strong emphasis on the defense of state sovereignty (when the war with the United States ended, he did not allow Santa Anna to enter his domain), seriousness in reports to the legislature, encouragement for the almost nonexistent mining industry, a concern for the different branches of public administration, attentiveness in the appointment of judges, and the prudent, effective, and forceful use of "the wide-ranging and unlimited faculties" that he requested and obtained from the state congress in order to combat "those who would disturb public order."[40] In some cases—civil unrest in Tehuantepec, for instance—his mere presence was enough to calm people down. Everything was done by decree, all in a moment, all powers concentrated into one—but within a legal framework. This new style could be seen in even the most trivial details. He would point out years later in his *Notes for My Children:*

> As regards bad habits there were some the sole purpose of which was to satisfy the vanity and pomp of the governors, such as having guards from the armed forces in their houses or wearing hats of a special kind at public functions. I abolished this custom as soon as I became governor, preferring to wear the same hats and suits that ordinary citizens wear and living in a house with no military guards or any kind of show, because it is my conviction that respect for those in government comes from the law and from honest conduct and not from special clothes or military display, things which are only proper for a king in a play.[41]

Something more than a new style was being tested in that remote state of the Republic: It was a new ethic of authority. With his almost idolatrous devotion to the law, his proven zeal for education (he founded fifty schools in rural areas, opened and subsidized new branches of the Institute of Arts and Sciences, encouraged the education of women), and his firm-handed concentration of power together with the seriousness that was the hallmark of his actions, Juárez seemed to be the exact opposite of Santa Anna. "His Most Serene Highness" was all frivolousness, thoughtlessness, rashness, impetuosity, and improvisation, with his emotional melodrama that fluctuated between tinsel, tears, and roaring laughter. The Indian from Guelatao—with his darkness of aged bronze, his look of a stone idol—seemed to have emerged from some ancient ritual drama. He was solemn, thoughtful, ordered, and impenetrable. Santa Anna gave a thousand or so dazzling speeches in which it is hard to find one sincere phrase or even a glimmer of authenticity beyond the authenticity of performance. In Juárez's speeches we sometimes have to allow for the rhetorical content, but coming from a man of absolute loyalties, his words, suddenly, could take on a special weight:

> I am a son of the people and I will not forget it; on the contrary, I will stand up for their rights and take care that they learn, that they grow nobler and that they create a future for themselves and abandon the path of disorder, vice and misery to which they have been led by men who only in words call themselves friends and liberators but through their actions are the cruelest tyrants.[42]

This was the man Santa Anna would never forgive for the 1848 decree forbidding him to enter Oaxaca. In retaliation, Santa Anna (during his final presidency) sent Juárez into exile. The two orphans, Ocampo and Juárez, met in New Orleans for the first time. These men could not have been more different, but even so they would form a holy alliance. Ocampo taught Juárez the principles of a moderately anticlerical position. On his part, Juárez would come to exert a strong influence over Ocampo. The latter had no faith in religion, while Juárez was all religion. Ocampo represented the spirit of the Reform, Juárez the Indian religious devotion needed to impose the spirit of the Reform. The teachings of Ocampo made Juárez accentuate his religious feeling for the law, his idolatry of the law, until it became almost dissociated from religion as such, almost creating, within Juárez's conscience, the separation of Church and State.

And as a person, this Indian represented something new and indefinable. A force not of nature but of the land and its history. He did not refuse power, as had every member of the creole dynasty from Iturbide to Santa Anna. He embodied it.

Most of the Liberals were not as "pure" as Ocampo. They called themselves *moderados,* and they *were* "moderates." They would have preferred a peaceful, harmonious coexistence with the Church.[43] The trial by fire for this position came in 1856 during the presidency of Ignacio Comonfort, the most moderate President in Mexican history. The tone of the majority in the recently summoned Constitutional Convention was neither radical red nor even Mexican rose but rather a pale pink. Both the President and the legislature wanted nothing more than to reconcile the Mexican political family on the basis of a twofold faith, in both tradition and progress, order and freedom. Gently, peacefully, the confusions between the spheres of Church and State would have to end. As the experience of the century demonstrated, the temporal had to free itself from the unproductive hands of the Church and pass into the active hands of civil society and a secular State (as well as into the invisible hand of the market). The sacred should return to its own proper compass: the intimacy of one's conscience and the interior of churches. Over these domains the Church should retain its indisputable sovereignty. But not beyond.

Before the delegates to the Convention had a chance to debate this issue, the first uprising supported by the clergy broke out in Puebla. The localized rebellion was put down without much trouble, and government reprisals were light—a partial takeover of Church property to which the ecclesiastical hierarchy reacted as if it had been total. In a moment of particular intensity, the moderate Convention refused to formally approve freedom of religion but did sanction "freedom of conscience" (which tacitly implied and tolerated freedom of worship). But immediately afterward, in the same text, the Convention voted to particularly "care for and protect" the Catholic Church "with just and prudent laws."[44] The moderation of the vote and the pro-Church addendum had no moderating effect on the attitude of the Church. The ecclesiastical authorities were as harsh in their condemnation of this proposal as they had been of the Law of Disentailment issued by the government a few months earlier. As a result of that law, the Church no longer held the title deeds to its rural and urban properties, but its former possessions still earned it an interest of 6 percent per year. The Law of Disentailment had actually turned the Church into an immense mortgage society and opened up possibilities for the creation of a vast new property-owning middle class. But its mortgage profits were no more soothing to the Church than the moderate "freedom of conscience" resolution of the Convention. No more than the frequent prayers to God within the Convention, the religious atmosphere of its sessions, or the sincere Christian faith of almost all the delegates. The Church accepted nothing. It discussed nothing. Its position was all or nothing. A moderate historian of the period expressed his sadness at the lost opportunity for reconciliation and negotiation: "The Church worked with tireless activity and its clandestine roles were countless. . . . They used every means possible to stir up hatred among the people for the government in power, to trouble consciences and inflame passions."[45]

The first fully liberal Constitution in Mexican history was promulgated on February 5, 1857. The delegates took their oath to it before a crucifix. Its first line mentioned God. Though it included some limitations on the power and privileges of the Church, its religious spirit was essentially no different from the Constitution of 1824 except for one point—it did not decree that Catholicism must be the exclusive religion of Mexico. The Church reacted with total rejection: "No one can lawfully take an oath on the Constitution."[46] Priests poured out anathemas and excommunications. (Only in the remote state of Oaxaca, Benito Juárez, who had again become governor, succeeded in persuading the archbishop to provisionally accept the Constitution with a Te Deum in the cathedral.)

To the cry of *Religión y fueros* ("Religion and special privileges!") the Conservative army and political leaders rose up—blessed by the clergy—against both the Law of Disentailment and the Liberal Constitution. Mexico began the Guerra de Reforma, the War of the Reform, a ferocious civil war that would last exactly three years, from January of 1858 to January of 1861.

During the first few months, the Conservative armies swept everything before them. But then the Liberal forces stiffened their resistance, and for two long, interminable years, a painful equilibrium was established. Each side steadily grew more radical. In one of the worst acts of the war, on April 11, 1858, in the town of Tacubaya near the capital, the Conservative army under Leonardo Márquez (known thereafter as "the Tiger of Tacubaya") massacred all their prisoners—commanders, officers, soldiers, even the doctors and medical students who were caring for the wounded. The "decent people" of the town, plus Church dignitaries and the military leaders, organized a huge public celebration of the victory. The city was filled with lights. Ignacio Manuel Altamirano, the great Liberal writer and editor, was there and wrote:

Light yourself up even more, damned city!
Light up your doors! light up your windows!
Light yourself up even more, you need light,
Party of the darkness! party of the cassocks![47]

While the Conservatives had abler generals, the Liberals—from their stronghold in the port of Veracruz—were led by the fierce will and authority of Benito Juárez and the "fiercely independent" ideology of Melchor Ocampo. The heartland of the country was the scene of numerous battles, yet, in contrast to the revolution of independence, humble people did not willingly participate in the war. Both sides used a draft to forcibly swell their ranks. But in a country of eight million, there were never more than twenty-five thousand men under arms. The War of the Reform was not a popular war in either sense of the word. The great mass of the people neither approved of it nor enlisted in it. It was a war between the ruling minorities.

In 1859 and 1860, as the bloodshed dragged on, the Liberal politicians in Veracruz initiated truly historic changes.[48] They issued the Laws of the Reform, through which, after three hundred years, the social, political, and economic position of the Mexican Church was irrevocably altered. These laws nationalized all ecclesiastical property without compensation; closed monasteries and convents; suppressed religious confraternities and monastic orders; made cemeteries national property; prohibited civil servants from taking part in any religious rites; set up civil courts to deal with births, marriages, and deaths; and permitted freedom of worship. Almost all the laws were formulated by Ocampo.

But the war had to be won before the Laws of the Reform could be put into effect. By 1859, both sides were in serious need of funds. The Conservatives were on the verge of mortgaging the country to Spain and the Liberals to the United States. Luckily for Mexico, neither deal went through. But while the treaty with the United States was still pending (it was eventually rejected by the U.S. Senate), the United States made a move that directly and critically aided the Liberals. As desperate on his side as the Liberals were on theirs, Miramón was trying to surround the port of Veracruz and cut off the Liberal capital. To complete his stranglehold, the "young Maccabeus" hired two Spanish steam-powered ships of war in Cuba, with the intention of using them to blockade the port. But the pro-Liberal diplomacy of Washington sent a U.S. corvette and two armed merchantmen speeding to seize the ships, "under suspicion of piracy." Though neither side knew it at the time, this ended the Conservatives' last best hope to turn the tide. Miramón retreated, and his armies began to lose their battles to generals directed from San Luis Potosí by Santos Degollado. A huge and total defeat at Silao, in the state of Guanajuato, ended Conservative hopes. In January of 1861, Juárez entered Mexico City.

The war was over, but the conflict went on. The resistance of the Church had been merciless, but so was the Liberal reaction after 1861. One historian coined the perfect phrase for this period: "The Pickaxe of the Reform."[49] Eminent Liberals literally picked up axes to destroy altars, church facades, pulpits, and confessionals. Scenes out of the French Revolution were reenacted. Images of saints were decapitated, shot full of holes, burned in public *autos-da-fé*; Church treasuries were robbed, archives were plundered, ecclesiastical libraries went up in flames. Bishops were stoned to death, and Church property was auctioned off. Nuns who had spent their whole lives cloistered were suddenly forced out of their convents. Ocampo ordered the expulsion of all Catholic bishops from the country with only two exceptions: the octogenarian bishop of Baja California (a desert) and the bishop of Yucatán (a historical island). "The government banishes the bishops," exclaimed the young Ignacio Manuel Altamirano (who had witnessed the massacre of Tacubaya). "It ought to hang them!"[50]

The Conservatives, defeated as an army but still present as scattered bands of armed men, responded with assassinations. Melchor Ocampo himself had resigned his ministerial position and returned to his beloved Hacienda de Pateo, where he was living with his daughters. He was fifty-seven years old, and he was satisfied. He had accomplished a great deal.

In May of 1861, a group of men with guns rode up to Ocampo's door and carried him off. Leonardo Márquez and Félix Zuloaga, now leaders of Conservative guerrillas, had ordered him brought to them. The group traveled for three days with their captive. In a small village, they let him write his will. "Before being shot . . . I declare that I recognize Josefa, Petra, Julia and Lucila as my natural daughters . . . and I adopt Clara Campos as my daughter . . . I die believing I did what I thought was good for my country." And after signing it and noting, "at the very place of my execution, the Hacienda of Tlaltengo," he added, at the bottom, "The will of Doña Ana María Escobar is in a notebook written in English, to be found between the living-room wall and the window of my dressing-room."[51] That will was a gift to his daughters. It offered them the identity of their mother (which they had not known), the same Ana María Escobar who had, like him, been taken in decades ago, as an orphan off the street, by Doña Francisca Xaviera Tapia. A squad of men shot him, and after the finishing bullet in the head, they hung the body of Melchor Ocampo from a tree.

Mexico was now on the brink of another war, that of the French Intervention, but it still had turned over the oldest page in its history: It had at last emerged from its traditions of theocracy. Juárez was reelected for another four years. He had given unto God that which was God's (perhaps a little less) and unto Caesar that which was Caesar's (perhaps a little more). Church and State were now divorced.

While Mexico was tearing itself apart in the drama of the Reform, far away in the north of Italy, in the small kingdom of Lombardy-Venetia, a ruler was trying to establish ideas of sovereignty and forms of democratic life similar to those of the Mexican Liberals. The moment for the unification of Italy had not yet arrived, but its outlines were on the horizon. The governor-general of Lombardy-Venetia was a kind of Austrian Iturbide, a liberal and romantic archduke whose admirable designs for the happiness of his subjects clashed with their most basic desire—to decide their own destiny.

Archduke Ferdinand Maximilian von Hapsburg had the misfortune to have come into this world two years after his brother Franz Joseph, who would become emperor of the dual monarchy of Austria-Hungary. While the firstborn prepared for the throne he would certainly occupy, the inclination of the second in line to the Austrian empire was to escape, to wander the seas, or his imagination, or the air. It was as if Maximilian had suspected that his brother's reign would last through the decades that in fact it did. As soon as he could, he visited Greece, Italy, Spain, Madeira, Tangiers, and Algeria, where he climbed the Atlas Mountains. At the tombs of the Catholic kings in Granada, he "proudly, eagerly and yet sadly" extended his hand "toward the ring of gold and toward the sword that had once been so powerful" and thought it would be "a beautiful and divine dream for a nephew of the Spanish Hapsburgs to flourish the latter in order to conquer the former."[52] He made that visit in 1854. The following year, as admiral and commander-in-chief of the Austro-Hungarian fleet, he traveled to Palestine; in 1856, to France, Belgium, and Holland; and in 1857, to the realm of his cousin, Queen Victoria.

A man of twenty-five by then, he already held a world of exotic landscapes in his memory and his imagination. That year he married Charlotte Amélie, daughter of Leopold I, king of the Belgians, and they took up residence at the castle of Miramar, which Maximilian had begun building two years earlier. It was a glittering limestone palace erected on a great rock overhanging the Adriatic Sea and including a small lake within its grounds. His office was an exact copy of the wardroom of his warship, the *Novara*. Maximilian wrote his *Memoirs* here, looking out to sea and soaring above it: "And if the hypothesis of the aerostatic balloons should one day become a reality, I shall devote myself to flying and I have no doubt whatsoever that it will give me the greatest pleasure."[53]

But on land, destiny was still against this blond, pale young man with lively blue eyes and a protruding jaw concealed beneath his carefully curled beard. Nothing was further from his mind than absolutism or devout Catholicism. He was a liberal, like his century. And nevertheless, the people of Milan privately respected him and publicly distrusted him. He was a foreign prince, an outsider. His plans and his works did him no good. He lived under the "constant humiliation of representing an indolent regime without a definite policy which reason in vain tries to defend."[54] In 1859, Franz Joseph responded to growing tension by sending reinforcements into the province, an action that in effect meant the removal of Maximilian. By doing this—wrote Maximilian—Franz Joseph ignored "decorum" and "the good name of an archduke." Was there nowhere in the world—that is, no monarchy in the world—for this prince of refined manners and his strong and ambitious wife, destined, like her husband—more than her husband—"to revive all the glories?" They seemed to have been born in the wrong place and the wrong century: "It is sad to see our splendid monarchy, so powerful in former times, sinking ever more steadily due to its incompetence, its errors and its incomprehensible course of action."[55]

In the "frightful" cold of December 1859, Maximilian left his wife in Belgium and sailed to far-off Brazil, which was ruled by his first cousin. He loved the adventure.

"Man is interested in the remote and the unknown," he wrote at his first sight of the coast of America, "and if there is a sign of life on any distant point, he is attracted to it. . . . It seems to me like a legend that I am the first descendant of Ferdinand and Isabela who since childhood has thought it his mission in life to tread on the continent that has attained such gigantic importance for the fortunes of humanity."[56]

At some stage during the crossing, he may have remembered the strange suggestion made by a group of Mexicans two years earlier, concerning a possible offer of the throne of Mexico. "Cela serait une belle position,"[57] he had remarked to his father-in-law. For the time being, Maximilian abandoned himself to the sensuality of a Brazilian woman, who (according to rumor) gave him a venereal disease from which he would recover in time. Sadly, it is said to have left him sterile. Despite the pleasures of the tropics, however, he could not forget the crumbling Austrian monarchy, now presided over by a "new Louis the Sixteenth,"[58] and the unstable balance among the European powers, endangering his property and his estates, including the idyllic refuge of Miramar. He was a prince in search of a kingdom.

The Mexican coterie in Paris was led by a professional exile, an ardent Francophile, the hacendado José Manuel Hidalgo, a friend of Eugenia de Montijo, the empress, a Spaniard from Granada. For two years, though for different reasons, both had been dreaming the same dream: to establish a monarchy in Mexico.[59] Suddenly, in 1861, events seemed to make this a serious possibility. The moratorium on repayment of public debts decreed by the Juárez government in June of that year began the cycle. In October, England, France, and Spain signed a Tripartite Convention in London designed to force Mexico into paying its debts and meeting other demands, not entirely unjustified in the case of the English and the Spaniards. Then, at the beginning of 1862, expeditionary forces from the three countries disembarked in Veracruz. Shortly afterward, their demands satisfied through diplomatic channels, Spain and England withdrew, leaving France in a position to carry out its true design—not the collection of exaggerated debts but the invasion of the country and the execution of a multiple project of reconquest. To Eugenia it meant revenge for Spain; to Napoleon III, a chance for France to return to North America, capitalizing on the Civil War in the United States.

Everything seemed to favor the foundation—or reestablishment, as its partisans would assert—of the Mexican Empire. The old creole monarchists made the suggestion to the empress of France. She knew how to reach her husband's ear, and Napoleon III hastened to pass the word on to the emperor of Austria-Hungary, who in turn ordered his ambassador in Paris to arrange for the suggestion to reach the most attuned ears in all the courts of Europe, those of his brother, the Archduke Maximilian, and his young Charlotte.

While Maximilian and Napoleon were carrying on lengthy negotiations about the conditions of financial, military, and diplomatic support for the new crown, the French army, after several major setbacks, was steadily advancing in Mexico. But the rout of General Lorencez's troops at Puebla—by the Mexicans under the command of General Ignacio Zaragoza on May 5, 1862—could have dampened Maximilian's enthusiasm, and there were some negative reactions in Europe. England refused to support the venture (Queen Victoria favored the vacant throne of Greece for Maximilian). Franz Joseph longed to distance his liberal and critical brother, but he claimed he was in no position to provide military or financial aid of any significance. All Maximilian's close relatives in the European nobility expressed their skepticism and fear. "They will murder you," warned Marie Amélie, Charlotte's grandmother and the wife of Louis Philippe of Orléans. His father-in-law, King Leopold, did not reject the idea but he urged him to obtain a "binding agreement" from Napoleon.[60]

It soon became clear that the whole project hung by a single thread: the support of Napoleon III. Maximilian played with the idea of stipulating his conditions: Napoleon would not be his "only patron," and Napoleon should not count on "his absolute submission."[61] But the truth was that his desire for the throne was so intense that he only listened to what favored his wishes and closed his mind to any effort or information that might have dissuaded him. At any and all cost, he had to strengthen his alliance with Napoleon. He had to believe that this remote country in peril was crying out for him, Maximilian, to come and rescue it.

He could not have been blind to the geopolitical risks that he was running. He was going to a distant land across the ocean, a country with a powerful and intrusive neighbor that would sooner or later emerge from its Civil War. He was dependent on a single patron, who might change his mind and whose position in an unstable Europe might alter at any moment. Nevertheless he decided that the fates would favor him. And not only fate but the desires of the most influential Mexicans and the votes of those more humble.

There were complications at the beginning of 1864 from his brother. Franz Joseph took a very hard line, demanding that Maximilian, once he became emperor of Mexico, should renounce his rights of succession to the Austrian throne. Weeks of tense and painful correspondence passed between them, involving both the family and other royal houses of Europe. Maximilian began to doubt the wisdom of his decision. Franz Joseph's pressure was an ominous sign. On his side, Napoleon insisted that they had negotiated the agreement Maximilian wanted and that things had gone too far for him to pull out now.

From Napoleon's point of view, much had been accomplished. Juárez and his government had abandoned the capital of Mexico in May of 1863 and headed north. A month earlier, following the siege of Puebla, which had lasted sixty-one days, the main body of Juárez's army had surrendered and been taken prisoner. It included twenty generals (among them Porfirio Díaz—who would escape to fight again—and Jesús González Ortega), three hundred officers, and eleven thousand soldiers. The thirty thousand men of the French army commanded by General Forey had occupied Mexico City in June. A Regency Council of three men—Salas and Almonte, both Conservative generals, and Archbishop Labastida—had been established. In July, an assembly of 215 "prominent men" voted to install a monarchy and appointed a committee to satisfy (after his death) the desire of Alamán: "We are lost, with no way out, if Europe does not soon come to our aid."[62] In October, Achille Bazaine, a very able general, was appointed to succeed Forey as commander-in-chief of the French expeditionary force. (When Forey returned to Paris, Napoleon would not allow him to speak to Maximilian and Maximilian surprisingly expressed no interest in meeting the one man who could give him the latest news from Mexico.)

In October, the Mexican committee met with Maximilian. Between November and February—while the Conservatives were fabricating lists of supporters for the future emperor—the advance of the French army had been inexorable. They took Morelia, Querétaro, Guanajuato, Guadalajara, Zacatecas. The month of March was passing, and Maximilian realized that it was too late for him to withdraw. Vacillation and insecurity were part of his character, but the last thing he would do was refuse this offer. Mexico meant redemption for his sad experience in Lombardy-Venetia and, above all, the accomplishment of the monarchic vocation that destiny had seemed to be denying him and his wife. The crown of Mexico, unlike that of Greece, was not "goods refused by half a dozen princes."[63] It was the promise of a great empire. With tears in his eyes, he received his brother at Miramar; with tears he renounced his rights to the Austrian throne and said good-bye to Franz Joseph. At last, on April 10 of 1864, the members of the Mexican committee arrived at Miramar with the signatures of support that Maximilian had demanded. The future emperor's voice trembled as he read, in Spanish, his positive answer: "With the vote of the prominent men of Mexico,"[64] he could consider himself chosen by the Mexican people.

The emperor of the French had generously provided the guarantees that Maximilian had always considered essential to the success of his mission. He therefore accepted the crown and would work for the liberty, order, grandeur, and independence of Mexico. As the Mexican imperial flag was waved in that reception room at Miramar, the members of the committee hailed the emperor and empress with their usual "Vivas!" The excitement was stronger than the emperor's fragile constitution; it all became too much for his nerves and he left Charlotte to preside over the celebrations.

The Treaty of Miramar, signed during those days, contained a series of stipulations about the money needed to finance the occupying army; the debts, claims, and obligations that the new government would assume; the salaries of the court and other items—all involving sums far out of proportion to the real economic state of the country. It had been drawn up on paper, for a hypothetical, poetic empire. This agreement did not go unnoticed by the watchful members of Juárez's cabinet, who followed each and every step in the adventure from their headquarters in Monterrey in the north of the country. José María Iglesias, one of Juárez's ministers and a highly competent jurist, analyzed the situation with devastating precision in his *Revista histórica*: "From our examination of the unforgettable Treaty of Miramar, it is perfectly clear that its stipulations cannot possibly be complied with . . . on the one side, there was perfidy, on the other sheer stupidity."[65]

The romantic dream began on May 28, 1864, when the warship *Novara* docked at the port of Veracruz. Strangely, in this city that had supposedly voted for the empire, there were no triumphal arches or welcoming crowds. The reception, according to a lady of Charlotte's court, was "glacial."[66] The empress wept, but Maximilian delivered a speech that denied reality: "Mexicans, you have wished for me; your noble nation, by a spontaneous majority, has chosen me, from today on, to oversee your destinies. I gladly surrender myself to this calling."[67]

On their way to the capital, a few provincial cities gave them a warmer welcome: 770 arches of entwined branches and flowers lined the stretch of road between Puebla and the pre-Hispanic city of Cholula, for the passage of the royal couple. Finally, after kneeling before the image of the Virgin of Guadalupe, Charlotte and Maximilian made their triumphal entry into Mexico City. Iglesias noted in Monterrey: "The Mexican empire is the product of an abortion. Feeble, rickety and ramshackle, it will have a sickly life and die an early death."[68]

Eight days after their arrival, they moved into Chapultepec Castle, to which they would give a new appearance evocative of Miramar. It was already, in itself, a metaphor for Miramar: a castle on a wooded hill. It was missing a lake (which they would create within the grounds) and marble furniture (which soon would arrive), and the ocean, which was replaced by a magnificent view of the Valley of Mexico: the shimmering lakes and canals, the countless church towers rising from the heart of the colonial city or surrounding towns and villages, and over them all the two majestic volcanoes crowned with snow. "It is the Schonbrun of Mexico," Maximilian enthused to his younger brother, Archduke Karl Louis, recalling the place where both had been born: "an enchanting pleasure palace built on basalt rock surrounded by the gigantic and famous trees of Moctezuma, which affords a view of such beauty that I can only compare it with that of Sorrento."[69]

In his daily life, the emperor soon began to imitate the Mexicans. He dressed in the style of the *chinacos*—the military Liberals—wearing a large sombrero, a short jacket, and pants that buttoned down from the knees. He moved around like any citizen, on horseback and without ostentation. He wanted to demonstrate his liberalness and his Mexicanness. He had not come to be emperor of the Conservatives but of all the Mexicans. He wanted to win the support of the Liberals because he was a liberal and because, as Leopold I had counseled him: "The Catholics, whether or not they want to be, have to be on your side."[70]

At the beginning of August 1864, while the occupying army was advancing toward the north and the west of the country, Maximilian went on his first trip through the historic Bajío: the granary of Mexico and the setting for the War of Independence. He visited Querétaro, Guanajuato, León ("the Manchester of Mexico"), and Morelia ("very

Maximilian von Hapsburg, Emperor of Mexico, 1866
(Private Collection)

Empress Carlota, 1867
(Private Collection)

liberal and for this reason really worth knowing").[71] At one meal, to the bewilderment of his fellow diners, he asked for the anti-Conservative song "Los cangrejos" ("The Crabs"), composed by the Liberal general Vicente Riva Palacio. On his way to Dolores, he stopped at the Hacienda de Corralejo, where Hidalgo was born, and ordered that a monument be erected on the site.

On September 16 he arrived in Dolores and delivered a speech at eleven o'clock at night from what had been Hidalgo's window. And then he gave the first official re-creation of the *grito*, the cry for Mexican independence.

On that same sixteenth of September, Juárez celebrated the anniversary of independence on a ranch in the state of Coahuila. He was preparing to march farther north and establish his government in Chihuahua. Iglesias wrote: "The deeply felt words of the President moved all those present. After he finished his speech, the soldiers sang patriotic songs, alternating them with popular dances."[72] Sung by those voices, the Mexican national anthem took on a new meaning. It was no longer the opera of war, as in the days of Santa Anna. It was the voice of war.

Beyond demonstrating how Mexican he was, Maximilian had to show that he was a liberal. He ordered that pictures of the heroes of independence should be painted and displayed in the National Palace. But there was only one real way he could prove himself to be liberal: by distancing himself from the Conservative party and from the demands of the Vatican.

"The worst people I have found in this country (he would say) belong to one of three groups: the judiciary, the army officers and the majority of the clergy."[73] To improve the legal situation, he ordered new codes drafted and the dismissal of incompetent judges. Toward the army, he took stronger measures: he sent Miramón off to Berlin to study artillery tactics and Márquez (who had butchered his Liberal prisoners at Tacubaya and murdered Ocampo) to Jerusalem, of all places, to found a Franciscan monastery. Instead of putting General Juan N. Almonte (who was the natural son of José María Morelos) into the key cabinet post of Foreign Relations, he appointed José Fernando Ramírez, a moderate liberal who in his spare time collected pre-Hispanic codexes and relics. With the clergy, he felt he had to draw the strictest lines: "The people in cassocks are evil and weak and the great majority of the country is liberal and wants progress in the most complete sense of the word."[74] It never occurred to him to draw any conclusions from this conviction. If all Mexico was liberal and its liberalism was legitimately embodied in Juárez, what was the role of His Majesty?

He went on dreaming. Surely he would be able to arrange a concordat with the Vatican on the basis of the Laws of the Reform. The nuncio whom Maximilian had so anxiously expected finally arrived, but he did not have the "good Christian heart nor the will of iron" that Maximilian had hoped for.[75] Or rather he used that will to transmit the concern of Pope Pius IX for his "defrauded" Church and to convey his total repudiation of this disturbing emperor's reformist ideas. Two weeks after the nuncio's arrival, during Christmas of 1864, Maximilian broke off relations with him and ratified the freedom of religion and nationalization of Church lands decreed by Juárez in Veracruz. Papal bulls and other documents arriving from Rome were to be evaluated by the Ministry of Justice and Foreign Affairs.

Among the members of Juárez's cabinet, these measures provoked joy and sarcasm. They marked the "most splendid triumph" of all the victories of the Reform. They demonstrated the soundness of Liberal principles, and the "reprehensible ingratitude" of the "so-called Emperor" toward "the authors of his ascent." A cool analysis of the internal and external factors at play here made it ever clearer to Iglesias that Maximilian would eventually yield: "Undeceived and repentant, he will abdicate and return to Miramar, or he will fall out with his few followers, thus putting an end to his ephemeral government which, up to now, has only been noteworthy for its inaction."[76] Five possibilities or some combination of them would precipitate his fall: a European conflict that would lead to a war in which the French would have to intervene; the withdrawal of the French troops because the imperial treasury could no longer maintain them; the death of Napoleon III; the reassertion by the United States of the Monroe Doctrine; and "the indefinite prolongation of the war waged by the Mexican supporters of Independence and the Republic."[77]

Maximilian foresaw no such destiny. On the contrary, he dreamed of expanding the Mexican Empire southward, into Central America, so as to include those territories that had formed part of the first empire ruled by Iturbide. Surely the advance of the French army into Nuevo León and Tamaulipas foreshadowed the final defeat of Juárez. The disorganized finances of the empire would stabilize. Mexico was not a country invaded by a foreign army. On the contrary, it was exactly what Maximilian wanted it to be. "I live in a free country, among a free people," he wrote to his younger brother, always comparing his own situation favorably to the decadence he had left behind: "If Mexico is backward in many respects, if it lacks comfort and material development, in exchange for that, and more important in my opinion, we are far more developed in social matters than Europe, and especially than Austria.

Here among us there reigns a healthy democracy, free of sickly fantasies after the style of Europe." [78]

"Carlota" (or, more affectionately, "Carlotita"), as she was known to her ladies-in-waiting, received visitors, and on Mondays she organized dances. For her, much more than for Maximilian, these soirées on the terraces of Chapultepec were a delight. But sometimes the sound of gunfire from the base of the palace hill would interrupt the festivities. For Maximilian, such outbreaks meant nothing. The year of 1864 was coming to an end with the best possible omens. In Chihuahua, where Juárez had established the constitutional government, Iglesias agreed with him: "1865 is born filled with a thousand promising signs . . . a future rich in hope announces to us the happy resolution of the second war of our Independence."[79]

Napoleon III soon realized that he had made a bad bargain. The results did not match his expectations. Perhaps "Juarismo" was, as "Carlota" wrote to Eugenia, "the most horrible form of demagogy," but the country was still resisting, still at war.[80] In February, Bazaine had forced the surrender of Porfirio Díaz in Oaxaca, capturing four thousand men and sixty cannons, but the cost had been immense. In regions apparently pacified, the struggle would start up again—with guerrilla warfare. On January 1, Juárez had stated, in a Declaration to the Nation, that the costly and chaotic rushing around of twenty-seven thousand French and allied soldiers only brought the day of victory closer. And the Juarista camp—always kept well informed by the press and by the diligent Mexican representative in Washington, Matías Romero—had for several weeks been hoping for and awaiting word of a great event. And then it happened: the final defeat of General Lee at Richmond. By the beginning of April of 1865, the Union had won the Civil War.

As Iglesias had predicted, the Union victory would be a critical factor in forcing the withdrawal of the French. Napoleon III was growing more and more disillusioned by the financial news from the empire, and he began to consider some measure of disengagement from the adventure. Maximilian, for his part, was not overly concerned. With persuasion and a demonstration of good faith, he felt the empire could attract the support of Lincoln's successor. There were also internal affairs that needed urgent attention, such as the new subdivision of the country into departments, syllabuses for the study of classical literature and philosophy, projects of colonization, and above all ministering to his true supporters: "The best are and continue to be the Indians; for their sake I have just passed a new law which creates a council concerned with aiding them, attending to their wishes, complaints and needs."[81]

"The Indians everywhere," observed a visitor to Mexico, "showed fanatic enthusiasm for the Emperor."[82] And they were right to do so. It was the first time since the colonial period that the villages and communities could rely on being heard by the authorities when they voiced their complaints on their own terms, based on their feelings of a divine right to the land rather than the liberal right of individual property.

In the area of the sugar plantations near the cities of Cuernavaca and Cuautla there had been killings, a few years earlier, of hacendados and Spanish administrators, of a sort not seen since the War of Independence. Maximilian knew of these tensions and tried to resolve them. He was aware that some of these pueblos (both Indian and mestizo) held property deeds that had been granted by the first viceroys of New Spain on behalf of the Spanish Crown, and that they retained surveys and maps marking precise boundaries, which had been violated by the haciendas over time.

The council created by Maximilian would eventually issue two significant decrees. One recognized the legal character of the pueblos so that they could defend their interests and reclaim rights to their land and water; the other prescribed not only the restoration of the land to its rightful owners but the donation of land to those villages in greatest need. Maximilian was not moved to these actions by any socialist impulse. It was the old paternal spirit of his royal ancestors.

The strangest thing of all was that Maximilian understood that his military situation was deteriorating as rapidly as his finances. Guerrillas and Juarista troops were renewing the battle in the states of Michoacán, Jalisco, Sinaloa, and Nuevo León. None of Maximilian's victories seemed to last. Nor was he totally blind. His complaints and letters about his situation were not those of a deluded or a stupid man. He had concrete and reliable information on his strategic situation, but he saw it from the wrong perspective and did not seem to draw the logical conclusions. Suddenly he came up with a solution that he was sure would turn things around. He would himself replace Bazaine as military commander, and he would himself take control of the Treasury. But more than anything, he had to follow his personal script. Thinking about an heir, for example. His sterility obliged him to make unexpected decisions, and he decided to adopt a young grandson of Agustín de Iturbide as the successor to the throne.

September 16, the anniversary of Independence, came around again. What counted most for Maximilian was still in place: his link with Napoleon—though the French emperor kept sending a stream of reprimands, criticism, advice, and veiled warnings—and also Maximilian's sincere desire to make his people happy. He had adopted them as he had the young Agustín. In his speeches, Maximilian now sounded like a Mexican patriot, like the men who were fighting him on the battlefields: "My heart, my soul, my work and all my loyal efforts are for your sake and for our beloved *patria*. No power in the world can divert me from accomplishing my mission; all my blood is now Mexican, and if God should allow new dangers to threaten our beloved fatherland, you will see me fighting by your side for independence and integrity."[83]

Iglesias thought Maximilian had no right to use the word *independence*: "To pose as its panegyrist hardly becomes him, since he has offered himself as the instrument for its destruction."[84] And he disapproved of the unveiling of a statue to Morelos in the Plaza de Guardiola, where Maximilian's praise for Morelos was "nonsense out of his mouth."

The following month, a harsh decree inspired by Bazaine belied the emperor's soothing promises. Anyone who "belonged to armed bands" could be executed by order of a military court.[85] This was equivalent to granting Bazaine discretionary powers over the life and death of the civilian population. The French general—who by then had married a young and beautiful Mexican to whose pleasures he was dedicating his growing leisure time—put the decree into effect with a lavish use of his firing squads. The measure did not daunt the Juaristas, but Maximilian decided that it should be tempered with mercy. After all, as he saw it, the Liberals were almost finished. Hadn't Juárez already left the country? Or at least he was at El Paso del Norte, almost across the border.

And then Maximilian's dreams opened into a whole new region of fantasy. He called upon Juárez "to come to me and faithfully and sincerely assist me."[86] Maximilian believed that Juárez was cornered on the American border, and that when he had once more extended his term of office, on December 1, 1865, the President had alienated many of his followers and collaborators. If he was the Republic, then the Republic consisted of only a handful of territory. One had to reach out to him. Maximilian would offer to make him chief justice of the Supreme Court.

Sad news arrived from Belgium. Carlota's father, King Leopold, had died. In his last letter to Maximilian, he had written: "What is needed in America is success; all the rest is only poetry and a waste of money."[87]

It was February of 1866. Maximilian spent some leisurely weeks in Cuernavaca, his favorite refuge. He had fallen in love with *la india bonita*, "the lovely Indian," a woman named Concepción Sedano. But on this occasion he came to Cuernavaca in the company of Carlota—

who told him about climbing "the Pyramid of the Seer" during her recent trip to the Mayan site of Uxmal in Yucatán—and they brought the child Agustín with them. They stayed in a spacious villa, surrounded by a flourishing garden and overlooking one of the most beautiful landscapes "I have seen on this earth." Describing the place to his old friend Baroness Binzer, he wrote:

> Just imagine the divine expanse of a wide valley stretching out before you like a golden robe and surrounded by range upon range of mountains piling up the most daring forms, tinted in the most amazing hues, from the purest pink, purple and violet to the darkest sky blue, some of them jagged and impenetrable rising rock upon rock as on the coast of Sicily and others covered with forests like the green mountains of Switzerland, and behind them all, standing out against the dark blue of the sky, the gigantic volcanoes with their snow-covered summits. Within that golden robe, just imagine in every season of the year—or rather throughout the year since there are no seasons here—an abundance of tropical vegetation with its intoxicating aroma, with its sweet fruit, and add to that a climate as benign as Italy in May and beautiful inhabitants, of friendly and honorable character.[88]

Among the countless springs, the "ancient mangoes," the "orange trees with their dense foliage," in "our comfortable hammocks" while "colorful birds sing us songs, we sway in our dreams."[89] The most deceptive of their dreams was to feel that they had really been adopted by the land of Mexico. Of the two words *Mexican Empire*, Carlota was fascinated with the noun and Maximilian with the adjective. "Life in Mexico is worth fighting for . . . the country and its people are much better than they are reputed to be," he wrote to a friend, "and you would be astonished at how well the Empress and I, completely Mexicanized by now, live among these people."[90] Swinging in the comfortable hammocks at the old colonial mansion built by the mining magnate Borda at the end of the previous century, they dreamed different dreams: she a dream of power and Maximilian a dream of love for the land that they had adopted.

Infernal news reached them in their paradise. On January 15 Napoleon III had decided to withdraw his troops from Mexico. The evacuation was to be completed within the year. International circumstances had forced the decision upon him: pressure from the United States, the Prussian threat, public opinion in France, the gross mismanagement of funds in Maximilian's empire. At the blow, Maximilian began to waver. Should he seek support from England? Should he abdicate? It was the moment that Carlota loomed large. She decided that she would travel to Europe. And would once again persuade Eugenia, persuade Napoleon, speak with the Pope. Maximilian would continue attending to the affairs of the empire; he would issue his Code of Civil Law. Before leaving, Carlota wrote down her thoughts for him. They clearly reveal her temperament, so different from that of her husband. All the experience of the kings of Europe, all the examples of her own line—and Carlota knew its history by heart—pointed to one lesson: Never abdicate!

> And now, I tell you: Emperor, do not let yourself be taken prisoner! As long as there is an Emperor here, there will be an Empire, even if it only has six feet of land. The Empire is nothing other than its Emperor . . . The Empire is the only way to save Mexico.[91]

Perhaps more clearly than anywhere else, the historical meaning of the Mexican adventure is revealed in these words of Carlota. The "biography of power" to which Maximilian and Carlota belonged was not the biography of Mexican power but the ancient biography of the crowns of Europe, which had begun a slow but irreversible decline since the French Revolution. In the nineteenth century, one by one the dynasties would vanish. Nation-states would emerge in Italy and Germany and republics in France and Spain. Austria-Hungary and England would be islands where the aristocracy would resist, with the dual empire menaced by disintegration from within and only England secure through its long-standing parliamentary system. Encased in this framework of decay, the princes without a throne suffered more than anyone from the pressures of history. When called upon to reign, they had to do so in a democratic and republican era, an age that was turning away from them. They felt as if they were statues modeled from life and set out in a wax museum. And this was why Carlota was furiously against abdication: She felt it as a defeat and dishonor to her lineage.

But what did Mexico have to do with this uniquely European drama? The Mexicans had other problems, above all the difficult integration of a political state and an economic market, the building of a nation out of a somnolent colony and its recent past of chaos, penury, and disintegration. Why did Mexico have to pay for the second pregnancy of Archduchess Sophie, the mother of Franz Joseph and Maximilian? Carlota never asked herself such questions. At the end of 1865 she had criticized Juárez for reelecting himself *in saecula saeculorum*, but she never saw the flaw in her own position. Nor did Maximilian. They had no doubts about the *legitimacy* of their mission.

Finally, in July of 1866, Carlota traveled to Veracruz and set sail for Europe. In the Liberal camps, the news spread like wildfire, accompanied by "Adiós, Mamá Carlota," a song composed by Vicente Riva Palacio:

Happy, the sailor
Sings in measured rhythm
And now he raises the anchor
With its singular glow.

The ship sails, over the seas,
Bouncing like a ball,
Farewell, Mamá Carlota,
Farewell, my sweet love.

From the distant shore,
Sadly looking after her,
The stupid nobility,
Reactionary and traitor,
In the depths of his breast
Foresees his defeat
Farewell, Mamá Carlota
Farewell, my sweet love . . .

As the chinacos—
Who already sing of victory
Retaining in their memory
Neither fear nor bitterness—
Cheer, while the wind
Whips the vessel on,
Farewell, Mamá Carlota
Farewell, my sweet love.[92]

"I suppose you will be very saddened," Juárez wrote ironically to a loyal governor, "by the departure of Mamá Carlota. . . . This hurried departure of the so-called empress is a clear symptom of the disintegration of Maximilian's throne."[93]

While Carlota's ship was crossing the seas and the French army was packing its bags to follow her, the Juarista troops were building up a momentum that they would not lose. Mariano Escobedo and several other Liberal commanders advanced from the north, Porfirio Díaz from the state of Oaxaca in the south, Ramón Corona in the west, and Regules and Riva Palacio in the state of Michoacán. The average age of these leaders was thirty-five. Practically all of them were veterans of the War of the Reform. Now they smelled victory. And it would be their victory, won not by the pen but by the sword.

In Europe, Carlota was received by Napoleon III before returning to Italy and Miramar. In Napoleon she saw Mephistopheles: Wheeling and dodging, he had determined not to send "a man or a centime more." He had broken his promise, he had dishonored France, he had abandoned them. During her nights she read the Apocalypse of Saint John. On September 9 she wrote a delirious letter to her "deeply beloved treasure," a letter that now only spoke of the luminous future awaiting them.[94] France, the United States of America, Juárez, the Liberals, the whole world would bow down before the sovereign:

> But everybody has to be clearly told: I am the Emperor, nobody needs a president, an emperor's son is not called a president, and a modern form of monarchy must be introduced with all the respect it deserves. They should bow their heads to you, since the republic is *une marâtre comme le protestantisme* [an evil stepmother like Protestantism] and monarchy is the salvation of humanity, the monarch is the good shepherd, the president the *mercenaire*, this says it all. As soon as the task of uniting the Mexicans has been achieved everywhere, few troops will be necessary once the rebellion has ceased and you can then stand before the world supported by your people. . . . The rest of the "*chinacos*" can be employed like the Garibaldini in Italy as a kind of militia or first line of defense for the national interest against foreign aggression. Only by keeping them busy at something will they disappear. . . . One has to take advantage of the elements available, but in this case it would be better if the French were to quickly begin their evacuation. If everything goes well, as it certainly should, immigrants will flock to Mexico from America and Europe and *you will have the most beautiful empire in the world. . . .*[95]

A few days after she wrote this letter, while visiting the Pope in the Vatican, Carlota showed signs of insanity. She was thought to be on the verge of death and she wrote her farewell to Maximilian. She seemed to be dying, but she did not die. After spending time in an asylum in Austria, she lived as a recluse in the twelfth-century Bouchot castle in her native Belgium. She would not die till decades later, in 1927, outliving Juárez, Maximilian, Porfirio Díaz, Franz Joseph, Pope Pius IX, Eugenia, Napoleon, Bazaine, and all the dramatis personae of her tragedy. But she would never know it. She went on talking about the Mexican Empire to a rag doll, which she addressed as Max, or else she would say that it had Max's heart. She kept it beside her always, like a treasure.

Ill and abandoned, Maximilian was ready to abdicate. He took a trip to Orizaba. By now he knew of his wife's insanity. Who would advise him now? Nobody and everybody. He issued a decree repealing the "Black Decree" of October 3, 1865, which had permitted summary executions. "Leave, leave that country," a loyal friend living in Havana advised him, "because in a few weeks' time it will become a theater for the bloodiest of civil wars."[96] By then, Maximilian was once again unsure. Now he would not go until he had left peace and order behind him in Mexico. The Conservative leaders Miramón and Márquez, who had returned from exile overseas, finally persuaded him to remain on the throne. The fierce cacique of the Indians of San Luis Potosí, Tomás Mejía, still loyally followed his emperor. They would all raise armies, they would change the course of destiny. Maximilian returned to Mexico City. But the drama of the Reform was really over. He was only its last caudillo, left on the stage to take his final curtain call.

On February 5, the tenth anniversary of the Constitution of 1857, the remaining French soldiers left Mexico. Bazaine tried to persuade Maximilian to accompany them. The emperor refused. A letter from his mother weighed upon his mind. She agreed "entirely" with Maximilian's decision to remain, since he would thus avoid the "appearance" of having been expelled. "My eyes are filled with tears. The Emperor has noticed and I think he has guessed the cause. . . . *And in spite of everything I am obliged now to hope you will stay in Mexico as long as possible and that you can do it with honor.*"[97]

Once again he wrote to his friends in Europe. "Now I am a general, in active service and in camp, wearing high boots, spurs and a wide sombrero. All I retain of my admiral's uniform is a spyglass, which I always carry with me."[98] He had almost infantile fits of hope; he referred with respect to "the bravery and manliness of the Liberal commander"; and he thought he had found an honorable solution to his personal drama. The State Council of the five *M*s (Miramón, Mejía, Méndez, Márquez, and Maximilian) collectively decided to march their Conservative army to Querétaro, which had been under siege for seventy days. But the reinforcements promised by Márquez never arrived, impeded through a defeat at Puebla by General Porfirio Díaz and through the selfish prudence of Márquez himself, who would die in his bed in 1913, some fifty years after his comrades. Up to this moment, Maximilian had let himself be carried along by excitements of the moment and apparent victories, but Márquez's failure to act broke his spirit. Desperately, he began to seek an encounter with "the redemptive bullet."

At dawn on May 15, 1867, with the imperialist forces themselves beseiged in Querétaro, a Colonel López, commander of the guard at the Convento de la Cruz, which was being used as Maximilian's headquarters, allowed a force of Liberal soldiers to make a lightning entrance and take the leading generals of the vanished empire prisoner. Maximilian escaped, on foot, and would get as far as the Cerro de las Campanas ("the Hill of the Bells"), where he tried to offer some resistance but then had no other choice but to surrender to the Liberal troops.

Certain details of this capture remain historically uncertain. Maximilian had made Colonel López his "compadre"; and he had given him a decoration on the very night before the dawn raid. Some versions of the story say that Maximilian himself had ordered López to make an arrangement with the commander of the Liberal troops, General Escobedo, in order to avoid more bloodshed.

It was to General Mariano Escobedo that Maximilian formally surrendered his sword. And as he sat in his prison quarters that night, he could hear—at a distance, in the camps—the Liberal soldiers singing "Adiós, Mamá Carlota."

Juárez, from San Luis Potosí, ordered the court-martials of Miramón, Mejía, and Maximilian. (Méndez had already been shot.)[99] All the prisoners were assigned lawyers of high quality—Liberal lawyers. Maximilian refused to attend his own trial. He denied that the court had any jurisdiction over him. At one point he thought of trying to escape. He could return to Miramar; he could write a history of his reign. Then he thought of Mejía, of Miramón, of the disgrace to his honor if he made such a move, and of the plain and simple fact that it would be impossible to hide anywhere. Was there a soul in Mexico who would not recognize his red-blond beard? He preferred to write several letters and telegrams to Juárez:

> I beseech you in the most solemn manner, and with the sincerity proper to the situation in which I find myself, that my blood be the last to be shed; and also that you dedicate that perseverance which has led to the victory of your cause . . . to the noble end of reconciling minds, and of finally achieving a firm and stable basis for the peace and tranquility of this unfortunate country.[100]

His detachment was sincere, but he still trusted that Juárez would show him mercy. European governments, the American representative, and even Garibaldi, the champion of Italian liberalism, appealed to Juárez. A seductive European princess threw herself at the President's feet. The wife of Miramón entreated Juárez to change his mind: "I appealed to his heart as a father and a husband . . . nothing would move that stony heart, nothing could soften that cold and vindictive soul."[101] Juárez would say, "I am not the one who has condemned them; it is the law, it is the people."[102]

Maximilian resigned himself. He would never recognize his responsibility in the quarrels of this "unfortunate country." His good intentions placed him, like a spoiled child of history, above and beyond responsibility. He wrote letters of farewell. He put everything in order, down to the smallest detail. To the physician who attended him, he commented: "I am happy: Altamirano has told me that the Liberal government will retain some of my laws."[103] A postponement of the death sentence raised his slender hopes for three days. Finally, on June 19 of 1867, on the Hill of the Bells overlooking the city of Querétaro, the emperor, at the age of thirty-five (together with Mejía and "the young Maccabeus") met the most Mexican of deaths: execution by firing squad.

While in Miramar Carlota was rambling on without end about the "sovereign of the universe,"[104] Juárez's black coach arrived in Querétaro. They say that he went down to the basement of the Convento de Las Cruces where the body of his enemy lay and commented, "He had short legs."

The tragedy of Maximilian and Carlota would feed the literary imagination of generations of playwrights, novelists, poets, and filmmakers in Europe and America. Franz Werfel wrote a play about them; Hollywood would do a film; Malcolm Lowry would conjure up their spirits in his great novel *Under the Volcano*, written in the same paradise of Cuernavaca where the emperor, swaying in a hammock, had woven his dreams. Paradoxically, both Carlota and Maximilian managed to gain in death what they had so desired in life: to become Mexican. The story of the doomed imperial couple would be remembered—more than anywhere else—in Mexico. And the dreamlike halo of their legend would bind them, within the fragile memory of the Western world, to the history of Mexico, affording them a place and a memory that many of their European contemporaries—princes, dukes, archdukes, and emperors—would never enjoy.

On July 15, 1867, after a four-year pilgrimage during which he had carried the Republic on his shoulders, Juárez entered the capital. A few weeks earlier, Mexico City had been occupied by troops under the command of the brilliant Oaxacan general, Porfirio Díaz—thirty-seven years old and with thirty-seven battles to his credit: "our Porfirio," as Juárez used to call him.[105] Although the citizens of the capital cheered the President as they had cheered Maximilian and Santa Anna, this time there was a conviction in the air that the country had really entered a new historic stage: its "Second Independence." The ideological and military debate around the two programs for nationhood proposed by Mora and Alamán was now over for good. Their last historical option—the importation of a European monarch—exhausted in the wake of their military defeat, the Conservative party of Alamán would disappear from the national scene to never (at least openly) return. Only the Liberals remained to consolidate the nation on the bases laid down by the Constitution of 1857 and the Laws of Reform. The progeny of Mora had triumphed.

The vengeful severity of the Liberals in 1861 seemed now to foreshadow a time of terror against the many collaborators of the empire. In some provincial cities, former imperial prefects were "brought to justice," that is, shot. But the new and broader stage of Mexican history took a different direction. When he arrived in the capital Juárez stated it simply, in a memorable manifesto:

> Neither in the past nor much less in the hour of total triumph for the Republic, has the government desired—nor should it desire—to be moved by any feeling of passion against those with whom it waged war. . . . We shall now put all our energy into obtaining and consolidating the benefits of peace. . . . Let the people and the government respect the rights of all, because among individuals, as among nations, peace is respect for the rights of others.[106]

Before the romantic aura of the imperial couple and their tragic end, Juárez's victory seems pallid, graceless, almost unjust. But Maximilian had never really *seen* Mexico. He could never perceive the obvious contradiction of his position as a liberal in a country already governed by liberals, nor come to terms with how he had been deluded by the old Mexican royalists. To face the facts could have led him, like Carlota, into delirium and madness. It was better to follow another course of insanity—to continue deceiving himself, to continue denying the evidence, to accept the calls to honor from his family, from his mother, from Vienna. It was better to die.

Juárez represented the opposite: the principle of reality. For him and for Mexico it was better to live. In his innermost conscience he knew—with a knowledge he did not have to constantly reexamine—that the country was tired of playing roles and simulating actions, of singing hymns of war and victory in the midst of the most humiliating defeats, of a glittering display of court ceremony in its national and provincial capitals while most Mexicans were still submerged in "torpid poverty." Through his countless letters, decrees, and laws, Maximilian had governed a kind of literary empire. Juárez was a man of his word rather than a man of words. He too had taken to writing while in exile, but his letters to his diplomatic representatives, governors, and military commanders had always dealt with practical matters. He did not play a role. He was not some sort of Napoleon, a "Napoleon of the West" as Santa Anna had been, nor was he, like Maximilian, a new and benevolent Charles V, a distant descendant of the Catholic kings. He was the President of a republic occupied by a foreign army.

Francisco Bulnes—the Porfirian-era historian and critic of Juárez—would say that Juárez knew no other language than that of his official role: severe, sober, irreproachable. But Bulnes was mistaken. Juárez was a public figure with a deep and tender private life. While he was in Chihuahua at the beginning of 1865, his wife, Margarita, his three daughters (Manuela, Felícitas, María de Jesús), his son-in-law (Pedro Santacilia), and his three sons (Benito, Pepe, and the newborn Antonio, whom he had never seen) went to live in a suburb of New York. For a long and painful period, he had no news of them. When he found out they were safe and sound, he began to write them often, giving them advice about all sorts of things that might come up in their daily lives. They should be careful with heaters, for example; they should not misuse this modern invention: "I believe that cold like hot weather, however terrible they may be, are conditions laid down by the laws of nature to conserve and invigorate man, plants and animals. And one must not contradict these laws if one does not wish to suffer for the sin."[107]

He worried first about the health and minimal material well-being of his family (he was scrupulous in his role of provider) and then about the education of his children. After cheering up his son-in-law (his "dear Santa") with "the triumph of our forces in Sinaloa against the French and the traitors" and rejoicing at the fact that "the spirit of the people is beginning to come alive again"[108] (all this written in January of 1865, when the republican cause was at its lowest ebb), he set down guidelines for the education of his own personal sheep: "I suppose Pepe and Beno are going to school. I beg you not to put them into the hands of a Jesuit or a minister of any other religion; they should learn to philosophize, that is, they should learn to investigate the why or the reason of things so that, during their passage through this world, they may take the truth for their guide and not errors and prejudices that sadden and degrade men and nations."[109]

This was the insight he had drawn from Ocampo, and it was now firmly embraced by a changed man. Around this time he received—from Matías Romero, his representative in Washington—news that overwhelmed him. His favorite son, Pepe, was gravely ill. Juárez did not delude himself. He wrote to his son-in-law:

> My dear Santa:
>
> I write to you still shocked by the terrible grief that is breaking my heart, because in the letter . . . I received from Romero last night he told me that my beloved son Pepe was seriously ill and, since he added that the doctor feared he might die, I am coming to understand that he says it is serious only to avoid immediately giving me the grim news of the child's death. But in reality my Pepito was already gone, he is already gone, isn't it so? And you will understand how much I suffer for this irreparable loss of a son who was my delight, my pride and my hope. Poor Margarita! she will be inconsolable. Give her strength with your counsel, so that she may withstand this cruel blow that misfortune has unleashed upon us. And take care of our family. You are their only shelter and my only comfort, while I remain totally unable to be with you. Farewell, my son, accept the feelings of the heart from your inconsolable father and friend. Excuse the crossings-out because my head is so confused.
>
> Juárez[110]

Never in any way neglecting his public duties, Juárez carried his pain within him: "I do not know how I can bear so much sorrow weighing me down." The loss of his "beloved son Pepe," and not knowing how his family was, how his wife was getting along—they were "very cruel punishment for a man like me who so dearly loves his family."[111] Nine months later, in El Paso del Norte, he was hit with the news of another death—his son Antoñito, whom he had never even known. Faced with this "new calamity for our family," he wrote to his son-in-law again: "You must imagine how much I have suffered and am suffering without even the comfort of being with you all so that we could console each other."[112] To his wife he wrote: "Even if words of advice mean little when it comes to natural feelings . . . [one must try] to accept."[113]

Justo Sierra, describing Juárez, would comment on how he demonstrated the incorrectness of the syllogism "inexpressive physiognomy, therefore impassive soul, therefore insensitive heart." Juárez suffered enormously and could not even count—like Maximilian—on an unconscious repertoire of dramatic roles within which to situate his own drama. It would have done him no good to see himself in the image of Cuauhtémoc or his Zapotecan forefathers or through any other metaphor of stoic suffering. He did not want to suffer and he did not identify with the defeated. He was tired of suffering. He was tired of the centuries of suffering. For him and for "his brothers," in whom he saw the profound incarnation of Mexico, he wanted a definitive victory. It was this will to salvation that saved him.

And he was helped by his friends. On March 21, 1865, his "wounded heart could not cope with any more emotion," but such friends as Lerdo, Iglesias, and Prieto and the men and women who were his neighbors in Chihuahua "made it a point of honor" to celebrate his fifty-ninth birthday. "They gave me a sumptuous luncheon," he wrote to Santacilia on March 23, "and today there will be a magnificent ball for the same purpose . . . during the luncheon there were toasts to Independence, to its defenders, to oppressed peoples, to the city of Chihuahua, to our family and to myself." One man in particular had stood out: "Our friend Guillermo was admirable with his lyre [i.e., with his poetry] and has played a very active role in everything that was done to celebrate my birthday."[114]

Guillermo Prieto not only loved and admired Juárez. He worshipped him. In Guadalajara, in 1858 toward the beginning of the War of the Reform, a squad of soldiers—from the local garrison that was believed to be loyal to the Republic—had burst into a cabinet meeting presided over by Juárez. As the President rose to face them, an officer gave the order to fire. Prieto, spreading his arms in the form of a cross, darted in front of Juárez and shouted, "Stop! Lower those guns! Brave men do not murder!" Amazingly, the soldiers listened and were shamed (and impressed by Prieto's bravery). They turned away and retreated from the council room.[115]

Prieto had later accompanied Juárez in 1859, on his long journey from Manzanillo to Veracruz via Panama and New Orleans. He had been a member of the government that had issued the Laws of the Reform and had come with Juárez to Mexico City, then gone into internal exile with him to Chihuahua. Now he consoled him "with his lyre."

In September of the same year, their friendship began to deteriorate. Conspiring with his most intelligent collaborator and advisor in the cabinet—the ex-rector of the Jesuit college of San Ildefonso,

Sebastián Lerdo de Tejada (the younger brother of the economist Miguel)—Juárez made the most difficult and contentious decision of his wartime government. He decided unilaterally to prolong his presidential term until the "cessation" of the state of war would make it feasible to call for elections.[116] According to the requirements of the Constitution, Juárez was to hand over power—at the end of his four-year term on December 1, 1865—to General Jesús González Ortega, the chief justice of the Supreme Court. Expecting that he would take over, Ortega had already laid claim to the office, but Juárez and Lerdo had prepared an arsenal of legal arguments and political contacts to prevent the transfer of authority. In essence, Juárez used the extraordinary powers given him by the Congress at the end of 1861, just before the European intervention. They had been granted "with no restriction other than that of protecting the independence of the entire nation, the form of government laid down in the Constitution and the principles and laws of the Reform."[117]

From then on, Juárez had used these special powers with remarkable vigor, boldness, and success. He had governed by decree, as he had in Oaxaca in 1850. With a decree on January 25, 1862, he had imposed the death penalty for anyone who, in the opinion of the republican authorities, furthered the reactionary cause. "As the leader of an endangered society," wrote Emilio Rabasa, one of the most acute political analysts of the Porfirian era, "he assumed all the power, arrogated to himself all rights including the most absolute, and before ordering an extreme measure, he would take care to issue a decree granting him the proper authority, so that the unlimited power he wielded was always based on a law." It was, in Rabasa's words, a "democratic dictatorship."[118]

Juárez still felt that the nation was in great danger. He would not only prolong his presidential term until he won total victory, but he also found an immediate way to deflect González Ortega's claim to the presidency by bringing charges against him for desertion of duties.

In October, while Juárez was preparing his maneuver, Prieto sent him a carefully worded letter in which he opposed the reelection and asked to be removed from his post as Administrator of Postal Services. Juárez replied with his typical severity: "I cannot accede to your request that I issue an order declaring that the work of the General Administration of Postal Services has been terminated, because I am not so naive as to give aid to the invaders by discrediting the administration of my country. . . . Nor can I tell you to leave, because I have no reason to tell you that, nor does the government reject you, nor are you a hindrance to it."[119]

Prieto was arguing for González Ortega, who had been a major force in the War of the Reform, only insofar as that general personified the law. To a friend he confided the motives for his protest. They were the same that, in the long run, almost every member of that generation of Liberals would voice against the President. Juárez, the defender of the law, had turned against it. He was behaving like Santa Anna. He was using every legal device he could imagine to retain the Presidential Chair.

> Juárez has been an idol because of his virtues . . . because his strength was justice. . . . Let us suppose that Juárez was necessary, illustrious, heroic and immaculate in power. Was it because of the man or because of his qualities? What would he be without these? . . . I can even suppose that Juárez's legerdemain will be successful. But is it honorable to be a party to it? Should one consent to such an escalation of power? Should we through tolerating this action authorize others of the same kind which would follow and soon? I, for my part, will not do it.[120]

In December, Juárez and Prieto broke off relations. Writing from El Paso del Norte, Juárez drily described the situation to his son-in-law:

> Shortly before I retreated to Chihuahua, Guillermo Prieto came to see me. . . . He told me that he loved me very much, that he was my singer and my biographer and that if I so wished, he would go on writing whatever I wanted. What do you think of that? I thanked him, pitying such weakness and not paying any attention to his falsehoods. . . . Anyway, this poor devil [is] out of the battle now. [He has] counted for something because the government made [him] count. Now we shall see what [he can] do on his own resources.[121]

That "poor devil" had saved his life, but in the judgment of the man who reserved all judgments for himself, Prieto's attitude was a challenge to the cohesion of the nation, which Juárez felt himself to represent and which, in fact, he did represent. To dissent, at that moment, was to desert, to defect, to put weapons into the hands of the enemy.

During the War of the Reform some years earlier, Juárez had acted in the same way—though much more severely—with Santos Degollado. This "saint" of the Reform had shown weakness (in Juárez's eyes) by approaching the English ambassador and asking him to negotiate a peaceful conclusion to the conflict; and he had suggested that a committee of ambassadors be set up to nominate a provisional President. There was no secrecy about the move. Degollado, who respected Juárez, had told him of his intentions. But Juárez had reacted harshly, dismissing Degollado at once as military commander of the Liberal armies and placing him under house arrest to await trial. There he would remain until Juárez finally agreed to let Degollado take up arms again, to attack the Conservative guerrillas who had assassinated Melchor Ocampo. Degollado died in the first encounter with them, fighting ferociously, "his head full of holes, one lung emptied by the thrust of a lance, and covered with bayonet wounds."[122] He had been, Prieto said at his funeral, "the sanctity of the Revolution."[123] Other Liberals stood up to praise and mourn him. But Juárez, who had already made his final judgment on Degollado, said nothing at all.

Under Juárez, the various states and regions learned a lesson that would always apply in the future. No regional cacique, no caudillo or general could truly oppose the center. Juárez inaugurated an era and an irreversible historical tendency, a fundamental centralism employing federal forms. But he had also given a powerful impulse to the creation of a *we* beyond localities, regions, or states—a *we* that was a nation.

When the Republic was restored in 1867, Juárez called for elections.[124] His only rival was the triumphant military caudillo during the War of the Intervention, Porfirio Díaz. Juárez won with 72 percent of the vote. During the speech that he made calling for elections on his arrival in Mexico City, he had clearly spoken of the need to amend the Constitution. Juárez thought the Constitution—certainly worthy of respect and veneration—was impractical. For a decade he had governed the country under the banner of the Constitution and at the margins of it—with practically unlimited powers in a regime that had suspended civil liberties. But tenacious defense of the Constitution as a symbol, as a cause, did not necessarily imply that it was untouchable. The time had come for change.

The major problem was the old quarrel between Congress and the executive power. It was natural enough after the experience of Santa Anna that the Constitutional Convention of 1857 should have granted absolute power to the legislature at the expense of the executive and the judiciary. But with telling logic and their impressive store of experience, Juárez, Lerdo, and Iglesias wanted to avoid the extreme of congressional hegemony over the executive, which had buried Morelos as well as Iturbide. And they felt that this imbalance of power was responsible for the disorder that ravaged the country during the twenties, when Mora and Alamán complained of "the dictatorship of the many." In the eyes of the Juaristas, Congress as created under the Constitution was no more than a permanent convention. Its relationship with the executive had to be modified so that the executive would be granted a power of veto, requiring a two-thirds majority to overturn it; the powers of Congress to call extraordinary sessions would be restricted; and, as in the United States, an Upper House would be created.

Juárez did not succeed in persuading Congress to allow these reforms, but their opposition did not hinder him. From 1867 on, the various pressures within the country called for the application of new, extraordinary powers and the suspension of civil liberties, powers that the President requested and received. While formally respecting the Constitution, he continued to govern without it. The victory had been won against the counterrevolution, against the "crabs." The Second Independence was secure, but the country had not yet reached that final and most precious objective: peace.

And the infighting within the Liberal party was a major obstacle to peace. A new generational struggle was tearing the party apart—disputes between the generation of intellectuals and lawyers who had drafted the Laws of the Reform in the safety of Veracruz and the generation of young army officers who had spent ten years on the battlefields, under arms for the Constitution, the Reform, and the Second Independence. Who were the real victors? Who could rightfully claim power? Nothing wounded the Liberal military leaders more than Juárez's decision to discharge tens of thousands of soldiers. It was true that the State could not afford the burden of an eighty-thousand-man army, which threatened to devour the entire, meager budget. But still the restless army commanders apparently found their subordination to "the scribblers" (*tinterillos)* intolerable. Soon the first revolution would break out, the first *pronunciamiento*, the first mutiny (*bola)* of the Liberal era against the "dictatorship" of Juárez.

Another continual threat to peace were highwaymen, bandits, and kidnappers. The country's roads were infested with these criminals born from the chaos of civil war. To deal with this problem, Juárez formed a special repressive force, the *rurales*, often composed, like a vaccine, of former bandits. And to complete the picture of internal violence, there were the Indian caciques and their warrior armies. Still ruling the Nayarit region like a separate empire was the fearsome Manuel Lozada, the *Tigre de Álica*. Farther north, in the fertile Yaqui valley of Sonora, the Yaquis revolted again, as they had in 1825, to defend "the valley God gave them," threatened by the implementation of Liberal laws.[125] In the far south of the country, in Yucatán, the Mayas stirred the fires of their implacable war against the white man. And in Chiapas, a new and bloody

War of the Castes broke out between the local creoles (the *coletos*) and the Indians.

Chiapas had already, in 1712, gone through an indigenous uprising fed by a complex and confused mixture of religion and racialism. "In no district of the Republic," noted an observer at the end of the nineteenth century," is there so firm a line as in Chiapas between Mexicans and Indians."[126] Throughout the rest of the country, for instance, the introduction of mules had made the pre-Cortésian reliance on human porters (*tamemes)* a rarity, but in Chiapas Indians continued to be used as beasts of burden.

Seventy percent of the inhabitants of Chiapas were pure Indians, of Mayan stock: Tojobales, Chontales, Tzeltales, and Tzotziles. In the rest of Mexico, especially in the Central Plateau, racial terms (words like *criollo* and *mestizo*) fell gradually out of use because of the accelerated process of *mestizaje*; but in Chiapas, the creoles and the mestizos were still *coletos* and ladinos. And both groups—even the priests and teachers among them—exploited the Indians.

It was yet another Mexican paradox that the President who had to contend with a major rebellion among the Indians of Chiapas was the Indian President, Benito Juárez. As had happened in 1712 (or in Yucatán in 1848), a symbolic religious affirmation—pagan, heterodox, critical of the traditional Church—was the flashpoint for the eruption of ethnic, religious, and social grievances. In Chiapas, the vehicle was not a virgin and her prophetess as in 1712, nor a talking cross (as in the War of the Castes in Yucatán), but "stones fallen from the sky," probably meteorites, which a man named Pedro Díaz Cuzcat saw as a divine message to the Indians, enjoining them to a violent liberation. With the aid of an Indian priestess and a focus on the stones, he organized an alternative religion. In the midst of offerings, incense, and flowers, he told the Indians: "You should not worship images that depict people who are not of your race." And so that they might have a proper lord to worship, he ordered the crucifixion of a young Indian on Good Friday of 1868.[127]

The prophet and the priestess were arrested by the authorities and sent to prison. But then a mysterious personage from Mexico City appeared on the scene, the engineer Ignacio Fernández Galindo, who stripped off his ladino clothes "and dressed himself in a black Indian tunic, a loincloth and a palm-leaf sombrero." He became the strategist and commandant of the rebellion. He advised the Indians "to make themselves the masters of all existing things." By the middle of 1868, this man was the leader of six thousand Tzotzil Indians, well organized and equipped not only with arms but with drums, trumpets, and provisions. In some villages, the Indians killed all the whites. It took the federal forces three years to control the situation. Ultimately, they shot the leaders of the rebellion and reestablished an uneasy peace. The final body count came to two hundred whites and eight hundred Indians.[128]

Juárez expended most of his time, resources, and efforts trying to suppress all these outbreaks. The generals who had risen against him had been defeated but only for a time, until another opportunity would come and a caudillo might appear to lead them. The bandits and the Indians, with their tawdry or sublime objectives, remained in a perpetual state of war.

Juárez stood for election again in 1871. Although the country was not fully at peace, it was difficult, at that moment, to justify his continuation in office. He had been the civilian caudillo of the Wars of the Reform and Intervention, but then the independence and integrity of the country had been threatened. The new situation was chaotic and discordant but from any point of view less menacing. New figures, new generations were knocking at the door, and the sixty-five-year-old President did his best to close it in their faces. He had been in power for almost fifteen years, and he seemed destined to remain in charge until he himself decided to leave. Responding to this situation, Sebastián Lerdo de Tejada, his friend, key advisor, and the last of his most faithful followers, decided to abandon him and to run against him for the presidency. He could count on supporters among the old reformists. Juárez's other opponent would be, once again, Porfirio Díaz, the idol of the young. Both hoped to defeat Juárez and would have done it if they had run together; but Juárez won with 47 percent of the vote. It was the most suspect of his victories. He had maneuvered Congress into amending the electoral system to favor his own candidacy and had promoted a string of electoral practices that jeopardized the freedom and secrecy of the vote.

The generation of the *chinacos*—the military Liberals—took up the cause of the Constitution against Juárez. They called him "the candidate of himself," the man who perceived power "as a right of conquest," "His Majesty Benito the First." One critic said, "Today it is not the Constitution that the government defends, but the Presidential Chair." Another went even further, asserting that "Julius Caesar was a greater man than Juárez and everybody blessed Brutus for killing him."[129] Not only the young distanced themselves from Juárez but also the old Liberals of the Reform and men, like Ignacio Ramírez, who had participated in the Constitutional Convention. Juárez was neither repealing nor violating the Constitution—he was draining its power.

Opposition to Juárez's reelection in 1871 led to reactions throughout the country against the man to whom the Colombian government had recently given the title *Benemérito de las Americas* ("Worthy of the Gratitude of the Americas").[130] In Oaxaca, Porfirio Díaz initiated a revolt on the basis of his *Plan de La Noria*. He had a simple, forceful slogan: "Valid Voting [*sufragio efectivo*], No Reelection." But his attempt failed. The federal troops were on the verge of total victory; Díaz had tried unsuccessfully to ally himself with the *Tigre de Álica*; and he was on the run in the mountains of northwestern Mexico when suddenly, on July 18, 1872, news came that shook the whole country. Juárez had died of a heart attack in the National Palace. Sebastián Lerdo de Tejada, the new President, declared an immediate amnesty, which Díaz reluctantly accepted. He would launch his next rebellion a few years later. In 1876 he led the revolution of Tuxtepec, which first brought him military victory and then, through an election, the Presidential Chair he had longed for. He would hold it for twice the number of years Juárez had.

Throughout the War of the Reform and the War of the French Intervention, Juárez had been the shepherd of his country. And he had continued to lead during the new period of civil strife, when death came to him at an opportune time not only for himself and his memory, but also for Mexico. He had never yielded power—not in Oaxaca, not in Mexico—nor was he ready to do so. If he was going to retain his hold on the Presidential Chair, he would have had to use more and more drastic means of repression against his own comrades. The Liberals would have consumed themselves, and Juárez would have been responsible. It would have been a sad thing for the balance sheet of his achievements to have become so soiled.

He had brought the country to the other shore. It is enough to compare the war of 1847 with that of 1862—the basic differences were due in large measure to Juárez. During the age of the caudillos and the *criollos*, Mexico was not a nation: It was an aggregate of regions and localities without national consciousness. In 1847 the "Mexicans," when they heard "the cry of war," had watched the American troops march by as if it were a parade, a theatrical performance of no concern to them. But the loss of territory, the real danger of national disintegration, the violence of the War of the Reform and the theological hatreds it aroused—all had helped to create a consciousness of national identity. Two additional and closely connected factors also played a decisive role: the legitimate, severe, and intelligent authority of Juárez throughout the entire process and the concomitant rise to power of the mestizos.

The enormous importance of Juárez in political terms was that he began to reestablish the order that had prevailed in colonial times but with entirely different premises: a state based on law, secular legislation and education, a market economy, and a general program of individual rights and genuine civil liberties. He had poured new wine into old bottles, succeeded in creating a legitimate basis of authority (within a country on the verge of disappearance), and guided it through two terrible storms, from which it had emerged transformed. Applying the instinctive knowledge of his forefathers, with religious fervor, he had transferred his ancient loyalties to the new political features of the nineteenth century: the law, the Constitution, the Reform. He lived through the years of his mandate like a shepherd called by God to lead a disorderly flock into emancipation, which he, in his personal life, had already attained—from the "torpid poverty" and ignorance of Guelatao, his birthplace, to the height of the Presidential Chair. He wanted to transmit this experience to his "brothers." And to a large degree he succeeded.

When Juárez died, Mexico was a different country. It was no longer a stage for the acting out of operas and plays. It had become a locus of history and reality. There was a new center in the old center. A new institutional order, *a new State*, revolved—like the earlier Aztec or colonial regime—around a new emperor, a new *tlatoani*, a sacred President, although the same tension continued to exist between the old and the new, the same crossroads where Morelos had been slaughtered. Juárez had created a fresh political legitimacy, which Porfirio Díaz would later consolidate. It was a powerful combination of tradition, legality, and charisma. Against the theocratic background of the past, a caudillo dressed in black represented the republican ideal of the future.

But Juárez achieved something more. He opened the door wide for mestizos to reach commanding positions in the life of the nation. During the second half of the century, the mestizos had certainly become the majority of the population, but the figure of the Indian Juárez would legitimize their position and their achievements. And though he had not brought peace to the country and he had not installed a true democracy and he did not respect federalism, he had still found a way not to break the law. Congress had acted, deliberated, stood in opposition. The judges of the Supreme Court, who were elected by popular vote, had done their duty with complete independence. The hundreds of newspapers of the period had been allowed complete freedom of speech. For the first time in its history, Mexico enjoyed full religious and civil liberties. The Constitution of 1857—for a brief shining moment—was put into practice. Never before had Mexico been closer to a democracy than it was in this period of the Restored Republic.

The target of this newly won freedom, often and with reason, had been Juárez's authority, but—even with all his severity and his resolution as a leader in war—he did not kill or repress his peacetime political opponents. He refused to give up power because he trusted no one else to wield it. He had lived too many years amid the chaos of Santa Anna, and he felt the Liberal victory was so fragile that it would be best for him to remain its constant shepherd, until a higher necessity arranged otherwise.

In Justo Sierra's searching biography, Juárez appears as a religious devotee of the entities sacred to him—the Constitution and the Reform, embodiments of the law. Just as Santos Degollado had transferred his religious faith of a primitive Christian to the cause of liberty, so Juárez had transmitted, in Sierra's view, his unyielding religious commitment as a Zapotec Christian to the presidential office.

But what was the shepherd's final goal? Who was Juárez really trying to save? Sierra might well have been right when he conjectured that "through the Constitution and the Reform, Juárez envisioned the redemption of the indigenous republic."[131] To save the Indians, "our brothers," from the clergy, from ignorance, from servitude, from "torpid poverty" was perhaps "his innermost and religious longing."

Juárez was not driven merely by a desire for power or his fervent attachment to the immutability of the law. He acted on the basis of an immediate, almost mystical connection with power. He believed that he represented a historical right over this land that none of his other political contemporaries possessed or even suspected. He did not invent a past, like the creoles. He did not hunt for phantom ancestors, like the mestizos. He was a son of this land and of this history, before there had been any Mexico or New Spain. Before 1821, before 1521.

At the farthest possible extreme from Santa Anna, he had infused the office of the President with a sacredness it had lost, the sanctity of an indigenous monarchy with legal, constitutional, and republican forms. And for this reason he was never to renounce power. He would die in power. Juárez and power were inseparable.

9

THE TRIUMPH OF THE MESTIZO

Three years before the Centenary Fiestas, Porfirio Díaz had been interviewed for *Pearson's Magazine* by American journalist James Creelman. They were sitting on the heights of Chapultepec Castle, looking out at the crystalline Valley of Mexico. Creelman would write, "There is no figure in the whole world who is more romantic or heroic than that soldier-statesman whose adventurous youth outshines the pages of Dumas and whose iron hand has transformed the warlike, ignorant, superstitious and impoverished Mexican masses, after centuries of cruel oppression by the greedy Spaniards, into a strong, progressive, pacifist and prosperous nation that honors its debts."[1]

From that scenic vantage point, the two of them surveyed the parallel biographies of the caudillo and his country. In the distance, eastward, the luminous Valley of Mexico was protected by the two volcanoes, the primeval couple: Popocatépetl ("Hill that Sends up Smoke"), who was sitting in mourning, according to the ancient Indian legend, because of the death of his Iztaccíhuatl ("Sleeping Woman"). Díaz had chosen that magical setting for the conversation. And the President, as he spoke, seemed to Creelman not only "the master of Mexico" but "the master of the history of Mexico."

Creelman wrote that Porfirio Díaz "moved along the terrace toward the inner garden, where a fountain set among palms and flowers sparkled with water from the spring at which Montezuma used to drink." And one could see "on the hillside of Chapultepec . . . white jessamine wreathing itself over rocks sculptured by the Aztecs."[2] It hardly mattered that Montezuma's fountain was a legend and the Aztec sculptures were material facts. Truth and myth melted into each other on this stage that had witnessed so many critical events in the life of Mexico. "To the rear was the pink-walled crumbling mill where Winfield Scott stood with his artillery in 1847."[3] And now Porfirio Díaz was leaving his mark on this theater of the nation. In the morning, he would swim and ride, surrounded by the same age-old *ahuehuete* trees that had watched the passage of *tlatoanis*, conquistadors, viceroys, emperors, and Presidents.

Besides the historical and political content of the interview, Creelman was subtly seduced by the visuals. Díaz spoke "from the heights of Chapultepec Castle" and "looked down upon the venerable capital of his country, spread out on a vast plain, with a ring of mountains flung up grandly about it." The President was "at the height of his career." The self-proclaimed patriarch of all Mexicans repeatedly "surveyed the majestic, sunlit scene below the ancient castle."[4] All of this seemed to be taken from a picture by José María Velasco, the supreme Mexican painter of the late nineteenth century. Always from a great height, with his eagle vision and scientific preciseness, Velasco had painted vast and motionless landscapes, barely touched by any human presence. But there had been a change in his approach since Díaz came to power. There were now signs of movement across the silent spaces of Velasco's mountains and canyons, not of people but of the physical symbols of progress—the railroad, ploughed fields, the distant smoke of a factory. Like the Velasco of the Porfirian era, Díaz too—from the heights of his political position, his age, his experience—obviously (Creelman was convinced) could envision at a distance the future of Mexico.

The old man told the American journalist the story of his life, presenting it as an epic (which in many ways it was), but there was another and more profound history captured in that life—the entire experience of *mestizaje*. Throughout his biography and the history of his time, the currents of the Indian and the Spaniard ran like two underground rivers. In Díaz himself they fused. Because of who he was, because of the long historical moment he dominated, he had become the quintessential mestizo within Mexico's biography of power.

Oaxaca had remained an indigenous sanctuary because it was a poor, rugged mountain region, with neither enough mines or even arable land to attract the conquerors away from other regions. For the Spanish Crown, it was an easy administrative decision to leave the local Indian nobility in place, with some small changes in status. They survived right up to the dawn of independence.

In the daily life of those "nations," respect was still paid to the union of religion and power. It had been that way since the era when the Mixtecs—warriors, workers of gold, builders, by legend descendants from an archer of the sun—had conquered the Zapotecs. It had been that way since the time when both nations had endured the yoke of the Aztec Empire. This Indian perspective, the fusion of religion and power, was the ideological context for the development of this area, across the nineteenth century, into what a later writer would call "the national factory of politicians and soldiers," the land of "the mystics of politics."[5] Juárez had come from its mountains, but it was in the capital, the city of Oaxaca, that Petrona Mori, a pure-blooded Mixtec Indian, gave birth in 1830 to Porfirio Díaz.

His father, José Faustino Díaz, died of cholera in 1833, leaving a widow with five children. Porfirio rapidly had to take over the responsibilities of his dead father. While his mother waited on customers at the Inn of La Soledad (the only one in the city), raised cochineal insects for their red dye, and supervised a primary school, her eldest son, Porfirio—no less enterprising and an heir to the practical skills of his father (who had been a blacksmith, mule-driver, and tanner)—fashioned chairs, school desks, and shoes, carved out rifle butts, and repaired pistols.

Although he would study at the Council Seminary and later at the State Institute of Arts and Sciences, his dream, very early, was to be a soldier. When he was twenty five, the currents of Mexican politics gave him his chance: He became the political chieftain of the mountains of Ixtlán, where he tried to organize a national guard.

During the War of the Reform, Captain Porfirio Díaz underwent his baptism of fire. He fought the first of the thirty-seven battles he would wage through the following decade; he suffered the first of his several wounds; he took part in the first siege of Oaxaca. President Benito Juárez assigned him the civil government and military command of the Isthmus of Tehuantepec where, as Díaz later recalled in his *Memoirs*, the situation was "extraordinarily difficult": "In my youth I had a harsh experience. When I commanded companies of soldiers, there was a time that for six months I received neither instructions nor assistance from my government, as a result of which I was compelled to think for myself and myself become the government."[6]

His strongest support came from a group of "fifty men of the nation of the Mixes from Santiago Cueva."[7] According to several chroniclers of the period, the Mixes were "notoriously idolatrous and superstitious, prone to contaminating the altars of their Catholic church with the blood of birds that they sacrificed to other gods,"[8] but they were also one of the most warlike nations among the many cultures of Oaxaca. Since pre-Cortésian times, they had withdrawn to their villages on the high mountains, like eagles to their nests, and their pride had not been broken by any other people, not by the Mixtecs, Zapotecs, Zoques, or even the Aztecs. When he needed more forces, Díaz could also count on a couple of hundred Zapotec Indians from a place that had long lived on a permanent war footing: the village of Juchitán. His *Memoirs* would recall the details of that "harsh experience" in which, through flattery, shrewdness, deception, and the strength of his character, he learned to command men who had war in their blood. "Don Porfirio Díaz," wrote the French traveler Charles Etienne Brasseur, who knew him then, "was the absolute master of the region and directed the valiant *Juchitecos*." When he first saw Díaz, Brasseur noted, "His countenance and his carriage vividly impressed me . . . he presented the most handsome indigenous type that I had ever encountered up to then in all my travels; I thought that he was the image of young Cociopij or of Guatemozín as I had often imagined him. . . . Well built, of outstanding distinction, his face showing great nobility . . . it would be desirable that the provinces of Mexico be administered by men of his character."[9]

Porfirio's political and military stint in Tehuantepec brought him not only an arduous apprenticeship but something more: living contact with his indigenous roots. The creole and mestizo population of the proud city of Tehuantepec inclined toward the "patrician" Conservative camp, but the Indians, above all those of the San Blas quarter, followed Porfirio as if they truly saw in him an incarnation of Cocicopij, Moctezuma's nephew who had been the last Aztec governor of Tehuantepec and whose memory was still vaguely retained, though enveloped by legend. In Tehuantepec the pre-Columbian past seemed more alive than the present. While the Compañía Luisianesa was building a railroad across the isthmus, the Zapotecs of Tehuantepec—believers in gods, saints, and protective animals (the *naguales*)—were slipping away to the grottoes of Condoy or the tunnels of Rayudija, or traveling to the ruins of Mount Guiengula, there to worship idols and tombs. For many visitors, these Indians seemed to have an aristocratic quality about them. The historian Juan B. Carriedo, in 1846, considered them "incomparably superior to the Indians of the rest of the Republic"—intelligent, affable, hardworking, cultivators of indigo, weavers of "extremely beautiful" silks, skilled and elegant horsemen. "They are the only Indians," in Carriedo's opinion, "whose women are beautiful."[10]

One of these women vividly impressed Brasseur as she had, for years, captivated those who knew her and those who did not. He met her in the house of Juan A. Avendaño, where she had come to play billiards with Governor Díaz and other important men of the town. In the mornings, she made cigarettes in the market or sold products in the neighborhood. "She mingled with the men, perfectly at ease, and she would take them on boldly at billiards with incomparable skill and assurance."[11] It was said that she was descended from ancient Zapotec aristocracy. Her name was Juana Catalina Romero. They called her Juana Cata. She was twenty-two years old. Brasseur went on and on with his description of her:

> a Zapotec Indian, with bronze skin, young, slender, elegant and so beautiful that she charmed the hearts of the white men, as had the mistress of Cortés, in another age. . . . I remember . . . that the first time I saw her, I was so impressed with her proud and haughty manner, with her superb Indian dress, so like that in which the painters represent Isis that I thought I was seeing that Egyptian goddess or else Cleopatra in person. . . . [12]

There were some who thought she was crazy, but the Indian people feared and respected her: "considering her a witch, deeply learned in herbs, charms, and potions, a sorceress who with just three words could make a rosebud blossom to kindle the passion of a lover, who could communicate with the *naguales* or with the spirits of Mount Rayudija."[13] It was months before Count Brasseur's fascination with her would fade.

But if it took Brasseur months to forget her (while still devoting long chapters to her in his *Voyage sur L'Isthme de Tehuantepec)*, there was another man who never would forget her: Porfirio Díaz himself. Juana Cata was certainly his informant and in all likelihood his mistress. According to a legend, erroneous perhaps but meaningful, Díaz succeeded in changing the route of the Trans-isthmian Railroad so that it would pass about six feet from the French-style chalet he had built for Juana. "When my grandfather visited her,"—Lila, Porfirio's granddaughter, would tell the story—"the engineer reduced his speed and whistled a signal; Juana Cata would half open her door and without the train even stopping, Porfirio would jump out as soon as the first step . . . came even with the running-board of the train."[14]

By the end of the century, that Indian woman—who had not learned to read and write till she was thirty, when she read her first book: a manual of etiquette—had become Doña Juana Catalina Romero, an unmarried matriarch and the undisputed authority in the isthmus, a calm but resolute *cacica* who settled disputes and removed political chieftains; owner of factories and sugar mills, exporter of sugar products that won international prizes; founder of a Marist college for young men and a college for women run by Teresian nuns; a traveler through Europe and the Holy Land, and the woman who reconstructed the old cathedral, cemetery, and *convento* erected by Cocicopij himself. After that reconstruction, she was venerated as a saint by the people of Tehuantepec. Every year, Doña Juana would host a lavish ball, which was frequently attended by President Porfirio Díaz.

Throughout his long stay in power, Don Porfirio never forgot that authentic "Doña Porfiria." Even in her old age, accompanied by her family, she used to visit the Castle of Chapultepec and sit at the same table with Díaz's wife, Carmelita. She and Díaz were united not only by the memory of a romantic love but by a kind of shared manumission. Both Porfirio and Juana Cata had fled their Indian origins, but in various ways they would continue to be influenced by the remote culture of their ancestors.

Those who dealt with the President perceived a force and reserve in his character, the habit of shielded "indirect speech" in his conversation, and many expressions and intonations in his language that seemed to stem from his Indian inheritance. His portraits, the chest covered with medals, show him with an almost white skin, like some hieratic American Bismarck. But when you saw him close up, other features stood out clearly. The old proverb came to mind: "When the Spaniard wanes, the Indian waxes."

Not only his personal and military biography reflected his Indian heritage. So did his personal style of governing and the attitude of those who obeyed him. Porfirio Díaz thought of himself as the total father of a family of minors and believed that this situation (which he intensely exploited) had its colonial origin:

> The Indians retain the traditions of an ancient civilization of their own. They can be found among lawyers, engineers, physicians, army officers and other professional men. [But they] care little for politics. They are accustomed to look to those in authority for leadership instead of thinking for themselves. That is a tendency they inherited from the Spaniards, who taught them to refrain from meddling in public affairs and to rely on the government for guidance.[15]

But in reality, it was an attitude that dated back much further. For the Aztecs—who were highly accomplished poets—their emperor (the *tlatoani*) was not strictly a god but was very like a god: "the light that like a ray of the sun gives us light, the mother who takes us on her lap, the father who carries us on his shoulders . . . the tree that throws a strong shadow and a strong circle . . . the heart of the city."[16] In lines like these, the poets presented the *tlatoani* as a total protective force.

Among the injunctions addressed to the new *tlatoani* during his elevation to the throne, Fray Diego Durán recorded: "You have to understand everything. You have to grieve for everything and be compassionate."[17] And as part of the same formal advice, Fray Bernardino de Sahagún listed: "Do your work calmly . . . it is right for you to assume the heart of an old man, of a serious and severe man . . . let your words be few and very serious."[18]

The *tlatoani*, the serious man, was expected to take on responsibility for the course of the sun and the cosmos in general—storms and earthquakes and eclipses. As a benevolent force, he represented the god Quetzalcóatl; he was father and mother and priest, the center of Aztec society. But throwing a shadow over all these responsibilities was the other face of the *tlatoani*, his role as mirror of that god whose name meant "Smoke in the Mirror" (Tezcatlipoca). This was a god of the instant, of "the here and now," and he had other names as well, all of them frightening: "Wind in the Night," "Enemy to Both Sides," "He Who Knows the People," "the Inconstant Creator." He was a warrior who could give a man honors and wealth and then suddenly strip them away. Sahagún quotes from a poem:

He thinks as he pleases,
does as he pleases . . .
He throws us here, and there.
We make him laugh; he laughs at us.[19]

The natural outcome of this cosmic individualization of capricious and punishing power in the god Tezcatlipoca was a fear of cosmic dimensions directed toward his earthly representative. Durán asked an Indian what Moctezuma looked like, and the answer was: "Father, I must not lie to you, I must not say what I do not know: I have never seen his face." Whoever raised his eyes to Moctezuma "as if he were looking at an ordinary man," Durán explained, "they then killed him."[20]

Clearly, nobody expected Porfirio Díaz to nourish the cycles of the sun and the moon, to avert eclipses or earthquakes, to fertilize the land and make the corn grow. His paternal and conciliatory style of exercising power, his "patriarchal policy" as he called it, derived more from his Spanish and Christian roots than from Aztec (or in his case Mixtec) ancestors.[21] But these descriptions of the *tlatoani* do reflect (to a degree) the style in which Díaz wielded his personal power throughout his long dictatorship and the manner in which his subjects, of all classes and ethnic origins, regarded and respected him. As with his Aztec predecessors so with him: Each new term of office "was like consecrating him as a god."[22]

And a god with both benevolent and fearful aspects. The sociologist Andrés Molina Enríquez would write: "In his hands, death has taken every form, prison has included every cruelty, physical punishment every horror, and practical punishment every sort, be it persecution, destitution, exile, servitude, indifference, contempt or oblivion. . . ."

Díaz believed that the only possible relationship between a government and its people was that of authority. The means toward maintaining it was, as Molina said, "to generate in all men that intimate atavistic fear which the term Majesty arouses."[23]

There was, of course, nothing particularly Mexican about the way in which the people obeyed their autocrat. But the sacred aura that accompanied their obedience was special, and it clearly had its roots in the Indian past.

There are at least eighteen possible translations into Náhuatl of the Spanish word *mandar* ("to command"), each with a different purpose or nuance. Included among them are: to command someone to do something, to command in return, to command by word of mouth, to command another, to command with harshness, to command others to enjoy themselves, to command whatever one wishes, to command like a prince. Díaz had only the one word, but when it came to applying power, many of these modes were thoroughly familiar to him, and for decades (as *pan o palo*, "bread or the bludgeon," benevolence or repression) he skillfully put them into effect.[24]

But Porfirio Díaz was a product not only of his Indian but also of his Catholic and Spanish past. In Oaxaca, facing the house where he was born, stood the parish church of La Soledad, which held one of the most venerated images in that pious city, a statue of the Virgin dressed in black silk trimmed with pearls and precious stones and credited with miracles. Not very far away stood the mysterious building where the rationalist and liberal Freemasons held their meetings. The local priest, José Mariano Galíndez, had preached a famous sermon against them, in the parish church on December 18, 1844, the day of Our Lady of Solitude: "We have allowed the most impious and licentious books to reach even here; they began to be read with a certain reserve but then spread at such a pace that they can now be openly sold in our bookstores. And so we come to lose our balance like children, letting ourselves be attacked by new doctrines. . . ."[25]

The home of these new, impious, destructive ideas, the sanctuary of modern philosophy, was the Institute of Arts and Sciences. The priests called it "the house of heretics" and its students "libertines." One of the teachers at this "house of heretics" was Benito Juárez.[26] Years later, Porfirio Díaz would become one of its "libertine" students.

The young Díaz, at the time he made his decision, was attending the Council Seminary as a candidate for the priesthood. It was difficult for him to move on to the Institute. Sole support of his family, a father to his several siblings, he had to convince his uncle, who was no less than the bishop of Oaxaca, José Agustín Domínguez. And he could not do it. This step, like so many others in Díaz's life, was taken forcefully. His uncle disowned him for life, but Díaz would make his move to the Institute of Arts and Sciences, where at one point he served as the librarian.

Although he would give private classes in Latin and register to study law, his vocation was not for letters but for battle. Benito Juárez, by then (in 1852) director of the Institute, gave an impetus to Díaz's military career by authorizing courses in military theory and tactics. Porfirio put his studies to good use when, five years later, he lived in the "warlike and hostile country" of Tehuantepec.[27] It was then he formed his personal army of fierce Juchitecos. Juárez himself, when governor of Oaxaca, had failed in his attempt to overcome and master the men of Juchitán. Díaz did something better: He won them over and he put them to use. During the War of the Reform, he would take part in twelve battles, was seriously wounded, created a secret police, suffered an attack of peritonitis, set up a munitions factory, and became expert in skirmishes, sudden attacks, and ambushes. But above all he grew skilled at managing men, at divining passions and ambitions so that he might make use of them. In 1860, President Juárez would refer to him, with respect, as "the man of Oaxaca."

During the War of the French Intervention, as commander of the Army of the East, Porfirio from the very first was at the forefront of combat and attention. His successes and reversals during the four years of the war read like legend. More than "the man of Oaxaca," he was becoming "the man of Mexico." In his *Memoirs* (published in 1892) he described his adventures: hairbreadth escapes, marches over mountains like Hannibal crossing the Alps, having to hide like an animal or an eagle, creating and leading new armies. Even when he was in prison, he never stopped fighting and resisting. The fruits of all this would come to him much later, in 1866, when his military success and reputation began to outstrip all the other generals.

"The good boy Porfirio," commented Juárez, who trusted him

Porfirio Díaz, 1864
(FINAH)

blindly, "doesn't date his letters until he captures a capital."[28] On April 2, 1867, in Puebla, he won his most important victory. It was the death blow for the empire. Before that, he had refused all sorts of offers meant to tempt him into an accommodation with the Emperor Maximilian. And on July 15, 1867, Porfirio delivered Mexico City, at peace and in admirable order, to Benito Juárez.

Twice, in 1867 and 1871, Díaz contended democratically for power. Tired of waiting and weary of the political maneuvers of the Juarista faction, which threatened to keep control forever, Porfirio and his broad coalition of lawyers and soldiers turned to the traditional path of Mexican politics: revolution. The first revolt, in 1871 against Juárez, failed. Díaz retired to a pleasant port city, Tlacotalpan, in the state of Veracuz, where he returned to the relaxation of woodworking—building chairs, desks and bookshelves—and fathering his sons, Porfirio and Luz. (In 1867 he had married Delfina Ortega Díaz, his niece, the daughter of his sister Manuela.) There he plotted a second coup, which would not fail, against Sebastián Lerdo de Tejada, Juárez's successor. Those who saw his triumphal entrance into Mexico city in 1876 understood at once that a new era was beginning. The novelist Federico Gamboa wrote:

> The chief, solemn, his head bare (his large embroidered sombrero was lying on the backseat of the vehicle) inclined half forward, responding . . . to the ovations. His chest jutting out, broad shoulders, strong neck; his head erect, virilely set upon his shoulders; his skin darkened by the sun; with black falling mustaches; his dominant, profound and determined gaze fixed on the horizon, the streets, the buildings; the serious set of his face lightly wrinkling his forehead; thick eyebrows; abundant loose hair; a restless and impressive nose exploring the smells of the surroundings, his forehead broad, his lower jaw prominent; his ears large and red. A perfect example of masculinity.[29]

Many years later, the sociologist Molina Enríquez (himself a mestizo) would describe the same scene in terms of his ethnocultural theory of the history of Mexico: "The new chief of the Nation, from then on, had to be a member of the mestizo element." But unlike the other mestizos of Mexican history (like Morelos, "the greatest of them all,"[30] or Porfirio's own Liberal and republican comrades), Díaz was already acting not as the chief of the Liberal party but as a patriot who *integrated*—into a single personality—the triple qualities of victorious soldier, upright administrator, and skillful politician.

Díaz came to power after having overcome—within his own biography—wars, rebellions, and generations. His trajectory had not been as long as that of Juárez (a pure-blooded Zapotec who in his childhood spoke no Spanish), but the deep veins of history along which he traveled were similar, from the indigenous sanctuary of Oaxaca into the full glare of the nineteenth century.

In his book *Los grandes problemas nacionales* (1909), which would have an enormous influence on the ideological program of the Mexican Revolution, Molina presented the history of Mexico as a process subject to rigid laws of evolution: The country had passed from Spanish to creole hands, from the creoles to the mestizos, but it had not been shaped into a homogeneous fatherland (*patria*). Still, there had been immense progress on the complex job of "integration," a trend toward forming the organic entity that Molina as a sociologist envisioned.

He believed that one man, Porfirio Díaz, had now crafted this work of integration, "without precedent in the history of humanity."[31] He pointed to *friendship*—understood as benefits to particular groups in exchange for loyalty—as Porfirio's primary tool. Molina coined the term *amificación* ("friendification") for the process in which he saw no danger but rather a blessing: "How complex has been the work of the estimable General Díaz, and how complex, of necessity, has been his responsibility! He is a unique man, who has had to govern and has wisely governed many distinct peoples within a single nation, groups that have been living in different periods of evolution from the prehistoric to the modern."[32]

Porfirio Díaz would boast that he had "preserved the republican and democratic *form* of government, defending the theory and keeping it intact."[33] Yet this preservation barely concealed the colonial past. While Díaz's harsh and majestic style recalled the figure of the Aztec *tlatoani*, the structure and operation of the Porfirian state (and its relation to its subjects) reflected features of the political model that Spain had transplanted to America. The proudly Liberal state of Don Porfirio, a tolerant layman in religious matters and openly progressive in economic affairs, was completely antiliberal politically.

Rather than criticize the "patriarchal policies" of Díaz, Andrés Molina Enríquez saw them as the natural political foundation of the country. The "coercive organization" and "obligatory cooperation" that Díaz introduced had been a return to Aztec and Spanish roots. In his concentration of power lay "the secret" of his political success, and it was "no more than the viceregal approach adapted to present circumstances."[34]

In a pamphlet published in 1897 (*Notes on the Policies of General Díaz*), Molina Enríquez compared Porfirio to the Roman emperor Augustus, who, in identical circumstances—when he recognized that governing with liberal and republican laws would lead to anarchy but also understood the sacred value of those laws—had found a similar way to solve his problem. A modern Augustus, Don Porfirio had respected all the constitutional forms, repealed no electoral laws, continued to hold regular elections, but had "demolished the apparatus of government and concentrated all the subdivided power into his own hands."[35]

As in the time of the Hapsburg viceroys, Díaz himself was both the central force and the main instrument of integration. And he had offered his "friendship" not only to each traditional ethnic and social group but also to Mexicans who had developed more contemporary associations. At the height of the famous strike at Río Blanco (1907), the workers did not try to solve their problems by turning to the courts, the press, or the Congress. In the hope of an impartial decision, they of course had to appeal to the man whose characteristics supposedly were:

> *solemn majesty and a lofty way of thinking*
> *gentle behavior, discretion and recollection,*
> *learning, prudence, common sense and comprehension,*[36]

as the early seventeenth-century poet Balbuena had described the ideal qualities of a viceroy, now presumed to be those of "El Señor Presidente." (The workers would eventually be answered with gunfire.)

A typical day in the life of Porfirio Díaz was very like that of a Hapsburg viceroy. Díaz handled the most varied problems of government: administration, the military, diplomacy, the Treasury; as well as legal, judicial, and religious matters at every level—federal, state, and municipal. But he had no time limits to his reign, no courts or church with any power to hinder him. The viceroys had always been besieged with petitions; and almost half the correspondence Díaz received during his many presidential terms consisted of petitions. But they were requests made not to a viceroy but to an absolute ruler.

Nevertheless, Díaz was usually noncommittal in his responses to petitions from his subjects. Not only Indians but creoles and mestizos of all social classes and economic levels asked him for everything under the sun: scholarships, loans, job promotions, pardons, protection from abuses, public employment, recommendations, pensions. The requests often came accompanied by a photograph and an enthusiastic declaration of loyalty. Though in practice he might do nothing, his answers were always courteous. "I shall try," he would write to them, "at the earliest opportunity." And he did invariably answer.

At the end of the nineteenth century, Federico Gamboa offered a fine, short synthesis of the resurrected monarchy:

> Every solution in public and private life has been left [in the hands of Don Porfirio]: from the learning of Greek and Latin to the uniform of the municipal coachmen; from the granting of divorces in cases of broken marriages to the matching of rich widows with foreigners sufficiently noble or royal; from border disputes between states to relations between neighbors and relatives; from the total figures of the harvest to the menus in each house; everything from the transcendental to the utterly trivial.[37]

As he unveiled—on one of the walls of the National Palace—a portrait of Charles III that was a gift from Spain for the Centenary Fiestas, Porfirio Díaz could feel satisfied that he had also continued what the Bourbons had begun. He had (for the good of the country) denied Mexicans the right to "think or give their opinions about the high matters of government" (temporarily, as it would only be during his lifetime).[38] He had opened up the economy both domestically (in 1894 he had abolished the *alcabala*, the internal sales tax in force since colonial times) and toward the outside world (building railroads to the border and encouraging foreign investment). The country was apparently on the road to impressive material progress, thanks to a new kind of "enlightened despotism," similar to the policies of the Bourbon emperors of the eighteenth century. The new public buildings of the Porfirian belle époque rivaled those of the viceroy Revillagigedo: the Palace of Fine Arts, the Post Office Building, the Legislative Palace, and (especially for the envoys) the small, luxurious mansions where they stayed during the Centenary Fiestas. In 1910 Mexico was doing its best to bridge the gap separating it from the advanced Western world. In some areas—certain fields of medicine, civil engineering, and the visual arts, for instance—Mexico was in no way inferior to the European nations. The elite of Mexico City had a cosmopolitan knowledge of philosophy, theater, and literature. In 1910 people read Nietzsche, Bergson, and William James and attended the plays of Shaw or Ibsen.

But fashionable intellectual theories (liberalism, positivism, evolutionism) were a verbal smokescreen hiding the true social realities of Mexico, at least from many of Díaz's supporters. Porfirio had re-created the Mexico of the Bourbons, but that viceregal Mexico—with its political absolutism and severe social inequalities—had supposedly ended with the revolution of 1810. Yet a hundred years later, despite the "peace, order and progress" of Porfirismo, the country was still 70 percent rural. One out of every two children died in their first year from whooping cough, malaria, yellow fever, and other contagious diseases. In 1900 there was only one doctor for every five thousand citizens. And in spite of the ten thousand primary schools, 84 percent of the population was illiterate. Mexico, as Humboldt had written in 1803, was still "the country of inequality."

A new type of small or medium-size rancher was flourishing in the north and west of the country. (There were more than fifty thousand of them, three times the number in 1877.) But Molina Enríquez had pointed to the continuing presence—in the heartland of the country, in "Old Mexico"—of the villain of history, the most characteristic residue of the viceroyalty: the hacienda. In 1876 there had been fifty-seven hundred haciendas; in 1910 there were more than eight thousand.

Not all of these haciendas deserved the harsh judgment of Molina Enríquez. Some were modern, commercially oriented enterprises, perhaps even geared to export trade, where peons were treated paternalistically but received wages and attention to their well-being. But most haciendas were different. As enormous as they were unproductive, they still retained a deeply feudal quality.[39] They were entities that consumed their own products, closed up within themselves in time and space, more aristocratic luxuries than business ventures. In the southeast, they were something worse. On the sisal-hemp plantations of Yucatán or the coffee plantations of Chiapas, the daily life of the peons bordered on slavery. In these regions, labor was scarce (in relation to the considerable need for it) and therefore valuable. The governmental authorities, local caciques, police, and landowners applied their own brand of justice to control the labor force (including private jails and widespread use of torture). The peons were paid not in cash but in products that the hacienda itself stocked in its "company stores." On these haciendas, exploitation had deep, centuries-old roots and entailed the lifelong and hereditary serfdom of laborers forever in debt to their "master."

In the cities, the gap between rich and poor was immense. Industrialists, bankers, important officials, Díaz's corps of intellectuals and technocrats (the *científicos*), and the richest landowners all lived amid marble, ivory, and tapestries in the old *palacios* at the center of the capital or in the French-style mansions built in the neighboring districts of Roma, Santa María, or Juárez. An increase in the number of government posts—and a degree of material progress—contributed to the growth of the urban middle class to about five hundred thousand throughout the country. This group constantly worried about the rising cost of living and their limited political influence. At the base of the urban social pyramid was the working class—nearly a million people. They inhabited densely crowded slums crisscrossed by dust or mud according to the season, among piles of garbage and the innumerable *pulquerías* (selling pulque, the cheapest of liquor), the customers reeling day and night among the dogs and the half-naked children.[40] So huge a social gap was a clear "seed of the violent storms that were to come" (as Justo Sierra had written—but as a description of the past: the social conditions of New Spain).[41]

Just as Porfirio, in his own person, had effectively integrated the Indian and the Spaniard, so *mestizaje* had in large part done the same for the people of Mexico. But many Indian communities continued (and still continue) to exist. About the Indians, Díaz used to say that they were "docile and grateful, all except the Yaquis and the Mayas."[42] Of course Porfirio knew that the Indians of Mexico (and the much larger body of mestizo peasants—the campesinos) were not really broken beyond any possibility of violent action. He had witnessed and fought in too many wars and revolutions, and he knew that when they begin, as he said, "they grow too fast to be stopped." And Díaz would sometimes speak of the revolutionary potential of the poor as the "sleeping tiger" of Mexican history—an animal that should not be awakened.

Among the purely Indian communities, the "indomitable" Yaquis had begun their war against the whites in 1825, to defend "the valley that God gave us." (And they had intermittently rebelled throughout the colonial period.) This episode in their long resistance to the central powers had lasted until 1909. In 1887, after the capture and death of their leader, Cajeme, Díaz had ordered that the Yaquis be sent money, animals, farm implements, clothing, and even an image of the Virgin of Guadalupe. "Spare no expense," the President had written to the military commander of the state. "We should not feel safe until we see every Indian with his goad in hand, walking behind a team of oxen and ploughing the fields."[43] The tranquility that he hoped for never came. The Yaquis not only spurned the offers of the government; they rejected any mediation from the Church. They demanded their autonomy and communal ownership of the land: "God gave all Yaquis the river, not a piece to each man."[44] At the end of the nineteenth century, the Yaquis and their chieftain, Tetabiate, swore "to die together rather than give up the land, even if we have to kill all the whites." In 1902, eight thousand soldiers entered their territory and began slaughtering their women and children. Soon afterward, Díaz launched a massive deportation of the Yaquis to Yucatán in the far south, at the other end of the country, where they were compelled to work, like slaves, on the sisal plantations.

But within the broad area of the high plateau at the center and west of the country (homeland for a multitude of races, seat of the ancient Mexican and Tarascan empires), in the eastern sierras overlooking the Gulf of Mexico (inhabited by Huastec and Totonac Indians), and in the indigenous Babel of Oaxaca with its twenty different nations, the pattern (among Indian and mestizo campesinos alike) was one of long periods of peace broken (though rarely) by brief bursts of violence. All these areas had two historic traits in common. They had been the scene of intense ethnic and cultural *mestizaje*. And they had been the zones where the process of conversion had been carried out with particular vigor and success by the various religious orders.

Across that vast landscape of nearly "complete repose," there had been, during the colonial period, only a few, brief moments when the tiger had awakened. The outbreaks were similar—spontaneous, collective, limited to specific communities without involving the surrounding areas, and quickly over with. Rather than rebellions they were local revolts, usually triggered by some injury or threat to the physical or psychic space of the community. The Indians or mestizos almost never questioned the legal authority of the viceroy or the sacred authority of the king. In the normal sequence of these revolts, the people would congregate in the local church, attack the local prison, chant "Death to the *gachupines*!" and "Death to bad government!" (local), and a few days later, once the wrong had been righted by the central authorities, they would return to their former state of repose.

After 1810, after 1910, throughout most of the country, the tiger would be awakened primarily through social rather than ethnic rage. The Indian (and Caste) army of Hidalgo had been, in part, a historical exception, while the southern territories of Yucatán and Chiapas—with their large, unassimilated, and oppressed Indian populations—were a territorial exception. The bloodiest Indian uprising in Mexican history had been the Caste War of Yucatán.[45] It began in 1848 and went through an acute stage of three years, then lingered on sporadically until the beginning of the twentieth century, when the Porfirian army—burning villages, destroying food supplies, forcing mass resettlements—reduced its final strongholds. That historical area (Yucatán and Chiapas), which corresponds to the land of the ancient Mayas in the southeast of Mexico, had two characteristics that ominously distinguished it from the rest of the country. It had only been superficially converted to Christianity; and *mestizaje* was almost nonexistent. A wall of distrust and resentment had grown up across the centuries between the whites and the indigenous population, who consulted their ancient chronologies and the cyclical prophecies of Katún, waiting for the right time to expel the intruders. Suddenly, the explosive mix of religious passion and ethnic grievance erupted. The Mayas wanted to assert their own religion (a syncretic combination of Christianity with ancient beliefs and ceremonies) and to be the absolute masters of their own land. Some of them, the *cruzoob*, followed the orders of caudillos who were also high priests; and deep in the jungle they worshipped a cross reputed to possess the power of speech and requiring human blood. Rage pent up for centuries was released as a will to revenge. Yucatán became the scene of mass sacrifices: "The women," a chronicle details, "were taken outside the city, stripped, raped, tied to stakes that were placed in the sun and then left in this state until nightfall, when the Indians returned and began their work of torture. All of them had machetes and with them they began to wound their victims, cutting slices from their arms, thighs and breasts."[46]

Although the development of the sisal plantations in Yucatán, during the closing decades of the nineteenth century, would bind a good part of the population in place, the *cruzoob*—who had retreated to the extreme southeast of the peninsula—continued fighting into the twentieth century, when Díaz finally overwhelmed them, exiling the rebels to the baleful National Valley (also known as the Valle de la Muerte) in the province of Oaxaca. The valley was a deep canyon in a site of great natural beauty where prisoners were forced to work with little food, many lashes, a workday from four in the morning till late at night, an average life expectancy after arrival of eight months, the sick left to die, and the dead often thrown into the swamps to feed the crocodiles.

But except for the far south and the Yaquis in the north, the major potential—with minor exceptions—for social resentment and violent upheaval at the bottom of the Mexican economic pyramid lay not in ethnicity (always a major factor in Peru, for instance) but in poverty. And especially in a problem as old as the history of Mexico itself: the issue of the land.

In a file buried in the General Archive of the Nation under the rubric "Lands Section," the inhabitants of Santo Tomás Ajusco, in the south of the Valley of Mexico, left testimony on the foundation of their village. The events took place in 1531, barely ten years after the Conquest. An Indian elder spoke to his "beloved children," explaining to them why they should obey the men of Castille and accept new limits to their lives. His duty was a sad one, like all the responsibilities of those who "protect the peoples" because it was "well-known how the defenders of the peoples are punished" like "the Great High Lord of Michoacán" whom the "envious," the "ravenous," those who "call themselves Christians," had burned alive:

Because they want only for they themselves to command,
Because they are ravenous for the metals of others and for the riches of others
And because they want to keep us under the soles of their feet
And because they want to make playthings of our wives and also of our daughters.

The ancestors had already predicted that "they would come from far-off lands to bring us sadness." And nevertheless, it was right that "we should be baptized," it was right that "we should yield to the men of Castille"—"to see if in this way they will not kill us," "to continue to see if in this way we will arouse their compassion."

Following this, the elder, using the old nomenclature (area of the live oaks, area between the nopal cactuses, the hill of the hare), described the new perimeters of their existence in detail. Then he added:

I think that if we hold this little bit of land perhaps they will not kill us.
What does it matter that the land we knew was larger?
But this is not done by my free will;
It is only because I do not wish my children dead,
let there be no more than this little bit of land
and let us all die on it
and also our children after us.
And let us work no more than this land
to see if then they will not kill us.[47]

Across the three colonial centuries, the peasants of Mexico used all the legal procedures within their grasp to defend the extensive or "little bit" of communal land left to them. Their natural enemies were the hacendados, who invaded their lands with cattle or tried to modify the boundaries. "You could have covered the entire territory of New Spain," Justo Sierra wrote, "with the dossiers of lawsuits about the distribution of land."[48] Even so, the very existence of the lawsuits and the relatively small number of uprisings proved that the Tribunal de Indios (the colonial "Court for Indian Matters") had stood the test of time. At the beginning of the War of Independence in 1810, there were four thousand pure indigenous communities remaining in "Old Mexico" alone.

But in no way was the tenacious defense of their "little bit of land" limited to communities still Indian. Other population groups as well, the pueblos, even if they were no longer purely indigenous, would often have old title deeds to their lands. The pre-Hispanic peoples, particularly those in the center of the country, had lived in sprawling settlements, the *altepeme* (plural of *altepetl*, a "territory," which came to mean the group of people occupying it). They would have a leader, called in Náhuatl—like the Aztec emperor himself—their *tlatoani* ("he who speaks"). Beginning at the end of the sixteenth century, the Spanish authorities modified this system. With the help of the friars and the *tlatoani*s themselves (who were now called caciques), the Spaniards gradually compressed the Indians into pueblos, which had plazas, straight streets, districts for each indigenous group according to their place of origin, and a small home assigned to each head of a family. Within walking distance, at the center of the pueblo, stood the church, which—built by the Indians, administered by their own "chiefs" along with the priests and dedicated to the patron saint of the pueblo or district—became their new symbol of collective identity. The importance of the local church and saint remained unaffected by the increase in biological *mestizaje* or the constant flight to Spanish towns, cities, or haciendas throughout the colonial period.

The disappearance of the Spanish protective state in 1821 meant that the Crown could no longer safeguard the rights of the older Indian communities or the land titles of the pueblos. "The Indian yearns for the return of the viceroy who would assure them of personal security and moderate taxes," noted Joel R. Poinsett, the first ambassador of the United States to Mexico. The new concept of individual property was bound to clash not only with the traditional Indian concept of communal property but with other legal formulations of colonial times, such as the "original sovereignty" of the Crown over the soil and subsoil of the kingdom of New Spain. And the poverty and exhaustion of the new Mexican state after the War of Independence immensely strengthened local and regional caciques, who built up little personal kingdoms, expanding their control over the communal lands of the communities. It did not take long to understand that the "repose" of the colonial period had come to an end. At first, only a few pueblos and Indian communities reacted violently to the new situation, but resistance grew during the 1840s. By the end of the war with the United States, their patience had almost run out. These groups seemed to understand that a crumbling nation-state—like Mexico in the fifties—could not guarantee their survival. Within the power vacuum, their only alternative was self-defense by force.

There were various disturbances in the heartland of the country. Peasants from the pueblo of Xicontepec, in 1848, drew the boundaries of their land across the patio of the Hacienda de Chichoncuac and occupied the neighboring Hacienda de San Vicente, where they erected new fences (*mojoneras*), reclaiming communal territory that they had lost. In October 1850, Indians from the municipality of Cuautla, similarly dispossessed, destroyed the stone wall raised by the owner of a hacienda. Although troops stationed in Cuernavaca received orders to move against these Indians, the soldiers refused to obey, arguing that "enraged by not having land to live on and convinced that their property . . . had been expropriated by the haciendas a long time ago, the people had lodged a complaint with the central government . . . and far from being heard, it had been forgotten."[49] In his report, the prefect of Cuernavaca wrote: "The word 'land' is the basis for unrest, the incentive for disturbance and an ideal means of inciting the masses."[50]

Francisco Pimentel, a nineteenth-century student of the "Indian problem," had written, "The communal system has made the Indian lose all feeling of individualism, of individual enterprise."[51] His apparent logic was that the Indians had to be returned to some presumed natural state, *before* communal ownership. This is exactly what the laws on disentailment of corporate property attempted to do in 1857, during the period of the Reform—with disastrous results for the Indian communities. The laws did not convert them into individual owners but only accelerated the forced sale of their properties, at ridiculous prices, to the hacendados.

Naturally enough, the Indians turned their backs on the Liberals. ("We have become the *gachupines* of the Indians," commented Guillermo Prieto.[52]) And the Indians had good reason to show "fanatic enthusiasm" for the Emperor Maximilian. Manuel Lozada ("the Tiger of Álica") as well as Tomás Mejía—the Indian cacique of the Sierra Gorda, in the northeast of Mexico—fought for the emperor. Maximilian showed particular sensitivity toward the problems of the Indians. He introduced decrees that recognized the juridic personality of the villages and ordered the restitution of lands usurped by the haciendas. Both laws would last the brief life of the empire. With the triumph of the Republic came new outbreaks of violence (among Indian communities as well as pueblos) around the issue of the land, until the iron, paternal hand of Porfirio Díaz gradually repressed them. In 1910 the old dictator must have surely assumed that the tiger was safely asleep.

But even while the Centenary Fiestas were in full swing, the villagers of Morelos were being threatened with the loss not only of their lands but of their lives. The modern sugar haciendas were sending armed men to seize village land and increase their acreage, in order to meet the growing demand for sugar on both the national and the international market. The pueblos were not slow to respond to this new threat of dispossession. In heated meetings, they claimed their "land and water," and like the Indians at Dolores in 1810, they shouted "Death to the *gachupines*!" (Many Spaniards owned or administered the sugar haciendas.) A distressed (and unsympathetic but prescient) young gentleman from the capital who happened to be in the area wrote to his friend, Francisco Bulnes: "I do not believe that the French Revolution had been prepared with as much audacity and equipment for destruction as is being done for this Mexican one. I am terrified! The speakers . . . without ceremony or shame, have raised the sacred banner of the poor against the rich."[53]

The tiger was stirring, in villages at the center of Mexico.

The Laws of the Reform had completely altered the position of the Church in the economy, in education, in social and political life. It had been a critical rupture, one of the most traumatic episodes in Mexican history. But Porfirio Díaz, with his "policy of reconciliation," had considerably softened its effects.

Díaz had been a pure Liberal but lacked the passions typical of a Jacobin. In fact he lacked almost every passion except patriotism and the will to power. He often said, "In politics, I have neither loves nor hatreds."[54] Perhaps his religious moderation owed something to the harsh death of his younger and only brother, Félix. When Félix Díaz had become governor of Oaxaca in 1870, he had implemented the Laws of the Reform with incredible fury. His Jacobinic rage was not limited to driving nuns from their convents, closing churches, or prohibiting the teaching of the Catholic religion in schools and seminaries. Determined to uproot the idolatry of the people, he roped the image of the patron saint of Juchitán (that village of warriors) to his horse and dragged it away, returning the saint days later with its feet cut off. The next year, 1871, Governor Félix Díaz joined his brother's first rebellion, against the Juárez government. When the effort collapsed, he tried to flee the capital but was pursued and captured by men from Juchitán. They took their revenge, torturing him (as they felt he had treated their saint) to death. They cut the soles of his feet, stuffed his genitals in his mouth, and made him walk on shards of glass and burning coals. Díaz, when he came to power, did not avenge his brother. He had learned a lesson, atrociously: the struggle over religion had to be resolved or at least soothed.

His political management of the Church was truly masterly. His intent was to end the discord, in contrast to his predecessor Lerdo de Tejada, a former Jesuit, who had gone so far as to expel the pious nuns of Saint Vincent de Paul from the country, triggering a rebellion of

Catholic peasants in the west. Díaz's formula was simple. He would not repeal the Laws of the Reform, but neither would he apply them. Through these laws, Díaz explained, "Party hatreds are maintained." Without them—or if they were allowed to exist only on paper—the unity of the *patria* could be achieved. As signs of good faith, Díaz took several private initiatives, heavy with significance from the standpoint of the Church. In 1879, when his first wife, Delfina Ortega, died in childbirth, he wrote a letter in which (of course unofficially and secretly) he renounced the Laws of the Reform. Thanks to this letter, his wife was allowed to receive the last rites of the Church. It was deposited (and stored in equal secrecy) within the archives of the Cathedral of Mexico.[55]

There were other, less dramatic acts of rapprochement. In 1881 Díaz succeeded in getting his friend, Father Eulogio Gillow, to bless his marriage to a girl of seventeen, Carmelita Romero Rubio (Díaz was then fifty-one). In 1887 the President in turn gave his blessing to the appointment of Gillow as the first archbishop of Oaxaca. (There was an exchange of gifts: from Díaz to Gillow, a pastoral ring with a large emerald set among diamonds; from Gillow to Díaz, a sumptuous piece of jewelry with representations of the victories of Napoleon Bonaparte.) Clearly Díaz was brought closer to the Church by way of his loves—and his interests. In his *Reminiscences*, Gillow himself put it this way: "Carmelita Romero Rubio was surprisingly at the heart of the evolution of General Díaz toward a policy of reconciliation with such profound consequences for the life of the nation."[56] When someone would denounce the secret existence of a convent, explained another writer of the time, "Díaz allowed his spouse to send an opportune warning to the nuns so that they might hide themselves in time; and when the District Judge would arrive to inspect the building, he would find that the denunciation had been unfounded, since there was not even a hint of a shelter for recluses but rather a college for poor children, who were taught by certain charitable women."[57]

By the end of the Porfirian era, the clergy had recovered part of its strength. Church ownership of property had grown (with lands and buildings registered in the names of intermediaries). The Church owned a number of schools, hospitals, and combative newspapers. The Virgin of Guadalupe had been crowned "Queen of Mexico" in 1895; the Order of the Sisters of Guadulupe had been created; the Jesuits had returned. But the President gave no concessions without receiving in return. The bishops supported "the pacifying work of Díaz,"[58] and in their Fifth Provincial Council of Mexico in 1896, they ordered the faithful to obey the civil authorities. For the rest, though "as a private citizen and head of a family"[59] Díaz declared himself to be a Catholic (apostolic and Roman), he still presided over the great Masonic ceremonies and maintained excellent relations with Butler, his principal Protestant minister. And when the moment of truth arrived, Díaz refused to sign a concordat with Rome.

It seemed the best of all possible worlds: the arranged, peaceful, friendly, mutually beneficial separation of the "Two Majesties" that had contended for the kingdom of New Spain: the Church and the State.

For the peaceful participants in the Centenary Fiestas, the problems between Church and Crown (or its later avatar, the State) seemed not a historic but a prehistoric matter. That Hidalgo, Morelos, and several other Insurgent caudillos had been priests, who had courageously defended the "only religion" and the *fueros* of the clergy, was a matter of little concern. It was also taken for granted that the great liberal thinkers of the nineteenth century should have come from religious colleges and seminaries. The important fact was that the Liberals had defeated the Conservatives who had fought to retain the privileges of the Church. Mexico had freed itself from the Crown in 1821 but had to wait until 1861 to free itself from the Church. After Juárez's knife had come the balm of Porfirio Díaz. The old problem had apparently been solved.

Mexican history had witnessed a chain of expropriations. In 1767 the Crown had expropriated the property of the Jesuits; in 1804 it had taken some of the Church's property; in 1821 the new independent state had expropriated the Crown; and in 1827 it had seized all the property of Spaniards living in Mexico (who were then expelled from the country). In 1861 the emerging Liberal state had then thoroughly expropriated the Church. With the defeat of Maximilian in 1867, "the Majesty" of the Church had lost its last bid for secular power. In a land where power had always had theocratic support, could the new state hold the loyalty of its people through abstract concepts like democracy, freedom, personal guarantees, and basic human rights? The Liberals felt a need to expropriate other aspects of the traditional power of the Church—its hold on the mind and spirit of the people.

Surely what was missing was a creed. The Liberals would have to create, as Justo Sierra wrote, "a patriotic religion which would . . . unify" all Mexicans.[60] And the official history of Mexico was created to serve this purpose. Liberals composed historical catechisms meant to arouse and reinforce a "holy love," a "deep devotion for the *patria*." Children (the new faithful) were to be taught "a great admiration for our heroes, making them realize that, due to the acts of our heroes, Mexicans now belonged to one big family."[61]

Strange as it may seem, one of the pioneers of this massive symbolic "expropriation" was that liberal Austrian in a foreign land: the Emperor Maximilian von Hapsburg. He had favored the civil beatification of the heroes of the independence movement, ordering their portraits to be painted and statues of Hidalgo and Morelos to be cast. If history could ever be fair, it would have reserved some place for Maximilian in the calendar of patriotic saints. The unfortunate emperor had shown that the most sensible way of persuading Mexicans to accept patriotic values was not, as so many thought, by destroying Catholic symbology but rather by expropriating and imitating it. He had been the first to teach the Liberals how to paint their own choir of saints.

During the long republican reign of Porfirio Díaz, Mexican Catholicism was what it had always been: a heaven filled with saints. But a parallel heaven of heroes and a corresponding hell of villains arose within the consciousness of the people. The images of the saints (on cloth and paper, in bronze, ivory, and stone) were venerated on the altars of churches and homes. When the Liberals came to power, idealized images of national heroes began to appear in the form of sculptures and busts, or neoclassical canvases in the style of David, the painter of the French Revolution. And they were distributed, like the saints, as educational prints for children. The priests would recount the "exemplary lives" of their saints during sermons and catechisms, while in speeches and books, public speakers and official historians would recall the heroic feats, famous sayings, and "exemplary deaths" of the Insurgents. To the holy calendar that hallowed the birth and death of the saints was added a new, patriotic calendar. For decades, Mexicans had celebrated Independence Day on September 16, but now there were new holidays: Constitution Day (February 5), Juárez's birthday (March 21), the victory against the French at Puebla (May 5), Hidalgo's birthday (May 8), Juárez's death (July 18), Porfirio Díaz's final victory in Puebla over Maximilian (April 2), and of course Don Porfirio's birthday (September 15). Relics of the saints and of the martyrs of the Church, bits of bone and scraps of clothing were treasured in chapels; urns holding the ashes of the martyrs for the fatherland were placed within monuments, and their clothing was displayed in special museums. Ever since colonial times, villages and cities had been named after one saint or another. The Liberals now renamed places throughout the country after their heroes: Dolores Hidalgo, Ciudad Juárez, Ciudad Porfirio Díaz.

On holy days, religious processions wound through the streets carrying images of saints and depictions of their lives. During the Centenary

Fiestas, there was a similar procession of heroes culminating in a kind of patriotic mass. The heroes would never replace the saints within the Mexican popular mind, but they would carve out a place for themselves and become the subjects of official anthems and popular songs, romantic novels and patriotic poems. The religion of the *patria* never supplanted Catholicism, but the fact remains that hero worship in Mexico assumed the peculiar form of beatification. In the collective imagination, the heroes of the fatherland would become lay saints.

For the Porfirian Liberals in 1910, this "patriotic religion" may have seemed like the natural culmination of the long, vast process of expropriation through which the "Majesty" of the State had prevailed over the "Majesty" of the Church. Would it end there? Apparently, there was nothing else to expropriate. And Díaz's "policy of reconciliation" seemed to permit both State and Church to prosper. The Church had lost its economic resources and been defeated politically in 1867, yet it was growing in social and spiritual strength. Between 1885 and 1910 the number of priests increased from 3,576 to 4,553. But Mexico no longer had 1.5 million people as in 1650. The population now was almost ten times larger. There was room for more priests and even a stronger Church. After all, Mexico was a secular country now.

But gathering in that Liberal heaven were the clouds of another storm. A new generation of young radicals was forming—Jacobins reborn, with the old anticlerical rage. Many of them—like their predecessors—were former seminary students with grievances against the Church; others came from the north, a border area that had developed in near isolation, on the margins of the religious history that had influenced the center and south. These frontiersmen (*fronterizos*) had played a decisive military role in the Liberal victory. They were liberal individualists in body and spirit, who felt an instinctive repulsion toward what they saw as the religious fanaticism of the rest of the country. In the eyes of the new radicals, Díaz had conceded far too much to the Church. For them, the pulpit and the confessional were not neutral institutions but the direct cause of Mexico's historical backwardness. A revolution led by these fanatics of antifanaticism, who detested religion with missionary zeal, could once again renew the struggle between the "Two Majesties."

The twentieth century was just beginning. It would be an era of strong states and imperious ideologies—ideas that on their arrival in Mexico could tempt new governments and ideologues to expel the Church from areas where it still exerted great influence: education, social welfare, and the minds of the men, women, and children of the nation. There *was* more left to expropriate for the State to become sole and supreme. A time was coming when the "Two Majesties" would confront each other again and the new religion would try to destroy the old.

In honor of Porfirio Díaz, this strange Liberal dictator, hymns and partitas had been composed, biographies and odes had been written. The nation regularly celebrated his birthday and the anniversary of his most important victory against the French invaders (April 2, 1867). With his name, they baptized cities, streets, markets, buildings, and children. You could see his solemn face and broad chest not only on the walls of government offices but also in store windows, on dinner plates, on packs of cigarettes and the boxes of matches used to light them. It was said that Tolstoy had called him a "prodigy of nature."[62] He was praised as "the good dictator," "the necessary man," "the irreplaceable president."[63] The Centenary Fiestas had been organized to mark the supreme moment of Mexican history. And he, Porfirio Díaz, considered himself its ultimate hero. It was he who completed the Holy Trinity of the *historia patria*—Hidalgo, Juárez, and Díaz.

Throughout its long decades of power, the "imperial presidency" of Porfirio Díaz had drawn political criticism, but Díaz should have noticed that as time went on these dissonant notes were also being sounded by his sympathizers and friends. On the last day of the nineteenth century, Justo Sierra, whom he admired, had written him a private and revealing letter, which Díaz must have ignored:

> Today, reelection means presidency for life, that is to say an elective monarchy in a republican disguise . . . there are supreme objections to indefinite reelection . . . it is important to realize that there is no possible way to ward off the risk of finding ourselves without the power to eliminate a crisis that could mean reverses, anarchy and a final harvest of international humiliation in the event that you should happen to die, from which may the fates preserve us though they, unfortunately, pay no attention to the desires of men. . . . Reelection also means that any preparation for the political future under your auspices and utilizing your unsurpassable condition of physical and moral strength is at present an unrealizable dream (and we all desire such preparation, even your most intimate friends, though to you they may say the opposite) . . . [and moreover it is] the inevitable impression among men of state and business in the United States, England, Germany, France . . . that there are no institutions in the Republic of Mexico—there is a man. On his life Mexico's peace, productive work and reputation depend.[64]

In 1904 Díaz did somehow have the humility to recognize his own mortality. For the first time, he named a Vice President. But in exchange he lengthened the new term of his endless presidency from four to six years. In 1907 came the Creelman interview. Díaz had one basic reason for conceding it, which had nothing to do with romanticism. The interview was meant to reassure United States public opinion and investors as to the political future of Mexico. After all, though many Mexicans (and he himself at times) thought the opposite, Porfirio Díaz was mortal.

The interview was published in the March issue of *Pearson's Magazine* with an imposing photograph of Don Porfirio. The article was titled "President Díaz: Hero of the Americas." And it contained Díaz's first direct answer to the burning question of the time. Would the old man reelect himself in 1910, for his eighth presidential term?

Díaz explained that one had to be cautious about returning to democracy as an absolute value. The abstract theories of democracy and the practical, effective application of them seemed to him "necessarily different" in a country like Mexico. Yet he firmly believed that the principles of democracy had taken root. The country's future was assured, thanks to those past decades of order, peace, and progress under his guidance. For twenty-seven years he had governed the Mexican Republic with such power that he might easily have set a crown upon his head. Now, at the height of his career, "this inscrutable leader, this astonishing man," wrote Creelman, "foremost figure of the American hemisphere and unreadable mystery to students of human government," made the most sensational announcement of his life: "I have waited patiently for the day when the people of the Mexican Republic would be prepared to choose and change their government at every election without danger of armed revolutions and without injury to the national credit or interference with national progress. I believe that day has come. . . . I will retire when my present term of office ends."

According to Creelman, "he folded his arms over his chest and spoke with great emphasis":

> I welcome an opposition party in the Mexican Republic. . . . If it appears, I will regard it as a blessing, not as an evil. And if it can develop power, not to exploit but to govern, I will stand by it, support it, advise it and forget myself in the successful inauguration of complete democratic government in the country. . . . This nation is ready for her ultimate life of freedom.[65]

Porfirio Díaz, 1905
(Collection E. Rincón Gallardo)

In Washington the interview was read with attention and moderate skepticism. When it was published in Mexico, it caused an immense commotion. The old dictator had to be taken at his word. It was time to begin forming the opposition party that Díaz would supposedly view as a "blessing."

The first serious political opposition to the Díaz regime had begun years before, in 1903. It was led by some young men who, like Porfirio, were from Oaxaca. But they were devoted not to power but to the denial of power. They were the anarchist brothers, Ricardo, Enrique, and Jesús Flores Magón. On February 5, 1903, the anniversary of the adoption of the Constitution, they hung a banner from the balcony of their new (and ephemeral) magazine, *Regeneración*. On it were the simple words: "The Constitution is dead." From then on Díaz pursued them relentessly, throwing them into prison in Mexico and eventually exiling them to the United States. But the seeds of a "new Reform" had been firmly planted. In San Luis Potosí, a new Liberal party was formed to seize the word *Liberal* from the Porfiristas and defend the principles of pure Liberalism, the political program of the Reform, and the 1857 Constitution.[66]

Because of their anarchist formation and their modern social sensibility, these new leaders had a different view of Porfirian progress. They did not deny that it existed, but they pointed to its immense costs in both the city and the countryside. They condemned the persistence, both feudal and servile, of the system of the haciendas, which had proliferated under Porfirismo. They also condemned the total lack of protective legislation for factory workers, the prevalence of child labor, the days of work from the rising to the setting of the sun.

In 1906–1907, members of this clandestine party infiltrated the workers' organizations of the mine of Cananea, in Sonora, and the large textile factory of Río Blanco in Veracruz. At both sites, strikes broke out, of a kind never seen before under Díaz. They were long and complex and they ended with the federal army moving in to break them with bayonets and bullets.

Don Porfirio then assigned a lawyer (one of his friends) to prepare a report on the social question. Apparently, the President-monarch, like a new Louis XVI, asked, "Is this a revolt?" The lawyer answered:

> The present movement is neither isolated nor restricted to the working class. . . . In addition to the hatred felt toward a certain circle (the *científicos*), there is the question of the workers and then the agrarian question. . . . It is certain that this is something serious,

> very serious. . . . The movements that have been initiated are precursors to those being prepared in other great centers of the country. . . . When the revolutionary idea is advanced enough to crest into action, the only way to dominate it is to lead it.[67]

Many years before, in 1891, Díaz had said, "I believe that when discontent with a government takes root, it gradually acquires so irresistible a strength that nothing is able to restrain it."[68] But Don Porfirio, at the height of his glory, did not imagine he would have to deal with a situation like that.

In the state of Chihuahua, in the north of the Republic, Francisco I. Madero, a young and wealthy landowner, had been struggling for democracy for years. In 1908 he wrote a book that became a best seller: *The Presidential Succession of 1910.* In it he made a passionate but objective assessment of the regime and rejected "patriarchal politics," to which he attributed "the corruption of the spirit, the disinterest in public life, a disdain for the law and the tendency toward deception, toward cynicism, toward fear. In the society that abdicates its freedom and renounces the responsibility of governing itself, there is a mutilation, a degradation, a debasement that can easily translate into submission before the foreigner."[69]

Madero made a prophecy: "We are sleeping under the cool but harmful shade of a poisonous tree . . . we should not deceive ourselves, we are heading for a precipice."[70]

Not that Don Porfirio would have agreed with him in the winter of 1907. Nevertheless, he had felt that it was the right moment to take the initiative and personally introduce the most revolutionary of ideas (but still an idea, not an action). Through the interview with Creelman, he had arranged to announce—before the world—his intention to retire.

The determined young Madero, to the surprise of Don Porfirio, had taken him at his word. In 1909 he formed the National Antireelectionist party and announced that he would be a presidential candidate in 1910. Soon he began a series of campaign tours, in the style of North American politics, all across the Republic. It was something totally new to Mexico. Don Evaristo Madero, Francisco's grandfather and the founder of the great business dynasty of the Maderos, commented, "It's the struggle of a microbe against an elephant."[71] But Francisco I. Madero did not flinch: He knew that microbes sometimes killed elephants.

Díaz viewed this attempt at political action with the same infinite annoyance and disdain he had felt when leafing through Madero's book. Poor "Panchito" had gone crazy. And as for the "really serious" opposition (floating the candidacy of General Bernardo Reyes, governor of Nuevo León and Díaz's eternal proconsul in the north), Don Porfirio, in 1909, had eliminated that threat in the most democratic manner he knew—by sending Reyes on an assignment into exile. No, definitely not. In spite of his declarations to Creelman, *Mexico was not ready for democracy*, nor was Díaz genuinely ready to leave power. In July 1910, the old dictator once again "sacrificed himself for the good of the nation."[72] A clearly fraudulent election gave him the victory over Francisco I. Madero. In the same month, from his prison in San Luis Potosí, Madero called on the people to revolt. And he set a date: November 20, 1910.

Meanwhile, Díaz had begun his eighth presidential term, which would end in 1916. Porfirio would then be eighty-six. But on the immediate horizon, there was an event of enormous significance: Díaz's eightieth birthday and the hundredth anniversary of the Mexican nation. Both biographies had to be celebrated together. Fundamentally Díaz thought they were one and the same.

And the Centenary Fiestas came to the provinces and the capital. And Díaz presided over the exaltation of the heroes and of himself. At the moment of greatest solemnity during the celebrations, the very day of "The Apotheosis of the Heroes," Don Porfirio, usually rather allergic to words—whether spoken or written—spoke and in his speech briefly summed up his interpretation of the history of Mexico: "The Mexican people, with vigorous energy and lucid insight, has moved from anarchy to peace, from poverty to wealth, from disrepute to respectability, and from international isolation . . . to the fullest and most cordial friendship with the whole of civilized humanity. . . . For the work of one century, no one will judge this to be a small achievement."[73]

He was right: It was no small thing. But Díaz was blind to half of history. Very soon, the loose threads of Mexico's past (the issues of the land and inequality and religion and lack of political liberty) would take over and fill the stage. Pulled along by these threads, the Mexican people would follow a path from peace to anarchy, from wealth to poverty, from international friendship to isolation. This too would be a final product of the nineteenth "century of caudillos."

Porfirio Díaz, the "hero of peace, order and progress" would end not like Juárez, at the height of his power and glory, but forgotten, far from his country, from the Oaxaca he pined for, far from Juana Cata, the Indian woman he remembered, crying out her name as he lay dying. When he passed away in Paris on July 2, 1915, he had been in exile since June of 1911. In Europe, he had been the object of considerable veneration, as the last great representative of a supposedly halcyon era. But like all the other antiheroes of the Manichean version of Mexican history, Don Porfirio—in the eyes of his own countrymen—would die without "glorious memory or honorable grave." And for him there was a heavier penalty. The remains of the others, even including Cortés, would find their final resting place in Mexico. Porfirio Díaz would not. In 1996, his bones, still banished, lie in a simple tomb in the Montparnasse Cemetery of Paris.

III

THE REVOLUTION

While the fireworks of the Centenary Fiestas were still bursting and the music and the rockets could still be heard, suddenly, like the spontaneous volcanic eruption of the entire Mexican past, everything was shaken and confounded by another uproar, not meant for entertainment—the formidable thundering of a new revolution. This would be no revolt out of an operetta like those of Santa Anna, or brief and almost bloodless, like those led by Porfirio Díaz himself. It would not even be like the War of the Reform, a war after all between elites, a war into which ordinary people, for the most part, were drafted rather than participating of their own free will. This time it would be "the greatest and most powerful of revolutions," the social revolution once prophesied by President Lerdo de Tejada, whom Porfirio Díaz had deposed long ago. Its first slogan, its verbal banner, was nothing new. Madero had taken it from one of Díaz's early manifestos: "Valid Voting, No Reelection."

Porfirio Díaz had believed that September 1910 would mark the centenary of a complete historical cycle, the consummation of Independence in progress, order, and peace. History fulfilled and frozen. He had not listened to the critical voice of Sierra in that letter written in the last hours of the nineteenth century nor (if he even read it) absorbed the closing phrase of Sierra's *History*: "The whole social evolution of Mexico will have been abortive and in vain, if it does not achieve the full goal of freedom." September 1910 would mark not a culmination but a beginning: the outbreak of a popular revolution very similar to the one that was being celebrated; a revolution that, like the one of 1810, would go on for some ten years; a widespread revolution that would lead to hundreds of thousands of deaths, a revolution led by new caudillos, some modern and some archaic, divided like their predecessors between the ideals of the future and the roots of the past, caudillos who were nationalists, democrats, anarchists, socialists, Jacobins, devotees of the Virgin of Guadalupe. In time, all these leaders would triumphally enter the new patriotic shrine of the nation: the pantheon of the Mexican Revolution. New epics, hagiographies, and catechisms would be written about them, recording their real and imaginary feats and the deaths of some of them—no longer in front of a firing squad but assassinated by traitors. This Revolutionary catechism would be combined with the history of the Liberals and melt into one: the Catechism of the Revolutionary Patria. Madero, Villa, Zapata, Carranza, Obregón, Calles, and Cárdenas would join Hidalgo, Morelos, Guerrero, and Juárez in the same heaven.

While in hell, the "traitors," the "sellouts," the "crabs," the "misled," the "reactionaries," the "bad Mexicans" would go on atoning for their endless guilt: Iturbide, Alamán, Santa Anna, Miramón, Maximilian of Hapsburg (and of Cuernavaca). They would now enjoy the company of a new and unexpected Liberal guest: Don Porfirio Díaz himself.

The Mexican Revolution was a vast historical readjustment in which the gravitational weight of the past corrected the rush toward the future. It had various stages and aspects and, in a very real sense, lasted for thirty years: from 1910 to 1940. The number of men actually under arms at any one time was never immense. Even at the most feverish point of hostilities (in May 1915), the armies numbered fewer than one hundred thousand men. The vast majority of the population of fifteen million fell into the category of "the peaceful" (*pacíficos*). But everyone's life would be profoundly changed by this Revolution.

Fighting never spread throughout the country. The principal military stages were quite localized. The state of Morelos, homeland of Zapatismo, and the Villista territory of Chihuahua were permanent theaters. There was action in the center of the country, in the west, and—somewhat less—along the Gulf coast. Mexico City lived in a continual state of apprehension, "with Jesus in the mouth," occupied in turn by warring armies who saw the city as their supreme prize.

The Revolution began with a modern democratic movement accompanied by an ancient claim to the land. Despite its initial triumph, this first stage unleashed an authoritarian reaction. The response to this counterrevolution stirred up military and social forces that, once they had won, failed to reach an agreement leading to the restoration of order. Dissension led to war and to a centrifugal splintering not very different from what the country had lived through during and after the War of Independence. The triumph of one faction returned the flood to its riverbed. Ideas and politics gradually began to replace bullets. During the final two decades of the process, Mexico was a laboratory for revolutionary changes under the auspices of the new State. At the end of the cycle, in 1940, the country had reestablished order, centered around a political system controlled by a "Revolutionary Family" rather than a single person, though personal government would continue to be a central feature of Mexican political life.

The Revolution led by Madero broke out on November 20, 1910, and in a matter of months spread to several areas of the country. The major centers of the insurrection were the states of Chihuahua and Morelos. Francisco I. Madero himself led the operations in Chihuahua, aided by men who were to become legends, like Pascual Orozco and Francisco Villa. The campesinos who followed Emiliano Zapata fought in Morelos. At the beginning of May 1911, Orozco and Villa occupied Ciudad Juárez, the most important customs post on the U.S. border, and through this occupation forced the government into negotiations that, by the end of the month, provoked the resignation of the dictator. "Madero has unleashed the tiger," said Porfirio Díaz in Veracruz before boarding the boat *Ypiranga,* which would carry him into exile.

A military coup carried out by General Victoriano Huerta would overthrow Madero. The "tiger" so much feared by Don Porfirio then truly awakened. A broad-based military movement to oppose the usurper was organized around the person of the governor of Coahuila, Venustiano Carranza, an elder of the Revolution. Between March 1913 and July 1914, various bodies of the Constitutionalist Army—so called because the movement tried to restore the constitutional order violated by Huerta—recognized the authority of Carranza as their commander in chief.

As long as the war was focused on defeating Huerta, Carranza was able to keep all the factions together. But once the usurper resigned and went into exile (on July 15, 1914) the Revolution was incapable of administering its own victory.

Loosely following the script of the French Revolution, the military chieftains met in a "Convention" (in October 1914) held in the city of Aguascalientes. Its purpose was to choose the new government and to define the future direction of Mexico. By then it was evident that Villa and Carranza were at odds with each other. The Convention produced a government that Carranza refused to recognize, immediately establishing his own government in the port of Veracruz. The leaders had to choose: whether to stand with Villa or with Carranza. At that moment, the Zapatista movement emerged from its base in Morelos and joined forces with Villa. Both threw their support to Eulalio Gutiérrez, the President named by the Convention.

Álvaro Obregón and Francisco Villa, two military titans, would confront each other—in the spring of 1915—in the Bajío, the Central Plateau of Mexico. With the crushing victory of Obregón, the government of the Convention fell apart, and the nationalist leader Venustiano Carranza became President.

The moment had passed for the three Revolutionary leaders, the caudillos whose purpose had been "liberation": Madero, "the Apostle of Democracy," with his *Plan de San Luis* (Program of San Luis) designed to liberate Mexico from dictatorship; Zapata, "the Caudillo of the South," whose *Plan de Ayala* was meant to return the land to the peasants, liberating them from the yoke of the haciendas; and Villa, "the Centaur of the North," who was like a blind force, acting not in response to a program but to an implacable, and often bloodthirsty, hunger for "justice."

The time then came for the chieftains, the jefes, who would try to direct the torrential action into manageable channels. One of them, Carranza, wanted a civilized Mexico under civilian rulers. The other, Obregón, wanted a civilized Mexico under a military government. For a while they worked together. Carranza convoked a Constitutional Congress at the end of 1916, and in February of the next year, a new (and genuinely revolutionary) Constitution was proclaimed in Querétaro.

Carranza held the presidency from 1917 to 1920. When he tried to make a civilian his successor, the powerful Army of the Northeast—under the apparent command of Adolfo de la Huerta (though the real chieftain was Álvaro Obregón)—rose against Carranza and defeated him. At the end of May 1920, the military leaders native to Sonora would take power and hold it for fifteen years.

Álvaro Obregón was President from 1920 to 1924. His attempt to maintain himself in power, directly and indirectly, would trigger a civil war among the Sonoran chieftains, and he would eventually be succeeded by two generals who were more truly "statesmen" (*estadistas*) than jefes. One of them was an austere primary-school teacher elevated to the rank of general by the Revolution, President from 1924 to 1928 and then "Supreme Chief" (*Jefe Máximo*) from 1928 to 1934: Plutarco Elías Calles. The other, who took office in 1934, was Lázaro Cárdenas, one of the youngest generals of the Revolution. When his term ended in 1940, the Mexican state had taken on a solid configuration: an all-powerful President elected every six years without possibility of reelection but with the right to name his successor from within "the Revolutionary Family," and a single (or almost so) party that served the Monarch-President in multiple controlling functions: social, electoral, and political.

Revolutions have been organized around ideas or ideals: liberty, equality, nationalism, socialism. The Mexican Revolution is an exception because, primordially, it was organized around personages. Each one called up a specific *ismo* in his wake: Maderismo, Zapatismo, Villismo, Carrancismo, Obregonismo, Callismo, Cardenismo. "Viva Madero!" proclaimed the slogan endlessly painted on the walls of the country. "Let us go with Pancho Villa!" shouted the horsemen of the División del Norte, who followed "the Centaur" and were moved by direct attachment to his person. "For my General Zapata!" the peasants of Morelos fought and died.

This charismatic element was less intense in the case of Carranza, the commander in chief of the Constitutionalist Army, or even the "Undefeated" General Álvaro Obregón, but the discipline and obedience of their armies was absolute and total. Between admiration and fear, ambition and faith, the Callistas were loyal to their *Jefe Máximo* just as the Cardenistas followed the most popular "Señor Presidente" that Mexico has ever known. The Mexican Revolution can hardly be reduced to the biographies of seven people, but without a knowledge of the specific lives of these personages, it becomes incomprehensible. For the experience of the nineteenth century was to repeat itself—power concentrated in emblematic figures.

There was something idiosyncratic, original, and even innocent in these men. They were not like the leaders of other revolutions who, in the name of humanity, defended abstract principles, broad ideological systems, prescriptions for universal happiness. The Mexican caudillos, jefes, and *estadistas* acted in accordance with modest Mexican categories. They did not consider universal history but the history of the fatherland (the *patria*). Except for Madero, they were not well read or educated, they had not traveled the world nor did they even completely know their own country, but only their own region, their own state, their own native soil. As with the Insurgent priests, a messianic attitude colored their actions: They wanted to redeem, liberate, bring justice, preside over the final coming of good government. The local histories from which they began, their family conflicts, their lives before rising to power, their most intimate passions—all are factors that might have been merely personal, though perhaps representative, if these were merely private lives. But they could not be in Mexico, a country where the concentration of power into a single person (*tlatoani*, monarch, viceroy, emperor, President, caudillo, jefe, *estadista*) had been the historic norm across the centuries.

10

FRANCISCO I. MADERO

The Apostle of Democracy

More glory in doing your own duty badly
than in doing another man's duty well!
Better to die doing what is right for you!
Shun doing what is right for another man.

—BHAGAVAD GITA

The first caudillo of the Mexican Revolution was not a peasant or a worker. He was the wealthy heir of one of the five richest families in Mexico: the Maderos. In the state of Coahuila, Francisco Madero's grandfather Evaristo, the founder of the dynasty, created the Compañía Industrial de Parras, originally involved in vineyards, cotton growing, and textile production but soon to become the center of a business empire: mining concessions; cotton mills in Coahuila, Nuevo León, the Gulf, Sonora, and Yucatán; cattle ranches; the Bank of Nuevo León; coal companies; rubber plantations; foundries. At the beginning of the twentieth century, the sun never set on the Madero dominions.[1]

Evaristo had married twice and in each case produced an enormous family. On October 30, 1873, Mercedes, the wife of Francisco—his eldest son by his first wife—gave birth, at the Hacienda del Rosario in Parras, to Evaristo's first grandson. He would bear the name of two saints who had founded religious orders, the first famous for action and the second for charity, Saint Ignatius of Loyola and Saint Francis of Assisi. And in Francisco Ignacio Madero's last name (*madero* means "wood"), there would be a suggestion of Calvary.[2]

Small in stature and delicate in health, Francisco Ignacio, at the age of twelve, entered the Jesuit Colegio de San José in Saltillo. His years there had a profound influence on his moral nature and his capacities for self-discipline. In his memoirs, he would later record some of his contradictory feelings: "I was deeply impressed by their teachings . . . but they made me encounter religion with somber and irrational colorations to it."[3]

He later studied briefly in the United States, spent some years in France, and traveled through western Europe. Neither the countries he visited nor their art and culture excited him during those years. His crucial experience was "the discovery of what has been most important to my life," the religion of Spiritualism.[4]

This doctrine—based on notions of the existence, manifestations, and meaningful communications from the spirits of the dead—was formulated in the United States during the middle of the nineteenth century but spread with a dizzying rapidity through France after a man named Allan Kardec became a convert and then its leading prophet and organizer. By 1854 there were more than three million practicing Spiritualists in the world and tens of thousands of "mediums" in Europe and America, who specialized in contacting the dead. By the time of his own death in 1868, Kardec had written several books and had founded the *Revue Spirite* and the Société Parisienne d'Études Spirites.

When Francisco I. Madero first leafed through the *Revue Spirite*—to which his father subscribed—the new faith, accepted by men as famous as the astronomer Camille Flammarion and the novelist Victor Hugo, was at the high tide of its expansion. Day after day, hundreds of pilgrims would visit Kardec's tomb or flock to the feet of his primary disciple, Léon Denis. One of these devotees was the young Madero. At the office of the Société in Paris, he bought all of Kardec's works. "I did not read his books," he would write in his *Memoirs*, "I devoured them, because his doctrines—so novel, rational, and beautiful—seduced me, and from then on I considered myself a Spiritualist."[5] Madero fervently began to attend Spiritualist sessions. He came to feel that he had a particular vocation for the role of "writing medium," someone whom the spirits would take by the hand and through him write their messages to humanity. Day after day, straining his senses, he tried to channel the spirits into his fingers but produced nothing of significance. Then one day he let his trembling fingers write, "Love God above all things and your neighbor as yourself."

Unlike many other devotees, for Madero the importance of Spiritualism did not lie in the chairs spontaneously tapping out coded messages, teapots walking across tables, or paintings coming alive, or in the movement's more literary concerns with the swarms of spirits roaming the world, the nature of Shakespeare's dreams, or the astral realms of Swedenborg. He was searching for ethical connections between Spiritualism and the Christian Gospels: "I have no doubts that the moral transformation I have experienced is due to my becoming a medium."[6]

Though he had finished business school in Paris, his father and grandfather decided to complete Francisco's education with a year at the University of California in Berkeley, where he improved his English and studied agricultural techniques. But there, once again, the important knowledge for Francisco was spiritual and moral. At Berkeley there were academics of the "progressivist school" trying to apply Christian morality to social problems. Not far away, at Stanford, there was a church of all faiths. The theosophy of Annie Besant—a mingling of Hindu ideas and Western influences—was a fashionable presence. (Besant was to become a leader in the early struggle for Indian independence.) The twenty-year-old Madero felt the impact of this whole complex of spirituality and public morality. He felt there was a general connection with the "revelations" he had received in Paris.

At the end of that year, 1893, he returned to take charge of one of the family haciendas at San Pedro de las Colonias. He was no longer a boy in poor health. He had worked hard to acquire physical strength and was now an excellent swimmer and dancer. Jovial, restless, hyperactive, he was full of projects for his hacienda: irrigation works, the introduction of American cotton, a soap factory, an ice factory, a meteorological observatory. He would write a pamphlet on the water rights of the Río Nazas that drew the praise of Don Porfirio himself. And he became a skilled administrator and businessman. By 1899 his purely personal capital had risen to the respectable sum of 500,000 pesos.[7]

Ever since his return to Mexico, he had also done charity work. It was part of the family tradition, but Francisco carried it to enthusiastic extremes. He had learned homeopathic medicine from his father and uncle, and from 1896 on, he could frequently be seen on the road, medicine bag in hand, visiting his peons to prescribe minute amounts of nux vomica, belladonna, lime charcoal, and many other potions, some of which he prepared himself. "It was a sight to see," said one of his friends, "how the indigent sick besieged him and how he would dole out relief for their pain, consolation for their troubles and money for their needs."[8] At his own house in San Pedro, where he lived in Franciscan austerity, he fed about sixty youngsters. His workers on the hacienda lived in clean houses, enjoyed high wages, received regular medical exams. Madero and Sara Pérez, who became his wife in January 1903, paid to support orphans and gave out scholarships. And they created schools, hospitals, and community kitchens.[9]

At the turn of the century, business and charity filled his days but not his nights. For years he had continued his Spiritualist experiments. In 1901 these took a new turn. He began to receive daily visits from the spirit of his brother Raúl, who had died by fire in an accident at the age of four. From then on, obeying the injunctions of "the spirit of Raúl," Madero began the constant practice of self-discipline, abnegation, and purification. "Raúl" was always urging him "to dominate matter" in favor of "questions of the spirit." Under this intangible rod, Francisco became a vegetarian, stopped smoking, and destroyed his private wine cellars. But the cleansing rites to which he subjected himself were directed toward an active rather than ascetic life. "You can have the only happiness there is in this world solely through practicing charity in the broadest sense of the word," the spirit of Raúl wrote with Madero's hand. "To be of aid" to others must be his mission and that of his family.

At the end of 1902, Raúl "suggested" that other spirits should be invoked. Before they could arrive, Bernardo Reyes, governor of the neighboring state of Nuevo León, violently crushed a political demonstration on April 2, 1903. The news deeply moved young Madero. Through his family, he was already familiar with Porfirio Díaz's methods of political control and manipulation, especially in the state of Coahuila. But now history had become immediate and tangible.

Months passed. And then the "spirit of Raúl" signaled a new direction through the hand of his living brother: "Aspire to do good for your fellow citizens . . . working for a lofty ideal that will raise the moral level of society, that will succeed in liberating it from oppression, slavery and fanaticism."[10]

It was a true illumination. The broadest meaning of charity took on, for Madero, the name of politics. "Great men"—"indicated" the spirit (and it was a premonition)—"spill their blood for the salvation of their fatherland." As his recently discovered vocation was taking form, Madero concentrated his energies for the first steps on the new "field of combat." In 1904 he was to fight a close electoral battle in his municipality. He founded the Benito Juárez Democratic Club, but he lost the election.

Along with his entrance into politics and his ongoing business responsibilities, he continued to be an active Spiritualist, reading, corresponding, and writing articles (on themes like God or the creation of the world). He would sign these articles with the name of his first alter ego, Arjuna, the hesitating prince in that classic of Hindu spirituality, the *Bhagavad Gita*, to whom the god Krishna teaches the significance of human action and the nature of the universe.

During the first months of 1905, Miguel Cárdenas, who was about to be reelected for a third term as governor of Coahuila, confided one of his concerns to the perpetual President, Porfirio Díaz: "I do not think it is far-fetched to feel that Sr. Madero, encouraged by the political passion that has overcome him and by the monetary resources available to him, could create some difficulties and even cause a scandal."[11]

He had reason to worry. A strong opposition movement had come into existence. Young Madero, who would soon begin to be branded as "crazy" and "unbalanced," supported the governmental candidacy of Frumencio Fuentes through organizing political clubs and financing a political newspaper (*El Demócrata*) and a satirical publication (*El Mosco*, "The Fly"). The President asked General Bernardo Reyes if it would be a good idea to jail Madero. Porfirio's proconsul for northern Mexico thought not but suggested instead that a strong cavalry squadron be stationed in the region and that the elder Francisco should be persuaded to calm his son down. The elections were held in relative peace in mid-September. The result was, of course and as expected, a victory for the official candidate.[12]

Despite the second electoral defeat of his political career, Madero did not lose faith: He published a manifesto in which he declared that the sovereignty of the State has always been "a myth" and complained of electoral fraud. The defeat did not quiet him. It made him more alert. Because he sensed that the curve of his democratic spiral would—in a few years—include the entire nation, he decided not to contest the result of the state elections. He wrote to his brother Evaristo, asking him to return from Paris to participate in "the great political battle shaping up for the future."[13]

His role and mission were becoming clearer. He would be a preacher, an apostle, a *medium* for the political spirituality he embodied, carrying his message of change—through the *word*—into every corner of the country. "Humanity has progressed," he wrote to a friend. "Little by little, tyrannies will be destroyed, and freedom, which brings Justice and Love along with it, will succeed in fulfilling the words of Him Who Died on the Cross."[14]

Beginning in 1907, "José," a more militant spirit, began to guide his steps. The notebook, intact after almost ninety years, in which Francisco Ignacio Madero, as medium, transcribes his messages from José recalls nothing so much as the writings of Saint Ignatius Loyola, the founder of the Jesuits. Each page is a struggle against the "yoke of the instincts," a demand for "gigantic efforts to conquer animality." José, like Raúl, uses guilt to influence Madero—and threats that he may abandon him forever. Yet the major spur is not fear but the hope of reward. If he were to succeed in dominating his "lower passions," then "we could accomplish something useful, effective and truly transcendent for the progress of your *patria*." And not only Mexico would benefit but Francisco and his wife. They would be able to produce the child they longed for.

The object was total self-control. "Burning prayers," "the saddest reflections," "the firmest resolves to purify oneself" had to follow each tiny descent into the mire of instinct. "Attempt to completely isolate yourself from the external world and close yourself into yourself, into your inner world where perfect calm reigns, and a profound, majestic silence." Another spirit arrived to second José and informed Madero that "all your actions must answer to a plan."[15]

The plan took shape. The spirits ordered him to preach (in a growing number of political letters) to sympathetic minds throughout Coahuila and the rest of the country. He began to write and publish more, in opposition newspapers that he often also financed. During that same year, 1907, Madero succeeded in taming his "lower nature" (which probably meant sexual abstinence). In October, the "spirit"—now convinced that his disciple and "brother" had triumphed over matter—ordained him, through a solitary and quixotic ceremony in an attic of his hacienda:

> Prostrate yourself before your God so that he may make you a knight . . . a member of the great spiritual family that governs the destinies of this planet, a soldier of liberty and progress . . . who fights under the magnanimous banners of Jesus of Nazareth.

That same month, the spirit warned him that the struggle was imminent and ordered him to "read Mexican history . . . so that you may begin your work as soon as possible." Through sustained effort and abnegation, "1908 will be . . . the foundation of your political career. . . . The book you are going to write will be the means through which your fellow citizens must learn your worth."

To prepare that book, Madero, at the end of 1907, entered a state of growing mystical tension. "Advised" by the relentless José, he got up even earlier, went to sleep late, gave up (with great difficulty) the "sacred siesta," ate little, drank no alcohol, avoided idleness and people, and prepared a detailed program of reading that included all five volumes of the standard collection of historical writings: *Mexico Across the Centuries*. On October 30, 1908, his thirty-fifth birthday, Madero—his book almost finished—took a message from José in the tones of the New Testament: "You have been chosen by your Heavenly Father to carry out a great mission on earth . . . for this divine cause you will have to sacrifice everything material, everything of this world."

Around the middle of November, he recorded an even more important communication:

> Your triumph will be a most brilliant one and of incalculable consequences for our beloved Mexico. Your book will cause a stir throughout the whole Republic . . . it will truly panic G.D. [General Díaz]. . . . You have to fight a shrewd, false, hypocritical man. You already know the antitheses you must use against him: against shrewdness, loyalty; against falsity, sincerity; against hypocrisy, candor.

It was signed with two initials: B.J. The spirit of Benito Juárez![16]

After sending his book to press, Francisco Madero went into solitary retreat for forty days and forty nights in the desert near the ranch he called "Australia."[17] As the new year opened, he wrote a letter to his father, giving his reasons why the book would be published no later than the twenty-fifth of that month:

> Among the spirits that fill space, there are some who are greatly concerned about the evolution of humanity, about its progress, and each time an event of importance is about to happen in any part of the world, a great number of them take on bodies, in order to save one or another nation from the yoke of tyranny.

He was certain that he was one of these spirits. "I have been chosen by Providence. Neither poverty, nor prison, nor death frighten me."[18]

Madero entered national politics, not with a manifesto or a "cry" but with the unusual gesture of a book. The first edition of *La sucesión presidencial en 1910* (*The Presidential Succession of 1910*) sold out at once. Dedicated to the Constitutionalists of 1857, to independent journalists, and to the "good Mexicans who will soon be known to the world for their integrity and energy," *The Presidential Succession* can be summed up in two almost medical (or homeopathic) formulas: a diagnosis of the Mexican sickness and a prescription for its cure.

For Madero, the Mexican sickness—a natural consequence of the militarism that had devastated the country throughout the nineteenth century—was absolute power in the hands of one man. Such power could not support any genuine progress, nor was there any infallible man who could wield it with balance. The example—Madero said—is before our eyes: In 1905, the small country of Japan, fortified (in his opinion) by democracy, had humiliated the mouldering Russian empire. The book offered various historical examples of absolute power, but none as telling as the Mexican czar himself. It was pure poison to recall, in 1909, that Díaz had come to power waving the banner of "No Reelection." In 1871 Porfirio the Liberal soldier had proclaimed, "Let no citizen impose his authority and perpetuate himself in power . . . this will be the last revolution." But the truth was, Madero argued, that General Díaz—a moderate, honest, and patriotic man for the most part—had been obsessed from then on by a fixed idea: to win and to retain power at any cost. His minions inisted that he was the "necessary man, the good dictator," but the balance sheet on his thirty years of administration—in the ways that most mattered—placed the Porfiriato solidly in the red.

On the slim credit side, Madero acknowledged material progress, at the cost of freedom; an advance in agriculture, though not without importing grains; buoyant industry, though monopolistic and subsidiary; and unquestionably peace—at the price of sacrificing all political life. The liabilities, on the other hand, were "terrifying": the slavery of the Yaqui people, the repression of workers in Cananea and Río Blanco, illiteracy, excessive concessions to the United States, and a ferocious centralization of politics. Economic, social, and political wounds translated into something worse. For Madero, absolute power had corrupted the Mexican soul, inciting "a disinterest in public life, disdain for the law, and a tendency toward cunning, cynicism, and fear."[19]

In language that even an opponent considered "virilely frank and accessible to all minds," Madero proposed a remedy, the restoration of democratic practices and the political liberty that makes all men equal before the law. There had to be a return to the Constitution of 1857. Or else "we are headed for a precipice."

A national Democratic Party would have to be organized under the slogan "Valid Voting, No Reelection" (*Sufragio efectivo, no reelección*). Díaz could run in a free election, retire to private life, or, as a compromise, continue in the presidency for one more term—at the end of which he would be eighty-six years old—so long as he held genuine elections for the vice presidency and for at least some governors and legislators.[20]

On February 2, 1909, Madero sent his book to President Díaz. Accompanying it was a firm, respectful, and noble letter in which he offered Díaz historical immortality in exchange for democracy:

> To develop your political program, principally based on the preservation of peace, it was necessary for you to assume the absolute power that you call patriarchal. . . . The entire Nation hopes that the successor to yourself will be the Law. . . . If I have taken the liberty of sending this to you . . . it is because I hope to obtain some declaration from you which—when presented to the public and rapidly confirmed by your actions—will make the Mexican people understand that it is now time they made use of their civic rights.[21]

In some region of the Milky Way, "Raúl" and "José" might well have smiled in satisfaction.

Madero then began the greatest practical lesson in democracy that anyone had ever attempted in the history of Mexico. The secret of "the Apostle of Democracy," as people were now beginning to call him, was a simple one. In confrontation with the mystique of authority represented by Porfirio Díaz, he would unfold the opposite mystique—of freedom. "I am above all," he would frequently repeat, "a dedicated democrat."[22] The Antireelection Center was founded in Mexico City in May of 1909. The following month, the first issue of *El Antireeleccionista* came out, directed by two intellectuals, Luis Cabrera and the young lawyer and philosopher José Vasconcelos, destined to become a major figure in the intellectual and political development of Mexico. By then Madero had sold much of his property, at a considerable loss, to obtain cash. He would use it to finance much of the antireelectionist work and initiate a series of long trips around the country accompanied by a small entourage.

In a number of places, large and small, Madero founded an antireelectionist club. He traveled through his home state in September, when he received the good news that General Reyes (briefly proposed as a candidate for the presidency) had disappointed his supporters by accepting an appointment from Díaz to a minor military post in Europe.

At every stop on his travels—despite harassment by the authorities ranging from the arrest of his sympathizers to denying him hotel rooms—Madero was cheered by thousands. A few days before his party's first national convention, the central government moved against the economic interests of the Madero family on a number of fronts. It took over the Bank of Nuevo León but had to pull back because the local public would only accept money issued directly by that bank. It put pressure on Madero's grandfather, the founder of the dynasty, and then accused Madero himself of "unlawful transactions in rubber" and issued a warrant for his arrest. The order was never carried out, at least in part because of intervention by the architect of Porfirian economic progress, José Ives Limantour, who was an old friend of the Maderos.[23]

In April 1910 Madero at last presided over the convention of the Antireelectionist party. In a speech, he warned against electoral fraud: "Force shall be met by force!"

Madero had not wanted a revolution. His goal was peaceful, electoral, democratic change. But the day before the convention he had met with President Díaz, and the interview seemed to have totally changed his mind. He felt he was dealing with a "child or an ignorant and distrustful rustic." He decided that "you cannot get anywhere with him." When Madero asked for guarantees, Don Porfirio answered that "he should trust in the Supreme Court," to which Madero responded not with an argument but with "a burst of laughter." He then wrote to a friend:

> General Díaz seemed to me totally decrepit. I found none of the qualities observed by his interviewers . . . he did not seem to me imposing, capable, or really anything. I learned about all of his projects, including those for the next two or three years, while he knows nothing of ours. . . . He must have been convinced that he had not imposed himself upon me and that I am not afraid of him. General Díaz has finally understood that yes! there are citizens manly enough to stand up to him. Porfirio is no brave rooster, but nevertheless we shall have to begin a revolution to overthrow him.[24]

The growth of antireelectionism was dizzying, the meetings ever more dangerous and intense. In Puebla Madero was acclaimed by thirty thousand people; by ten thousand in Jalapa; in Veracruz he spoke of regaining the rights of individuals, the liberties of municipalities, and the autonomy of states. In Orizaba, where Porfirio's army had killed strikers at Río Blanco, he delivered—to an audience of twenty thousand workers—one of the defining speeches of his social politics, anchored in classical liberalism:

> You desire freedom, you desire that your rights be respected, that you be allowed to form strong organizations, so that united you

> can defend your rights . . . you do not want bread, you want only freedom, because freedom will allow you to win that bread.[25]

Madero was reinvigorating the liberal ideals that many Mexicans had fought for in the Wars of the Reform and the French Intervention. Some came to believe that he would put an end to taxes, prefects, and authorities. "The extraordinary part of those apostolic orations," his faithful friend Roque Estrada remembered, "was a tremendous enlightened sincerity and a profoundly felt faith in the cause."[26]

At the beginning of June 1910, he began what would be his last speaking tour. In Monterrey the government decided to arrest him. The action was stupid, counterproductive, and too late. It only made his apostle's halo glow more brightly. Don Porfirio may not have been directly involved in the decision, or, if he was, it was a sign of his diminished faculties.

From his prison in San Luis Potosí, Madero—in a feverish rhythm—continued to pour out letters. He inspired everyone with the same courage: "All of you may be sure that I will not weaken for one single moment." Nor did he weaken, even when Díaz was reelected at the beginning of July. So as to leave no legal expedient unexplored, his party submitted a huge and detailed memorandum on electoral fraud to the Congress. But the massive evidence died there.

Through the influence of his father with the governor of the state (and the posting of a substantial bond) Madero was allowed to move around the city on horseback during the day, accompanied by guards. On October 4 he galloped away from the guards to take refuge with sympathizers in a nearby village. They put him on a train heading north the following morning. He was then smuggled across the American border, hidden in a baggage car for the last stage of his journey by sympathetic railway workers (a group that would later play a key role in the Revolution). That same month, in San Antonio, Texas, he issued the *Plan de San Luis*. Madero had written it in prison, with the participation, among others, of Ramón López Velarde, who would become one of the great Mexican poets of the century. The document called for revolution, Madero to become provisional President, a general refusal to acknowledge the authority of the central government, the restitution of land to villages and Indian communities, and freedom for political prisoners: "Fellow citizens, do not hesitate, even for a moment! Take up arms, throw the usurpers out of power, recover your rights as free men!"[27]

The patriotic rhetoric did not cloud Madero's practical instincts. The Revolution that would begin on November 20, 1910, already had its plan of action, its formal representatives in each locality. At night Madero may have invoked friendly spirits, but in the mornings he wrote to New York asking for information about his stock market shares in rubber. The code was: 1 share = 100 Winchester rifles.[28]

On Sunday morning, November 20, 1910, ten men, including a guide, accompanied the leader of the Revolution to the border at the Río Grande. Madero's uncle Catarino was supposed to be waiting there with four hundred men. But when Madero arrived, he found nobody at all. Eventually his uncle appeared, with a force of only ten. It would be impossible to attack Ciudad Porfirio Díaz and begin the Revolution with just twenty men. Madero decided to go into hiding. A few days later he traveled incognito to New Orleans with his youngest brother, Raúl (who had been given the same name as the brother lost in 1887).

There had been uprisings "for Madero" in Chihuahua, Sonora, Tamaulipas, Coahuila, and Veracruz. But the exiles (including Madero) knew little about them, and the hoped-for revolution seemed to be a fiasco. But to Madero's mind, it could not fail. After that interview with Díaz, he had understood that a revolution was inevitable. His democratic campaign—essential to the success of the coming revolution—had then prepared public opinion for it, justifying the resort to arms. Madero had seen the enthusiasm of the crowds. The Revolution could not fail.[29]

The most extraordinary proof of his faith can be found in the letters he sent from New Orleans to Juana P. de Montiú (his wife's pseudonym). In one of them, he told her that he was sleeping well, that he never missed his siesta, that he read in the library, exercised at the YMCA, went to the opera, and felt confident about the outcome of the struggle. As never before, Madero needed to unite his strengths and his faith. He found the necessary power in his own character but also in a new reading of the *Bhagavad Gita*. In his long sessions at the New Orleans library, he took careful notes and titled them "Commentaries on the *Bhagavad Gita*." In the *Gita*, the god Krishna speaks for eighteen chapters to Prince Arjuna (who sometimes briefly questions or answers). Arjuna—who is the leading warrior of the Pandava brothers—hesitates to begin the final battle (of the Indian epic, the *Mahabharata*) against their cousins and enemies, the Kurus. He sees his relations and friends in the opposing army and hesitates at killing his kinsmen. To incite him to combat, Krishna speaks of the unreality of death, the transience of the world of the senses, the necessity for renunciation, and a paradox that particularly impressed Madero (as it later would Mahatma Gandhi when he led the long struggle for the independence of India):

> And in truth I say that action is superior to inaction. . . . Hear my words, O prince! in truth I tell you that whoever acts according to *dharma* [the proper order of the world, the duties of one's station in life], without desire for the fruit of his action, renounces his action at the same time as he performs it.

In his commentaries, Madero wrote that the paradox of action-inaction was solved by the key word: *duty*. "'Renouncing the fruit of our actions' should be understood to mean that we should not perform any meritorious action in view of the reward we expect from it but because we consider it to be our duty."

In a letter to his wife, Madero said that he had come to identify even more with Arjuna, his old pseudonym, and went on:

> I have the feeling that my life is NOT in danger. But if something happens, I will go to my grave satisfied that I have done my duty. I send you my Bhagavad Gita. Guard it carefully along with the notes I drew from it. . . . Know that I always carry your love within my heart.[30]

In February 1911 Madero finally entered Mexico with 130 men. Shortly after, he personally led an attack on Casas Grandes, where he took a wound in the arm. For a couple of months he directed the revolutionary operations, though not always in an orderly and coordinated manner. He knew a little about the struggle taking root throughout the Republic, but even more strongly he foresaw that it would spread. The localized revolutionary juntas on the United States border operated freely. The American government did little to stop the flow of arms. A pro-Madero lobby in Washington began to have its effect. The sabotage of telegraph lines and railroads hindered the movements of a federal army less fierce than it had been painted. Between February and March, the number of armed encounters doubled. By April, the insurrection had spread to eighteen states, including Morelos, south of Mexico City, where the leader was Emiliano Zapata. In New York, in Washington, and on the border, the Díaz government sat down to negotiate.

In an address on April 1, Porfirio struck a blow that, in other circumstances, in another age, and faced with a less fervent opponent than Madero, would have been masterly. He took up as his own the banners of the Revolution, including the "interesting" point of the restitution of lands to the dispossessed, and he completely replaced his long-standing cabinet. With their conditions fulfilled, no further reason existed for "sadly mistaken or perversely deceived Mexicans" to refuse to lay down

Francisco I. Madero and his wife, 1911
(FINAH)

their guns. But Madero did not retreat. A promise from the administration was not a "sufficient guarantee." He demanded the resignation of President Díaz and Vice President Ramón Corral.

The talks went on. The full revolutionary leadership signed a document detailing fourteen points that they required for an agreement. The items included pay for the revolutionary troops; the release of political prisoners; and the Revolutionary Party's right to name the Secretaries of War, Education, the Interior, Justice, Communications, and Public Works. Díaz's resignation was not mentioned but it was clearly understood to be necessary.

At that moment, Madero began to waver. He signed the document, repented on the following day, and soon afterward repented having repented. Faced with the possibility of Díaz's resignation, he sensed the moment of triumph. For the first time he felt the need to lead the country rather than the opposition. But Madero only understood leadership as magnanimity. He insisted that, even if General Díaz was asked to resign, it should be done "in a form that would not cause him pain, to see if in this way we can manage to avoid more spilling of blood."

On May 7, in a manifesto to the nation, Díaz admitted that the November rebellion "had gradually been spreading," and—a decisive move—implied that he might resign "when his conscience told him that . . . he would not be turning the country over to anarchy." Whatever may have been his intentions, the manifesto strengthened the revolutionary cause.

On the following day, before Ciudad Juárez, Madero hesitated again. He wanted a cease-fire, but his troops, under the command of Pascual Orozco and Francisco Villa, were uncontrollable. On May 10 the city fell to the revolutionaries.[31] Three days later, Orozco and Villa demanded that Madero execute General Navarro, the federal commander of the town. Madero refused, then quickly left his office. Minutes later, "covered with sweat and asking for a drink of water," he arrived at the house where he was staying. Dr. Vázquez Gómez asked him: "Where are you coming from, so agitated?" and Madero answered: "I have just taken General Navarro and his guards down to the riverbank. They wanted to shoot them. . . . I took them to the Río Grande by car and from there they crossed to the other side."[32]

On May 21, the Treaty of Ciudad Juárez that ended Madero's Revolution was finally signed. The President and Vice President would resign before the end of May. Díaz's Minister of Foreign Affairs, Francisco León de la Barra, assumed the interim presidency, for the purpose of calling general elections. The troops would be demobilized in each state so that peace and order could be restored. Four days later, Porfirio Díaz presented his resignation.

Madero's star had reached its zenith. The enthusiasm of the people seemed to predict every imaginable success for the Apostle of Democracy. On his way to Mexico City, he was cheered at the train stations. Wherever he passed, he heard applause, *vivas*, the pealing of bells, the explosions of rockets.

On the morning of June 7, 1911, a strong earthquake shook Mexico City. "When Madero arrived/even the earth trembled" (*cuando Madero llegó/hasta la tierra tembló*) a folk poet would sing in a ballad.[33] In the afternoon Madero made his triumphal entry. He was received by a hundred thousand euphoric people, a fifth of the capital's total population. Two magic words painted on the walls and imprinted on everyone's mind summed up the moment: "Viva Madero!" The wife of a foreign diplomat preserved this image of the thirty-seven-year-old liberator in her diary: "Madero has a pleasant and spontaneous smile. There is something in him of youth, of hope, and of personal goodness."[34]

But in the end, that fiesta of freedom would be deceptive. Within the victory itself, for Madero, were the seeds of defeat. He had spent his entire political life fighting against absolute and personal power, promoting democracy and freedom understood as the absence of coercion, as equality before the law. To the man whose fixed idea was to be free of power came the necessity now of exercising power. He faced a dilemma like that of Morelos nearly a hundred years earlier (Madero had discussed it in *The Presidential Succession*): whether to remain a military caudillo or to sanction a power outside himself. Like Morelos—who had died because he gave precedence to the Congress of Chilpancingo—Madero yielded primary power to the Constitution of 1857.

His duty—his dharma, as he might have said—had been to free the people of Mexico and give them the chance to govern themselves. At this stage of play, it was not his move: It was the nation's turn, it was the turn of every player in the democratic chess game: judges, legislators, governors, journalists, the voters of the capital and of every state and town. He had been uncompromising as a liberator; it was right (in his eyes) that he should be liberal in power. Consistent with his vision of the world, he had "reestablished the supremacy of the law." He had done what he had urged Porfirio Díaz to concede. He had chosen "the law" as the worthiest successor to the dictator. It was up to the people now (and especially the political classes) to do their part.

Luis Cabrera later wrote about Madero: "The first duty of a surgeon is to clean out the rot before he closes the wound." But Madero was an apostle, not a surgeon. The only imposition he would permit himself was never to impose. He put all his "faith in the capacity of the people to govern themselves with serenity and wisdom." To be an autocrat like Porfirio—even just a tinge of it or under some disguise—would have seemed to him (if he ever considered it) moral suicide.

He made two fatal political mistakes. His first was allowing Francisco León de la Barra to serve as interim President and permitting Porfirian "representatives"—formally "elected" in 1910 but of course chosen by Díaz—to continue sitting in the Legislature. These concessions left partially in place a "Porfirismo without Don Porfirio," which would be hostile to Madero. His other grave error was demobilizing the revolutionary army. It dispirited the men and seemed to deprive the Revolution of some of its legitimacy (and Madero of important military support).

According to the German ambassador, Von Hintze, the goal of interim President Francisco León de la Barra was to undermine the legitimacy of the future Maderista regime.[35] His greatest achievement was the rupture between Madero and Emiliano Zapata, leader of the peasant revolution. The Caudillo of the South had trusted that Madero would carry out his commitment—in the Program of San Luis—to restore lost lands to the villages. Madero, after sending his troops home, was trying to solve the problem in a slow, studied, and peaceful way. Halfway through 1911, Madero went to Cuernavaca and Cuautla to meet with Zapata.

There he was received with affection and with hope. Madero urged Zapata "to continue to trust me as I trust you," and Zapata answered, "I will always be the most faithful of your followers."

But any agreement between those two was an explosive mix for those who wanted a return to the Porfiriato (under some other Porfirio). The landowners and the governor of Zapata's state of Morelos appealed to President de la Barra. In the Senate and the Mexico City press there were hysterical outcries about "the ferocity of the Attila of the South." But the real Attila was in command of the federal troops stationed in Morelos. His name was Victoriano Huerta.

Zapata still had most of his men under arms. Madero made a formal request to De la Barra for "ample powers" so that he could personally persuade Zapata to send his soldiers home. Three days later, Madero returned to Cuautla. As a sign of good faith he brought his wife with him. From there, he sent word to the President that the demobilization would begin as soon as certain reasonable demands were met. Zapata's most important requests were that Raúl Madero become commander in chief of the army in Morelos and that Huerta's troops be pulled back to Cuernavaca.

Madero felt that it was "difficult to overcome their distrust . . . which seems justified given the attitude assumed by General Huerta, who without express orders is advancing on Yautepec."[36]

Not only the Zapatistas rejected Huerta and had no faith in De la Barra. Madero's own mother, Doña Mercedes González Treviño, wrote to him during those days:

> This is to tell you . . . that you should get rid of the federal troops. Don't just keep on contemplating things. Impose yourself a little on that De la Barra because, if not, we are going to have to fight. . . . Huerta . . . must be removed . . . [He is] arranging the counterrevolution.[37]

Huerta marched on, attacking Yautepec, firing at the president of the town council when he tried to wave a white flag. De la Barra sent a message to Madero (on August 20, 1911) saying that the federal advance would be halted, if Zapata disbanded his troops within twenty-four hours. Madero insisted that Huerta be removed and defended Zapata: "The information you have received on the excesses of Zapata is greatly exaggerated. . . . I know what is being said about Zapata in Mexico City and it is not accurate. The hacendados hate him because he stands in the way of their continuing abuses and threatens their undeserved privileges." De la Barra, "the White President," mused that it was "truly disagreeable to deal with an individual [Zapata] from such a background." He insisted on the demobilization and defended Huerta as "a loyal and upstanding soldier."[38]

On August 25 Madero gave up arguing. He wrote a bitter letter to De la Barra. The President had been wrong to send Huerta, a man who "provokes hostilities instead of calming them," into the state of Morelos. He insisted that Zapata's requests be met as soon as possible: "If you do not fulfill what I offered in your name, with the agreement of your Council of Ministers, I will be held up to ridicule and, not only that, they will think that I came to betray and deceive them and I cannot resign myself to that."[39]

He left Morelos promising Zapata that the demands would absolutely be met when he, Madero, became President. But although he threatened in private, he did not attack De la Barra in public. Unfortunately for him, he was then doubly branded by the Zapatistas: He had not won them their demands, and he had "come to betray and deceive them." Only in this way can we understand why the Zapatista movement broke with him so quickly, in November of 1911, just a few days after he would finally take office.

But the erosion of his political prestige had continued through the four long, ambiguous months of the interim government. The winner of the Revolution had—for a time—renounced the rewards of his victory, and through his renunciation, he himself became the first to undermine the Revolution. It did him no good to justify his action on the basis of the constitutional legitimacy he thought he was protecting. Most of the revolutionaries did not agree with him. They felt confused, disillusioned, even betrayed. The old regime, maintained almost intact, saw a chance to fill the vacuum and prepared its forces to reverse the Revolution at its own chosen moment. Even before he became President, Madero had failed in the eyes of enemies and allies.

All the strength and wisdom he had put into the Revolution seemed to turn against him when the moment of command came. From his letters to De la Barra, it is clear that Madero knew every movement his enemies were making across the political map, but he believed that he could gradually undermine their influence by displaying the serene outlines of his own spiritual map. In 1911, he published, under his new pseudonynm of "Bhima" (one of Arjuna's brothers in the *Mahabharata*), a *Spiritualist Manual* in which he discussed politics purely as a product of morality: "There can be no doubt that if all good men were to cast off their egoism and get involved in public affairs, the people would be governed wisely and the most worthy and virtuous men would be those occupying the highest posts; and it follows that these men would then do good and accelerate the evolution of humanity." [40]

Strangely, poignantly (and surely unknown to Madero), Bhima in the *Mahabharata* is the archetypal representation of brutal, thoughtless violence in behalf of his just cause (and a relentlesss avenger, who swears and carries out a vow to drink the blood of a mortal enemy). One tiny touch of the real Bhima might have saved Madero and Mexico from a whole world of cruelty. But for Madero, it was not him but Providence itself that had triumphed. It was the time to decree perpetual peace, constitutional order, and universal brotherhood. And his term in office would sometimes seem like a party of one at which the President "is always smiling, invariably smiling." Porfirio had lost. He, Madero, had won. His virtue and his goodness were now at the service of humanity.

Madero became President in November 1911, through the freest election in Mexican history. He governed for fifteen months with so many problems that, in retrospect, his term seems like a near-miracle of survival.

First of all, he was no diplomat. He governed by feeling, not calculation. He always appeared to withdraw from reality or to pass beyond it. The cabinet he formed was unstable and inefficient, made up of antagonistic elements united in the name of some ideal reconciliation. The Senate—still with the same membership as under Díaz—tenaciously opposed him, discrediting and paralyzing his attempts at reform. From 1912 on, the Lower House had a Maderista majority but it was nevertheless dominated by venomous oratory against the executive. There were governmental problems in eleven states. Yet nothing was so irresponsible (and persistent) as the assaults of the press. The newspapers of Mexico City poured out jokes, nicknames, cartoons, rumors:

> President Madero was criticized for being short; for not having grave gestures and a hard gaze; for being young; for loving and respecting his wife and parents; for giving speeches; for being a vegetarian; for not being afraid; for greeting Emiliano Zapata with an embrace . . . for not being an assassin; for studying Spiritualism and being a Mason; for going up in the air in an airplane; for dancing; and of course for choosing Pino Suárez as his vice president.[41]

It was a cruel paradox that the press—for whom freedom of speech is both its support and its reason for existence—would, throughout Madero's presidency, implicitly request a return to Porfirian silence. Some of his people suggested that the President revive the *Ley Mordaza* (the Gag Law), but Madero always refused to put any limits on the freedom of the press, "so necessary"—he had said in August 1911—"for it to fulfill its lofty mission." Gustavo, Madero's influential brother, affirmed that "the newspapers bite the hand that took off their muzzle." Madero—in this respect always consistent—chose to maintain the democratic ideal in the hope that impure reality might some day rise to its level.

Beyond political opposition, Madero had to confront—aside from the Zapatistas—three especially serious rebellions: by Bernardo Reyes, Pascual Orozco, and Félix Díaz. In an interview with Madero in July 1911, Reyes had promised to contend for power using democratic means, but his subversive maneuvers soon became an open secret. On December 14, 1911, he crossed the northern border into Mexico. But for him it was late. During the first decade of the century, a broad sector of the country would have responded as one man to Reyes, but after his repeated and, at times, undignified submission to Díaz, and later to the Maderista triumph, no one supported him. Eleven days after his frustrated rebellion, the former proconsul of the northwest surrendered in Linares, Nuevo León. Porfirio Díaz, the devotee of authority, would have had him shot; Madero, the devotee of liberty, locked him up in the Santiago Tlatelolco prison of Mexico City.[42]

In March 1912, the rebellion of Madero's former general Pascual Orozco broke out in Chihuahua, a revolt with no other platform than his own resentment and with no greater popular support than that of Don Luis Terrazas, the largest landowner in the country. In an early battle, the rebels defeated the federal army under the command of General José González Salas, who, fearing the barbs of the press, committed suicide. He was replaced by Victoriano Huerta, who defeated Orozco's *colorados* in three major battles. In September 1912, Orozco fled to the United States, but Huerta could not savor his victory. He had quarreled with President Madero about the supposed insubordination of Francisco "Pancho" Villa. Huerta had ordered Villa's execution, but Madero commuted the sentence. On September 15, the evening before Mexico's Independence Day, Huerta, drunk as usual in a cantina (*El Gato Negro* in Ciudad Juárez), bragged to his officers: "If I wanted to, I could make an agreement with Pascual Orozco and I would go to Mexico City with twenty-seven thousand men and take the presidency away from Madero." Told of Huerta's boast, General Ángel García Peña, the new Minister of War, stripped him of his command.

But only a few days later Madero promoted Huerta to major general. On this occasion he was not moved by kindness. Though Huerta had often deceived him, he felt a need to win the general over. Perhaps this gesture would finally convince Huerta of Madero's goodwill.

In October 1912, a revolt led by the "uncle's nephew," as Félix Díaz was known, broke out in Veracruz. His program was as reactionary as his last name: "to reclaim the honor of the army trampled by Madero." A few days later, Díaz had to surrender and was locked up in the prison of San Juan de Ulúa. This time Madero thought he might have to order an execution. But then the press and "high society" mobilized to defend "uncle's nephew." They called Madero a despot and a tyrant for daring to think of executing "poor Félix" (as Porfirio tersely said in Paris, when he heard of the abortive attempt). Madero still seemed disposed to carry out the execution, but democracy, against which Félix had risen, saved his life through the separation of powers. The Supreme Court abrogated military jurisdiction over the case, and "the nephew" ended up being sentenced to prison. And so, once again, Madero failed before both his allies and his enemies. Friends of his government called him weak and indecisive. Díaz's supporters did not even give him credit for his mercy.

In spite of having overcome these and other rebellions, Madero by the end of 1912 found himself politically isolated. When Cuban ambassador Manuel Márquez Sterling had arrived in Mexico, he was told:

> You have come at a bad time . . . and soon you will have to see the government in pieces and Madero perhaps on a ship to Europe. He is an apostle whom the upper classes despise and the lower classes distrust. He has deceived us all! He does not have an ounce of energy . . . he has the mania to proclaim himself a great democrat! Don't shoot . . . ! Do you think that a President can govern if he does not shoot, if he does not punish, if he does not make himself feared, if all he does is invoke laws and principles? The world is all lies. How can Madero claim to govern us with the truth? If within the Apostle there were a Don Porfirio, hidden and taciturn, Mexico would be fortunate.[43]

In reality, the global situation within the country was much less alarming than the picture painted by Márquez Sterling's nostalgic informant. The Mexican people, who had poured themselves out in arms and in votes for Madero, had not welcomed any of the rebellions. Even in the sullen domain of Emiliano Zapata, the peasant revolution was giving ground before the humanitarian and democratic policies of Felipe Ángeles, the new military commander of the area. Business was going on as usual in the country, bank assets were growing as was external trade, yet little by little reality was being infected by rumors, distortions, and an atmosphere of mistrust artificially created by the press. In 1912, during Orozco's rebellion, United States ambassador Henry Lane Wilson was behaving on the basis of this kind of hysteria when he urgently asked his government to evacuate American "refugees" by sea. Though somewhat doubtful of Wilson's warnings, the U.S. State Department sent a boat, the *Buford,* to the coast of Sinaloa with a crew of five hundred. The "refugees" from this "anarchic country" came to a total of eighteen. The incident triggered the ridicule of the *London Times*: "The only refugees collected so far seem to be people wanting to travel gratis to San Diego."[44]

The hysteria also concealed the achievements of the regime. A month after becoming President, Madero had created a Department of Labor and sponsored the first convention of the textile industry, which regulated and humanized work in the factories. He was the first Mexican President to legalize labor unions and the right to strike. During his term, the Casa del Obrero Mundial (the "House of the World Worker") was created, an organization with anarcho-syndicalist connections that soon came to play a fundamental role in the consolidation of the Mexican labor movement.

Madero's experience as a highly capable agricultural administrator led him to take a number of initiatives in agrarian policy. He reorganized the system of rural credit, created experimental agricultural stations, planned the conservation of forest resources. And the idea of achieving agricultural reform through expropriations and the restoration of land to agricultural communities was legitimized during the Madero era.

There were real achievements in other areas of social and economic policy: the creation of industrial and elementary schools, new railroad concessions in the southeast, work on new and important highways. The policy of safety inspections was instituted for roads and highways and bridges, while the foreign oil companies had to accept new taxation policies.[45]

But though there were many material changes, the most important were political. Madero was scrupulously respectful of the separation of powers. He never interfered with the courts, and he was in favor of the broadest possible representation in the legislature (where for a short time a Catholic party had a voice and a vote). Universal, direct voting was introduced everywhere. He wanted to give a "political personality back to municipal governments" and initiated the firmest and most clear-cut decentralization policies in recent Mexican history. He always took great pride in the consistency between his original revolutionary program and his actions as President: "The Federal Executive . . . has respected the law . . . the bitterest enemies of the Revolution, those who fight it in the political arena, must confess that thanks to this movement they condemn today, they can exercise rights consecrated by the Constitution that could rarely be exercised in earlier periods."[46]

All of these achievements were real, but they were accomplished by a man with a mysterious incapacity for the art of politics, for the clockwork of ends and means. The very force of his personal coherence was to prevent him—like other apostles throughout human history—from achieving the goal he fervently represented.

The period between the beginning of the coup against Madero and his eventual death are remembered as the "Tragic Ten Days." It is part of the nature of an apostle that his Calvary may be better remembered than his work, though in a certain sense, perhaps, his Calvary *is* his work. The *Decena Trágica* is the best-known episode of Madero's life and movement. All the central images are engraved in Mexican memory.

On Sunday, February 9, 1913, General Manuel Mondragón—with a group of armed men—left military headquarters in Mexico City on his way to the prison of Santiago Tlatelolco, where he freed Félix Díaz and

Bernardo Reyes. Meanwhile, the young cadets of the Military College, by order of the conspirators, seized the National Palace in the Zócalo. But General Lauro Villar, a Madero loyalist, passionately harangued them, and the boys surrendered their position.

Unaware of this turn of events, General Reyes thought he could freely enter the National Palace and was shot dead at the gate. But unfortunately for Madero, Villar was wounded in the exchange of gunfire. Madero rode down on horseback from his residence at Chapultepec Castle, escorted by cadets from the Military College. Díaz and Mondragón then took control of the Ciudadela (the city's main military garrison), where there was enough ammunition for them to resist a long siege.

Madero yielded to Victoriano Huerta's pleas and pathetic professions of loyalty and named him commander-general of the city in place of Villar. The capital lived through days of turmoil and anguish. By February 11 there were more than five hundred dead and wounded. A continual exchange of artillery fire between federal troops and rebels threatened all life on the streets.

But observers began to notice something strange. Huerta sacrificed men, but he avoided trying to capture the Ciudadela. Díaz and Mondragón were also willing to lose men, but their shells never damaged any key targets. Few knew that a secret arrangement had been forged under the protection of the United States ambassador, Henry Lane Wilson.[47]

The ambassador, who had previously served in Chile and Belgium, had been assigned to Mexico in 1910 and immediately established excellent relations with Porfirio. Lane Wilson had always hated Madero. His reports to the U.S. State Department are consummate fabrications of arrogance, calculated lies, and hysteria. President William Howard Taft distrusted him. Nevertheless, the ambassador moved from a campaign of slander to direct intervention. On that day of February 16, he wrote to his German counterpart, Von Hintze:

> General Huerta has, since the beginning of the Díaz uprising, been in secret negotiations with Díaz; he would come out openly against Madero if he were not afraid that the foreign powers would deny him recognition. . . . I have let him know that I am willing to recognize any government that is able to restore peace and order instead of Mr. Madero's government, and that I shall strongly recommend to my government that it grant recognition to any such government.[48]

Ambassador Wilson was at the very center of the plot. He turned part of the diplomatic corps against Madero, floated (on his own responsibility) unfounded threats of military intervention, and warded off any possible armistice. To him, Madero was a "fool," a "lunatic," whom "only resignation could save." "The situation"—he commented to the Cuban ambassador—"is intolerable: I will put it in order." And he had to do it soon. On March 4 Woodrow Wilson would become president of the United States. His strong democratic convictions would make him very unlikely to support a government created through a military coup (*golpe*). The balance could then tip in favor of Madero.[49]

For his part, Madero remained confident. He continued to be, above all, a man of faith. Besides, by February 16 he had received a telegram from President Taft that conveyed his concern but officially discounted any danger of intervention. A few days later, with the telegram in hand, he responded to the senators who had—like the diplomats—unsuccessfully asked him to resign. "It does not surprise me that you come to demand my resignation, senators named by General Díaz and not elected by the people! You consider me an enemy and would take pleasure in my fall." But he was not prepared to resign: "I would die, should it become necessary, fulfilling my duty."[50] It was then that he confided to his friend José Vasconcelos: "After this blows over, I will change the cabinet . . . the responsibility will now fall upon you young people. . . . You will see, this will be resolved in a few days, and then we will renew the government. We must triumph, because we represent the good."[51]

On February 17, his brother Gustavo discovered by accident that Huerta was negotiating with Félix Díaz. Gustavo personally took Huerta prisoner at gunpoint, and the general, at two in the morning, was brought before Madero. The President listened to Huerta's entreaties, as he denied his involvement in the plot and promised to arrest the rebels within twenty-four hours. It was a key moment. And Madero made a suicidal decision. In spite of Huerta's previous commitments to Porfirio Díaz and Bernardo Reyes, in spite of the disrespect and mockery Huerta had shown him in Morelos in 1911, despite the fact that his own mother had warned him against the "counterrevolutionary" Huerta, despite the arrogant threats of Huerta at Ciudad Juárez, despite rumors that Huerta had earlier met with Félix Díaz in the pastry shop *El Globo,* despite—at that very moment—the confirmation of his arrangements with the rebels, Madero freed Huerta, personally returned his pistol (at Huerta's request), and granted him the twenty-four hours he requested to demonstrate his loyalty. He then reprimanded his brother Gustavo "for being carried away by his impulses."

Why did Madero listen to Huerta? Perhaps, as Vasconcelos believed, because on the eve of an unjust defeat, saintly men can fall victim to a kind of paralysis.[52] Perhaps as a challenge to the Providence that had always smiled upon him. Or to turn the other cheek, or to love his enemy, or perhaps to perform his first open and deliberate act of sacrifice. The answer is not to be found within the world of politics.

Huerta and his ally, General Blanquet, closed their circle of treason. On February 18, Blanquet arrived at the National Palace with a squad of men and arrested the President after a bloody shoot-out. Madero slapped him in the face and shouted at him, "You are a traitor!" Blanquet responded calmly, "Yes. I am a traitor."[53] Meanwhile, Huerta had invited Gustavo Madero to a meal in a downtown restaurant. There he asked, in an offhand way, to see Gustavo's gun, and when Madero's brother handed it to him, he pointed his own weapon at Gustavo's heart and told him that he was now a prisoner.

Gustavo—who had one glass eye that had earned him the nickname of *Ojo Parado* ("Still Eye")—and the quartermaster general of the National Palace, Adolfo Bassó, were taken to the Ciudadela. The Cuban ambassador, Manuel Márquez Sterling, in his book *The Last Days of President Madero*, describes what happened there:

> Jeers, insults, angry shouts mark their arrival. An individual named Cecilio Ocón is the judge who interrogates the defendants. Gustavo rejects all the accusations of his enemies and invokes his privileges as a legislator. But Ocón, after condemning him, along with Bassó, to execution, slaps Gustavo brutally: "This is how we respect your privileges!" Félix Díaz intervenes and they lead the prisoners into another section of the Ciudadela. But the mob of soldiers, full of courage, follows them in a frenetic, screaming chorus. Some of them mock Gustavo, others swing their iron fists against the defenseless man of politics and they torment him and they provoke him. Gustavo tries to strike out at the worst of them. And a deserter from the twenty-ninth battalion, Melgarejo . . . pierces Gustavo's only good eye with his sword, blinding him at once. The mob breaks into savage laughter. The disgraceful spectacle has amused them. Gustavo, his face bathed with blood, weaves and staggers, groping his way; and the ferocious audience accompanies him with bursts of laughter. Ocón takes him to the room where he is going to be shot. Gustavo, concentrating all his energies, pulls away from the murderer who is trying to force him along. Ocón,

rabid, tries to grab him by the lapel of his coat. But his adversary is stronger than he is. The pistol finally ends the fistfight. More than twenty barrels discharge against the dying martyr who shudders out a final sigh, on the floor. "He is not the last patriot," shouts Bassó. "There are still many brave men behind us who will know how to punish these infamies!" Ocón, with his clouded gaze and unsteady walk, points a finger and says, "Now, that one."

The old sailor, ramrod straight, walks to the place of his execution. One of the executioners tries to put a blindfold on his eyes. For what? "I want to see the sky," he says, in a strong voice; and raising his face toward the infinite spaces, he adds, "I can't find the Great Bear . . . Ah yes! There it is, glittering . . ." and then, saying his farewell: "I am sixty-two years old. Let it be remembered that I died like a man." He unbuttoned his overcoat to show his chest and he gave the order, "Fire!" as if he wanted to overtake Gustavo on the threshold of another life, beyond the Great Bear.[54]

With the President and Vice President José María Pino Suárez imprisoned, Ambassador Wilson lost no time in concluding the *Pacto de la Embajada* (the "embassy agreement") with Huerta and Díaz, according to which both would in succession assume the presidency of the Republic. In the words of the German diplomat, "Ambassador Wilson planned the coup. He himself boasts of it." When he learned of Gustavo's murder, Pedro Lascuráin, Madero's Minister of Foreign Relations, volunteered to obtain the resignations of Madero and Pino Suárez. Madero (ignorant of Gustavo's death) thought that by accepting he would stop the bloodbath and save his family from any harm. Serenely he wrote out his resignation. It was his first and last gesture of weakness, an understandably human action and only for the moment not that of an apostle.

According to the Constitution, if the President were to resign, the Foreign Minister was to become interim President. Lascuráin served as President for forty-five minutes and then resigned in favor of Huerta, who could thus claim to have preserved constitutional forms.

"Huerta will not keep his word," Madero said to Márquez Sterling. The train that was supposed to take him to Veracruz, there to meet a cruiser that would carry him to exile in Cuba, "will not leave at any hour."[55]

Despite the pleas of Señora Madero, Henry Lane Wilson did not lift a finger to save him. On February 19, the ambassador wrote to Washington: "General Huerta asked my advice about whether it would be better to send the ex-president out of the country or to place him in a mental asylum. I replied that he should do what would be best for the peace of the country."[56]

Suspecting he might be slaughtered, though still unaware of his brother's death, Madero had the fortitude to joke with Márquez Sterling on the night of February 21 when the Cuban visited him in his captivity. That night the ambassador saw him sleep "a sweet sleep."

The following night would be his last. General Felipe Ángeles was with Madero and Pino Suárez in the quartermaster's room of the National Palace, from which they were taken to be murdered. Based on Ángeles's testimony, Manuel Márquez Sterling wrote:

> That afternoon of the crime, the government had set up three camp beds in the jail, with mattresses, an addition meant to deceive them into believing they would be there for a long time. By now, Madero knew of Gustavo's martydom and, in silence, he contained his grief. Around ten at night, the prisoners went to bed. . . .
>
> Don Pancho, wrapped in his blanket—as Ángeles reports—hid his head. The lights went out. And I think he was crying for Gustavo.
>
> A few minutes later, an officer named Chicarro came in with Major Francisco Cárdenas and ordered Madero and Pino Suárez to accompany them to the Penitentiary. His face streaked with tears, "Don Pancho" embraced the faithful Ángeles and got into the car that would carry him to his death.[57]

Several months later, a representative of the British Foreign Office sent his government a detailed account of the assassinations. At five in the afternoon on that day, a British citizen with a car-rental business had received a telephone message from the rich Mexican landowner Ignacio de la Torre, General Porfirio Díaz's son-in-law. The message said that the Englishman should send a big car to De la Torre's house as soon as possible. The order was filled, the car driven by a Mexican chauffeur. After a long wait, he was told to take the car to the National Palace, and at 11:00 P.M., Madero and Pino Suárez were brought out and put inside. The car was escorted by another vehicle in which there was a unit of *rurales* under the command of a Major Cárdenas. For some months, this officer had headed a group of men assigned to protect the hacienda of Sr. Ignacio de la Torre, near Toluca. The major was a man who felt much admiration and warm personal affection for Porfirio Díaz, and he had sworn to avenge his overthrow.

The automobiles drove along a winding road toward the penitentiary and went past the main entrance to the farthest end of the building, where they were ordered to halt. Some shots were fired, which went over the roof of the automobile, and Major Cárdenas ordered the two prisoners out of their car. When Madero stepped out, Cárdenas put his revolver to Madero's neck and killed him with a single shot. Pino Suárez was taken to the wall of the penitentiary and there executed.[58]

An unconfirmed legend says that when he left the National Palace, Madero was carrying his "Commentaries on the *Bhagavad Gita*." Did he find any consolation in the mystique of detachment that Krishna had preached to Arjuna? Or was his final "fall" beyond his understanding?

Porfirio Díaz had always feared a revolution that would arouse Mexico's violent past. "We must not awaken the tiger." This horrible crime woke the tiger to a violence seen before only during the War of Independence. The old social and economic grievances of the Mexican people had of course precipitated the Revolution. But during the long, painful, and revelatory civil war that was about to begin, there was—aside from the drive for revenge—an element of historical guilt for not having prevented the sacrifice of Francisco Ignacio Madero.

The young philosopher Antonio Caso, who helped to carry his casket, first called him San Francisco Madero. A time was coming, as had happened before in history, when a society would grow and mature bearing the burden upon its shoulders of the death of a just man.

11

EMILIANO ZAPATA

The Born Anarchist

The land will be our own possession, it will belong to all the people—the land our ancestors held and that the fingers on paws that crushed us snatched away from us.

—MANIFESTO ISSUED IN NÁHUATL BY EMILIANO ZAPATA, 1918

Even before the Spaniards came to Mexico, the small district of Anenecuilco, in the heart of the present state of Morelos south of Mexico City, had already felt the burden of economic subjugation to centralized power. It is listed in the Mendocino Codex among the settlements paying tribute to the Aztec Empire. After the Conquest, Anenecuilco ("Place Where the Waters Swirl") became a village, and in 1579 it was forced to defend its independent status—which it did well—against attempts by the Marquisate of the Valley (Cortés's immense territory) to absorb it into other units or compel the people to work on projects unconnected with their village. The identity of Anenecuilco was threatened again in 1603, when the authorities tried to dissolve the village along with two neighboring communities into a single entity. Ahuahuepan and Olintepec, the other villages, yielded and disappeared. Anenecuilco did not.[1]

In 1607 the Spanish viceroy Luis de Velasco assigned the village a grant of land. Later in the same year the newly constituted Hacienda del Hospital took that land away. It was a loss the village would never forget.

During the seventeenth and part of the eighteenth century, Anenecuilco barely survived. In 1746 it had only twenty families, who defended their legally inherited lands from a triple menace: the haciendas of Cuahuixtla, Hospital, and Mapaztlán. In 1798 the village formally requested more land and protested the ruling of the royal Audiencia in favor of the hacendado Abad, owner of the Cuahuixtla estates. By the end of the century its population had grown. The 1799 census registered thirty-two Indian families "with everything including a village chief." A witness recorded, in 1808, that the Indians of Anenecuilco—some of whom bore the last name Zapata—"leased lands from the Hospital Hacienda because theirs were not sufficient for their needs." The villagers kept going to court, asking for that grant they had received and lost in 1607, requesting other land and water rights taken from them by the haciendas. But the colonial period ended without any decisions.[2]

Anenecuilco made its tiny contribution to the War of Independence. Its small church sheltered and saved the life of one of the Insurgents closest to José María Morelos—Francisco Ayala, who had married a woman from the village with the maiden name of Zapata.

The pueblo reopened its case against the Mapaztlán Hacienda in 1853, when the people again requested documents from the National Archives. In 1864, they turned to the Emperor Maximilian and directly asked him for their lands. Maximilian, sensitive to protests against the often troublesome haciendas, ruled in favor of Anenecuilco. Unfortunately for the village, the empire disappeared.

In 1874 the village chief of Anenecuilco, a man named José Zapata, wrote to the hero of the War of the French Intervention, General Porfirio Díaz, who had visited the region:

> The sugar mills are like a malignant disease that spreads and destroys and makes everything disappear so that, with an insatiable thirst, it can take over land and more land. You saw this when you visited us . . . and you promised to fight and we believe, or rather we are certain, that you will do it. . . . This is only a reminder, so that you will keep this problem in mind and not forget. . . . "We will not rest until we get what belongs to us." These are your own words, General.[3]

Two years later, the government of Lerdo de Tejada overthrown and the Porfiriato launched on its long life, President Díaz—who seems to have responded politely to one or more earlier letters—received another, even more hopeful and assertive communication from the people of Anenecuilco:

> Since we have regularly been sending you letters to remind you, we believe you have not forgotten us, though your last answer was on the 13th of January of last year, we know it is because you have so much to do. . . . We hope and trust that someday justice will take charge of our problems, with devotion we hold on to the papers that someday will prove that we are the only true owners of these lands. . . . We are not criticizing, we only want to be sure you have not forgotten us. . . . The sugar mills [are] more and more tyrannous and heartless. We do not want to commit any violent act against them, we will wait with patience until you give us the signal to begin our fight.[4]

The letter also mentioned that José Zapata had died. For his only response (there was none in practical action), Díaz wrote on the blank margins of the letter: "Answering you in the same terms as always. I am with you and will help you to the utmost. I am sorry about the death of Señor Zapata, who was a faithful servant and an able friend."

In 1878, Manuel Mendoza Cortina, the owner of the Hacienda of Cuahuixtla, affirming that "justice for the poor has already gone off to heaven," made another move to dispossess Anenecuilco, this time of their water. One of the village leaders, Manuel Mancilla, began talks with him, in secret, trying to reach a mutual agreement. When they discovered what was going on, his neighbors cut off his head. They threw the corpse on the road, near the Hill of Flints (*Cerro de Pedernales*). And he was buried as a traitor outside the boundaries of the pueblo.

For Anenecuilco, the Porfirian regime was no peaceful interlude. In 1883 the residents taxed themselves to buy guns. In 1887 the eastern end of the village was destroyed by their archenemy, Mendoza Cortina. In 1895 the Spaniard Vicente Alonso Pinzón, the new owner of the Hospital Hacienda, occupied the pueblo's grazing lands, killed some of their animals, and put up barbed wire fences.

At the beginning of the century, Anenecuilco, for the umpteenth time, took the legal road again. Yet another request was sent off to the National Archive for copies of their title deeds, and they requested the opinion of Francisco Serralde, a famous lawyer. After closely examining the deeds, Serralde wrote: "The deeds fully cover the 600 varas of land given to the natives of Anenecuilco by decree and by law." In 1906, offering the opinion as evidence, the villagers appealed to the governor of the state, who called a meeting (which led to nothing) with representatives of the Hospital Hacienda.

A year later, Anenecuilco was still waiting. When Porfirio Díaz paid a visit to the nearby hacienda of Tenextepango, owned by his son-in-law Ignacio de la Torre, a village delegation went to see him. Forty years had passed since they had written him their hopeful letters. Porfirio promised yet again that he would intercede. The governor warned them that they needed "relevant title deeds" to validate "their alleged rights." Nothing happened.[5]

One of the most politically active villagers would remember that 1909 was "the heaviest year." In June the administrator of the Hospital Hacienda decided to make a decisive move (unaware he was stepping toward the edge of the precipice): He refused even to rent land to Anenecuilco. In September the new village leader, a young man named Emiliano Zapata, carefully studied the community records, many of them in Náhuatl, the language of his ancestors (for which he used a translator). Again the village, in his person, went to a lawyer for support and a local social activist for advice in preparing a claim. The hacienda administrator answered Zapata with words that would take on a tragic celebrity: "If the people of Anenecuilco want to sow their seed, let them sow it in a flowerpot, because they will get no land even on the barren slope of a hill."[6]

In April 1910, the tone of a letter from Anenecuilco to the governor of Morelos was not combative but pleading, almost imploring: "We are willing to recognize whoever the owner of said lands may turn out to be, whether it is the pueblo of San Miguel Anenecuilco or someone else, but we wish to sow on said lands so as not to suffer, because the sowing is what gives us life, from which we draw our livelihood and that of our families."[7]

In May the village played its final card, a new, contemporary letter to President Porfirio Díaz. He replied that he had again referred the matter to the interim governor of the state. In Cuernavaca, the state capital, the governor then received a delegation and asked for a list of the people directly harmed by the hacienda's actions. Just two days later, the villagers sent him the information, introduced by a paragraph crystallizing the intact historical memory of the pueblo:

> List of the persons who have annually sown their seasonal crops on the fields . . . which were . . . included in the grant of lands to our village on September 25, 1607 by the viceroy of New Spain, today Mexico, as shown on the map attached, and of which property the Hospital Hacienda has deprived us.[8]

Halfway through 1910, Emiliano Zapata decided that the situation was critical, and he made the decision postponed for centuries. On his own initiative and at his own risk, he occupied and distributed the long-disputed land.

Soon afterward, President Díaz ordered the heirs of Vicente Alonso Pinzón to formally return the fields to Anenecuilco. In December of 1910, Zapata broke down the stone walls and carried out a second distribution of former Hospital land. In Anenecuilco, they held a rodeo to celebrate the historic recovery. Three centuries after its issue, the land grant of Viceroy Luis de Velasco went into effect.

Emiliano Zapata's true homeland was not Mexico or the state of Morelos, but the village of Anenecuilco. Linguistically and ethnically, it was no longer a purely indigenous community, but in deeper areas of its being, it was still very much Indian. Generation after generation—with a growing indifference to the accidents of other histories not its own—the pueblo had continued to claim its viceregal rights to the land, as if 1607 were always only yesterday. And its motives were more than economic. During the centuries of their legal battles with the haciendas in the courts of government after government, the people of the community had shown that the issue was dignity as well as possession. In one of the cases brought during the colonial era, the Indians of Anenecuilco said that they wanted a decision "even if they had to give up the lands that the Cuahuixtla and Hospital Haciendas would return to them."[9] The pueblo kept demanding that its enemies—the haciendas—recognize its right to exist as it had always been, sanctioned by ancient authority reaching back beyond the Spaniards. Zapata not only knew the whole history of that small universe. He was its embodiment. Everything else was abstract, alien.

His parents were Cleofas Salazar and Gabriel Zapata. They had ten children. Emiliano, the second youngest, was born on August 8, 1879, with a mark on his chest in the form of a hand. The first pants he wore were decorated with coins, and his uncle Cristino Zapata told him they were like the clothes of the famous *Plateados* ("Adorned with Silver"), the elegantly dressed bandits who had ravaged the state of Morelos in the second half of the nineteenth century. (Uncle Cristino had fought against them.) His father's other brother, Chema Zapata, gave him a souvenir—a muzzle-loading rifle—from the time "of the silver."[10]

Emiliano went to elementary school in Anenecuilco, where his education included the rudiments of bookkeeping. At sixteen he became an orphan but was far from helpless. He managed to buy a team of ten mules and used them to carry corn from the ranches to the towns. For a time he hauled lime and bricks for construction work on the nearby Hacienda of Chinameca. Besides his business as a mule-driver, he was a successful farmer. "One of the happiest days of my life," he once confessed, "was when I made around five or six hundred pesos from a crop of watermelons I raised all on my own." Zapata was always proud of earning an independent living.[11]

This self-sufficient farmer was no drunkard—though he did enjoy his brandy—nor a carouser—though he relished the San Miguelito Fair every September 29—nor a gambler (though he always carried his pack of cards). But he was surely "very much the lover" (*muy enamorado*). Many years after his death, the old women of Morelos still sighed as they remembered him: "He was so brave and so handsome." His sister recalled how "especially seductive and charming 'Miliano was with women." He took great pride in his huge mustache: It separated him from "girlish men, bullfighters, and priests."[12]

What everyone—not only women but all who knew him—found most attractive about Zapata was his character as the *Charro of Charros*—the finest of horsemen. Emiliano appeared in bullrings astride the best cowboy (*vaquero*) saddles, riding the best show horses. He loved to compete in rodeos and races and to fight bulls from the saddle. His *charro* clothes and style were in their own way classic, balanced, free of affectation or extreme elaboration. One of his secretaries, Serafín Robles—"Robledo" as the Jefe called him—once described it: "General Zapata's dress until his death was a *charro* outfit: tight-fitting black cashmere pants with silver buttons, a broad *charro* hat, a fine linen shirt or jacket, a scarf around his neck, boots of a single piece, Amozoqueña-style spurs, and a pistol at his belt."[13]

Robles asserted that in the entire south "there was no other *charro*" like Don Emiliano Zapata: "He could speed away like a flash of lightning . . . on horseback he flew . . . he was a bull rider, roper, horse breaker . . . he would lance the bulls, place the *banderillas*, fight the bulls from horseback and also on foot."

His gifts as a *charro* brought Emiliano not only amorous and aesthetic admiration but economic rewards. He was hired to train the finest horses belonging to a nearby and very well-connected hacendado, Don Porfirio's son-in-law Ignacio de la Torre, who would later organize the murder of Madero.

But he was more than a great horseman or energetic lover. Zapata was the living memory of Anenecuilco. And he steadily added to that memory, learning from what he saw and experienced. Between 1902 and 1905 he had quietly been involved in the dispute between the neighboring village of Yautepec (where he had relatives) and the Atlihuayán

Hacienda. Jovito Serrano, the leader of the Yautepec villagers, had gone to the lawyer Francisco Serralde (who had earlier given advice to Anenecuilco and had also defended opponents of President Díaz). Serralde's impression of the conflict was clear and farsighted. He wrote to Porfirio: "If the Supreme Court does not give these men justice, you may be sure . . . that there will soon be a revolution." The President agreed to see Jovito Serrano, and Zapata was a member of the delegation. Soon afterward, he learned that Serrano had been deported to Quintana Roo. Nothing more was ever heard of him. Zapata watched, recorded it, remembered.[14]

The year 1906 produced a major event for Zapata's intellectual education. The teacher Pablo Torres Burgos settled in Anenecuilco. He gave no formal classes but began supporting himself by selling vegetables and cigarettes and trading in books. Emiliano Zapata was among the friends who had access to his small library, where the liberal *El Diario del Hogar* (*The Home Journal*) and the anarchist *Regeneración* punctually arrived, the best newspapers of the anti-Díaz opposition. Soon afterward, a similar intellectual miracle occurred in Villa de Ayala. The teacher Otilio Montaño came to live there and began to teach and propagate, with fervor, a still more incendiary literature: the works of the Russian anarchist prince Kropotkin. Zapata admired Montaño so much that he asked him to become his compadre.

In 1908 Emiliano Zapata left Anenecuilco for romantic reasons. Perhaps remembering the feats of the *Plateados*, who carried their lassos "for the women they liked," Zapata kidnapped Inés Alfaro, a Cuautla woman, with whom he set up house and had a son, Nicolás, and two daughters. The father of Doña Inés denounced the action to the authorities, who forced Emiliano to enroll in the Seventh Army Battalion. But he was only there for a while, since we know that he actively participated, one year later, in the political campaign against the official candidate for state governor.[15]

In September of that same year, the inhabitants of Anenecuilco named him President of their Defense Committee:

> After the meeting was over, the elders took Emiliano aside and gave him the documents they were in charge of, and that were the same that have been passed down to us. . . . Emiliano accepted them and, together with his secretary Franco, began to study them. Franco spent eight days with Emiliano in the church choir room, reading the documents and trying to untangle the rights laid down in them. During those days they gave up all their other work and only went out to eat and sleep. It was in this way that the future caudillo

> drank down the deep waters of the suffering of his village and closely bound himself to the destiny of his remote Indian ancestors. . . . Because he wanted to know the meaning of the Aztec characters on the traditional map, Emiliano sent Franco to the town of Tetelcingo, where they still spoke Náhuatl and preserved many Indian customs. . . . Franco went to see the local priest, who was an Indian from Tepoztlán, land of the great Náhuatl translators. The priest was able to decipher the indigenous names and Franco brought the information back to his pueblo.[16]

Midway through 1910, Zapata carried out his tiny local revolution and took back the land lost in 1607. At the end of the year, he planted his watermelons again and was gored in one of his many bullfights. Within the village, buried in a secret place and secure inside a tin box, rested the viceregal land grant and deeds, maps, ownership claims, notebooks full of legal disputes and verdicts. Later, when he set out for the great Revolution, Zapata entrusted the box to his faithful Robledo: "If you lose them, compadre, you will be hung out to dry on a *cazahuate* tree."[17]

Without the Maderista whirlwind, that small rebellious move in Anenecuilco might have passed unnoticed even in the history of the region. But Madero's Program of San Luis included a promise to the pueblos—a return of lands seized by the haciendas. For Zapata and his intellectual friends, Torres Burgos and Montaño, those words were music from heaven. When the Revolution broke out, the men of Anenecuilco and some nearby villages decided to send Pablo Torres Burgos as their representative to Madero in San Antonio, Texas. Addressing the people in Yautepec, Otilio Montaño shouted: "Down with the haciendas! Long live the people!"[18]

The hour had come for Emiliano Zapata to begin *his* revolution. A young lawyer from the capital, one of the first outsiders to join Zapata's uprising, described him during those first days as he gathered his men in the town of Jojutla. Octavio Paz Solórzano, enraptured, would write:

> He ordered his men to gather in the Main Plaza, to begin their march. He was in the middle of the plaza on horseback . . . surrounded by some of his leaders, when suddenly a shot sounded. At first no one realized what had happened . . . but Zapata had felt his hat tilt; he took it off and saw that there was a hole in it. . . . They looked toward the town hall and saw the man who had fired rushing away from one of the balconies. This happened in less time than

Emiliano Zapata, 1912
(FINAH)

> it takes to recount it. Those nearest Zapata were about to charge the town hall when Zapata shouted: "No one move!" and, without the slightest hesitation, he rapidly directed the magnificent horse he was riding toward the door of the hall. . . . He rode up the steps of the building, before the astonished gaze of those present at the scene, who then saw him appear behind the balconies, carbine in hand, riding through the rooms of the Governmental Palace. After he had checked all the offices without finding anyone . . . he had the horse descend the steps and reappeared on the plaza, perfectly calm, to the admiration of his troops and of the large crowd watching him, on that spirited dark chestnut horse that was a gift from Prisciliano Espíritu, the priest of Axochiapan, and he had the cigar in his mouth he would never abandon even in the thick of battle.[19]

A few weeks later, Torres Burgos was surprised by federal soldiers while taking a siesta from which he would never awaken. Zapata, suddenly, was the leader of the Revolution of the South.

He received word of a challenge from the hated Spanish administrator of the Chinameca Hacienda, a man named Carriles. The Spaniard was reported to have said that if Zapata were "so brave and so much a man, we have thousands of bullets and enough guns waiting to welcome you and your men as you deserve." Paz remembered how "Zapata's eyes flashed with rage," and he made the decision to attack Chinameca—his first military action—not out of any preconceived plan but for honor.

The result was "a frightful slaughter" and for the living, a ballad (*corrido*) composed by Marciano Silva, an old southern singer who had joined Zapata's army:

> *There came the terrible Zapata*
> *with justice and the right;*
> *he spoke in the voice of command: "Bring an axe*
> *and break down this gate!"*
>
> *The earth trembled on that day*
> *as Zapata entered.*
>
> *He assembled them all—the hour was eleven!*
> *and he made them kneel down before a rock.*
> *"Kiss this cross and blow bronze bugles*
> *and you shout 'Death to Spain!'"*

Long live General Zapata!
Long live his faith, and his ideas!
Since he is ready to die for the patria,
so am I, for the nation![20]

He took on a load of supplies at Chinameca and then marched on. As he moved through towns and villages, his army steadily swelled. Why did they follow him? One of the reasons, caught in the ballad, was a kind of historical cyst that echoed the War of Independence fought a century before: hatred for the Spaniard. Constancio Quintero García of Chinameca recalled: "I joined because of that fear of the Spaniards; now they were going to brand us like animals." Espiridión Rivera Morales, also of Chinameca, explained: "We were sowing just a little bit of corn on the hillsides, because the Spanish bastard had taken all our land." Others came to Zapata because they yearned for freedom and justice: "We had more rights in the mountains, on horseback, as free men, than staying there, because the *rurales* were after us, charging us for living, for the hens, for the pigs . . . and Zapata gave us rights. . . . We had to follow him! That's the reason. I couldn't take the injustice any longer." Others went with him simply because they had nothing: "Eighteen others came with me from my village; all of them farmed only the stony slopes of mountains. The mine workers never joined us; they were sons of bitches. They didn't come because they were in with the gringos . . . because they were paid good wages."[21]

There were many others in the region who did not join him. The majority remained *pacíficos* ("peaceful"), as the Zapatistas called them, including many peons who stayed loyal to their hacendados. Zapata generally respected their choice, but his deputies often did not. Sometimes peons would join the Revolution and then quickly desert to return to work on the haciendas. If Felipe Neri, one of Zapata's generals, caught such men, he would cut off part of an ear, "so that I may recognize you."

By May 1911, there were only two government strongholds left in the whole state of Morelos: Cuautla and Cuernavaca. The city of Cuautla was defended by a famous cavalry regiment: the "Golden Fifth." Zapata asked them to surrender peacefully, but the mayor of the city refused.

The fighting went on for days. Sometimes a wall would collapse, leaving the enemies face to face, and they would then struggle with all their strength and determination for control of the piles of earth and bricks behind which they could shield themselves and continue to fire.

At close quarters they would sometimes not even shoot but try to brain each other with the butts or barrels of their rifles:

With gasoline they bathed
the railroad cars where soldiers
were fighting and lit them without
attending to the screams they heard . . .

All of the poor soldiers were
turned into nothing but ashes
and their remains were flung away
across the fields of Cuauhuistla.[22]

On May 17, Felipe Neri captured the Convent of San Diego in a furious assault and suffered the injury that perhaps may explain his vocation as *mochaorejas* ("clipper of ears"). A bomb flung at the wall of the church bounced back to explode at his feet, seriously wounding him and leaving him deaf for life. Two days later the Zapatistas were in control of Cuautla.

In Mexico City the old dictator received the news with real alarm. He knew that the "guerrilla fighters of the south were brave." Later in exile he would say, "I was calm until the South rose." Six days later he resigned.

On June 7, 1911, Emiliano Zapata met Francisco Madero. It was at a luncheon where Madero was surrounded by flatterers, and it left a bad taste in Zapata's mouth. Soon after, Madero visited Morelos, an area he had neglected during his presidential campaign. He showed himself equally generous to hacendados and revolutionaries, and it made Zapata uneasy. He could not understand why Madero would listen to those who criticized the violence of the Zapatistas in the capture of Cuautla. Had there been or had there not been a Revolution? Soon the Mexico City press, incited of course by the hacendados, began a campaign to discredit "the bandit" Zapata, who was expected to launch an insurrection at any moment. Then, once more, Madero invited Zapata to come to Mexico City.

The two caudillos met at Madero's home on June 21. The young Gildardo Magaña would remember how Zapata—courteously, using images and parables but also decisively—explained the reasons for *his* revolution. There was tension in the air. The Caudillo of the South relaxed things by walking toward the President and pointing to a gold chain Madero wore on his vest:

> Look, Sr. Madero, if I—able to do it because I'm armed—take your watch away from you and keep it, and if, as time goes on, we happen to meet again, and both of us are carrying the same weapons, would you have the right to ask me to give it back?
>
> Of course—Madero said to him—and what is more, I would ask for compensation.
>
> Well that is exactly—Zapata ended up saying—what has happened to us in the state of Morelos, where a few hacendados by force have taken over village lands. My soldiers (the armed peasants and all the villages) insist that I tell you, with all due respect, that they want you to move immediately to restore their lands.[23]

For a time, Zapata trusted Madero. He sincerely wanted to retire to a private life and enjoy his forthcoming marriage to the woman who would become his only legal wife: Josefa Espejo. But first he had to leave Morelos under the command of a man like Raúl Madero—or any soldier for that matter except certain federal generals, most especially not Aureliano Blanquet or Victoriano Huerta. Meanwhile, interim President De la Barra, that relic of Porfirismo, was pressuring Zapata to demobilize his army.

A process began that could not be stopped: a sad breakdown of trust. Time and time again, addressing Madero directly or indirectly, Zapata repeated the words *trust* and *loyalty* to imply the opposite: betrayal. Madero understood and made another visit to Morelos, in honest good faith, believing he could speak effectively to both Zapata and the Porfirista government he had foolishly left in place.

When he arrived in Morelos on August 18, Madero made a speech in which he called Zapata "most honorable general" (*integérrimo general*). Despite everything, they still seemed to believe in each other, but they were moving within a closed frame created by the hacendados, the capital press, the racist views of President De la Barra, and the zeal of General Victoriano Huerta—a pure-blooded Huichol from Nayarit and a veteran of the wars against Yaquis and Mayas—who was advancing on Yautepec "to weaken Zapata until I can hang him." Four days after he arrived, Madero realized that the central government was paying absolutely no attention to him and he left, only offering Zapata promises to be met "when I come to power."[24]

During the interim presidency, Zapata had to endure military and verbal assault. And it was the words that enraged him most. It would sicken him to hear the federal *pelones* ("short-hairs" because of their military haircuts) say that his men were "cow-eating bandits."

In the Lower House, José María Lozano went much further: "Zapata is the new Attila, a Spartacus, the liberator of slaves, the man who promises wealth for everyone. He is a total danger to society. He is the underground rising with a will to sweep the surface clean. . . . Zapata is no longer a man, he is a symbol."[25]

When Madero became President, he and Zapata made one last effort to reach an agreement. Zapata's conditions could not have been more reasonable. He asked for the withdrawal of a federal general from Morelos, the naming of Raúl Madero to replace him, and then he also made a mild reference to the problem of the land, asking Madero "to issue an agrarian law that would try to improve the conditions of workers in the fields." Madero made a decision he would later repent. He warned Zapata "to surrender to good judgment and leave the country . . . your rebellious attitude is doing serious harm to my government." After that, there would never be anything else between them but anger and regret.

"Victories in which those who have been defeated are those who win" could mean only one thing to Zapata—betrayal. He was a man of absolute convictions. He could not see Madero's reluctance to distribute the land, or his failure to impose himself on De la Barra and Huerta, as anything but treason in the biblical sense of the term, the sin of Judas that includes all other sins. In Zapata's final letter to the President, he wrote: "You can begin counting the days, because in a month I will be in Mexico City with 20,000 men and I will have the pleasure of coming to Chapultepec . . . and hanging you from one of the tallest trees in the forest."[26]

This failed encounter between two men of faith would be one of the most tragic moments of the Revolution. Madero himself, in the final hours of his life, admitted as much to Felipe Ángeles.[27] Perhaps by then Zapata's attitude made more sense to him, and he understood that it was the view of a man who had been waiting for centuries.

"I can pardon those who kill or steal," Zapata used to say, "because perhaps they did it out of need. But I never pardon a traitor."[28] He lived obsessed by treason. In the *Plan de Ayala* drawn up by Zapata and Otilio Montaño and signed on November 25, 1911, in the small mountain town of Ayoxustla, the word *betrayal* (*traición*), referring to Madero, is directly used five times (and implicitly, with great harshness, at various other points).

But the *Plan de Ayala* was far more than an attack on Madero. It was Zapata's attempt to provide an ideal program for *his* revolution, rooted—as was everything Zapata did—in his own biography, in the history of Emiliano Zapata and the life of Anenecuilco. To his secretary, Robledo, he explained the basis (as he saw it) for the Program of Ayala:

> My ancestors and I—within the law and in a peaceful way—asked earlier governments to return our lands, but they gave us no attention and no justice. Some of them were shot with some excuse like claiming they tried to escape (*la ley fuga*); others they shipped off to the State of Yucatán or the Territory of Quintana Roo, from where they never returned; and others, like myself, were forced into military service through the hateful "draft" system. That is why we now claim our rights with arms, since we will not get them in any other way. Because you must never ask, holding a hat in your hand, for justice from the governments of tyrants but only pick up a gun.
>
> I spent three days getting my ideas clear and then I communicated them to my compadre Montaño so that he could give them a form, resulting . . . in the Program we wanted.[29]

The *Plan de Ayala* was original in a number of ways, but its primary goal was to "begin to continue" the Revolution that Madero "had gloriously begun with the support of God and the people" but that "he did not bring to a fortunate conclusion." There were three central articles. The *Plan* called for the restitution to "the pueblos or citizens who hold the proper deeds" of "lands, mountains and waters usurped by the hacendados, *científicos* or caciques." Furthermore, a third of the "lands, mountains and waters" monopolized by these owners would be expropriated—but with compensation—so that the pueblos and citizens could make use of them and "improve in every way . . . the lack of prosperity and well-being among Mexicans." Those who directly or indirectly "resisted this Program" would have their holdings nationalized, and two-thirds of their former wealth (which could have remained theirs had they peacefully agreed to the first two articles) would then be applied "to paying war indemnities—pensions for widows and orphans of the victims who fall in the struggle for this Program."[30]

The Program was not an argument for extreme radicalism. The original *Plan de Ayala* did not envision the disappearance of the haciendas. They were to give back what they had taken—against the law—and they were to accept reasonable reductions in their holdings for the good of all Mexicans. The dream of Zapatista redemption was to create a mosaic of small autonomous holdings whose owners would be united by a strong sense of community. For the Zapatistas, the *Plan de Ayala* was the means to this goal and it would always retain the character of Holy Writ, of a messianic promise.

From then on, the Zapatista revolution is the history of a war without mercy "against everything and everybody." As its leader would say, "Revolutions pass, revolutions will come, I will go on making mine."[31]

The rebellion was at first amorphous and scattered, but it took on shape and strength in response to the attacks by federal soldiers. Each side had its extremists: the government in the person of General Juvencio Robles, who burned villages and moved whole populations to "resettlement locations."[32] On the Zapatista side, there was the southern leader Genovevo de la O, who blew up trains full of people.[33] And then the Madero government decided to change its tactics. Felipe Ángeles, the new chief of operations, immediately put an end to Robles's savagery and refused to broaden the war, in spite of the dynamite attacks against the trains. He said that he thought "the attitude of the Zapatistas is justified." They did not want "the flourishing orchard of Morelos to become their hell."[34] In the major cities, elections were held that showed a clear will toward legality and change. Slowly, through peaceful means, space was being opened for the ideas of agrarian reform. The Zapatista movement began to weaken, without enough arms or supplies. For a time, Zapata moved his men out of Morelos and retreated to the district of Acatlán in the state of Puebla.

The movement was jolted out of its decline by the fall of Madero. Briefly, it seemed that Zapata might consider the possibility of an agreement with Huerta in exchange for official recognition of the *Plan de Ayala*, but any accord between them was really impossible. After all, Huerta could not be trusted. Hadn't he and his allies "betrayed and assassinated their leaders"?

The war flared up again, rising to a new level of terror. Generals Cartón and Robles hanged Zapatistas from the trees and telegraph poles, forced peasants into their army, took civilian hostages, returned to the process of "recolonization" and the looting and burning of villages. The ferocious campaign only increased support within his state for Emiliano Zapata.[35]

In 1914 the balance shifted. Zapata's forces took Chilpancingo in March, where he had the pleasure of executing General Cartón:

Long live the forces of Chilpancingo!
Death to Huerta! Death to Cartón!
Many weapons firing in celebration
were heard when Cartón's life ended.[36]

Swiftly advancing, Zapata occupied Jojutla, Jonacatepec, Cuautla. The federal soldiers—damaged as well by their defeats in the north at the hands of Obregón and Villa—fled the state. And so did the hacendados of Morelos—this time for good. In June 1914, foreseeing the end of the government of Victoriano Huerta and the imminent triumph of the

Zapatista soldiers, circa 1912
(FINAH)

Revolution, the Zapatista movement published its ratified text of the Program of Ayala with the primary objective of elevating "the portion concerning the agrarian problem . . . to the level of constitutional law."

At heart, the Zapatista movement never stopped being an island unto itself. This was both its strength and its weakness. The Zapatistas felt they had been betrayed by Madero. And on their own, they had beaten back the murderous assault of Huerta's army. They saw no reason to trust anyone or to give up their resistance.

Venustiano Carranza, after the fall of Huerta, ran straight into this village wall of Zapatista suspicion. He dispatched a whole series of emissaries to seek agreement with Zapata. None of them met with any success.

To the first envoy, Zapata confided: "I see dangerous aspirations in Carranza." With another emissary, General Lucio Blanco, he let the opportunity slide to establish a relationship with a man who could have been a noble and natural ally. Other men were sent who were true revolutionaries, strong partisans of agrarian reform. Zapata flatly rejected them all.

Manuel Palafox, recently become Zapata's close advisor (an able, pragmatic man and a great administrator), warned the final delegation that Carranza's only option was to renounce the presidency, let the Zapatistas participate in the selection of a new government, and—literally—"submit" himself to the *Plan de Ayala*, "without changing so much as a comma." Carranza of course would not accept these conditions, and in September 1914 he broke completely with Zapata. The Caudillo of the South responded with an agrarian decree much more radical than the Program of Ayala. The nationalization of enemy possessions would for the first time extend to urban property and—also for the first time—forms of ownership were to be established that recalled the Aztec *calpulli* (shared communal ownership of land). In embryo, this was a call for the *ejido*, an institution that would later develop out of the Mexican Revolution and involve the collective ownership and cultivation of fields (not merely the joint right to grazing land, for which the term was used in Spain).

Much more enigmatic and significant than Zapata's transitory relationship with Carranza was his connection with the Convention of Aguascalientes (when the generals gathered to decide on a post-Huerta government) and especially his contacts with Francisco Villa. It might be assumed that the popular roots of the Convention would have finally opened up the possibility, for Zapata, of a national agreement. But it only half happened.

To begin with, Zapata himself did not go to Aguascalientes. Nor did any of the principal leaders of the Revolution of the South. Only his intellectuals attended. Many of them were anarchists who had joined Zapata out of a true populist vocation, not through curiosity or opportunism. One of them, Soto y Gama, collected his moment in the sun (or fifteen minutes of fame). Before the astonished audience of generals (who miraculously did not open fire on him), he ripped up the "rag" of the national flag, since one had to destroy—according to the anarchist gospel of Kropotkin or even the peaceful Tolstoy—all abstractions that oppressed the people.[37]

Between Zapatismo and anarchism, a profound connection did exist. "Rebellious peasants are natural anarchists."[38] They want freedom for their village, perhaps a world of independent villages. The state for them is an evil, something to overcome, a condition to transcend. And the rhetoric of anarcho-syndicalism connected and exalted peasants and intellectuals. The primary slogan of Zapatismo, "Tierra y Libertad" ("Land and Liberty"), was a contribution of the Mexican anarchist Ricardo Flores Magón (who took it from Russian socialist Alexander Herzen).

Although the Convention broke with Carranza and, in principle, accepted the Program of Ayala, its alliance with Zapatismo was to be brief.

On December 6, 1914, the forces of the Convention moved into the capital. Zapata and Villa, riding in together, led their troops. At one point, Villa's military cap fell off and Zapata swung down the side of his horse to scoop it up from the ground and hand it to Villa, without ever slowing his pace.[39]

Mexico City had trembled like a defenseless maiden before the imminent assault of the Zapatista "hordes." When they finally arrived, these hordes behaved like peaceful flocks of bewildered peasants, who carried—as a symbol of their struggle for the permanent and traditional—the same standard as the armies of Hidalgo a century before: the Virgin of Guadalupe. A terrified young man-about-town remembered years later how he had been approached by the Zapatistas, not to rip out his heart—as he feared—but with this polite request: "Young master, would you let us have a little paper money?" They wandered the streets like lost children, knocking on doors to ask for food. Dressed in coarse white cotton, with their Franciscan sandals, enormous straw hats and cartridge belts, carrying their machetes, they did not look like soldiers, nor did they want to. They were peasants out of place.[40]

The confusion the Zapatistas felt at occupying the capital and exercising a power they did not want or understand was shared by Zapata

Zapata and Villa entering Mexico City, 1914
(FINAH)

Zapata with Villa sitting in the Presidential Chair, 1914
(FINAH)

himself. At this moment, when he was at the height of his power, Zapata's "natural anarchism" showed its generous and tragic content. He hardly visited the city. He stayed in a small dark hotel, one block from the railroad station. Two days before entering Mexico City, on December 4, he had met with Francisco Villa in Xochimilco, on the outskirts of the capital. Many hoped this meeting would produce an agreement that would change the history of Mexico. Villa, as an observer recalled, "was tall, robust, weighing about 180 pounds, with a complexion almost as florid as a German, wearing an English [pith] helmet, a heavy brown sweater, khaki trousers, leggings, and heavy riding shoes." [41] He looked like what he was: a soldier, the caudillo of the powerful División del Norte. In contrast, Zapata "seemed a native from another country," [42] with his slender face, his dark skin, his *charro* outfit complete with huge hat, useful for shading him from the sun and hiding his expressions, absurd as military dress. He was what he looked like: a peasant in arms.

We have a shorthand transcription of their dialogue. It is perhaps the only time when we can truly hear Zapata. Two sections are especially striking. In the first, he shows the self-sufficient, local, peasant nature of Zapatismo, down to the country metaphors he uses. For men who want power he has only contempt, and he equates power with the city and its arid landscape of sidewalks:

> VILLA: I don't need government jobs because I wouldn't know how to manage them.
>
> ZAPATA: That's why I warned all my friends to be really careful, and if not—they get the machete (*laughter*). . . . Well, I think we won't be taken in. We've only loosened the rope on them, looking out for them, looking out for them, over here, over there, to keep after them while they're grazing. . . . The men who have worked the hardest are the last to get any good from those sidewalks. Just nothing but sidewalks. And I'm speaking for myself: when I walk on a sidewalk I feel like I'm going to fall.

In a second significant section—during which the voice of General Serrato is also heard—Zapata explains to Villa the importance of distributing the land:

> VILLA: Well, we should give the people these bits of land they want. Once they've been distributed, there will be some who'll try to take them away again.
>
> ZAPATA: They feel so much love for the land. They still don't believe it

when they're told: "This land is yours." They think it's a dream. But after they've seen other people drawing crops from these lands, they too will say: "I'm going to ask for my land and I'm going to sow there." More than anything, this is the love the people feel for the land. Generally everybody gets their living from it.

SERRATO: It seems impossible to them, that it had really happened. They do not believe it. They say: "Maybe tomorrow they will take it away."

VILLA: You'll see how the people will command and that they're going to see who their friends are.

ZAPATA: The people know that they want to take away their lands. They themselves know that they can only defend it themselves. But they would die before they give up the land.[43]

Villa talks about "bits of land they want," Zapata about "*the land*." The huge difference in attitude between the warrior and the guerrilla was caught in the famous photo that shows a euphoric Villa sitting in the Presidential Chair next to a surly and suspicious Zapata, always wary of a bullet perhaps springing out of the camera instead of the flash of a bulb. A Zapatista witness to the scene remembers: "Villa sat in the chair as a joke, while Emiliano stood to one side, and he said to Emiliano: 'Now it's your turn.' Emiliano said, 'I didn't fight for that, I fought to get the lands back, I don't care about politics.'" And later he said, "We should burn the Chair to end ambitions."

Both were sons of the people, but they were very different men. And so were their attitudes and intentions. One was savage and boisterous, the other mystical and taciturn. One fought for the sake of the fight, the other for the Program of Ayala. The city of Mexico, the National Palace, the Presidential Chair, the powers of government were symbols, for Zapata, of deception, the centuries of assault on the lands and dignity of Anenecuilco. Thus his physical aversion to politics. And why he would always keep repeating statements like: "Whoever tries to tempt me with the Presidency of the Republic—and already there are some people hinting at the offer—I'll *wipe them out*!"[44]

Octavio Paz has considered, with clarity, the historical destiny of this friction between Zapatismo (which encompassed the life and death of his own father) and the two entities that were alien and even opposed to it: the city and the State: "In the inhuman context of history, particularly at a revolutionary moment, Zapata's attitude had the same meaning as Hidalgo's gesture before Mexico City. He who refuses power, through a fatal process of reversion, will be destroyed by power. The episode of Zapata's visit to the National Palace illustrates the nature of the peasant movement and its later fate."[45]

* * *

While the troops of Venustiano Carranza were fighting and defeating Pancho Villa's forces in different areas of the country, Zapata, in Morelos, was enjoying a breath of peace. The miraculous parenthesis had in fact already opened at the beginning of 1914, with the defeat of Huerta. It would last till the end of the following year.

Zapata used the time to put the anarchist utopia of *his* revolution into practice. The transition from peasant life to guerrilla life and from there to his ideal world was a smooth one for Zapata. During the campaign against Madero's federal soldiers, and above all against Huerta's *pelones*, the profile had been drawn of a peasant society that even in war was faithful to itself; an army divided into small units, decentralized, respectful in its relations with the villages, mindful of its Indian roots, and devoted to religion.

"Some followed the chief, others another leader, well . . . there were various leaders," recalled a Zapatista veteran. They usually operated in units of from thirty to a few hundred men under the command of the most energetic guerrilla, sometimes even a woman who would be called the *coronela* or the *capitana*. They rode the small local horses or moved on foot or on muleback. Most of their weapons—even some cannon—were taken in lightning raids on their enemies.[46]

This condition of dispersal was a natural continuation of prerevolutionary life in Morelos, where the true political cell was the village. Moving in small groups made it easier for the guerrillas to attack by surprise or from concealment, to vanish into the countryside and to gather supplies from the villagers. And the people would offer not only tortillas but ammunition and information.

Another notable feature of that war of wandering villages was its bias toward indigenous values and consequent respect for the Indians. In a chronicle of the town of Milpa Alta compiled years later, we can read the testimony in Náhuatl by an old Indian woman:

> The first that we knew of the Revolution was when one day the great Zapata arrived from Morelos. And he stood out for his beautiful clothes. He wore a broad hat, and leggings and he was the first great man who spoke to us in Náhuatl. . . . His men wore white: white shirts, white pants and sandals. . . . All of his men spoke Náhuatl (almost like us).. . . When those men came into Milpa Alta, you could understand what they were saying.
>
> Señor Zapata stood up in front of his men and spoke to all the people of Milpa Alta: [this must be a confusion since we know that Zapata did not speak Náhuatl] "Join me! I have risen up; I have

risen in arms and I bring the men from my region with me. Because we don't want our father Díaz to take care of us. We want a much better President. Rise up with us because we don't like what the rich pay us. It's not enough to clothe or feed us. I also want all the people to have their land; then they can sow and harvest corn, beans, and other grains. What do you say? Will you join us?"[47]

Religious devotion was another vital aspect of Zapatismo. Besides the Virgin of Guadalupe on their banners, the guerrillas "wore their most beloved saints on their hats to protect them." And in Zapatista territory, priests were not persecuted. On the contrary: Many did their part for the cause. The priest of Tepoztlán had translated the Náhuatl documents of Anenecuilco; in the first days of the uprising, the priest of Axochipan had given Zapata the beautiful horse he rode off to war; a priest from Huautla had typed the first copy of the *Plan de Ayala*.[48]

Zapata would set up his headquarters in the village of Tlaltizapán. By now he was a living legend in Morelos. "Here even the stones are Zapatista," said one of the faithful. He was "our defender," "our savior," the "Chief himself," the "scourge of traitors."[49]

As Otilio Montaño had once envisioned, Tlaltizapán became the "moral capital of the Revolution." Every day Zapata would hear petitions—"let us water our fields with the water from that hacienda"; "give us ten pesos to take care of our needs and if God wills he'll give us corn and we'll pay you back"; "keep my old lover away from me because I'm afraid to go out in the street and someday he'll shoot me"—and then he would issue his orders and go off to lounge in the plaza toward evening with some of his leaders, drinking, smoking a good cigar, discussing fast horses and fierce fighting cocks. His nights he spent with a woman in the village, and he fathered at least two children there. In fact there were rumors that he lived with not just one woman but three sisters "under the same roof and in the greatest harmony." (He was known to display such amorous democracy before and after that period of peace. We know of at least twenty of his women and seven of his children.) In his spare time, he still loved to perform on horseback and encounter the bulls.[50]

Around him, in the state now controlled by the Zapatistas, a peaceful revolution was developing. The *hacendado* class had disappeared, and Morelos was really an independent territory, a constellation of small communities like those envisioned by Kropotkin. An attempt was being made to resurrect an ancient concord, mythical or long lost.

Direct, local democracy was practiced. The distribution of land was carried out in accordance with the customs and needs of each community.

The Zapatista leaders were forbidden to impose their will on the villages. There was no state police, no formal, vertical chain of command—political or ideological—of any kind. The Zapatista army—formed of the people, a true "armed league of communities"—served the social order, which was democratic and civilian.[51]

If the goal was a return to peasant origins, the conditions "in the beginning" had to be understood. For Zapatismo, there could be no true return without comprehending the clear contours of rights over the land. A new and authentic map was needed. To draw it, a generation of young agronomists arrived in Morelos to reexamine and establish the boundaries of each village. The "young engineers" (*ingenieritos*) had to respect the viceregal deeds some of the villages possessed as well as the opinion of the elders. For them it was a lesson in living history. Addressing one of the agronomists, Zapata commented: "The villages say that this stone wall is the boundary; you're going to draw the line along it for me. You engineers often care a lot about straight lines but the boundary is going to be this stone wall, even if you have to work for six months measuring all the ins and outs."

It took months to "recover the map" and restore lands to the one hundred villages of Morelos. Meanwhile, Zapata set up four sugar mills and tried to persuade the peasants to cultivate commercial crops instead of corn and beans. But his concern was more protective than progressive, more moral than economic: "Now that there is money, we must help those poor people who have suffered so much during the Revolution. It's very right that we help them because we still don't know what they may have to suffer in the future. . . . We have to make these sugar mills work, because now they are the only industry and source of employment in the state."[52]

What really was his personal utopia? Soto y Gama records a conversation Zapata had with Enrique Villa:

> —Emiliano, what do you think of communism?
> —Explain to me what it is.
> —For example, all the people of a village farm . . . their lands together and then they distribute the harvest equally.
> —Who makes the distribution?
> —A representative, or a council elected by the community.
> —Well look, as far as I'm concerned, if any "somebody" . . . would try to dispose of the fruits of my labor in that way . . . I would fill him full of bullets.[53]

For Zapata, "land and liberty" were distinct ideals but inseparable and equally important. And so anarchism—as preached to him by some of his intellectuals—"did not entirely displease him," although he was not convinced that it was any better than the only program he believed "would make the Mexican people happy": the *Plan de Ayala*.

Yet the roots and shape of *his* ideal world were always older than any modern program. Once when he was asked about "the first and final reason for his rebellion," Zapata had someone bring him the dusty tin box that held the documents of Anenecuilco's struggle for the land. He leafed through them and then he said: "For this I fight."

"This" was the land. Zapata fought religiously for the land that was, to the Zapatistas as to all peasants in traditional cultures, "the mother who nourishes us and cares for us" (as Saint Francis wrote in his *Canticle of the Beings*). The word *patria*, in his manifesto issued in Náhuatl to the Indian people of Tlaxcala, becomes "our beloved Mother the Earth, which is called the Fatherland."[54]

During the 1915 truce, instead of building up external alliances, he isolated himself even more, plunging deeper into the search for an ancient order to the point of wanting to reconstruct it from the memory of the elders. His war was reactive and passive, not active and willed. It was a resistance in place that eventually would wear itself out. Zapata would not leave his home though power was there, outside, threatening him. He was not looking for a useful map by which he could advance but for a mythical place, for the breast of his mother earth and her constellation of symbols.

Zapata's peaceful moments ended with the crushing defeat of Villa by General Álvaro Obregón. President Carranza could now concentrate on destroying Zapatismo. In August of 1915, the Zapatista revolution began to die. It was not a collapse but, in the words of John Womack, "a ragged, bitter and confused giving way."[55] The process had many aspects, almost all of them painful. First there was the murderous onslaught of the federal armies. General Pablo González wanted to finish off the Zapatistas "in their own dens." In Jonacatepec he took 225 civilian prisoners and shot them all. He executed 283 people in cold blood when, in June 1916, he took Zapata's headquarters at Tlaltizapán. The Zapatistas shifted their command center to Tochimilco, at the foot of the Popocatépetl volcano. Then González and his army of 30,000 men stepped up the pressure, adopting and multiplying the methods of Juvencio Robles: fire and looting, execution and mass deportation, with one new element, the riotous destruction of as much property as possible in order to impoverish the people.

Zapata retreated before the pressure and began another guerrilla war. In October 1916, he decided to mount a spectacular offensive. He carried out isolated but effective attacks on water pumps and streetcar stations near Mexico City. The impact on public opinion was enormous. By the end of November, González began to withdraw his forces. By the beginning of 1917, the Zapatistas had won back their state.[56]

When his troops entered Cuernavaca, Zapata wrote to Octavio Paz Solórzano, his representative in San Antonio, Texas:

> They have left Cuernavaca unrecognizable. The houses have no doors. The streets and plazas have become dung heaps. The churches are wide open. The images of the saints have been destroyed and stripped of their garments. And the city has been abandoned because they took away all of the "peacefuls" (*pacíficos*) by brute force . . . we found only three families in hiding.[57]

One of the most striking historical facts in that new chapter of violence, which lasted from the end of 1915 until the end of 1916, was that it was accompanied by great legislative creativity on the part of the five members of the Zapatista intellectual junta. It was as if they were responding to Carranza by feverishly drawing up law after law and delineating the ideal country that they might have governed.

They issued an agrarian law and legislation on labor accidents, general education, public assistance, the administration of justice, a press law prohibiting censorship, and a truly remarkable law authorizing plebiscites and with a deeply democratic cast to it: "Throughout the history of nations there have always been traitors to the cause of the people, men who, seeing the people forget to exercise their political rights, take them away and along with it their civil rights."

But none of these laws ever had a chance to move from mere paper into the context of a functioning state.[58]

Perhaps even more painful than the ferocious war or the unusable laws was the breakdown of Zapatismo from within. It was hard enough to justify—to the *pacíficos* in Morelos—any continuation of the struggle, given that the enemies were no longer Porfiristas but themselves revolutionaries. To make things far worse, the Zapatista leaders began to disagree, to quarrel, and finally to begin killing one another. It was in part the negative consequence of their dispersed organization, and it was fatal to the movement. Antonio Barona killed Felipe Neri for daring to try to disarm ten of his men. Then he killed Francisco Estrada and Antonio Silva, till the ferocious Genovevo de la O "wiped him out" in return by dragging him to death behind a horse through the streets of Cuernavaca. There were far too many more.[59]

The accidental death of one of the fighters closest to Zapata—Amador Salazar—through a stray bullet was a heavy blow for the caudillo. The Zapatistas had reestablished their headquarters in Tlaltizapán, and there Zapata had Salazar buried, dressed as a *charro*, in the truncated pyramid that he had erected to accept the remains of his comrades in arms. But his most shattering loss occurred in May 1917, when a court-martial issued the sentence of death to Zapata's compadre, Otilio Montaño, coauthor of the *Plan de Ayala*, who had given him Kropotkin to read years ago and had studied the Anenecuilco documents with him in 1909. He was accused of being the mastermind of a conspiracy against Zapata. It was said that there were incriminating letters, that it was not the first time that his revolutionary zeal had weakened, and that Zapata himself had spotted him prowling around his house. The judges produced no proof against Montaño, nor did they open the trial to the public. During the trial and execution, Zapata disappeared from Tlaltizapán.

Before his death, Montaño dictated a will in which he stated: "I am going to die, there is no doubt about it, but there where justice is done, there I will be waiting for you sooner or later." Montaño, who was above all a religious being, was denied the last rites. He resisted being executed with his back to the firing squad, but they forced him to turn around. He spread his arms out wide and declared, "in the name of God," that "he died an innocent man." Later in the day, they took the corpse to Huatecalco and hung it from a *cazahuate* tree, with a sign on his chest warning: "This is the fate met by traitors to the *Patria*."[60]

For a long time, perhaps always in some sense, Zapata had felt delusions of persecution and a fixation on betrayal. Time and time again he would repeat his favorite statement: "I can pardon those who kill and steal, but I never pardon a traitor." Now Soto y Gama was ordered to write a "decree against traitors . . . a damned race that should be rooted out without hesitation. There must not even be a seed left of traitors." Soto y Gama diligently drew up the decree.[61]

A month after Montaño's death, as if fate were enforcing the law of an eye for an eye, Zapata's older brother Eufemio died a hard death. Eufemio Zapata was a terror to the peasants but especially to drunkards, because he had taken it into his head to break the people of that particular bad habit. One day he beat an old man unconscious with the stick he always carried, berating him for his drunkenness at every blow. The man's son got a rifle and shot him, then dragged the dying Eufemio to an anthill (of fierce, stinging ants) and rode out of town.[62]

More and more suspicious, Zapata grew steadily more isolated. In August of 1917, they brought him the head of Domingo Arenas, the

Indian caudillo (and agrarian activist) who had led his Tlaxcalans into Zapata's ranks in the now remote year of 1914. Arenas had gone over to Carranza, then had left him and was putting out feelers about a possible return to Zapata. But he was a "traitor," and Zapata was very happy to see his head. Yet it afforded him only a brief change of mood.[63] "Everything made him angry . . . many leaders were afraid to go near him." The air was getting harder to breathe. Pablo González returned to Morelos and along with his chief henchman, Jesús Guajardo, began to apply his usual ruthless methods. Even nature turned against the Zapatistas. There were outbreaks of typhoid, malaria, and dysentery.

Zapata began to feel as if he were being strangled. He now had a single obsession—the need to make alliances. There was no revolutionary or even counterrevolutionary leader he did not approach: Félix Díaz, Manuel Peláez, Francisco Villa, Felipe Ángeles, Álvaro Obregón, and, in utter despair, Carranza himself. All without success. A sign of his extreme unease was his manifesto to the nation of April 1918, when he called for a kind of "popular front" without even mentioning the *Plan de Ayala*.

The cold professionalism of Pablo González and "the pacifying operation of Spanish influenza" closed the circle ever tighter on the Zapatista guerrillas, who now numbered no more than a few thousand. In August 1918 they lost Tlaltizapán again and again retreated to their refuge of Tochimilco at the base of Popocatépetl.[64]

The man who had always feared and hated traitors died as the victim of a deceit carefully orchestrated by Colonel Jesús Guajardo and his superior, Pablo González. Rumors—not completely unfounded—of a disagreement between the two officers had reached Zapata's camp. In a mood of renewed optimism, Zapata wrote to Guajardo and invited him to join the rebel side. González intercepted the letter and used it to blackmail and incite Guajardo, who saw this as an opportunity to redeem himself and demonstrate his loyalty.

He sent Zapata a positive answer. As mistrustful as ever, Zapata asked him to shoot Victoriano Bárcena—a Zapatista who had accepted amnesty from the government—along with Bárcena's men. To prove his sincerity, Guajardo executed them all.

Partially satisfied with this guarantee against treason, Zapata arranged a meeting with Guajardo. The colonel gave him a present, a handsome chestnut horse named As de Oros ("Ace of Diamonds"). Then came the next step. Guajardo agreed to give Zapata twelve thousand cartridges. They would have to meet at the Chinameca Hacienda, where Zapata had delivered construction supplies during his years as a mule-driver. It had later been the scene of his first battle, against the Spanish administrator's private forces.[65]

On the morning of April 10, 1919, Zapata surrounded the hacienda but waited outside with his men. He was still suspicious. His deputy Palacios went inside to confer with Guajardo, who repeatedly invited Zapata to lunch. Finally, around one forty-five in the afternoon, Zapata agreed to enter the hacienda. An eyewitness—Major Reyes Avilés—described what happened:

> [Zapata commanded]: "We're going to see the Colonel and I want only ten men to come with me." He rode toward the gate of the hacienda. Ten of us followed him as he had ordered, while the rest waited full of confidence under the shade of the trees with their carbines in their cases. A squad, lined up in formation, seemed ready to pay him honors. Three times the bugle sounded the salute of honor, and as the last note faded away and as our commander in chief appeared at the threshold of the gate, the soldiers who had presented arms fired their rifles twice at point-blank range in the most treacherous, cowardly, and vile manner, not giving us any time to reach for our pistols, and our unforgettable General Zapata fell, never to rise again.[66]

As always happens when the heroes of a people die, extravagant stories began to spread. The corpse exhibited in Cuautla was said to lack a small wart on its face or the birthmark of a hand on its chest, and so it was not the body of Zapata. Or the little finger Zapata had once injured was intact and unscarred on this corpse. People swore they had seen him at night, mounted on As de Oros. Much later there were reports of an old man glimpsed behind the walled-up door of a house in Anenecuilco—it had to be Zapata.[67] Nineteen years later there was somebody who said: "I saw his body. The one they killed was not Don Emiliano, but his compadre Jesús Delgado. How could I not recognize him, I who soldiered to his orders and earned these stars!"[68]

At the age of eighty, another Zapatista veteran offered another story: "It wasn't Zapata but his compadre who died in Chinameca, because the day before he received a telegram from his other compadre, the Arab. Now Zapata is dead, but he died in Arabia; he sailed to Arabia from Acapulco."[69]

There were times when Zapata had foreseen his own death, but he had struggled to stay alive, always watching his back and lashing out at "traitors." Yet perhaps more than any other figure of the Revolution, his death would be a transfiguration. The up-and-down details of his human biography would fall away and something purer would emerge, destined for a far longer life. Octavio Paz would later write of him: "Along with Morelos and Cuauhtémoc he is one of our legendary heroes. Realism and myth meet in this melancholy and passionate figure of hope, who died as he had lived: embracing the land. Like the land itself, he is formed of patience and fecundity, silence and hope, death and resurrection."[70]

As with all the peasant revolutions of the twentieth century, the Zapatista movement and its caudillo inexorably moved toward their tragic end. But the symbol of Zapata, "with all the beauty and visual poetry of popular images"[71] would live on, powerfully, within the minds of the Mexican people. In a *corrido* that depicts Zapata as a child, Emiliano makes a promise to his father and then the singer makes another, broader promise whose emblematic overtones resonate in the newspapers of yesterday and today:

"I will make them return the stolen lands
and I will quiet your pain.
This is an oath, not boasting and bluster,
I give you my word of honor."
It is the beginning of a long history
I undertake to relate to you.
Register it, all of you, and let your memory
never permit it to be forgotten.[72]

12

FRANCISCO VILLA

Between Angel and Iron

Dice Don Francisco Villa:
De nuevo voy a atacar,
me han matado mucha gente,
su sangre voy a vengar.

Said Don Francisco Villa,
"I will attack again!
They've killed many of my men.
I will avenge their blood!"

—"The Ballad of the Battle of Celaya"

Of all the provinces of northern New Spain, none suffered as much as Nueva Vizcaya from the long war against the "barbaric Indians." The Tobosos and Tarahumaras rebelled against the Cross and the sword during much of the seventeenth century. Later came a nightmare that would outlive the colonial period and ravage the north of Mexico until the very end of the nineteenth century: the Apaches. Between those nomadic centaurs and their Mexican opponents there was not only a war to the death, but a grim, shared escalation in the modes of killing. This ferocious and pitiless environment, "always unsettled," as the chronicles described it, was the school of life for the man whose epic incarnates a profound area of the Mexican soul, its darkest and most vengeful anger, its most innocent aspirations for the light: Francisco Villa.

He was probably born around 1878 in San Juan del Río, in the modern state of Durango. His father, Agustín Arango, was a sharecropper and an illegitimate son of Jesús Villa. Agustín died young, leaving his wife and five children to fend for themselves. Doroteo Arango was the eldest. He never went to school and began to support the family by working on the El Gorgojito ranch, which belonged to the López Negrete family. It is likely that the owner, his son, or the administrator of the ranch tried to rape Martina, Doroteo's sister. Her brother defended her with gunfire and fled to the nearby canyons with their terrifying names: the Canyon of the Devil, the Canyon of the Witches, the Canyon of Hell. He was soon caught and jailed but avoided summary execution on the pretext of a supposed escape attempt (the *ley de fuga*) by wounding his jailer with a stone pestle and then actually escaping. Around 1891 he became a bandit.[1]

According to Dr. Ramón Puente, who was his secretary for a time and later wrote a perceptive biography of him, Villa alternated between episodes of banditry and long periods of civilized existence, as if his life as an outlaw were always directed toward a new beginning, toward the hope of salvation. Puente concedes that Doroteo Arango had learned to rob and kill with the bandits Antonio Parra and Refugio Alvarado, alias "the Hunchback," but denies that he changed his name to Francisco Villa in honor of a bandit with the same name. He says it was a search for family ties, a return to his legitimate last name, that of his grandfather. At the time popular legend places Villa in the mines of Arizona or on the railroads of Colorado, Puente describes him setting up a butcher shop in Hidalgo del Parral. Supposedly he was married to Petra Espinosa (after abducting her) and enjoyed a good reputation. Around 1910 he moved to Chihuahua and was then, according to Puente, still working as a butcher, leading a quiet life but feeling wronged by society, by the government and the laws that oppress the poor and drive them toward crime.[2]

The anti-Villa version omits the periods of civilized peace and denies the episode of the outraged sister. For his detractors then and now, Villa was no more than a murderer.[3] We may never discover the truth about Villa's life before the Revolution, but there is testimony that can help us get closer to it. The first stories John Reed sent to American newspapers contain facts and conjectures that Reed himself did not include in his *Insurgent Mexico*.

When Reed came to Mexico in 1913, he interviewed witnesses to

Villa's raids between 1900 and 1910—the clerk of the Parral town hall and the Chihuahua police chief among others. He supplemented his information with a careful reading of old local newspapers. "His misdeeds," Reed wrote, "have no parallel with those of any other celebrated person in the world."

Between 1901 and 1909 Villa is certainly known to have murdered at least four people. He can be credited with participation in ten premeditated incidents of arson, innumerable robberies, and various kidnappings from farms and cattle-raising haciendas. In 1909 when, according to Puente, Villa was an honest butcher, the real Villa and his gang burned the town hall and archives of Rosario, in the district of Hidalgo. During that raid, Villa picked up the seal that he would later use to brand his cattle. In May 1910 he showed up on the San Isidro ranch posing as "H. Castañeda, cattle dealer." After looting the place, his band killed the owner and his young son. Then in October Villa and his men—his compadre Urbina among them—robbed the Talamontes ranch in the Jiménez district of Chihuahua. But from the beginning of that crucial year, Villa had established connections with Abraham González, chief of the Madero antireelectionist campaign in Chihuahua. In July, Claro Reza, one of Villa's compadres, denounced him to the authorities. When Villa learned of it, some reports say he entered Chihuahua and stabbed Reza through the heart. Other versions set the scene outside the cantina Las Quince Letras where Villa, without bothering to dismount, draws his pistol, cuts down Claro, and trots out of the city. Another version places the events on the Paseo Bolívar. Reza is walking along with his girlfriend, and Villa waits for him while eating an ice cream, then turns to face him, shoots him, and strolls away without anyone daring to follow.[4]

All this sounds like a Western movie: the grim outlaw and his band laying waste to dusty ranches—spirited horsemen, nervous, sinewy cattle, rapid bursts of gunfire, and endless chases. But we should not forget that we are always dealing with a Mexican reality. The old military colonies established during the eighteenth century in the north of Mexico to combat the Apaches—and enlarged by President Juárez in 1868—lost their fighting edge after the final defeat of the Indians in 1886. They had accumulated an inertial force of death (both confronting it and dealing it out) across the centuries, and now that force had nowhere to go. Then, during the Porfirian period, a new predator appeared, intent on seizing land and cattle from these colonies—the hacienda. When regulations were established on the sale of cattle—for the benefit of the haciendas—rustling began to spread. The result was a new culture of violence.

Segments of the rural poor and middle class saw the expansion of the haciendas, in land and cattle, as a global injustice. Who were the bandits? The government and the haciendas who promoted the idea of "cattle enclosures" or the ranchers who had freely grazed their cattle on the land since colonial times?[5] This situation in the Mexican north favored the rise of "social bandits," whose reputation would be that of robbing the rich to give to the poor. Reed himself was able to verify the magnanimous side of Villa, who was already the subject of "many traditional songs and ballads celebrating his exploits—you can hear the shepherds singing them around their fires in the mountains at night, repeating verses handed down by their fathers or composing others extemporaneously."[6] In his sympathetic articles, Reed called Villa "the Mexican Robin Hood."[7]

Villa would never deny his career as a bandit, but it is probable that before the Revolution he had acted for reasons other than—or at least complementary to—mere personal gain. If he had been a poet, he might have written, like his predecessor, Sinaloan bandit Heraclio Bernal:

Harmful without knowing it
I have been to society,
But I have always wanted
To really belong to it,
Yet I never could.[8]

Villa was a bandit out of a movie but also an avenger, a righter of wrongs. The qualifications neither cancel nor lessen the weight of the criminal term, but they give it a social shading and—when the time came—they opened up the possibility of a revolutionary mission.

His time did come, shortly before he "settled accounts" with Claro Reza. In the Palacio Hotel of Chihuahua, Francisco Villa met Francisco Madero. Tearfully, he told Madero the story of his life, his adventures and his crimes, he gave reasons for them, he confessed. Madero showed absolute confidence in him. The Apostle of Democracy forgave Villa's past and absolved him. The Revolution, for Villa, seemed like "something that was going to redeem him and redeem his class, the race of the poor." His nineteen years as a bandit had given him a knowledge of every slope and corner in his territory and taught him "more than one trick." Now he would be able to use "this knowledge for the cause of the people."

The Maderista revolution began to reveal Villa's tactical genius. In his first battle, he deceived General Navarro's troops into believing he had much larger forces by setting up a number of hats on sticks. With a small group of loyal and well-armed men he distinguished himself in the battles of San Andrés and Santa Rosalía, and the capture of Ciudad Juárez. Pascual Orozco, the major military commander during that first stage of the Revolution, considered him a "good guy." Both of them protested Madero's decision not to execute General Navarro, the federal commander of Ciudad Juárez. According to the El Paso *Morning News,* Villa threatened Madero at gunpoint, to which Madero answered: "I am your chief; if you dare to kill me, shoot!" Villa then wept and begged forgiveness, but deep down he thought that Madero should "hang these *curritos* [Spaniards]." For the time being, he offered an example with his personal, point-blank execution of José Félix Mestas, a sixty-year-old former Díaz official. Despite this and other violent incidents, Madero—after the triumph of the Revolution—compensated Villa with fifteen thousand pesos, which he used to open a butcher shop.[9]

The goodness of that man who had forgiven him everything, including a threat against his life, made a deep and lasting impression on Villa. In 1912 Orozco urged Villa to join his rebellion against Madero but only impelled him to enroll again as a Maderista. This time he was part of the federal troops. Leading a brigade of four hundred horsemen, Villa put himself under the command of General Victoriano Huerta, who gave him the rank of honorary brigadier general. Quickly Villa learned the arts of war, the formations, the strategies. Huerta admired Villa's ferocious charges and began to fear him. In Jiménez he used a trivial pretext to accuse Villa of insubordination and had him court-martialed. Villa was sentenced to death. Before the firing squad lining up with their rifles, Villa threw himself to the ground, sobbing and begging. Miraculously, Raúl Madero arrived on the scene just in time to save his life. A telegram from the President had commuted his sentence to imprisonment: yet another debt Villa owed to his savior.

In the penitentiary in México City, he met Gildardo Magaña. The young Zapatista taught him to read and write and informed him of the *Plan de Ayala*. In June 1912, he was moved to the prison of Santiago Tlatelolco, where General Bernardo Reyes—who had been there since his aborted rebellion—taught him the rudiments of civics and national history. In December 1912, Villa planned his escape. A small file and great shrewdness did the job. He then embarked on a long trek that brought him to El Paso, Texas, in January 1913.

After the assassinations of Madero and Pino Suárez, Villa met with the Sonoran revolutionaries José María Maytorena and Adolfo de la

Huerta in Tucson. They gave him a modest amount of arms and supplies for the rebellion. In April 1913, followed by seven men, with a few pack mules, two pounds of sugar, and a little salt and coffee, he entered Mexico to avenge the death of his redeemer.[10]

In the middle of 1913 there would have been no way to predict the victory of the Constitutionalist Armies under the command of the *Primer Jefe* ("Top Chief"), Venustiano Carranza. Although Carranza had named Manuel Chao his commander in the state of Chihuahua, it was Villa who was really in charge. Within a few months his force had grown from eight men to nine thousand. At the end of September 1913, when Carranza retreated to Sonora after a failed assault on Torreón (the nerve center of the railroad system), Villa formed his famous Northern Division (División del Norte). Within a few days he had seized Torreón, allowing him to move his troops quickly and efficiently by train. Around the middle of November he made an unsuccessful attempt on the city of Chihuahua but along the way accomplished his first dazzling exploit when he took Ciudad Juárez. It was his access not only to a customs post on the U.S. border but on to a greater stage: Mexican history and, at times, worldwide celebrity.

While some of his soldiers began a skirmish at the outskirts of the city of Chihuahua (to draw off the enemy), most of the soldiers, led by Villa, intercepted and emptied two coal trains at the Terrazas station. The men boarded the cars, and with cavalry following them, headed for Ciudad Juárez. At each station Villa took the telegraph operator prisoner and requested instructions from the base at Ciudad Juárez, pretending to be the federal officer in charge of the trains. He would keep telling them it was impossible for him to continue south as commanded, and they kept agreeing, ordering him to steam ahead north toward them. On the night of November 15, 1913, while the federal soldiers were sleeping soundly or relaxing in the gambling casinos, a silent signal—a flare—launched the attack. In no time Villa's troops had taken the barracks, the arms depot, the international bridges, the racetrack, and the casinos. In the United States, newspapers and public opinion marveled at the achievement. One American general compared it to the Trojan War.[11]

A week later, Villa looked down on Tierra Blanca and said, "I like those plains for a big battle." It would be fought from the twenty-third to the twenty-fifth of November. Five thousand federal soldiers halted their trains in the middle of a plain, where they were surrounded by soft sand. Six thousand Villistas watched them from the mountains. Villa had assured them of water, bread, horse feed, munitions, and machine guns by keeping railroad communication open to Ciudad Juárez. A well-equipped hospital train would attend to the wounded.[12]

Villa's cannon began to fire from the plateau, and then he himself led the mass cavalry charge. The federal troops lowered the angle of fire for their heavy guns, but it did them no good. Villa cut off their retreat, and they were trapped, completely at the mercy of the revolutionaries. Blood soaked the soft sand. There were a thousand dead and immense spoils. Soon they were singing the "March of Tierra Blanca" in all the plazas of the state of Chihuahua. The headlines in America filled eight columns: PANCHO VILLA RIDES TO VICTORY![13]

The federal army pulled out of the city of Chihuahua. There was a brief period of anarchy—violence and looting. Then Villa entered the city and on December 8 assumed the governorship of the state. He would hold the post for a month. On January 10 he took Ojinaga, the last federal garrison in Chihuahua.

On January 17 he had a conversation by telegraph with Carranza. The tone was cordial. After greeting his "esteemed leader" with "respect and affection as always," he gave him a tangible sign of his loyalty. He had "not hundreds, but millions of cartridges" and thirty-eight cannons, all at the disposal of the *Primer Jefe*. "If you were not with us," Villa added humbly, "I don't know what we would do." For his part, Carranza responded amiably by announcing to Villa that he had chosen him to be one of his "principal collaborators" for the coming campaign in the south.[14]

Around that time came a predictable event. Villa—that figure out of a movie—attracted the attention of Hollywood producers. On January 3, 1914, the Mexican Robin Hood, the fierce horseman so like those of the Old West, the "future peacemaker of Mexico," who always respected American property, signed an exclusive contract with the Mutual Film Corporation (for twenty-five thousand dollars) to film the exploits of the División del Norte. Villa agreed to schedule his battles during the day, to bar any cameramen not with Mutual, and if necessary, to simulate combat. Mutual would supply food and uniforms. Thousands of feet of documentary film were shot, along with some fiction footage. Raoul Walsh—later to become a famous American director—played the young Villa in the film titled *The Life of General Villa*. In 1967 Walsh would recall: "Day after day we tried to film Villa riding toward the camera but he spurred and whipped his horse so hard that he went by at ninety miles an hour. I don't know how many times we repeated: 'Señor, slowly please, slowly.' In the mornings we succeeded in postponing the executions from five to seven o'clock so that there would be good light."[15]

Francisco Villa, 1912
(FINAH)

On May 9, 1914, *The Life of General Villa*, with Villa himself appearing in several scenes, opened at the Lyric Theater in New York. He must have found the script moving. Two lieutenants rape his sister; he kills one but the other escapes; Villa declares war on humanity; the Revolution breaks out in the north; Villa captures city after city, arrives in the capital, finds the lieutenant, strangles him, and—the happy ending—becomes President. His legend circled the globe: Pancho Villa, superstar.[16]

But his military career was even more successful. In March 1914, he set out for the south with an impressive army—sixteen thousand fully equipped men and a hospital train able to handle fourteen hundred wounded. To his great good fortune, he had at his side a man whom he would come to revere: Felipe Ángeles, not only a master of the art of war (and a friend of the martyred Madero) but also an expert in mathematics and ballistics and above all in understanding human nature.

At the beginning of April, in one of the most intense battles of the Revolution, Villa captured Torreón again. It was an achievement not of tactics but of frenzy, of waves of blood and cannon fire. Reed would write: "Villa is the Revolution. If he died, I am sure that the Constitutionalists would not advance past Torreón in a year."[17]

In April, San Pedro de las Colonias fell:

In the old cemetery
the battle was so harsh
that there were dead below
and more dead above.

Easter Monday in the night
the federals retreated,
escaping with their wounded
on trains and on beasts.

They went by way of Saltillo
through the arid desert,
leaving their wounded behind them,
and their dead on the roads.

But all their officers—
they drove away in cars,
some with their mistresses
and others drinking toasts.[18]

In May the battle of Paredón began.[19] Another witness and chronicler, Vito Alessio Robles, describes it with awe: "A hurricane of horses whips past our flanks. It is a magnificent spectacle. Six thousand horses in a cloud of dust and sunlight. . . . The battle is over without our artillery having had the occasion to fire a single shell."[20]

In Paredón, Felipe Ángeles interceded with Villa and saved the lives of two thousand prisoners; he would do it many more times.

Villa had now become a radiant military caudillo, whose men were blindly devoted to him. Rafael F. Muñoz described what must have been a general feeling in the División del Norte: "They surrounded the cities no matter how large they were; they flooded the cities no matter how sprawling. They would move launching shouts of enthusiasm amid the streams of blood. They fell seeing others advance. Before clouding over forever, their eyes would be dazzled with victory."[21]

And then—for the first time—serious friction surfaced between Villa and Carranza. They had a meeting in Chihuahua that was disastrous. Rodolfo Fierro, Villa's favorite gunman, had just shot down William Benton, an English rancher—with no provocation, though Villa had previous troubles with the man. It was an incident with possible international overtones, but it was hardly the only one dividing Carranza and Villa. Carranza could not stand Villa's arbitrary willfulness. He thought him uncontrollable. Villa disliked the *Primer Jefe*'s cold ambition, the shifty look behind his glasses. How different from Madero! he must have thought. Carranza was not a friend: He was a rival.[22]

But the open break between the two came on the eve of the decisive battle of Zacatecas. Carranza sent a message to Villa, ordering him to detach units under Generals Natera and Arrieta to attack the garrison. Villa refused. "It's just sending men to the slaughterhouse," he informed Carranza by telegram. The Division was accustomed to triumphing together. Carranza called Villa a man without discipline, and Villa exploded: "Who asked you to stick your nose into my territory?"

Though Villa resigned "to avoid any suspicion of ambition" and Carranza accepted his resignation "with regret," it was Villa who prevailed. Ángeles wrote a letter of mass resignation, cosigned by eleven of Villa's generals. Carranza had put their backs against the wall and by doing so had freed them. Without any authorization from Carranza, they confirmed Villa as their commander in chief and marched, more united than ever, on toward Zacatecas. On June 23, after eleven days of a battle to be remembered, Felipe Ángeles wrote: "And finally, the serene descent of evening, in the full certainty of victory that comes smiling and tender to caress the forehead of Francisco Villa, brave and glorious soldier of the people."[23]

Not even Villa's most dogged detractors could deny that the defeat of Victoriano Huerta—in the way it proceeded—would have been unthinkable without the energy of Francisco "Pancho" Villa and his División del Norte. He was the "arm under arms of the Revolution" (*el brazo armado de la Revolución*).[24]

Did Villa have a vision of utopia? The answer is uncertain. A firm no, if by *utopia* one means an organic program like the *Plan de Ayala*. But yes, if we take into account his brief period as governor of Chihuahua. Somewhere along the way, Villa had discovered the outlines of his own earthly paradise and put it into effect with the rapidity and decisiveness of a cavalry charge.

His first measure was to confiscate the property of the Chihuahua potentates who were enemies of the Revolution. Creel, the Terrazas, Falomir "had to pay their debts to the vengeance of the people." The Villistas used menace and torture, acted on tips and denunciations. They seized the wealth they could see and found much of what was hidden. But Villa did not use the funds for his personal profit. The confiscations were "to guarantee pensions for widows and orphans, defenders of the cause of justice since 1910." Some of it was also used to set up the State Bank of Chihuahua. An initial capital of ten million pesos guaranteed the paper money that the bank issued, circulation of which was required by law. Throughout 1914 at least, the Villista currency was regularly quoted on the market. His best collateral was not the cash in the safes but the word and strength of Villa.[25]

"Socialism—is it a thing?" Villa once asked Reed.[26] Although he had no idea what that "thing" was, his utopia had some moderately socialist traits. Reed called Villa's government "the socialism of a dictator." And in some sense it was. Assisted by his able secretary, Silvestre Terrazas, Villa showed himself to be a rigorous administrator. He lowered the cost of basic necessities, organized their rationing and distribution, made abuses and extortions punishable by death, and put his entire army to work on electrification, streetcars, telephones, the water supply, and the slaughterhouses of the capital.[27]

One of the most personal facets of *his* socialism emerged with children. Villa loved them—his own and those of others. He took in hundreds of homeless children and paid for their education. During his brief government he hired teachers from Jalisco and opened many schools, which he would attend—as if he were another pupil—during celebrations or competitions. His educational proposals included a military university for about five thousand students and an elementary school for every hacienda.

His chief preoccupations were his soldiers ("my good boys"), children, and his "race of the poor." He would speak vaguely of the land as the people's heritage or as an individual enterprise, but never with the religious sense of the Zapatistas. In the Arcadia of his imagination, life would be lived in the countryside surrounded by school desks and rifles. Mexico would be an immense and fertile military academy.[28]

John Reed met Villa for the first time at the end of 1913:

> He is the most natural human being I ever saw, natural in the sense of being nearest to a wild animal. He says almost nothing and seems so quiet as to be almost diffident. . . . If he isn't smiling he's looking gentle. All except his eyes, which are never still and full of energy and brutality. They are as intelligent as hell and as merciless. The movements of his feet are awkward—he always rode a horse—but those of his hands and arms are extraordinarily simple, graceful and direct. They're like a wolf's. He's a terrible man.[29]

The greatest writers of the Mexican Revolution tried to convey their own impressions of Villa. The words "wild animal" (*fiera*) or "feline" appear in many descriptions. For Martín Luis Guzmán, "his soul, rather than of a man, was that of a jaguar"; for Mariano Azuela, his "curly head of hair [was] like that of a lion"; and for Vasconcelos, he was "a wild animal who in place of claws had machine guns and cannon." His eyes were the most disturbing feature of that wild animal. Vasconcelos and Puente remembered them "as being bloodshot"; for Rafael F. Muñoz "they stripped souls bare"; Mariano Azuela saw them "glowing like coals."[30] But it was Guzmán who best understood their expression: "His eyes were always restless, mobile as if overcome with terror . . . constantly anxious . . . a wild animal in his lair, but an animal that defends itself, not one that attacks."[31]

This characterization of a Villa always on the alert to defend himself fits his earlier life as a bandit on the run, pursued, fearful, sleeping whenever he had the chance, moving at night and resting during the day, sexually unbridled, artful, always alert to strike a sudden blow or attack at dawn. An animal hounded by his own lack of trust:

> I have seen him rifle in hand, throw a blanket over his shoulders and disappear into the darkness to sleep alone under the stars. In the morning he invariably reappears from a different direction, and during the night he slips silently from sentinel to sentinel, always

> alert. . . . If he would find a sentinel asleep, he would kill him at once with his revolver.[32]

Two implements were essential to him: his horse and his gun. It was impossible for him "to navigate," as he used to say, without a horse. You cannot visualize a sedentary Villa, or a Villa moving on foot. A horse allowed him to pursue or to flee; it was the prologue or epilogue to death. And death itself was the pistol. Martín Luis Guzmán compared his pistol to his gaze:

> The mouth of the barrel was a foot and a half from my face. I saw the feline glow of Villa's eye shining above the sight. The iris was like a tiger's eye gem, with infinite, microscopic points of fire. The golden grooves that spread from his pupil became fine bloodshot lines through the white of his eye until they disappeared under the eyelid. The evocation of death sprang more from that eye than from the black circle that ended the barrel. And neither blinked in the least: they were fixed; they were one piece. Was he aiming the barrel so that his eye could fire? Was he aiming his eye so that his barrel could fire?[33]

It was not Villa's eyes that aimed the gun, but his entire nature. Yet that wild animal was also a feeling, plaintive human being, compassionate to the weak and tender with children, cheerful, a singer, a dancer, a complete teetotaler, imaginative, and a great talker. In the classic sense of the word, he was a centaur.

The English consul, Patrick O'Hea, who was harshly critical of Villa, once wrote: "If I were asked to define Pancho Villa, my answer would be, 'which one?' Because the man changes to the rhythm of his successes or failures."[34] But O'Hea found multiplicity where there was, in fact, a kind of duality. Martín Luis Guzmán saw an almost mystical element in Villa's nature that was the deepest key to his immense acceptance among the people: the duality of the hero who embodies, at once, vengeance and hope, destruction and mercy, violence and illumination.

In contrast to the look of him, Villa had a thin voice. Once Puente saw him give an unjust command, retract it, change his mind, show his doubts, start to cry—feeling himself helpless within his ignorance. And the man who never stopped "once his hand had touched the handle of his gun" was also seen to "tremble in the presence of books as if they were something sacred."[35]

One expression of his duality was Villa's attitude toward women. According to Soledad Seáñez, one of his last and most beautiful wives,

"Francisco was terrible when he was angry but very tender when in a good mood." Villa often respected the forms of love, from conquest to separation, in an almost gallant, gentlemanly, and paternal manner. His abductions were not entirely animal: He wanted the women to love him, he courted with imagination, and dozens of times he agreed to marry, although once the union was consummated he would tear up the records.[36]

In many other cases, his behavior was atrocious. Once he intercepted a letter full of complaints in which his young lover Juana Torres called him a bandit and other insults. He made her read it out loud to him, spitting in her face at each epithet. A number of women gave themselves to him willingly, but there were others who certainly had no choice.[37] He was not a man likely to take no for an answer. When John Reed asked him about the many stories of his having raped women, Villa answered, "Tell me, have you ever met a husband, father or brother of any woman that I have violated?" He paused. "Or even a witness?"[38]

But the decisive biographical proof of Villa's duality can be found in the two men closest to him, equidistant and extreme extensions of his nature: Rodolfo Fierro—"iron" like his name—and Felipe Ángeles, a soldier on the side of the "angels."

Fierro was a murderous animal, nothing more, but marked by a "sinister beauty." He was a tall man, taller than Villa. He would speak in a soft voice, without threats or rhetoric, and his gestures and manners were civilized enough. But he killed like he breathed, with as little effort and no less naturally. Villa's soldiers called him "the Butcher" (*el Carnicero.*)[39]

He once stationed himself with a pistol in the middle of a corral. Three hundred prisoners, ten at a time, were pushed through a gate and told to run. Anyone who could reach the far side and clamber over the fence would be allowed to live. Fresh pistols were handed to him, and for two hours Fierro never stopped firing. Only a single prisoner managed to cross the sand and hurl himself over the fence, in the twilight, to live. Then Fierro felt a cramp in his trigger finger and let his attention slide away to massaging his sore muscle.[40]

John Reed would describe Fierro involved in his vocation: "During two weeks that I was in Chihuahua, Fierro killed fifteen inoffensive citizens in cold blood. But there was always a curious relationship between him and Villa. He was Villa's best friend; and Villa loved him like a son and always pardoned him."[41]

O'Hea remembered how the man enjoyed killing defenseless people, alleged to be spies or critical of Villismo. It was Fierro who had cut down the Englishman Benton because he thought he was reaching for a gun, when it was really a handkerchief to wipe the sweat from his forehead that Benton wanted. ("Just a misunderstanding," commented Fierro.) This "great handsome animal," this perfection of murder in mind and in body, was one of Villa's possibilities—his instinct for killing. "I only know," wrote O'Hea, "that this man, with his wandering gaze and his cold hand, is evil itself."[42]

But Patrick O'Hea also conceded that Villa had another side, to which pure men were attracted. Each of the revolutionary factions drew a different type of intellectual. Those who moved toward Zapatismo, for instance, were anarchists or Christian mystics. The democratic idealists were drawn to Villa. Like him, they detested "politicians" and "ambitious men" and the followers of Carranza, who were generally more pragmatic people. Almost all of the Villista intellectuals had been loyal to Madero and saw in Villismo, led by themselves, the chance to establish a continuity with the enlightened liberalism of the nineteenth century.[43]

The idealists were more concerned with education—another tie to Villa—and with democracy than with the distribution of land or the problems of labor. They came to Pancho Villa with the same attitude as that doctor of the Enlightenment, in the eighteenth century, who confronted the *enfant sauvage*. They would have to teach him all that should and should not be done—to reconstruct knowledge from the beginning of human time. Reed understood the problem: "The whole complex structure of civilization was new to him. You had to be a philosopher to explain anything to Villa."[44] As for Villa, he actively sought the help of these men. He would learn passages from the Constitution by heart and lost no opportunity to proclaim—with sadness, with humility—his intellectual helplessness: "It would be bad for Mexico if an uneducated man were to be president."[45]

Although Villa never completely accepted the dictates of his teachers—he would sometimes even call them "freeloaders"—he must have seen in each one of them a potential Madero. With his redeemer dead, he needed to believe in a man who could unite purity with authority. He found him in Rodolfo Fierro's polar opposite: General Felipe Ángeles.

In Villa's eyes, Ángeles was the complete and ultimate man: a soldier *and* a scholar. According to Martín Luis Guzmán, Villa felt a "superstitious admiration" for Ángeles. He knew of the hours Ángeles had lived through with Madero and Pino Suárez; he had heard about his benevolent treatment of the Zapatistas; he admired his love of music and books, his honesty, his sensitivity to the just causes of the people, his sense of mercy. He must have listened many times to words like these from Angeles:

Felipe Ángeles, 1915
(Fondo de Cultura Economica [FCE])

> The Revolution was to liberate us from the masters, so that the government would return to the hands of the people itself, so that in each region they may elect honest, just, sensible and good men whom they know personally and oblige them to act as servants of the popular will expressed in the laws, not as their overlords.[46]

This was the voice of Madero, speaking through a very different man: a consummate artillery expert, a technician of war. He was an armed angel, who carried a sword in one hand and the scales of justice in the other. Villa valued him so highly that he himself baptized one of Ángeles's sons so as "to have the confidence to call him his compadre." A few months later, at the Convention of Aguascalientes, Villa would propose Ángeles as a candidate for the presidency. This strong Madero, this military Madero, was Villa's other possibility—his luminous space of peace.[47]

There were times when one word—*justice*—served to merge both sides of Villa. Villa wanted "a justice as clear as the light of day, a justice that even the most ignorant could apply." A justice as convincing as the words of Madero. "He is hostile to the present," Reed wrote, "to laws and customs, to the distribution of wealth . . . to the system. . . . More than almost anything else, what he wanted was to apply his awesome justice, a justice of extermination, of relentless revenge." There was, for instance, his differing approach to three kinds of prisoners during the Constitutionalist war against the Huertistas. He considered the common federal soldiers, the "short-hairs" (*pelones*) to be men who had been drafted into the army and believed that they were fighting for the *patria*. Frequently he would just let them go after a battle. But toward the troops known as *colorados*, federal irregulars, many of them former rural policemen (*rurales*) who had fought for Pascual Orozco in his rebellion against Madero—men who were Huerta's toughest, cruelest, and most loyal troops—Villa's judgment was that they were peons volunteering to fight against the cause of liberty, and therefore evil. When he captured them, they would be shot. As would the federal officers, "because they were educated men and ought to know better." Like the Horsemen of the Apocalypse, Villa did not impart justice; he imposed it. He was not an arbitrator but a stern judge. Before opponents like the *colorados*, Villa showed the coldness of iron and the heat of an avenging angel.

In July 1914, Victoriano Huerta went into exile. The victorious Constitutionalists immediately had to face a critical priority: reconciling Villa and Carranza. Villa had wanted to march on to the capital from Zacatecas, but Carranza had different ideas. He cut off the coal supplies to the Villista trains and gave Álvaro Obregón's Army of the Northwest the honor of triumphantly entering Mexico City on August 15, 1914.

A month earlier, in Torreón, representatives of Villa and Carranza had signed an agreement. It was a waste of paper. In August an old animosity flared up between Villa's compadre José María Maytorena—the governor of Sonora—and the Carrancista military commander, Plutarco Elías Calles. Obregón went to Chihuahua for the purpose of calming down the antagonists and trying to reach a compromise between Villa and Carranza. The two caudillos were face to face.

"Look, my good dear friend [*compañerito*]," Villa said to Obregón, "if you had come here with troops, there would have been *a lot of gunfire*. But since you're coming alone, you can feel safe. Francisco Villa will not be a traitor. The destiny of the *patria* is in your hands and mine; if we get together we'll tame the country *in less than a minute*, and, since I am a humble man, you'll be the President."[48]

Obregón prudently evaded Villa's proposition, while listening to him and thinking: "He is a man without much control over his nerves." The two of them then put together a list of proposals for directing the political future of the country and sent it off to Carranza. The *Primer Jefe* accepted some of them but considered the general content so important that it could only be given a fair hearing at a national convention attended by all the revolutionary generals.

A movement of troops by the Carrancista general Benjamín Hill revived the crisis in Sonora. Again Obregón went to Chihuahua. More iron than angel this time, Villa received him with suspicion. On September 16, 1914, from the main balcony of the city hall, they watched a military parade. Obregón knew that the display was meant to impress him. And it did. He counted fifty-two hundred men, forty-three cannon, and tens of thousands of "little Mausers." He also sensed that Villa did not trust him, that the caudillo of the División del Norte "could erase him from the book of the living." Any pretext would serve. And since Villa wanted a pretext, he found one.

"General Hill," he shouted at Obregón, "thinks that I can be played with. . . . You're a traitor and I'm going to have you shot right now." When Obregón did not explicitly agree to order Hill's withdrawal, Villa sent for a firing squad. At this point, Obregón accomplished the first major defeat in Villa's career. It was a victory and defeat of the mind. Seeing how agitated Villa was, Obregón spoke to him with a poise that disarmed him and with arguments that confused him. He did not confront Villa directly but used Villa's own impetus (as if it were a judo match) to disable him: "Since I put my life at the service of the Revolution I have considered that it would be my good fortune to lose it. . . . [By shooting me], you would be doing me a favor, because that death would give me a status I do not have; and in that case you would be the only loser."

Villa began to waver. An hour later he called off the firing squad. Now more angel than iron, he broke into tears and said: "Francisco Villa is no traitor; Francisco Villa does not kill defenseless men, and certainly not you, my good dear friend, who are now my guest."[49]

Obregón breathed easier but showed no emotion. Carefully concealing his apprehension, he stayed a few more hours in Chihuahua and was escorted away by the Villistas José Isabel Robles and Eugenio Aguirre Benavides, whom he knew well. Villa would change his mind twice more before Obregón could safely leave the state, with his life and his subtle triumph.

Obregón's second psychological victory over Villa occurred during the Convention of Aguascalientes. It was a matter of coolness and maneuver to reach and establish a stronger position. While Villa hovered outside the city with his troops and only came in to sign the agreements with his blood on the flag, Obregón took part in the debates and won many allies in both the Villista and the Zapatista camp. When the Convention disavowed Venustiano Carranza and named Eulalio Gutiérrez provisional President for twenty days, Obregón shrewdly did not come out for Carranza. He rode above the currents, allowing for the flow of time.

But there was no way to stay neutral. Pancho Villa would not give up until he saw "the tree of Don Venus" fall. Carranza announced that he would only resign if a strong government could be formed to carry out the social demands of the Revolution. Eulalio Gutiérrez was forced to tip the balance and change the situation from difficult to impossible by naming Villa commanding general of the Army of the Convention. Some of the other generals tried to arrange the reconciliation that would have saved many thousands of lives. And at one point Villa even proposed an unbelievable solution: He and Carranza should both commit suicide.

Eventually the factions realigned themselves into two groups. Obregón moved firmly to the side of Carranza. The possibility of civil war loomed large. A city newspaper published a cartoon of a terrified mother—the Revolution—who had just given birth to twins, one with the face of Venustiano Carranza, the other of Pancho Villa. From the door of the hospital room, the poor father—the People—exclaims in horror: "If I can't handle one, what am I going to do with two?"[50]

Several months later, the new war began between the "twins" (*cuates*) of the Revolution. The Army of the Convention marched on Mexico City. Although Carranza had the support, among others, of Francisco Coss in Puebla, Cándido Aguilar in Veracruz, Francisco Murguía in the state of Mexico, and the noted generals Pablo González and Álvaro Obregón, the Convention could count on Zapata, Villa, and various other generals who controlled the center, north, and west of Mexico. It was at this key moment that the two most popular caudillos of the Revolution—Villa and Zapata—met in Xochimilco.

In the agreement they signed there, Zapata and Villa were attempting to set down solid foundations for their victory, but the underlying and vehement theme of their conversation is really the opposite: defeat. Villa made his feelings very clear: Power was for others. "I don't need government jobs. . . . I understand very well, that we ignorant men fight the wars and the cabinets have to put it to use."

All Villa and Zapata could do was "find people" who could "put those jobs to use" but on the condition "that they don't bother us" because "this ranch is too big for us; it's better out there, away from here."

Villa wanted to retire after setting "the people on the road to happiness." He spoke of his future and peaceful "sweet little ranch" and of his "nice cabins" but admitted that there was still "a lot to do" in the north. He was not particularly interested in who might be chosen for the cabinet and felt his own mission was simply "to fight very hard." The word *pelear* ("to fight") appears nine times in the conversation. After the meeting ended, Villa and Zapata went into the dining room, where they were served a Mexican-style banquet. And afterward, Villa delivered these words from the mouth of an "uneducated man": "When I can really see where my country is going, I will be the first to pull out, so that you can see we're honest men, that we've worked like real men of the people, that we're men of principle."[51]

Those words delivered at the height of his power are more than a revelation: They are an omen of his third psychological defeat. He had *beforehand* agreed that he would submit to "the cabinets" if "they didn't bother him." *Beforehand,* he had renounced political power. He was moved not by "natural anarchism," as Zapata was, but rather by his lack of self-esteem and his ignorance. Politics were for dishonest, ambitious men, people with no principles. Following directly from this vision came his destiny: to battle on, to battle on blindly, or until the coming of a new Madero in whom he could believe. Again, a life to be lived between extremes: the angel or cold iron. But if, *beforehand,* he renounced the intermediate ground of political action, Villa had no hope of defeating a rival like Álvaro Obregón in any sphere other than war.

Two days after the Agreement of Xochimilco was signed, the soldiers of Villa and Zapata entered the capital. There was the famous scene of Villa playing the joke of his life—between bursts of laughter—by sitting himself down in the Presidential Chair. Everyone else laughed too (except Zapata, who could only see the Chair as a symbol of centralized and hated authority) because they knew that it meant nothing, it was all for the photographers, to see how it looked, to see how it felt.[52]

Paying no attention to the fate of the "cabinets of the Convention" (an ephemeral matter of "politics"), Villa enjoyed his stay in the capital. He went to banquets, courted cashier girls, flirted with the actress María Conesa, ordered Fierro to kill young David Berlanga for daring to criticize him. (Berlanga is remembered for smoking a cigarillo before dying with so steady a hand that the intact ash never fell till after he was shot.) And Villa sent needy city children off to study in Chihuahua, shed floods of tears at Madero's grave, and renamed a street, the Calle de Plateros, after his savior Madero.[53]

While his elite corps, the *Dorados* (the "Gilded Men"), sang "Jesusita en Chihuahua" and "La Cucaracha"—

(With the beard of Carranza,
I will make a scarf
to be worn on the sombrero
of his father Pancho Villa)

—he planned his final campaign against the forces of Carranza. Pancho Villa did not know, and would never know, the degree to which his psychological defeats had prepared the ground for his later, conclusive losses.

On the eve of the great battles of the Bajío—the fertile central basin north of Mexico City—Villa could count on the open sympathy of the American government. President Wilson thought that he was "the greatest Mexican of his generation." And Villa never let an opportunity pass to thank the "wise" President Wilson for his goodwill and for his decision to avoid war with Mexico. He would almost always agree to any request from George Carothers, the local American consul. Very rarely would Villa meddle with American interests, and he always maintained a functional friendship with General Hugh Scott, the commander on the border. Nevertheless, the crucial question of gaining American recognition would be settled not in the meeting rooms of polite diplomats but on the fields of the Bajío. Not with "politics," Villa would say, but by "pouring gunfire" down on his "dear good friend" Obregón.

Beyond Villa's own psychological formation, other factors contributed to his military defeat. One of the most important was Zapata's failure to cooperate with him, and another was the dispersion of the Villista troops. They had to fight on three fronts: the west-to-northwest region from Jalisco to Baja California, a north-to-northwest zone from Coahuila to Tamaulipas, and the Huastecan region running from San Luis Potosí to Tampico. At the beginning of 1915, half the country was the theater for a civil war that would have lasted longer if Obregón had not decisively defeated Villa in the Bajío: at the fierce encounters of Trinidad, Resplandor, Silao, Santa Ana (where Obregón lost his arm), León, and the two battles of Celaya.

In Celaya, Villa wanted to use the same tactics of assault across open ground that had served him so well at Tierra Blanca, Paredón, and Torreón. Felipe Ángeles advised against it, primarily because of the terrain. Surrounded by irrigation ditches that would allow the enemy to entrench themselves, Celaya was not a good setting for Villa's cavalry charges. Nothing could have been further from the soft sand dunes of Tierra Blanca. Since they had missed a chance to attack Veracruz, Ángeles advised bypassing Celaya and moving on north. But Villa was impatient and refused to listen to him.

On April 6 he launched his first charges against Celaya. Obregón seemed to be in great danger, and the Carrancistas took heavy losses. "At this hour we have about two thousand casualties," read Obregón's telegram to Carranza. "Assaults of the enemy are coming very fast. Be assured that while I still have a soldier and a cartridge I will do my duty." But by one in the afternoon the following day, Villa had made more than thirty cavalry charges. None of them had broken through. Then Obregón ordered his reserve cavalry to attack in a pincer formation. Villa's exhausted forces were caught by surprise. "The Villistas," Obregón reported, "have left the field strewn with bodies . . . more than a thousand dead have been found."

A week later, Villa—literally—charged again. Conscious of Villa's "crude and entirely impulsive character," Obregón spread his soldiers over a wider area but kept to the same basic strategy. He resisted the charges from the trenches, then had his soldiers fake a retreat and, at exactly the right moment, sent his reserves sweeping into action to surprise the enemy. Villa lost ten thousand men—four thousand dead, six thousand prisoners—as well as thirty-two cannon, five thousand weapons, and a thousand saddled horses. For a brief time he would recover a measure of his strength, but at León his rosary of defeats continued. Once more Felipe Ángeles warned him about the terrain, but Villa ordered charges, the all-out assault technique of the División del Norte, with the *Dorados* riding last to be sure that no one turned and fled against orders. But at León his men fell in grim numbers before his old enemy, General Benjamín Hill.

After yet another loss at Aguascalientes—the final encounter between Obregón and Villa—the army and their leader, dazed and decimated, turned north, intending to double back and attack again from the west. Villa blamed his defeats on lack of ammunition, insufficient reinforcements. But it was not really the truth. Obregón had defeated Villa militarily with the same techniques he had used for those psychological battles in Chihuahua. He would not confront assault with assault, or even provoke an attack. He had let Villa come and fall into the logic of his own momentum, until Obregón could strike the finishing blow at the moment his enemy was most exposed. Imitating the mental tactics he had used in Chihuahua, he now employed a kind of judo of armies.[54]

Villa's military fate was sealed in these battles. The Villista currency lost value at a dizzying rate: from fifty cents to the dollar in 1914 to five cents after Celaya. The money would become a collector's item. Food shortages and inflation scourged the Villista territories. In August 1915 Carothers informed his government: "Villa is bankrupt and is taking over everything . . . to collect funds." Even the American sympathies for Villa and all his diplomatic compliance with them could not mask his financial and military defeat.

More painful to him than defeat and bankruptcy must have been the desertions. One by one, his lieutenants left him or betrayed him. Several of them joined Carranza. Tomás Urbina, his compadre and former comrade in crime, threatened to rebel. As much a butcher as Fierro but with a broader and more intelligent amorality, Urbina presided over a real economic empire. At his Hacienda de las Nieves, Reed wrote, "the town belongs to General Urbina, people, houses, animals and immortal souls . . . he and he alone wields the high justice and the low. The town's only store is in his house."[55] Villa sent men to take him by surprise, then reluctantly turned him over to Fierro "to dispose of as he wished."[56]

On September 11, 1915, Felipe Ángeles himself abandoned Pancho Villa. And on October 14, while marching toward Sonora, Rodolfo Fierro met a death worthy of his life. He was riding his horse and wearing a vest that was extremely heavy because it was stuffed full of gold coins. His horse began to sink in quicksand at the Casas Grandes

Lagoon. The story goes that he screamed for his men to throw him ropes, promising gold to everyone, but that they heaved the lassos short, intentionally, and the Butcher was swallowed up slowly by the mud.[57]

On October 19 the American government, "totally disillusioned" with Villa, officially recognized Carranza. Villa must have felt this as the greatest of betrayals. For almost five years he had made every effort to respect the United States. Often he had yielded to Scott's requests, to the advice of Carothers, to the initiatives of Wilson or the State Department. Unlike Carranza, he had almost always said yes to them, and now they repaid him with a stab in the back. His response was brutal and threatening: "I emphatically declare that there is much I have to thank Mr. Wilson for, because he relieves me from the obligation of giving guarantees to foreigners and especially to those who had at one time been free citizens and are today vassals of an evangelical professor of philosophy. . . . I take no responsibility for future events."[58]

Villa's last real military campaign was his attack on Sonora at the end of 1915. Perhaps because both his steep rise and his fall had been so sudden, he still did not feel himself defeated. In Agua Prieta—from November 1 to November 3—his cavalry charges broke against the barbed wire and cannons of General Plutarco Elías Calles. There were only three thousand men left in the División del Norte. Obregón took the execution of the death blow directly into his own hands. He reduced the last Villista strongholds in Sonora. Then Ciudad Juárez and Chihuahua surrendered to him. At the beginning of 1916, Villa the warrior became a guerrilla.

There are many legends and perhaps even more interpretations of Villa's assault on the American town of Columbus, New Mexico. It has been credited to German machinations againt the United States or Villa's belief that he had discovered a plot by Carranza to make Mexico a protectorate of the United States. With Pancho Villa almost anything is possible, but to attribute rational realpolitik to his actions is going more than a little too far. Most likely his attack on Columbus was motivated by human passion, for revenge. Before the battle of Agua Prieta, at the end of October 1915, he had told an American reporter: "The United States recognized Carranza . . . paying me back that way for the protection I guaranteed their citizens. . . . I'm finished with the United States and with Americans . . . but, by the life of God, I can't believe it."[59]

Knowing he was lost, he had again become a wild animal, just as he was before the Revolution—but now without hope and with rancor, an animal betrayed. Since the end of 1915, he had turned steadily more bloodthirsty. In San Pedro de la Cueva Villa had all the men shot en masse and with his own pistol killed the priest, who was on his knees, clinging to him and begging for mercy. In Santa Isabel his troops shot down American miners. He would later burn some victims alive and execute old men. Now as never before, he distrusted everyone. He would disappear at night, always sit with his back to the wall, never eat until one of his men first tasted all his food. He organized surveillance patrols and spy missions. The greatly reduced number of followers he had left now called him "the Old Man." They did not lose faith in him, but they did lose their identity. If Villa was now an outlaw, what were they? Revolutionaries or bandits?[60]

At dawn on March 9, 1916, Villa attacked the small border town of Columbus. Rafael F. Muñoz puts these quite believable words into his mouth: "The United States wants to swallow Mexico; let's see if it doesn't get stuck in their throat!" The assault went on till noon. There were fires, rapes, looting of banks and businesses. The attackers stole a number of guns and horses and killed some civilians. Before reinforcements could arrive to engage him, Villa galloped away, well satisfied. The land of the gringo invaders of Mexico had been invaded. He could never have considered the enormous danger he was bringing upon his country.[61]

General John Pershing, later to become a hero of the First World War, entered Mexico in command of the "Punitive Expedition," for the sole purpose of capturing Pancho Villa. The Carrancistas were also looking for him, with similar passion. Months passed, but Villa could not be found.

Only a few of his men knew where he was. Villa had taken a serious wound, in his right leg, during an encounter with federal troops. He was hiding, high in Chihuahua's Sierra of Santa Ana, in the Coscomate Cave. They had trekked Villa up there on the back of a burro. Every jolt meant intense pain. Then they had raised him into the cave with ropes and covered the entrance with branches. He was there for six weeks, with a few pounds of rice and sugar. Badly set, the leg from then on would be stiff and shorter than the other. To walk evenly, Villa would have to wear a specially made shoe.[62]

But even when he came down from the cave—a difficult and painful descent—none of his enemies found him. The Punitive Expedition, one of the most expensive hunts ever mounted for just one man, was a spectacular failure. Day after day the military report was the same: "I have the honor to inform you that Francisco Villa is everywhere and nowhere." The experience of his nineteen years as a bandit had never proved more useful. According to Muñoz, Villa said: "No one can follow me on horseback or on foot, nor on the plains nor in the mountains. They won't take me alive, not even with a trap . . . [I'm] just like the wolves."[63]

But the wild animal did more than defend himself. On September 16, 1916, he raided the city of Chihuahua with only eight hundred men, captured it, and held it for a few days. There followed months and years of fruitless skirmishes as well as vicious murders. In December 1918, Felipe Ángeles came to him again, after almost three years of separation. The federal government thought that Ángeles was only back to fight, but he had a more noble intention: to offer Villa his own Maderista mission.

He was coming not as a soldier but as an apostle. The Revolution was drawing him back like a moral magnet, like an inescapable destiny. He could not tolerate exile and idleness in the United States. Fearful of an American invasion, he was seeking to unite all Mexicans and he was returning to Pancho Villa not to advise him on artillery methods but to preach a respect for life, a spirit of "reconciliation and love."

Ángeles wrote to a friend: "I am going to work with ignorant and savage people, whom the war may perhaps have made worse. I am going to reach the humanitarian and patriotic qualities within them."

Villa and Ángeles moved along together for five months. After skirmishes and small battles, Ángeles succeeded in saving the lives of hundreds of prisoners. But Villa wanted to repeat history. When he launched a ferocious and lavishly bloody attack on Ciudad Juárez, Ángeles left him for good. Villa tried to persuade him to stay, but Ángeles had decided to carry his message elsewhere.

He would have no success with other caudillos. Eventually he found himself on the run, alone, in mountains he did not know and where no one knew him. A man named Salas agreed to hide him and then turned him over to the government. Ángeles would be court-martialed in the Theater of Heroes in the city of Chihuahua.

"The trial of Felipe Ángeles," Puente wrote, "is one of the most sensational trials of the Revolution; his judges are also his most merciless enemies, spurred more by partisan zeal than a broad spirit of justice." Ángeles took over his own defense and gave one of the most moving speeches in Mexican history. With self-control, with vehemence and clarity, he expressed his credo. Like Madero, he did not preach hatred, "because hate sits badly in one's soul" but spoke for "the opposite passion, love."

These seemed like words Madero never had time to say. And as Ángeles defended himself, he defended Villa: "Villa is good at heart; circumstances, men, injustice have made him bad."[64] Why should he not go to Villa? To whom else but Villa could he preach goodness? "I blame the current condition of Villa and his men on the governments that have not had compassion for the disinherited and have turned them into wild animals."[65]

His speech was a democratic, instructive, and egalitarian sermon in which he defended the Constitution of 1857, public education, and the new socialist currents. The public acclaimed that former soldier of Porfirio Díaz who truly identified with the poor and the oppressed, that strange military and scholarly Quixote for whom even the ferocious Zapatista general Genovevo de la O had felt affection. It was from the people that Ángeles had learned his disdain for "statesmen with frozen hearts." His message and his person seemed no longer of this world. He had been betrayed by his own comrades, but he found room in his response for only "three words: purity, love, hope."

The jury condemned him to death. Carranza refused to commute his sentence. During the trial, Ángeles had leafed through Renan's *The Life of Jesus*. He died believing, word for word, in the final statement he put down on paper: "I know they are going to kill me, but also that my death will do more for the democratic cause than all the gestures of my life, because the blood of martyrs fertilizes great causes."[66]

The "Corrido of Felipe Ángeles" says this about his death:

The clock sounded the hours. It was time for the execution.
"Prepare your weapons well and shoot me in the heart.
Don't feel sadness for me, because with a man like me,
you do not fire at his head, you fire at his heart."[67]

It was as if the man who had embraced Madero in his last hour could have died in no other way. And their new embrace was final, total, a union in martyrdom.

Now without his angel or his man of iron to guide or guard him, Villa decided on a last cinematic action—in order to make his presence felt. He crossed the Bolsón de Mapimí River and attacked Sabinas in Coahuila. Carranza was now dead and the Sonoran generals had taken power. The interim President, Adolfo de la Huerta, was a man Villa respected. He had not forgotten that De la Huerta, along with Maytorena, had provided him with money for his eight-man invasion of Mexico in that distant April of 1913. General Eugenio Martínez began talks with Villa that would end in an agreement to surrender.

On July 28, 1920, the last 759 Villistas laid down their rifles. They were awarded a year's wages, and their leader was given the Hacienda of Canutillo. On the triumphal journey to his new hacienda, those left alive of his former comrades in arms came out to embrace him. Journalists harassed him with questions on the meaning of the armistice. With his arms around the shoulders of the federal generals Martínez and Escobar, Villa answered with the most ambiguous and poetically suggestive of his jokes: "You can say that the war is over; that now honest men and bandits walk together."[68]

It must have seemed like a dream to him. It *was* his old dream of retiring, as he had told Reed, "raising cattle and corn . . . among my *compañeros* whom I love, who suffered so long and so deeply with me." At times it seemed that peace was smiling upon him. His *Dorados*, like children, took classes in reading and writing. Villa himself "charged" into work the way he had led those legendary cavalry attacks. He repaired tractors, replanted fallow land, set up schools, and spent time with his young son. In this peaceful place of retirement, at a prominent location, Villa set up two images, two presences: a bust of Felipe Ángeles and a portrait of Madero. His martyrs.

He wanted peace but was denied it as if he had no right to it. Physical and mental pain assailed him. The fracture in his leg gave him trouble. He grew terribly jealous over his last wives. (The number of his marriages had become legendary. He had scattered "little Villas" everywhere.) He dared to criticize the "profiteers" of the Revolution, even though he himself opposed the distribution of land in his area. And he began to indulge in habits that were bad for him: tobacco and anis. The former abstainer became almost an alcoholic.

Nothing tormented him more than the fear of an attempt upon his life. But he was careless enough to meet with Adolfo de la Huerta and to offer his help in any future contest between De la Huerta and the Sonorans, Calles and Obregón. Word of the meeting was passed on to the government, and they began to plan his assassination.

A group of people rented a house in the town of Parral near Villa's hacienda. It was a building everyone who entered or left the town toward the northwest had to pass. There they lived for three months, waiting for the chance to surprise their victim at a moment when he might be vulnerable. The ever-alert Villa finally—and only in part—relaxed enough to give them that chance.

He attended a christening (as the godfather) in the town of Río Florido and then drove back to Parral, where he stayed for several days to deal with some business. On the morning of July 20, 1923, he decided to return to Canutillo. His wife was at the hacienda, pregnant with their second child, and she had told him before he went away that she felt a vague fear that this would be their last good-bye.

Villa and his men left their hotel at eight in the morning, an hour when children were on their way to school, but the town "had an oddly mysterious quality." There were no police on duty, and the soldiers of the garrison were on review outside the town, though it was late in the month for this kind of military ritual. Despite this last unusual detail, Villa was not disturbed by the overall scene. What had happened to his famous astuteness, his eternal suspicion? Time had passed, quiet time, and blunted them.

The car was full of people; Villa was driving, and on his right sat Trillo, his secretary, who had decided—in order to economize a little—not to bring along the entire force of Villa's bodyguards, saving the expenses for fifty men and fodder for fifty horses. An elderly candy seller standing by his stall gave the prearranged signals. The vehicle was coming and Villa himself was at the wheel.

The car turned the corner to meet with a massive blast of gunfire. All the occupants of that mysterious house—the bales of hay piled on the doorstep and the constant flow of armed men had made it look like a barracks—fired at once on the Dodge. The car veered out of control and crashed into a tree. Nearly everyone was killed, almost instantly. The moans and the cries died away. One of the attackers rushed out to put a bullet in Villa's head. His lifeless body was already doubled up near the driver's door, the right hand still reaching for his gun. His skull was full of holes, and in the autopsy it was hard to find his heart. It had been turned to pulp by the expanding bullets used by his assassins.[69]

He was buried the next day. Across Mexico, the people mourned him because in that life they saw a metaphor for their own. It was the most complex of metaphors, encompassing ignorance and aspiration, brute courage and tenderness. And it was a metaphor too of stern justice, of revenge.

Three years after his death, robbers broke into his grave and stole Pancho Villa's skull. Did they want the bones of an angel or a souvenir of cold iron?

13

VENUSTIANO CARRANZA

Nationalism and the Constitution

The Latin-American nations still need strong governments able to maintain law and order over undisciplined populations, prone at any moment and on the slightest pretext to disorders.

—VENUSTIANO CARRANZA

Liberty and sovereignty were never abstract terms for the men of Coahuila. During the Hapsburg era—as Nueva Extremadura—the province had been politically dependent, against the will of its people, on neighboring Nueva Vizcaya and Nueva Galicia. Later, in Bourbon times, Coahuila reluctantly submitted to the dictates of the Intendency of San Luis Potosí. When independence came, the new state of Coahuila y Texas could at last enjoy a sovereign existence. Two decades later, in 1836, it would suffer the painful loss of its northern region.

For the Coahuilans, the independence and then American annexation of Texas left a memory of two wrongs—the interventionist power had seized their territories, and the central government had been unable to defend them. This twin historical trauma reinforced the Coahuilans' old vocation for self-government (and distrust of outside powers).

Besides this feeling for liberty and sovereignty, the inhabitants of Coahuila also showed a strong frontier identity. It was evident in their physical courage and their resolution—like medieval knights—to defend themselves against the "barbarians" (the Toboso, Lipan, and Comanche Indians), but also in the more subtle characteristic of preserving Spanish culture in ways as various as wine growing or the forms of city government. Precisely because they were frontiersmen, living in a zone by definition under threat, they felt the *values* of the center with greater depth and urgency.[1]

One of these frontiersmen, Jesús Carranza Neira, had fought against the Indians and on the Liberal side in the War of the Reform. A mule-driver and rancher, Don Jesús became, during the War of the French Intervention, the primary conduit for information between Juárez and his generals in Coahuila. In 1866 he even gave Juárez a personal, interest-free loan to help him support his family in exile. When the Republic was restored, Don Jesús Carranza, now the patriarch of a family of fifteen children, received a grant of lands that secured his personal fortune.[2]

As time passed, all of Don Jesús's children would learn the history of Benito Juárez, that austere Zapotec in a black Prince Albert coat who carried the *patria* like a tabernacle in his carriage. But one of the children in particular enshrined the example of Juárez within his memory. He was the eleventh child of Don Jesús: Venustiano, born December 29, 1859.

Venustiano Carranza studied at the Fuente Atheneum, a famous liberal school in Saltillo, and in 1874 entered the newly founded Escuela Nacional Preparatoria (the National High School) in Mexico City. While he was there, Porfirio Díaz led the rebellion of Tuxtepec; Juárez's successor, Lerdo de Tejada, fell from power; and the triumphant armies of Díaz entered the capital. Venustiano wanted to become a doctor, but a serious eye disease cut short that possibility. The young Carranza decided to return to Coahuila and raise cattle. In 1882 he married Virginia Salinas, with whom he had two daughters. (Decades later, he would begin another family and have four more children, all sons.) In 1887, at the age of twenty-eight, he took his first political office as municipal president of Cuatro Ciénegas.[3]

A proud liberal individualism—typical of the northern ranchers but acute in Coahuila—and the Juarista affiliation of the Carranza family were perhaps the principal factors in his decision to participate in the events of 1893, when Coahuila lived through an early foreshadowing of the Revolution. Before the impending reimposition of Governor José María Garza Galán under the usual Porfirian pretext of reelection, three hundred ranchers in Coahuila armed themselves and offered resistance. Among them were the eldest sons of Don Jesús, Emilio and Venustiano Carranza. Díaz immediately reacted by turning the problem over to

Bernardo Reyes, his man in the north (especially for the states of Nuevo León, Tamaulipas, and Coahuila). Several times Porfirio expressed his suspicions, not entirely unfounded, that the real instigator was his old opponent Evaristo Madero—using both the occasion and the Carranzas. With Reyes serving as an intermediary, Díaz, despite his conjectures, granted an audience to Venustiano, who explained to him in detail the causes and justifications of the movement. The issue—more than anything else—was Coahuilan resentment against the reimposition of the governor, their independent will to be heard in matters that affected their state.[4] Wisely, Díaz understood that he would gain more by replacing Garza Galán. He knew that Coahuila had always been an unstable entity and later wrote, in a letter to Reyes: "If it is a fact that we cannot win any of them over, let us not risk losing them, because sooner or later civil war is bound to break out in that state . . . and we must cultivate the little we have going among them."[5]

With the triumph of the movement against Garza Galán, Venustiano Carranza achieved a more personal victory: He consolidated a friendship with Bernardo Reyes, to whose political influence he already owed the municipal presidency of Cuatro Ciénegas. He would fill that position again for 1894–1898 and move on to legislative appointments (in formal terms "be elected," but of course no one was really elected under the regime of Don Porfirio Díaz). In 1904 Governor Miguel Cárdenas—another Bernardo Reyes man—recommended that Díaz support Carranza for senator: "Señor Carranza's background, his love of order and the other qualities he possesses, as well as his devoted and single-minded support for your administration, are a sure guarantee of his loyalty."[6]

But his loyalty was really to Bernardo Reyes. Venustiano was part of a suspicious and somewhat frustrated generation that saw in Reyes the seed for a renovation currently blocked by Porfirio's technocrat advisers, the *científicos*. Yet his relative distance from Don Porfirio did not carry him to the extreme of sympathizing with the democratic convictions of his fellow Coahuilan Francisco I. Madero.

When the elections for governor were held in 1909, Carranza obtained the President's consent to declare his candidacy. He had already served a short term as provisional governor. He could rely on support across the political spectrum, from Governor Cárdenas to Francisco Madero, who strongly endorsed Carranza's candidacy. Don Evaristo Madero, the greatest magnate in the state, called him "honest and vigorous." Almost everyone agreed with the special attention his program gave to municipal autonomy and judicial independence. Only one supporter was missing: the great elector himself. It may have been the now remote events of 1893 that moved Porfirio Díaz to support the opposing candidate, who took office in December 1909. It was then that Carranza, resentful toward the President, approached that "person of no political significance," Francisco I. Madero.[7]

In January 1911, Carranza met with Madero in San Antonio, Texas. Madero—in February—named him provisional governor of Coahuila and commander in chief of the Revolution in Coahuila, Nuevo León, and Tamaulipas. Speed had never been one of Carranza's virtues, even less so now that he had turned fifty. The insurrection he was to have led was postponed. Some thought that he was still loyal to Reyes. Madero grew impatient but did not lose faith in him. With practically no military actions to validate him, Carranza nevertheless took part as Madero's Minister of War in the talks begun at Ciudad Juárez on May 3, 1911, to negotiate the end of the Porfiriato.

The discussions were held in a cottage on the outskirts of the city. (The revolutionaries called it their "*palacio nacional.*") The Porfirian representatives were bargaining over the conditions for the resignation of Díaz and Corral when Venustiano Carranza suddenly spoke up. As a student in Mexico City he had witnessed the Tuxtepec revolution. Better than any of those present, he knew the nature of revolutions in Mexico. From his experience, he drew, beyond his arguments, a prediction:

> We representatives of the will of the Mexican people cannot accept the resignations of . . . Díaz and Corral because that would be recognizing the legitimacy of their government. . . . A revolution that makes concessions is a revolution lost. The great social reforms that our fatherland needs can only be achieved through decisive victories. To win, truly to win, revolutions must be relentless. What do we gain from the resignations of Señores Díaz and Corral? Their friends will remain in power; the corrupt system that we are fighting against today will remain. An interim government will be a vicious, anemic and sterile prolongation of the dictatorship. Alongside this rotting limb, the healthy element of the Revolution will be contaminated. There will come days of mourning and misery for the Republic and the people will curse us, because—for the sake of a sickly humanitarianism—we will have spoiled the fruit of so much effort and so many sacrifices. I repeat: when a revolution makes concessions, it commits suicide.[8]

The Maderista revolution did not listen to Carranza, and it compromised by allowing an interim government. The consequences would follow in time exactly as Carranza had predicted.

On June 3, 1911, a polite but guarded Carranza received Madero in Piedras Negras. For a short period he became the provisional governor of Coahuila. In August 1911, Carranza resigned in order to "put the power of the vote into practice"—to contend electorally for the governorship that Díaz had refused him and that the Maderista revolution finally allowed him to win at the polls.

Carranza was governor for a year and a half. During this brief time, he began to transform the judiciary, the legal codes, and the tax laws. He proposed legislation on accidents in the mines, and moved against "company stores" (which peons were compelled to use), commercial monopolies, alcoholism, gambling, and prostitution. His government invested 375,000 pesos in education and—among other educational initiatives—opened nine night schools. But his proposals on education met with more success than his attempts to regulate mining interests or labor conditions. And it was then he realized that the great foreign interests had to be controlled through laws not at the municipal or regional but at the national level.

Although in practice he did not push for it enough, Carranza toyed with the old Spanish concept of municipal autonomy, which had a long tradition in the federalist state of Coahuila. His experience in Cuatro Ciénegas had convinced him that the political progress of Mexico could only come from the bottom up, from the "school of democracy," which might take the form of the autonomous municipality. He believed in the worth of small communities, and—in the face of the large haciendas and mining operations—he promoted small-scale agriculture and mining. But he also showed a distinctly patriarchal conception of politics. With our freedoms achieved, he had said to his fellow citizens in 1911, "all we need do is to enlighten the people, to teach them—with dedication, with concern and with love—how to make sensible legal use of their freedoms, and to guide them until they understand public problems."[9]

Carranza had clear ideas, but he was not an idealist. He knew the times were not right for peaceful reconstruction, and he recognized that there were potential threats. He put considerable effort into maintaining security forces, sometimes by reenlisting former Díaz *rurales* (against Madero's wishes). By September 1912, a growing distance between Carranza and Madero had become public and flagrant. Carranza's primary interest was in defending the sovereignty of his state. And he was observing, with immense concern, the deterioration of the presidential image, foreseeing that his warning at Ciudad Juárez on "the suicidal revolution" might very soon come true. Madero had long considered

Carranza "vindictive, spiteful and authoritarian . . . a phlegmatic old man who asks one foot for permission to move the other."[10]

While the apostle prepared for martyrdom, the "old man" Carranza, acting in a way not at all phlegmatic but clearheaded and full of guile, established ties with the governors of San Luis Potosí, Aguascalientes, and Chihuahua and ensured the loyalty of such future stars of the Revolution as Cesáreo Castro, Francisco Coss, and Pablo González. During the Tragic Ten Days, he sent young Francisco J. Múgica to offer Madero refuge in Coahuila. Nothing took him by surprise. The "old man" had the benefit of his years. He had lived through much history and he had listened to and read about much more.[11]

Many who knew him commented on Carranza's particular interest in history and its lessons. For Luis Cabrera, Carranza was a true "practical encyclopedia of the history of Mexico." His golden age was the period of the Reform; his most beloved figure, Benito Juárez. "Juárez was for him," José Vasconcelos wrote, "all of human greatness above and beyond universal geniuses." Carranza had almost no experience abroad, but he had supplemented his physical stasis with considerable literary travel. His favorite European history was that of France, its politics—in a conservative, classical version—and also the record of its social customs. But his greatest passion of course was the history of Mexico, which he read voraciously.[12]

At the end of February 1913, confronted with the success of the military coup and the death of Madero, Carranza thought he saw the echo of an earlier chapter in Mexico's history. In 1858 "the reaction" had deposed President Comonfort—a moderate incapable of exercising his constitutional powers. The War of the Reform had begun between Liberals and Conservatives. Benito Juárez had won that war and during its course issued the Laws of the Reform, which would profoundly alter Mexican life. He had incarnated not only an authority that resisted but an authority that legislated. Then Juárez had to confront the French invasion and the empire of Maximilian. The lesson that he had taught through his actions in this second stage would be equally enduring: national sovereignty as a supreme value.

To Carranza, the moral was clear. The new conservative reactionaries, headed by the usurper President Victoriano Huerta, had overthrown the constitutional President. What was needed was a new Juárez invested with legitimate powers to defend the banner of constitutionalism and, in due time, propose new Laws of Reform. In the overflowing river of Mexican history now running backward toward the days of the earlier Reform, foreign powers—especially the United States, more arrogant than in 1847—

would try to catch whatever fish they could. Again Mexico would have to fight for its sovereignty (this time with no allies) against Europe and against the United States. Both had, after all, turned their backs on President Madero.

Carranza adopted his historical chart from Juárez and at least some authoritarian methods from Porfirio Díaz. But from Madero, Carranza had learned practical lessons about all that he should *not* do. Resounding within him were his own words at Ciudad Juárez. The Revolution had not been "relentless." The interim government had been "a vicious, anemic and sterile prolongation of the dictatorship." An "unhealthy humanitarianism" had "spoiled the fruit" of the Revolution. Madero had made concessions and "a revolution that makes concessions commits suicide."

Things would be different now. Everything would serve the principle of authority, as Carranza would move toward assuming a national role. For the purpose, he could rely not only on his historical wisdom but on certain natural attributes. For one thing, his age. In 1913 he was fifty-three, by far the "old man" of a revolution begun by men who were twenty or thirty years younger. His height and bearing also were of help to him. The Spanish writer Blasco Ibáñez would later describe him as "majestically tall, robust and strong despite his years."[13] When John Reed met him in 1914, he saw "a towering figure, seven feet tall it seemed" (he was really six feet four inches) and compared that "vast, inert body" to "a statue."[14] A third component was his beard (someone called it a "flowering beard"), which won instant respect from the skeptical Martín Luis Guzmán:

> The way he stroked his beard with the fingers of his left hand—which he would place under the snowy cascade, palm outward and fingers curled while he raised his face slightly—announced tranquil habits of reflection from which one could expect—at least I then imagined—nothing violent, nothing cruel. "Maybe," I thought, "this is not the genius, not the hero, not the great altruistic politician that Mexico needs, but at least he does not usurp his title: he knows how to be the *Primer Jefe*.[15]

Carranza almost had the instincts of a modern advertising man in the means he would use to elaborate his image. It was a presentation meant to inspire obedience and order without coercion. Few Mexican politicians—and of course even fewer Mexican revolutionaries—have paid as much attention to formal appearances. On launching the Constitutionalist revolution—the second chapter of his own Juarista epic—

he took care that his title, *Primer Jefe* of the Constitutionalist Army, should be matched by his dress. If his condition was dual—a civilian *and* a revolutionary—Carranza had to be and appear both. He put together his own uniform: a northern-style wide-brimmed gray felt hat, a twill coat with no military insignia but bearing the gold buttons of an army general, riding breeches, patent leather boots, and side-buttoning pants of Saltillo leather.[16]

Another important characteristic of his style was a certain slowness of manner. There was something naturally deliberate in his voice, in his gestures, and, according to Luis Cabrera, even in his intelligence. But this slowness was also a way of stalling for time. Carranza lacked Díaz's political instincts but made up for it by letting events simmer and by filtering problems and people. It was almost impossible, for example, to have a face-to-face interview with Carranza. John Reed had to first submit a questionnaire to Carranza's secretary, Isidro Fabela—one of the men closest to the *Primer Jefe*—for censoring. Carranza's knowledge, his years, and his deliberate process of thought and action had made him a stubborn man.[17]

Behind that "huge mask" (as Reed described him) who was trying to become the reincarnation of Juarista authority, there lay a rustic and paternal inflexibility. His psychological and internal rhythm was not that of the Revolution but of the countryside, the tempo of the ranches, made up of cycles and fatality. This deliberate personal rhythm, as well as his resources and stratagems, his sense of authority, and, above all, his interpretation of the Reform, would mark the modes and substance of the Revolution. He wanted to be another Juárez, to command like Don Porfirio, and to avoid the errors of Madero. To a certain extent, he succeeded. And he also succeeded in leading and, to use his favorite word, "channeling" (*encauzar*) a revolution that had far more complex and powerful undercurrents than he himself suspected. No one in Mexican history lived the transition between the nineteenth and twentieth centuries as fully as Carranza. He was a man who was a bridge. Like the Liberals of the Reform, Madero had desired the pure empire of *the law*. Before and after Madero, militarism had meant and would mean an almost pure empire of *the act*. Carranza lived the tension between acts and laws: old and new acts, old and new laws. His biography is, without a doubt, the most complex of the Revolution.

Carefully following all the proper forms, Carranza—at the end of February 1913—obtained a mandate from the Coahuila legislature to rebel against the government set up after the murder of Madero. On March 4 he openly broke with Huerta and a few days later suffered his first military defeat. While retreating toward Monclova (later to be the site of his temporary government), he stopped at the Guadalupe Hacienda, where a group of young officers were preparing the famous *Plan de Guadalupe*. In it they disavowed Huerta and chose Carranza to be "the *Primer Jefe* of the Constitutionalist Army." Carranza accepted the designation and promised to call a general election upon the triumph of this revolution.

The signatories—Francisco J. Múgica, Jacinto B. Treviño, and Lucio Blanco, among others—were expecting a new version of Madero's Program of San Luis and the inclusion of measures like the distribution of land. Carranza's model, however, was not Madero but Juárez: "This Revolution must only be for the purpose—and everyone should know this—of restoring constitutional order, without drawing the people, through deceptions, into a struggle which is bound to cost much blood and then generate, should it fail, larger revolutionary movements. The social reforms the country needs must be carried out but should not be promised in this program."[18]

A few days later, a delegation from Sonora headed by Adolfo de la Huerta visited Carranza in Monclova and endorsed the Program of Guadalupe. From the very beginning, Sonora would be the primary garrison of resistance against the federal troops. It was a remote and powerful state whose middle classes had produced a number of natural leaders: Álvaro Obregón, Benjamín Hill, Salvador Alvarado, Juan Cabral, Plutarco Elías Calles.

The struggle had barely begun. Always faithful to the guidebook he had drawn from history, Carranza took the first measures of war. He decided it was proper to repeat Juárez's extremely severe decree of January 25, 1862, which would be used to pass judgment on Huerta, his "accomplices in military mutinies," and the "supporters of his so-called government." The law prescribed the death penalty for, among others, those who had rebelled against legitimate institutions and authorities or had made attempts against the life of the Supreme Leader of the Nation. It was a legal justification for executing prisoners of war.[19]

The early stage of the rebellion lasted from March to August of 1913. Carranza divided the Republic into seven operational zones, of which only three actually functioned: the northwest, under the command of Pablo González; the center, under Pánfilo Natera; and the northeast, under the orders of Álvaro Obregón. In July, Monclova fell into Huerta's hands while the rebels tried unsuccessfully to take Torreón. In August, the *Primer Jefe,* aware of the vulnerability of his position, decided to return to his Sonoran stronghold. Again he was reminded of Juárez's long marches in 1863 from the capital through the northern deserts before taking refuge in Chihuahua, choosing never to leave Mexican soil. Although he could have reached Sonora more easily by way of the United States, Carranza preferred to cover the three hundred kilometers from Piedras Negras to Hermosillo by way of Torreón, Durango, southern Chihuahua, the western Sierra Madre, and northern Sinaloa. He refused to set foot in the United States. It was a matter of dignity—and style.

On September 14, 1913, he met Álvaro Obregón in El Fuerte, Sinaloa. Obregón observed him and remarked: "He is a man of details." When Carranza arrived in Hermosillo he set up his government with eight sections parallel to those of Huerta. On September 24, in the chambers of the town council, he delivered one of the most important speeches of the Revolution.[20]

It began with a long historical reflection. Mexico had to reverse the course of four centuries: "three of oppression and one of internal struggles which have been driving us into the abyss." During the Porfirian dictatorship—an epoch like that of Augustus or Napoleon III "in which everything depended on one man"—the newspapers deceived the public by writing about progress when all that really grew stronger was the submissiveness of the national soul. Carranza did not mention Madero by name, and he qualified the importance of the Maderista slogan. In his view, the struggle transcended the ideal of "Valid Voting, No Reelection," just as it went beyond the Program of Guadalupe: "Once the armed struggle called for by the Program of Guadalupe is completed, México will have to embark upon the formidable and majestic task of the social struggle—the class struggle, whether we ourselves like it or not—and oppose the forces who oppose it. New social ideas will have to gain the respect of our masses."

Carranza was not a social revolutionary. Only by understanding this can we understand the words "whether we ourselves like it or not." But with his sense of historical necessity, he already half understood that the "Revolution was the revolution," an almost telluric movement, as the writer Luis Cabrera had called it, which men can in the best of cases channel but never cut off. The proposals he included in that speech must be understood from this perspective. They were as personal to him as his agricultural metaphors: "The country has been living in illusion, starved and luckless, with a handful of laws that are of no help to it; we have to plough it all up, drain it and then truly construct."

This immense program of rectification would require, as Carranza announced for the first time, the drafting of a new constitution. Other no less decisive measures would be the founding of a national bank and laws to be decreed in favor of peasants and workers, drafted by their representatives. But the most important message of the speech dealt with sovereignty, the primordial value for anyone from Coahuila. The loss of Texas had not been forgotten: "This fratricidal struggle [also] has as its goal . . . respect from powerful peoples for those that are weak."

To the personal and political gifts that validated the legitimacy of his leadership, Carranza with that speech added another: He was now an ideologue of the Revolution. Military victory was to be followed by a period of social reforms, a new constitution, other new laws and institutions, and a different attitude that "would shake off international prejudices against us as well as our eternal fear of the colossus of the north."[21]

Carranza remained in Sonora until March 1914. There he heard of the first scintillating victories of Villa and of the advances of González and Obregón. Without ever leaving the territory of the nation, he moved on to Ciudad Juárez. He would remain there until the complete triumph of the Constitutionalists over Huerta, in July 1914.

During the Constitutionalist revolution—while Obregón and González moved their armies south and Villa was winning victory after victory—Carranza had to play an especially difficult double role. Besides attending to the economic administration of the war, he had to maintain the unity of the Constitutionalist Army under his command and spar with foreign countries, especially the United States. On the chessboard of internal unity, his principal opponent was a wild animal, Francisco Villa; on that of foreign affairs, a moralist, the American president Woodrow Wilson.

Although at the beginning their relations were almost cordial, Carranza and Villa never really understood each other. The sense of authority that the *Primer Jefe* claimed for himself was incomprehensible to the fierce warrior. The problems Villa caused with foreign governments began to pile up: He had herded the Spaniards of Chihuahua like cattle and confiscated their property, then permitted the murders of the Englishman Benton and the American Bauch. In April 1914, Villa arrested the governor of Chihuahua, Manuel Chao, a Carranza loyalist. For the *Primer Jefe,* it was the last straw. He called Villa in and lectured him, even refusing to listen to his words until Villa agreed to free Manuel Chao. It was a brave thing to do because he was in Villa's domain of Chihuahua and surrounded by Villa's troops. Yet as happened in this case, his sense of authority could easily become authoritarian. Without weakening his firm stand but with a little less zeal and a little more sympathy, Carranza might have gotten Villa's more willing consent. But Venustiano Carranza was not a man for subtleties. His idol Juárez had also been criticized for authoritarian zealousness.

To Carranza, the confrontation simply seemed another obvious les-

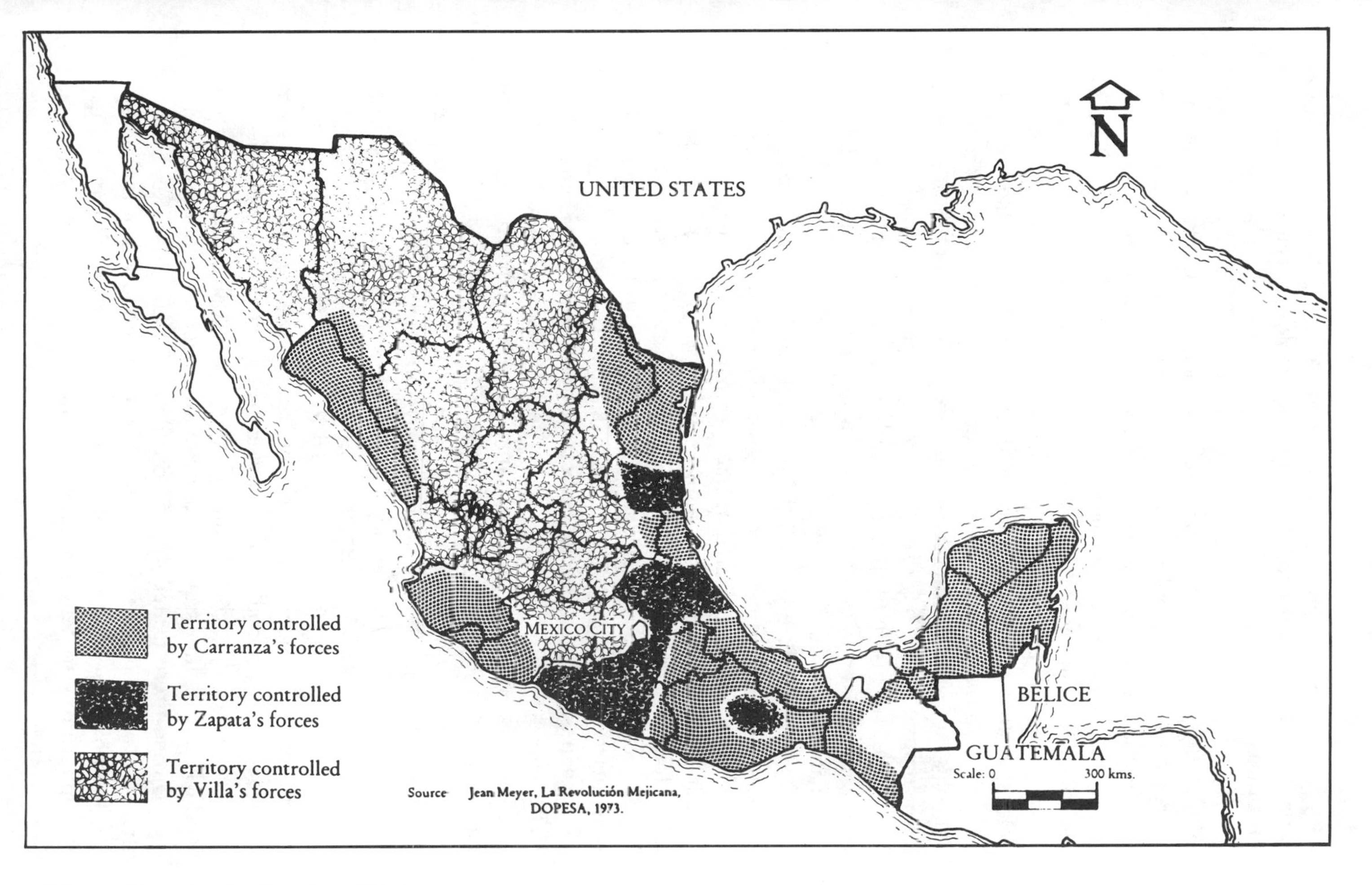

Military Situation in October 1914

son. Better to sin doing too much, like Juárez or Díaz, than doing too little, like Madero. But from then on, the break between Villa and Carranza was inevitable.[22]

While the soldiers fought on their battlefields, Carranza emulated Juárez on the battlefield of diplomacy. Juárez had contended against France and, to a much lesser degree, England and Spain. Carranza's opponents were England, Germany, and, more than any other, the United States. He knew that absolute protection of national sovereignty was a necessary condition for the triumph of the Revolution, and in defense of that sovereignty he used all the knowledge he had inherited, learned, or possessed through his personal character. For Woodrow Wilson—his American counterpart—Carranza was always full of surprises, an incomprehensible bag of tricks, a man indifferent to good intentions. But at the conclusion of their long and complex relations, which lasted through highs and lows over a period of seven years, both would end up reasonably victorious.

On the American side, all the moves involved the same tactics that Porfirio Díaz had used on home ground. The strategy of the State Department was to alternate menace, threat, and violence with moral preaching, reconciliation, and support. Like the good Coahuilan he was, never forgetting the loss of Texas, Carranza used the tactic of distrusting all their moves as fraudulent. His premise—this time more Porfirian than Juarista—was very simple: Even if a biblical apostle occupied the White House, Mexico could expect no good from the United States. "The danger is in the Yankee who lies in wait for us," Don Porfirio had said in Paris. And how could Carranza forget the sinister role of Henry Lane Wilson in the martyrdom of Madero?

When President Wilson's first official representative came to Mexico in November 1913, Carranza made him wait ten days before receiving him with cold formality. He was not impressed by any American good intentions (to recognize his government), nor would he agree to compromise with the reactionaries and form a provisional government. In February of the following year, after the murder of Benton by Villa's *pistolero* Fierro, Carranza would not allow the Americans to act as intermediaries and made it clear to Simpich, the American consul, that all disputes were to be worked out with him, Carranza, and not any other revolutionary leader.[23]

During those heated days of the Benton affair, John Reed for the first time met Carranza. Reed offered him the goodwill of the newspaper for which he was a correspondent. Carranza thanked him and used the opportunity to deliver a diatribe against the United States and perfidious Albion:

> To the United States I say that the Benton case is none of your business. Benton was a British subject. . . . Why should they not come to me? England now has an ambassador in Mexico City who accepts invitations to dinner from Huerta, takes off his hat to him, and shakes hands with him! . . . If the United States intervenes in Mexico upon this petty excuse, intervention will . . . deepen a profound hatred between the United States and the whole of Latin America, a hatred which will endanger the entire political future of the United States![24]

The intervention was not long in coming, although for different reasons. On April 21, 1914, the marines landed at Veracruz. The form was that of "the bludgeon," but the purpose was really "bread" for the Constitutionalist forces. Wilson wanted to implement a blockade against the Huerta government. The marines were there to seize arms shipments coming in from Germany. Carranza understood this but still demanded the immediate withdrawal of the marines and threatened to declare war. Mexican sovereignty had been violated, and Venustiano Carranza was no more willing to accept it than he had been to save himself many miles and muscle pains by taking a shortcut through U.S. territory in his long march of August 1913. The marines stayed for months. And when U.S. Secretary of State Bryan, always more extreme toward Mexico than Wilson, later threatened not to recognize the Constitutional government after Huerta had fled, Carranza showed his continuing displeasure by not even bothering to respond.[25]

On August 15, 1914, Obregón—representing the Constitutionalists—signed the Teoloyucan treaties with the last representatives of the government risen to power on the body of Madero. Carranza refused to make any concessions, to change even a word. Five days later, the *Primer Jefe* arrived at the capital. As always, Carranza carefully staged his arrival. He came on horseback, flanked by his general staff, and began the journey in Tlalnepantla, eleven kilometers from the National Palace, so that he could cross much of the city and accept the enthusiasm of crowds that numbered close to three hundred thousand. He must have remembered Juárez's entry into the capital on January 11, 1861, after the final defeat of the Conservatives. History, in his view, had repeated itself with surprising accuracy.[26]

According to the *Plan de Guadalupe*, the overthrow of Victoriano Huerta should have meant the triumph of Constitutionalism and, at least in theory, the end of the Revolution. But in fact it had hardly begun. Venustiano Carranza was the *Primer Jefe* of the Revolution but not the only jefe. Two caudillos of the people refused to submit to his authority: Pancho Villa and Emiliano Zapata. Peace depended on the difficult task of winning them over. Seen from the perspective of the future, the discord between Carranza and the two caudillos seems natural. Aside from being Mexican, they had nothing in common.

Social class, culture, one could even say civilization profoundly divided Carranza and Zapata, who broke off all connection with the President on September 5, 1914. It was the old conflict between north and south, the liberal Mexico of the nineteenth century and the ancient Mexico of the viceregal and pre-Hispanic eras. With Villa, the problem was more narrowly political. A "lover's quarrel," Álvaro Obregón called it, with obvious exaggeration but pointing to something real. It was more a difference of passions and personalities than of belief or ideology.

Villa had a whole litany of complaints against the *Primer Jefe*, ranging from their confrontation over Governor Manuel Chao to Carranza's refusal to grant Villa the honor of entering Mexico City at the head of his troops (and refusing even to promote him to major general). Although Villistas and Carrancistas signed the Treaty of Torreón on July 8 and each group agreed to recognize the other and to call a convention of generals to decide the political future of Mexico, Carranza already knew that "the lover's quarrel" would end in divorce. By September he had written to the governor of San Luis Potosí, Eulalio Gutiérrez: "If we cannot come to a peaceful agreement and the armed struggle begins—not because we desire it but through the force of circumstances—we want to be prepared."

In that September of 1914, the caudillos were playing poker and the stakes were high: the political future of the country. Everything seemed unsure. The leaders were swept by contradictory feelings—a disinterested will toward agreement and a suspicion-ridden need to form alliances in an underground fight for power. Villa was the first to break his poker face: On September 23 he disavowed Carranza. At that moment, power belonged to no one, and almost no one was loyal to anyone but himself.

The Convention of Aguascalientes met on October 5. Until then the quarrel had been between individuals and individual personalities: Carranza against Villa, and, oscillating between them, a swarm of generals more or less Villista, Carrancista, or independent. Once the Convention had begun, the conflict would be more than political. Legal and moral issues would count; it would be a confrontation of legitimacies. What was the lawful locus of power in Mexico? Would it be the sovereign Convention of Aguascalientes represented by 150 of the most distinguished generals of the Revolution (soon to be joined by Zapatista civilian intellectuals) or the *Primer Jefe* of the Constitutionalist Army, who held the executive power in accordance with the Program of Guadalupe?[27]

Though he did not directly participate in the sessions of the Convention—he was neither a general nor a Zapatista anarchist intellectual—the young philosopher José Vasconcelos formulated the juridic defense of the Aguascalientes Convention. Since February 1913, true popular sovereignty, Vasconcelos wrote, had rested with the citizens rebelling against the usurpation, the members of the Constitutionalist Army, "which is the army of the sovereign people." Article 128 of the 1857 Constitution (which was still in force at the time) referred to the moment when the people would *recover* their freedom by defeating an anticonstitutional government. And what would be the vehicle of this recovery, fully endorsed by the Constitution? The army in rebellion. All the jefes at Aguascalientes, instinctively and without exception, agreed with Vasconcelos.[28] And the Convention called for Carranza's resignation.

In a message that he sent to them on November 23, 1914, Carranza declined the invitation to come to Aguascalientes. Although he was surprised by the urgency with which the assembly asked him to resign and declared that his retirement must not clear the way for a "regime only apparently constitutional," he agreed that he would leave office and, if necessary, the country, provided that three conditions were met: 1) the establishment of a preconstitutional regime "that would take charge of carrying out the social and political reforms the country needs before a fully constitutional government is reestablished"; 2) the resignation and exile of Villa; and 3) the resignation and exile of Zapata.

A week later the Convention's joint commissions of war and of the interior (including generals Álvaro Obregón, Felipe Ángeles, Eulalio Gutiérrez, and Raúl Madero) in principle accepted Carranza's conditions, but their statement did not convince the *Primer Jefe*. When the Convention named Gutiérrez provisional President—for only twenty days until his position could be ratified—its members sent an ultimatum to Carranza, calling for his immediate resignation. Four days later Carranza answered with a long telegram, from Córdoba in Veracruz state, where he had transferred his government to forestall any aggression against him. His reasoning was neither philosophical nor legal but practical, and, to a certain extent, historical.

He would not accept the orders of the Convention nor would he resign until his three conditions had been fully met. Villa was still in command of his División del Norte and was beginning to meddle with the command of other divisions; Zapata's power had not been diminished but exalted by the Convention; and as to his first condition (the need for a legitimate preconstitutional regime), his reasons, although more complex, were no less clear-cut. However much revolutionary legitimacy the provisional President might possess, what kind of power could he effectively wield, given the way in which he was chosen?[29]

From his careful readings of French history, in which he always preferred the classic to the romantic version, Carranza had learned to distrust "assemblyism." The experience of the Restored Republic in Mexico also confirmed, in his view, the historical incompetence of deliberative groups. Beyond the political and military threat represented by Villa and Zapata—whose attitudes, no less than that of Carranza, provoked the final split and the civil war, the *Primer Jefe* was concerned not so much with abstract legitimacy as with concrete responsibility and efficiency.

In his opinion, only he, and not the assembly, could guide the "formidable and majestic social struggle" into its proper channel. This act of assertion against the revolutionary assembly "by the man entrusted with the Executive Power" was a decisive moment in Mexican history and foreshadowed the future growth of presidential authority.

What would have been the political structure of Mexico if Carranza had yielded to the Convention? Perhaps more democratic, perhaps more anarchic and fragile. We will never know because the eventual winner would be Carranza. He had no doubt that he represented (and served) the real interests of a revolutionary state that still had to be shaped. All his actions were based on that conviction.

The State that arose from the Revolution was, in many of its important features, a creation of the *Primer Jefe*. He felt Juárez would have acted no differently. He continued to follow in the steps of Juárez, later by proclaiming the laws of the new Reform, but at this particular moment through moving his government, as Juárez had, to the city of Veracruz.

At the end of November 1914, after negotiating the unconditional withdrawal of the American occupation forces from the city, Carranza set up his government in Veracruz. He would be there until October of the following year, when the military situation shifted in his favor. It was not so at first. The Carrancistas dominated the outlet to the Gulf, the entire southeast, and a good part of Tamaulipas and Veracruz, but the rest of the country was controlled by the unstable alliance of the Convention,

Villa and Zapata. In April 1915 Obregón defeated Villa in the Bajío; in May Pablo González began the final campaign against Zapata; in July Francisco Lagos Cházaro—the last Convention president—surrendered, and in August the Constitutionalists entered the capital. The diplomatic recognition of the Carranza government by the United States in October 1915 was a simple acceptance of military victory.

But military actions alone could not keep the Revolution alive. It needed political movement and social reform. Carranza formed his cabinet, of civilians and soldiers from the professional middle classes, and on December 12, 1914, he began to fulfill the promises he had made in his speech at Hermosillo.

He issued his "Additions to the Program of Guadalupe," the beginning of his "formidable and majestic social struggle." The second article of this document pointed toward the legislation to come, *his* Laws of Reform.

Earlier, in Veracruz, the city where Juárez had instituted a law legalizing civil marriage, Carranza had legalized divorce on the symbolic date of Christmas Day, 1914. His will toward symmetry with Juárez was a matter both of detail and of basic vision. When he had said (at the beginning of 1915), "Today the social revolution begins!" he meant a revolution to be achieved *through laws*. The presentation of his social program, in Article 2 of his "Additions," shows a degree of continuity with the constitutional liberalism of the nineteenth century. Although it discusses restoring lands to communities and the breakup of large estates, it indicates no intention at all to create a new`system of property or preach a social gospel. The insistence on themes like municipal autonomy, the independence of the judiciary, and equality before the law are also clear signs of this surviving Liberal emphasis. As one of his cabinet members said, he wanted "to make the Revolution constitutional."

From the announcement of the first reforms at the beginning of 1915 until the oath to the new Constitution in Querétaro on February 5, 1917, Carranza's preconstitutional government would fight a battle on many fronts, at least as complex as the military campaigns. "Ploughing it all up," as he had said in Hermosillo, required a hard look at the problems rooted deep in Mexican life and history: the problem of the land, the inequalities between rich and poor, the question of sovereignty over natural resources, the relations between Church and State, and the structure of political power. The initiative for reforms would come directly from Carranza, or from his generals, or sometimes from social pressure. For its leaders and for society, those two years of the preconstitutional regime—1915 and 1916—were a time of historic experimentation.[30]

* * *

Indirectly, symbolically, the agrarian reform put into motion by the law of January 6, 1915, was the achievement of Zapata. In Porfirian times it had often been said that there was no land problem in Mexico. In *Los grandes problemas nacionales* (1909), the prophetic Andrés Molina Enríquez had warned how serious the problem was, and he had offered solutions like those later urged by the Zapatistas in their *Plan de Ayala*—the restitution of lands to communities and the division of large estates. For many years Molina's warning had been a cry in the wilderness. Suddenly, at the dawn of the Maderista revolution, Zapatismo proved how wrong the nonbelievers were. There was not only a land problem—there was an unresolved historical grievance, the old complaint of the peasants against the Liberal era that had denied their culture, encircled their lands, threatened their ancient way of life.

For a time Carranza also thought that the problem "had been exaggerated," but he gradually yielded to the evidence and to pressure from his aides. In September 1914, when talks with Zapata collapsed, Carranza declared that he was ready to legalize the agrarian reforms of the Program of Ayala "not only in Morelos but in any state where these measures are required."

In Veracruz, Luis Cabrera, an intellectual who was one of Carranza's principal advisors, put the final touches to a new agrarian law, issued in early January 1915. The term *ejido* by now referred to village lands communally owned and exploited, but the outlines of the institution were less precise than they would later become. The Carrancista agrarian law saw the *ejido* as a reparation for past injustice, not as a new system of ownership. Lands that were their rightful inheritance would be returned to pueblos that had lost them, and new acreage would be created by expropriating land adjacent to the villages. On paper, the mechanism was simple. Pueblos would send their requests to the Local Agrarian Commission, where decisions would be made on restitution or a grant. If the decision went in the petitioner's favor, a special executive committee would define the boundaries and draw up a provisional title deed. In each case, a National Agrarian Council would make the final decision and the office of the President would issue the final deeds. Those losing land would have the right of appeal.

In practice the mechanism turned out to be restrictive, complicated, and slow. The beneficiaries of the law were to be "the pueblos," but the law did not precisely define them. The social fabric of the Mexican countryside included figures that the law ignored: partners owning small ranches, tenant farmers, agricultural peons, laborers living in shanties on the haciendas. Carranza had hoped for the peaceful submission of reality to the law, but violent reality, in many areas, went beyond the law. There were land invasions, forests cut down, conflicts, confiscations. On June 11, 1915, Carranza felt obliged to issue a manifesto to the nation in which he declared that no land would be confiscated "which had been acquired legitimately from individuals or governments and which did not constitute a special privilege or monopoly."

The National Agrarian Council took more than a year to get established, and when it finally did—on March 8, 1916—it worked at a snail's pace. On September 19, 1916, to the despair of some radicals and many villages that had been dispossessed or were in need of land, Carranza suspended the provisional grants. A month later, on the basis of deeds dating from 1801, the Commission made its first definitive restitution of lands, to the village of Iztapalapa. Before the promulgation of the new Constitution it was to confirm only two more, both on the outskirts of Mexico City. It was a meager harvest.[31]

With labor problems, Carranza's practice of both approaching and distancing the issues, of legal initiatives and practical restraint, was similar but rockier. He proposed to institute and improve the reforms he had introduced while he was governor of Coahuila. One of his first decisions in Veracruz had been to modify the Liberal Constitution of 1857 so that his government could legislate on labor issues, and he had set up a Commission on Social Legislation, whose job it was to study various international labor laws and adapt them to Mexican conditions.

Carranza and his aides in the preconstitutional government had tried to lure the peasants to them with the siren of the agrarian law. Toward the workers, their approach was more direct. They negotiated a political agreement. The idea actually came from Obregón, who was already showing his genius not only for war but for politics. He himself approached the leaders of the Casa del Obrero Mundial ("House of the World Worker"). As an earlier token of confidence, he had already given them the Santa Brígida Convent and the Colegio Josefino as sites for their offices. In spite of its anarcho-syndicalist origin opposed to any relationship with power, the Casa (in Veracruz on February 17, 1915) signed an agreement with the Constitutionalist Army that would have far-reaching effects. In exchange for a governmental promise to support the demands of the working class in the future, the Casa pledged "to take up arms" in defense of communities already controlled by the Constitutionalists and also to fight against "the reaction," which meant not the vanquished Huertistas but the forces of Villa and Zapata. Six "Red Battalions" were immediately formed. In their ranks were more artisans—carpenters, typesetters, bricklayers, tailors, stonecutters—than industrial workers. Almost three thousand men began to mobilize for war at the general headquarters of the Casa in Orizaba. Among them was the painter José Clemente Orozco. Only a few days later, the Red Battalions would move into action.[32]

But this connection between the Carranza government and the working class proved to a be a deceptive honeymoon. After the military victory of constitutionalism, when the workers tried to use their right to strike (on various occasions toward the end of 1915), Carranza's government began to react with growing harshness.

In January 1916, when the railway workers called a strike in solidarity with the Orizaba textile workers, Carranza drafted the railroad workers into the army. Around the same time, the electrician Luis N. Morones organized the Federation of Union Workers of the Federal District (FSODF), with the objective of returning to the anarcho-syndicalist tradition. A few days later, on January 13, the Red Battalions were summoned to Mexico City, where they were dissolved. And only days after that move, Carranza ordered his governors to prohibit workers' rallies and to arrest labor leaders whose "activities tend to upset the public order."

The enormous economic problems of the preconstitutional government, the decline of the currency, and the constantly increasing prices touched off a wave of strikes. In May 1916, the electricians, streetcar workers, and telephone operators went out on strike in the capital. Benjamín Hill, who was military commandant at the time, threatened the FSODF strikers with "severe punishment," but he postponed confrontation by paying them off with a new currency known as "the unfalsifiable" (*infalsificable*). It was not long before it began to lose value. The workers demanded to be paid in gold and opposed layoffs, which had begun to be widespread.

August 1916 would be the most crucial month. At the headquarters of the Mexican Union of Electricians, a massive assembly called for a general strike. On August 1 Carranza made a brutal move against the "ungrateful working class." He resuscitated the law of January 25, 1862, under which strikers could be condemned to death. The following day the Casa del Obrero Mundial was closed forever by the police. Carranza would always go on believing that the workers had been "denying the sacred recognition of the fatherland . . . of the principle of authority . . . of every system of government." On their side, the labor organizations would draw a lesson from the confrontation, one that would make them wiser. It would be best to wait for the coming of a President with the political and social sensitivity to institutionalize agreement between government and labor.[33]

* * *

Of deepest importance to Carranza was legislation to defend or reclaim the natural resources of Mexico. At the beginning of 1916 he said, "The present Revolution aspires first and foremost, aside from its pursuit of social welfare, to preserve above all the integrity of the nation and its independence."[34] But the country could never be whole or independent if it reverted to the Porfirian tendency of free-and-easy concessions to foreign investment, especially in oil and mining.

This particular battle was waged on several fronts: against the oil companies, the mining companies, and also governments that defended them, centrally the United States. The most arduous combat was fought over oil. Porfirian legislation gave a landowner the right to combustible and bituminous deposits in the subsoil. Thanks to this law (and almost no taxation), two rival companies had accumulated immense or immensely productive properties: El Aguila—Lord Cowdray's English company—and Mexican Petroleum, which belonged to the American magnate Edward Doheny. The former operated in the zone of Poza Rica and Papantla, the latter in Tampico.

On January 7, 1915, Carranza issued a decree from Veracruz that suspended all oil exploitation until a regulatory law could be issued. The engineer Pascual Rouaix set to work on planning such a law and soon issued an opinion that justice required "the restoration to the Nation of what belongs to it, the wealth of the subsoil, the coal, the oil." In May 1915, Carranza sent Rouaix to observe the operation of the American oil industry—its refineries, fields, and laboratories—with the idea of creating a national Mexican petroleum company.

The power of foreign oil interests would cap this budding nationalist well for some years. From the end of 1914 until 1920, General Manuel Peláez, feudal lord of the Huastecan region where the companies operated, charged fifteen thousand dollars a month to protect the oil companies against all interference from the center. The only options open to Carranza's government were measures like increasing export taxes—Peláez did not control the ports of entry—restricting exploitation through temporary permits, or endorsing the famous Calvo Doctrine (named after its Argentine originator), which put Mexicans and foreigners on an equal legal footing. Perhaps most important, the government could alert public opinion to the justice of its true objective: regaining control of the subsoil for the good of the nation.

The mining companies—80 percent of them in American hands—presented a different, though no less serious problem. In the area of mining interests, the Porfiriato had not been as generous as with oil. Continuing the viceregal tradition, it had not granted landowners the automatic right to noncombustible minerals in the subsoil. This favorable circumstance was of at least some help to Carranza. When he first raised the taxes on foreign mining interests (March 1, 1915), the United States government and the companies argued that the increases were unconstitutional and demanded their repeal. Two months later, Carranza granted concessions to small mining companies and imposed taxes and license-expiration dates on the large companies. U.S. Secretary of State Lansing insisted on the repeal of the decree but only secured some temporal extensions. On the ultimate expiration of mining licenses, Carranza stood firm. He tightened the screws on August 15, 1915. From that moment on, foreigners could not defend their rights through diplomatic means—they had to use the Mexican courts. Once again Lansing thundered: He "emphatically disagreed" with the new decree, he insisted on the validity of diplomatic recourse; in his view these were "confiscatory taxes." Carranza turned a deaf ear; he would delay some deadlines and tax payments but would not reverse his course. Not once did he repeal a nationalist decree.[35]

There was one ancient historical process that Carranza would have preferred not to touch, a question he did not mention in his Hermosillo speech. It was the relationship between Church and State, which he, like many others, thought was no longer an issue. Almost forty years of Porfirian conciliation seemed to have succeeded in smoothing down the rough edges of reactionary and Jacobin confrontation. Suddenly, in 1914, the first symptoms of renewed anticlericalism began to surface. Much of the violence of the Carrancista armies was directed at the Church—its men, its open or hidden property, its symbols, and its very existence.

The entire Republic came to be the stage for a bizarre and theatrical performance. In contrast to the Zapatista devotion to the Virgin of Guadalupe, evident in their images, the crosses they wore, their banners with the image of the Virgin, in contrast also to the Villistas who were cautious in the face of religion, the Carrancistas gloried in acts of premeditated and jubilant sacrilege. They drank out of chalices, paraded wearing priestly vestments, built fires in confessionals, shot up sacred images, converted churches into barracks, carried out mock executions of the statues of saints. In the state of Mexico, they banned sermons, fasting, christenings, masses, confessions, and even kissing the rings of priests.[36] The painter José Clemente Orozco witnessed and shared in some of these scenes:

> When they arrived in Orizaba, the first thing they did was attack and loot the town churches. The church of Los Dolores was emp-

> tied and we installed two flat presses, several linotypes and machinery for a graphics workshop in the nave. . . . The church of El Carmen was also raided and given to the workers of La Casa del Obrero Mundial as living quarters. The saints, the confessionals and the altars were turned into firewood by the women, for cooking, and we dressed ourselves in the altar hangings and the vestments of the priests. We all left decorated with rosaries, medallions and crucifixes.[37]

It was often the Carrancistas from the north, especially Sonora, who swung the axes in the churches. Faced with the anticlericals, Carranza did not tear his hair out but neither did he remain impassive. He knew that the politically minded portion of the clergy shared responsibility for many of Mexico's historical misfortunes, but he was not a partisan of extremes and excesses. In August 1916 he halted an avalanche of confiscations from the Church by issuing a decree that centralized power (in the Ministry of Finance) over the use, preservation, and maintenance of ecclesiastical property.[38] He shrugged his shoulders before the anticlerical fiesta, but he put limits on the *business* of anticlericalism. When the real anticlericals took power in the twenties, the country would endure not theater but civil war: the crusade of the Cristeros.

In September 1916, Carranza sounded the political clarion call for that decade and many more to come. He convoked (as he had said he would in Hermosillo) a Constitutional Convention (*Congreso Constituyente*) to be held in Querétaro in the Bajío north of Mexico City. "The liberal spirit of the Constitution of 1857 will be scrupulously respected; we only want to purge it of its defects because of the contradictions and obscurity of some of its precepts, because of the gaps in it and because of the changes that past dictatorships have made with the deliberate intention of denaturing its original and democratic spirit."[39]

Carranza's concern was legitimacy and continuity, just as the Constitution of 1857 had maintained a continuity with the federal Constitution of 1824. He expected that he and the Constitutional Convention would develop the charter for a strong and balanced state, with a much stronger and more efficient executive power (but one that would not yield to dictatorial temptation) and a less overbearing legislative power than had been sanctioned by the Constitution of 1857. He foresaw a judiciary whose independence would be guaranteed by giving judges lifetime appointments. And he felt there should also be provisions—at the other end of the political scale—for municipal autonomy.

Carranza himself felt no need to incorporate his social reforms into the Constitution. They could be handled through issuing laws that would immediately go into effect, as Juárez had done with the Laws of the Reform, and then they would follow their own historical course, their own process of maturation. He was confident that the new Constitution would validate his concept of authority and would respect his psychological rhythms. He was right about the first point, wrong about the other. He believed that the Convention would concentrate on "purging" the political defects from the Constitution of 1857. And that the representatives would in effect approve the changes Carranza had made in the structure of state power. But, to his considerable surprise, they would accelerate historical time by introducing new and radical social reforms into the very text of the Constitution.

Carranza thought that the final chapter of his "second Reform" would neatly end in Querétaro. Instead it was the beginning of the Mexican social revolution.

Frank Tannenbaum—the great American friend, scholar, and critic of the Mexican Revolution—wrote in 1932:

> The Constitutional Convention of 1917 was the most important single event in the history of the Revolution. . . . Whatever may be said about this Convention and all possible things both good and bad have been said, it once and for all set a definitive legal program for the Mexican Revolution, and laid the legal foundation for all of the conflicting currents of the last fifteen years.[40]

Venustiano Carranza—unaware of what the Convention had in store for him but convinced of its historical importance—staged one of his characteristic triumphal processions from Mexico City to Querétaro. At eight in the morning, on November 18, 1916, "with a great sense of drama and history," he left the National Palace in Mexico City on horseback, accompanied by fifty other riders, to make the long day's journey to Querétaro.[41]

Carranza installed his retinue, prepared his speeches, and on the afternoon of December 1, when the delegates had already assembled for the first day of discussions, he made his entrance, with an impressive escort, into the council chamber. A young audience was waiting for him, one that included workers, professionals, small businessmen, journalists, and teachers. They listened to his speech respectfully but not submissively. Carranza was speaking to them from another century. They were impatient and romantic; they believed in "the law" only as a direct lever for revolutionary change. They did not represent the Reform; they represented the Revolution. After his speech, many of the delegates poured into the local taverns, asserting they had never felt greater need for a tequila.[42]

As the Convention proceeded, a consistent pattern emerged. The representatives close to Carranza, the *bloque renovador* ("renewal faction"), would offer a mild modification of some aspect of the 1857 Constitution. They could be more or less assured of 85 conservative to centrist votes. The more radical delegates (and there were 132 of them) would challenge the Carrancistas, always with a more far-reaching proposal. So it went with Article 27, the centerpiece of agrarian reform. The renewal faction argued that nothing more was needed than the 1857 text. The object was to take over and distribute enough land to establish a pattern of small private ownership. All that was required was proper administration.[43]

Immediately, Pascual Rouaix formed a voluntary commission or "core group," to study and modify the project. The sociologist Andrés Molina Enríquez joined the meetings. He was not himself a delegate to the convention, but he was welcomed for the contributions he had already made to Mexican history and sociology. Drawing on the insights of his *Los grandes problemas nacionales*, Molina guided the new law in a very different direction from the Carrancista nonproposal. For him, the solution could not be found through respecting the Liberal Constitution with some slight retouching but in returning to the spirit of colonial legislation. The individual should not be the ultimate owner of the land:

> The Nation, like the king before it, has full rights over lands and waters; it only recognizes or grants individuals direct domain and under the same conditions as in the colonial era. The right of property thus derived permits the Nation to retain within its power everything necessary for its social development, and to regulate the overall state of property as well as permitting the government to solve the agrarian problem. Upon Independence, inadequate civil legislation was adopted, which made provision only for freehold property . . . leaving the Indians without support or protection. The evil was exacerbated by the Reforma and reached a peak during the Porfirian regime, ignoring the existence of the communities.[44]

As finally approved, the new Article 27 returned "the original ownership of lands and waters" to "the Nation"—as the colony had entrusted it to the Crown. The Nation (i.e., the body of all Mexicans) had the right "to transfer control to individuals, creating private property." Obviously, this capacity to create private property implied the complementary right to revoke it in the name of the public interest. And the Nation also had the right to regulate private property in the interest of the people, while communities that had "none or not enough land and water" could take them from nearby large landholdings. Legally at least, from this moment, the era of the vast landed estates (*latifundios*) and haciendas had ended.

And if the State (as representative of the Nation) was to be the ultimate owner (and guardian) of the land, it would also own what lay beneath the surface. Article 27 gave the Nation—in place of the king—direct control over all minerals or substances in the subsoil including oil. Carranza had not proposed any more radical measures than the Calvo Clause, legally equating Mexicans and foreigners in any dispute before the courts. The young radicals went far beyond him, specifying that only "native-born or naturalized Mexicans" could have full property rights in Mexico. The government could grant such rights to foreigners, but they would have to recognize that they were in effect Mexican nationals with respect to their property rights. They could not make any legal appeals to their own governments, as foreign owners in Mexico had been doing for decades.

The young radicals had now revolutionized the concept of property both in Mexican and in international law. Only the Bolshevik Revolution would go further. Within the classic liberal scheme that respected individual property as an end in itself, the Mexican government had always negotiated from a position of weakness with foreign oil or mining companies. The principle of retroactive rights was always invoked against any assertions of ownership by the Mexican government. With the new version of Article 27, there was no question now of retroactive rights because the Nation was seen as having *always been* the owner of its soil and subsoil.[45]

All these provisions would set into motion fundamental economic and social changes as well as confrontations with powerful forces inside and outside of Mexico. The colonial past had been summoned into the twentieth century, to exert its power over the future.

A similar process—Carranza arguing for almost no movement at all; the radicals charging in, taking over, and charging ahead—created the entirely new Article 123, on labor relations. In his speech of December 1 (with its indirect benefits for the bartenders of Querétaro), Carranza had discussed labor problems from a Liberal rather than Revolutionary perspective. He had spoken—benevolently though in general terms—of specific improvements in working conditions. His faction then proposed small changes in the Constitution, dealing with the individual rights of workers rather than any notion of a social class. They made no mention of the right to strike. The young radicals brushed them aside and created a text with subterranean, perhaps unconscious echoes of the social Catholicism proposed by Pope Leo XIII in his encyclical *Rerum Novarum*. Among its foremost points were an eight-hour day, the abolition of child labor, provisions to protect working women and adolescents, required holidays, a reasonable salary to be paid in cash, profit-sharing, the establishment of boards of arbitration, and compensation in case of dismissal.

Article 123 was, of course, also inspired by more recent socialist trends. But its formulation—in February 1917—preceded the Russian Revolution. It was perhaps the first twentieth-century assault on liberal notions of adequate social progress. Among the delegates to the Convention was a man named Estebán Baca Calderón, a survivor of the attack against the strikers of Cananea by Porfirio Díaz's army. When he raised his hand to vote yes, he must have felt that those lives had not been sacrificed in vain.[46]

Once it became clear that Carranza and his partisans would not be able to moderate either the decisions or the energy of the Convention, words like flames (and stones) were to be expected when the delegates confronted the ancient issue of the Church.

"The clergy is the most dismal, most perverse enemy of the fatherland!" shouted Francisco J. Múgica—who had once been expelled from the Zamora Seminary—in a session discussing the always potentially explosive relationship between Church and State. For these new and angry Jacobins, the Church was a den of thieves, outlaws, con men—a many-headed monster that devoured Mexicans, and especially the Mexican woman, through aural assault at the confessional booth. A man named Recio from Yucatán proposed that confession be constitutionally prohibited, while delegate Alonso Romero elaborated a multiple image of the woman at confession as an adulteress, the priests as satyrs, and the husbands—who would allow their wives to pour the secrets of the home into the licentious ears of priests—as pimps.

Taking into account the small detail that nearly all Mexicans were Catholics, the Liberals close to Carranza recommended prudence and realism. Carranza himself offered some tolerant words: "The customs of a people do not change overnight; for a people to stop being Catholic, the triumph of the Revolution is not sufficient; the Mexican people will continue to be just as ignorant, superstitious and attached to their ancient customs until one educates them."

But once again, in the articles on religion, 3 and 130, the Carrancistas would be defeated. The anticlerical spirit of the 1917 Constitution went far beyond 1857. The new Constitution would refuse the Church recognition as a legal entity, deny priests various rights and subject them to public registration, forbid religious education, and prohibit public religious ritual outside of churches. And all churches would officially became the property of the nation.

Curiously enough (just as with Article 27, on the land), a critical attitude toward the Liberal Constitution was based in part on colonial patterns and origins. The state was reclaiming, for its own use, the royal domination (*real patronato*) exercised by the kings of Spain (and especially by the Bourbons), leaving the Church free only in the areas of doctrine and private devotion. But the new patronage would be enforced by a state that now had the constitutional means to be aggressively antireligious—whenever it chose to be—and there could be no recourse to Rome.[47]

The Constitution was a great effort to impose order on chaos, a bridge toward a new national life across the proliferation of death. Throughout Mexican history (but especially in its most violent days—the Conquest, the Wars of Independence and the Reform, the great Revolution), death itself, death as an ultimate value and source of value, has been as much a protagonist as the Virgin of Guadalupe. From both the Indian and the Spanish traditions comes a sense of the enormous importance of the moment of death, its potential for the transfiguration of life and the living. The great deaths—of major and minor figures, of heroes and even of human devils—were remembered through the ages, especially when they were shaped with classical courage and dignity. Over and over again in the history of Mexico, a character (often intensely moved by love for his people) when captured or trapped by his enemies, with nothing more to offer in life, makes a gift to "Mexico" of a memorable death, an iconic legacy for the future generations.

If a figure in Mexican folk song (or modern ranchero ballads) shouts, perhaps drunkenly, on the verge of mortal combat, *La vida no vale nada!*, he is not asserting, literally, that "life is worth nothing" but that death—how someone dies—has immense worth. There is a colloquial phrase for this attitude: "*hombrearse con la muerte*"—"to face death like a man" or, more precisely, "to push death around."[48] The individual death of a major historical actor is of course the purest moment within this aesthetic of death, but there was room, on the stage of enduring memory, even for the chaotic hell of mass death within the Mexican Revolution—transfigured by the sensibility of writers and visual artists but also as received by the Mexican people. José Clemente

Venustiano Carranza with children, 1917
(FINAH)

Orozco, the greatest Mexican muralist and equally powerful when he chose to use words, describes how:

> the tragedy tore up everything around us. Soldiers moved along the iron tracks to the slaughterhouse. Trains were blown up. . . . The people grew accustomed to slaughter, to the most merciless selfishness, to satiating their senses, to pure, unconcealed animality. Small towns were attacked and every kind of outrage committed. The trains coming from the battlefields . . . emptied their loads of tired and wounded soldiers, battered, exhausted, sweat-soaked, ragged. . . .
>
> Farce, drama and barbarity. Clowns and dwarfs scurrying after lords of the knife and the gallows in intimate conversation with the smiling madams of brothels. Insolent commanders, bold with alcohol, demanding everything at the point of a gun.
>
> Shots exchanged in dark streets, at night, and then screams, curses, insults that could not be forgiven. Windows shattering, dull blows, crys of pain, more gunfire.
>
> A parade of stretchers with the wounded wrapped in bloody rags and suddenly a savage peal of bells and the thunder of shots. Drums and bugles sounding a reveille drowned out by the multitudes cheering for Obregón. Death to Villa! Viva Carranza! *La Cucaracha* chorused to the rhythm of bullets. Riotous celebrations of the victories of Trinidad and Celaya while the unlucky Zapatista peons taken prisoner would fall before Carranza's firing squad in the courtyard of the church.[49]

And for others, the blood and fury of the Revolution would sound the deeper notes of a religious return to origins. Long after those years, in one of the most important books of the Mexican twentieth century—*The Labyrinth of Solitude*—Octavio Paz would write:

> . . . the Revolution is a search for ourselves and a return to the mother . . . a going to extremes, an outburst of joy and abandonment. . . . And with whom does Mexico commune in this bloody fiesta? With itself, with its own being. Mexico dares to be. The revolutionary explosion is a prodigious fiesta in which the Mexican, drunk with himself, at last comes to know, in a mortal embrace, his fellow Mexican.[50]

That outburst of redemption and grief, that *hombrearse con la muerte*, that multicolored parade and swarming theater, was a huge and tragic gesture for all Mexicans, for the common people as for the middle classes, but those who would legislate and administer and govern felt the need to interpret, to align the significance of events within a rational context. Destruction and death could not be only a matter of chance. There had to be meaning and finality in the flow of events. Civil war had to become social philosophy, and the Constitution of 1917 was the moment of transfiguration.

For Carranza it may have been like waking up in the morning and seeing a beloved child miraculously grown and mature and with a stubborn mind of its own. His new Laws of the Reform had been taken out of his hands, expanded and incorporated into the Constitution of 1917, making it a legal and doctrinal production very different from—and even contradictory to—the 1857 Constitution. But though the Carrancista delegates had not been able to impose their social moderation, they would succeed in introducing their own political alterations to the old Liberal Constitution—or rather the structural changes desired by the *Primer Jefe*. The executive power would be immensely strengthened and the legislative power greatly weakened. There would be no Vice President. (And judges would be chosen for life—to assure their independence—while the institution of municipal autonomy would be fortified.)

The best men among the radical delegates were moved by a profound humanitarianism, by the will "to put the welfare of the majority before that of the minority and by the belief that this could not be achieved without the initiative and the active support of the Revolution now become a government." They distrusted the Liberal laws because, in their view, these laws had almost always served to disguise privilege and oppression. So that justice might be done, so that the spilling of so much blood might be justified, so that the weak would be protected, the radicals felt they had to go to the heart of things. And they listened to the voice of Andrés Molina Enríquez, who spoke for a return to Mexican traditions. In his view, only the colonial, patriarchal, corporate, and traditional State had cared for the welfare of the weak. The radicals framing the future looked consciously or unconsciously toward the past.

While the supreme values of the radicals were social justice and material equality, the ideology of the Carranza faction is less simply presented. They wanted full national sovereignty, municipal autonomy (a strong tradition in Carranza's state of Coahuila), individual liberties, and the strengthening of State powers. But there was a genuine conflict between these last two objectives because Carranza distrusted the "Liberal utopianism" of the 1857 Constitution.[51]

"In order to be free," Carranza maintained, "it is not enough to want it but one also has to know how to be free." [52] Despite his connections to nineteenth-century Liberalism, his personal concept of democracy was in part opposed to classic constitutional liberalism: "Democracy . . . cannot be anything other than the government of noble, profound and serene Reason. . . . Democracy . . . must not seek the majority in partisan compromises of whatever origin or shielded under whatever name but in the representation of all classes and all legitimate interests."[53]

The word *vote* appears nowhere in his definition. The idea of parties is rejected. For Carranza, as for all postrevolutionary rulers since that time, the paternal and omnipresent State was in itself the embodiment of democracy.

The radicals had returned ownership of the land, waters, and subsoil (once the property of the Spanish Crown) to the Nation of all Mexicans (and their representative, the State), and the State had also been granted unlimited power to control the religious life of the country. Without being radicals, Carranza and his group had, to a certain extent, also returned to the colonial past. Strengthening the executive power (and weakening the legislative), they made the President almost omnipotent. Placed above social groups but "representing" them all, the new State—with more explicit power than Don Porfirio could ever have imagined—assumed the responsibility of paternally guiding the nation toward progress, prosperity, justice, and equality. The involuntary union of the radical and the Carrancista programs in the Constitution of 1917 gave birth to twentieth-century Mexico. A purely democratic revolution launched to overthrow a dictatorship ended up by creating an equally authoritarian regime. This marriage of the twentieth century with the colonial past was consummated in Querétaro when, on February 5, 1917, after two months of passionate debate, the new Constitution was proclaimed.[54]

In May 1917, the *Primer Jefe* became the constitutional President of Mexico. During his three years of ceremonious government, there was no military or social peace, none on the international front, none in the minds of Mexicans. The Revolution was not over.

The economic panorama could not have been more disastrous. For the revolutionary leaders and intellectuals, 1917 would be remembered as the year of the Constitution. For most of the Mexican people it would go down in history as one of the worst years of the century, "the year of hunger." The government, with an enormous debt of almost 750 million pesos and no internal or external credit, could barely do more than meet its insatiable military budget. Unemployment was rising, and the country was in ruins. The inequitable structure of Porfirian progress had collapsed. Crops went unharvested, railroads had been destroyed, cattle were exported to buy munitions, mines and industries closed, banks failed, capital was stolen or hoarded. The cities were short of water, food, coal. Black markets were flourishing everywhere. The worldwide epidemic of influenza as well as typhus and other diseases were spreading through the country. The nation's agriculture was close to catastrophic. Subsistence crops failed; for lack of funds, grain could not be imported; hunger was rampant. The campesinos often had to eat bran that might be sold to them mixed with sawdust. Or with the cows gone, they had to overcome a great Mexican revulsion against killing and eating horses. Elsewhere they ate earth.[55]

The country was still a huge rebel encampment. In the mountains of Morelos, Zapata was carrying on *his* stubborn revolution. Félix Díaz was active in the state of Veracruz, hoping to bring back the era of his uncle Porfirio. In Oaxaca, there were Guillermo Meixueiro and José María Dávila, also inspired by memories of Porfirismo and by a tradition of local autonomy. (They called themselves Soberanistas—partisans of their own sovereignty.) In the Huasteca, General Manuel Peláez continued to protect the American oil companies against any interference from the national government. In San Luis Potosí, the enemy was the Cedillo brothers and in Michoacán, Inés Chávez García, a hyena somehow born with a human body. Through Chihuahua roamed the government's greatest nightmare—the rebel of legends, Pancho Villa.

In his three years—almost to the day—of constitutional government, Carranza succeeded only in eliminating a few minor leaders and two luminaries: Felipe Ángeles—betrayed and sentenced in a rigged trial—and Emiliano Zapata, lured, betrayed, and shot down. When he had them in his hands, Carranza showed no mercy to his enemies and perhaps thought that he was following in Juárez's footsteps. But Juárez, when the Republic was restored in 1867, had decreed a broad and generous amnesty. In this area, the student lagged far behind his teacher. History would blame him for it and with reason.[56]

In international affairs, Carranza showed real courage, especially toward the United States. From beginning to end he defended Mexican interests with an obstinacy that at the time seemed like blindness or inflexibility but in retrospect proved to be a fortunate mixture of firmness and flexibility. During those three years of internal and external torment, various groups in U. S. government and business openly fought to withdraw recognition from Carranza's government, pressing for its overthrow and even for military intervention. On the other hand, President Woodrow Wilson almost always took a mild and conciliatory position. He prided himself—and rightly so—on never having used major military force against Mexico. And Carranza showed considerable skill in exploiting the disagreements between powerful groups and individuals in America. He divided and he prevailed.

He would not let himself be provoked, and he did not provoke others. The rewards were considerable. He never had to abrogate or revise the terms of his legislation on foreign interests. He introduced new rules of the game—favorable to Mexico—in international relations with governments, companies, and individuals. And he did all this during the First World War, from which the United States was to emerge more powerful than ever.[57]

Even his limited flirtation with Germany—as a counterbalance to American influence and pressure—did not lead Carranza to lose his sureness of touch in foreign affairs. He hoped for aid from Germany before, during, and after the World War, but not to the extent of renouncing Mexican neutrality or embarking on risky adventures that might provoke the United States or its allies. There was the incident of the famous Zimmermann telegram, sent from the German Foreign Office to the German ambassador in Mexico but intercepted and published by the British Secret Service:

> We intend to begin unrestricted submarine warfare on the first of February. Nevertheless, we will try to keep the United States neutral. In the event that we do not succeed in doing so, we will make Mexico a proposal of alliance on the following basis: joint command of the war, a common peace treaty, generous financial support, and an understanding on our part that Mexico is to reconquer its former territories in Texas, New Mexico, and Arizona. The details of the arrangement we leave to Your Excellency. . . .

The telegram was apparently the culmination of a series of attempts to involve Mexico in a war with the United States (and prevent the Americans from entering the war in Europe). None of the German machinations tempted Carranza into making any foolish moves.[58]

He was equally prudent domestically on religious issues, refusing to put the new constitutional Articles 3 and 130 into effect. His governors in some states were impatient. When the governor of Jalisco insisted on the official registration of priests, a widespread and effective protest closure of stores and businesses was organized. The state government yielded, and peace returned. Immediately—to prevent recurrences—Carranza proposed amending the constitutional articles on religion.

He wanted to substitute an entirely new Article 3 for the Constitution's prohibitions on religious education. Carranza argued—and perhaps for the first time he was speaking from almost a pure nineteenth-century Liberal position—in favor of "freedom of conscience" and the need to respect the decisions of parents "concerning their children's education."

He proposed to leave Article 130 in place but eliminate some of its most incendiary requirements, like the right of state legislatures to restrict the number of priests and the clause requiring priests to be Mexican-born. If the anticlerical elements of the new Constitution were insistently imposed, there was "a serious risk of prolonging the irritation characteristic of the struggles over religion which have proved to be so grim in the Old and in the New World."

To his disappointment (and to the future anguish of Mexico), these proposals found no echo. The state legislatures and two-thirds of the National Congress rejected them.[59]

But Carranza, as constitutional President, not only invoked nineteenth-century Liberalism in his sensible opposition to the social dangers of anticlericalism—he also put brakes on agrarian reform and on the labor movement. The young radicals who had defeated him had not convinced him. The new articles already formed part of the Constitution, but they could be modified or only partially enforced, given the powers that the same Constitution had conferred on the President. Carranza thought that much of the new revolutionary legislation was erroneous, and he tried to oppose and alter it. But more powerful currents than he could ever dam were flooding over the Liberal island Don Venustiano Carranza now firmly recognized to be his territory. And it was too late to build new bridges.

On the eve of the 1920 presidential election, Carranza was, more than ever, a man whose time had passed. A new ideology and a new attitude rooted in both the past (the moral values of the colonial era) and also looking toward the future (under the influence of modern socialist ideology) had left him far behind. The country was still in chaos, and to make things worse, much of the new political class was treating public office like private property. "The Old Man doesn't steal, but he lets them steal" became a popular saying. And a new verb, *carrancear,* was coined, meaning "to steal." Despite his personal honesty, many ordinary Mexicans began to associate Carranza with corruption.

And then he decided to open fire on the last of his battlefronts. Perhaps remembering Juárez's defense of civilian government in 1871 against "the odious banner of militarism," he determined that he would add a new page to the history of Mexico. It was an addition both desirable and sadly premature—putting an end to military domination of the government. The shrewd old man felt that, in Álvaro Obregón, a new Porfirio Díaz was waiting for his chance at the presidency. To the Spaniard Blasco Ibáñez, who was then writing his book *El militarismo mexicano*, Carranza said: "The evil of Mexico has been and is militarism. Very few civilians have become President. Always generals and what generals! . . . This has to end, for the good of Mexico. I want a civilian to succeed me as President, a modern and progressive man who will keep peace in the country and expedite its economic development. It is time that Mexico began to live like other nations."[60]

But the time was right in moral, not political terms. And politically Carranza blundered in two serious respects. He turned away from Obregón—the military victor of the Revolution—and then chose for President a man nobody knew, Ignacio Bonillas, the Mexican ambassador in Washington. At the time, there was a popular song—current throughout Latin America—about a poor shepherdess abandoned and wandering, ignorant of her parents or where she was born. All she knows is that her name is Flor de Té. The malicious Mexican public immediately baptized Carranza's candidate: "Viva Bonillas! Viva Flor de Té!"

The real power in Mexico at the time was not Carranza but the tight-knit group of generals from Sonora who—ever since the Huerta coup and the *Plan de Guadalupe*—had played a decisive role in the Revolution. They all saw Obregón as their leader, the undisputed caudillo who had defeated Villa, "the man most popular and most feared." And Obregón made his intentions clear: "If I don't get to be elected President, it will be because Don Venustiano does not want me. But before the bearded old man can rig the elections, I will rise against him."[61]

In April 1920, the Sonorans issued their *Plan de Agua Prieta*, repudiated Carranza's government, and renewed the Revolution on their own.

Martín Luis Guzmán hit on the perfect adjective to describe the fall of Venustiano Carranza: *ineluctable* ("irresistible").[62] It was the last in a series of defeats that had begun much earlier at the Constitutional Convention in Querétaro. The new generations—whether radical intellectuals or Sonoran generals—were not about to accept the admonitions of the old patriarch. Carranza fully realized this but would not back down. He returned to his Juarista map of action and issued a "clear, categorical," and dignified statement directed to History, his favorite muse:

> Those who think I am capable of giving in to the threat of an armed movement, however large and powerful, are completely mistaken. I will fight as long as it takes and with all possible

> means. . . . I must leave it settled, affirmed and established that the principle of public power should no longer be a prize for military caudillos whose revolutionary merits do not excuse their later acts of ambition.[63]

It was not a matter of saving lives, not even his own life, but of salvaging principles. His duty was to preserve, and he would act in obedience not to strategy but to destiny. His end was "irresistible" because he freely assumed it: "My only goal is to defeat the rebels by arms or to die fighting in the contest. . . . Since the year 1913 I have been living on borrowed time."[64]

Rebel pressure at the beginning of May made him decide to leave the capital and again go to Veracruz. On his last night in Mexico City, he reread one of his favorite books, an eighteenth-century French biography of the Emperor Justinian's general Belisarius, who had led thirty years of brilliant campaigns for his emperor but then (according to this legendary version, which seems to have confused him with Boethius) fell from favor and was blinded in prison, where he stoically accepted his fate. "Whoever completely gives himself to the fatherland must think of her as poor," Belisarius says, "because that which he offers because of her is priceless."[65]

In the stoic virtue of that Roman general submitting himself to fate, Carranza found his final image and consolation. But when he left the next day by train, he was still the elaborate traveler—filling sixty railroad cars with his people, with weapons, with his files and possessions, and with the entire National Treasury in bars of gold.

He was moving now with an almost deliberate inertia, no longer impelled by an instinct for survival but by the will to leave a legacy, for history: "History will recognize the patriotic motivation of my actions and will judge them. I proceed as I think is my duty for the good of the country."

Carranza's nomadic government—strung out in its long line of railroad cars—took losses for the first time on the outskirts of the capital, at Villa de Guadalupe. It would not be the last attack. "Each kilometer awakened new fears, each station greater threats."

On May 14 there was a bloody gunfight in Aljibes Station with the forces of Guadalupe Sánchez, who like the great majority of the generals had turned against the constitutional regime. In the midst of the shooting, General Urquizo reached the platform of the presidential car where Carranza, seated, tranquil and "without fear," watched the disorder and the panic. Over and over Urquizo tried to persuade him to leave the train and escape. Carranza refused. He was challenging fate. "He did not move from the chair in which he was resting; not a muscle moved in his face . . . some projectiles ricocheted balefully off the train . . . and others off the gilded handrail of the platform."

At last, yielding to the pleas of General Murguía, Carranza descended and without any hurry mounted a fresh horse. (Two days earlier, his had been killed under him.) Attacked on all sides, the caravan fell apart: "It looked like the remains of a shipwreck." Since the route to Veracruz was cut off, Carranza's depleted party decided to change direction. They would head north. Only a handful of generals accompanied him: Mariel, Murguía, Barragán. In the face of this disaster, Carranza remained unruffled. He was all calm, fortitude, order. Now he was even more aware that he was playing a leading role in a true historical drama. Juárez was no longer the model. Carranza himself was the hero.[66]

On May 20, after six days of a long, difficult march, the caravan of a hundred men—many of them civilians—crossed a river and passed through settlements to reach an area controlled by General Rodolfo Herrero, a former rebel who had, a few weeks earlier, accepted amnesty from the Carranza government. Herrero was elaborately obsequious and persuaded Carranza to continue on to the remote village of Tlaxcalantongo, there to spend the night and await news from General Mariel, who had gone ahead to reconnoiter the way north.

At one in the morning, Herrero announced that his brother had been wounded in a nearby village and that he had to go to him. Carranza's people were immediately suspicious, but Carranza quoted history: "Let's say what Miramón said in Querétaro: 'God be with us for these twenty-four hours.'"

Carranza could usually fall asleep anywhere, but that night he lay tossing and turning in the corner opposite the door. Around three in the morning, a messenger arrived with a note from Mariel, which said that the following day's route was clear and safe. Carranza remarked, "Gentlemen, now we can rest." Twenty minutes later, suddenly, in the midst of the darkness and the rain, shouts were heard and a burst of gunfire.[67]

"Crawling like reptiles along the muddy ground," Herrero's men had reached the spot outside the hut exactly opposite from where Carranza was sleeping. Then they had opened fire. Days later, Carranza's Minister of the Interior, Aguirre Berlanga, testified:

> Immediately after the first shots, the President said, "They've broken my leg, I can't move," and I answered, "What can I do for you, sir?" but he did not answer, and I don't know whether he heard my words because the gunfire continued, intensely, as also the shouts of "Death to Carranza!" "Come out, you old bearded goat!" "Crawl

> out on the ground!". . . and they entered . . . and pointed their guns. . . . Señor Carranza was not dead yet, but he never spoke again, only a death rattle.[68]

There is evidence suggesting that once Carranza's leg was broken, in those minutes of uproar and rain and darkness, the President realized he was lost and chose to die by his own hand. He put on his glasses. He picked up his Colt .45 revolver. Steadying the pistol between the index finger and thumb of his left hand, he pointed the muzzle at his chest. He managed to fire off one or more shots. The death rattle sounded, and death came to him.[69]

In the Casa de Carranza Museum today are the bullets taken from the body of Francisco I. Madero after his murder. Carranza had kept them in his home. Perhaps because they were the symbol of a fate and a passive denouement he had always hoped to avoid.

He had said more than once, when he left on his long train that moved so slowly for so short a time, that he would return to Mexico City either victorious or dead. Whether or not he died by his own hand, he never surrendered to his enemies, and his death had a dignity that could well be compared to that of Miramón in Querétaro.

Women threw themselves down before his open coffin, on the way to the Dolores Cemetery of Mexico City, crying out, "Our father is dead!" The term said a great deal. That tenacious old man had, for a long time, presided over the Revolution like a strict and intelligent father. The remnant of another sensibility, another era, a man of the nineteenth-century Reform, Carranza channeled a social struggle that without him might have spilled over into endless chaos. A bridge between the centuries, he carried Mexican history toward a shore not to be found on his chart, too complex for his vision but also more sensitive to human suffering.

14

ÁLVARO OBREGÓN

Death and the General

Sólo venimos a dormir,
sólo venimos a soñar:
No es verdad, no es verdad que venimos a vivir en la tierra.

We only came to sleep,
We only came to dream:
It is not true, it is not true that we came to live on the earth.

—AZTEC POEM

"I had so many brothers and sisters that when we ate Gruyere cheese, only the holes were left for me." This joke, one of the endless stream that Álvaro Obregón used to recount or invent, gives a fair idea of his origins. At one time, there would have been enough cheese for everyone in the household of Francisco Obregón and Cenobia Salido. But beyond their having eighteen mouths to feed, politics and nature had done them harm. In 1867 the Liberal government had confiscated a large part of the family estate because of the sympathy Don Francisco's business partner had shown for the empire of Maximilian. A year later, a devastating flood came close to confiscating the rest. By 1880, the year Don Francisco died and Álvaro, the last of the children, was born, the

Hacienda of Siquisiva—the family home in the northwestern state of Sonora—was well on its way to ruin.[1]

The youngest child was brought up there by his mother and three of his sisters, teachers by profession—Cenobia, María, and Rosa—who were never to leave him. As a child he learned all the jobs on a farm and spent much of his time in the company of Mayo Indians, men noted for their courage. One of his brothers, José, ran a school in Huatabampo where Álvaro received his elementary education. Books and poems attracted him, but necessity and his innate energy took him out of school and launched him into life.

During his adolescence he became a jack-of-all-trades. He was good at whatever he put his hand to and tried to learn something about everything he could. By the age of thirteen he was his own boss: He had begun raising tobacco and he started a small cigarette factory. With members of his family, he formed an orchestra of which he became the conductor. He learned photography and carpentry, and it was not long before he discovered his great talent for mechanics. In 1898 he became a lathe operator at the Navolato sugar mill and then the ace mechanic at the Tres Hermanos mill owned by the Salido brothers, his maternal uncles who were prosperous hacendados in the Mayo River region.[2]

In 1903 he married Refugio Urrea, and in 1904 he began to sell shoes door-to-door and also tried his luck as a tenant farmer. Though at first he suffered some losses due to sudden floods, he was in a position, by 1906, to buy a small farm from the federal government that he humorously named La Quinta Chilla ("The Poor Man's Manor").[3]

By 1907, when Álvaro Obregón was twenty-seven years old, his financial future began to look up. The chickpeas he was raising at La Quinta Chilla were gradually raising him out of poverty. But nature was hostile to him in another, deeply tragic way: In 1907 two of his four children, including the eldest, died very young, and, in that same year, his wife also died.[4]

In 1909, Obregón, a widower with two small children—now cared for by the three older sisters who had cared for him—made his first great splash: He invented a chickpea harvester that he was soon manufacturing on an assembly line and selling to all the local farmers in the Mayo Valley. When Obregón took his first trip to Mexico City in 1910—to attend the Centenary Fiestas—he could feel a legitimate pride in his growing prosperity.

The hard work required to repair his family's battered fortunes surely dominated Obregón's early life, but it did not make him a bitter man. On the contrary—he was known as a lighthearted, genial, romantic, and very likeable person, with a creative and alert mind. His great mechanical ability was complemented by a truly amazing memory. He could remember the order of an entire deck of cards arranged at random after seeing them once, and he became the most feared poker player in Navolato, a professional reader of minds. The sugar-mill owner finally paid him not to play.

But this hardworking and expansive man, in whom everyone recognized extraordinary mental powers, nourished a darker side, an undercurrent that turned his thoughts toward death. Something in his childhood, perhaps the loss of his father when he was only a few months old, struck him so profoundly that he spent the first five years of his life in silence. The child who said nothing and recorded everything was repressing a tension that finally surfaced when a friend of his mother disparagingly called him a "monkey," to which the young Álvaro answered with his very first words: *vieja loca* ("crazy old woman!"), already combining humor with aggression.

Despite economic need, the early years passed for him with more joys than pain, but when he was an adolescent, he had a strange and sorrowful experience that would leave a heavy mark upon him. When his older sister Rosa described it, she seemed to share her brother's torment:

> When Álvaro was about fifteen, he was working on our brother Alejandro's hacienda, located some thirty leagues from Siquisiva. They both slept in the same room . . . and one night Álvaro woke up frightened and moaning . . . Alejandro woke up, and when he saw what was going on, he was startled by it and asked him what was wrong. By then Álvaro was fully awake and he told him he had just had a horrible dream, in which he had seen our mother dead. This had a terrible effect on Alejandro but nevertheless he tried to comfort Álvaro and calm him down, telling him it was only a nightmare and that he should go back to sleep. But now Álvaro couldn't sleep and he lay awake the rest of the night. At dawn they heard a horse galloping toward the hacienda. What had happened? It was nothing less than a man sent to bring them the sad news that our mother had died in Huatabampo that very night. Álvaro never forgot that nightmare. Each time he told the story, he became terribly nervous and . . . [5]

Rosa Obregón did not go on, but that "and . . ." trailing off reflectively suggests the obsessive presence of death in the life and dreams of Álvaro Obregón.

And when he was twenty-nine years old, in 1909, the year before

Madero began his revolution and the year Obregón invented his chickpea harvester, he wrote a poem that seems to be partly modeled after the great medieval Spanish poem "Coplas por la muerte de su padre" ("Stanzas on the Death of his Father") by Jorge Manrique. Obregón's poem is hardly a masterpiece, but it does have a certain rough power. He titled it "Fuegos fatuos" ("Empty Fires"), referring to the phosphorescent glow sometimes seen over swamps due to the decay of organic matter, called *ignis fatuus* in Latin, "will o' the wisp" in English, "foxfire" in the swamp country of the American South. As in Manrique, the poem speaks of the emptiness of life and of souls after death, all leveled and equal:

. . . humanity in its affliction

drags along so much fruitless vanity,
forgetting the tribute
that it will have to render to the cemetery.

There everything is the same; the boneyard
is the same as Calvary;
and although they may come from different conditions,
the men and the women, the old and the infants,
in the dark nights,
foxfires, unreal, they all pass on together.[6]

"I'm used to fighting with the elements of nature: the frosts, the wheat rust, the rain, the winds that always come when you don't expect them. How could it be hard for me to defeat men, whose passions, intelligence and weaknesses I understand? It's easy to switch from being a farmer to a soldier."[7] Obregón was speaking in 1912, when he first turned soldier, on the side of the Madero government against the rebellion of Pascual Orozco. But in 1910, when Madero rose against Porfirio Díaz, it had not been so simple for Obregón to shift from farming to soldiering. His new prosperity, his small children, and even a fairly open sympathy for Don Porfirio dissuaded him from joining the initial struggle. His nephew Benjamín Hill and almost all his future comrades in arms would always reproach him for being a latecomer. But he would blame himself even more. In 1917 Obregón had the courage to make a full confession in the prologue to his self-exalting and not very readable opus: *Ocho mil kilómetros en campaña* ("The Eight-Thousand-Kilometer Campaign"):

> . . . The Maderista anti-reelectionist party was divided into two groups: one composed of men who answered the call of duty, who abandoned their homes and broke every connection with family or occupation to seize a rifle, or a shotgun or the first weapon they could find; the other of men who listened to the call of fear, who could not come up with any weapons, and had their children, who would be left orphans if they should perish in the struggle, and they had a thousand other ties that Duty cannot sever when the phantom of fear takes control of men.
>
> I bore the shame of belonging to the second of these groups.[8]

In April 1912 came another chance. This time he did not miss the opportunity. To fight for Madero against Pascual Orozco, he rapidly collected three hundred men and joined the Fourth Irregular Battalion of Sonora, under the command of General Sanginés.[9]

In a matter of days, Obregón astonished his superiors. Disobeying orders—as Porfirio Díaz had in battle—he devised tactics to lure the enemy into traps, surprise assaults, encircling maneuvers that won him rich spoils and automatic promotions. Victoriano Huerta himself said when he met him: "Let's hope this commander is a good omen for the fatherland!"[10]

At one point in the struggle, Obregón found a military outlet for his mechanical ingenuity. Disregarding his superiors who believed in collective trenches, he conceived and, in a way, invented the idea that each soldier should dig his own foxhole, which would be cheaper, faster, and safer. It would be some years before this tactic would be widely adopted by armies throughout the world.[11]

Colonel Obregón had now crossed a boundary. Death was no longer a shadow that might pass through his mind or a wake of family misfortunes trailing behind him. It was there before him, every day, in war. Death was almost his friend now, in the open, beyond war or peace. Some of the feeling shows in a letter to his son Humberto, then five or six years old:

> By the time you get this letter I will have left for the northern border with my battalion, summoned by the voice of the *Patria*, which is in this moment having its guts torn up and there cannot be a single Mexican who does not come to its aid. I am only sorry that you are far too young to accompany me. If I have the honor of dying for this cause, give thanks for being an orphan and you will be able to call yourself, with pride, the son of a patriot.[12]

In December 1912, Colonel Obregón resigned from the army and returned to his farm. He enjoyed his country peace for two months. In February 1913, the barracks uprising overthrew President Madero. Without a moment's hesitation, Obregón offered his services to the government of Sonora. They made him chief of the War Department with an immediate mission. On March 6, 1913, he left Hermosillo with orders to capture the three main strongholds in the northern part of the state: Nogales, Cananea, and Naco.

Between March 1913—when he fought his first battles on the border—and August 1914—when he entered Mexico City in command of an undefeated Constitutionalist Army—Obregón deployed his great natural gifts in a more demanding enterprise than the cultivation of chickpeas. Fortune followed his every step, but it was a fortune armored with calculation and observation. It took him only a few days to drive the federal armies out of their garrisons. His goal was then to move south and capture the port of Guaymas. But there as elsewhere Obregón would never order headlong charges like Pancho Villa. He drew the enemy out, away from their base of operations, forced them to protect their lines of communication and meet him in skirmishes, undermining their morale before he would make the decision to confront them in open battle. Showing his good judgment once more in this instance, Obregón did not attack Guaymas: he laid siege to it with a fraction of his forces and continued his advance toward the south.

In May 1913, he won the battle of Santa Rosa through a carefully timed encirclement of the enemy forces. A few days later he captured three hundred men and all their artillery in Santa María after first carrying out a meticulous study of the terrain and geometrically planning blockades and maneuvers to cut off their escape. By now he was commanding officers (like Benjamín Hill) who had been with Madero from the very beginning, but they formed up in ranks, though not always willingly, behind the legitimate authority of Obregón.[13]

On September 20, 1913, Venustiano Carranza named Obregón commander in chief of the Army of the Northwest with jurisdiction over Sonora, Sinaloa, Durango, Chihuahua, and Baja California.[14] Two months later, in an exemplary display of his ability to command and deploy men and to make use of artillery support and the nature of the terrain, Obregón took Culiacán. During the battle he was wounded in the leg, but he joked about the wound. The bullets, he felt, did not take him very seriously. "They wounded me, yes, but my wound couldn't have been more ridiculous: a Mauser bullet bounced off a rock to slap my thigh."[15]

In Obregón's area of responsibility, there was a lull of almost five months then in the fighting, a time that Carranza dedicated to the political and military organization of the Revolution. Obregón used the period, among other things, to distance Carranza from Felipe Ángeles, the only military commander who might overshadow him.

The soldiers of Pancho Villa followed their caudillo through conviction or a fondness for the man, because of his charisma, and sometimes also because they wanted the spoils and trophies of war. Obregón's troops did not serve him for any more magical motives than hard, cold cash. The Sonoran revolution was, more clearly than the others, an organized enterprise. His soldiers depended more on their salaries than on plunder. Among them, the Yaqui Indian battalions stood out for their bravery. Obregón had been at home with Indians since his childhood (and he had recruited Mayos to help him get elected municipal president of Huatabampo in 1911). Now he had managed to enlist thousands of Yaquis in exchange for an agreement. After the victory, they expected their immemorial demand for the lands "God had given them" to be satisfied. Obregón was not the only Sonoran leader to recruit Yaquis, but more of them fought for him and he knew how to treat them and how to make the best use—militarily and psychologically—of their legendary military prowess.

In April 1914, he began to march again. While Villa dazzled the national and foreign press with his wild cavalry charges, Obregón advanced along the Pacific Coast, using the more effective weapon of his ingenuity. In May 1914, on the Topolobampo coast, when the federal gunboat *General Guerrero* threatened his troops, a remarkable event occurred. For the first time in the military history of the world, a plane—the *Sonora* piloted by Alberto Salinas of Obregón's army—attacked the gunboat, flying (at a height of three thousand feet) eleven miles out to sea.

In May 1914, Carranza, already at odds with Villa, ordered Obregón to speed up the southern campaign. Obregón blockaded Mazatlán, as he had done at Guaymas, and continued on to Tepic, where he cut the Guadalajara-Colima railroad, isolating both of the besieged ports. At the beginning of July he simultaneously threatened positions to the south of Guadalajara and defeated the federal troops in Orendáin. There were eight thousand enemy dead, and his army captured sixteen artillery pieces, five thousand rifles, eighteen trains, and forty locomotives. A few days later, Obregón—now wearing the eagle of a major general—was sweeping forward toward the capital. He arrived in Teoloyucan on August 11, when he signed (on the mudguard of a car) the treaties that ended the regime of Victoriano Huerta. Five days later, at the head of eighteen thousand men, he entered Mexico City.[16]

Marching beside the ranchers and miners in their felt hats, khaki pants, and brown leather leggings were the Yaquis, "to whose bravery the Sonoran rebels undoubtedly owed their victories." Some entered playing small drums, while others still wore the clothes in which they had left Sonora:

> the same short cotton pants, the same sandals, the same embroidered shirts, the same ribbons tying back their hair. Some had gotten used to boots and others had adopted the ten-gallon hat, without of course changing their outfits. They all carried Winchester 30/30 rifles and wore several well-filled cartridge belts; and nobody had left their bows, quivers, slings and blowguns behind—frightening objects for the urbane inhabitants of the capital.[17]

Mexico City, although it quivered to the rhythm of the Yaqui drums, was much more afraid of the menace from the Attila of the South, Emiliano Zapata. And many considered it a relative blessing that a "white man" should be the first to "take the city." Some even compared Álvaro Obregón to Hernán Cortés. But the hopes of the well-dressed citizens soon went up in smoke. Obregón came not as a messenger of peace but of revenge, for the death of Madero.

Unlike Villa, who had a purely irrational relationship with death—the death of others, not his own—Obregón seemed to have made a painful pact with death from the very beginning. He had not abandoned "the delights of home," as he used to call them, because of any jubilant delight in joining the *bola*, because of profound democratic or social convictions or even through pragmatic calculation. It was as if a destiny he could not avoid had torn him away from those "delights."

When he decided to join the struggle as a leader he had, in full awareness, prepared himself beforehand for the possibility that he would lose his life. He did not play with death, but he faced it like a bullfighter, with indifference and disdain. In his western campaign he had taken a number of risks, some of them reckless. He carried no weapons, did not move when a grenade fell only a few yards away, made some dangerous river crossings when he might have avoided them or waited for calmer waters.

The fascination with death—as theme, as goal and value and obsession—shows even in the language of his public manifestos, where the word *blood* comes easily to his imagination. The Huertistas were "a pack of dogs" and "let us satisfy their thirst for blood until they choke on it." Or when he ended his stay in Mexico City with the words, "A people cannot shed too much blood in defense of their freedoms."

He had once said to Carranza: "All of us who take part in this affair do it for patriotism and to avenge the death of Señor Madero."

When he heard this, Carranza must have been disturbed. For him, the struggle was much more than a vendetta, much broader than a "matter of patriotism." It was a historical cause like that of Juárez, involving the entire nation, its sovereignty, its laws, its internal order, and its destiny. Not even Zapata fought for revenge and a vague patriotism, but "for the land." Other revolutionary jefes justified their involvement with idealistic, social, or moral theories. But to Obregón, the Revolution was not a matter of theories but of war, of the dance with death. And the murder of "Señor Madero"—the apostle whose memory bothered his conscience because he had not joined him in his first, brave uprising—required repayment.[18]

Three days after his arrival in the capital, Obregón went to the French Cemetery to pay his respects to Madero. Standing by the grave, facing some of the men who had been Deputies during Madero's presidency but whom he believed had behaved like cowards when Madero was slaughtered, he handed over his pistol to María Arias—a woman who had publicly protested the events of February 1913—with these provoking words: "I give my pistol to María Arias, the only man to be found in Mexico City at the time of Huerta's coup."[19]

It was as if by punishing the cowardice of Mexico City he could atone for his own initial doubts of 1910 and cleanse himself of his error. The city had apparently been chosen to pay the immense moral cost of his close personal involvement with death. For this reason—beyond the objective responsibility of the capital, which did in fact exist—Obregón had come to make the "sadly celebrated Mexico City" suffer for its sin, and within it most especially the clergy, the bourgeoisie, and resident foreigners. For the work of retribution, his memory was his greatest ally. He remembered an attack upon him published (in the city of Tepic) by the newspaper *El Hogar Católico* ("The Catholic Home"). Like many Constitutionalist leaders, Obregón had the mistaken idea that the clergy had been an important source of support for Huerta, and he therefore decided to impose a fine of half a million pesos on the Church, to be paid to the Revolutionary Council for Aid to the People. Later, he jailed and then expelled Paredes, the Vicar General (the chief ecclesiastical judge) and 167 other priests from the capital.

It went worse for the rich men of the city. Obregón remembered that businessmen like Pugibet, who owned the El Buen Tono cigarette factory, had applauded Huerta as the savior of the *patria*. In response, he imposed special taxes on capital, real estate, mortgages, water, pavements, sewers, carriages, rented and private automobiles, bicycles, and on and on. Businessmen who were hoarders—wartime monopolizers of vital supplies—were given forty-eight hours to deliver 10 percent of their merchandise in staples most urgently needed (corn, broad beans, oil, lard, tallow candles, coal) or else their entire stocks would be confiscated.

He was no easier on foreigners. The special taxes applied to them just as they did to Mexicans. A group of foreign businessmen called a special meeting in the Hidalgo Theater to protest these taxes. Suddenly they were confronted with General Obregón, who warned them that they would not be exempt from Mexican law. Outside the building he had stationed a triple cordon of soldiers, with loaded guns. Obregón informed the foreign businessmen that they were being assessed half a million pesos and that they would also have to pay a kind of moral tax. He required them to sweep the streets of Mexico City.[20]

In September of that year, he briefly left the capital to make his visit to the lair of Pancho Villa. At no time did he show greater bravery or a cooler hand for poker at the highest possible stakes. Of his own free will, he stepped between the jaws of the wolf, not once but twice. He came with a proposal for reconciliation but also to see his potential enemy up close. He observed him, studied him, weighed him. Villa paraded his military might, and Obregón took advantage of the display to photograph it within his memory. At Villa's first threat to shoot him, Obregón staked his life, and manipulated his opponent. The game went on for days without his ever lowering his cards or his gaze. When the American consul in Chihuahua offered to clear the way to El Paso for him, Obregón refused out of a sense of dignity—and recklessness. Villa let him go but then made him return. And Obregón asked one of his two guardians (and friends), José Isabel Robles, to intercede with Villa for him, but only to avoid "being insulted and affronted." He wanted Villa "to shoot me without humiliating details."

Again Villa let him go and again tried to get him back. This time, without doubt, he was going to finish off his "dear good friend." Obregón would have met his end with dignity during the poker game, but now that he had already left the table, he must have thought that this back-and-forth process was becoming too undignified. When he heard of Villa's new order to intercept his train, he decided to get off, and to the question "What will you do now, General?" he answered, "Die killing."[21]

With the help of the goddess of luck together with Eugenio Aguirre Benavides and José Isabel Robles—by then great admirers of his courage and willing to partially deceive their commander, Villa, and shuttle Obregón from station to station and train to train—his life was saved this time. He did not have to "die killing." Death continued not taking him seriously.[22]

Beginning in October 1914, when he attended the Convention of Aguascalientes, Obregón would formulate his own consistent political and military program. But there is no reason to doubt his goodwill toward reaching an agreement between the various factions. Obregón did not want to trigger a civil war. And so, astutely, he chose to remain neutral, squarely in the middle for more than a month between the two prevailing currents of legitimacy: the Convention and the preconstitutionalists of Carranza. When the moment of truth came in mid-November (and Obregón would opt for Carrranza), he had nevertheless made some good friends in the ranks of the Convention and had won some Villistas and even more Zapatistas over to his side. His apparent indecision perhaps shows only his great sense of opportunity. He had not wasted time torturing himself over ideas and convictions. But he had closely observed his future enemies and he had tried to guess their next moves.

Early in the civil war, Obregón carried off a real stroke of genius in military and political recruitment. If in Sonora he had enlisted the Yaquis by offering them land in return for collaboration, why not try the same with a group analogous to the Yaquis in the urban environment: with the workers? Making a decision that would have great historical consequences, Obregón went beyond the wishes of the *Primer Jefe,* who distrusted the working class. Through his promises to La Casa del Obrero Mundial, he succeeded in forming and enlisting the anarchist Red Battalions to fight against Villa and Zapata. Though Carranza, after his victory, would disband the Red Battalions and repress working-class movements, the step—at this moment of the civil war—had implications for his and future Mexican governments. The Red Battalions not only brought considerable military support to the Carranza forces; with them came something even more important—a halo of legitimacy from the proletariat, the class that according to socialist ideology would inherit the future.[23]

At the beginning of 1915, Obregón easily defeated the Zapatistas at Puebla. But he had no illusions. He knew that his chief enemy (and the generals who obeyed him) had control of most of the country. Pancho Villa was now moving past Guadalajara toward the Bajío. Felipe Ángeles occupied Saltillo and from there dominated the northeast. Calixto Contreras and Rodolfo Fierro were in the west, while Tomás Urbina was roaming Tamaulipas and San Luis Potosí. Obregón respected the enemy's strength but also noted their fragmentation.

He knew that if he took Villa on in the Bajío, the Caudillo of the Northern Division would have to fight almost a thousand miles away from his primary supply bases. Obregón had always thought that Villa must be defeated in the center of the country. Ángeles, aware of the danger, wore himself out trying to persuade Villa not to take the bait.

At the beginning of April 1915, Fortunato Maycotte, Obregón's aide, had repaired the railroad lines across the zone where the battles would be fought. The supply line from Veracruz remained open and safe from Zapatista attacks. The moment had come for the first battle of Celaya.

Villa's tactic, as always, would be the brutal charge. Obregón tried to economize his strength and his resources. His strategy, as always, would be to draw the enemy out and resist his assaults. On April 6 the battle began.

Villa's ferocious attacks would cost Obregón substantial losses of men, and once again he would be dancing with death. His telegram to Carranza, at a moment when things looked most bleak, did not fail to include an embrace offered—if necessary—to death: "I will consider it my good fortune if death should surprise me as I throw a blow in the face of its fatal onslaught." Good luck and military genius favored him. Villa wore out his troops with their charges against Obregón's entrenched army, and he was routed by the encircling counterattack of Obregón's reserves. One week later, Villa came at Celaya again. Obregón (with reinforcements of five thousand men) just did the same thing, receiving the charges from a strong defensive position, biding his time, and decimating the enemy in a lightning counterattack by well-rested forces. In this second battle of Celaya, he lost two hundred men (dead, wounded, and prisoners), Villa ten thousand.

Although these rivals would confront each other three more times—in the Bajío (Trinidad and León) and later in Aguascalientes, the Villista army—physically and in spirit—had been wounded to death before Celaya. At Trinidad, their attempts to attack from all sides were frustrated by Obregón's strategy of forming a giant, uneven rectangle, with his forces in defensive position along lines of between sixteen and twenty-two kilometers. His generals kept pleading to break out and take the offensive. "Don't lose the shape," Obregón would answer, and his forces maintained it until the order came for the crushing counterattack.[24]

Obregón knew he was winning the war, but in some strange way he also seemed to feel the wounds he had inflicted on his victim. Perhaps he had never before so vividly experienced the dizzy heights of victory and the emptiness of life. Once again he wrote poems:

I have run after Victory
and I won her
but when I found myself beside her
I felt despair.

The glows of her insignia
illuminated everything,
the ashes of the dead,
the suffering of the living.

In another poem, he tried to write about the clear light of dawn, the flight of birds, the colors of the landscape, mountains that "meditate" and even the fragrance of flowers, but all of it contrasts with a single obsessive image:

But man, the fool, does not even so much as notice
how near to him is the eye of the rifle of death.

At the beginning of June, Obregón set up camp at the Santa Ana del Conde Hacienda in Guanajuato. Paying no attention to danger, he went walking with some of his subordinates toward the front-line trenches. Artillery shells began to fall around them:

> We were some twenty-five yards away from the trenches . . . when we felt the sudden explosion of a shell that threw us all to the ground. I sat up before I fully realized what had happened and then I could see that I was missing my right arm, and I felt sharp pains in my side, which made me suppose that it had also been torn by shrapnel. I was bleeding so much that I was sure that it was useless to prolong a situation from which all I could gain would be delayed and anguished dying, a painful spectacle for my comrades. . . . With the hand that was left to me, I drew my pistol . . . from my belt and shot at my left temple, trying to finish the work that the shrapnel had not; but I failed because the pistol had no bullets; my aide, Captain Valdés, had emptied it the day before when he cleaned the gun. . . . Lt. Colonel Garza, who had already gotten up and had kept his wits about him, realized what I was trying to do, ran toward me and grabbed the pistol away, after which he, Colonel Piña and Captain Valdés carried me off

> . . . and leaned me up against one of the patio walls, where my officers thought I would be less exposed to enemy fire. At that moment, Lieutenant Cecilio López . . . arrived and drew a bandage from his knapsack and they wrapped up my stump.[25]

On that morning of June 3, 1915, Álvaro Obregón, having had his fill of valor and drowning this time in his own blood, decided to put an end to the vanity of living but did not succeed. The index finger pulled the trigger; chance denied him the bullet.[26]

At the end of his "8,000-kilometer campaign" (85 against Orozco, 3,498 against Huerta, and 3,644 against Zapata, Villa, and the men of the Convention), another side of the vanity of things surprised him, a sweeter and more soothing countenance—that of fame. Suddenly he understood that not very far away, almost within his reach, was the Presidential Chair. No caudillo could equal him, not even the *Primer Jefe,* to whom he would remain loyal for now, while well aware that he could leave him at any time without the slightest damage to his own prestige. He was the strongman of Mexico, the man who had won the Revolution. In 1917, he was only thirty-seven, the same age as Porfirio Díaz at the triumph of the Republic in 1867. And as Díaz with Juárez, he felt the victory was more his than Carranza's.[27]

For Obregón, the parallel with Díaz was not entirely unconscious, but as a practical man he was not guided by an awareness of history. Unlike Díaz, Obregón did not immediately resign his command, nor did he show any sign of losing interest in politics. As Minister of War in Carranza's preconstitutional government, he continued his work as a military manager. He collected statistics on various aspects of the military, reorganized medical management and services, opened a staff college, a school of military medicine, the Department of Aviation and a school for pilots, and he placed munitions factories under the direct control of the military. His goal was to create a thoroughly professional army.[28]

At the beginning of 1917, the new Constitution was on the table in Querétaro. It was the high point of Carranza's career, but Obregón, with good political sense, decided to steal some of the audience. The Convention gave him an opportunity to distance himself publicly from Carrancismo—though not as yet from Carranza—and to adopt a halo of ideological daring. During the debates on key issues, such as Article 3 on religion or Article 27 on national resources, Obregón took up residence in Querétaro, where he received visits from radical legislators. Even Andrés Molina Enríquez consulted with him. Invariably, Obregón supported the most extreme measures. No risk intimidated him: not the United States, not civil war. To his aura of undefeated victor and martyr, he added that of the most radical caudillo of the Revolution.[29]

Soon after the formal swearing of the oath to the new Constitution, Obregón, who had just taken a second wife in María Tapia, resigned as Minister of War and retired to his Quinta Chilla, no longer of course a "poor man's manor." As Porfirio Díaz had grown his sugarcane and waited, so Obregón raised his chickpeas. But unlike the caudillo from Oaxaca, Obregón showed no impatience. "I have such good eyesight," he would joke years later with his friends, "that from Huatabampo I managed to see the Presidential Chair."[30]

He began to overeat, probably in reaction to the physical problems caused by the loss of his arm. He put on weight, his hair turned gray, he began to look bloated. As he became more obsessed with his health, he would visit hospitals in the United States and take careful mental note of every day-to-day change in his body. Five years after he lost his arm, he looked like an old man. But he was only forty.

As he grew fatter, so did his wallet. In a few years, La Quinta Chilla expanded from 180 to 3,500 hectares, primarily planted with chickpeas. In 1917 Obregón founded a Cooperative Agricultural Society that soon united all of the chickpea growers in Sonora and Sinaloa. He led efforts to raise the quality and price of their product. Nor did he have to wait long for results: In 1918 the price of chickpeas doubled and General Obregón earned fifty thousand dollars.

He was genuinely happy to be working the land once again. Obregón had built his life with his own labor, and he felt proud of having won this battle as well. During that calm of two years—always glancing out of the corner of his eye at the Presidential Chair far to the south—Obregón swiftly built up a small empire: He bred cattle, exported leather and meat, bought mining stocks, opened an import-export agency, and employed fifteen hundred men.[31]

It was his happiest time. Everybody wanted to be with the brave and attractive victor, the charming talker with his penetrating intelligence. Naturally enough he began to show signs of narcissism, but it was an egotism without excessive passion, without solemnity, because deep down Obregón did not take himself seriously. He had not lost his sense of the vanity of everything human but, blessed by fortune, he was living on the sunlit side of that conviction—with humor. He loved fun and jokes, even acting the fool. And he was able to play the entire keyboard of comedy (except for irony, which was never part of his style).[32]

He could tell story after story. He could create absurd situations, applaud—and rapidly memorize—other people's jokes, and he had the great gift of being able to ridicule himself. He could use his prodigious memory to create impossibly comic situations. The poet José Rubén Romero (in the presence of Obregón when he was already President) once recited a sonnet he had recently composed. Obregón said to him, "That's a poem I've known for a long time," and immediately recited it back. The poor poet, turning various colors, hemmed and hawed and said, "How can you have read it? I've never published it." To which the President of the Republic answered, "Not you, but the man who wrote it! I read it in a magazine and liked it so much that I learned it by heart," and he recited it again from the beginning.

When the poet was saying his good-byes, Obregón told him not to go away embarrassed. "It's just that I have the ability to memorize everything. As you were reciting your sonnet, I learned it by heart."[33]

Jokes and death (or both of them together like the smile on a Mexican cardboard skull) are ways to escape the tensions of life and come to terms with the emptiness of human existence. Toward life, Obregón was capable of anything except taking it seriously.

By the beginning of 1919, Álvaro Obregón began to reap a very special harvest: the unanimous popularity that would carry him to the presidency. The main opposition to his possible election was Carranza. In the eyes of the Old Man, Obregón did not understand the great national problems, had no program of government, and worst of all, lacked the qualities necessary to govern. He was a military man, and the President intended to put an end, once and for all, to militarism. In January Carranza publicly stated that he was against "premature launchings" and "political effervescence."

Obregón thought that popular pressure would make the Old Man yield. It did not take him long to undeceive himself and take the offensive. In June he declared his own candidacy in a message to the nation. And in August he made a secret agreement with the top leaders (Grupo Acción) of the CROM (Regional Confederation of Mexican Workers), a newly formed umbrella labor organization under the leadership of Luis Napoleón Morones. In it he agreed that if he became President, he would, among other things, create an autonomous Department of Labor, designate a Minister of Industry and Commerce sympathetic to the CROM, and issue a new Labor Law. Here again he repeated the strategy that had served him so well with his Sonoran Yaquis and the Red Battalions.

In November he made a series of grand tours. The tone of his campaign was triumphalist but with reason. If he had triumphed over nature, over rain and wheat rust and winds, and then over Orozco,

Huerta, Zapata, Villa—surely he could displace his ex-chief Carranza, even perhaps winning his consent?

He threw himself completely into the campaign, traveling, speaking, organizing support. He moved across the nation the way he had led armies, village to village, city to city. He would leave his first-class railroad car to walk back and campaign in second and third class. His popularity grew not only because of his speeches on morality and honest government but through rubbing shoulders with the masses, with all his fellow citizens.

His maneuvers worked for him and so did—just as in battle—the attacks of the enemy. The Senate suspended his military rank, which, far from doing him harm, helped him. Now he could present himself as what he had been in the beginning: a civilian whom destiny had transformed into a soldier. He could retain the advantages of a military aura without its drawbacks.[34]

In January of 1920, the ship of legality represented by President Carranza had sprung many leaks, and as happens in such cases, almost everyone abandoned the boat. Carranza had principles on his side; Obregón and the Sonoran dynasty had specific interests and youth and power on theirs. A meeting of governors was orchestrated to discredit Obregón, but it had no effect.

The tensions built up, and Carranza began to move toward direct repressive action. The government wove a plot against Obregón, coming up with a minor officer who claimed Obregón had approached him to join an armed uprising. Disguised as a railwayman, Obregón escaped on a train, to the south. A sympathetic state government awaited him in Guerrero. The Chief of Military Operations for the state was Fortunato Maycotte (his old lieutenant at the battles of Celaya, who had supervised the all-important task of repairing the railroad lines). Obregón snapped to attention before him and said, "I am your prisoner," but Maycotte answered him with "No, you are my commander."[35] There in Guerrero, on April 20, in the town of Chilpancingo (where Morelos had convoked his Constitutional Congress), Obregón accused Carranza of trying to impose an unpopular candidate—Ignacio Bonillas, the civilian Carranza had chosen to succeed him—and of financing the campaign with public funds. From that moment he was putting himself at the disposal of "the Constitutional Governor of the Free and Sovereign State of Sonora in order to support him in his decision and to cooperate with him until the High Powers are overthrown." The governor was Adolfo de la Huerta, who would be the formal head of the revolt, though Obregón was its brains and real commander.

On April 23 the Sonorans issued their Program of Agua Prieta. Carranza fled from Mexico City toward Veracruz, once again like Juárez but without his luck. On May 20 he would die in Tlaxcalantongo. It was one of the fastest military takeovers in Mexican history. At the end of May, Adolfo de la Huerta, commander in chief of the Liberal Constitutionalist Army, would become the interim President of the Republic.[36]

Along with the years of fighting, Mexico had passed through a revolution no less profound and perhaps more far-reaching: a revolution of ideas. There were new conceptions about property, the agrarian problem, the relations between workers and employers, the political role of the Church, the economic function of the State. Although unhappy with much of the new legislation, Carranza had recognized that the 1917 Constitution would be a watershed in the life of the nation, like Juárez's Laws of the Reform. In Querétaro, Obregón had supported the radical articles of the Constitution, but he had made no mention of them (or of the Constitution) in his manifesto to the nation on June 1, 1919, announcing his presidential candidacy. Unlike Carranza or the legislators he himself had supported, Obregón was not interested in ideologies. His social and political ideas were eminently practical.

For Obregón, at least in his manifesto, there was only one basic problem in the country. Seeking power and wealth, the revolutionary caudillos had become vehicles of reaction. There was a danger that these new material interests could impede "the advanced principles of the struggle, and, above all, the principle of valid voting (*sufragio efectivo*)." The peace and achievements of the Revolution were at risk because "they did not allow the country to liberate itself from its liberators."

This impressive phrase would spread like a whirlwind, but the manifesto went still further. To "liberate the country from its liberators," Obregón proposed a "course that will break with all formulas and patterns." Unwittingly imitating Napoleon III, he called for a kind of national plebiscite (about himself). He offered no social platform (because that would be nothing more than "rhyming prose") but instead a moral and political goal: to purify the government and defend the freedom of the vote—which he certainly assumed, in this campaign, would lead to his election. As for the economic problems of the country, his basic concern was to give foreigners the confidence and guarantees to invest in Mexico. The manifesto ended with a call to all Mexican citizens to form the new Great Liberal Party.

Obregón's campaign speeches would show a similar pragmatism. He had a very poor opinion of the great landowners. But it was based not on their record of abuses, injustices, or exploitation, but rather on something worse in his opinion—lack of productivity. Their backwardness, their routine minds, their protectionist zeal had stripped them of any possibility of competing on the international market. They were, in short, bad businessmen.[37]

From this insight flowed Obregón's solution for the problems of the countryside:

> It is indisputable that true equality—however much we might wish for it or do wish for it—cannot be realized in all the breadth of meaning of the word, because in the struggle for life there are men who are more vigorous, more intelligent, better conditioned and prepared physically and intellectually than others; and they undoubtedly are those who must gain greater advantages from their efforts in the struggle for life. But it is necessary—and this we should be able to accomplish—that those on high feel more affection for those below; that they do not consider them as merely elements of labor at their service but as coworkers and collaborators in the struggle for life.[38]

The social role the government should play must be limited to "achieving a balance of the factors of production . . . guaranteeing the rights of the workers but preserving capital. . . . [Government must be] the pointer on the scale." In relation to the United States, Obregón thought that the attitude of the Mexican administration should change: "We will respect the rights of each and every Mexican and foreign citizen in our Republic; and when we prove with deeds that we know how to follow this policy, we will have the right to demand for ourselves as well the respect of all the other nations on earth."[39]

This was his conciliatory fishing pole, trying to attract the investment of "honest capital" from American business.

Throughout his campaign, Obregón did not speak a single word of ideological radicalism. Some detractors began to identify features of another time in his ideas. He wanted to pacify the country and modernize agriculture and industry. He had said his government would be more interested in administration than in politics. Had he not confided to Lucio Blanco in 1914 that they—he and his entourage—would be "the new *científicos*?" But there was no need for any clever attempts at analysis. Obregón himself had already said it: "Don Porfirio's only sin . . . was to grow old."[40]

Adolfo de la Huerta's brief six months of interim government, which preceded the presidency of Obregón, was more important than has sometimes been recognized. Among other things, the interim President calmly, through negotiations, at last brought widespread peace to the country. Villa, Pablo González, Félix Díaz, the remaining Zapatista chieftains, and other generals across Mexico—one by one they laid down their guns. The Revolution now had no armed enemies to confront. Many exiles—who had fled at different moments—began to return to Mexico. One of them was José Vasconcelos—to become the best choice De la Huerta made for any important position and an invaluable legacy for the government of Obregón.

When Vasconcelos came back to Mexico in 1920 and assumed the position of Rector of the National University, he was thirty-eight years old, two years younger than Obregón. During the final years of the Porfirian regime, he had been part of a group of philosophers, writers, and humanists known as the Ateneo de la Juventud ("the Athenaeum of Youth"). By 1920 some of them were already becoming well known: the writers Pedro Henríquez Ureña, Alfonso Reyes, and Martín Luis Guzmán, the philosopher Antonio Caso, the painter Diego Rivera. Vasconcelos had been a Maderista from the earliest hour, had rebelled against Huerta and had been Minister of Education in 1914 during the brief government of Eulalio Gutiérrez. In 1915 he was forced into exile and left for Europe, South America, and the United States. With the fall of the Carranza regime, which he hated, Vasconcelos returned with an almost messianic project for the education of Mexico.

As Rector, he invented the motto *Por mi raza hablará el espíritu* ("The spirit will speak for my people"), but few recognized the quasi-religious overtones of his words and the dimensions of his project. Obregón—who became President in December 1920—listened to him and supported him. Only a few days after taking office, Obregón created the Ministry of Public Education and federalized its sphere of action. And he asked José Vasconcelos to head the new governmental agency. Vasconcelos could then apply his skills and his sympathies to accomplish changes in Mexican education and culture that still, in many ways, live on and influence the Mexican present.

In education it was a new dawn, and a major effort went into the construction of schools: rural, technical, elementary, and also specifically Indian. As if they were soldiers or missionairies on a modern crusade, hundreds of teachers were sent into the remotest corners of the country. In Teotihuacan, for example, the anthropologist Manuel Gamio opened a "comprehensive school" designed to educate peasants without uprooting them from their culture.[41]

The young American scholar Frank Tannenbaum—then at the very beginning of his long and fruitful love affair with the Mexican

Revolution—entitled the first article he ever wrote on Mexico "The Miracle School." It dealt with the process of social rehabilitation at one of these new schools in the Colonia de la Bolsa, a Mexico City slum. In the midst of an environment of "bums, tramps, thieves, pickpockets and burglars," the "miracle school" was introducing practical instruction for children and adults, teaching methods of hygiene, how to cultivate gardens, and various crafts and skills that Tannenbaum felt were as important as literacy because they would strengthen the sense of community.[42]

But the creation of schools was only one facet of the new cultural energy. Another, equally brilliant, was the development of the fine arts. During Obregón's term, Mexico experienced a true rebirth of national values, a varied and broad-based return to its origins: the indigenous and Spanish past, the colonial and provincial traditions. In 1921, under Vasconcelos's initiative, Diego Rivera, José Clemente Orozco, Jean Charlot, Fermín Revueltas, David Alfaro Siqueiros, Roberto Montenegro, and other muralists took over the walls of venerable colonial buildings to express the social gospel of the Revolution.

Daniel Cosío Villegas, one of the intellectual squires of Vasconcelos on that cultural and educational crusade, would long afterward write nostalgically about the time:

> Then there really was an evangelical environment in which to teach your neighbor how to read and write; then there was really a feeling in the breast and in the heart of every Mexican that educational action was as urgent and as Christian as satisfying thirst or staving off hunger. Then began the first great mural paintings, monuments that aspired to catch for the centuries the anguish of the country, its problems and its hopes. Then there was faith in books, and in the timeless masterpieces of literature; and books were printed by the thousand, and thousands were given away. Founding a library in a small remote town seemed as important as building a church.[43]

Unfortunately, the age was not entirely favorable to the new educational gospel of Vasconcelos. By 1923 the uncertainty hovering over the coming presidential election unsettled public life and in many ways impeded the difficult job of organizing the educational system. Vasconcelos would resign. But the "pact" between Obregón and Vasconcelos had turned out to be immensely fruitful. The rural schools had spread and multiplied. Vasconcelos's Ministry had formed a generation of artists and writers who would in turn educate new generations. A first impulse had been given to the foundation of a new publishing industry.

But more than anything a road to creativity and reconstruction had been opened that soon would inspire a number of national commitments.

Other aspects of the Obregón regime were less luminous. A harsh and stormy political environment made everything doubly difficult. On each one of the social and political fronts opened up by the Constitution of 1917, a climate of violence existed that Obregón could not always moderate.

Great tension, for instance, was generated by the attitude of the powerful CROM. Its leader, Luis N. Morones, wanted all the clauses of his 1919 agreement with Obregón to go into effect one after another. Once an anarcho-syndicalist, Morones's stated philosophy now was "Blessed are the idealists, for theirs is the kingdom of all the disasters." Even though Morones and his Grupo Acción already had representatives in important administrative posts, they tried to broaden their public influence, and in great measure they did. By 1923, the CROM, with a heavy use of violence, had almost overwhelmed its most serious competitor, the anarchist CGT (General Labor Confederation). By the end of Obregón's term, it was clear that the CROM had broken free of presidential control. It would strengthen its position even further by signing a new and stronger agreement with the future President, Plutarco Elías Calles.[44]

In social policies, Obregón kept to the letter of the Constitution more closely than Carranza. During his term he distributed 921,627 hectares of land, almost five times more than had been done under Carranza and De la Huerta together. Article 123 on labor relations was still not really enforced, but Obregón did not oppose its most important provisions. In the Federal District (*Distrito Federal*), businesses were slowly coming around to the idea of Sundays off with pay; local boards of arbitration sometimes functioned; and the right to strike was respected, provided that the workers involved were part of the CROM. (Otherwise, as happened with an attempted railroad strike, the police or the army moved against them.)[45]

The relationship with the Church was also marked by an atmosphere of tension. When Pope Pius XI was installed in 1922, Obregón sent him public congratulations and a private message emphasizing the "complementarity" of the revolutionary and the Catholic programs. But for the most part the Church was very far from resigning itself to Articles 3 and 130 of the new Constitution, and some bishops resisted the distribution of land or the secular unionization of workers. The clashes between Cromistas and members of the ACJM (Young Mexican

Catholic Action) began to appear daily in the newspapers. The most serious incident between government and the Church happened in Guanajuato. Father Filippi, the apostolic delegate, went to a hill in the city to consecrate Christ as King in the open air. The people prostrated themselves at his feet, but religious services outside of the precincts of a church were illegal. The government invoked Article 33—which permitted expulsion of foreign priests—and threw him out of the country. Nevertheless, it was clear to anyone who could think calmly that Obregón did not share the anticlerical ideology of Plutarco Elías Calles, his former Minister of the Interior (who had just resigned to prepare for the coming elections). And in 1924 the Pope dispatched a new apostolic delegate to Mexico.[46]

Obregón had a driving obsession to be recognized by the American government. When in 1921 he organized Centenary Fiestas to commemorate the Independence of Mexico, there was of course a touch of Porfirian nostalgia but also a diplomatic motive: to show how many nations had already recognized his regime and call attention to the isolated position of the United States. He did not yield to crude U.S. threats, when they tried to make the repeal of Article 27 (returning original ownership of "lands and waters" to the Nation) or a treaty of friendship and trade the necessary conditions for recognition. But he did make some concessions that were no less important. In 1921, the Mexican Supreme Court ruled in favor of Texas Oil, setting a precedent of nonretroactivity in the application of Article 27; in 1922, Finance Minister De la Huerta signed the De la Huerta–Lamont treaty, in which Mexico recognized a debt of $1.451 million to the International Committee of Bankers. But the United States government demanded even more. And when the Bucareli Conferences ended in 1923, Obregón had obtained two things: recognition by the United States and a reputation as an opportunist or, more precisely, a "sell-out" (*entreguista*).

In essence, the Bucareli Agreements contained two provisions. In the first, the two countries committed themselves to forming two joint claims commissions: one to pay compensation for American losses during the Revolution, the other for mutual damages since 1868. In the second agreement, the Mexican executive power agreed not to apply Article 27 retroactively with respect to oil and to indemnify Americans in cash for all expropriated land not earmarked for *ejidos*, or even land taken for this purpose in lots larger than 1,755 hectares.[47]

Obregón's defenders would argue that the President had not modified Article 27 directly nor restricted the freedom of future governments to reinstate and enforce its regulations. For others he would remain "a sellout."[48] But there was an urgency in Obregón's need for recognition at that moment, and it was beyond economic considerations. No uprising against his government would now be able to count on American arms. The year 1923 was turning dangerous.

Obregón's nephew, the Sonoran general Benjamín Hill, and other liberals had formed an opposition party, the Liberal Constitutionalist Party (PLC). Shortly after Obregón became President, Hill would die under mysterious circumstances. (He fell seriously ill and—though attended by Obregón's personal physician—died after a banquet offered him, in the interests of concord, by Obregón and De la Huerta.) Until 1922, the PLC fought a battle as splendid as it was unsuccessful for decentralization, valid voting, and the total separation of the three governmental powers—key features of Madero's original program. They envisioned a true parliamentary regime with full democracy for Mexico. But Obregón (with the help of the CROM and other groups in the national Legislature) nullified the influence of the PLC, though it had supported him for President.

In 1923, when it was clear that Obregón favored Calles to succeed him, another opposition group, the Cooperativist Party, proposed De la Huerta as a candidate and began to argue for some of the Liberal Constitutionalist principles. A sector of public opinion felt that Obregón was repeating the political error for which he had attacked Carranza. He was imposing his own candidate on the country.

Rather than give up power, the Sonoran dynasty disintegrated. "The hecatomb was upon us and the political life of the country took a frightful leap backward," wrote José Vasconcelos. A leap into war. With the death of Murguía and Lucio Blanco in 1922 and the assassination of Villa in July 1923, three of the greatest real or potential threats to the regime had disappeared. Nevertheless there remained a long line of generals who dreamed of sitting in the Presidential Chair. At the end of 1923 came the De la Huerta rebellion. More than half the army fought against Obregón. Some generals joined the rebellion only in the hope of someday occupying that Chair, others through a democratic sense of shame.[49]

Obregón announced that, for him, the rebellion was a stroke of good luck. "This little job was beginning to bore me."[50] While the rebels built up strength in Veracruz and Jalisco, the President marched on the Bajío. Many of his former comrades fought against him. A group of them—Salvador Alvarado, Manuel Diéguez, and others—massed their troops in Ocotlán, Jalisco. Fortunato Maycotte (Obregón's former devoted aide and compadre) with Cesáreo Castro confronted the President himself in Esperanza, Veracruz. The battle at Ocotlán was ferocious and to the death. General Roberto Cruz, one of the most violent men of his time, later wrote: "No one showed any mercy. . . . First they fought over the river and then over every inch of land as if it were their final stronghold. How many dead there were scattered everywhere! I saw many bodies. I also saw heads swimming in the river . . . like huge fruit fallen from who knows where."[51]

The rebels were crushed. Their remaining forces retreated all the way to Yucatán and some of the leaders even farther, into exile. Obregón was back at his old work. He was risking his own life and disposing of others, including the lives of his former friends. Salvador Alvarado was shot down at point-blank range by a subordinate who betrayed him in the town of Frontera, Tabasco, not very far from Yucatán, where he had once conducted a short-lived socialist experiment. Manuel Diéguez, the old warrior for social reform, escaped with his men from Jalisco and crossed Guerrero, Oaxaca, and Chiapas before being taken prisoner. He wired Obregón reminding him of the days they had fought together; Obregón responded that his lack of pride was comparable only to his fear of death. He had Diéguez shot.[52]

Obregón's compadre, Fortunato Maycotte, was hunted down like a deer. He fled across Puebla, Morelos, Guerrero, Oaxaca, the Mixteca, pursued by land and by boats at sea. In Puerto Escondido, driven by thirst, he emerged from hiding and was taken prisoner. Obregón ordered him executed. The President had never been a merciful man. Now less than ever.[53]

Seven thousand men died in the rebellion. At the end of his term, Obregón confided to an admirer: "I am going to leave by the front door of the National Palace, bathed in the esteem and affection of my people." Perhaps by then he was mistaken. He had taken office "bathed" in nearly universal support, but four years later opinions of his government were divided. Obregón may have secretly admitted to himself that the transition from soldier to President had been more difficult, less surefooted than moving from farmer to soldier.

To his credit, he had encouraged educational reform, made some advances in fiscal and monetary policy, achieved a somewhat tense modus vivendi with the Church, and modestly supported the demands of workers and peasants. But against him, his enemies pointed to his dealings with the United States, his political centralization, the suppression of parties in the Legislature, and a failure to live up to the democratic promises in his manifesto of June 1919. And there were those who felt that he was modeling himself even more closely after Porfirio Díaz, that he would choose stand-ins for President and in effect continually reelect himself at least until, like Díaz, he had to commit the sin of growing old.

At the end of his presidential term, Obregón returned to the farm again, this time to Náinari—his property in the Yaqui region—to devote himself again with enthusiasm to agriculture. And he kept himself even more busy by expanding his small business empire into a vast enterprise. With some not entirely legitimate help from the brand-new National Bank for Agrarian Credit, he bought much of the land belonging to the American-owned Richardson Company and expanded his affairs until they included—aside from his normal harvest of chickpeas and cotton—the irrigation of the Yaqui Valley, a rice mill in Cajeme, a seafood-packing plant, a soap factory, tomato fields, an automobile agency, a jute-bag factory, and more. His business growth was financed not only by Mexican bank loans but also with money borrowed from willing California banks.[54]

It was his last period of humor and happiness, of peace at the side of his wife and his small and grown children, within a quiet family life that for him was a pure and ultimate value. To a visiting ambassador, who saw him dressed as a farmer and asked whether he was wearing a disguise, Obregón said: "No, ambassador. There, in the Presidency, it was there I went around in disguise."

The reply itself was also, of course, a disguise. He still had such good vision that he could keep his eye on the Presidential Chair from Náinari.[55] But he both wanted and did not want to occupy it again. He confided to a friend: "Before they called on me to feed me to the cannons, now they call on me to feed me to the crises." The Presidential Chair attracted him not for the power it would give him—even less so for any programs of social and economic reconstruction he could put into effect—but for the aura of duty and sacrifice surrounding the position.

By April 1926, nobody was blind to the intentions of the great *Manco de Celaya* ("the one-armed man of Celaya"). He spent more and more time in the capital and was a frequent guest of President Calles at Chapultepec Castle. "Calles is not the problem," the old Liberal Antonio Villarreal wrote to José Vasconcelos, "It is Obregón. You cannot imagine the ambition there is in that man! Don Porfirio was a joke in comparison!" After several stormy sessions, the Chamber of Deputies and the Senate finally, in October 1926, opened the way to the reelection of Obregón.

Álvaro Obregón, 1921
(CONDUMEX)

But this time the Sonoran's presidential ambitions ran up against a wall of generalized unpopularity and two opposing candidates: first General Arnulfo R. Gómez and then later his own "little finger," Francisco "Pancho" Serrano (a close friend who had been beside him when he lost his arm to the Villista artillery barrage).

From October 1926 to April 1927, Obregón (one more time) experienced the familiar taste of war. Commanding fifteen thousand men, he himself led the final military defeat of the Yaquis to whom—in exchange for their blood—he had once promised freedom in their own land. It was the end of a hundred years of war. He called it, "a brilliant opportunity to close out this shame for Sonora."[56]

He began his presidential campaign in May 1927. Most of the army supported him as did the National Agrarian Party, but the powerful CROM and a good part of public opinion were against his reelection. In the end he would force the CROM to bow to him and would ignore public opinion. But first he had to get rid of his rivals. Gómez, who saw what was coming, told the French diplomat Lagarde that Obregón was "mentally unbalanced," bordering on "megalomania," and that he would try to force the issue. "It is a fight to the death in which one of us must die."[57]

The first to die was neither Gómez nor Obregón but Serrano. Gómez and Serrano made a decision for preemptive action and conspired with General Eugenio Martínez to arrest Obregón and Calles on October 1 at the Balbuena airport. But Martínez betrayed them. Serrano fled to Cuernavaca, where he was warned that an arrest was imminent. Unsure of whether to trust the warning, Serrano took no precautions. General Fox took him prisoner, along with a group of his political sympathizers, then riddled them all with bullets. Arnulfo Gómez called for an insurrection, and troops from five garrisons responded, but they were nowhere near enough. Gómez himself fled to the mountains of Veracruz, where he was discovered hiding in a cave and shot the same day.

The Sonoran civil war seemed to have no end. As the bloodbath went on, very few words were spoken about the ideals of the Revolution. The political violence of this period seemed much more naked, painful, cruel, and arbitrary than the social violence of 1910 to 1920. There had been purpose then in the Revolution, meaning, even a kind of poetry. The struggles now were just a string of murders.

Obregón was aware of this degradation. Strictly speaking, he lived it as a descent. Already red with blood, the stage was beginning to turn black, and in a sense Obregón began to summon up his spirits, the vague presentiments of his own death, as he had done before his great battles. Perhaps he thought of all the many dead, his friends, his enemies, and the friends turned enemies. Not long before, Jesús M. Garza—the lieutenant colonel who had taken the pistol from his hand in Guanajuato—had ended a long struggle to live with the great love of his life, whom he had married but could not live with in peace because alcohol and the war had maimed his soul. The man who had refused to let Obregón kill himself was now dead by his own hand. Death was moving ever closer to Obregón.

In November 1927, a Catholic engineer, Segura Vilchis, made an attempt on his life, hurling a bomb at Obregón's car. The General did not flinch and, smiling, went on to enjoy an afternoon of bullfighting, his favorite diversion. A few days later, Segura Vilchis and the Pro brothers, one of them a priest alleged to be part of the plot, died before a firing squad. In January 1928, the French diplomat Lagarde noted cautiously in a report: "This is what will happen if Obregón reaches the Presidency alive." People in the street said similar things. In Orizaba, a Cromista stronghold, there was another attempt against his life. And one night, he was at the house of his former secretary, Fernando Torreblanca, when shots were heard. Obregón said quietly to his son Humberto: "They were not for me."[58]

Now he spent very little time in conversation, he told no more jokes nor showed any of his old garrulousness. Although he was only forty-eight, he looked far older. The end was near at hand, and he somehow knew it, was perhaps waiting for it. "I will live until someone trades his life for mine."[59]

Héctor Aguilar Camín, a scholar of Sonoran history, gives us one of the last images we have of Obregón, in May 1928 at his Náinari Hacienda:

> In the blazing heat of May the undefeated General—one-armed, graying and now reelected President—is doing his books and sending messages from a small office decorated with a giant ear of corn harvested with a farmer's pride on his own land. Outside, his farm dogs are barking and howling obsessively and unusually. Obregón asks his chauffeur to quiet them, and the chauffeur goes out to quiet them but the dogs keep on barking. He orders that they be given food, and they give it to them, but the dogs will not stop barking. "Give them fresh meat," shouts the general through the window, but fresh meat does not calm them either. After an hour of listening to the dogs barking, the last caudillo of the Mexican Revolution, unnerved and anxious, thinks that he sees a formal omen of his own fate in the tenacity of the pack of hounds. "I know what those dogs want"—he says somberly to his chauffeur—"they want my blood."[60]

On July 17, 1928, his personal secretary wrote him a warning note in red ink on his schedule, letting him know there was a rumor of a conspiracy against his life. Obregón either did not see the note or did not want to see it. He had lived all his life on the alert. One more warning may have meant little to him. At La Bombilla Restaurant, a group of his supporters were giving a banquet in his honor. Someone commented: "Look at the general! What can he be thinking about? It looks as if he's staring into infinity." His assassin approached him and showed him the beginning of a portrait. It was a good likeness; the man had talent. Obregón agreed to let him continue the sketch. Moments later, as musicians were singing the light music of *Limoncito*, José de León Toral—a fanatic Catholic—opened fire, trading his life for that of the undefeated victor of the Revolution.[61]

Shortly afterward, Toral was judged and executed. He was so different from his victim—thin and dark and trembling. Almost a shadow. Obregón, in 1909, had already written the epitaph for this fatal exchange of lives:

and although they may come from different conditions,
the men and the women, the old and the infants,
in the dark nights,
foxfires, unreal, they all pass on together.

15

PLUTARCO ELÍAS CALLES

Reform from the Roots

His birth was disorderly, and this perhaps explains his passionate love of order, the inviolable, of what should and should not be done.

—THOMAS MANN, *MOSES*

In the state of Sonora, Elías was not just another name. It was the emblem of a great landowning dynasty, with many branches, marked by wealth and prominence dating back to the eighteenth century. The Liberal colonel José Juan Elías Pérez—a great grandson of the founder—died in 1865 of wounds suffered in combat against the French troops of the Emperor Maximilian. His wife, Bernardina Lucero, was left with eight children.[1]

The eldest son, Plutarco, barely sixteen, had to take on the responsibility of managing a family inheritance that was shrinking every day due to neglect, Apache raids, and steady losses to cattle rustlers. Soon the family had to give up some of its ranches and then lost more territory through inability to exploit it (the government taking it over as unused land) or by sale through necessity to the American-owned Cananea Copper Company. The responsibilities were too heavy for the young Plutarco. He began a steep decline into depression and despair.

He would never marry. In Ures, in 1872, he fathered a son (Arturo) with Lydia Malvido. Four years later in Guaymas he had a relationship with María de Jesús Campuzano, who gave him two children: Plutarco, born on September 25, 1877, and María Dolores. He would soon abandon Guaymas and settle in San Pedro Palominas, leaving his children to the care of God (in whom he did not believe). For him, all family responsibility was like an avalanche that had descended on him and with which he could do nothing. Except escape. Alcohol was his refuge and his primary commitment.[2]

One of his brothers, Alejandro, who had become a highly respected administrator of properties in Guaymas, took on the responsibility of baptizing young Plutarco a year after his birth—in the absence of Plutarco senior. His mother died when he was three, and he was raised by his maternal aunt, María Josefa Campuzano and her husband, Juan Bautista Calles. They took him from Guaymas to Hermosillo, where he received his primary education. His father showed no interest in the child. Juan Bautista Calles, a dealer in liquor and groceries, came from a family of schoolteachers. It was he who cared for and educated Plutarco, who as a mark of gratitude took Calles for his last name.[3]

Beginning in 1887, the Porfirian governor of Sonora, Ramón Corral, had carried out a profound reform of education. He had brought in expert teachers from the famous (and ultramodern) teachers' training schools of Jalapa and Mexico City. The objective—drawn like so many currents of the time from Paris—was a form and content based on "positive" reason and science. But in Sonora, the pedagogical initiatives of the government in favor of secular education also had more immediate motives: the recent and aggressive competition from Catholic education.[4]

In 1888 young Plutarco Calles took the special courses for teachers at the new academy founded by Governor Corral, the Colegio de Sonora. He observed firsthand the educational conflict between the government and the Church. He was, by conviction, jubilantly atheist. ("When I was an altar boy as a child," he would recall decades later, "I stole the alms to buy candy.") The profession he had chosen, of schoolteacher, was one of the most prestigious in that remote and sparsely populated society, where there had been no real contact with European humanism or the Enlightenment. In Sonora, at so great a distance from the true literary and spiritual currents of the century, pedagogy seemed the beginning and end of human knowledge. It was a kind of lay religion—clear, disciplinary, rational, and methodical—defending an abstract and almost literary scientism, alien to the test of experience but with a certain rigor in its approach to forming the human personality.[5]

One of the first products of this new pedagogy in Sonora was Plutarco Calles, who in 1893 became an inspector for the Public Education Council in Hermosillo, a teacher at a school for boys, and, the next year, a children's aide at the Colegio de Sonora.

During his first year at the *colegio* he met Adolfo de la Huerta, who was, like him, from Guaymas. They had an exchange that already shows us something about Plutarco Calles:

> —I'm told, Plutarco, that you're from Guaymas.
> —Yes, I'm from Guaymas.
> —From which family?
> —From mine.[6]

In 1897, when he was twenty, he returned to Guaymas, where a year later he would begin to teach. He had come back to see his father (who would always move like a ghost through his life and mind), and he had now adopted his real last name, but without giving up the Calles. That year, his father took him away from Guaymas—"because the son of a bitch spends too much time in love"—and brought him to Arizpe. Soon Plutarco returned to Guaymas, disoriented and now a bit of an alcoholic like his father.

He felt himself assaulted by "chaos and pain," triggered by his illegitimacy and the disorder of his life, both seen as the fault of his father. To society at large, Plutarco Elías Calles was illegitimate because his father never married, but he was even more so in the eyes of religion. Denying the authority of religion would at least in part be an attempt to negate his own illegitimacy.[7]

In 1899 some of the chaos began to lift. Plutarco turned away from his father's path and moved down a different road. He married Natalia Chacón (by civil rite only), and a year later she bore the first of many children. For a couple of years he worked unsuccessfully at a series of jobs. He tried farming and failed at that. He became the manager of a flour mill, which after four years was seized by a bank. Calles then tried to become a small entrepreneur in Guaymas, setting up "Elías, Smithers and Company, Dealers in Fodder, Seeds and Flour." In the warehouse of that store, the city's Maderistas, with whom Calles sympathized, held meetings. Spiritualist gatherings also took place there, attracting local devotees. In April 1911 the business folded. Then, for a brief time, Elías Calles weakened and turned to alcohol. But it was no lasting despair. He was not the kind of man who would relinquish hope and action. It might be said that he had no choice. By 1911 there were four children to support. But something stronger drove him—the urge to build, with pedagogic discipline, a life directly opposed to that of his father.[8]

In September 1911, at Agua Prieta on the American border, Calles, having recently opened a small general store that sold a little of everything—machinery, groceries, wines—took up a new profession.[9] The governor of Sonora, José María Maytorena, made him a justice of the peace (*comisario*). To his considerable though not very successful experience as a businessman, teacher, and occasional farmer, he now added a job with political and police responsibilities, like the sheriffs of Arizona across the frontier. He had to maintain order, administer justice, and manage the local customs station.[10]

When Madero fell and was murdered, Calles did not hesitate. Indecision was never one of his traits. He moved his family to Nogales and quickly began to coordinate the recruitment of volunteers across the border in Douglas, Arizona. On March 5, 1913, in command of a small regiment that recognized "his capacities as an intellectual best prepared to organize and direct all the forces," Calles reentered the country.[11]

The Sonoran army in rebellion against the Huerta government was commanded by Álvaro Obregón. Calles—assigned the rank of lieutenant colonel—occupied Agua Prieta with no trouble and on March 16 moved into his first action, an attack on the town of Naco. Obregón had canceled the assault, but his telegram arrived too late. The Huertista garrison was able to repel the attack. Calles then installed himself in Nogales to organize a supply of arms, while Obregón commented, "Calles stays out of danger."[12]

In August 1913, Governor Maytorena returned—from a leave of absence—to the state and his political position. The leaders of the movement that had immediately rebelled against Huerta considered his absence "for reasons of health" an act of indecision and even of cowardice. With the arrival of Maytorena, Calles was about to resign his command when Venustiano Carranza came to visit Sonora. Calles would get to know him, and a current of mutual sympathy developed between them. They had similar temperaments: closed, thoughtful, tenacious, disciplined, and vigorous. Over the next year, a confrontation began to build between Maytorena (more and more inclined to ally himself with Villa in defense of the sovereignty of his state) and Calles—now a full colonel and loyal to Carranza.[13]

Near the end of September 1914, just before Villa and Carranza would go to war with each other, Calles (who had been visiting the

Primer Jefe in Mexico City for the Celebrations of Independence) returned to Sonora and passed through Agua Prieta. At a distance, he saw his father, sitting in an armchair and talking—between sips of mescal—to a boy who recalled, years later, what old Plutarco had said:

> He said . . . "There's Plutarco over there . . . he thinks he's a big shot because he's a colonel. They screwed Don Porfirio and now they think they're going to win the Revolution . . . but they won't win anything, they'll get slapped around . . ." I remember Colonel Calles walked by very fast . . . and greeted his father with a wave of his arm from a distance. Don Plutarco waved back and slowly began to get up from his chair when he saw the colonel hurrying away. I noticed that he was moved . . . he almost started to cry and then dropped back to his chair and a swig of mescal.[14]

Calles was hurrying toward war. Sonora was to become one of the first battlefields between Villistas and Carrancistas. In the same town of Naco he had once failed to capture, Calles and Benjamín Hill, with only eight hundred soldiers, awaited Maytorena's assault. He came on October 1, with ten thousand men including Yaquis, and besieged the town for 107 long days. Calles and Hill had done a flawless job of preparation. They had set up all their services below ground, raised barbed wire fences, dug trenches, attended to every detail: supplies, transportation, arms, telephones, electricity, water. On the north they were protected by the American border. Maytorena attacked from the south, the east, the west. The city resisted, and in the end Maytorena—though he had far more troops—had to lift the siege and retreat.[15]

Calles, now a brigadier general, remained in Agua Prieta training brigades and seeing intermittent action against Maytorena's forces at various points in Sonora. On August 4, 1915, Carranza named him interim governor and military commander for the state of Sonora. Although he immediately turned to the work of government, which was (as he now already knew) his true vocation, he still had to face a final military problem, much more daunting than Maytorena's Yaquis.

On November 1, 1915, an American reporter asked Pancho Villa:

> —General Villa, will you attack Agua Prieta?
> —Yes, and the United States, if necessary . . .
> —When?
> —I'll decide that.
> —How many cannon do you have?
> —Count them when they're roaring.[16]

Villa himself led the attack on Agua Prieta, and Plutarco Elías Calles defeated him. The official report Calles sent on November 4 to inform the high command of his victory is a model of order and intellectual clarity: general ideas, the enemy situation, details of the terrain, an analysis of his own initial means and choices and alternatives, how he executed his plans, the orders he gave, maps of the actions, and a summary of the results. Villa's eighteen thousand men had broken against carefully placed mines, barbed wire fences, ditches, and trenches aligned by Calles, who commanded only a fourth as many men. On the outskirts of the town 223 Villistas lay dead. Calles wrote to Obregón: "The chief of the attacking forces did not carry out his pompous promises of the evening before."[17] Days later, in the small town of San José de la Cueva, Villa shot all the men—including the priest—in cold blood. Among them were several named Calles. Although the Sonoran never forgave the offense, his victory at Agua Prieta had closed a chapter.[18] For Calles it was the end of the armed Revolution and the beginning of his personal revolution, his pedagogical dictatorship.

On the very day he became interim governor of Sonora, Calles announced his far-ranging *Programa de Gobierno*, proof that he had taken advantage of his intervals of inactivity to think out "the revolution of ideals and the reforms leading to progress" that he would now offer his people. Like the good teacher that he basically continued to be, he would first of all reform public education: opening schools in all towns with more than five hundred residents, requiring mining companies and industrial concerns to create their own schools, and establishing, under his auspices, a system that would include scholarships, libraries, reading rooms, teachers' colleges, and adult schools. Like the good justice of the peace that he had been, he would reform the legal system, promoting new civil and penal legislation. Though he not been a very good farmer, he would reform agriculture, "the main element of national wealth due to the abundance of rivers and the richness of the land." He would push for better wages in the fields and the division of the great landed estates. And he would create an official agricultural bank for the state of Sonora. Nor was all of this enough for him. He would also open new roads, preserve the old ones, encourage economic competition for the benefit of the consumer, propose a new tax system, create welfare institutions, encourage habits of cleanliness through public addresses by the governor himself, and—furthermore—encourage workers to form mutual-aid associations.[19]

Four days after taking office, he dropped his first bombshell: a decree that, if it had been enforced to the letter, might have pushed his father into the grave two years before his time (or saved him to live more sad and desperate years). The decree prohibited the importation, sale, and manufacture of "any amount" of intoxicating beverages. The penalty was five years in prison, but to demonstrate how serious he was, Governor Calles ordered the execution by firing squad of a poor drunk in Cananea.[20] Among the avalanche of decrees, there were others with clearly moralizing intentions—he prohibited gambling and authorized the police to arrest not only the gamblers but also spectators—and some that were genuinely moral: a grant of amnesty to the followers of "the criminal Villa" and an order to shut down *planchas* ("planks"), the torture rooms in prisons.[21]

The process of moralizing Sonora was then applied to history and to politics. In 1916, Calles decided to go after another great enemy of modernity: He took away the citizenship of "wandering tribes and those of the Yaqui and Mayo Rivers as long as they persist in retaining the anomalous organization they have today in their settlements and towns." In another decree, he seized—for the people of Sonora—all possessions of those who had given material or moral support to Orozco, Huerta, or the government of the Aguascalientes Convention. In order to purify public services, in May of 1916, he had a questionnaire distributed among public employees with such questions as: "What role did you take in the political struggle of General Plutarco Elías Calles against José María Maytorena when the aforesaid military officer was in the capital of the state trying to combat the treason of the Maytorena faction?"[22]

All the promises in his Program of Government were translated, almost immediately, into decrees. Besides ordering the creation of a vast school system, he established a register of property and published an Organic Law of State Courts. A minimum wage was enacted for day laborers and hacienda peons, and all sources of production (mines, industries, lands) not being presently exploited within the state were declared available to be used for the public good. A carefully prepared Revenue Law for the coming year was issued on the last day of 1915. Public education was assigned a huge 22 percent of the budget. When in May 1916 Calles took over the state Office of Military Operations for a couple of months and left Adolfo de la Huerta as interim governor, he already had fifty-six decrees to his credit. He had issued almost six per month.[23]

A year later, on June 25, 1917, he took office as the formal constitutional governor of the state of Sonora. (It was the year Plutarco senior died in Agua Prieta. He ended his days living in a hotel, trying to hide bottles of liquor under his pillows from the nurses, who would always find them and take them away from him.)[24] Plutarco junior—with a brief hiatus in 1918—continued to govern Sonora until May 1919, when Carranza named him Secretary of Industry, Trade, and Labor. The following January he resigned to join Álvaro Obregón's electoral campaign.

He had shown himself to be as much a man of actions as of words and laws. After enacting his Labor Law, for instance, he put down all agitation with an iron hand, expelling several sympathizers of the anarchist IWW from his state and ordering the execution of an old social activist, Lázaro Gutiérrez de Lara. Against the Yaquis, "a deadly hindrance to the progress of the state," he fought relentlessly, but he showed the same determination with the American Wheeler and Richardson companies when they monopolized unused land. One of his most radical measures, in fact without precedent anywhere in the country, was the expulsion of every single Catholic priest from Sonora.[25]

But his most personal work was not destructive, it was pedagogical. He inaugurated the Teacher's Training School, organized a congress on education, opened 127 primary schools, and, in the same preconstitutional period, conceived of a project that would remain very close to his heart: the Cruz Gálvez Vocational and Arts Schools for orphans of the Revolution. They were to be named in honor of one of his lieutenants, "fallen for the *patria.*" In August 1917 he circulated a statement throughout Sonora, titled "For the Salvation of My Race," signed by himself, in which he explained the meaning of this project and asked Sonorans to add their labors to his, so that it might be realized. He wanted to construct two large buildings—for young women and for young men—with funds raised by public subscription. The personal motivation is clear enough: The schools were to prevent and heal the abandonment that he himself had suffered. In his Decree Number 12 announcing the creation of the schools, he underlined that they should shelter "all orphans in general," without any political prejudices.

By 1920 the schools had their buildings. There were 468 boys (all of them living at the school) and 396 girls. Calles sent his own daughters to school there. Boys learned carpentry, farming, typing. The girls had an orchestra, the boys a band. When they addressed the governor, all of them, male and female alike, called him "Papá Calles."[26]

Calles's activities in Sonora were a political laboratory for his presidential programs. He was the teacher in power. Using all the practical experience he had accumulated—in his illegitimacy and abandonment, in his life as a teacher, businessman, agricultural worker, administrator, and justice of the peace—he was already trying to reform society from the roots.

* * *

Plutarco Elías Calles was Minister of the Interior during most of President Álvaro Obregón's term as President. In temperament, he was almost the exact opposite of his chief. To Obregon's expansiveness and joviality, Calles was a counterpoint of introversion and weighted reflection. His deep voice inspired respect. He was and looked strong, even-tempered, and inflexible. He sometimes smiled but rarely laughed. His face, like his reasoning, was symmetrical.

An old Porfirian—the normally dyspeptic historian Francisco Bulnes—calmly noticed that "General Calles has the good physique of a dictator."[27] The French ambassador to Mexico called him "realistic and cold, with a clear temperament and a firm will."[28]

One of his strongest qualities as a politician was silence. There was even a rhymed popular saying about it: *En el hablar es parco/Plutarco*, "Plutarco doesn't say much." Silence and his gaze. Calles did not look at people; he drilled into them. Someone said, "With one look, he knows your biography."[29]

In different circumstances (from those of Mexico in the years 1920 to 1928) Calles and Obregón might have been at odds. But history and politics—more than mutual sympathy or friendship—united them. Obregón never really understood Calles. Probing souls was not his specialty. Deep down, he was scornful of him, because the yardstick he used for measuring men was strictly military. Calles was for him "the least military of all the generals." And Calles judged men fundamentally by their minds and their plans. He would not have dared to publicly issue a judgment on Obregón, but he must have thought that he was a man without a program and he would have thought the less of him for it. Despite all this, they always needed each other. Though the only profound resemblance between them was that both were devoted family men, they always maintained forms and displayed the appearance of cordiality.[30]

In the middle of 1923 Calles retired for a time to the Soledad de la Mota Hacienda in Nuevo León, the property of his son, Plutarco junior. There he relaxed by opening a school and teaching the first class. He knew he had been chosen to succeed Obregón, and he was isolating himself to reflect on his future government. (Such introspective retreats would be a constant feature of his later years.) Despite the expressions of allegiance that began pouring in, he was still—at the age of forty-six—troubled and pursued by the illegitimacy of his birth. There were rumors in Sonora that he was of Muslim origin, or more specifically that he had Syrian blood. They called him "The Turk." (For many Mexicans, all Middle Easterners were "Turks.") Calles never responded to the rumors, perhaps to avoid giving the matter too much importance or letting himself be provoked. Or, more likely, because he did not want to reveal, at the heights he had reached, the social and religious stigma of his birth.[31]

Before assuming the presidency, between August and October of 1924, Calles traveled to Europe with some of his family. He had a medical purpose—an old leg wound from the siege of Naco was troubling him—but he also had an important political goal. He wanted to study the governmental and economic order of Europe, especially in social-democratic Germany. Up to then he had considered himself a socialist. His trip to Europe would make him more of a social engineer.

In Germany he spent time at industrial plants and cooperatives and asked for an example of every tool. In the evenings he would leaf through translated documents such as "The Reorganization of Rustic Estates in Prussia for Their Better Exploitation," "Domestic Careers for Prussian Peasant Girls," or "Agricultural Cooperatives and Rural Credit in Europe." In Hamburg, Calles also made use of his self-instructional visit to declare that Mexico would open its arms to European immigrants, expressly including Jews. His message crossed the German border to reach small Polish cities. Since the United States had sharply reduced its immigration quotas, hundreds of Jewish families responded to Calles's invitation and came to settle in Mexico. It was the origin of the country's Eastern European Jewish (Ashkenazi) community.[32]

From Germany he went to France, where Edouard Herriot, the radical socialist prime minister, received him with honors. Perhaps he would have liked to travel to the England of Ramsay MacDonald, but he made do with detailed information about the labor movement. His goodwill toward this movement was so deep that years later, when there was a great strike in England, he sent the coal miners $200,000 from the Mexican government. After Europe he went to the United States, where he visited President Coolidge and attended a banquet organized in his honor by the American Federation of Labor and addressed by Samuel Gompers. It must have been in the United States that he acquired a book he devoured: *The Profits of Religion* by Upton Sinclair. Its first lines were eloquent: "This book is a study of the Cult of the Supernatural, from a new point of view—as a source of income and a shield for privilege."[33]

"The Turk" took over the Presidential Chair on December 1, 1924. Guided by his considerable and arduous experience of life, having meticulously observed the European social order, he was sure he had learned important lessons. Now he would try to apply them.

* * *

Plutarco Elías Calles, 1925
(Fideicomiso Archivo Plutarco Elías Calles. Fernando Terreblanca)

"In my judgment and I say it in good faith," Calles explained a few days after taking office, "the revolutionary movement has entered its constructive phase." And this was true of the first two years (1925–1926) of his government. His intent was to repeat on the national level—but now in a broader, richer way—his achievements as governor of Sonora.

There were a series of "battlefronts." One of the most important (and dramatically innovative) was the area of banking and finance. Its commander was his Minister of Finance, Alberto J. Pani, his close collaborator a brilliant young man named Manuel Gómez Morín, who had been Undersecretary of Finance during Obregón's presidency and financial representative of the Mexican government in New York City.[34] Their first success (on January 7, 1925) was a new General Law of Credit Institutions, which revised the Porfirian law of 1897. Five days later, the National Banking Commission was founded.[35]

On September 1, one of the dreams cherished by all Mexican governments since the days of Porfirio Díaz became reality: a central bank (modeled on the American Federal Reserve), the Bank of Mexico. It began its operations surrounded by widespread suspicion, with very few associated banks and troubled by a plague of generals flowing into its offices to request direct loans. Gómez Morín, president of the bank's Administrative Council, knew that the first years would be crucial, and he chose to follow a conservative policy. He issued very limited amounts of money and allowed the Bank of Mexico to operate for a time as just one more commercial bank, building up its credit and gradually persuading other banks of the advantages of association. He granted some personal loans and exceeded the limits fixed for loans to the government, but along with these deviations he considerably advanced the consolidation of the bank. It became a solid institution, destined to endure.[36]

On February 1, 1926, Calles's government took another step on the reorganization of loans for social purposes: the founding of the National Bank of Agricultural Credit. Gómez Morín, its creator, thought that the new establishment was "one of the greatest things accomplished during the Revolution." Calles saw in it the ideas he had brought from Germany on cooperative societies and the so-called Raiffeisen Fund for agricultural credit. The institution was designed to operate as a bank that would finance regional and local agricultural societies, with the object of promoting widespread decentralization in agriculture.[37]

In 1927, 378 local societies with seventeen thousand members had been established. The National Bank of Agricultural Credit functioned dynamically but was attacked by the same disease as the Bank of Mexico: Revolutionary generals. (With the combat phase—though not the violence—of the Revolution essentially over, the term *Revolutionary* was by now taking on a privileged, ennobling, and already "sacred" aura.) Unfortunately, the Bank of Agricultural Credit did not defend itself as well as the Bank of Mexico. So-called courtesy loans began to flow like a great river toward various generals but especially toward one—Álvaro Obregón.[38]

On the battlefront of transportation, there were two major campaigns: to improve the railroads inherited from the past and to construct highways as a legacy for the future. Pani had considered privatizing the railroads, since they accounted for some $400 million of the national debt. Yet the problem was not so simple. There were far too many employees—which was the major financial drain—but the railroad workers would also violently resist layoffs. There would certainly be losses not of money but of lives. The restructuring of the railroads, as first planned, had to be canceled.[39]

Elsewhere on the railroads—in construction—the Calles government scored a major achievement, completing the South Pacific railroad that connected Nogales, Hermosillo, Guaymas, Mazatlán, Tepic, and Guadalajara. And in the campaign of the highways, almost everything went off smoothly. At the beginning of his term, Calles had set a goal of ten thousand new kilometers of highway. Though the figure proved totally unrealistic, seven hundred new kilometers were constructed in the first two years of his presidency, most of it financed by a gasoline tax.[40]

And so it went: new irrigation projects, the expansion of the network of schools (with its great personal importance to Calles) and especially his Central Agricultural Schools, which he saw as an important method for "modeling the new Mexico" from the "raw material" of the peasants. There would be modernization of the army, public health laws and campaigns, housing projects, the enthusiastic encouragement of sports, a war on alcoholism. A new body of laws, the Calles Code, was elaborated during these years, though most of it did not go into effect until later. One clause completely abolished any distinction between legitimate and illegitimate children. Calles was reforming from his roots.[41]

The most significant innovation in all this activity was a great expansion in the economic role of the State. The Porfirian regime had already intervened in the economy, but not in any social sense. "Who should be able to reach out a hand to the poor?" Calles asked a reporter, and answered himself—"Only one agency: the government."

Mexico had no social class that could, through its own efforts, draw the country toward material progress. The State had to take the initiative—creating banks, reservoirs, roads, schools, laws, and institutions for the benefit of society. But society had not been consulted as to the utility of that vast program of State action. Under Calles, the government was becoming the only judge of the interests of society, an institution that would treat the Mexican people—just as Díaz had—like a minor child under the protection of its governing father.

In his foreign policy, especially toward the United States, Calles turned his back on Obregón's Bucareli Agreements and tried to return to the positions of Carranza, favoring strict adherence to Article 27 of the Constitution. Various initiatives on oil were discussed in the Legislature. The most radical was proposed by Morones, who was both the leader of the CROM and a very influential Minister of Industry, Commerce, and Labor. Pani and Aarón Sáenz, Minister of Foreign Relations, made the mildest proposals. All the initiatives—however weak or strong—were opposed by the oil companies. The American ambassador, a hawk named Sheffield, thought that Mexico would be, or already was, the second Bolshevik country on earth: "Soviet Mexico." On June 12, 1925, American Secretary of State Kellogg issued the first threat against Mexico to be sounded in the Calles era: "This government will continue to support the government of Mexico only while it protects American lives and interests and fulfills its international compromises and obligations. The Mexican government is now being tested before the world."[42]

A year later—with the Mexican government and the American oil companies no longer talking to one another—Calles issued the new oil regulations. If only the opinions of the oil companies, resident Americans, or Ambassador Sheffield had counted, the United States would have invaded Mexico. But along with intellectuals, sympathetic journalists, and the Democratic Party in Congress, other forces were putting a different kind of pressure on Washington—U.S. banks and broader business interests were in favor of a peaceful solution.

On November 10, 1926, *The New York Times* announced that the moment had come to break off relations with Mexico. Along with the quarrel over oil there was now an international clash: Mexico and the United States supported opposite sides in Nicaragua. The United States preferred the conservative Díaz, while Mexico inclined toward the liberal Sacasa and not only with words, as the Mexican general Escamilla Garza remembered: "General Calles sent two expeditions to Nicaragua, one by the Pacific and one by the Atlantic. I was in command of three boats. . . . We moved along the coast to avoid the gringo boats. . . . I had taken on a crew of 500 men, most of them Mexican. After 56 fights and skirmishes, we made the Tipitapa agreements with the Americans."[43]

As if to put meat on the plates of those who were convinced of an inevitable "Soviet Mexico," Alexandra Kollontai, the first ambassador from the Soviet Union, arrived to open relations. Her first words were: "There are no two countries in the world more alike than modern Mexico and the new Russia."[44]

In January of 1927, the government canceled the permits of oil companies unwilling to comply with the new regulations. President Coolidge announced that Mexico—already on probation—might be treated the same way as Nicaragua.

Said *The New York Times*: "These are some of the most serious words in the diplomatic vocabulary. They are never used officially unless it is a formula of the most extreme warning, and generally as the prelude to an ultimatum, a breaking of relations or war itself."[45]

Mexican diplomacy reacted not with bravado but with astuteness: It proposed international arbitration by the International Court of The Hague and harvested sympathy in the American Congress, where Mexico could count on the support of Senators Borah and La Follette.[46]

Arturo M. Elías, the President's half brother and a very active consul general in New York, reported that, on the advice of those friends of Mexico, Carlos and Ernesto (authors Carleton Beals and Ernest Gruening), Mexico should make a declaration in defense of the right of each country to support the government in Nicaragua that it considered most democratic. The tactic worked. Only one American in forty—commented *The Washington News*—wanted war with Mexico. Kellogg stopped calling the country Bolshevik. In mid-January 1927, in his telegrams to Calles, Consul Elías used a premature word: *triumph*.[47]

But in March of that same year, Manuel Téllez—the Mexican ambassador to the United States—rushed back to Mexico with disturbing information. There were rumors that the United States was going to send the marines into Mexico. Calles made a public threat, a "conflagration" of the oil wells in Tampico that "would light up the sky all the way to New Orleans." But just when everything seemed lost, tensions lifted, thanks in part to a dramatic maneuver of counterespionage. Within the American embassy, men working for Morones laid their hands on documents that mentioned a future invasion. A few days later, Secretary of State Kellogg—caught in a political crossfire—admitted that three hundred documents had been stolen and at the same time lowered the voltage of the entire situation. Only the oil companies themselves continued to defy Calles. When pushed too far, the President would shut down a well. Sheffield was soon replaced. It did seem like a victory for Mexico. And it was, but only in part.[48]

Calles understood that he had to move carefully. He could not apply the oil regulations too rigorously. He could not, for instance, make them retroactive. On September 29, 1927, Calles and Coolidge would inaugurate a direct telephone link. At the end of October, a new ambassador, Dwight Morrow, arrived with specific instructions from Coolidge: "Keep us out of a war with Mexico." He knew—as his famous journalist friend Walter Lippmann had told him—that what was motivating Mexico was not Bolshevism but rather something different and found throughout the world: nationalism. His ideas, his tactics, and (perhaps more than anything in Mexican eyes) his attitude would be like Sheffield turned inside out. He would try to reach agreements rationally and courteously, avoid arrogance, identify a little with the people and culture of the country, showing a combination of shrewdness and respect.[49]

Almost effortlessly, Morrow—who came from the investment firm of J. P. Morgan—moved into the center of Mexican economic and political life. He worked closely with the new Finance Minister, Luis Montes de Oca. He studied the budgets and negotiated a total restructuring of the foreign debt, so that Mexico could invest and grow. In the oil conflict, he reached an agreement with Calles so that judgments favorable to the companies could be recognized as a precedent, and he tried to amend the most sensitive articles of the Oil Law. As a gesture of goodwill, he brought his son-in-law Charles Lindbergh, "the ambassador of the air," to visit Mexico City.[50]

Calles had made the United States withdraw its threat of invasion. He had conjured away the scarecrow of "Soviet Mexico." And he had reduced the harshness and hysteria of American diplomacy at a time when America's president was not a man with a social conscience (like the Roosevelt to come) but Calvin Coolidge, for whom "the business of America [was] business." Calles had to make some concessions, especially on oil. But he won as much as he could in 1927, a year when he had to deal with two serious problems. There had been a sharp decline in international prices for Mexico's principal products: oil, industrial metals, and silver. And within the country a crisis was developing over the old sore point of religion.

When Calles took office in 1925, there had already been a decade of continuous agitation around this long-unresolved problem. And on this front, 1925 would not be a peaceful year. On February 21, under the auspices of the CROM, the Mexican Apostolic Catholic Church was created, under the leadership of a "Patriarch Pérez." Inaugurating a Mexican church separate from Rome was not a new idea. The bold thing was to actually attempt it at that moment, when the powers of Church and State were confronting each other with unprecedented tension in every area touched by the Constitution of 1917. There were peasants who at the advice of priests rejected the lands they were given and *agraristas*—supporters or armed militants of official agrarian reform—who demanded apostasy in exchange for a parcel of land. The partisans of a growing Catholic unionism sometimes fought in the streets with the CROM. But the greatest conflict of course was over the anticlerical Articles 3 and 130. The Church would not recognize them. Nor would it forget all the insults and attacks it had suffered during the Revolution.

The Callista state was determined to decree and impose strict fulfillment of the Constitution. In 1925 the tension between the "Two Majesties" was building up in many states of Mexico. In May, a woman named Jáuregui, said to be an unbalanced fanatic, made an attempt on the President's life. Nevertheless, even though the National League for the Defense of Religion (a primarily middle-class organization) had already been created, no one clearly saw what was coming.

The open break came in 1926. All through the previous year, Calles had kept hoping that his governors would carry out the requirements of the Constitution, but the laws were applied too loosely for his taste and he finally decided to take more severe measures. In January he asked Congress for extraordinary powers to amend the penal code and introduced more regulations of the forms of worship. On February 4, Archbishop Mora y del Río—in the newspaper *El Universal*—was quoted as attacking a number of provisions of the Constitution. The League for the Defense of Religion applauded him, but the President did not. He said to one of his aides: "This is a challenge to the government and the Revolution! I am not prepared to tolerate it. Now that the priests are making these moves, we must apply the law as it stands." Although Mora y del Río denied that he had been accurately quoted in *El Universal* and added that the objections as reported were not valid under present conditions, Calles had made up his mind. He ordered all governors immediately to apply Article 130. Church schools were closed; foreign priests were expelled; there were riots, demonstrations, fighting in the streets.

The Vatican advised moderation, but the bishops were divided in their opinions between resistance and passivity. In a pastoral letter they asked that the government of Calles follow in Carranza's footsteps and agree to amend Articles 3 and 130. Calles responded directly to Mora y del Río: "I want you to understand, once and for all, that the agitation you are provoking will not be enough to alter the firm resolve of the Federal Government. . . . There is no other way . . . than to submit yourself . . . to the law."[51]

On July 2 the Calles Law was officially published, amending the penal code. It included punishments for crimes related to religious teaching and worship. Article 19, the most inflammatory, made the official registration of priests obligatory before they could exercise their ministry.

The response was immediate. The League organized an effective economic boycott in several states. The bishops published a joint pastoral letter announcing the suspension of masses, to begin the moment the Calles Law officially went into effect. Ambassador Sheffield wrote to his Secretary of State: "The President has become so violent on the religious question that he has lost control of himself. When this topic has been dealt with in his presence, his face turns red and he has hit the table to express his hate and profound hostility toward the practice of religion."[52]

The conflict continued to build. Calles declared that "naturally" he did not intend in the least to soften the amendments and additions to the penal code. He was confident—and told Lagarde—that "each week without religious ceremonies will cost the Catholic religion 2 percent of its faithful."[53] He announced that "we are at the point when the lines will be drawn forever across the fields. The hour when we will unleash the final battle is approaching; we are going to know if the Revolution has defeated the reaction or if the triumph of the Revolution has been ephemeral."[54]

He really believed this. For Calles, the battle against religion, the crusade for his version of secular enlightenment, had apocalyptic importance. But not only for him. On July 31 the faithful crowded into churches across the country. It was the last day of masses.

On August 21, Calles held a long meeting with Leopoldo Ruiz, bishop of Michoacán, and Pascual Díaz, bishop of Tabasco and secretary general of the Mexican episcopate. At that point, the attitude of the bishops was conciliatory and even somewhat humble. But Calles was strident, lacerating; his brief responses were filled with imperatives. His vision of the role of the clergy in the history of Mexico was absolutely negative, with no relief, and he would therefore not concede a single point to the bishops, nor would he throw them, as they asked, "a plank to save us." The meeting went at once to the heart of the problem: the enforcement of Articles 3 and 130, and in particular the thorny matter of the official registration of priests:

> CALLES: Can the legitimate representative of the people, which is the government, require anything less than knowledge of who administers its properties? . . . Without any relaxation or hesitation, you will have to submit.

RUIZ: Against the dictates of our conscience.
CALLES: The law is above the dictates of conscience.
DÍAZ: The law . . . can be amended . . . with your help . . .
CALLES: I am not the one who will solve the problem; that is within the competence of Congress and I must honestly say that I am in absolute agreement with this law that you wish to amend, since it satisfies my political and philosophical convictions.[55]

Calles went on to say that he thought "in all sincerity" that the Mexican clergy had always placed itself on the side of the oppressor, that the Catholic missionaries had for centuries done nothing to help the poor. The bishops offered arguments to the contrary, but Calles immediately rejected them.

Then they tried to persuade him to follow a Porfirian mode and at least throw some kind of public mask over the application of the law requiring the registration of priests. After all, there were mayors who already felt they had the right to name or remove local priests. But Calles refused even this. And he concluded: "I am going to show you that there is no problem, since the only one you could create would be to commit yourself to a rebellion and in that case the government is fully prepared to defeat you."[56]

As they left, the bishops told him they had no intention of fomenting a rebellion. But they had no need to. Calles had hoped to suppress the people's "fanaticism" by cutting it off at the roots. A huge sector of the peasant population of "Old Mexico" contradicted his illusions in armed rebellion. For them, the "cause" was clear: They were fighting to bring back masses, they were fighting to defend religion. Their war cry was *Viva Cristo Rey!* ("Long Live Christ the King!"). They would become known as the Cristeros and their war as the Cristiada.

The War for Christ would last three years. It would be a savage war, on both sides, pitting groups of mounted guerrilla fighters against a federal army that had recently been modernized by its commander, General Amaro. The modernization had followed the European model, concentrating on infantry supported by artillery and aircraft, and it had left the federals weak in cavalry, an outmoded arm in Europe but essential for fighting in the territory of the Cristiada (especially strong in such west-central states as Jalisco and Colima), where there were few highways and entire regions inaccessible to ground troops. It would be a crucial advantage for the Cristeros.

The initial mass uprisings of January 1927 (encouraged but neither supplied nor strategically organized by the League for the Defense of Religion) were repelled by the federal troops, who inflicted heavy losses.

Calles thought the uprising might be over in a couple of months. But without any real central organization and (highly unusual in Mexico) without a caudillo, the peasants fighting for their religion became bands of guerrilla warriors, capturing weapons from their enemies and waging a ferocious war like that of the Zapatistas (and more effective). The National League for the Defense of Religion had abandoned its hopes of directing the Cristiada and moving it toward the overthrow of the government (to be followed by its own rise to power), but in 1927 they succeeded in hiring an unusual figure to supply military expertise and leadership for the Cristiada. General Enrique Gorosticta was a Mason and a liberal agnostic, with considerable experience at war and few practical opportunities under a government he detested. He joined the Cristiada as a mercenary officer but would become closely connected to the soldiers of faith who followed him, offering them critical leadership and technical expertise until his death in battle in 1929, shortly before the end of the Cristiada.

The rebellion eventually spread to thirteen states across the center of Mexico. In many areas, the federal army would sally out of the cities to repress a Cristero attack. The federals would hang men, burn villages, sometimes shoot priests (about 90 of them were executed during the war) but then would pull back to their urban garrisons, leaving the countryside to the mounted guerrillas. By the time the war ended, close to 50,000 Cristeros were in arms. Another 25,000 had died in combat. The Cristiada would cost upward of 70,000 lives, a precipitous drop in agricultural production (a decline of 38 percent between 1926 and 1930), and an internal migration (to the cities) of 200,000 as well as an external emigration (mostly to the United States) of 450,000 people. (The establishment of a major Mexican presence in California dates back to these years.)

As early as March 1927, Calles had begun to consider a compromise. He continued to believe that the Cristeros were only a minority, but he could not ignore the costs of the war. Obregón intervened and tried to find a solution that would bring peace without making the government lose face. In June 1927 Calles let several military leaders of the rebellion out of prison and permitted worship in private homes. He fell ill under the pressure and resorted to a folk healer (*curandero*), showing a less than rationalist streak, which would come out again in his old age. To fill his cup to the bitter brim, a great sorrow came to him then: the death of his wife.[57]

Ambassador Dwight Morrow arrived in October 1927 and quickly made a judgment on the situation: "The country is upside down. The poor have almost nothing but the consolation of the Church and there will be no lasting peace or progress until an agreement is reached." With the support of the American government, he immediately took charge. He himself drew up an agreement between Calles and Father Burke—an American Catholic with full powers from the Vatican to undertake negotiations. At the end of their meeting, in March 1928, Calles said to Burke: "I hope your visit marks a new era for the life and the people of Mexico."[58]

The end of the war was delayed until June 1929. When Obregón was assassinated in July by a Catholic militant, the peace process was postponed. Rome, meanwhile, denied that the Pope had ever given his blessing to the Cristeros. But with or without Rome, they continued their war: The "cause" was still there, for there was no celebration of mass. Finally, in June 1929, a peace agreement was concluded.[59]

While it is true that Calles did not invent a conflict that had deep historical roots, he was surely its primary catalyst. His entire biography points ahead toward this war. In his files he kept love letters written in Aguascalientes in April 1913, by Bishop Ignacio Valdespino to a Sonoran lady, and other documents that implicated priests in love affairs. For Calles, as for many Revolutionaries, such things were the definitive proof of the historic debasement of the clergy.

What he really would have liked was "to eradicate" the Catholic faith from Mexico. He did not want a purer Christianity, closer to its origins. Nor was he a romantic atheist insisting that God did not exist. Calles neither enjoyed blasphemy nor was he drawn to inventing counterliturgies. He was a teacher from Sonora, a state of fiercely independent and often anticlerical frontiersmen, and he did not understand, respect, or forgive the center of the country, "Old Mexico," where men—as he used to say—were not "real men." He was a priest of the faith of "progress and evolution," an imperious and apparently rational reformer driven in reality by a blind and irrational passion: to deny original sin—the sin of his own origins.

Calles's extremely complex presidential term (1924–1928) had many political protagonists: the generals, the union leaders of the CROM, the *agraristas*, the cabinets. There were eight thousand political parties in the country. The verbal and physical battles between them spun an almost insoluble tangle at national, state, and local levels. Pacts, ruptures, threats, confrontations, campaigns, embezzlement, light-handed theft, bravado, shoot-outs, riots. The political life of Mexico turbulently overflowed the formal boundaries of legislatures, cabinet meetings, or the press and spilled into bars, brothels, and casinos.

The dominant figures were always the generals. Every single one hoped to become President. Most of them probably preferred Obregón to Calles, but most of them also would line up behind either if breakaway generals tried a coup. Some—like Lázaro Cárdenas—were creatively independent men; others—like Claudio Fox, Roberto Cruz, and Eulogio Ortiz—born killers. And there were the feudal caciques, like Caraveo in Chihuahua or Cedillo in San Luis Potosí. The closest man personally to Obregón was "Panchito" Serrano and to Calles, Arnulfo R. Gómez. But both would oppose Obregón's reelection, and Calles and Obregón would swallow the bitter pill and destroy them.

The CROM leader and Minister of Labor, Luis Napoleón Morones, was another typical personality of the time. His wild parties in a mansion in the Tlalpan neighborhood, his enormous diamond rings, bejeweled women, and luxurious black Packards are only part of the story. He was a true popular leader and a great speaker. He fought political battles—which often turned into gunfights—against the military, the Church, parties (like the National Agrarians) who supported Obregón, the railroad workers affiliated to the Communist Party, the anarchist CGT, and the press. His capital sin may well have been his aspirations for the presidency of the Republic instead of being satisfied with the presidency of the workers, but beyond Morones himself, it was the CROM under his domination that laid the pattern and groundwork for the later emergence (in 1936) of the most powerful labor organization of twentieth-century Mexico: the Confederation of Mexican Workers (CTM).[60]

With words, with actions, and with pistols, all these leaders fed the image of *México bronco*, "untamed Mexico." It was perhaps the natural aftermath of the Revolution, but a deeper explanation lay elsewhere. Without political institutions to resolve the problem of the presidential succession in a peaceful way, democratically or otherwise, the country was condemned to an alternating pattern of violence. If each presidential term lasted four years, the cycle would repeat itself as it had with Obregón and now with Calles: two years of work and two years of restlessness and bloodshed. To make things worse, in 1926 everything pointed to the return of Obregón and the sacrifice of a principle that had sparked the Revolution: Madero's slogan of "No Reelection."

As the days before the election dwindled down, the killings mounted. Beginning with the death by poison (in April 1926) of General Ángel Flores—an early opponent of Obregón's reelection (and a projected rival)—the generals began to fall. After the aborted rebellion of Gómez and Serrano, 25 of them would soon be dead along with another 150 people shot down in preelection gestures. According to General Roberto Cruz (chief of the Police Inspector's Office for the Federal District and so in a sense the lord high executioner for *México bronco*), "It is possible that Calles has let himself be influenced by Obregón." Without Obregón's advice, the President might have at least exiled some of the rebels. On the other hand, there is no doubt that Calles ordered the rapid death of the Pro brothers—one of them a priest—when they appeared to be implicated in the engineer Segura Vilchis's attempt on Obregón's life.[61]

"What an iron character Calles had," the killer general Roberto Cruz recalled in 1961. "There was no one in the government, absolutely no one who refused to obey him or would even so much as stand up to him on a matter of principle, not a single person who would resist any of his decisions."[62] Martín Luis Guzmán wrote that Calles "was not bloodthirsty (in the sense that he did not take pleasure in killing), but at the same time it did not bother him and he disposed of the lives of others with supreme indifference."[63]

After the assassination of Obregón on July 17, 1928, anything could have happened, the most likely perhaps a coup under the pretext of a supposed complicity of the Calles regime in the killing of the great man. Although the murder of Obregón had long been a possibility, the public reaction was surprise, confusion, sometimes hysteria. And Mexico was not at peace. In the west, the Cristeros were still waging their revolution. It was by no means clear whether the death of Obregón was or was not the end result of a widespread conspiracy. The country was tense and troubled.

Calles kept cool. His special gifts stood out as never before: severity, reflection, consistency. Each step that he took, or permitted, was a wise one. At first he let the anger of the Obregonistas flow freely but without allowing it to touch him personally and deflecting it toward two natural targets: Morones, the CROM patriarch, whom many thought might be responsible for the assassination; and the assassin Toral himself, whose interrogation Calles left to his victim's friends. Meanwhile Calles called a meeting of the thirty most prominent generals, to request their support and to propose that the interim President be a civilian. Although by that time some Sonoran dissidents were already plotting a rebellion, the coup attempt was delayed by the speed and effectiveness of Calles's actions.

At this highly delicate moment for the country, Calles hit upon the idea of introducing an ambitious political reform. In order to present it to the nation, he waited until his final Report to the Legislature (the *Informe*) on September 1, 1928, when he invested his proposal with a particular gravity and solemnity, speaking clearly and forcibly in his harsh, deep voice.

The last caudillo had died with Obregón: "There is no personality of indisputable stature, with a firm hold on public opinion and enough personal and political force to merit general confidence through his mere name and prestige." It was a misfortune, but also in a way a blessing:

> I do not need to remind you how the caudillos—perhaps not always intentionally but always naturally and logically enough—have hampered the appearance, formation and development of other national power options, to which the country could resort in moments of internal and external crisis, and how they sometimes delayed or blocked, even against the will of the caudillo himself—in that same natural and logical way—the peaceful evolutionary development of Mexico as an institutional country, in which men may become, as they should be, mere accidents with no real importance beside the perpetual and august serenity of institutions and laws.[64]

He noted that he could have "if . . . conscience had not forbidden it, disguised a resolution to continue in office under the claim of benefiting the public." He had not done so, and "at the risk of redundantly emphasizing this solemn declaration," he stated that "never, for no motivation and in no circumstances, will the current President of the Republic of Mexico come to occupy that position again."

On this total condemnation of reelectionism—which stamped a final seal on the Maderista ideal—Calles based his new project: "an opportunity . . . to pass once and for all from the historical condition of 'the country of a man' to a nation of institutions and laws."

The second part of the speech is rarely remembered by historians. In it, Calles spoke of inviting "the political and clerical reaction" into the Legislature in order to begin "the battle of ideas," without which the "Revolutionary Family" could run the risk of losing its vitality and degenerating into a conflict of factions. "The reaction" could function as a "moderating tendency." Their "presence in both houses"—concluded Calles—"would not endanger the hegemony of a Revolution that has already triumphed within the consciousness of the public and so can open itself up to contention, from which the nation would ultimately benefit."[65]

Calles was—at least in words—calling for the end of the age of caudillos. And he was depriving future aspirants of any right to the role. It was the beginnning of a significant change in the structure of Mexican political life.

He also had to settle another tricky problem. Obregón—the President-Elect—was dead. On December 1, 1928, an interim President would have to succeed Calles, until another election could be called. Calles's next right move was to support Emilio Portes Gil for the job. He was a lawyer from Tamaulipas, a young but experienced politician with a radical reputation and no history of close ties with Calles, or with Obregón either. All in all, an unobjectionable candidate with an additional advantage for the Obregonistas who were the political force of the moment—he was a bitter enemy of Morones. Portes Gil would sit in the Presidential Chair from December 1, 1928, to February 5, 1930.[66]

With the problem of the interregnum solved and the reaction to the assassination cushioned, Calles could now unleash his great institutional project: the creation of the National Revolutionary Party (PNR). At first he was president of the Organizing Committee for the projected party but then gave up formal direction of the committee to operate as a referee. During the three months between this resignation and the first convention of the PNR in Querétaro (where the future presidential candidate was to be nominated), the party united its forces, identified itself with the nation, and elaborated an open, classless ideology including "radical action, centralized organization and even moderate evolution." To compete with the Obregonista Aarón Sáenz, who thought he had the nomination salted away, Calles pulled an "old revolutionary" out of his sleeve, the engineer Pascual Ortiz Rubio from Michoacán, who was then the Mexican ambassador to Brazil. Although Ortiz Rubio, a contemporary of Calles, had an impeccable record dating back to his commitment to Madero, he had been out of the country for many critical years and had no personal political base. All that being so, the competition was waged entirely within the PNR and at the last moment Calles decided in favor of Ortiz Rubio. The real winner was not the former ambassador to Brazil but the new institution, the PNR, which now, for the first time in Mexican history, as a party determined the succession.[67]

One more coup attempt—the long-delayed Sonoran conspiracy—surfaced around the same time. Led by Gonzalo Escobar, it was one of the last rebellions of the generals. The relationship Calles had developed with the United States was of considerable help to the government of Portes Gil. The Americans refused to deal with the rebels in any way or let them purchase any kind of arms, while they freely sold weapons and ammunition to the government. Federal forces took only a few weeks to crush the rebellion, leaving a thousand dead rebels on the field and two thousand wounded. Forty-seven generals were killed or forced to resign their commissions. For Calles and his project of institutionalization, it was a great victory. In his first military test away from the protective shadow of Obregón, Calles (though Portes Gil was officially President) had shown that he could triumph not only with reason but with force. In his *Informe* to the Legislature on September 1, 1928, he had promised to be "the guarantor for noble and disinterested conduct by the Army," and he had warned the generals against the "inexcusable and criminal" act of trying to take power by means other than "those specified by the Constitution." When the "antipatriotic and disloyal" act occurred, against which he had preached, the guarantor had then played his part. He had become, more than ever before, the "strongman of Mexico."[68]

In June 1929 Calles (settling into his role as "referee") gave another important speech, in which he referred to the "political failure of the Revolution." In his judgment, the Revolution was now victorious in the economic and social spheres but "in the specifically political field, on the terrain of democracy, with respect to the vote and the basic purity of individuals or groups of electors, the Revolution has failed."

It was up to the party (and not the executive branch) "to correct the errors that the Revolution has committed in the area of politics." But not only the party—the opposition as well. The Revolution should go to the two houses of the Congress and open them, as his September message had already suggested, to "the reaction." Above all, the vote should be respected, as should "all legitimate triumphs of opponents in politics."[69]

The moment was approaching when these sentiments would be put to the test. The only opposition candidate was José Vasconcelos, who was already exhibiting—across the entire country—one of the most impressive and broadly inclusive democratic campaigns in the history of Mexico. With his "battalions" of university students and with the sympathy of the middle classes, intellectuals, and even workers in the northeast, Vasconcelos was trying to return to the Maderista roots of the Revolution and open the way to democracy. In his speech of 1928, Calles had invited the opposition to begin a battle of ideas in the Congress, but he had also seemed to suggest that power would continue to belong "to the Revolution." For Manuel Gómez Morín, that young man who had participated in (and helped create) the economic reconstruction programs of Calles's initial years and someone who knew the President well, the speech meant an opportunity to found a party that could confront the PNR, from the very beginning, by taking part in "the struggle of ideas." Calles was initiating access not to power but to competition—unequal, but competition all the same—with power. Gómez Morín tried very hard to persuade Vasconcelos to found such a party, but unfortunately for the political life and future history of the country, he was unsuccessful.[70]

In mid-1929, Calles announced his belief that "the future of Mexico is guaranteed" and went off to Europe for almost five months. Meanwhile, at a moment when the Cristeros were about to establish a connection with Vasconcelos, President Portes Gil signed the agreement with Rome that put an end to the war. (Roberto Cruz would sum up the pact with humor: "Let them violate the Constitution but only a little. And let us look stupid, but also only a little.")[71]

The battle now was political, focused on the campaigns of Ortiz Rubio and Vasconcelos, during which the PNR went through its initiation into the technology of electoral fraud, a "science" that later became its highly refined specialty. Throughout the whole country, the party extended its political machinery, linked from that moment on to the government; and then it moved—first with words—against the followers of Vasconcelos. *El Nacional Revolucionario*, the party organ, proclaimed that "a country cannot be governed with literary teachings . . . the PNR does not distribute *The Iliad* [as Vasconcelos had done] . . . but rather 35 million hectares." A hail of insults was directed at "the pedant" Vasconcelos and "the intellectuals, homosexuals, bourgeois, students, feminists and fanatics" who followed him. Actions would soon follow words: strong-arm men breaking up meetings, assaults, and finally murders. The young Vasconcelista Germán del Campo was shot dead in the street; and months after the elections, in March 1930, there was a bloody massacre of Vasconcelistas in the town of Topilejo, on the outskirts of the capital.[72]

In November 1929, Pascual Ortiz Rubio had been declared the victor in elections that he most probably would have lost had they been honest. Vasconcelos went into exile disillusioned with a people that had not followed the example of the men of 1910 to defend the freedom of the vote. He would always believe that his defeat was due to "an antinational alliance" between "proconsul" Dwight Morrow and Plutarco Elías Calles.

Calles remained the head of the PNR until August 1931. Although his enemies attributed his permanence in power to pure ambition, Calles's motivation was more complex. He had created a political organization that needed time to develop. Tumult in the Legislature would be preferable to the coups and assassinations of the twenties. Calles was moving from "referee" to his long, unofficial but profoundly powerful role as the *Jefe Máximo* ("Foremost Chief") of Mexican politics. From the house in the Colonia Anzures where the "humble citizen" Calles lived, one could see the Castle of Chapultepec, and people began to say maliciously, "The President lives there, but the man who gives the orders lives across the street." But this time, popular wisdom was not quite fair; the man who gave orders may have lived "across the street," but Ortiz Rubio did not simply let himself be ordered around.

And in September 1932, when he resigned his office (with certain achievements behind him like a new Labor Law and the inauguration of the Chapultepec Zoo, which would delight generation after generation of Mexicans), Ortiz Rubio would leave with dignity. In his memoirs, he wrote bitterly of the "thinly veiled dictatorship" of Calles. But yet, in those years, the country had made advances toward an institutional life. The "nonreelection" of the President was now a permanent feature of Mexican politics. The *Jefe Máximo* gave orders, but not as a tyrant. He governed within a framework of broad ideological tolerance, through an institution—the PNR—that in fewer than four years had come to dominate the legislative, parliamentary, electoral, and in general the whole political life of the country. Clearly it was not democracy, but it was closer to it than all the previous Revolutionary regimes except for the government of Madero. Thanks to the PNR, Mexico avoided the militarist destiny of almost all Latin America. Right up to its revamping in 1938, the PNR was a civilized conclave of generals who resolved their differences without drawing their revolvers. It softened and contained violence—until violence could fall out of fashion.[73]

The PNR was here to stay. But for it to "solidify"—as young general Lázaro Cárdenas had written to Calles in October 1930—it required a "personality who could exert authority over both politicians and soldiers."[74] Calles, however, was not about to leave the scene gracefully. Any President who wanted to become fully independent would have to earn the space for that freedom of movement.

After Ortiz Rubio, checkmated on all sides by Calles, resigned "with my hands clean of blood or money," Calles chose not a President but an administrator for the final two years of the six-year term. Abelardo Rodríguez had made a fortune smuggling liquor into the United States during Prohibition. When Calles—at one point in 1931 when he felt his power threatened—had moved out of his very public shadows and taken over the portfolio of War in Ortiz Rubio's cabinet, he had summoned Rodríguez into his office and made him an Assistant Minister of War. As interim President (from September 1932 to December 1934), Rodríguez, in his own words, left "politics to the politicians." Sometimes this division of labor cost him his personal and presidential dignity, but Rodríguez was almost always prepared to pay the price.[75]

Normally, during the *Maximato* (the period from 1928 to 1934 when he was the undisputed *Jefe Máximo*) Calles kept a physical distance—sometimes considerable—from the precincts of power. He traveled to Europe twice for long periods and would regularly retire to his various ranches or other businesses. Distance served to emphasize his actual power. Between 1930 and 1935, prominent Mexican politicians practiced two sports: pilgrimage to the haciendas of the *Jefe Máximo* to request his "advice and guidance," and golf, to which General Calles had become addicted in Cuernavaca.

Occasionally, when he wanted to—as in 1931—he would hold a cabinet post for a time. More often, he took charge of official commissions that called for a strong hand. As chairman of the Commission for the Reorganization of the National Railroads, he proposed and carried out measures as severe as the dismissal of ten thousand workers and a general reduction in salaries. For a brief period he was director of the Bank of Mexico, which he turned into Mexico's first true central bank, while also pushing through measures that included a new General Law of Credit Institutions.[76]

But more important than his jobs were his opinions on the great national problems. With the collapse of Morones and the CROM and the strengthening of the General Confederation of Mexican Workers and Peasants (CGOCM) under the Marxist leader Vicente Lombardo Toledano, the labor movement had lost direct political influence but now had a growing social base and increasing ideological radicalism. When pressure from the workers' unions began to rapidly mount in 1934, Calles condemned what he called "agitation." But he was not taking the new union leadership into account. He was in control of the political apparatus (the cabinet, the party, the army), but he had neglected the union cadres and—even more so—the membership. He would soon pay for his error.

Between 1924 and 1928, social policy had been a central element of his government. During the Maximato, Calles paid little attention to it. His obsession was the consolidation of political reform. Besides, at fifty-seven years of age, he felt exhausted. He had liver and spinal problems that kept him in bed for many days. In November 1932, his second wife died after a painful battle with cancer, leaving him with the sad conviction that his life would never again be whole. And so, at the height of his glory, he began to feel that his hour was over—his hour and that of the Revolution. Symbolically, a commission was created to collect funds for converting the unfinished structure of Díaz's old legislative palace into the Monument to the Revolution.[77]

One internal battlefront remained open for Calles: the religious question. During those days he tightened his connections with the most fanatic of the antireligious fanatics: Tomás Garrido Canabal—the governor of Tabasco. He could no longer order the registration of priests, but there was still Article 3 of the Constitution, on the education of the young, and he had something to say about its importance. In the pious capital of Jalisco, on July 20, 1934, Calles delivered his famous *Grito de Guadalajara*:

> The Revolution is not over. . . . We have to enter a new phase, that I would call the period of psychological revolution: we must enter and conquer the minds of the children, the minds of the young, because they do and they must belong to the Revolution . . . [and the Revolution must] uproot prejudices and form the new national soul.[78]

It was the inner voice of the teacher again—and for the last time—trying to reform the psychological makeup of the Mexican people from the roots, through their children.

But other projects were under way, to fill the "minds of the children" with the content of another revolution, quite different from Calles's Sonoran anticlericalism. Since the beginning of 1931, a sector of the unionized left, represented by Lombardo Toledano and several academic and official groups, were proposing the adoption of "socialist education." In October 1934, shortly before the new President, Lázaro Cárdenas, was to take office, Article 3 of the Constitution was amended to read: "The education imparted by the State shall be socialist and in addition to excluding all religious doctrines, shall combat fanaticism and prejudices, to which purpose the schools shall organize their teaching and activities in such a way as to create in our youth a rational and exact concept of the universe and of social life."[79]

Among workers, agrarian activists, intellectuals—and within the government itself—a different attitude toward politics was making great strides. Its most prominent feature was the new socialist idealism (and dogmatism). It was not only a Mexican tendency, but a sign of the times. After the 1929 Depression and Hitler's rise to power, the Western world was hungry for new beliefs that could explain the crisis in the capitalist system and fortify opposition to nazism. Socialism and often communism satisfied this craving for the advent—within a reasonable and inevitable time—of a classless society.

"Within the governments of the Revolution," wrote Lombardo Toledano in October 1933, "there have been and there are men who believe that the Revolution has not yet taken place and that it must be carried out." Calles had a different concept of the Revolution. He agreed that its goals—the social and economic, political and moral betterment of the people—were far from completed, but he distrusted wide-ranging ideological formulas. He was a harsh and violent reformer but not a revolutionary in the Marxist sense. For him, politics was still more a matter of character than ideas; and where ideas counted for him, they were not ideologies. The new men knocking on the doors of power thought otherwise. They were the new priests of socialism.[80]

Lázaro Cárdenas would make use of this new politics, this new ideology, to displace his old chief. He did so because he partly shared these beliefs, but also because they provided him with a social power base—of workers and peasants—and a reinvigorated revolutionary legitimacy that Calles could now only combat with the weapon of criticism.

When Calles chose Lázaro Cárdenas for the presidency, he was surely aware that this was no Abelardo Rodríguez, that he was selecting a man who might not give him unconditional support. But he also knew that Cárdenas was the "most solid" of the Revolutionary generals. Despite his youth—in 1933 he was barely thirty-seven—Cárdenas had a long military career to his credit, fighting against Zapata, Villa, Peláez, the Yaquis, Carranza, and De la Huerta. And also a respectable political career as governor of Michoacán, President of the PNR, Minister of the Interior, and Minister of War and the Navy. The "Old Man" knew that Cárdenas had shown some independence during the government of Ortiz Rubio, but Calles made his choice—for a dynamic succession. He preferred a measure of risk and possibility over candidates who would only represent submission. Besides, the Old Man, even though he was not so old, felt as if he were. He had lived on the crest of the wave for more than twenty years, and in a state of perpetual tension since infancy. The passing of time and the taste for power did not rejuvenate him, as it had rejuvenated Porfirio Díaz. He was tired.[81]

Cárdenas took the oath of office on December 1, 1934. Calles flew to Los Angeles for an operation and returned to convalesce at his El Tambor Hacienda on the Pacific Coast. Political pilgrims arrived with news from the capital. He was told of renewed religious confrontation, which pleased him, but other news did not: He disapproved of the wave of strikes and the declarations by some members of Cárdenas's cabinet that Mexico was on its way to the dictatorship of the proletariat. Cárdenas was treating him with reserved respect, asking him for advice on financial matters and maintaining that he was needed in the capital. But for Calles the President's public pronouncements were just as alarming as the labor agitation: "We must fight capitalism," said Cárdenas, "the liberal capitalist school that ignores the human dignity of the workers."

Finally, in May 1935, the Old Man returned to Mexico City. He made a statement of open support for Cárdenas: "I am totally optimistic with respect to the general situation of the country, and have complete faith that the government of the Republic will duly solve any problems that may arise." A few days later, in a much publicized interview, Calles attacked the labor unions: "The strikes hurt capital less than they do the government, because they are closing the doors to prosperity." The interview triggered a crisis. Cárdenas asked for the resignation of his cabinet—until then at least partly composed of Calles supporters—and made a statement about political agitation by resentful groups seeking to promote themselves and sow division.

Cárdenas said he felt that the wave of strikes was the natural result of a legitimate struggle of interests and that "reasonably solved, in a spirit of equity and social justice," they would contribute to a strengthening of the national economy. In reality, he was in favor of them. The final attack of the President on the slipping *Jefe Màximo* was not long in coming: "I think I am justified in asking for the Nation's full confidence and for the revolutionary group to adopt the necessary serenity in order to continue collaborating with the Executive branch in the difficult task it has assumed. . . . I exhort all men of the Revolution to meditate deeply and sincerely on their duty."

This indirect statement marked his break with Calles. On June 16 the Old Man complained that his words, delivered with "the best of intentions, for the good of the country and the government, have been given a twisted interpretation." And then he announced: "I retire forever from politics." In the Zócalo of the capital, huge contingents of workers roared their support for Cárdenas. The Legislature witnessed a dizzying change of allegiances: the Callistas of yesterday almost instantly became the Cardenistas of today.

There was still one episode to come in this bloodless drama. At the end of 1935, Calles returned to Mexico City. He had come back to publicly defend his record, now being subjected to a campaign of vilification. He called his return an act of dignity. For four months there was tension—and some violence—between partisans of Calles and Cárdenas. And then Cárdenas made a historic decision. He would order his predecessor and rival into exile, but he would not kill him.

On April 9, 1936, Cárdenas shipped the former *Jefe Máximo* out of the country. A few days before, the journalist José C. Valadés had interviewed Calles, in bed at his Santa Barbara Hacienda. Calles declared himself a sworn enemy of communism, criticized the Spanish Republic, and gave his interpretation of Marx: "For Marx, the individual does not exist, and therefore, freedom does not exist. . . . Marx makes the individual a cog in a great machine he calls the State. The State rules, the State commands, the State dominates; for the State, man is nothing."[82]

Unfortunately Calles could not claim to have been a defender of freedoms. He had been a great reformer but not a liberator. And he had recently been reading *Mein Kampf* with interest and respect. In his opposition to the rise of Marxism, he had begun to slide slowly toward the opposite precipice of Fascist sympathies.

Calles spent five years in San Diego, California, finally in true retirement. During the Second World War, in March 1941, President Ávila Camacho thought that the time had come for national unity and invited Calles to return to Mexico. Days later Calles appeared on the balcony of the National Palace, flanking Ávila Camacho, along with all the other former Presidents including Cárdenas.

He spent his last four years, like Voltaire's Candide, cultivating the garden of his Palomas Hacienda in Cuernavaca. There he became the successful farmer that he never was in his youth, planting flowers and fruit trees. He played golf, often alone. He continued to speak with the tones and inflections of a teacher. What did he think about during his long and melancholy morning walks? It is difficult to know, since he did not confide his final opinions to anyone. About his nights, on the other hand, Calles's archives—lovingly maintained and ordered by his daughter Hortensia—conserve some surprising information.[83]

From the middle of 1941 until his death in October 1945, Calles attended, once a week, the Mexican Circle of Metapsychic Investigations. The Spiritualist sessions were regularly frequented by political personalities such as the former labor leader Luis Morones and the future President Miguel Alemán. On December 26, 1943, the "spirit guide" of the circle communicated a message through one of the living:

> General Plutarco Elías Calles was and continues to be a patriot. He has never been as well prepared as he is now. He understands better than ever the problems of the country and those of humanity. Day by day, the hour draws near in which our poor and unfortunate fatherland will turn to his experience and wisdom. No one could help the fatherland better than this man of strong character, perfected by the years, without egoism or vanity.[84]

Two months before he died—on October 19, 1945—he said that he believed "in a Supreme Being." His nineteenth-century rationalism—put between parentheses when he sought the medical help of a *curandero* during the Cristiada—withered away completely now at the end. He wanted to reform his own personal history and that of his country from the ground up, eliminating impure and disorderly elements, imposing "order, the inviolate, what should and should not be done." He had wanted to banish faith and enthrone reason. At the end—in the darkness of the human condition—he found himself groping for some kind of faith, reaching for hands around a Spiritualist table, trying to interrogate the dead.[85]

16

LÁZARO CÁRDENAS

The Missionary General

The good king does not exclude the poor or the forsaken from his palace, he listens to the complaints of everyone, he does not dominate his subjects as if they were slaves, he guides them as if they were children.

—FATHER JUAN DE MARIANA, SEVENTEENTH CENTURY

All the earlier caudillos of the Revolution had made a choice, as adults, to join the violent currents of their time. The man who was destined to culminate and complete the revolutionary cycle entered the whirlwind when he was only an adolescent and almost against his will. "I think that I was born for something. . . . I always live with the sure sense that I will win fame. But how? I don't know," he had written in 1912, among the first notes of a diary he would continue throughout his life. In a dream, he saw himself "on a stormy night in the mountains, at the head of a large body of soldiers . . . freeing the *Patria* from the burden oppressing it." But in his day-to-day life, he was already, at only sixteen, the head of a large family.[1]

His father, Dámaso Cárdenas—a much beloved storekeeper and pharmacist in the town of Jiquilpan at the heart of the state of Michoacán[2]—died that same year, leaving his mother, Doña Felícitas, with eight children. As the eldest son, the young Lázaro had to leave school and go to work at the La Económica printshop, where the Revolution came to him and Jiquilpan in June 1913. The town was occupied by anti-Huerta forces who had their manifesto printed at La Economica. A company of *rurales* reoccupied the town and broke up the printshop, stealing the press and burning its files.[3] His mother feared for Lázaro's life, and he agreed to leave town, on foot, for the "hot country" of Apatzingán, where his mother's brother was the manager of a hacienda. (A hundred years earlier, in that same impoverished region, the priest Morelos had issued the first Mexican Constitution.) It would be no small surprise for Doña Felícitas when, not long after, she found that her beloved Lázaro had taken refuge not only with his uncle but with the Revolution.[4]

General García Aragón, a former ally of Emiliano Zapata, was operating in the area. One day Lázaro displayed his handsome, left-handed penmanship to the general and recounted his experience as a printer, office worker, and clerk. It earned the boy a commission as a first lieutenant (*capitán segundo*) in charge of correspondence.[5] Two months later he had his first taste of war and had to escape from Huertista forces riding double on another man's horse. More war followed but also his first contact with what he would later call "the agrarian meaning of the armed struggle" and with the Indian communities of the Meseta Tarasca, who were fighting against foreign companies that were cutting down their forests.[6]

A year after the Revolution (and its enemies) had knocked on his door at the printshop, the dreams Lázaro had set down in his diary were dimly beginning to take on form. But out of devotion to his mother, a sense of responsibility as head of the family, and because he lacked the temperament of a warrior —though never in the least a coward—Lázaro tried again, as 1914 opened, to keep himself out of the Revolution. He spent time as a civilian in Jiquilpan and then in Guadalajara (some of it in hiding because the Huertistas still had his name on their local blacklist). But the call was too strong for him. On June 23 he served as liaison between two revolutionary chieftains, one of whom—Zúñiga—he had known in Apatzingán. Decisive battles were about to be fought in the region of Jalisco between the Constitutionalist Army and the federal troops of Huerta. Now nineteen years old, Cárdenas joined the Revolution for good. His mother—when she saw Zúñiga in a violent rage that almost led him to shoot down a priest—begged Lázaro, "with tears in her eyes . . . 'Don't you do that!'" It did not alter his decision, but it was a message he would not forget, a plea against the bloody excesses of war.[7]

When Álvaro Obregón signed the Treaty of Teoloyucan, which would send Huerta into exile, Captain Cárdenas was one of the anonymous witnesses. In the civil war between Carranza and the Convention of Aguascalientes, the young officer was loyal to his leaders, first on the side of the Convention (where he felt uneasy) and then as part of the Carrancista forces. He fought against Zapatistas and Villistas, and on March 28, 1915—as a lieutenant colonel not yet twenty years old—he first shook the hand of General Plutarco Elías Calles. Chance had brought them together, but currents of mutual sympathy flowed between them immediately. The rigorous and lucid former teacher from Guaymas was always in search of disciples.[8] The young man in command of the Michoacán-Jalisco Twenty-second Regiment of Cavalry had seen his father die when he was sixteen, and already, just three years later, death had taken all the men who had brought him into the Revolution. He was a soldier ready for a new father. A week later—after he had led his three hundred men in a successful attack—Cárdenas noted in his diary, with satisfaction and admiration: "On the road the general made me a present of his black horse . . . and nobly freed the prisoners."[9]

In mid-September, the young lieutenant colonel was ambushed at Santa Bárbara. He held out against eight hundred infantry for more than three days without rest or resupply. Cruz Gálvez died in this battle, the officer after whom Calles, when governor of Sonora, would name his schools for orphans of the Revolution.[10]

As a reward for this display of valor, Calles promoted Cárdenas to colonel and acclaimed him as a "brave leader." In private he called him *chamaco* ("the kid"). At the battle of Agua Prieta, Cárdenas commanded the first line of resistance against Villa himself and played a vital part in General Calles's victorious defense.[11] At the end of that year, with the enemy defeated in Sonora, Cárdenas could leaf with filial pride through a *Revista Ilustrada*, whose principal story narrated "the admirable defense of Agua Prieta." Three photographs illustrated the story: Álvaro Obregón, Plutarco Elías Calles, and Colonel Lázaro Cárdenas.

In March 1918 the young colonel left for Michoacán. He had asked for permission to fight the bandits who were devastating his native state. On the way he stopped in Guadalajara, where Doña Felícitas was on her deathbed. "She had breath enough to wait for my arrival," he would write in his diary.[12] The loss of his mother put fury into his fighting against the three plagues of Michoacán: Jesús Cíntora, who operated in the Balsas region and Tierra Caliente; José Altamirano, marauding through the cities of the center, and, the bloodiest of all, Inés

Chávez García, who used a knife to kill prisoners and took pleasure in watching his thugs rape women. Unfortunately for Cárdenas, his anger and courage did not translate into victories. He was showing himself to be a truly valiant but reckless commander, capable of taking on too much and putting his men and himself into danger. Finding himself trapped, with one bullet left in his pistol, by the bandit Altamirano, the young officer was about to shoot himself rather than be captured.[13] One of his officers stopped him, and at that moment a bandit tried to grab him by the neck. Cárdenas shot him in the head with that last bullet, and then the men around him managed to fight off the bandits, who rode away to celebrate their victory. But on that same day, the exhausted Cárdenas and his officers reformed and resupplied their company and attacked Altamirano at his ranch, in the middle of a drunken party. Though the attack saved the lives of some prisoners, the bandit chieftain still escaped. And Cárdenas did not learn caution. Nevertheless, by the end of 1918, both Altamirano and Chávez had been hunted down and killed.

In 1920 he followed his former Sonoran commanders in their revolt against Carranza. In mid-May, he learned that President Carranza's battered column had entered the area of his authority. On May 20, the day before he turned twenty-five, Brigadier General Lázaro Cárdenas marched toward the mountains of Puebla to intercept Carranza. But the flooding El Espinal River blocked his way. While Cárdenas was waiting for the waters to recede, Carranza was betrayed and met his death at Tlaxcalantongo. "The kid" was lucky, as he had been before in dangerous situations he had himself created through youthful recklessness, and as he would be many times in the future. He had been freed of direct involvement with the death of Carranza. He never had to confront (or discover how he would have faced) the venerable President.

In 1922 Cárdenas was still constantly on the move. Now he returned to Michoacán, where the governor was Francisco J. Múgica, the vociferous and radical anticlerical of the 1917 Constitutional Convention. An old friend of Cárdenas, Múgica (and his family) had been helpful to Doña Felícitas after the death of Don Dámaso. But now Múgica was in direct conflict with the central government. His radical measures—prominent among them the beginning of land redistribution, a fierce anticlericalism, and a very advanced labor law—had almost led to civil war. De la Huerta, the Minister of Finance, suggested "the resurrection of Lázaro"—that is, making Cárdenas governor—as the only way to solve the problem, but President Obregón had other ideas. In 1923 Cárdenas received orders to take Múgica into custody and bring him to

Mexico City. While on the road he was surprised by a telegram from Obregón: "I acknowledge your report that General Francisco J. Múgica was killed while trying to free himself from his guards."[14] The order was clear, but Cárdenas could not accept it. He never acknowledged receiving the telegram, and he allowed Múgica to escape.

At the end of 1923, the rebellion of De la Huerta erupted against Obregón's choice of Calles to succeed him in the Presidential Chair. Obregón sent Cárdenas to harry the rear guard of one of the most brilliant generals of the Revolution: Rafael Buelna, "the Gold Nugget" (*el Granito de Oro*), who was fighting under the command of General Enrique Estrada. On December 12, Cárdenas advanced with his two thousand horsemen on the trail of Buelna, who, with his greater experience and military skill, set up a trap. Obregón's instructions had been clear: Harass, do not attack. But at Huejotitlán, Buelna maneuvered Cárdenas into a position where he was exposed to assault by a rested infantry column. Cárdenas and his men fought bravely for every inch of terrain, but the infantry wore them down and Cárdenas took a serious wound in the stomach. He lost a great deal of blood during eight hours of fighting without any medical attention. The battle lost, he was prepared to die and sent a message to Buelna, taking full responsibility for the encounter, offering him his life, and asking only that his men be spared.

But Buelna, on his own initiative, ordered Cárdenas to be carefully carried from the top of a hill and brought to his headquarters. (A telegram later arrived from his commander, Estrada, with similar instructions.) He was taken to Guadalajara and put in a hospital. Buelna and Estrada knew of Cárdenas's honorable nature, his aversion to bloody extremes, his unstained record, his youth. Everybody liked him. Though the defeat infuriated Obregón, General Calles sent "an affectionate greeting full of satisfaction at the knowledge that you are well." During his convalescence, he learned that the devout women of Jiquilpan, thinking he was dead, had "organized public prayers for his resuscitation."[15]

Soon afterward, fate exchanged the roles and in Colima Cárdenas found himself with the power of life and death over Estrada. Of course he did not kill him, and he paid his debt by clearing the way to exile. He would have done the same for the unlucky Buelna (who would die in battle in 1924), had he fallen into his hands. Did Obregón know of Cárdenas's merciful behavior, so different from his own? We do know that he considered Cárdenas "correct but incompetent."[16]

On March 1, 1925, knowing that the impending Oil Law could lead to unpredictable reactions from the American oil companies, President Calles appointed his faithful "kid" as Chief of Military Operations in the Huastecan region. He would hold that post for three years.

Soon after he arrived, he was given news that raised his spirits: His friend General Múgica, who had taken a leave of absence from the army, was coming to work a small oil concession in the area. He would arrive in Tuxpan in the middle of 1926.

For Cárdenas, Calles had been a military and political maestro. He respected the older man's strength and the clarity of his goals but especially his radical reformism as governor of Sonora. He had seen Calles ponder and implement an avalanche of decrees. But busy with the rhythms of war, Cárdenas had lacked an ideological teacher. He found one in Múgica.

Eleven years older than Cárdenas, Múgica was a small, agile man, restless and physically strong. In his studies as a day student at the Diocesan Seminary in Zamora, he had reached the level of theology, but "justifiable causes" obliged the rector, Leonardo Castellanos, to expel him. After having assimilated Christian socialism into the very marrow of his bones, Múgica had at some point decided to move from Christian belief to secular socialism. The Revolution was a godsend for him. He had been one of the signers of the *Plan de Guadalupe* calling for revolt against the usurper Huerta; he had participated in the first act of land distribution (carried out by Lucio Blanco on the Los Borregos Hacienda owned by Félix Díaz); he was part of Carranza's government in Veracruz; he experimented with radical measures in Tabasco, and—his finest hour—he was, along with Andrés Molina Enríquez, the ideological soul of the radical articles in the 1917 Constitution. As an anticlerical radical, he surpassed the Sonorans (not an easy achievement) and made himself troublesome enough for Obregón to have wanted him dead.[17]

Now Múgica had the opportunity to repay Cárdenas for saving his life. He did it by imparting a persuasive lesson: "socialism as the appropriate doctrine for resolving the conflicts of Mexico." Around 1926 and thanks to Múgica, Cárdenas read some of Karl Marx.[18] And his chief of staff, Manuel Ávila Camacho, gave him books on the French Revolution. But no book could compare with the privilege of having one of the primary ideologues of the Mexican Revolution at hand to advise him. They traveled widely together and became even closer, mutually respectful friends. With Cárdenas, Múgica did not share his passion for Baudelaire nor probably the more literary pages of his diary, precise and beautifully written notes on the natural life of the Huasteca.[19] He did admire what he called Cárdenas's "amorous anarchy," and they were both drawn to similar women, "slender, light-skinned, gracefully moving." And whenever he could, Múgica would preach to his pupil on the ravages of religion.

Yet Cárdenas learned his greatest lessons during that time from direct observation of the behavior of the oil companies. With the passage of the years, he would vividly remember one scene: "When we met General Múgica in the oil fields of Cerro Azul and Potrero del Llano, we were detained at the gates of the companies that had closed off the roads and it was only after waiting an hour that their guards arrived to open up the way for us. And this happened to the Military Zone Commander himself!"[20]

The companies would boast of having "powerful friends," and they felt themselves to be on "conquered territory." They would cheat the Mexican tax system by using underground installations connected to the port. They had left nothing good in the places they exploited: not one school, not a theater, not a hospital. Only wasteland. A few days after Cárdenas arrived in the zone, they had tried to bribe him with an offer of fifty thousand dollars and a luxurious Packard driven right up to his door.[21] Cárdenas and Múgica discussed all this and more during their tours of the area. And Cárdenas began to toy with the idea of expelling the oil companies from Mexican soil and abolishing the existence of that "state within the State."

None of the major battles of the Revolution had been fought in Michoacán. But during the eighteenth and nineteenth centuries, the state had been a center first for the great battle between liberal and conservative ideas for the conscience of the Mexican elite and then a bloody ground for the Wars of Independence and the Reform. "Two-faced Morelia, heroic in your masses, reactionary in your elite," wrote Múgica in 1927, somewhat inexactly in his qualifying phrases but pointing, with truth, to the intense duality evident not only in the city of Morelia but in the history of Michoacán—Ocampo and Munguía, Mora and Alamán, Hidalgo and Iturbide. In the 1920s the old antagonists within the state had taken on new faces. The most important confrontation was between Jacobins, now anticlerical socialists, and, on the other side, Catholic groups concerned with social questions—as a result of Pope Leo XIII's encyclical *Rerum Novarum* (issued in 1891). They were zealous opponents, heads and tails of the same coin, arguing questions of this world with theological hatred.

Rerum Novarum—the Church's attempt to answer and redirect the rising tide of socialist ideology—had called for fair salaries, workers' savings banks, mutual societies, and the division of agrarian properties. In 1913, in the small and pious city of Zamora, a national conference was held of the Confederation of Workers' Circles of Mexico. The Catholic organization had been founded the previous year. Speeches calling on workers to claim their rights alternated with attacks on liberal individualism and socialism. "The excessive abundance of democracy has made the Mexican people drunk," a prominent priest asserted, but its harmful effects could not be compared with those of the "monster that is driving its claws into the heart of the *Patria* . . . the terrible scourge . . . the cancer: socialism . . . [represented by] teachers with pretensions of redeeming humanity."[22]

The other side, that of the "monster," was no less active, evangelical, and intolerant. During his brief term as governor of the state, supported by a small army of young radicals but with no social base for support, Múgica had attempted to distribute land and enact an advanced labor law that regulated everything from fieldwork to housekeeping. In December 1922, a leader associated with the Communist Party, Primo Tapia, founded the League of Communities and Agrarian Unions of Michoacán. Its goal was radical agrarianism.[23] Tapia would be murdered in April of 1926, but radical agitation—in both countryside and city, agrarian and union—went on until it encountered and was transcended by a much more serious and generalized upheaval: the Cristiada. In Michoacán, the quarrel between the "Two Majesties," Church and State, was not only a question of ideas but of fierce social bases that supported them. The Church spoke of its "flock," the State of its "masses."[24]

Onto the field of this old and intense ideological war stepped Major General Lázaro Cárdenas, barely thirty-two years old in 1928, to begin his campaign as the only candidate for governor of Michoacán. Cárdenas the candidate—and thereby already governor elect—had the feeling of a loyal son toward the ideals of the Revolution and his paternal mentors from the first Revolutionary generation: Calles and Múgica. But he was also a young father of the Revolution; he had grown up in it and was now responsible for fulfilling its ideals.

And in the future governor of Michoacán, this sense of duty came with a feeling of the legitimate and almost unlimited right to command. He had been an officer in charge of men since he was sixteen. He had fought against Huertistas, Carrancistas, Zapatistas, Villistas, Yaquis, bandits, a whole rainbow of rebels. But this familiarity with blood would not translate into either a blind will to kill or a taste for brutal impositions. Cárdenas might well agree that the religious "fanaticization" of the Mexican people was a shame, but he was no Múgica living obsessed with the ghost of his own Catholic past. He had a different style—compounded of the likeable good-fellowship of his father, the gentleness of his mother, the Indian patience of his Aunt Ángela. And his vision of social problems came to be in some degree different from that of his mentor, Múgica: less profound but more serene, more balanced, with a wider range. There was no will in him to persecute priests. He was a firm and martial reformer like Calles, a convinced idealist like Múgica, a relentless manipulator of the masses—all of it framed within a temperament marked by genuine humanity and communicated almost sweetly. In short, he was the perfect politician.

As governor, Cárdenas had learned from Múgica's errors (and made his mentor proud of him and perhaps nostalgic for his own unsuccessful term of office). Laws and words and arguments alone would not serve. From his first days in power, in 1929, the young governor began to create political shock troops. Young teachers (a group here as elsewhere central to radical Mexico) and various members of the Communist Party and of Primo Tapia's disbanded Agrarian League would help in the political and ideological structuring of a new organization. Its slogan would be "Union, Land, Work." Governor Cárdenas was its first, honorary president. The Michoacán Revolutionary Confederation of Labor, which would soon become the powerful CRMDT, was born.

The organization developed an agrarian and labor program that was slightly more progressive but very close to the propositions expressed in Articles 27 and 123 of the Constitution and in the Labor Law of Michoacán approved during Múgica's term as governor. The CRMDT favored a broad solution to the problem of the land, medical care and schooling on the haciendas, and such steps as an eight-hour day and a minimum salary. But the end of the program statement pointed beyond the Constitution of 1917: "Only a transformation of the existing capitalist system will provide the worker his emancipation from the condition of pariah."[25]

In the cities, however, the CRMDT membership consisted more of small-business employees and odd-job and white-collar people than industrial workers. There were lottery-ticket sellers, chauffeurs, waiters, shoeshine and errand boys. The teachers, grouped into the State Bloc of Socialist Teachers of Michoacán, took an important leadership role. Women and young people were also represented by their respective blocs, but the real core of the CRMDT, the true political shock troops, were the "agrarianists" (*agraristas*), many of them organized in bands with access to weapons.[26] Four years after it was founded, the organization had four thousand agrarian committees and one hundred thousand members. It was the first mass organization in Mexican history *brought into existence* by a government and vertically linked to it. The form would have an important future (and also was an echo of the viceregal past, of the corporativist organization of society under the Crown).

The state government financed the CRMDT with funds that were never listed in official records. Union halls were provided, and the state of Michoacán would "regularly furnish transportation in a train of up to 14 cars for the movement of all state delegations." Hotels were unnecessary. Union representatives were put up in the town halls.

Other organizations were in effect forbidden to represent workers and peasants. The CRMDT was "the only institution that responded to the yearnings of the Michoacán workers." Its members came to occupy 95 percent of popularly elected offices, from municipal presidents to minor judges. And in 1931 the state government banned the formation of unions within the growing number of communal land communities (the *ejidos*), leaving the field completely open to the CRMDT. The one legitimized union then dedicated itself to the "organization and ideological transformation of the peasants to encourage them to request lands . . . supporting the legislative measures of General Cárdenas."

The governor began to fashion a messianic image for himself, but one that was in tune with his character. The CRMDT would agitate, manipulate, stir things up. And suddenly Cárdenas would appear with the soothing balm of his presence. Or the towns and villages would express their needs and Cárdenas would come and hear them. He went everywhere. And he would never dress up as a peasant but always wore a dark suit. A serious man, a man to be respected. And a man who would always listen, he himself, in person.

His paternal and compassionate presentation of power was in the mold of a benevolent priest, or a missionary father—one of the friars of the spiritual conquest who introduced not only Christ but European fruits and vegetables and crafts, a master of both spiritual salvation and the material well-being of a community. But the governor's paternalism also had another side—a refusal to accept criticism, a swollen sense of pride. It was the same inclination for the bold and stubborn leap forward that had led him into ambushes and the shadow of death as a young officer. He intervened in everything and would often treat local legislators like common soldiers, instruments for carrying out his orders. And it was very hard for him to acknowledge mistakes, "though with time he could come to accept them."[27]

Electoral freedom or the division of powers at any level were not important to him. (During his governorship, the same CRMDT people would be steadily reelected.) What truly mattered was the revolutionary and protective mission of the State: "It is not possible for the State, as an organization of public services, to remain inert and cold. . . . It must assume a dynamic and conscious attitude. . . ."[28]

The 1917 Constitution had provided for such a State on the national level, but Cárdenas—in Michoacán—was practicing something different from the authoritarian though economically laissez-faire state foreseen by

Carranza or the moralizing and military visions respectively of the Sonorans Calles and Obregón. In its political logic, though not in its social aims, it was closer to the "integral paternalism" of Porfirio Díaz, but the radical shades and structures Cárdenas gave to it were something new. Perhaps these innovations were only imaginable in a region with religious roots and tensions as profound as those of Michoacán. It lay in the heart of "Old Mexico," where the Church had made its presence most vividly felt, where for centuries (during and after the viceroyalty) it had been trying to shape its flock into an integrated totality of secular and spiritual, local and regional, country and urban existence. The new political structure that Cárdenas was creating within the State would compete with that other Majesty but would also, even if unconsciously, imitate it: the State as a Counter-Church.

The Tarascan Indians, intact as a people since the days of Vasco de Quiroga—the benevolent priest of the spiritual conquest who had taught them the practical means for survival—recognized the priest in Cárdenas. As they had called Quiroga Tata Vasco ("Dear Father Vasco"), so they would christen the young governor Tata Lázaro.

Besides the governor himself (the executive power) and the united, single front of workers (the CRMDT, his political shock troops) a third, "sacerdotal" limb—the teachers—completed the structure of Cárdenas's first-created government. Just as the Church gave great importance to its schools and seminaries, prayers and homilies, so the State was to vigorously pursue a social education that would permit "children to become true human beings, men of enterprise and of action."

The Michoacán government devoted almost half its modest budget to promoting education and—through issuing a regulatory law—soon made dozens of businesses and haciendas open up their own schools. It was necessary to "socialize the schools," to imbue children and adults with feelings of brotherhood and solidarity, to put aside—Cárdenas said—"useless and overly refined knowledge transmitted dogmatically and cruelly." The teachers were to become agents of change, carriers of "the new revolutionary ideology."[29]

To "modify the spiritual attitude of individuals, to displace fanaticism once and for all," Cárdenas concentrated his educational drive on the old Cristero zone: Coalcomán, Apatzingán, the "hot country." The job of the young teachers—modeled after the "traveling players" (*saltimbanquis*) of Vasconcelos's educational campaign in the early twenties—was not to distribute copies of *The Iliad*; their principal assignment was to "defanaticize" and "de-alcoholize." They would often use methods recalling those of the missionary fathers, relying on aesthetic appeals, especially theater—little plays with a message. And their instruction would include soap-making, how to preserve fruits, the encouragement of athletics.[30]

Ideological training was central. Besides being teachers, they were to be agents of revolutionary change and were expected to be familiar with cooperative and union issues. "We gave a course in advanced civics," one of them remembered. "That way we began to organize, or rather advise the peons on how they should organize themselves and ask for land and they went around forming *ejidos*." The teaching centers were "foci for ideological fermentation" that distributed socialist propaganda, in large editions financed by the state government. And when the teachers went into haciendas, they often carried guns. (The hacendados and their armed guards might very well not be ready to listen quietly.)

The war against illiteracy and the spread of technical education were much more successful than the campaigns to defanaticize and dealcoholize. The village of Zurumútaro—where the teacher joined the community in burning the statues of their saints—was a rare exception. Much more common were reports like that of the *maestra* discouraged by the poor results of the campaign against drunkenness in her area: "The people were very fond of alcohol, and most of them were sexually loose and had a great number of adulterous affairs. You were always seeing them drunk and with somebody else's woman."[31]

There was also the problem of teachers not sufficiently revolutionary, neutral toward religion or unconcerned with communicating revolutionary ideology. The CRMDT decided on the need for ideological purification in the teacher-training schools. A "purifying commission" was created to exclude trainees lacking "an advanced ideology." In the last *informe* of his term as governor, Cárdenas (politely) excommunicated such lukewarm participants in the educational wars: "Within society, in contrast to this type of teacher who has not gained the influence or received the consideration that his ministry deserves, the social guide must appear who advances, with courage, into the struggle."[32]

The first months of Cárdenas's government had coincided with the final months of the War of the Cristiada. Even during the military confrontation, Cárdenas had tried to persuade and pressure the Cristeros, to offer them amnesty rather than hang them. In December of 1928 he had secured the peaceful surrender of the Cristero leader Simón Cortés, and in the town of Aguililla he had persuaded Father Ríos to go up in an airplane and superintend the surrender of rebel troops.

For Cárdenas, anticlericalism was important only as a means "to emancipate the workers and their families from the claws of confessional fanaticism" so that they would be free to participate in and benefit from his agrarian and social programs, "with full spiritual liberty." But he was not about to be pushed around by the Church either and—in 1932, late in his governorship—responded to clerical agitation against his agrarian programs by amending the state constitution. A maximum of three ministers "of any religion" were to be allowed within a single district (Michoacán then had eleven districts.)[33]

His personal attitude remained tolerant. If a priest carried "divisive" activities too far, the governor would simply go to his friend, the Bishop of Morelia, Luis María Martínez, and have the man transferred. And there is a description of Cárdenas arriving—by chance—at the town of Tzinapan while almost the entire population, including the municipal president, were attending an elaborate mass performed by a priest who—according to the constitutional amendment—had no right to officiate in that town. The whole population went into a state of high anxiety "at the presence of a man who was considered an 'anti-Christ' by the fanatics."[34] Cárdenas summoned the priest ("the people surely expected that [the governor] would skin him alive") and urged the man to request the proper permission in the future, offered to expedite the process for him, but also asked that he, as an educated man, respect and value the laws and teach that respect to his flock. At the sight of the priest coming out of the meeting all smiles, the townspeople cheered their governor.

Unfortunately, that governor could not be everywhere. There were many among Cárdenas's political shock troops for whom "defanaticization" mattered more than anything. And at times blood ran in the streets. At Cherán, for instance, an Indian municipality, *agraristas* of the CRMDT poured into town for a conclave, choosing the deliberately provocative time of Holy Week, and hung the central square with their red and black banners marked with the hammer and sickle. At one point they insisted that the village leader—brought almost by force from the office of the government political party, the PNR—swear a solemn oath to their flag on the podium they had erected in the square. When he refused ("my only flag is the tricolor") someone put a pistol to his chest. Suddenly, at the outskirts of the crowd, a voice cried out, "Don't swear it or you're a dead man," and the Catholics began to fire from all sides. The clash between defanaticizing fanatics and Catholic fanatics resulted in more than thirty deaths and many wounded.[35]

One of the radical teachers of the CRMDT—Salvador Sotelo, who taught in the village of Ario—used to ring the church bells to summon the people to "socialist sacraments": "Receive the honey that the laboring bee, symbol of the worker, extracts from the nectar of the flowers so that your life may be pleasant."[36]

Ten years later he said he felt betrayed and abandoned. The CRMDT was gone—dissolved by order of Cárdenas. The old priest had returned to Ario. The church bells only rang for mass.

All the governments of the Revolution between 1917 and 1928 had distributed 131,283 hectares to 124 pueblos in Michoacán. During only four years as governor (September 1928 to September 1932), Cárdenas shared out 141,663 hectares among 181 villages. During his term he issued laws on the use of unexploited lands "to alleviate the pressure of requests" and authorizing land expropriations for the public good.[37]

While Calles, the *Jefe Máximo,* was saying in Mexico City that the *ejido* had failed, Cárdenas was insisting: "There has been no failure with the *ejido*; what is wrong is that the peasants need better means to cultivate the land. . . . The *ejido* . . . will be the basis for the prosperity of the country."[38] The *ejido*—as Cárdenas would conceive of it—had two forms. Essentially based on pre-Hispanic ideas of collective land ownership, the *ejido* was an agricultural unit that was the property of the state. In its most common form (and the only one currently functioning, though under altered conditions) the individual members of the *ejido* (*ejiditarios*) held a provisional title to their individual portions of land, which they worked as family farms though they were not, ultimately, their private property. (Within each such *ejido*, there would be some areas and services shared in common—like water sources or land for grazing cattle.) Cárdenas would also introduce—under the influence of Lombardo Toledano—"collective" *ejidos,* similar to the Russian *kolkhoz*, where the land was to be farmed communally.

An *ejido* would be governed by a committee (*comisariado*) elected by the general assembly of all *ejiditarios.* Under no circumstances (according to rule) could *ejido* lands be sold, legally seized, or automatically inherited. Only the government—for violations of *ejido* regulations—could cancel an *ejiditario*'s right to his land. This was the tool that Governor (and later President) Cárdenas would apply profusely in his attempt to solve Mexico's ancient problem of the land.

The first to oppose the agrarian policies of the governor were, of course, the hacendados. Most of them had no good arguments, but they had ample means: shrewd lawyers, protective legal shelters for fertile land, paid hoodlums and tame peasant unions, territory spuriously listed as divided or else divided among family members as a preventive measure. When the Chamber of Commerce, Agriculture and Industry asked Cárdenas to stop the distributions in 1930, he replied—as always courteous but firm—that there was still much to be given away and he advised "the owners to make it easier for the government . . . and to accept the fact that there is no other solution to the agrarian problem in Michoacán or in the entire Republic."

The clergy was generally as hostile (or even more so) than the hacendados, primarily because of agrarianist anticlericalism. Father Trinidad Barragán of Sahuayo publicly prayed to God that "the earth swallow up the *agraristas*." But there were priests who had more rational, less totalizing objections to the programs. Father Federico González, for instance, of San José de Gracia, was not opposed to land distribution (since 1926 he had been distributing land—sold to peasants at moderate prices and long-term payment—on behalf of María Ramírez Arias, the owner of El Sabino Hacienda), but he had strong doubts about the worth of the *ejido*. He felt the institution would decrease productivity and that the members of an *ejido*—not really feeling themselves owners of the land—would not improve it and would be tempted to deforest and misuse it. Father González did not consider *agraristas* impious or evil but felt that economically and socially a system of small landownership would be preferable and far less socially divisive.[39]

In some areas, the distributions in their favor were opposed by the peons, shanty residents, and tenant farmers themselves. This must have been the hardest (and in his eyes strangest) opposition for Cárdenas to confront. In the town of Zacán, a witness recalled the day when "the government people" came to divide the land: "We hadn't asked for this thing of the *ejido*, nor did we know what it was. When the government people came, we thought they were looking for Cristeros again and we didn't believe anything they said."[40]

At the end of 1913, when Cárdenas was eighteen years old and on one of his first forays for the Revolution, he had been present at a discussion between his commanding officer, former Zapatista García Aragón, and Casimiro López Leco, the Indian cacique of the Cherán district. It was the time of his first real contacts with Indians, and on that day he learned about the grossly one-sided contracts that the mountain communities had signed with American entrepreneur Santiago Slade. Under pressure and threat from the Porfirian prefects, they had leased away their immense wealth in forests for ninety-nine years at an absurdly low price.[41] Twenty-five years later, when Cárdenas became President of the Republic, he would free that wealth from foreign hands and return it to its rightful owners.

Cárdenas would always have a special connection with the Indians. It was a genuine love. He himself felt that it was based on the tenderness he felt for his godmother, his Aunt Ángela. She was mute and her silence had dramatically accentuated her Indian qualities. As governor of Michoacán, he freely gave his time to hearing Indian leaders and ordinary people, counseling them and trying to settle their differences. Near the end of his term, he wrote to Múgica from the Indian town of Paracho, in the heart of the Meseta Tarasca: "Here I am, a Patriarchal Court of Law, presiding under an arch in the plaza. I am sorry that I cannot spend more time here. I would gladly stay a year." His political aide (fluent in Purépecha) Victoriano Anguiano—the son of a wealthy Indian cacique—was there with him in Paracho and wrote: "What impressed me most was the strict paternal commitment with which he rejected any attempts to kneel to him or kiss his hand, which . . . the village elders wanted to do in recognition of his authority. He would take them by the hand with cordial energy and raise them up so that they looked him in the eye."[42]

Around this time, the Cultural Center in Carapan had been attacked with stones and gunfire by Indians, who were afraid that it represented an assault on their religion and ancient customs. When Cárdenas arrived, the plaza was full of brilliantly dressed women—in spotless, glowing primary colors—and the "stoical and reserved" men who reminded Anguiano "of the splendor of Tariácuri in the Purépecha Empire." With Cárdenas beside him looking out at the crowd, Anguiano began to speak:

> The Governor ordered me to explain to them that they had been deceived by those who had said those missionaries of culture were going to deprive them of the Catholic religion . . . that the education to be given to adults and children was so that they might live better, in less unhealthy and impoverished conditions. I began my address in Spanish, but soon realized that it was like pouring water in the ocean. Then I began to explain things to them in our sweet and melodious Purépechan language and the effect was magical. Their faces were transformed by expressions of confidence, looks of comprehension and smiles of recognition for a fellow Indian. And of course they understood and accepted my explanations; and their reserve, their doubt, their distrust—with which people descended from pre-Columbian cultures always view the mestizos, the whites and their actions—changed to ingenuous joy and complete confidence. Afterward, the Indian dances began with their multicolored costumes and the melodies sounded, with their mournful gaiety. And in the midst of that fiesta of color, simplicity and the spontaneous surge of pure, great hearts, we ate together by the streams sheltered by the foliage of the trees.

But for those Indians, it was not only the sweet Purépechan language that dissolved "their reserve, their doubt, their distrust." The man beside Anguiano mattered to them, the sincere straightforward gaze of "Tata Lázaro," who looked them in the eyes.

Cárdenas's intensely active term as governor of Michoacán was interspersed with high national positions. It was an extraordinary apprenticeship, at a very young age, one that no other revolutionary general could approach. He learned to navigate through all waters before becoming the supreme pilot. The comments of Gonzalo N. Santos—cacique of San Luis Potosí, brilliant humorist and raconteur, smooth and effortless murderer—are perceptive: "Professional Cardenistas paint Cárdenas as a Saint Francis of Assisi, but that's the last thing he was. I have known no other politician who was better at hiding his intentions and feelings than General Cárdenas . . . he was a fox."

Cárdenas was President of the PNR from November 1930 to August 1931. At once he began to *act*—reorganizing the party newspaper (*El Nacional*), creating the National Sports Confederation, inaugurating an anti-alcoholism drive, traveling to Oaxaca in order to personally offer his consolation and assistance to the victims of a powerful earthquake. His goal for the PNR was to give it "stronger character as a popular organization." In the dance for power between Calles as the *Jefe Máximo* and President Ortiz Rubio, Cárdenas inclined toward the President but avoided quarreling with Calles. Partly because of this tilt toward Ortiz Rubio, he resigned the presidency of the party and almost immediately took over the vital administrative and security position of Minister of the Interior in the national cabinet. He would be there for less than two months (August 28–October 15), putting his greatest efforts into reconciling the *Jefe Máximo* with the President. It was a futile attempt. In the middle of October 1931, Cárdenas returned to Michoacán, keeping his impressions of the Maximato to himself.

For two months after the end of his term as governor, Cárdenas was in a brief political limbo. From the heights of a powerful governorship and top positions in the party and the cabinet, he had fallen to being the military commander of Puebla. "He had no money to rent a house," recalled Gonzalo N. Santos, "he was practically exiled from Michoacán and his closest partisans hounded and humiliated by General Calles. . . . But Cárdenas knew how to accept it like a good Tarascan, to pretend and to play 'the tame house cat.'" His only consolation during those days, certainly no small thing, would be the love of Amalia Solórzano, the beautiful young woman from Tacámbaro whom he would marry in September 1932.[43]

Lázaro Cárdenas *(left front)*, 1932
(FINAH)

Nevertheless, he had time to support the *ejido* distribution on the great Atencingo sugar hacienda. And on January 1 came his salvation: He was named Minister of War and the Navy in the cabinet of President Abelardo Rodríguez.

Then came the "Revelation" (*destape*). Against all odds (and against the advice of some of those closest to him) Calles chose Lázaro Cárdenas as the next candidate of the PNR, that is, as the next President of Mexico. He may have done it for several reasons, but one of them was surely personal affection. Pascual Ortiz Rubio had once heard the *Jefe Máximo* say, "I love Cárdenas like a son." Between the "Revelation" in June 1933 and his acceptance of the PNR candidacy in December, Cárdenas shared long days with Calles on the *Jefe Máximo*'s ranches. He seemed simply to be accepting the designation, acquiescing in Calles's decision. But there were small things that bothered the Old Man: Cárdenas did not join him in his pleasures, neither drinking nor partying. In the long run, would he follow him in ideas and actions?

Only the tour Madero made before the Revolution can be compared—in extent and thoroughness—to Cárdenas's "campaign," which was really of course a process of display and a presentation of ideas by the future President. He began on January 1 in Michoacán, where he explicitly declared that as President he would "do what I did on taking over as governor of Michoacán." He would create "a single workers' front." In Yucatán he warned those willing to listen closely enough: "The basic agrarian principles will soon be put into effect in this state. . . . The lands should be given to you so that you yourselves [the peasants] can continue cultivating sisal." In Veracruz he stressed, once again, his other key ideal: "The division among the workers of Veracruz is unfortunately well known. . . . To uniting them—and all the workers of the country—I will tenaciously dedicate myself."

In June he began to tour the northern states. In the north he said and wrote little. The land of the Sonoran generals, and of Carranza and Pancho Villa, felt alien to him. He noted only such things as the desperate conditions of the Tarahumara Indians, the needs of the migrant workers. When—in Coahuila—the Callista general Eulogio Ortiz showed him the beautiful hacienda "with which the Revolution had done him justice," Lázaro Cárdenas noted, from the heart of his collective, gregarious, anti-individualist being: "Our people present a mosaic of criteria. We will try to fuse them into one." The hacienda was in the cotton-growing area of La Laguna, to which, in 1936, Cárdenas would return with a purpose less pleasurable for General Eulogio Ortiz.[44]

In October he could breathe again. He was back in Jiquilpan. The distances he traveled—the figures laid out in railroad miles and planes and boats and horses—were enormous, but much more striking was the simplicity and sincerity of the message presented again and again: "Create a unified front of workers," and "Carry out the grants of land to which the pueblos have their right." His intentions were clear: to extend his experiment in Michoacán to the whole of Mexico.

Symbolically predicting changes to come, the newly inaugurated President Lázaro Cárdenas made several minor initial decisions. He installed a direct telegraph line so that the people could send their complaints directly to him. He opened the doors of the National Palace to all the visiting groups of peasants and Indians who wished to see him. He moved his official residence from the sumptuous Castle of Chapultepec to the more modest residence of Los Pinos;[45] and he made this remark to Luis L. León, director of the official newspaper, *El Nacional*: "Look, Luis, it would be helpful if from today on, every time *El Nacional* mentions the name of General Plutarco Elías Calles, they try to omit the title, '*Jefe Máximo* of the Revolution.'"

But if he wanted to bring about changes that were more than symbolic, it was absolutely necessary to ensure the loyalty of the army. Cárdenas decided to execute a very careful surgical operation, composed of many small moves. Allies of the Sonorans were promoted into high positions with no power, distanced from the centers of power, or eased into retirement. Officers with old histories of opposition to the Sonoran generals—former Villistas, Zapatistas, Carrancistas—were incorporated into a growing body of pro-Cárdenas commanders and filtered into sensitive military positions throughout the country. Well before Cárdenas moved against Calles, the military strings of the country were firmly in his hands.[46]

There were steps to gain greater control over the judiciary, the Legislature, and the press. Cárdenas was not a man at ease with the division of powers. He put through a law eliminating lifetime appointment for judges, limiting their terms to six years (the same length as the presidential term) and undercutting their independence. The Legislature suffered only one blow, but it was a bruising one: A group of Callista deputies and senators who opposed Cárdenas's labor policies were compelled to resign for "incitation to rebellion and seditious maneuvers." The press enjoyed great liberty throughout the Cárdenas period; but at the beginning of his government, Cárdenas put through some changes in the laws that could serve as weapons against it if necessary: notably a provision that allowed the authorities to prohibit the transport of publications "that denigrate the nation or the government" and the establishment of a state monopoly on newsprint.

But the fundamental weapon of change was public, strident, and continually dramatic: countrywide labor agitation. First there was the surging importance of the CGOCM (the General Confederation of Workers and Peasants of Mexico) led by Marxist intellectual Vicente Lombardo Toledano. Then came the establishment of an even more solidly based group—the Federation of Workers of the Federal District headed by the famous "Five Little Wolves": ex-milkmen Fidel Velázquez and Alfonso Sánchez Madariaga and three former chauffeurs: Fernando Amilpa, Jesús Yurén, and Luis Quintero. These organizations—along with the railroad workers, the oil workers, the miners, electricians, telephone, telegraph, and transportation workers—began or supported (and in some cases displayed their power by only threatening to join) an unprecedented wave of strikes. There were more than five hundred of them throughout the country between December 1934 and May 1935.[47] Cárdenas's Minister of Public Education, Ignacio García Téllez, went so far as to say that Mexico was moving "toward the dictatorship of the proletariat."

But the President himself had other goals. He wanted to kill three birds with one stone. Through his open approval of the strikes, he was shepherding the working masses toward the unification he had so clearly foreseen during his campaign. (In February 1936 the CTM—Confederación Mexicana de Trabajadores—would be founded.) And the agitation would also provide an excuse for changing the rules of the game between workers and owners, favoring the weaker party under the protective shelter of the State. Then, at the right time, the great labor mobilization would provide an impregnable base for the President when he was ready to rid himself of the *Jefe Máximo*.

It was clear to anyone who had observed Cárdenas's term in Michoacán that he understood the masses and their leaders and knew how to manipulate them. But the Sonorans and their courtiers had not paid close enough attention, and now they would look in the wrong directions until it was too late.

For the freedom to carry out his program and to progress toward the Mexico he wanted to bring into being, "the kid" had to eliminate the Maximato and eventually send his old political mentor across the border into a comfortable exile. In June 1935 he used Calles's antilabor statements (which he had perhaps manipulated Calles into making) to break the Maximato, through his own response, which indirectly (but clearly enough as a signal) called Calles to account for "perverse intentions" against the rightful attempts of the workers "to create a more solid economic situation."

The working class poured into the streets calling for the head of Plutarco Elías Calles. Cárdenas immediately formed a new cabinet—without the Callistas—and then proceeded to purge the governors. Between 1935 and 1936 he declared their powers void, annulled elections, and forced the governors of fourteen states "to go on leave." On April 9, 1936—to enthusiastic support—he sent Calles into exile. But Calles would die in his bed. He would be the first of the major Revolutionary caudillos to avoid that final meeting with the bullet.

The purge was a major turning point in the history of Mexico. The change in political leadership encouraged many others: an end to the military hegemony, the end of legislative blocs, still greater centralization into the hands of the executive, a taming of the other branches of power, the rise of mass politics and a corporate state that had already taken shape during Cárdenas's administration of Michoacán. It was also—in the person of many Cardenistas—a change of generations. Coming onstage were those who had only witnessed the Revolution. The men who had fought it were mostly retired or dead.[48] But not the leader of the new generations, "the kid" who had mounted the whirlwind at the age of sixteen.

Lázaro Cárdenas was a man very sensitive to symbols. When he felt himself firmly in control, he noted in his diary on February 8, 1937:

> Today I issued a pardon for all political, civil and military prisoners (numbering more than 10,000), who have taken part in rebellions or riots under past administrations. The intent of this law is to eliminate divisions between Mexicans and at the same time, give the country greater confidence.[49]

Adolfo de la Huerta and the son of Porfirio Díaz, along with hundreds of other exiles of the Revolution, returned to Mexico. One of them, Rafael Zubarán Capmany, commented: "Cárdenas has the heart of Madero and the character of Carranza. I have spoken at length with him. He is a man."

In February 1936, Cárdenas made it very clear that he was principally concerned with Articles 27 and 123 of the Constitution (on land and labor) and not the anticlerical Articles 3 and 130: "The government will not repeat the mistake committed by previous administrations in considering the religious question a preeminent problem. . . . It is not the government's job to promote antireligious campaigns." The teachers, who in Michoacán had been the spearhead of "defanaticization," now were given different orders: "From here on, there should be no antireligious propaganda in the schools. We must concentrate all our attention exclusively on the great cause of social reform." The persecution of religion did not magically end, but it lessened dramatically during the Cárdenas presidency.[50]

Cárdenas had sounded the retreat for the anticlerical programs he had in part supported as governor. He continued to promote, as President, what had been the other side of "defanaticization": general education, but he did it in a minor key. During his entire term, the great debate on "socialist education" filled the air: public discussions, doctrinal texts, university agitation, speeches by Lombardo Toledano, threats, homages to Lenin, the anniversary of the Russian Revolution elevated to a national holiday in the calendar of the Ministry of Education; confusion among teachers, parents, and students; doubts about what exactly would be the "rational and precise meaning of the universe" to which the new socialist Article 3 of the Constitution cryptically referred; many meetings, the founding of the Workers University, workers dressed as students, students dressed as workers, more speeches by Marxist leader Vicente Lombardo Toledano, miles of ink and hurricanes of talk. In contrast, the National Palace housed the man who had already been nicknamed the "Sphinx of Jiquilpan," not exactly a silent man but not a bombastic one either. A man who kept his own council and chose his words carefully. He could contain religious persecution, but he could not control all the chatter or the drive toward "socialist education." Yet fundamentally he saw it as a minor concern. And he himself did not join in.

Along with the new President came a new style of power. A wide-ranging front-page interview amply illustrated with photos, on September 8, 1935, presented Cárdenas just as he wanted to appear and (almost) as he was. "Neither gambling, nor night life . . . seduce our president." He did not go to parties, he did not like protocol, "the dawn never caught him in bed." He was tireless, vital, on the move, he had no guards accompanying him. He had closed casinos and prohibited jai alai, but he enjoyed "country sports." Never, not even when he took office, did he put on a tuxedo. He almost always wore a dark suit. And he was extremely courteous.

There was a refinement to the image, alien to the cruder, *norteño* quality of the Sonorans. The rumors of his love affairs did him no harm. There was some criticism of the fact that he favored his younger brothers, but that after all seemed a venial sin. And more than anything, his image was that of a man who helped people, tranquilly, gently, with delicacy and tact.

* * *

For "the Revolutionaries of yesterday" (northerners, whether Carrancistas, Obregonistas, or Callistas) the *ejido* had been only one, limited means toward land reform. Cárdenas, the "Revolutionary of today," saw it differently:

> By the mere fact of requesting *ejidos*, the peasant breaks the economic tie to the landowner, and in these conditions the role of the *ejido* is not to produce the economic equivalent of a salary . . . but rather—by its extent, quality and the way it is used—it should suffice for the total economic liberation of the worker, creating a new economic-agricultural system . . . to replace the regimen of salaried field-workers and eliminate agrarian capitalism in the republic.[51]

The *ejido* was to be not only a means toward justice—after centuries of exploitation—but a method of economic improvement. The institution would not only surpass the hacienda morally, it would leap past it in production. And—at a deeper level of Cárdenas's agenda—agrarian reform would destroy the hacienda and the political power of the hacendados, many of them Revolutionary generals, "to whom the Revolution had done justice."[52]

Before Cárdenas, the process of agrarian reform had focused mostly on the central grain-growing region of the country. He would extend its reach throughout Mexico and vastly broaden its resources. There would be an Ejido Credit Bank, services of all kinds available from the Agrarian Department, irrigation, communication, even government-sponsored sports for the new *ejiditarios*. An entire program of redemption.

Between October 1936 and December 1937, Cárdenas concentrated his time on what he called his "agrarian workdays." The confrontation with the *Jefe Máximo* was over and done with. There were day-to-day political details in "the city of gossip and selfish interests and corruption," but as much as he could, he would go out into the country where "everything is pure" and where he would be able to "live with the needs and anxieties of the people so as to more easily find a way to heal them."

The nature and course of the Agrarian Reform reflects—more than any other of Cárdenas's projects or achievements—the complicated nature of the man, his staunch virtues and his self-deceptions, his capacity for innovation and his dogged stubbornness. It was a program carried out with boldness and recklessness, inspired by truly noble feelings of benevolence but imposed from above in a paternalistic and even arrogant manner, a project for economic benefit that had destructive economic results, a gift (genuinely received) of a greater sense of dignity to the peasant but accompanied by an insistence on generalized assumptions with the consequent failure to evaluate the very specific (and varying) conditions of the Mexican agricultural mosaic. A closer look at each individual case—perhaps sacrificing speed for accuracy—might have led to a much better record in making the proper *local* decisions for the good of the peasants themselves. (The Agrarian Reform, for instance, was in general more successful in the north of the country, where modes of behavior encouraged a more cooperative attitude toward working together, than in the *tierra caliente*, the "hot country," where customs were much less favorable.)[53]

And the progress of the reform would also be conditioned by a basic conflict of attitudes. Cárdenas was reviving—to a degree—the stance of Las Leyes de Indias, decisions from on high for the good of the humble. The peasants, on the other hand, would very often be suspicious of their new master "without a face"—the State. And with their essentially conservative nature, they would turn to a proverb like *Más vale malo por conocido que bueno por conocer*, "The evil you understand is better than the good you have yet to know."

In October 1936 Cárdenas took his first major agrarian step on the national level—the division of the La Laguna haciendas in Coahuila. Until then, no one had dared to touch or even to think of touching the truly modern agricultural regions of the country. On the Laguna complex, the best land consisted of 220,000 irrigated hectares, where a small group of large and medium landowners produced cotton. They included Revolutionary generals—Pablo Quiroga, Miguel Acosta, Carlos Real, and Eulogio Ortiz. Three large foreign companies controlled most of the economic movement in the area. The hacendados—in seventeen years of operation—had earned 217 million pesos and reinvested very little of it. The key to the nature of the business in this region of lakes lay in the unpredictable behavior—droughts or flooding—of the Nazas River. Only substantial capital could risk planting without absolute assurance of a harvest, which is why the hacendados always distrusted the official plan to create the Palmito reservoir. They were afraid (with good reason) that the government—once the water supply was assured and regulated—would expropriate them.

The CTM and the Communist Party had done union organizing and political work in the region. The groundwork had been laid. When Cárdenas arrived (in his famous Olive Presidential Train), he would stay for close to two months, rejecting all contact with the owners and personally overseeing the distribution of the land. When his turn came,

General Eulogio Ortiz shrugged his shoulders and, in words that would be remembered, he said, "The Revolution gave me the land and the Revolution takes it away from me." Cárdenas noted in his diary: "He should have said: during the Revolution I acquired it and today I return it to the people."[54]

There were many emotional moments. On November 10 Cárdenas exclaimed: "All those who have worked the land for wages . . . are going to be sure of their own place within the *ejido*!" Ten days later, on the anniversary of Madero's first day of the Revolution, a former soldier of Pancho Villa turned in his old gun for an iron plough. Cárdenas was deeply moved. In a photograph from La Laguna, he stands with his left hand in his pocket, his right hand crossed over his chest, confetti in his hair and a boyish smile on his face, while a group of peasant women surround him and a small boy almost leans back into his arms. The final figures would be 447,516 hectares distributed among 37,753 *ejidatarios*.

Decades later, in 1961, Cárdenas would admit: "Lands like La Laguna and other regions were distributed even against the wishes of peasant leaders themselves, who would have preferred to continue fighting for unions on the haciendas." But Cárdenas was not interested in encouraging peasant unions. He wanted to free the peasant from the hacienda system, and yet there would be a major problem—which would have its repercussions and which Cárdenas did not talk about or perhaps did not honestly see. It was the dominating fact that under the *ejido* system—whether individual or collective—the individual campesino did not really own the land. The State was the owner, a new and all encompassing "hacendado."[55]

When he returned from La Laguna, Cárdenas was riding a wave of emotion. He had taken the first, great step in his national agrarian program, and as he noted in his diary, he felt ready to confront one of the worst national abuses, a long-standing sore in the Mexican south: "At the end of 1937 we will move on to totally resolve the agrarian problem in Yucatán, that has dragged on for long years and we must finish with it so that we can save the indigenous people—the majority of the peons in the sisal-growing zone—from their misery."[56]

In Yucatán, the hacendados—the self-described "divine caste"—had violated the agrarian laws for years. Assault and murder by strong-arm men were standard techniques for control, and they would send their peons to provoke and attack members of the local *ejidos*, while refusing to obey the laws requiring them to rent out their equipment, moderate their harvests, or renew exhausted fields.

On August 1, 1937, Cárdenas arrived by train in Mérida, the capital of Yucatán. On the third, accompanied by Governor Palomo, he spoke from the balcony of the Mérida Literary Institute to a huge crowd from the Popular Alliance. Thousands of Mayan peons waving the national tricolor and the red flag of revolution, carrying banners with agrarian slogans, listened to him with fervor. "Ninety years after the beginning of the last tragedy of the Mayan race, the Revolution has come to offer you, through the sisal fields, some minimum compensation for the blood spilled in your struggle for the land!"[57]

It was a belated response to all they had suffered in the War of the Castes. And for the benefit of the hacendados, Cárdenas announced that his decision was irrevocable.

On August 22, before a committee of peons from the municipality of Abalá, Cárdenas and Governor Palomo began their work at the Temozón Hacienda. Accompanying the men were the women—barefoot, with their golden skin—and the children squatting on their heels. The flags and banners read: TEMOZÓN EJIDO COMMISSION; UNIFIED FRONT FOR WOMEN'S RIGHTS; MILL OF NEXTAMAL; THE MENA AND SOSA WOMEN'S LEAGUE. The head of the Agrarian Department announced: "The President of the Republic and the Governor of the State . . . hand over possession of the *ejidos* in this region of Abalá, including the town of Temozón." At that instant, 23,500 hectares of sisal fields and 66,700 hectares of fallow land were entrusted to 8,404 heads of families in 70 *ejido* communities in Abalá, Tixkobob, Muna, Ekmul, Dzidzantún, Seyé, Tekantó, and other communities with their melodic names in the language of the great Mayan civilization. The grants went on for weeks. Twenty-three days later, close to 65 percent of the Yucatecan sisal fields were in the hands of those who had been working them for over a century.

A year passed, and by then the peasants were thundering against the Agrarian Department and the governor. What had happened—they implicitly flung in Cárdenas's face—to the 35 million pesos he had promised them? Where were the early payments on their harvests? There was a great deal of "poverty, hunger, unrest and discontent." The President lamented: "What has Yucatán asked for that we have not given them?"

Much of the fault was in the planning. Of the 272 *ejidos*, only 10 really received proper land and planting allotments. All the remaining 262 were in one way or another deficient, and it was difficult and uneconomical for them to produce and refine their sisal. In a broader context, the individual *ejiditario*—due to the organization and administrative structure of the centralized, overarching *ejido* organization—

President Cárdenas *(center)* with the Indians, circa 1936
(FINAH)

"The Sisal-Growers of Yucatán"—often came to feel that he was merely working for another great landed estate. And the Ejido Bank, full of corruption, could be as rigid and impersonal as the hacendados had been.

The Sphinx of Jiquilpan was careful not to confide his impressions even to his diary. It would appear that he continued to believe that the mode of distribution had been correct and just, and that—if there had been errors—they could be ascribed to a weak revolutionary spirit in the officials assigned to get the new system working.

His critics called the Yucatán land reform a failure and blamed Cárdenas for "the experiment carried out on the living flesh of a people, an impulse perhaps born of generous sentiments, but guilty of carelessness and lack of foresight."

Perhaps they were forgetting that the conditions before the experiment were also desperate, and that for centuries, long before Cárdenas existed, a conjunction of history and nature had been conducting a cruel experiment on the living flesh of the Mayan people.

When he had made his first automobile trip through Yucatán with Governor Palomo, Cárdenas had given a banknote to a beggar who had jumped up on the running board. Cárdenas had said to him: "Take half and give the other half to someone who needs it as much as you do." In his second and last visit as President, there were more (and more frequent) small acts of charity. At least "he could do that."

Yet another set of circumstances prevailed in two haciendas of his own native state of Michoacán, conditions that might have called for a more careful procedure than instant expropriation. In 1938 the President wrote a letter to his old acquaintances, the Cusi family, of Italian origin, owners of two haciendas that were in many ways exemplary. The Lombardía and Nueva Italia haciendas were very well run, and the peons had neither registered complaints nor shown antagonism toward the landowners. Cárdenas recognized that the Cusis had been "good hacendados" but warned them that their lands would have to be redistributed. The hacienda as a form could not continue to exist. (And in this insistence, Cárdenas was fulfilling a legal requirement of the 1917 Constitution, only partially or symbolically honored by previous Revolutionary governments.) In November the peasants received the haciendas lock, stock, and barrel: land, buildings, machinery, cattle, lemon orchards. This time the expropriation could not fail: These were the perfect haciendas freed from their hacendados. In few other experiments did Cárdenas put greater personal faith.

In order not to dismember the productive unit, 1,375 peasant owners of 32,196 hectares were integrated into five *ejido* communities and a collective society for *ejido* loans. Shortly afterward, the inhabitants of one of the communities petitioned: "We wish to be responsible for our success or failure. We are not at all convinced that we can prosper inside the society."[58]

Cárdenas rejected the petition. For a time—while the first harvest cycle lasted—the *ejiditarios* were ecstatic. At the time of the distribution, the fields had already been sown: The good results and the light work fooled them. Soon afterward, nonexistent profits were distributed. Loans that were not repaid piled up. (In 1944, Cárdenas would persuade the government agencies involved to write off all those loans.) Corruption surfaced within the *ejidal* society and within the bank. A witness remembered: "No one really knew which way to go . . . they expected us to know everything overnight; how to sow, clean canals, care for the cattle and everything else. On the hacienda a peon went out to sow or to water, to scare off ducks or to carry loads, but he did not do everything. Before, one knew what to expect, but afterward it all turned into a mess."[59]

The profits from the first years—real or fictitious—were spent in the most extravagant way: a stadium, or a swimming pool that would later be left unfilled. In 1941 the five communities separated to become "responsible for their own success or failure." Cárdenas would remain in contact with the region until the sixties were well under way. And he would never emotionally accept the fact that this community of *ejidos*, the apple of his eye, the experiment perfectly prepared to prosper in collective brotherhood, would turn out far different from his hopes.

But there were some far more successful divisions and assignments of the land, even in a context of recalcitrance from those who were to benefit. In the Mexicali Valley of Baja California, three American companies had created a closed economic circle—renting much of the land, controlling the water, binding farmers to them with loans. Cárdenas went there to break this control, and according to an eyewitness, when he first began to divide the lands, the local peasants were suspicious and unwilling to accept the grants. "He had to give some lands to hairdressers, waiters, *croupiers* from the gambling tables."[60] But when he visited the zone two years later, no matter who may have been the first recipients of the land, he was cheered as a popular hero. The *ejidos* in this area had asked for their lands to be divided into individual farms, and the government had agreed. "Look at him, touch him," said the mothers to their children. "He gave us the land."

The same messianic image he had created in Michoacán now was being spread through the length and breadth of the country. It was a projection—and an honest one—of compassion, inspired by an almost obstinate faith in the goodness of his work. But obstinacy was often not enough for the realities of Mexico.

For the Indians in general (beyond distribution of lands), Cárdenas devised a crusade for health, education, and bread. In December 1935 he founded the Department of Indian Affairs. But much of the effort never left the planning stage, and the projected brigades of specialists were never assembled and dispatched in large enough numbers. Progress came in small doses, or else things would refuse to change or they would change for the worse.

In June 1939, for instance, he visited the Yaqui Indians. He offered to construct storage buildings for them, bridges and houses, and to mediate between the eight villages in their jurisdictional disputes. (He only refused their requests for new churches.) But on the issue of the land, a report published some years later (1943) by the Banco Agrícola showed that the Yaquis by then had jurisdiction over 400,000 hectares but only farmed 2,000. A schoolteacher living in the region explained:

> Ex-President Cárdenas's plans were . . . intended to benefit the Indians on the right bank of the river as well as the mestizos and creoles on the left bank. But it did not happen that way. The men of Cajeme . . . have prosperous collective *ejidos*, because they were given loans and canals were constructed. . . . The Indians have only the land. They have no canals. They get no loans. . . . The men of Cajeme are healthy, happy, they live off their work, they have nice houses, they eat well. The Indians are all sick, they live off what they steal for much of the year, they are frequently drunk, they live like animals, in dark hovels.

"It is not true," Cárdenas said, "that the Indian is a hindrance to his own improvement or indifferent to progress. If he only infrequently shows happiness or sadness, hiding the secret of his emotions like a sphinx, it is because he is accustomed to being forgotten." Cárdenas visited them, he praised and honored them, and he offered them limited state assistance with scant results. In the end he would resign himself to giving the only thing that depended directly on himself: his attention, his ear, his presence.

From the economic point of view—national, regional, local, ejidal, individual—the gigantic operation of sharing out the land did not come close to fulfilling the hopes of President Cárdenas. The sudden increase in public spending, a deficit of more than 40 percent in the budget for 1936–1937, and an overdraft of 87.6 million pesos against the Bank of

Mexico all contributed to a rise in prices, which was also fueled by a pronounced drop in agricultural productivity.

Yet in political terms, the reform was a resounding success. During his term, Cárdenas distributed 18,352,275 hectares among 1,020,594 peasants. The hacendado class disappeared off the map, and the word *hacienda* was relegated to the history books. The profiteers of the Revolution who had "done themselves justice" by taking over "beautiful estates" now had to find another way to earn their living. On the other hand, though the "master" (*amo*), the "boss" (*patrón*) had disappeared, he had been replaced by an immense bureaucratic network stretching from the commissioner of an *ejido* to the offices of the Agrarian Department. Clearly Cárdenas had not foreseen or desired this result. His ideal, as Frank Tannenbaum wrote, was different: "a Mexican nation based on the autonomous and independent government of the pueblos in which each individual will be assured of his own *ejido*, will be free of exploitation and will actively participate in the problems of his community."[61]

But in Cárdenas's grand design there was an implicit assumption (and a significant error)—his faith in the purity and impartiality of the authorities. The *ejido* linked the peasant to the State much more than to the land. And paternalism could often turn into subjection. The peasant would run the risk of becoming not a free man but an element of political capital.

The CTM (Confederation of Mexican Workers) was not the CRMDT. In Michoacán, the working class had barely existed. In the Mexico of 1936, the nation's working class did not need the President to invent it. The 350,000 members of the new organization had at least two decades of experience in union battles. The relationship between the CTM and the government would be much more equal than the pattern in Michoacán.

The most immediate benefit the workers gained from their connection with the State was a new way of dealing with their employers. When in February 1936 Cárdenas personally went to the industrial city of Monterrey—to support with his own presence the strike against the glass industry—he was not there to take over factories and put them into the hands of the workers. His goverment was not Communist (as the factory owners kept screaming), nor was it financed by Russian rubles. The government was encouraging political "solidarity" strikes, but the whole display had a specific purpose—it was geared to establishing the new rules of the game between "the factors of production" (capital and labor) as well as between them and the government. They were rules that would essentially remain in effect for decades (and even to the present day) and they were set down—like an unofficial constitutional amendment—in the famous Fourteen Points published in *El Nacional* on February 12 of that same year.

The Fourteen Points called for one national union (it would be the CTM); specified the need for cooperation among government, labor, and industry; required industry to "always consider the demands of the workers within the margin offered by their economic possibilities"; and named the government "arbitrator and regulator of the life of society." The objective was to create a definitive, though not entirely subservient "company union," but the company was a single and vast one—the government of Mexico.[62]

The CTM had its ups and downs during the early years. The Communists retreated before the forces of Fidel Velázquez and the other "little wolves." In June 1936 the miners and metalworkers left the organization, and a year later the electricians followed. By 1937, according to data from the Communist Party, only 44.3 percent of the original members remained in the CTM.

But in August of that year, Cárdenas—that is, the State—began to collect repayment: "We ask of the labor organizations . . . consistent cooperation in seeking agreements before going out on new strikes. . . ." It was necessary to give "precedence" to the problem of Yucatán and to show solidarity with the peasants. One by one, the groups that had deserted began to return. The Communists received a visit from the international leader Earl Browder, who, in the face of rising fascism in Europe, ordered "unity at all costs."[63] It was the height of the politics of the "popular front." Cárdenas not only believed in it and stimulated it, he also embodied and capitalized upon it for the sake of the State.

For the State, not for the CTM. In a fundamental decision at the very beginning of its existence, Cárdenas—in February 1936—prohibited the CTM from organizing peasants, offering as his reason that "it would incubate seeds of disintegration." In August 1938, he created, on the spot, an organization to be called the National Peasant Confederation (CNC), directed by the reliable and honest Graciano Sánchez. (In the same corporative spirit he promoted the official integration of the Chambers of Commerce and Industry.) And in order not to put all his eggs into the one CTM basket, he also kept alive labor organizations like the anarchist CGT and the old CROM.[64] His practice of balance tipped to an extreme with the generous asylum he offered to Leon Trotsky, hounded through much of the rest of the world by Stalin's anathemas and secret police.

The painter Diego Rivera had requested the asylum at the end of 1936, and Trotsky landed in Tampico in January 1937. The Communists and the CTM shrieked in outrage. The Communists would of course continue to protest, but the CTM would yield. And the Cardenista state continued to solidify as the representative force for the entire life of the nation.

On March 17, 1939, Cárdenas wrote a letter to the mine workers. After giving them a summary of the agrarian policies of the government—from La Laguna to Baja California—he also gave them a warning:

> Why the mine workers, who are on a higher level economically and socially . . . should choose this moment to agitate the working masses with the apparent intention of maintaining an important sector of the working masses in a state of constant unrest . . . is incomprehensible. . . . I hope . . . it is only a matter of isolated actions by mistaken individual elements who obey movements alien to the workers' movement or else it is an issue of antipatriotic commitments by sectors that do not understand our problems.

With his usual courtesy, he was ordering them not to go out on strike. In 1936 strikes had been neither "antipatriotic" nor "mistaken." Now in 1939, the State—taking the same protective stance as supreme arbiter and judge—explained that they were.

The comparatively "hard line" taken with the workers (from 1938 on) had much to do with national priorities defined at the center. But a specific internal labor process (though not finally resolved till 1940) also had its influence: the failure of the experiment at having workers themselves run the railroads. Cárdenas tried it. It was part of the spirit of continual experimentation that characterized his "new deal" as much as it did Roosevelt's in Washington and much of the rest of the world in the 1930s.

Antonio Madrazo, Executive President of the National Railroads, had sent him an alarming memorandum in February 1937. Continual union demands were paralyzing the railroads, which in 1908 had been nationalized administratively but not economically. (The railroads were still expected to make an operating profit, which belonged to the shareholders.) Out of every peso earned, 55 centavos now went for salaries; efficiency had declined; the union frequently threatened to impose "the penalty of expulsion" (from the union and so from their job) against workers who worked too hard. Cárdenas made a decision: full expropriation in June 1937. He created the Autonomous Department of National Railways, as a nonprofit national patrimony. Less than a year later, in April 1938, he turned over administration and operation of the enterprise to the Union of Railroad Workers.

There followed—between August 1938 and March 1940—six serious passenger-train crashes and hundreds of minor accidents. Agustín Leñero, Cárdenas's private secretary, sent the President a telegram reporting that "almost all of the accidents can be attributed to worker error due to lack of discipline." The experiment was a failure and would be finally ended in December 1940. If it had worked on the railroads, Cárdenas might have moved toward greater direct workers' control of industry. The failure would impose a retreat and an open search for a new way. The result would be industrial capitalism under the protection, regulation, and arbitration of the State.

It was during this period that the labor movement was officially assigned a place within the ruling party. Cárdenas reorganized the PNR (renaming it the PRM—Party of the Mexican Revolution) in line with the goals he had expressed when president of the party in 1930–1931. What had been primarily a conclave of generals was transformed into a party of the masses: labor, peasants, bureaucrats, the military, all forcibly affiliated not as individuals but through their respective organizations. It was a process of integration, a movement toward the future in terms of rationalizing, controlling, and applying power, for the purpose of "uniting these disparate groups so that they do not act anarchically"; but it was also a strong call from the Mexican past, the voice of the integral State, of the corporativist model.[65] In form it echoed Porfirio Díaz but perhaps more strongly the ancient theocratic-political unity of the Hapsburg colonial period. Thirty years after the beginning of the Revolution a new political father was returning to an ancient, persistent form of government: power as an ordered, architectural structure.

Cárdenas had not forgotten his experiences in the Huasteca, with his friend Múgica during the mid-twenties, confronting the arrogance of the oil companies. From the beginning of his term, he showed signs of a hardening attitude toward them. On September 1, 1935, he declared that the Oil Law of 1925 "was not in harmony with the basic principle of Article 27 of the Constitution."[66]

The pressure then built up for nearly three years, event after event, in a slow steady crescendo. In 1936 a law was issued on expropriations for the benefit of the people. American ambassador Josephus Daniels received guarantees from the President that it would not be applied to mines and oil. In the middle of the year, the eighteen thousand members of the Mexican Oil Workers Union successfully forced the companies to sign their first collective bargaining agreement. And Cárdenas created a government entity called the National Oil Administration.[67]

Then, in 1937, a new confrontation broke out between the oil workers and the companies. The President and Múgica (now his Minister of Communications) and Finance Minister Suárez repeatedly denied that the government had any intention of nationalizing the oil companies. They pointed out that Mexico—now at the high point of its program of Agrarian Reform—needed the income furnished by the American oil companies. But Cárdenas wrote (to himself) in June 1937: "The entire oil industry should also come into the hands of the State so that the Nation can benefit from the riches of the subsoil that are now shipped away by foreign companies. To accomplish it, we will follow another procedure."[68]

The Council of Conciliation and Arbitration declared that there was an economic conflict between the companies and the union, and so, in accordance with labor law, a committee of experts was chosen. The committee issued a 2,700-page document with forty conclusions unfavorable to the companies. They would have to pay the 26 million pesos their workers were demanding, not the 12 million pesos the companies were offering.

The companies were no longer confronting the Oil Workers' Union (or the CTM). They were face to face with the government. Cárdenas explained to Ambassador Daniels that, from now on, the State would take part in all decisions on wages (and of course new taxes) within the oil industry. After all, the report had shown massive manipulation of their account books by the companies, huge hidden profits, and inflated expenses. The companies put their lawyers to work in an attempt to refute the report and meanwhile raised their offer to 20 million pesos.

Meetings continued to be held. The government insisted on its figures. In November 1937, Standard Oil issued a hard-line statement: "We cannot pay and we will not pay." They had made a simple calculation—the Mexican government would not dare go any further. Mexico needed American personnel to manage the industry as well as the markets and resources that only the companies could offer. To general astonishment, the government then made an unprecedented move, canceling a concession that Standard Oil had received in 1909. The taboo against retroactive cancellations—one that neither Carranza nor Calles had ever felt strong enough to overcome—was broken.

On December 18 the Federal Conciliation and Arbitration Council ruled on the appeals of the companies. They had to pay the full 26.3 million pesos to their workers and hire eleven hundred new employees, with whom they would have to sign contracts.[69] The companies cried miscarriage of justice. For them, it was the "most extremist contract that had ever been given to workers in any industry in any country." On the twenty-ninth they asked for protection from the Supreme Court of

Mexico. The atmosphere was building toward crisis. The reserves of the Bank of Mexico plummeted. One of its official balance sheets was not even published.

More meetings. Cárdenas refused to reduce the 26 million pesos required from the companies, but to Mr. Armstrong, their official representative, he suggested that he might make other concessions. In a heated meeting on March 8, 1938, he offered a firm commitment to prevent any further problems, once the full payment had been made to the workers:

> —And who will guarantee that the increase will not be more than twenty-six million?
> Cardénas answered—I will guarantee it.
> —You? (*general smiles from the company representatives*)
> Cárdenas rose to his feet.—We are finished.

And they were. On the next day Cárdenas noted in his diary, "I am optimistic about the attitude that the Nation will assume if the government is obliged to take radical action."[70] And on the following day, Cárdenas called in his old friend Múgica, reminded him of those months together in the Huasteca a decade earlier when two generals of the Revolution stared at the unyielding countenance of the oil companies and their state within the State. In an act of poetic justice, he entrusted Múgica with formulating a manifesto to the nation that would explain—at the right moment—his reasons for taking the oil and then would request the support of the people.[71]

Politicians and financiers—including the companies themselves—thought that, at most, the Mexican government might send troops to occupy the refineries. But now everything rushed along. On the fifteenth, the Federal Commission pressured the oil companies to pay and on the next day declared them officially in contempt of court. ("They will not dare to expropriate us," said Armstrong.) On March 18 itself, the companies met with the President and finally agreed to pay the 26 million pesos but objected to any other commitments. It was much too late.

That very night, at ten o'clock, Cárdenas addressed the nation by radio from the National Palace, announcing "the far-reaching decision of the Government of Mexico to recover the oil wealth that foreign enterprises have been exploiting . . ." And he requested "the support of the people not only for the recovery of the oil . . . but also for the dignity of Mexico, that foreigners think they can ridicule after having obtained great benefits from our natural resources. . . . With will and a small amount of sacrifice from the people to resist the assaults of the interests affected, Mexico will come through gracefully."[72]

Two hundred thousand people flooded the Zócalo, cheering the President. Long lines that would become legend formed at the Palace of Fine Arts as people from every level of society united to offer whatever they could—from the jeweled earrings of society women to chickens and turkeys fresh off the farms—to help pay the country's debt of honor, the compensation to the oil companies. *Corridos* were composed and sung in the streets:

On the eighteenth of March, the day of the great sensation!
He nationalized the oil then! The Chief of our Nation! . . .
And so Mexico is giving the world its great lesson!
History is being redeemed through our Revolution! . . .
They wanted to make a joke of the laws of our free nation
without noting how they were born from the roar of cannon![73]

On Sunday, the President had gone with a few close friends to the Nevado de Toluca, where he swam alone in the icy waters of an old crater. "Swimming in the volcano," wrote his secretary, Raúl Castellano, finding it a perfect metaphor for the events that Cárdenas had just lived through. The oil companies engineered a wide and effective commercial boycott against Mexico, which for a time was obliged to sell its oil to the Axis countries or engineer complicated exchanges. There were embargoes, shortages of spare parts even in industries that had nothing to do with oil, campaigns of slander, writers paid to spread an image throughout the world of "a Mexico that stole whatever it could get its hands on." The British government sent an insulting note that provoked Mexico to suspend relations. Tension continued with the U.S. government, but for Washington the risks of a confrontation were greater than any possible benefit. On December 7, 1941, when World War II began for the United States, the quarrel over the oil effectively ended. The attitude of both presidents, Roosevelt and Cárdenas, had contributed—as much as the international situation—to a settlement. Frank Tannenbaum wrote:

> It is typical of Cárdenas that throughout so much commotion, he has known how to keep his head. He did not hurl epithets at the American people, did not daily denounce the American government, did not insult the Secretary of State, did not ridicule the president of the United States. Quite to the contrary, he remained friends with Josephus Daniels and once noted: "I was very lucky to be

President of Mexico when Roosevelt was President of the United States."[74]

Article 27 of the Constitution had finally been fulfilled in both letter and spirit. Mexico—from that moment—was more truly itself.

In his book *Homage to Catalonia*, George Orwell mentions the good Mexican bullets he used to keep in reserve for the moment of battle against Franco's Fascists. On September 15, 1936, when Cárdenas gave the traditional *Grito* of Independence in the Zócalo, he added: "Long live the Spanish Republic!" These were not just words: The desire would take on form as aid during the war and as an offer of refuge after the defeat of the Republic. On June 7, 1937, a group of war orphans arrived in Mexico. Two years later, thirty thousand Spanish Republican refugees disembarked. There were pro-Fascists who complained of the terrible "Communist" immigration. But the influx was not only an act of mercy. It brought immense cultural and economic capital to Mexico, as the country welcomed the cream of Spain's intellectual and scientific elite. One of the many institutions formed in collaboration with the refugee Spaniards was the Casa de España en México. Daniel Cosío Villegas—Cárdenas's ambassador to Portugal in 1936–1937—had recognized the great opportunity, not only for beneficence but for intellectual gain offered to Mexico by Franco's decimation and uprooting of Spain's greatest minds and sensibilities. He became secretary general of the Casa de España. (Its director was the writer Alfonso Reyes, now returned to Mexico after many years of self-imposed exile.) The Casa de España would develop into a famous school of higher education, El Colegio de México. Cárdenas's Mexico became—and Mexico continues to be—a country of asylum for the persecuted of other lands. Cárdenas said: "There is no antipathy or prejudice in our country against any country or race in the world. . . . Distinctions or persecutions of any part of the population are contrary to the spirit and the laws of my government. Among us, any North American is welcome, black or white, Jew or Catholic, all we ask is that they obey our immigration laws."[75]

Every action Cárdenas took in foreign policy was inspired by this moral attitude: his condemnation of the Italian Fascist invasion of Ethiopia; the censure of Japan in the Sino-Japanese conflict; his condemnation of Hitler's Austrian *Anschluss*; his order to Mexico's permanent delegation to the League of Nations in Geneva to condemn the Nazi persecution of the Jews; his protest against the German invasion of Czechoslovakia, Belgium, and Holland and against the Soviet invasion of Finland.[76]

The feeling for liberty that Cárdenas projected in his foreign policy had concrete and palpable repercussions on national political life. The journalist José Alvarado wrote: "During the six years that he was at the National Palace, his work was freely discussed throughout the entire Republic and his regime was the object of rude and violent attacks." He never took any reprisals.

He wanted his term to be different from the bloody years of the Sonorans. And he succeeded. The death of General Saturnino Cedillo—the leader of the last rebellion by the military in the history of contemporary Mexico—was not due to the President. When Cedillo's revolt, possibly fueled by foreign oil interests, had begun, Cárdenas had personally gone to San Luis Potosí to try to persuade the general—who was the cacique of that region—to surrender.[77] His deputy, Pedro Figueroa, remembered his instructions: "The general does not want us to kill anyone. He wants no bloodshed. This rebel movement could have ended in an hour, but it lasted a few days because General Cárdenas had given express orders to free the prisoners and, not only that, give them money to go back to their homes and their jobs."

He was infuriated when Cedillo was killed in January 1939, under uncertain circumstances, in the mountains of San Luis Potosí.

And to his generosity in offering political asylum, to the unrestricted freedom of speech that he permitted, in the end he added encouragement for another right, and one that was critical for peace and justice in Mexico: respect for freedom of worship.

> In all countries there are various tendencies for and against religious beliefs and practices, but with respect to Mexican law and government, there is complete religious liberty in our country. Throughout Mexico (except in Tabasco where for many years there have been no churches), the churches are open and religion is freely practiced. The Catholic Church gave its support to the Mexican government on the recent oil expropriation. The Church took this step, because it understood that all the people of Mexico enthusiastically backed this measure, so necessary for the general welfare of our country.

One of the greatest paradoxes of that six-year term was the coexistence of a corporativist state with the broadest civil liberties. This would continue to be one of the central and to a degree fortunate paradoxes of Mexican life. For a paternal politician like Cárdenas, the coexistence was understandable. He was the good father, dominant yet tolerant. And he was willing to encourage the natural freedom of his children.

* * *

On December 1, 1940—the day on which he was succeeded as President of Mexico by Manuel Ávila Camacho—Cárdenas wrote in his diary: "I put my effort into serving my country and I gave my greatest commitment to the people most in need. I eliminated many privileges and in good measure I distributed the wealth that had been in the hands of the few."[78] Word for word he was writing the truth. But the political legacy of his presidency was more complex than the benevolence of his intentions (or some of his accomplishments).

During his last two years—as his presidency wound down and the sexennial problem of the succession approached—he was subject to a wave of criticism. Much of it was laughable or worse—conservative rantings about his having given the country away to Russia, absurdities from some sectors of the Church comparing him to Calles, and even (from the brilliant economist Manuel Gómez Morín, for instance, whose other criticisms were often perceptive) an attack on Cárdenas for his stances at Geneva—in support of Jews and other victims of nazism and fascism—and his welcome to the Spanish exiles, all of which Gómez Morín described as "permitting Mexican functionaries to operate as agents . . . for factions that are foreign to us."

But the man who had created Calles's economic program in the twenties also had things to say more worthy of attention, when he founded the National Action Party (PAN) in 1939, primarily in opposition to Cardenismo. Gómez Morín recognized Cárdenas's "honorable intentions in cases like agrarian collectivization and bureaucratic syndicalism" but felt the results were "lamentable, or in the best of cases, nonexistent." To his mind, the Agrarian Reform was inspired by "a false and artificial concept of struggle," and the labor organizations had been "poisoned with politics and by ends, tactics and objectives that were not theirs." The greatest sin of the regime, he concluded, was its "mental and moral confusion."

Yet there were other critical observers, like Manuel Moreno Sánchez (formerly a judge in Michoacán and critical of the impunity with which the CRMDT had been allowed its range of violence) who were able to recognize the true paradoxes in Cárdenas's record, to feel the breadth of the man and as a result both judge him firmly and still be drawn to respect him:

> In Cárdenas there is at the same time the appearance of a tolerant and fair President and the shadow of a cacique who wants to solve everything without thinking about the structure of the Constitution, or the division of powers, whether legislative or judicial. . . . He

> was always on the side of the poor and he had never been poor. . . . He is certainly a destroyer, because every revolutionary is born to destroy; but if he was unable to produce accomplishments leading to material development, no one can deny him his principal accomplishment: the moral transformation of the humble. With Cárdenas many pariahs have learned that they are men.

After Calles's creation of the PNR (which Cárdenas had turned into the mass-based and much more corporativist PRM), the problem of the presidential succession seemed to have been resolved, though not through any prospect of democratic elections. The party—which meant the outgoing President, more or less consulting with other party leaders—would choose the new President. The generals would keep their pistols holstered.

In his final two years of office he lowered the pressure on the Agrarian Reform, was much less receptive to labor agitation, and showed no interest in nationalizing any further industries. Yet the approaching end of his term in power—as in all societies, democracies included—opened him up to attacks from his opponents and controversy among his supporters looking toward the future. As he had done in 1932 in Michoacán, Cárdenas—going by his own reading of the political circumstances—selected a moderate military man (in this case one of his faithful followers: Manuel Ávila Camacho) rather than Múgica, who would have seemed like the natural, radical "Cardenista" successor.[79] But the elections would be neither clean nor peaceful, because General Juan Andrew Almazán (a long-ago Zapatista) ran as an opposition candidate and—like a lightning rod—attracted a range of anti-Cárdenas resentment—with his speeches against socialist education, the *ejido* ("a new colonial land grant"), Cárdenas himself, and the leaders of the PRM. He was in favor of municipal autonomy, small individually owned farms, and also (like Cárdenas of course) spoke in favor of the Allies against the Axis in the European war, which may have partly lost him at least the active support of the far-right Sinarquista movement (founded in 1937), which completely rejected the Mexican Revolution. Yet, in the middle classes, among labor groups like the oil workers and railwaymen resentful of the CTM, and also some old revolutionaries (*cartuchos quemados*—"spent bullets") including Diego Rivera—there was considerable support for Almazán against Ávila Camacho. But the brand-new machinery of the PRM was not going to lose its first elections. The CTM and the army diligently collaborated in the manipulation of the ballot boxes. "From the countryside," wrote Luis González, "a million votes were cast by ten thousand people (in favor of the official candidate)."

The Electoral College had the nerve to publish final figures that read: Almazán: 15,101 votes; Ávila Camacho: 2,476,641.

Cárdenas would be succeeded by a general but one with a strongly civilian temperament. On December 1, 1940, a new era—or rather the six years that were an introduction to it—opened in the political life of Mexico.

As Cárdenas's presidency faded into history, his reputation and prestige grew. His youth (he was only forty-five at the end of his term) continued to serve him, as he lived on through thirty more years of Mexican history. He became an icon, a kind of moral *Jefe Máximo,* the only true living Mexican Revolutionary, the moral conscience of the Revolution.

He continued to spend time with the common people. Sometimes he would become a judge who mediated disputes between *ejiditarios* and small property owners. Others took him for something more: Believing him a miracle worker, an old man brought his paralyzed son and in a supplicating voice, said: "Cure him for me, sir, with your miraculous hands, so that he can help me at my work."

Under the shade of the mesquite trees he would promise to build a school, introduce irrigation, send a teacher, open up the dirt road to the neighboring town. He continued attending to his flock, like a village priest for all villages. He would arrange to arrive unannounced in pueblos, so that the peasants would not "sacrifice their chickens" to honor him.

In some places, his Jeep would be the first car the inhabitants had ever seen. "Bring us the fire that doesn't burn, Tata." And Tata brought the electricity to many towns. At an even more personal level, in his own house in Mexico City, he took in and cared for eighteen children.

He would roam the roads of Mexico for the rest of his life, offering the common people his time, his wishes, and his hopes. He would remain the missionary father, motivated by compassion as had been the policies of his administration (with their mixed and uncertain results). And he would remain an influential and vital figure in Mexican political life, across the terms of Presidents with vastly different perspectives and moral values.

IV

THE MODERN STATE

By 1940, the word *Revolution*—always written with its capital letter—had assumed an official and definitive meaning. It was no longer the revolution of one caudillo or another. The Revolution was now the name for a single, permanent, all-encompassing process. It involved not only the armed struggle of 1910 to 1920 but the Constitution of 1917 as well as the process of social transformation and creation of institutions based on that program. But for the common people of Mexico, the Revolution had another meaning. It was an event that transcended the human order of things. It was natural or divine like earthquakes or droughts, a cataclysm of cosmic proportions and terrestrial origins, an eruption beyond the limits of history and much nearer to them than history. For better or worse, it had changed everyone's life. From either perspective, "before" and "after" were to be measured in Mexico by the Revolution.

Into the holy shrine of the *patria*—where the Aztecs, the Insurgents, and the Liberals had found their rest and enduring memory—the new secular saints, the caudillos of the Revolution, came riding in force. The patriotic calendar of holy days would expand to register the dates of their births and deaths and great deeds. The names of cities and towns and streets and schools would be changed to honor them. The grudges and hatreds that had separated them in life, or even led them to kill one another, now seemed mere accidents before the founding myth that united them—the Revolution, generous mother, who reconciled them all.

In its first decade (1910–1920), most of the blood had been shed in the direct physical confrontations of factions at war (and the repercussions on civilians). The lives lost in combat numbered 250,000. Related causes (typhus, influenza, hunger) took another 750,000. Much of the Porfirian elite (politicians, intellectuals, priests, soldiers, businessmen) vanished into death or exile. There was a moment, in 1915, when there were almost no professors for the University. Mines, factories, and haciendas closed. The banking and monetary system fell apart. Very few cattle herds remained. The railroad system—Don Porfirio's pride and achievement—suffered damage from which it would never recover. Only the oil sanctuary of Veracruz was preserved intact.

The second stage of the Revolution (1920–1935) witnessed far more complicated violence—ethnic, political, religious, and social. The Sonorans, "untamed" (*bronco*) men, fought ferociously against the Yaqui Indians and against the Catholics of the Cristiada (who were for them "Old Mexico") and also fought endlessly among themselves. The goal of the ethnic violence was to end once and for all the centuries-old resistance of the Yaquis. And the Sonoran generals would accomplish it, more savagely successful than even Porfirio Díaz (such was their repayment for the invaluable support they had received from these courageous Indians during the Revolution). Meanwhile, the civil war among the Sonorans had thrown the country back a century in time. Each region had its revolutionary caudillo turned cacique, a new "lord of lives and haciendas" with his dream of reaching the Presidential Chair.

Calles had closed the violent decade of the twenties with two solutions destined to endure: the formation of the PNR—the party of the State—and a final agreement with the Church to end the Cristiada. But social violence continued in many parts of the country. Before Cárdenas came to power, deputies from different blocs within the PNR would still sometimes open fire on one another in casinos, brothels, or on the very floor of Congress. Even at the height of Cardenismo, members of the old CROM fought in the streets against the new, socialist-oriented unions. By the mid-thirties, all this combat had an ideological basis: a confrontation between the Sonoran concept of the Revolution (focusing on economic progress promoted by the State, oriented toward private property, antireligious, and bordering, at moments, on fascism) and the Cardenista conception (concerned with social justice under the supervision and protection of the State, oriented toward collective property, tending to adopt socialist doctrines, and bordering, at times, on communism). Cárdenas's victory over Calles and the clarity of his social choices (if not his results) halted the killing. But the elections of 1940 would not be peaceful. There would be dozens killed and hundreds wounded. The stereotype of Mexico as the "land of the gun" continued to be well earned, though in comparison with the horrors occurring in the world that year, the violence of Mexico would look like child's play.

Along with bullets, the Revolution had brought its material achievements. Anyone who had lived during Porfirio's Centenary Fiestas—and people forty and over were old enough to remember—could appreciate the immense changes. There were now thousands of schools, innumerable public services, a broad network of roads, and extensive irrigation projects. The hacienda had disappeared. The governments of the Revolution—and Cárdenas most of all—had redistributed almost 15 percent of the national territory (around 26 million hectares) among 1.81 million peasants. In parts of the country individual ownership prevailed, but by 1940 half the rural population formed part of the new class of *ejiditarios*. There were 20,000 *ejidos*, of which 900 were collective. Criticized by many, and sometimes by the peasants themselves, the *ejido* had brought about an authentic revolution in the ownership of land.

In 1910 the slender working class of Mexico had known what it was like to strike but not as a right. It had been an exceptional act of disrespect and daring, which sometimes would end in the flow of their own blood. By 1940, thanks to Article 123 of the Constitution and the labor policies of Calles and Cárdenas, the workers held a position in society that was not only legal and legitimate but visible and strongly influential. And the CTM was beginning to integrate the entire Mexican working class. There had been a genuine revolution in the rights of labor.

In the struggle for the subsoil of Mexico—which other Presidents had waged but not won—Cárdenas had achieved the first great nationalist triumph of Mexico in the twentieth century, through reclaiming the oil (while Díaz, back in 1910, had in contrast opened all the doors of Mexico's wealth to foreign investors). There were many more social achievements. Had they been worth all the suffering? Had there been more construction than destruction? The dead had no voice, and for the people—the majority of whom had primarily *undergone* the Revolution—this government had grown from the Revolution and therefore it had the right to rule. Among the Revolutionaries themselves—participants and sympathizers, generals, professional men, and intellectuals, the old and the new devotees—questions were asked, judgments made, but even dissidents felt themselves to be part of the Revolution. It was the others, the bad Revolutionaries, who had betrayed her, violated her, not fulfilled her possibilities, led her astray and corrupted her. In 1940, the Revolution, that immense promise, remained alive and in force.

But the Revolution was now the State, with a legitimacy that came not from the ballot box but from the legendary bullets of the Revolution. Adapting Madero's famous slogan, "Valid Voting, No Reelection," the Revolutionary State had tranquilly blanked out the first phrase but scrupulously observed the second. Porfirio Díaz had been "reelected" seven times. It was hard to imagine any President now—after the assassination of Obregón—daring to try to reelect himself.

And the new Revolutionary State had a strength that came from its being as ancient as it was recent. Especially under Cárdenas, at a culminating point in its development, it combined the spirit of the missionaries with that of the Crown and the more hierarchical features of the Church—protective, corporative, and paternal. Though it expressed (in its essential functioning) the Spanish tradition of linkage between the "Two Majesties," the Revolutionary State—as molded by Cárdenas—had most of all supplanted the Church. The corporate State would offer its benediction to everyone (organized in groups with varying degrees of dependency); the State would protect and oversee and judge (nowhere more clearly than in the Agrarian Reform); the State would teach the truth, in its schools and, as the Church had always done, in the visual gospel of painted walls. The health and welfare of Mexicans were now primarily the responsibility of government. So often seen as the enemy of the Revolution, the Catholic Church would continue to leave its lasting legacy on the State produced by the great upheaval.

The Agrarian Reform—as the distribution of land—in practice developed into a state instrument of political control over the peasants, but it had clear Christian roots. Some saw it as the Mexican correlate of the biblical multiplication and sharing out of the bread. The Marxist leader Lombardo Toledano published a pamphlet, in 1921, titled *The Distribution of Land to the Poor Is Not Opposed to the Teachings of Our Lord Jesus Christ and the Holy Mother Church.* He lavishly quoted the Old and New Testaments, the fathers of the Church, and the encyclical *Rerum Novarum* of Pope Leo XIII to demonstrate that the agrarian distribution was only a continuation of the teachings of Christ. The pamphlet was illustrated by Diego Rivera: an Indian peasant behind his team and plough; beside him with her babe in arms, his wife; illuminating the scene from the heavens and surrounded with splendor, Jesus Christ.

End of the Ballad, Diego Rivera, 1928
(Jorge Rodríguez, Editorial Clío)

In his murals, Rivera would depict the Revolution as the coming of a brilliant light. It had been preceded—in Rivera's ultraofficialist version of Mexican history—by the unremitting darkness of the *Conquista* and religious and social slavery (all of which had descended upon another time of light, the earthly paradise of the Indians). With the new illumination came the sacred distribution of the land, the rural *maestra* teaching children to read while a soldier stands guard over the scene, the workers on strike and invincible. Zapata appears either as a beautiful living figure in white peasant clothes, as pure and noble as the white horse whose reins he holds, or dead, his body whole in its shroud, below the earth, germinating like the corn.

With greater subtlety than Rivera, perhaps because he himself had been a direct witness of the great revolutionary spree, the other great muralist, José Clemente Orozco, created the Adam and Eve of Mexico—Cortés and La Malinche, nude, hand in hand, a dead Indian at their feet. (The original sin of Mexico—the mestizo family born not from a simple union but from a complicity in bringing death.) A Franciscan missionary embraces an emaciated Indian, thus saving him, but his embrace is hypnotic; through his eyes he transmits the new faith, with fanatic concentration. A Hidalgo very like the real one—almost insane, seized by his "frenzy"—cleanses history with fire: His torch sweeps across a heaped-up mass of men and corpses stabbing one another with knives. And Orozco paints the Revolution itself as he had lived it—"farce, drama, barbarity." Light, in the murals and paintings of Orozco, is not Rivera's bright and pure luster of a Christian Heaven transformed into the paradise (or testing ground for triumphant saints) of the Revolution. It reflects the shifting shadows of history and often, more precisely, the red and black background glow of Purgatory or of Hell.

In Rivera, the human condition is no longer a fundamental issue. History has altered it; the world has been transformed once again, into a greater and better and more beneficent stability. The self-image of the new State and its decisions on the meaning of history are given their classic, sacred representation in his murals. And sentimentality already speaks through color and form—though at a high level of skill in the master, much less so in his many followers. In a sense, the Revolutionary State saw itself as a mural by Rivera—and would cling to its belief in ever less likely circumstances.

Only glimmers of tragedy and despair (mostly in the past and now overcome) appear in the work of Rivera, but for Orozco the

The Trench, José Clemente Orozco, 1922–1926
(Jorge Rodríguez, Editorial Clío)

vision of death—with its most tangled and arbitrary assertions—is at the very center of his view of the Revolution. The bodies fall majestically in the trenches, like statues come to life only to die. A statement of ideology sprouts skulls. The knives and the guns are not arranged—as with Rivera—in pleasing Italianate masses. They tear and blast the painted people and disturb any easy sense of balance in the viewer.

The culture and art of Mexico would follow Orozco rather than Rivera. The great novelists of the Revolution, writers of novelized autobiography—Martín Luis Guzmán, Mariano Azuela, José Vasconcelos—wrote, as Orozco painted, the agony of the Revolution, depicting its dark truths and not an artificial curtain of light. There would always be official artists and voices, but the real culture of Mexico would be a critical culture, more than anything an image in negative of the State. (And it would be so in Mexico because—with minor and transitory exceptions—freedom marked the world of art, culture, and thought.)

At the end of its own process of gestation through these centuries of history, incorporating past modes of power within itself and solidifying to resist a legacy of violence, the Revolutionary State faced, in 1940, a world swept by an ocean of war. Throughout the Revolution, Mexico had been turned in upon itself, meditating on its wounds and hopes, focused on redeeming the past. The pendulum of Mexican history was about to swing again—as it had after the wars of the nineteenth century and in the same direction—toward the future.

In the political life of Mexico the presidential succession had always been a trial by fire. Because he chose the wrong candidate in 1920, Carranza provoked a rebellion and paid with his life. By selecting Calles instead of De la Huerta in 1924, Obregón triggered a civil war among the caudillos of Sonora. At least in part because he slighted Madero's precept of "no reelection," Obregón was assassinated in 1928. Calles witnessed the parade of three Presidents chosen by himself until the fourth, Cárdenas, sent him into exile. In 1939 it was Cárdenas who had to exercise the most delicate of presidential decisions. His choice baffled many partisans of the Revolution, and not only the radicals who wanted Múgica but observers as prudent and intelligent as the intellectual Daniel Cosío Villegas:

> I was disappointed not in Cárdenas' term of office but in his choice of successor. When I realized that Cárdenas was supporting Ávila Camacho, unquestionably conservative in tempera-

> ment and political tendency, I knew that the Revolution was going to change.... Cárdenas could have found a man who would have continued his work, not curbed it. But the turn toward Ávila Camacho represented a change of direction.[1]

The act of "revealing" the new candidate, for Cárdenas, marked a real yielding of power. He could retain moral power—and he did, an immense moral influence right up to his death in 1970—but he, who had liquidated the Maximato, was not going to become a new *Jefe Máximo*. He felt that he had carried the social program of the Constitution to its ultimate consequences, but with realism and humility he weighed the tensions that this very process had created. His political instinct told him that he could not stretch the string of the reforms any tighter, at the risk of breaking it and plunging the country into a new civil war. He would write, years later:

> General Múgica, my very dear friend, was a widely known radical. We had avoided a civil war and we were enduring, because of the oil expropriation, terrible international pressure. Why choose a radical when I had already left a revolutionary instrument in place?... When we left power, the workers were organized; the peasants too and the Agrarian Reform were on the march ... the members of the army had been incorporated into the party of the Revolution. Was this, or was it not an instrument of progress with which Mexico could continue its liberation? I cannot make judgments about what happened afterward; I feel I was perfectly justified.[2]

And so he came to the conviction that Mexico needed peace. Peace to assimilate the fever with which the country had lived for thirty years, to decide how it would use its "revolutionary instrument," and above all, peace because external circumstances required it. "I wanted peace for the country, and in the period of conflict that the world had begun to experience with the Second World War, it was Manuel Ávila Camacho who could insure it."[3]

The fiercely contested elections of 1940 must have made Cárdenas even more certain of his diagnosis. The country's middle classes (and some other groups) supported Almazán, the leader of the Revolutionary Party of National Unification, who promised to end the social and ideological agitation of the Cardenista era. The forces organized by "the revolutionary instrument" supported Ávila Camacho. PRM gangs provoked bloody street fighting. On Election Day there were at least 30 dead and more than 150 wounded. Cárdenas at one point was ready to accept the victory of Almazán, but "the instrument" closed ranks and decreed the victory of the official candidate by an unbelievable margin. Almazán left for Havana and, to the displeasure of his followers, shortly afterward announced that he would not lead a new revolution.

The elections had brought the country to the edge of another abyss. Where would Múgica have led it?—Cárdenas might have asked, with relief, when on December 1, 1940, he yielded the presidential sash to his lifelong lieutenant, the likeable and conciliatory Ávila Camacho.

17

MANUEL ÁVILA CAMACHO

The Gentleman President

Hidden away in the Sierra Norte of Puebla, near the state of Veracruz, in a region of natural abundance and beauty surrounded by a variety of Indian cultures, rich in fruit trees and crops ranging from beans to coffee, endowed with a rich copper mine discovered in 1890 by the Italian businessman Vincenzo Lombardo Catti, the city of Teziutlán was a little mirror of Porfirian progress. A popular almanac reported in 1897: "Teziutlán is a commercial market of importance, very often visited by traveling salesmen from businesses in this country and abroad. . . . It depends on a group of businesses that handle significant capital and sell on a large scale in the principal markets of Europe and the United States."[1]

The numerous family of Manuel Ávila Castillo and Eufrosina Camacho Bello grew up, at the beginning of the century, within the middle class of that "internal port" with its fifteen thousand inhabitants. Following the eldest, Maximino, born in 1891, came eight more children. The second-oldest boy, Manuel, was born on April 25, 1896, the year that Porfirio Díaz was reelected for the fourth time, and two years before the railroad came to Teziutlán.

Maximino and Manuel completed their primary studies in the Liceo Teziuteco, a lay school that taught practical subjects and was becoming well-known in the region. It was attended by children of the

French and Italian immigrants who had established themselves in the nearby "hot country" of Veracruz and of course by the wealthy of Teziutlán, like Vicente Lombardo Toledano, the eldest grandson of Don Vincenzo Toledano, whose family passed their summer vacations at the Lago de Chapala in the remote state of Jalisco, a fashionable tourist site during the Porfirian period. The young Lombardo would accompany his father as they cruised that lake in one of the two family steamboats or—back home—hunted deer, boar, and other large game on the outskirts of Teziutlán.[2] The life of his fellow students, the Ávila Camachos, was much more modest. Their father, Manuel Ávila, formerly an overseer and administrator of haciendas in the Veracruz area, had a mule-driving business and would take Maximino on the extended trips required by his profession, while Eufrosina cared for the rest of the family.

The Revolution broke the patterns of life in Teziutlán. The Lombardo fortune sank headfirst and would remain only an ambiguous memory for Vicente, later to become one of Mexico's leading socialist intellectuals. The Ávila Camacho family lost their father in 1916. Maximino, as the eldest, took charge: He continued working as a mule-driver but also as a cowboy on various haciendas, as a mailman and even as a salesman for Singer sewing machines.

And then the two eldest sons went off to the Revolution. Bearing the burden of her children, Doña Eufrosina, like many another woman of the time, revealed surprising survival skills. Between 1915 and 1916 she ran a store in Puebla, later moved to Mexico City, where she bought passenger buses, then lived in Morelia and later in Sayula, Jalisco. Her itinerary was not capricious: She was following Maximino and Manuel. Finally, in the middle of the 1920s, she returned to her beloved Teziutlán, where she would live across from the Iglesia del Carmen as the faithful devotee she was of that Virgin. She purchased property in the center and outskirts of the city and later acquired a dump truck that would be rented out for the construction of the road from Teziutlán to the coast. The economic balance of the Revolution had not gone badly for her, but she had lost her husband and would see two of her sons die—Miguel, of influenza in 1918, and Eulogio, the youngest, who would be killed in 1932 during agrarian turmoil in the region. It is said that when Doña Eufrosina died in 1939, she made her son Manuel swear that he would never persecute the Church.[3]

The violent times were well suited to Maximino's personality. A friend of his youth remembered him as a "very mischievous boy, restless, sarcastic, emotional in his own way and above all adventurous."[4] In 1912, troubled with the problems of the family, he was bold enough to write a series of letters to President Madero. In March he offered the President his services. He wrote that he could form a group of volunteers within fifteen days. In June he asked the President to assign him "some commission . . . even one extremely dangerous." When he received no positive answer to his requests, he informed the President in November that he had resolved to enter the Military College, because he had "a genuine desire to be a soldier." His final communications were anguished: He no longer was asking for employment, but only "a small amount each month to support my little brother and pay for his education."[5]

Apparently Maximino spent some months at the National School for Officer Candidates. Close to the time of the Tragic Ten Days, he deserted and then joined the forces of General Gilberto Camacho in the Puebla region. Between 1913 and 1920 he participated, on the Constitutionalist side, in a number of military actions in Puebla, Oaxaca, and Veracruz, rising to the rank of cavalry colonel. In 1920 he joined the rebellion of Agua Prieta and became part of General Lázaro Cárdenas's brigade. And in 1924 he took part in the defense of Morelia against the forces of De la Huerta and was promoted because of his "heroic behavior." Years later he commanded troops and fought with considerable brutality against the Cristeros.

His official record of military service did not mention other "paramilitary" activities, such as his participation in the 1929 mass murder of student followers of Vasconcelos in Topilejo. During the early 1930s Maximino was already a powerful force in the politics of his state. In economic alliance with the millionaires William Jenkins and Axel Wenner-Gren, he would amass a decent fortune, worth two to three million pesos at the time and composed of cattle and farming ranches and magnificent country manors. (At his death, his wealth would be far greater.) In league with the skillful Puebla politician Gonzalo Bautista, he would become governor of the state in 1937. His close friend, a man no less fearsome, Gonzalo N. Santos, the cacique of San Luis Potosí, gave this description of him: "The governor of the state, Major General Maximino Ávila Camacho, was in command in Puebla, I mean in command and not just governing, because he commanded the military, the finance ministry, the telegraphs, the mails, the administration of the railroads, and the diocese."[6]

Santos—who was the least saintly of men—said that Maximino was "ferocious." And the pious souls of Puebla agreed, trembling at the sound of his name. Fathers with attractive daughters shuddered at the thought that Maximino might abduct them. And workers had even more to fear. Their strikes were put down by force over and over, and in

1937 they were forbidden to display those "filthy rags" (the red and black flags of the anarchists). Witnesses say he not only reprimanded his subordinates in the Puebla government, he sometimes struck them with a whip. When Maximino snapped his fingers, the rich men of Puebla ran to him as fast as they could move. "You don't fool around with Maximino," said Santos.

Manuel was a faithful copy of his brother—in reverse. Though his official record of army service does not list him as formally joining before 1919, there are credible indications of previous military activity at his brother's side. But in contrast to Maximino, Manuel showed no passion for the military life. Although he seems to have participated in actions against the armies of the Convention and especially, at the beginning of 1915, in the capture of Puebla by the illustrious Sonoran general Salvador Alvarado, his contributions were primarily administrative. Thanks to some modest training as a bookkeeper, he became Secretary of the Local Agrarian Commission of Puebla and paymaster for the Division of the East. In 1919, in the Huasteca, he would make the acquaintance of a young general, a good-natured and humanitarian man of his own age who would become a real brother to him: Lázaro Cárdenas.[7]

From then on, Manuel became Cárdenas's right hand, the Chief of His General Staff. Wherever the chance of war took Cárdenas, Manuel followed him: the Huasteca, preparing to hunt Carranza; Michoacán, saving the life of the radical Múgica; on the Isthmus of Tehuantepec, in Jalisco and the Bajío. Nevertheless, during the rebellion of De la Huerta in 1924, he was not to be found with Cárdenas at the forefront of battle but was installed in a hotel in Morelia. And there, functioning as the paymaster for the army, he was taken prisoner by General Enrique Estrada, who defeated Cárdenas and, immediately afterward, saved his life. To the relief of Doña Eufrosina, who was then living in Morelia, Estrada showed the same generosity toward her son Manuel.[8]

Halfway through the 1920s, Manuel had problems with the high command of the army. In the judgment of an investigatory committee, he could not prove that he had ever joined the Revolution. They claimed that there were no records either of his earlier activities or of his participation in the struggle or in the army. The committee recommended that "he not be recognized as possessing any military status at all." Perhaps Cárdenas's intervention with Calles may have gotten his comrade out of a tight spot. Under any circumstances, when the War of the Cristeros broke out, the officers of the army had other tasks more pressing than investigation of dossiers.[9]

Now in command of troops, according to his official service record, Ávila Camacho fought against the Cristeros. But the word *fought* is in his case ambiguous. His weapon was persuasion. It was well known that in Atotonilco el Alto and Sayula "he won the affection of his enemies because he was noble and magnanimous with spies whom he allowed to leave the military zone and settle in another city."[10] It was in Sayula that he met Soledad Orozco and would marry her in a religious ceremony.

People used say that Ávila Camacho was "a very kind man." He not only prevented his men from abusing the Cristeros; he had in fact reached a kind of modus vivendi with them, to avoid confrontations and to gain time. In April 1927, very early in the war, when his troops took the town of Pihuamo in Michoacán, he summoned the Cristero chiefs and offered them amnesty. For three years he persisted in his proposals of peace until the arrangements between the Church and the government favored the end of the Cristiada.

When the Cristero leaders received orders to demobilize their troops, they decided to send José Guízar Oceguera to confer with Ávila Camacho:

> Guízar Oceguera requested that the Cristero soldiers of Cotija be allowed to bear arms when attending "a mass in honor of the Virgin of San Juan del Barrio," a petition which was, to his surprise, accepted by General Ávila Camacho; he also asked that officers and leaders be allowed to keep their pistols, arguing that they might need them and this too was conceded; he pleaded that the combatants not be deprived of their horses, a plea that was accepted; he asked that ten pesos be given to each soldier, to which General Ávila Camacho responded that he had neither the money nor the authorization to do it. When Guízar Oceguera explained that it was advisable to give them the money in order to prevent them from committing robberies because of their need and suggested that the money could be obtained "from the administrator of the hacienda of Santa Clara and the rich men of that region who would benefit from peace," the suggestion was implemented by General Ávila Camacho.[11]

During the 1930s, Calles, the *Jefe Máximo,* gave him the command of the Twenty-ninth Operational Headquarters, located in Tabasco, territory of the most frenzied anti-Catholic ever born in Mexico, governor Tomás Garrido Canabal. Compared with that cacique, Plutarco Elías Calles was a model of Catholic piety. Garrido Canabal had a standing order that his lieutenants and public servants salute him martially each day with the cry "God does not exist!" to which he would respond, "Nor has he ever existed!" It is said that he named one of his sons Luzbel—the Spanish for Lucifer. (Graham Greene's novel *The Power and the Glory*—about a priest on the run secretly administering sacraments—was based on those years of defanaticizing fanaticism in Tabasco.) It was a matter of course that this possessed individual should clash with the temperate son of Doña Eufrosina. On one occasion Garrido Canabal tried to bribe him, and another time, pistol in hand, he tried to arrest two young men in the very offices of the military headquarters. "How could I deliver them over for him to slaughter them?" Manuel remembered years later. "I said to him, 'Whatever the reason for this persecution, it's shameful for you as governor to act like this.'" Yet again, Manuel saw persuasion as having an effect: "I feel sure that, through several later conversations, I managed to initiate some slight change in Garrido."[12]

When his friend Cárdenas became the presidential candidate in 1933, Manuel Ávila Camacho's lucky star ascended to heights that it is likely, deep within himself, he did not passionately desire. He became the Chief Administrative Official of the bureaucracy, later Assistant Minister of Defense and, from December 1937, Minister of Defense. His official service record lists his request, in early 1939, for an unlimited leave of absence in order to devote himself to affairs of "a private and political nature." The affairs were the process of competing for the presidency of the republic.

"Manuel is a piece of steak with eyes!" Maximino is supposed to have shouted in his office in the state capitol in Puebla, when a good friend, the *pelón tenebroso* ("the bald man in the shadows"), Gonzalo N. Santos, told him that General Cárdenas and an important group of governors were supporting his brother for the presidency. "I'm the eldest in the family, I've formed them since they were children. When Manuel was little I made him break a wild burro and gave him a peseta for it. . . . I made him a soldier when I had already been a military man for years. . . . I am the governor of Puebla, which used to be a nest of scorpions and now I have it perfectly under control. . . . Here no voice counts but mine!"[13]

Maximino, who held the rights of the firstborn in the Ávila Camacho family, felt that he had the same rights over "the family of the Revolution." Nevertheless that family chose, instead of the fierce *cacique*, the man who in time would become known as "the Gentleman President."

"He was a man who was a hundred percent prudent, in his actions, in his decisions," recalled General (and lawyer) Alfonso Corona del Rosal half a century later. He had met Ávila Camacho not on the field of battle but on a polo field, polo being the President's favorite sport and one in which he was a highly skilled defensive player. From then on, the personal qualities of the President caught the general's attention: "very stable, he wasn't changeable, he listened, he had patience, he didn't make sudden decisions, they were always sensible ones."[14] Other testimonies confirm his politeness. He would get up to receive visitors and "always wrapped his wishes in courteous expressions." He dressed discreetly, without pretension, in a jacket or a sport coat. He traveled by auto with a single aide. "He was very much in love with his wife, Doña Soledad," recalled Adolfo Orive Alba, the young director of the National Commission for Irrigation who used to accompany him in visiting villages where they would discuss plans for small-scale irrigation. The marriage produced no children but a love that caught Orive's attention: "For the general to meet a fashionable actress of the time—María Félix, Sofía Alvarez, Dolores del Río—would be like his becoming acquainted with a photograph. They pleased him but at a distance, like the faces of strangers."[15] (The fuller truth would seem to be that he was not a totally faithful husband but that his love for her was always central and compelling.)

At the extreme other end of the planet stood Maximino. He had "the presumptions of a dictator." Now that his brother was President of the Republic, Maximino one fine day—with his pistols—took over the Ministry of Communications, drove the Minister out, sat down in his chair, and telephoned Manuel to inform him of the administrative adjustment. It was an appointment the President had previously agreed to, and prudently (like a man maintaining control of a fire in a fuel depot) he permitted the unorthodox changing of the guard. Maximino in this incident concisely revealed—to the entire city of Mexico—his three most ominous qualities: arrogance, headlong haste, and soaring pride. He could be volcanic, irritable, and a great nuisance but also passionate and feeling. His romantic affairs with celebrated actresses were numerous and widely appreciated. Around 1945, the Spanish popular singer and flamenco dancer Conchita Martínez was a regular at his home. "Maximino," remembered Santos, for whom being "a drunk, a carouser and a gambler" were not defects but virtues—"had become infatuated with her and when her husband showed signs of 'not understanding,' Maximino sent some men to beat him up and kick him out of the country, leaving the beautiful Spaniard for himself."[16] Incidentally, Maximino was twice married and had at least eleven children in and out of matrimony. One of his sons, Eulogio, wound up living with Manuel and Soledad.

And Maximino had a passion for dressing well. The tailor who had served him since he was governor of Puebla, a Jew born in Warsaw, recalled years later:

> The General arrived in his enormous black Packard and parked right in front. It was my siesta time, but I—of course!—raised my gate and attended to him. Several *pistoleros* flanked him. A woman came in the car, it was the lady bullfighter Conchita Cintrón. He would lay his pistol down in the dressing room (it would have thrown off the fitting) and then, with his index finger, he would run through all the English fabrics on the shelves. "I want one gray suit, one light, one dark, one striped . . ." It could be ten suits at a time. When he would try them on, he would embrace me saying, "Ask me for anything, Maestro, maybe a gas station?" Of course I never asked him for a thing. He paid me well, and he behaved well with me.[17]

One day Maximino forgot his pistol with the inlaid gems on the handle. The good tailor almost had a heart attack. What to do with that "thing"? Immediately he telephoned "My General," and Maximino sent some of his *pistoleros* to collect it. To the tailor's son, who delivered that particular batch of suits, Maximino gave a huge tip, which lasted the boy a month. That was Maximino.

The two Ávila Camacho brothers shared what could be called a ranchero culture. They owned ranches, they liked ranchero songs, jokes, food. Maximino loved the bulls. It was said—an unlikely legend—that he had traveled to Spain as a young man with the intention of becoming a professional bullfighter. He was the owner of the Cuatro Caminos bullfighting ring, north of the capital, not far from La Herradura ("The Horseshoe") ranch, his brother's private residence. Manuel, on the other hand, adored horses. He would tell touching anecdotes about the horses in his life. One of his closest friends, Justo Fernández from Veracruz, recalled: "His fondness for horses was shared by all the military leaders of that period, because the Revolution was made on horseback and the railroad was called the iron horse. . . . Because he had been a cavalry dragoon, once he became President he gave an important impetus to the sport of horse racing. . . . At the end of his life, he owned [among other horses] two Arab stallions and a mare, La Aurora."[18]

Through his initiative and on land that had belonged to La Herradura, the Hipódromo de las Americas was constructed during his term of office. He would watch the races from a presidential box built specially for him. While "the ferocious" Maximino was firing off shots

Manuel Ávila Camacho (*right*) with his brother Maximino, 1939
(FINAH)

at his wild parties, perhaps the greatest pleasure for his brother, the Gentleman President, was to take a leisurely ride across his ranch: "There he was with his spouse, watching the agricultural work being done. . . . It pleased him to see, at five kilometers from La Herradura, what irrigation work could be done, what spring, small stream, small river could be put to use."[19]

Another trait that linked the brothers was their weak health. There was diabetes in the family. Manuel had a heart attack during his presidential campaign and two more while President. In early 1945, when the presidential race was just beginning, in which he intended, one more time, to make his rights of primogeniture prevail, Maximino "surrendered his soul to the Lord," apparently due to a massive heart attack. Earlier that day, he had spoken at a public function in Atlixco with many guests. Since suspicion is a Mexican weakness, there were some who wondered whether something more than seasoning had been added to the food.

At the beginning of the 1940s, before a newsstand in Mexico City, any citizen could point to the two most read and most independent daily newspapers: *Excélsior* (with its three editions) and *El Universal*. Beside them were *El Nacional* and *El Popular*, which were left-wing and nationalist. Both of these were official journals though under different jurisdictions, the first representing the government, the second the CTM. Among the magazines *Hoy* stood out prominently, along with the more superficial *Todo*, *Futuro*, *Sucesos*.

Until the invasion of Czechoslovakia in 1939, *El Universal* showed moderately pro-German sympathies, sincerely believing that the Munich Agreement had rectified territorial abuses created by the Treaty of Versailles and avoided a war in which "twelve, fifteen or twenty million of the youth of the European nations would fall on the field of battle."[20] For that newspaper, an additional advantage to the agreement was the isolation of the Soviet Union, a country of "Asiatic" origin whose revolution had been the work of "Jews." Like many other voices of the West, *El Universal* came to the truth late. With the Nazi invasion of Czechoslovakia in March 1939, it lost its hopes and was not surprised at the Hitler-Stalin pact. When in 1941 Hitler launched Operation Barbarossa against Russia, *El Universal* declared, "Perhaps the salvation of humanity lies in the war now tearing apart the totalitarian powers."[21]

The principal edition of *Excélsior*, through the later 1930s, maintained an attitude similar to that of its competitors. Nevertheless, there were editorial commentaries in favor of democracy as "a just form of government marked by tolerance, courtesy and true peace." But its critical emphasis was directed against "the tragic farce of communism" rather than "fascist dictatorship." After Munich, *Excélsior* editorialized, "As soon as the movement for the Sudetenland found new strength through the titanic energy of Hitler and the incomparable discipline of the German people, those who were oppressed in Czechoslovakia felt that they could count on a supremely powerful protector and they demanded their rights."[22] Six months afterward, *Excélsior* corrected, though belatedly, its enthusiasm for the "inestimable benefit" of Chamberlain's policies: "Czechoslovakia has disappeared from the map. . . . The democracies now have only one duty . . . to go to war, if necessary, against fascism and against communism, in defense of their sovereignty as free nations."[23] Its opinion would be no different after the Hitler-Stalin pact: "Bolshevism and fascism are two birds of a feather and it would be ingenuous to believe that when it comes to totalitarian issues, the two totalitarianisms would not come to an agreement."[24]

The "Late News" edition of *Excélsior* frequently included a section called "Perifonemas," which was widely read and discussed. Its anonymous author was the poet Salvador Novo, who wrote on April 4, 1939:

> [Hitler is] the warrior to whose voice the world listens. His words merit careful meditation; they batter like the blows of a hammer but are as clear and luminous as the most refined thought, free of demagogy but endowed with the most authentic patriotism, free of cowardice but endowed to the marrow with a will toward peace, a lasting peace.[25]

El Popular on the other hand—with its left-wing compass—read the Führer more accurately. When Czechoslovakia was invaded, it editorialized: "For months, since Munich, we have said that the fascist program of conquests and absorptions was not fully completed with the annexation of the Sudetenland. The betrayal by Chamberlain and Daladier peacefully gave Hitler the right to tear Czechoslovakia apart. . . . Nazism is on the march toward other lands on other continents."[26]

And *El Popular* made a prediction that could not have been more cogent: "The cruelest war in history is being readied."

Among the weekly political publications of the time, none was as popular as the magazine *Hoy*. It was less moderate than the daily newspapers in its pro-Axis sympathies. José Pagés Llergo published various reports from Germany and Japan that became famous. "I had the honor," he recalled somewhat later, "of being the first journalist in the last three years to speak with Hitler." (He was lying or—to be kinder—immensely exaggerating. A photo was snapped of Pagés not far from

Hitler because he had managed to approach him as part of a crowd at an airport, getting near enough to throw a question at the Führer, which was never answered.) In a long review of an anti-Hitler book, Pagés compared Hitler to Napoleon and Julius Caesar and remarked: "Hitler is the man of destiny for Germany . . . a mind of prodigious perception . . . like Napoleon he is an utter realist, he hates fictions and farces. . . . despite his bursts of anger, his excessive egoism and his physiological anomalies, Hitler's answers are violent but brilliant . . . his glance makes you shiver . . . he is a visionary of genius."[27]

The attitudes of the political reviews toward the Axis oscillated between sympathy and outright support. At the latter extreme stood *Timón* ("The Rudder"), an ephemeral journal financed by the Germans and directed by no less than the old and embittered José Vasconcelos. Whatever inner process had brought one of the most extraordinary men of Spanish-speaking culture to embrace the cause of Hitler continues to be a mystery. Perhaps when so explosive an ensemble of antagonisms was in play (anticommunism, anti-Americanism, anti-Semitism, antiliberalism) the result could only be a commitment to hate. "Hitler's strength does not come from the barracks," wrote Vasconcelos, "but from the book [*Mein Kampf*] inspired by his penetrating intelligence."[28]

But ideological prejudice leading to blindness in the face of reality was not found exclusively in journals of the right. The intellectual left also showed marks of fanaticism, as in the case of the political journal *Combate*. Narciso Bassols, the director, had represented Mexico at the League of Nations, where he condemned Mussolini's invasion of Ethiopia. On his return to Mexico, he wrote articles denouncing the Munich Agreement as an "infamy" and a "capitulation." But suddenly the picture turned complex. How was one to interpret the Molotov–Von Ribbentrop nonaggression treaty? Bassols, like almost all the left, justified and supported it. And even though Leon Trotsky, in Mexico itself, was revealing many of the grim facts about Stalin's purges and mass murders during the late thirties, neither Bassols nor the rest of the Mexican left would publicly accept or assert the truth.

Confronted with the facts, the Cardenista government had retained much greater serenity, a spirit of justice, and a sense of reality. Mexico protested against the Italian invasion of Ethiopia, supported the Republican cause against Franco more than any other country, stood up for the suffering Jews at Geneva, condemned the Japanese invasion of China, and was the only country besides the Soviet Union that rejected Hitler's *Anschluss* of Austria. The agreement on this issue with the Soviets did not imply collusion. At least in this whole area of policy, the individual human being was much more important to Cárdenas than ideologies and governments; and so he gave asylum to Trotsky and to many persecuted by various nations. On May 22, 1940, Cárdenas wrote in his personal diary: "Germany is developing a most active process of propaganda and is attempting in every other way as well to win people over to its cause. Its campaign of expansion, like any assault on whatever country, is in conflict with the feelings of the Mexican people."[29]

To situate itself beyond fascism and communism, in a realm of relative freedom, Mexico required above all an attitude of prudence and caution. As the war went on, Cárdenas was able to verify the wisdom of his decision on the succession—these were, precisely, the specific virtues of Manuel Ávila Camacho. It was not for nothing that they had shared the military life, hundreds of days and nights facing the enemy or his threatening shadow. The anxieties of the Revolution had created their friendship. The anxieties of the Second World War would consolidate it.

Although when the war began the Mexican government had declared itself neutral, by 1941 it was clear that its stance was definitely opposed to the Axis powers and that Mexico was openly collaborating with the United States. The government of Ávila Camacho suspended commercial relations with Germany and withdrew all Mexican consuls assigned to Germany, France, and Holland. At the same time it recognized, in Mexico, the accredited diplomatic representatives of the countries that had been invaded, which meant nonrecognition of the German conquests.

With the Japanese attack on Pearl Harbor, Mexico took more drastic measures against the Axis. It broke off all diplomatic and consular relations and supported continental solidarity by giving permission for all boats and naval ships of any American country to anchor in Mexican waters and ports, requiring only prior notice to the authorities. The same decree authorized the executive power, in case of obvious emergency, to permit the forces of other American republics to cross the territory of the nation.[30]

At the same time, President Ávila Camacho made the first preparations for military collaboration with the United States. On December 10 the Military Region of the Pacific was created, uniting a number of military zones from north to south. General Lázaro Cárdenas was assigned command.

Mexico entered the war on May 23, 1942. It was the first time the country had ever involved itself in a war outside the Americas. The pretext was the least of it: a German attack in the Gulf of Mexico on the oil tankers *Faja de Oro* and *Potrero del Llano*. Consistency and realism were the fundamental reasons: to continue the sensible international trajectory of the country in the 1930s and to support the United States, which had recently entered the conflict.

A survey by the magazine *Tiempo* reflected the indecisiveness of public opinion, but little by little the strong pro-German feeling that was especially predominant in the middle classes (a reflection of traditional anti-Yankeeism) began to move toward moderate sympathy for the Allied cause.[31] It was the first time since the United States had initiated its Good Neighbor policy that a glimmer of authentic Pan-Americanism ran across the continent, beginning with the meeting between Franklin Roosevelt and Ávila Camacho in the Mexican city of Monterrey. And then came the support of Hollywood, with Walt Disney's sympathetic cartoon characters: Mexican *charros*, the dancing ladies of Rio de Janeiro, and Argentine gauchos—along with the discreet introduction of likeable Americans into Mexican films. But nothing helped the Allies more than victory—achieved with a dash of Mexican participation. Most dramatically, in the middle of 1944, Mexico sent a group of pilots into the Pacific war. This was Squadron 201, which returned at war's end to a hero's welcome. They had fought in the Philippines, especially over Luzon, and later over Formosa. Five of their planes had been shot down, and they had received an official message of appreciation from General Douglas MacArthur. The Mexican cinema proceeded to make films about the exploits, real and imagined, of these descendants of the "illustrious lineage of the eagle warriors."

Before Mexico had made its official decision, the war had polarized opinions in the press. The internal front of Mexico was also divided into irreconcilable factions. In the western states, the old wound of the Cristiada not only did not heal, it was opened again with the rise of Sinarquismo, an armed movement (pro-Nazi in its sympathies) whose object, supported by tens of thousands of peasants, was to finish off "atheistic and communist tendencies." At the other extreme, with a small number of militants but a similar fanatic zeal, the Communist Party ceaselessly denounced "fascist and Falangist tendencies" and took the credit for the death of that "objective ally" of "Nazi fascism" that Leon Trotsky had supposedly been. The religious intolerance and extravagant dogmatism of Article 3 of the Constitution, which since the amendment of 1934 prescribed "socialist education," were an unpardonable insult in the eyes of the Church and the faithful of the middle and upper classes. For good measure, in the politicized circles of these classes, the bloody elections of 1940 had left a trail of disenchantment: They distrusted Ávila Camacho (Joke: How does Ávila Camacho resemble the income tax? In that he's a goddamn imposition!), but they were contemptuous of former presidential candidate Almazán. When he returned to the country, they booed and shouted insults at his image in the newsreels.

The workers' movement was also living through a time of ideological agitation. Beginning in the 1920s, the unions had been converted into a crucial element of power and legitimacy, but in the 1940s their political destiny was uncertain. The CTM (its member unions were normally the official ones) was gradually consolidating its power, but unionism of the moderate left was equally strong and sought a greater margin of independence with respect to the government. And there were also unions of anarchist and Communist inspiration that operated with efficiency, boldness, and confrontation. The situation was equally tense in the countryside. While the peasants who belonged to *ejidos* were learning how to live under the new conditions of the Agrarian Reform, the small and medium landowners (and the still substantial number of large landowners) were seeking legal and extralegal methods for defending their patrimony.

An early sign of conciliation on the part of Ávila Camacho was the plural nature of his cabinet. There were Cardenistas, Callistas, and a rising group—the Avilacamachistas. Particularly important in the last faction was the Minister of the Interior, Miguel Alemán Valdés, who was only thirty-six years old.

Another sign of the changing times was the personal willingness of Ávila Camacho to set limits on his own power. When, in the context of the war, he decreed a suspension of constitutional freedoms, he immediately proceeded to regulate the situation through a permanent statute permitting review by the Attorney General. In 1934 Cárdenas had serenely suppressed the lifetime appointments of Supreme Court judges, ordering their terms to coincide with the presidential six years. In 1941 Ávila Camacho announced an amendment that reestablished lifetime appointments. In his choice of judges, he asked for and followed the advice of Luis Cabrera, well known as an opponent of Cardenismo. The selected judges showed themselves to be honest and even sometimes independent.

He took his most decisive step toward national unity in September 1942, when he invited all the ex-Presidents to appear with him, together, on September 11. Upon a large reviewing stand constructed for the occasion in the center of the Zócalo, Abelardo L. Rodríguez, Pascual Ortiz Rubio, Adolfo de la Huerta, and Emilio Portes Gil took their places while, to the left and the right of the president, respectively and appropriately, Lázaro Cárdenas and Plutarco Elías Calles filled out the line. "How are you, My General?" Cárdenas is supposed to have asked and Calles to have answered, "Very well, My General." Three years later, when Calles died, Cárdenas would write in his diary, "When he returned from exile, General Calles greeted me with nobility." The Gentleman President had achieved a small miracle.[32]

For Ávila Camacho, "National Unity" was more than just a pleasant phrase or a state of mind to maintain in confrontation with the Axis powers, or a balancing of his cabinet, or even the gist of his symbolic gesture toward the former Presidents—it was the guiding principle of his six-year term. No former President of twentieth-century Mexico had dared to publicly declare a religious affiliation. Perhaps to honor the final wishes of Doña Eufrosina (combined with his own inclination and political astuteness), Ávila Camacho in 1940 had confessed to the journal *Hoy*: "I am a believer . . . but to be Catholic is not to be clerical nor fanatic. I am a Catholic by origin, in moral feeling."[33] No one read the qualifications, only the affirmation. "The sentence conquered a great deal," his niece remembered.[34] For the left, the declaration was a mistake: One should not be tolerant with those who "were creating a counterrevolutionary consciousness in children and young people."[35] But public opinion, as a whole, received it with relief: It represented the principle of reconciliation between the Revolutionary State and the Catholic Church. Religious processions and manifestations returned to the streets, and schools of a clearly religious nature began to operate with a certain degree of freedom.

Sinarquismo, with its secret organization and its mystico-military theories, struck some blows, like the temporary "takeovers" of Guadalajara and Morelia in 1941, symbolic actions that were a bit more than just spectacle. To slow the advance of the Sinarquistas and to gain their cooperation in time of war, the government granted them the colony of María Auxiliadora in Baja California. At the end of his term, the conciliatory disposition of Ávila Camacho achieved something more: channeling the Sinarquista movement toward political action by registering it as a political party. Nevertheless, hard-line Sinarquista cells continued to function within the Church. And on April 14, 1944, a lieutenant named De la Lama y Rojas, a Catholic extremist, shot at President Ávila Camacho and would say about his failure to kill the President: "I was not able, sadly, to complete my mission." (Maximino would seem to have accomplished his. The failed assassin was shot dead while in police custody, probably on Maximino's orders.)[36]

But extremist religious rejection of the government of the Revolution was on its way out of fashion. Ávila Camacho had begun a trend toward healing the great division. Not only the political left considered this new policy of conciliation mistaken or at least excessive. Solitary democratic and liberal voices also feared the reappearance of the

Church in the political life of Mexico. In an essay that would trigger verbose responses, Daniel Cosío Villegas, near the end of Ávila Camacho's term, warned about the danger of the right wing coming to power. In those circumstances, "the velvet and emaciated hand of the Church would show itself as it really is, with all its lust for power, with the incurable obscurantism with which it faces the problems of the country and the real men within this country."[37]

An additional proof of good faith (and of faith) on the part of the government was the amendment of the controversial Article 3 of the Constitution, dealing with education. It was done late in the regime (December of 1945), and the ideas and activities of the National Action Party (PAN)—founded in 1939 by Manuel Gómez Morín—significantly contributed to it. For Gómez Morín it was a personal battle. Between 1933 and 1934, at the time when the State was trying to impose socialist education, Gómez Morín had been Rector of the University (the UNAM in Mexico City, the country's premier university). It had been a time of intense radical pressure, when the government went to the extreme of cutting the University budget and threatening to close the institution, which it considered "elitist, superfluous and reactionary." While Gómez Morín carefully managed the limited public funds remaining and engineered a successful campaign for private funds, dozens of professors (with the support of thousands of students) defended freedom of instruction by continuing to teach on voluntarily reduced wages.

Perhaps the most important debate of the Mexican twentieth century had been staged at that time between the philosopher Antonio Caso and Vicente Lombardo Toledano, the former defending liberty of thought and research, the latter proposing the adoption of socialism as scientific truth. Pressure from the professors would force Lombardo Toledano out of his position as director of the University Preparatory School. And the government would eventually give in and restore the University's funding.

Now, eleven years later, Gómez Morín's political party, the PAN, launched a final battle on this issue—in the press and in the national Legislature. They were supported by the Church and the *Unión Nacional de Padres de Familia* (the National Association of Fathers of Families). The final emendations to Article 3 did not come up to PAN's expectations (they would have gone to the opposite extreme, prohibiting the requirement of lay education), but they abolished the clauses on education necessarily being "socialist" and "defanaticizing." Education was instead to be "democratic and national."[38]

* * *

National unity included the world of culture. Thanks to the good offices of the Gentleman President, José Vasconcelos had agreed to become director of a beautiful official library, the Biblioteca Nacional. In addition, the *Maestro de América*, his collaboration with Nazi Germany now a thing of the past, joined other distinguished Mexicans (the painters Orozco and Rivera, the musician Carlos Chávez, the cardiologist Ignacio Chávez, the writer Mariano Azuela, and others) in the foundation, in 1943, of the Colegio Nacional, modeled after the Collège de France as an institution through which the government recognized its sages. As a natural corollary, the government established the National Prizes for Arts and Sciences in 1946.

In that climate of internal harmony, there arose a number of institutions of study and scientific investigation, as well as journals and publishing houses, almost all of outstanding quality and destined to endure. Their founders belonged to the Generation of 1915—also known as the "Generation of the Seven Sages." Gómez Morín and Cosío Villegas were members of this group, men who had been of student age during the latter days of the Revolution and had studied under Antonio Caso, the only member of the Ateneo de la Juventud who had remained in Mexico. Cosío Villegas had founded the publishing house Fondo de Cultura Económica, and the economist Jesús Silva Herzog, since 1942, had been publishing the important journal, *Cuadernos Americanos*. And in 1943 a new and high-level institution of learning, the Technological Institute of Monterrey, was founded.

Catholic culture experienced a notable renaissance, after almost a century of retreat. Two humanists, the brothers Méndez Placarte, launched the review *Abside*; Gómez Morín created the publishing house Editorial Jus, and Father Ramón Martínez Silva established the Universidad Iberoamericana. In the scientific sphere, Luis Enrique Erro, in 1941, founded and directed the Astronomical Observatory of Tonantzintla.

President Ávila Camacho could even feel himself to be a legitimate member of the Generation of 1915, thanks to his creation of the Mexican Institute for Social Security. Gómez Morín had conceived of it in 1927, but only now did it become reality. These and other moves established the basis for a new health and social-welfare system.

In a less Olympian area, the workers' movement, the Gentleman President tried in vain to tighten the knots (as he himself put it) of "love and harmony." Unity, in the strict sense of the word, seemed impossible. The cost of living, which in the cities sometimes reached critically high levels, provoked workers' protests. There were also other, ideological factors at work. The orthodox left of the Communist Party followed the directions of Moscow. The CROM continued to exist, as did the anarchist CGT. On the unionized left functioning within the CTM and still allied with the government, Ávila Camacho's "*paisano*," Lombardo Toledano, had great influence, but the time he spent involved with the Workers' Center for Latin America (the CTAL) favored the growing power, since 1941, of Fidel Velázquez, a moderate leader with the ideology of a chameleon, who had already shown his dexterousness by removing the Communists from the top leadership of the CTM in 1937.

With Lombardo Toledano's help, Ávila Camacho achieved some important agreements. In December 1941 the CTM "requested" the President "to indicate the tasks proper for us . . . in order to resolve in the best way possible the crisis that weighs upon our people."[39] The "best way" was the creation, in June 1942, of a National Workers' Council, which decided "to avoid confrontations" and "give its total support to the government." On April 7, 1945, the CTM signed an agreement between workers and industry (*Pacto Obrero-Industrial*), which prescribed cooperation between owners and workers to achieve "the industrial revolution" in Mexico.

On occasion the President intervened directly to calm minds and conciliate interests. It happened with one of the most conflict-ridden public enterprises in the country: the National Railways. The issue as always was whether to comply with the wishes of the pugnacious Association of Railroad Workers or to modernize the enterprise. Under Ávila Camacho, the confrontation was merely postponed. When union elections among three feuding groups (independent, pro-Communist, pro-government) led to the victory of the independents, loser resentment drove the President to sit them all down at a table and apply his well-known method of persuasion. He got them to form an Executive Coalition Committee, but it never really functioned. An attempt on the part of management to adjust the collective contract encountered the natural resistance of the independent leaders. The next President would have much more disagreeable surprises in store for the Association of Railroad Workers.[40]

Seen from whatever angle, the demilitarization of Mexico was a phenomenon as admirable as it was mysterious, especially when contrasted with the military background of Mexican history. With the well-known exceptions of two brief interregna (the Restored Republic of 1867–1876 and the fifteen months of Madero's presidency), the country, ever since Independence, had been primarily governed by soldiers. Between 1910 and 1920 it had gone through a revolutionary experience unique in Latin America—also the work of generals who did not easily accept giving up power. To support the 35,000 cavalrymen and 31,000 infantrymen who fought the Cristeros during Calles's presidency, the government expended 30 percent of its budget, more than triple what it spent on education (9 percent). Many other Latin American countries managed to maintain a healthier symmetry: like Argentina (17 percent and 20 percent) or Uruguay (15 percent and 14 percent). But from that time on, the relative proportion of the budget dedicated to the military steadily declined. Cárdenas lowered it to 19 percent, Ávila Camacho, despite the war, to 14 percent, and future presidents would bring it down to as low as 6 percent.[41] Along with this decrease, the number of positions assigned in each cabinet to military men kept shrinking. How can we explain this peaceful transition from military to civilian power?

An important influence—as opposed to the Argentine, Peruvian, or Chilean armies—was the fact that the postrevolutionary Mexican Army lacked a tradition. For the sons of the upper classes in those South American countries, most of them creoles, a military career was not the necessary means for access to power but a mark of social prestige that was usually transmitted from generation to generation. In Mexico the wave of change in the nineteenth century had altered this situation. The creole army of Iturbide and Santa Anna had little glory to boast of; it had lost the war with the United States, it had lost the War of the Reform, and it had become an accomplice to the French Intervention. Its Liberal adversaries were not career military men but newcomers with military talent, most of them mestizos like Porfirio Díaz. These soldiers had garnered the historic victories of the War of the French Intervention—May 5 (1862) or April 2 (1867). When Díaz came to power, along with many of his old comrades-in-arms, the process of forming a professional army began again and went on across the decades of the dictatorship. In the final Porfirian years, when the veterans had died, the young generals perfected themselves at the French military school of Saint-Cyr, maintained a sumptuous academy in Mexico, and allowed themselves the luxury of patenting various weapons. It was an army organized more for show than for action. Madero preserved it almost intact, and this single decision cost him the presidency and his life. With notable exceptions like Felipe Ángeles, the high officials of this army surged to power with Victoriano Huerta and were liquidated by the Constitutionalist revolution.

The new Revolutionary generals, some of them military geniuses, were once again newcomers as soldiers, and they behaved accordingly. Former schoolmasters, farmers, public employees, and bookkeepers, they dreamed of dismounting from their horses, removing their epaulets, and taking charge of a civilian government. Some of them found more pleasure than others in the profession of soldiering, but when the military adventure was over, both sorts—the born civilians like Manuel Ávila Camacho and the natural soldiers like his brother Maximino—had become the undisputed rulers of the country.

Obregón was another natural soldier. He felt enough strength as a military man to violate the primary dogma of the Revolution—no reelection—and try to emulate Don Porfirio. Nor did he shrink at eliminating many of the generals who had supported him through the years of war, including some of his closest lieutenants. If he had achieved reelection, he would have probably remained in power for one or more decades, establishing a new militarist tradition. His assassination changed the picture.

Calles was a natural civilian. His specific genius was for politics, and though he had been a brave and capable soldier, he felt comfortable in a suit. His historic speech on the passage from "the era of the caudillos to the era of institutions" set the country moving, in moderate installments, toward the solution Carranza had wanted to impose prematurely in 1919. The mere fact that the generals, weakened by then but still powerful, were willing to lay their pistols on the table and agree to ordered rules for accession to power—that by itself was a notable advance in self-control. It was a step preceding full civilian government.

Cárdenas moved farther in the same direction. A thoroughly honorable military man but without great luster or talent, he had always disapproved of the classic barracks behavior of his brother officers: the drinking bouts, brothel escapades, arbitrary arrests and kidnappings, shootings and robberies during the days of the armed struggle. Though fearless (and much too reckless) in his battles, he found violence of any kind repugnant. When he came to power, he further speeded up the political subordination of the officer class.

When Cárdenas ordered them en masse into the PRM, the officers complied without a word of protest. The docility of the military was due not only to the gradual victory of a civilian trend among the caudillos. There were also demographic reasons. Many of the representatives of the Revolutionary generation (born between 1875 and 1890) had died in the Revolution or in the successive "cleansings" of the 1920s. Others were living in exile. The rest remained but with their energies considerably reduced by weariness and age. The very fact that President Cárdenas was much younger operated as a symbolic limit to the ambitions of the oldest generals, the "spent cartridges." Although the more radical among them resented the fact that Cardenas had "dug up" a "little man" who was "an upstart" and made him President, most of them merely kept quiet or else lined up behind him. They did not support the opposition on the left (Múgica) or the right (Almazán).

Cárdenas's military generation had entered the Revolution late and held lesser commands, or else they had joined in the 1920s and seen action against the various "disloyalties" of those years. Many had become politicians and were recognized as such, but once they had exchanged the military tunic for a civilian suit, it was difficult to change back again. A typical case was General Juan Andrew Almazán himself, who had long been a prosperous builder of roads. When his election bid was defeated with a heavy use of fraud, it was his civilian followers rather than the military ones who were prepared to take up arms for him. But, though he exposed himself to public scorn, Almazán did one more service to civilian rule in Mexico—he renounced the road of violence.

During the presidency of Manuel Ávila Camacho, the shelving of the military was completed. In his inaugural address, the President, with total clarity, marked the new direction: "Any attempt on the part of politics to penetrate the precincts of the barracks deprives civic life of a guarantee and provokes division among the armed forces. . . . We must enhance our armed forces as an immaculate defense for our institutions."[42] By a new ukase of the President, the officers left the PRM, returned to their barracks, and were ordered not to meddle in politics.

To ensure that "the army would always favor the government, defend its institutions, and remain loyal," the government began to explore a whole slew of methods. On land from La Herradura ranch donated by Ávila Camacho, the army began to construct a modern central headquarters and a central military hospital. The government supported the creation of schools of military medicine and engineering, and special schools were opened for "children of the army." A savings fund was created to serve as a basis for the founding of the Army and Navy Bank. Another, less honorable type of aid for high-ranking officers came through granting them succulent concessions, the management of gas stations, for example.

"To face the bulls of Jaral, you need horses from the same place," went a popular proverb. It was the generals of the Revolution—those for whom the saddle was not a definitive home—who faced and tamed the generals of the Revolution. Men like Calles, and especially the tandem of Cárdenas and Ávila Camacho, on their own initiative exchanged the politics of weapons for the weapons of politics, and even more extraordinary, they passed these weapons down to a generation of civilians. Their greatest lesson was in setting limits to their own power, a willingness to leave the historical stage at the right time. Their successors, the university graduates, benefited from their decision, but when the moment arrived, they would lack the vision and stature to apply it to their own tenure in power.

Some generals established themselves strongly in states or regions and constructed local domains (*cacicazgos*) run with an iron hand. The State did not oppose them but made use of them for its own benefit. "You obey me, I protect you," was the golden rule. Cárdenas had no other choice but to use "the bludgeon" against the cacique Cedillo, but President Ávila Camacho was only moved to dispense "bread" to such local rulers. In this, as in many other ways, the governments of the Revolution adjusted to the Porfirian tradition.

A truly perfect example of this process was Gonzalo N. Santos, who inherited the *cacicazgo* of the Cedillos in the state of San Luis Potosí. A founding member (Card No. 6) of the PNR, professional organizer of victorious electoral campaigns by hook or by crook, soldier, deputy, senator when necessary, Santos had accumulated tens of thousands of hectares under the protection of the Revolution. He had twelve thousand armed men, "*his* Huastecos," under his direct command. At the end of his life, he did an extremely rare thing for a Mexican politician. He produced his own memoirs—cheerful, macabre, and very well written. And so we know—from his own testimony—that his professional killers accompanied him wherever he went. If a priest preached against him, he calmly had him shot. If he liked a piece of land, he would approach the owner and ask him to set a price. If the owner refused or seemed doubtful, Santos would throw up his hands: "Let his widow decide!"

In exchange for impunity within his own territory, Santos contributed significant services "to the Revolution." In the elections of 1929, a newspaper in his capital city reported, "When more than 2,000 people had collected, two trucks full of PNR policemen began to fire on the crowd with Thompson machine guns." There were dead and wounded, but Santos, who directed the operation, described it as "firing into the air to drive away the quail." During Ávila Camacho's campaign, Santos took on various jobs, from obtaining the commitments of old Revolutionary leaders to dispersing "quail" with his Thompson. On the day of the 1940 elections, he covered himself with glory. In his memoirs, he describes "the bloody elections of 1940" with absolute honesty and sincerity, as if it had been a military campaign:

> We moved against the polling station with our guns out and since there were some shots fired at us we responded with energy, firing

away freely. We grabbed all the ballet boxes full of votes for Almazán and we broke them and we carried away all the papers and records and ballots and so on, and we left the polling-station table in pieces and drove away all the Almazán supporters.

At another polling station, he discussed procedure with the poll-watchers:

I said to the poll-watchers, "When you transcribe the voting list and fill up the box at voting time, don't discriminate against the dead, because they're all citizens and they have their right to vote."[43]

When his compadre Ávila Camacho came to power, Santos collected in cash and in services:

"Look, compadre" . . . I must have said to the President, "so that you don't think I'm a man who has to be coaxed about everything, I'm going to ask you to buy my *Tres Filos* ranch, in the municipality of Tamuín. It's 17 thousand hectares that I bought from some little old American ladies who inherited it and live in the United States. Let the Ministry of Agriculture and Cattle-Raising buy 10 thousand hectares and the other 7,000 I'm going to give you as a present, dividing them into small properties." "Set me a price, compadre," Don Manuel said to me. "Five hundred thousand pesos," I answered him. He said to me, "Go right over to Marte R. Gómez and he'll pay you at once." And so it was.[44]

Around 1943, he collected in services. Santos was about to become governor of his state for three years, a term specified in the Constitution. But his plans were more ambitious. In the presence of the Minister of the Interior, Miguel Alemán, he listened to the judgment that Alemán's secretary, Rogerio de la Selva, had prepared in response to a request Santos had made:

"You can't be elected legally for six years, because the Congress does not have time to amend the Constitution since it's going to be elected on the same day you will be." I answered him, with Alemán right there: "These are very small problems for a man of character. There will be two calls for elections in San Luis Potosí. First the deputies will be elected and then will come the election of the governor when the deputies have already amended the Constitution to

> lengthen the term." Alemán said to De la Selva, "Don't quibble any more or keep Gonzalo waiting, 'we'll go torture the Constitution' and let his term be six years." And so it was. The Constitution did not complain about having been tortured.[45]

His "mandate," as he comically called his absolute power, did not only last through his six years as governor. In San Luis Potosí, not a fly moved without Santos's permission. From his ranch, El Gargaleote, Santos ordered "whoever followed him and whoever followed the man who followed him." Who had the guts to oppose him? He became known (the most famous of his nicknames) as the proverbial "Dark Chestnut Horse—that dies before it gets tired."

Unlike Maximino, Santos lived on for many years, tightly controlling his feudal domain, but his very eccentricity, like Maximino's, was a sign of anachronism. In postwar Mexico, in a world of planes and automobiles, the old-fashioned regional caciques had begun to be museum pieces. (Though in some areas—especially the most backward, like the state of Guerrero—somewhat more modernized but very real caciques still exist.)

In the harsh words of the intellectual Miguel Palacios Macedo, the peasants were "the political herd" of the Revolution. It was a sad sight to see them being bused in from their villages, before the elections of 1940, to parade in support of Ávila Camacho. "Long live the man we said before!" some of them shouted, not even knowing the name of the candidate. After the massive and indiscriminate distribution carried out by Cárdenas, the agrarian problem had worsened into the most painful and urgent of the country's dilemmas.

Although he trusted in the *ejido* as the means for the economic rescue of the country, Cárdenas had fundamentally conceived of the distribution of land in terms of a religious humanitarianism—like a vast ceremony of moral restitution to the Mexican poor. Little by little he himself had become aware that merely handing out land was not enough to ensure even minimum prosperity. What he never could (or wished) to see was the hidden legal and political flaw in his conception. The *ejiditario* could not legally sell, mortgage, rent, or traffic in any way with his land. If he stopped cultivating it for two years, he would lose it. His fate rested in the hands of the *Ejidal* Committee, which could take his allotment of land away at any moment, even by inventing an accusation. The *ejido* had freed the peon from the hacienda, but its magnanimous protective spirit had reproduced, in practice, the same paradoxical inequalities that existed in the colonial *Leyes de Indias*.

Critically needed in the Agrarian Reform was a strategy that might have carried it out in stages, first in the areas that produced crops for industry, then in the poorer, grain-growing areas. Time had not been spent on the needed agricultural research. No efforts had been made to evaluate the changes in crops and techniques that would best counteract the unfavorable natural conditions of Mexican agriculture. There had been no consistency in the way the Reform operated, among other reasons because "the allocations had not been dictated by prudence or necessity but rather by the urge to be seen as the greatest distributor of lands."[46]

For the founder of the PAN party, Manuel Gómez Morín, the greatest problem lay in the political conception of the *ejido*. At the root of protection was control. To maintain control of the peasant population in order to perpetuate its own power, the goverment denied ownership of the land to the peasant. The "ejidal rights" of the peasants meant little in practice: If they did not legally own the land they worked, they were naturally not moved to invest in it. The Ejidal Committees controlled the land, and they dealt out favors in exchange for political support, always in behalf of the ruling party. And one vital mode of control was access to loans. The Agricultural and Ejidal Banks were not really banks. They used money from the federal Treasury to exert control over the peasants; and they also, in the process, annually lost a fifth of their investments.[47]

The government of Ávila Camacho recognized some of these problems and from the first tried, partially, to turn off the faucet. This President distributed only 5 million hectares, as opposed to the 17.89 million Cárdenas had apportioned. But while the paperwork became slower and more difficult for the functioning of the *ejidos*, the assignment of private allotments and certificates of immunity to smaller independent landowners flowed with greater ease. And like his former Sonoran chiefs, Ávila Camacho returned to an emphasis on irrigation as the key to agricultural development in a country as arid as Mexico. At the end of the war, the government assigned 15 percent of its budget to irrigation, and during his six-year term, thirty-five major irrigation projects were completed, most of them in the potentially productive zones of the center and north. The total acreage with new or improved access to water amounted to 549,129 hectares, more than all previous governments together had accomplished. Stimulated by this impetus and by the expansion of foreign demand, the private agricultural sector, the rancheros, increased their production of commercial crops like fruits and vegetables.[48]

Opinions on the problem of Mexican agriculture continued to

divide Mexicans. The old Revolutionaries would always live with the conviction that the *ejido*—whatever its possible limitations—had liberated the peasant. If it had not made him richer, at least it had given him greater dignity. "The hacendado," Marte R. Gómez responded to a critic of the *ejido*, "made his work-gangs of peons walk out to the fields singing the Alabado [the Hymn of Praise to the Eucharist] . . . they had to kiss the hand of the lord of the hacienda and call him 'father' or at least bow deeply to him, sombrero in hand. . . . debts were inherited from generation to generation . . . there was no possible escape, because he who was born in the hacienda, though legally not its slave, was surely its servant."[49] The Agrarian Reform had fundamentally altered these colonial customs.

Those critics of the Agrarian Reform who argued for individual ownership of land allotments were met with paternalistic arguments. What would the peasants do with their lands? They would sell them to the first bidder, it would encourage a return to the great estates; the peasants would become day laborers again on what had briefly been their land. The Agrarian Reform had been no panacea, but any return to the past seemed worse.

But in the judgment of Cosío Villegas, who had studied agricultural economy at Cornell, the Agrarian Reform, that sacrament of the revolutionary creed, was in a state of crisis: "The truth is that it was in the worst possible situation. It had been harsh enough in its destructive performance to arouse all the hatred and rage against it from those who had suffered . . . but in its constructive aspect, its success had not been clear enough to maintain the indomitable faith of those who had hoped it would bring happiness on earth to ten or twelve million Mexicans."[50]

Ávila Camacho's presidency marked the beginning of a long, arduous process: the rise of the National Action Party, the PAN. Their democratic credentials were suspect, given the closeness between its members and the Catholic hierarchy, the sympathy many of them had shown toward Franco's regime, and their initial lack of enthusiasm for Mexico's entrance into the war. But with all its limitations, PAN proved to be a legitimate descendant of the two great movements toward democracy of the Mexican twentieth century: Madero's Revolution and José Vasconcelos's campaign to break the governmental monopoly of the ruling party.

Gómez Morín had worked since 1927 to organize an independent political grouping. PAN, at its inception in 1939, impartially united Maderistas and Huertistas, intellectuals and men of finance, but its basic membership consisted of independent professional men from the middle classes of the capital and provinces. Economic autonomy would be decisive for the survival of the party. "We all had to continue working," Gómez Morín explained. "It was a principle that . . . everyone considered his political work as one more feature of the daily agenda: attending to the children, going home, working and *doing politics*."[51] It was a kind of apostolic mission. The predominance of the PRM was so complete (it had never lost a race for President, Senator, Deputy, Governor, Municipal President) that even to think of power seemed impossible, inconceivable: "When we founded the National Action Party, we said that it was no job for a day but an ongoing fight forever."[52]

From the very beginning, PAN's internal functioning was democratic: open discussions, voluntary and individual membership, decisions by majority vote. Its basic platform, among other points, opposed the anticlerical articles of the Constitution, supported "municipal autonomy," called for hereditary family ownership of the *ejidos* and breaking the connections between the benefits of the Agrarian Reform and loyalty to the PRM. And the party stood for financial accountability in government and the freedom of the vote.[53]

In 1943 PAN participated for the first time in legislative elections with fifty candidates and deputies. Contending that twenty-one of them had won, the party tried to defend its case before the Federal Electoral College, which was due to begin its hearings on August 4. On that day the representatives of PAN discovered that the iron door of the hearing room was closed—to them. The PRM members showed their credentials and flowed inside. There followed the "battle of the pushes." While the PRM deputies elbowed their way in, the PAN members were shoved outside. Shortly afterward, the Electoral College awarded all seats to the PRM. None of the Panista candidates served in the Thirty-ninth Legislature.

In 1945, in the city of León, PAN had conducted a vigorous campaign and was convinced that it had won at least some minor positions. The PRM declared all its own candidates victorious. There was a massive demonstration in the main plaza. (Seeded through the demonstrators, there seem to have been government provocateurs, trying to precipitate violence that could justify an assault on the protestors.) Federal troops opened fire on the crowd. The count was twenty-six dead and some thirty wounded.[54]

The National Action Party immediately sent a note to the Supreme Court of Justice, asking for its intervention. On January 7, 1946, the Court for the first time in its history did intervene, voting 20–1 to investigate the events. On the following day, Ávila Camacho set in motion a process to remove the state governor. Nothing else of significance would happen, but at least it had not all been completely swept under the rug.

Of particular importance for the political history of the country would be the initiatives on electoral reform presented by PAN in 1942 and 1945. The objective was to create independent agencies within the government to supervise elections. The CTM argued that "the present electoral law should continue in effect since it guarantees the existence of revolutionary regimes that favor, as part of an evolutionary process, the exercise of civil rights." Nevertheless, the Ministry of the Interior would on December 6, 1945, send to Congress a new federal electoral law, incorporating the PAN proposals, such as the requirement for strictly maintained membership lists for political parties, and the creation of an independent Federal Commission for Overseeing Elections. The only reservation was that the new electoral agencies, without exception, continued to be the exclusive monopoly of the party in power.[55]

In 1946 PAN put up 110 candidates, of whom 4 were declared winners. In the Legislature these deputies presented nearly eighty different initiatives, dealing with the central bank, education, monetary policy, the need for protecting small landowners against expropriation, the organization of democracy, the electoral system, the reorganization of the railroads, irrigation: "Many of these initiatives were not even debated," Gómez Morín recalled. "In the whole three years we didn't succeed in getting them debated. Some served as the basis of initiatives presented by the government, more or less changing the wording . . . we didn't oppose them; it didn't matter to us who presented the initiative, only that things got done."[56]

Gómez Morín himself was one of the losing candidates. He had run in the district of Parral, in his home state of Chihuahua. PRM hooligans broke into his meetings, turning portable phonographs up to maximum volume. (In other places they used large drums, mariachis, animals, claxons, and eventually gunfire.)

Cosío Villegas would direct some of his acid criticism at the PAN. Gómez Morín, an old friend of Cosío's since their years at the National Preparatory School, had the "undeniable merit" of having been "the first person to shake up the political apathy characteristic of the Mexican," but aside from this moral triumph he granted him almost nothing. The Panistas lacked any popular attraction, they belonged to the upper froth of society, they circled in the shadows of the churches, they represented "plutocratic and very transient interests . . . [and were without] principles or men."[57]

PAN's intimacy with the Church had aroused the Jacobin within Cosío Villegas. And it is true that, within PAN, there was an authoritarian current barely distinguishable from Sinarquismo, but there was more to the party, and most important, it did have men and it did have principles. The Panista lawyers in the Legislature had fleetingly awoken the most servile arm of the government from its sleep. Clearly, Cosío Villegas, great reader that he was, had not read their numerous initiatives.

The military alliance with the United States favored a settlement of the lingering oil controversy, a return to the U.S. practice of buying silver from Mexico (suspended because of the oil expropriation), and the almost total cancellation of interest due on Mexico's long-standing debt to the U.S. Banking Commission, which dated back to 1921. The Mexican economy began a qualitative leap forward, especially in the industrial sector. Protective tariffs and government stimuli, through loans and tax exemptions, helped the process along.

The country's new mission was to manufacture what it previously had to import. While Europe and the United States were settling the destiny of the world, Mexico—a country that traditionally supplied raw materials and precious metals—showed a modest surge in areas like metal products, textiles, manufactured foods, chemicals, domestic electrical appliances, furniture, and the construction industry. At the same time, new banks were founded to compete with those established during the Porfirian era, as well as large department stores like El Puerto de Liverpool. Due to the arrival of substantial capital seeking protection during the war, Mexico accumulated an important reserve of funds, but it could not be used to import tools of production or indispensable spare parts. One of the negative effects of this situation was an inflation of almost 100 percent through the five years of the war.[58]

It was a great age for publishing, radio, the Mexican cinema, and for the expansion of tourism. The Fondo de Cultura Económica became the most important publishing house in the Spanish-speaking world. The exiled writers and thinkers from Spain—fleeing Franco's physical and spiritual destruction of Spanish intellectual life—produced a wealth of translations from European literature and thought. These were circulated throughout Latin America, where the Fondo (and other Mexican publishers) created a true revolution of cultural communication.

Beginning in the forties, Acapulco became the honeymoon spot par excellence. Agustín Lara, the romantic composer of the century in Mexico, wrote his famous song "María Bonita" during those years (in praise of his wife, the actress María Félix): "Remember Acapulco, lovely María . . ." Not far from the city's old airport was the Papagayo Hotel, property of Juan Andrew Almazán (some called it his "consolation prize"). A North American investor named Blumenthal—who carried on business in the luxurious Ciro's Bar of the Hotel de la Reforma in Mexico City—opened another, magnificent hotel in Acapulco, the Casablanca, which after the war began to attract Hollywood movie stars. From then on tourism, like so many other Mexican functions, had a caudillo, in this case Miguel Alemán, Minister of the Interior. He saw the activity as a gold mine for the development of the country and also, of course, for his own.

Mexico became synonymous with its music and its actors. The Latin American market bought everything in advance from the country that was experiencing its "golden age" of cinema. The previous decade had prepared the ground. The major themes of Mexican cinema had then been explored by the caudillo of film, Fernando de Fuentes: the bitter remembrances of the Revolution (*Vámonos con Pancho Villa*, "Let's Go with Pancho Villa" of 1933; *El Compadre Mendoza* of 1934), the very successful genre of ranchero melodrama (*Allá en el Rancho Grande*, "There on the Rancho Grande" of 1936), the early exploration of urban themes (*La Casa del Ogro*—"The House of the Tyrant" of 1940). There were already the beginnings of a star system: the brothers Soler, Joaquín Pardavé, and above all, Mario Moreno "Cantinflas," who in 1940 made the film that defined him as a comic actor, a true Mexican contribution to the picaresque art of the world: *Ahí está el detalle* ("There's a Small Problem There"). But substantial investment was lacking, and that came with the war.

The Office of the Coordinator of Inter-American Affairs for the American State Department favored support for the two media (radio and cinema) considered strategic in a country of few readers like Mexico. Starting in 1942, capital, equipment, agreements, exchanges, and grants for technicians flowed into Mexico from Hollywood or Washington. Twentieth Century–Fox donated sound equipment to Clasa Studios; RKO supported the construction of the Churubusco Studios (founded by the radio magnate Emilio Azcárraga Vidaurreta in 1944). The impetus from abroad encouraged Mexican money, both public and private, to flow into the industry as well. With the help of the Banco Cinematográfico (founded in 1942), various companies were created in which not only the cinematic impresarios Wallerstein and Ripstein invested but also Maximino Ávila Camacho himself. And eighty-five feature films were made in 1945, compared with twenty-nine in 1940.

With the country's entrance into the war, the nationalist imagination was unleashed: Movie Mexicans discovered nests of Japanese or German spies or died in battle. *Escuadron 201* (1945) invented battles for the famous air squadron to arouse patriotic feelings; a Spanish exile (played by Angel Garasa) died in the air defending the new fatherland, which had years before received him with open arms.

The wartime reductions in the film industries of Europe and the United States favored the consolidation of a Mexican star system, creating legends in the Spanish-speaking world. While the tormented Lupe Vélez killed herself in Hollywood, Dolores del Río returned to Mexico, leaving behind her lover, Orson Welles, to play romanticized Indian roles that seemed to have stepped out of a mural by Diego Rivera. But the most celebrated actors of the period—film lovers in *El peñón de las animas* ("The Rock of Souls") of 1942 and husband and wife ten years later—were María Félix and Jorge Negrete. After her role in *Doña Bárbara* (a novel by the Venezuelan Rómulo Gallegos, which he himself adapted for the screen), the gorgeous María would be known everywhere in the Spanish-speaking world as "La Doña." And Jorge Negrete, the "Singing Charro," would be received like a god in Madrid, Havana, or Buenos Aires.[59]

Jorge Negrete was President Ávila Camacho's favorite *charro*, elevated to a national prototype. But that thundering affirmation of ranchero Mexico, those songs in which the Singing Charro exclaimed, "I am Mexican, my land is fierce!" or "Jalisco, don't back down!" that macho who boasted that he must "eat that prickly pear even if it stabs my hand," that Juan Charrasqueado who was "a drinker, a carouser and a gambler," that supremely noble and "always the lover" corporal who, mounted on a fine mare, serenaded the landowner's daughter with his lusty friends, that Quixote who accompanied by his own Sancho—the actor *el Chicote*—roamed the village fairs singing and kissing girls and tipping his sombrero "bordered with silver" and then waged his gunfights in the cantinas—that character stood for the past, not the future.

Urban Mexicans, even those of modest means, no longer identified with the image of the *charro*, but at the same time they loved that vanishing vision of themselves; and they gave him a long and romantic *adiós*. The sport of cockfighting, which for centuries had been the most important popular diversion, lost its place in those years to wrestling. This too was part of the changing picture. Cantinflas ceased being the rascal from the provinces to become the unpunished invader, the "upstart," the "poor bastard," the "scum," who coming from the impoverished outskirts of the city establishes himself at the center of urban institutions in order to disrupt them. Jorge Negrete himself would end his days, in 1953, as the leader of the Actors' Union, a role very distant from the bucolic ranchero evenings of his movies.

The countryside was going out of style. Mexico was still far from being an urban nation; 65 percent of all Mexicans actually lived in the country. Nevertheless, among the ruling elites, a fundamental mutation was occurring, a change of paradigm, from the rural to the urban, from the peasant to the worker, from the agricultural to the industrial. The process was evident in politics and in business. The old Revolutionary elite had owned agricultural enterprises: Obregón planted chickpeas, Calles had sugar mills, Cárdenas agricultural ranches, Ávila Camacho cattle ranches. The young generation rising to power came from the provinces, but they had acquired a different economic viewpoint. Some of their money came from the urbanization of ruined haciendas.

Significantly, during those same years, a film genre of nostalgia for the age of Porfirio came into fashion. A well-known comic actor, Joaquín Pardavé, played "Don Susanito Peñafiel y Somellera, friend and servant to Don Porfirio." Around him the plot would reconstruct the Porfirian environment—the literary and artistic bohemia, the variety theater, the music of the Spanish operettas, the modes of dress, and the life of the court. "Ay, what times, Don Simón!" said the words of a catchy song. Enchanted with the new genre, the old rich (who were few) and the new (who were many) began to assume their social positions with an ostentatious naturalness that had not been seen in Mexico for many years—sumptuous dances "in black and white" at the old Country Club founded by the English colony of Mexico City in 1905, huge bets on the jai alai contests, new social sections in the newspapers. The moral climate was changing, and more and more it was coming to resemble the regime that the Revolution had deposed.

Was Porfirian Mexico rising from its ashes? No, it was "the Revolution" descending from its horse to enter an automobile (preferably a Packard) and once in it, coming to think that, after all, "order, peace, and progress" were not such insignificant objectives. The ruling elites of the country—most of them unconcerned with ideologies—approved of the change. They argued that the country could not stagnate. Perhaps for this reason even the Communists supported the presidential candidacy of the man whom Vicente Lombardo Toledano had baptized "the Cub of the Revolution": Miguel Alemán Valdés.

Months before the "Revelation" of the candidate, Gonzalo N. Santos, the Dark Chestnut Horse, had received the assignment of flattering Maximino and persuading him that Alemán was "the right man." The "ferocious" Maximino thundered against his brother, claimed his rights to occupy the Chair beloved by all, swore to kill that "pretty boy Alemán"; but all this was (perhaps) countered by the mysterious banquet at Atlixco. Maximino would rest in peace, and so would many Mexicans.

Other members of the cabinet aspired to "the Big One," but the Gentleman President, with courteous means, dissuaded them. General Miguel Henríquez Guzmán, a former comrade of Cárdenas, wanted to continue the military line for at least another six years, but Ávila Camacho was determined to pass on his power to a civilian. Only the Minister of Foreign Relations, Ezequiel Padilla, broke discipline and launched his candidacy outside the official party. On their part, PAN offered to make Luis Cabrera a candidate, but he refused because of his age.

For the first Sunday of July in 1946, when the elections would be held, the official party had a new name: It was no longer the PRM but the PRI, *Partido Revolucionario Institucional.* Once again there were street brawls and deaths, but nowhere near as many as in 1940. The official propaganda, which painted Padilla as an ally of the United States, had its effect: His popularity was in no way comparable to Alemán's. The final figures were: Alemán 77.9 percent, Padilla 19.33 percent.[60]

It was in November 1946, in the midst of the generalized euphoria created by the impending arrival in office of "the Cub," that Daniel Cosío Villegas wrote "The Crisis of Mexico." He published it the following year in *Cuadernos Americanos,* the famous Mexican review of ideas that circulated throughout Latin America. The essay was the dissonant chord in the fiesta. Cosío Villegas announced nothing less than the death of the Mexican Revolution.

Cosío spoke of the Revolution as a historical fact and a national program that had corresponded "genuinely and deeply to the needs of the country" but was now on its deathbed. One by one he examined each of its original intentions—the impulse toward democracy, the commitment of the people to social justice and economic improvement in country and city, the nationalist zeal, the crusade for education—explaining in each case how they had been adulterated or abandoned.

Madero had destroyed Porfirismo, but he had not constructed democracy; Calles and Cárdenas destroyed the great landed estates, but they did not create a new Mexican agriculture. And Mexico retained almost nothing of the original apostolic spirit of the new education. Cosío Villegas argued that the Revolution should not become a closed historic cycle. "The only ray of hope—very pallid and distant, to be sure—was that the Revolution itself would purge its men and reaffirm its principles." Mexico had found its way in the Revolution. To abandon that way was not only a mistake, it was suicide: "If we do not reaffirm the principles, but only sweep them out of sight; if we do not purge the men, but only adorn them with Sunday clothes or titles . . . of lawyers! then there will be no regeneration in Mexico and, as a result, regeneration will come from outside, and in no very great length of time the country will lose much of its national existence."[61]

They let the dogs loose on Cosío Villegas. Someone called him "the gravedigger of the Revolution." The private secretary of the new President warned him not to publish anything like that again.

In public, the Minister of the Interior spared his life: "In Mexico, señores, no person holding unorthodox beliefs will be persecuted."[62] This pardon, disdainful of his ideas and of the anguished warning he had extended, would hurt him "more than prison or death." From Renan's *History of the People of Israel*, Cosío copied down a passage: "Narrow minds always accuse those who see clearly of desiring the misfortunes they reveal. The duty of Cassandra is the saddest that can settle upon the friends of truth."

Shortly afterward, he asked the Rockefeller Foundation for a grant to study the transition from the Porfirian regime to the Revolution, so as to help him understand the transition from the Revolution to "Neoporfirismo."

But as the six years came to an end, the happiest man in Mexico (after Miguel Alemán himself, the Cub of the Revolution), was President Ávila Camacho. His personal style had produced good results: peace abroad, order within, progress in the cities. A sensible, conciliatory policy had improved relations with the Church. Like the old founder of a family business who yields control to his son, the university graduate, Ávila Camacho was now bequeathing power to his symbolic son. "What a fine thing it is that university graduates can now become President!" he said to an associate on the next-to-last day of his term. "I belong to the army, and I love it very much. But for Mexico the era of the generals is over now. I am sure that civilians will successfully do their duty."[63]

Ávila Camacho and Alemán "were very similar men," recalls a friend of both, the intelligent Veracruzan politician Marco Antonio Muñoz. "They had a very strong friendship: They spent long periods of time riding on La Herradura Ranch; their wives—Doña Cholita and Doña Beatriz—were good friends, and Sr. Presidente Ávila Camacho moreover was a compadre of the *Licenciado* ["Graduate" or "Lawyer"] Alemán. I understand that he and Doña Cholita baptized the *Licenciado* Alemán's youngest son, Jorgito."[64]

And as a symbol of the new age, the General and the *Licenciado* also confronted each other on the field—of a golf course.

18

MIGUEL ALEMÁN

The Businessman President and the System

On May 17, 1947, a new play opened in the Teatro de Bellas Artes in Mexico City: *El Gesticulador* ("The Imposter") by Rodolfo Usigli. The public welcomed it with great interest, but the government reacted violently. Some of its performances were canceled, and critics were paid to attack the play in the press. Though Usigli, with premonitory awareness, had written the work in 1938 and published it in the literary review *El Hijo Pródigo*, no director had dared to put it on the stage. Their fear was well founded. Using the real language of politics for the first time in Mexican literature, the play proclaimed not only the death of the Revolution—which Cosío Villegas was documenting at that very time—but its transfiguration into the lie of its perennial, institutional existence.

The central character is César Rubio, a mediocre professor specializing in the history of the Mexican Revolution, who has moved to the north of the country in the hope of improving his luck. He is accompanied by his wife, Elena, and his children, Julia and Miguel. Shortly after they arrive in their new city, they happen to be visited by a professor from Harvard University who is doing research for his biography of a genuine precursor of the Revolution: a general who was also named César Rubio. During a conversation with his American colleague, the professor suddenly decides to claim that he himself is the real César

Rubio. Knowing that the general died by assassination and that almost nobody left alive remembers him, the professor convinces the "gringo" that he is telling the truth. He, César Rubio, disillusioned by the moral direction of the Revolution, in a final act of abnegation, had made a decision to disappear into anonymity until now, thirty years later, when he has decided to reveal his true identity.

The story makes the *New York Times*. The political forces in the state approach Rubio reverently. A former combatant, almost blind, recognizes him with deep emotion. The public acclaims him. He immediately receives proposals to run for governor of the state against General Navarro, a typical Revolutionary cacique who has entrenched himself in power. While his wife looks on powerless—though in private she threatens to expose the farce and leave the city—César the professor is transfigured. He knows that he is not the real César Rubio, but he *feels* that he is or should be. The mask and the real features fuse. He loses his individual identity. The lie turns as real as truth. And he feels that his lie will revitalize the truth of the Revolution. He tells his new followers who listen to him entranced:

> The great politician comes to be the beat, the heart of things . . . he is the hub of the wheel; when he breaks, or rots, the wheel, which is the people, shatters; he separates everything that should not work together, he binds everything that should not exist apart. At the beginning, this movement of the people whirling around the one creates a sensation of emptiness and death; later the one discovers the function of this movement, the rhythm of the wheel that would not work without a hub, without the one. And he feels the only peace that comes with power—which is to move and make others move, on time and with the flow of time. And because of this it happens that the politician, in Mexico, is the greatest creator and the greatest destroyer.

Apart from Elena, who sadly submits to the lie, there is only one other person who knows the truth: his opponent, the corrupt General Navarro. Unseen by them, the son Miguel overhears Navarro speaking to his father with contempt: "I don't know how you've been shameless enough, brazen enough to begin this farce. . . . Your first name is César and your last name is Rubio, but that is all you have in common with the general." Yet the professor is not intimidated:

> It may be that I am not the great César Rubio. But, who are you? Who is everyone in Mexico? Wherever you look, you see imposters,

> impersonators, fakes; assassins disguised as heroes, bourgeois disguised as leaders; thieves disguised as deputies, ministers disguised as wise men, caciques disguised as democrats, charlatans disguised as lawyers, demagogues disguised as men. Who asks them to account for themselves? They are all hypocritical imposters.

Navarro has a secret of his own—he is the man who assassinated the real General Rubio. The professor reveals that he—the historical researcher—knows of Navarro's crime, and the general then realizes that he cannot publicly denounce the imposture. Each one is aware of the other's lie. The people must choose between them in an election, but before that can happen, Navarro's hired gunmen assassinate Rubio. Navarro pretends to be shocked, and he announces to the people:

> César Rubio has fallen at the hands of the Reaction while defending the ideals of the Revolution. I admired him . . . I was ready to withdraw in his favor because he is the governor that we needed. (*Murmur of approbation.*) But if I am elected, I will make his memory more venerated than any, ever, the memory of César Rubio, martyr of the Revolution, victim of the conspiracies of fanatics and reactionaries. . . . The capital of the state will bear his name, we will build a university in his honor, a monument that will carry his memory forward to future generations. (*A burst of applause interrupts him.*)

"You're disgusting!" Miguel in another scene says to Navarro, "and for Mexico you are a bloodsucker! . . . But that's not what matters to me. . . . It's the truth that matters and I will speak it! And I will shout it out!" He swears that he will find proof that his father was not a hero and that Navarro is a murderer. But Navarro is not disturbed. "*Viva Navarro!*" shout the people at the sight of him. "No, boys, no!" he answers, "*Viva César Rubio!*" Miguel says only, "The truth!"

The truth is that the Revolution has died, but the reality is that it cannot die. The wheel of politics that "connects everything" around the politician "that is its hub" needs the Revolution in order to continue moving, to construct and to destroy. It is because of this that the professor and the assassin in their disguises complement each other. Both feed the myth of the Revolution: One dies reincarnating it and the other—his murderer—honors it. Only Miguel "wants the truth so that he can live"; only he "hungers and thirsts for the truth" and cannot breathe in this "atmosphere of lies," but his destiny is to leave the picture, to flee in the final scene:

> He covers his face with his hands for a moment and seems about to break down, but he straightens up. Then, in desperation, he picks up his suitcase. . . . The sun is blinding. Miguel leaves the scene, fleeing from the very shadow of César Rubio, which will pursue him all his life.[1]

Beginning in 1940 and with the passage of time, the goals of the Revolution had been wiped away without being resolved. In itself, the indignant official reaction against the essay of Cosío Villegas proved the accuracy of his diagnosis. He called for "a purging of men and a reaffirmation of principles," but the practical resolution reached by the "Revolutionary Family" was something quite other. It did purge its men, opening the door to a new generation of young lawyers, but they, once in power, radically changed the principles of the Revolution, on occasion for the better, and pretended that, by doing this, they were actually carrying them out.

"Politics connects the one with everything that is originally there," César Rubio says—after his transfiguration. "He knows the cause and the purpose of everything. At the same time he knows that the one cannot reveal them. He knows the price of a man." The President is the owner of truth and of men. He decrees what, and who, is or is not Revolutionary. And so, in 1946, the party that brought the new President of Mexico to power had changed its name and given itself the luxury (or blundered into a Freudian slip) of adopting a name that of itself implied a contradiction in terms and, therefore, a lie, but a lie accepted as truth: the PRI, *Partido Revolucionario Institucional* (the Institutional Revolutionary Party).

At the hub of power, in 1946 as always, as in the time of the *tlatoanis* or the viceroys, stood the new leader, the President. Around him whirled the wheel, that is, the people. The hub moved and moved the people. He was "the greatest creator and the greatest destroyer." Like César Rubio, Miguel Alemán took on the role of a Revolutionary who was dead, but he ended by believing and feeling that thanks to himself, and in him, the Revolution was alive. The wheel of power became a fiesta of disguises. Some dressed up through cynicism; others were not even aware that they were wearing costumes; some sincerely believed that they were not. The wheel grew until it included almost the entire country. The Revolution had died, but in the Zócalo the crowd shouted: "*Viva la Revolución!*"

Pursued by the shadow of his father, the false revolutionary César Rubio, Miguel the son would dedicate his life to the search for truth.

There were Mexicans who would join him in that search. Pursued by the shadow of his father—the true Revolutionary Miguel Alemán González—the new President, Miguel Alemán Valdés, would install the regime of a sham Revolution: the "Institutional Revolution." Most Mexicans, of their own free will, would adjust to it. Very substantial groups would prosper under its shade. The real history of how this lie became reality is more dramatic than that of César Rubio himself.

"The Cub of the Revolution" was not merely a well-chosen campaign phrase. In the case of Miguel Alemán Valdés, it was the stamp of his personal destiny. He was the son of General Miguel Alemán González, one of the first precursors of the Revolution, a man who had done almost nothing else in his life but continually make the Revolution.

The year was 1906. The broad stage was the state of Veracruz. "Only Veracruz is beautiful," goes one of the infinite number of sayings current among the Veracruzanos as they thank God for the grace of being born in an area gifted with every imaginable natural opulence. The smaller stage was the tropical region in the southern part of the state. The zone of Los Tuxtlas is near the basin of the Coatzacoalcos River and its tributary, the Papaloapan. It is surrounded by lakes where children could catch *mojarra* fish with their bare hands, and it is filled with luxurious vegetation, some of it put to their own uses by the celebrated sorcerers of the area. Los Tuxtlas is known not only for its beauty, its coffee and tobacco plantations, but for the tendency of its men to "rage against" their neighbors at the slightest provocation—or even a little before. The specific stage was the town of Sayula. There lived Alemán González, a character out of a novel.

Born in 1884, Alemán was the nephew of a powerful cacique of the region but had no other particular good fortune. He had been a telegraph operator and a supervisor of track-laying gangs for the Interoceanic Railway, but after a while, he could rely on a small, steadier income from a grocery store. He found his true passion in devouring the revolutionary literature of the banned Mexican Liberal Party and conspiring against the government of the dictator Díaz.[2] A Oaxacan worker named Hilario C. Salas had introduced him to the anarchist movement of the Flores Magón brothers and would very soon participate with him in the rebellions of the nearby Popoloca Indian communities against the government, triggered by the grant of their lands to the Pearson Company of England. Acayucan, where Alemán's parents lived, was the nerve center of the conspiracy. The uprising began on September 23, 1906. Although they could count on sympathizers in various towns of the region, the revolutionaries failed completely.

Some of their leaders ended up in the dungeons of the old prison of San Juan de Ulúa.

Both Miguel Alemán González and his wife, Tomasa Valdés, had been married before. He would have two sons with her: Miguel, the elder, born in Sayula in 1903, and Carlos. Antonio, a son from his first marriage, lived with them. The sons went to a school with the barest of furnishings. They had to sit on soapboxes (and remembered the brand name, Octagon).[3] When they had to travel to nearby villages, or to Almagres, where there was a train station, they went on horse trails by mule. The youngest son, Miguel, would recall many years later that his sandals (*huaraches*), made from very rough leather, used to hurt him a great deal. "To make them less painful, to soften them a little, I would urinate on them."[4]

Four months before the Maderista revolution broke out, Alemán entrusted his family to his parents in Acayucan and went off again to the gunfire with his friend Cándido Aguilar. Shortly afterward, Doña Tomasa moved with her sons to the home of her maternal grandparents in Olutla. It was a town that claimed to have been the birthplace of La Malinche, where Doña Tomasa's extended family ran small businesses catering to farmers and cattle herders and manufactured "black soap." A teacher from the town remembers that "they made this soap from the fat of cattle . . . it was a very gentle soap that everybody liked very much."[5] Meanwhile, in Ciudad Juárez, the triumphant Madero revolution raised her husband Miguel to the rank of lieutenant colonel.

Returning to Acayucan as commander of the local Corps of Volunteers, Alemán enjoyed—or rather endured—a few months of peace until the Huerta coup returned him to his natural environment: the world of war. In 1913, together with Salas, who was now his commander, he endorsed "the proclamation of Los Tuxtlas" and joined the Mixta Morelos column. Miguel, his ten-year-old son, helped his mother deliver milk on horseback throughout the municipality, but occasionally he also helped his father. Together with his brother Antonio, the boy used to steal bullets from the federal soldiers quartered in the area. Inserted in milk jars, discreetly hidden among mangoes and guayabas, the provisions reached his father in the mountains.[6]

Under the command of Cándido Aguilar, Alemán in 1914 rose to the position of military commander of Puerto Mexico and took part in various military actions in Puebla and Tlaxcala. In 1915, while the family was sheltered in Orizaba, Alemán had his share of glory: He participated with bravura in the great battles of Celaya. Obregón entrusted him with the defense of the telegraph line and railway from Hidalgo to Querétaro and the important military command of the fortress in

Pachuca. In 1916, already promoted to general by Carranza, he fought Peláez, the cacique in the pay of the oil companies. Suddenly, an order from his superiors transferred him to Tuxtepec, in Oaxaca, where he reported against his will. He considered the move a humiliation. Some connect him after that with Félix Díaz, who was continually plotting his own revolution. Mysteriously, for a time, General Alemán was without a command, but in 1919 he became Chief of the Civil Guard in the Córdoba region.

By 1920 he had moved with his family to Mexico City. The Sonoran generals distrusted him and kept a vigilant eye on him. They never forgot that he was a friend of Cándido Aguilar, Carranza's son-in-law who was living in exile in San Antonio. And Alemán did go to San Antonio, there to share the exile of his friend, but in 1922 he reappeared in Mexico, not in the capital but in the sierra of Veracruz, risen in arms against the government at the side of the faithful Carrancista Francisco Murguía. When the De la Huerta rebellion broke out, with Veracruz its principal stronghold, General Alemán continued to fight the government but without formally joining those particular rebels.

His son Miguel had interrupted his college preparatory studies in Mexico City to be near him and to help him if possible, accepting whatever dangers were involved. But the two had a definitive conversation in which his father made Miguel see "the usefulness of returning to my studies and choosing an occupation more stable than the military."

In 1925 General Alemán was elected deputy to the local legislature in San Andrés Tuxtla. It seemed to be his farewell to arms, but stability was not to be his. This time his antireelection sentiments were directed against Obregón, and his commander was General Arnulfo R. Gómez, who had been one of Calles's trusted men. In Jalapa, he met his son Miguel, who was finishing up his work in the Faculty of Law and Jurisprudence. And then a destiny that he seemed unable to avoid began to play itself out for the general. Gómez was murdered, and Alemán was stripped of his rank. He lived on the run, in the mountains. In March 1929, he joined the national rebellion of General Gonzalo Escobar.

On March 9, General Alemán assaulted and captured the state capital. The federal troops counterattacked, drove him out, and pursued him. Alemán, the university graduate (*licenciado*), would remember half a century later, "My father never even considered the possibility of surrender, although it appears that his enemies offered him the guarantee of a pardon."

The end came on March 20: "They surrounded him here, near San Juan Evangelista, in a place called 'the Thicket of Aguacatillo,'" an old teacher from Acayucan recalls, ". . . there they surrounded him and opened fire but they didn't take him alive. He first killed the general who was with him, Brígido Escobedo, and then he killed himself."[7]

While General Alemán was away at war, his family, protected by Cándido Aguilar, had been sheltered in the Hotel Español of Orizaba. His mother ran a small store, and the general's cub finished his primary studies, then moved on to the *secundaria*, the next level of the Mexican educational system. Accustomed since childhood to earning his living, the boy put his basic mercantile skills into practice: He organized the buying and selling of empty bottles and was the first pupil in the history of his school to start a business in marbles. He was behind his fellow students academically, but an uneven development was the natural consequence of the gypsy life Don Miguel had dealt them.

In 1920 the young Miguel entered the National Preparatory School in Mexico City. In various ways, the cub would exert authority among and over his fellow students. He was older than any of the others, by five years in some cases. But he also had an exuberant temperament. Like the land from which he came, an opulence ran through his blood; he was congenial, a friend to others, joyful, hyperactive, always eager to start new things. He loved his father but resented the permanent anxiety at home and the constant need to defer his studies. And in fact, in 1923, he had no choice but to interrupt them again, because his father needed him.

The months in 1923 when he worked as an employee of El Águila Oil Company in Coatzacoalcos were not a waste of his time. Alemán learned English and moved from the company post office into the department of explorations. Then, after making a dangerous trip in which he tried to persuade his father to accept an amnesty, the cub returned to Mexico City and began his studies at the University in the Faculty of Law and Jurisprudence, then headed by Manuel Gómez Morín.

Among his teachers were some of the "seven sages" of Mexico: Gómez Morín himself in public law; his old preparatory-school teacher Lombardo Toledano now teaching industrial law; Alberto Vásquez del Mercado in business law; and Daniel Cosío Villegas.

In those University halls where the maestros of the Generation of 1915 lectured, a new generation was forming. Most of its members had no memory of the Porfirian era. They had been born to the Revolution. Intellectual sons of the men of 1915, biological sons of the Revolutionary generation, they were heirs in a double sense, and in public life they would act as such. The general's son, Miguel Alemán, Jr., would become their leader and representative figure.

Alemán wanted to make the Revolution (as his father's generation had) but in his case neither with weapons nor in opposition but rather through achievements and within the regime (just like their teachers of the Generation of 1915). From the time he first came to the University, he aimed at integrating his friends into a compact group. The specific "us" of those young men congealed with surprising speed, helped by the fact that all of student life took place within a small area in the center of the capital. "Alemán was well known for his Veracruzan temperament and his love of joking. He specialized in bathing people; he would throw water from the second floor onto passing students and sometimes professors. He was a very likeable guy."[8] He seemed to have a very clear sense of his generational destiny, and he knew how to transmit it to his fellow students. The list of Alemán's friends was endless, and all through his college years he was refining and sharpening his leadership skills.

In September 1927 he made a proposal to those old friends who had come to the University with him from the National Preparatory School. They would form a group, in order to create a special connection among them. An agreement was drawn up and signed, and Grupo H-1920 came into existence:

> We are ready, and we swear it by what is most sacred, to be of aid to each other in the tremendous struggle of life and not to spare an atom of strength in raising up any one of us to whom fate has been adverse or who is seen to be at some particular moment in urgent need of aid. Many of us, and we are certain of this, will come to fill prominent places in our social and political life; they will continue to be obliged to aid those of the group who need it. Group H-1920 will consist only of those men who already formed part of it in the year 1920, while studying at the National Preparatory School of this capital.
>
> The members of this group are obliged to lend aid, through whatever means may be required, when asked by any other member. He who being able to help refuses to do it—after the group has considered and voted on the case—will be expelled from the group and punished according to the decision of the majority of the members of the group.
>
> Whoever does not wish to continue being a member of the group will explain his reasons, according to which his resignation will or will not be accepted at a special meeting of the group; but if his reason for resignation is to avoid helping the others, he will be severely punished, and there will be the additional aggravating cir-

cumstance of his having proved false to an agreement sealed with honor.[9]

This group was being formed not to serve the country or their fellow men, but to serve themselves. Friendship, so understood, as a pact of mutual assistance, would be the standard of Alemán's behavior.

It was a time when student agitation was changing and intensifying its tone. The National Center of Antireelectionist Students was formed, which supported Generals Arnulfo R. Gómez and Francisco Serrano against Obregón. The name of Miguel Alemán, Jr., appeared on the list of founders. His struggle coincided with that of his father, who was about to begin fighting again in the Huasteca region of Veracruz. But the young Miguel's connection with the opposition would be fleeting. In the following year, when the campaign of Vasconcelos awoke the civic soul of the middle classes, Alemán, already a lawyer, kept his distance from him. "Our generation," he recalled in his old age, "maintained itself between extremes, adopting the revolutionary ideals but assigning them a meaning more in accord with the pacification the country so desired."[10]

Knowing that his father the general could need him at any moment, and already in large part responsible for his mother, who had been left alone, and his brother Carlos (a sad and retiring boy who would study dentistry in Guadalajara), he completed the work for his law degree in 1928. His thesis, "Professional Illnesses and Dangers," was the product of good fieldwork in the mines of Pachuca's Real del Monte, where he developed silicosis. Two weeks later, Obregón was assassinated. In the following year, his father the general killed himself in the Thicket of Aguacatillo. The words of the general resounded in the ears of the new lawyer: "Go back to your studies, choose a profession more stable than the military." Wasn't this a prescription that could be applied to the entire country?

Now a lawyer, he spent some time in Las Choapas, Veracruz, litigating in behalf of the Oil Workers' Union. With his father dead, he then returned to the capital and worked in the office of his maternal uncle, the politician Don Eugenio Méndez. At the beginning of 1931, he married Beatriz Velasco, the daughter of a wealthy family in Celaya, and entered the government as a junior lawyer in the Ministry of Agriculture and Development, where later he became Director of the Department of Forests. Then he left the bureacracy to practice law again, opening an office with (of course) friends—Gabriel Ramos Millán, Manuel Ramírez Vásquez, and Rogerio de la Selva. His specialty was labor law, then called industrial law. Alemán defended workers and represented unions: the miners of Real de Monte and the oil workers of El Águila. He also used his good connections with Veracruzan politicians to make contacts with the transportation unions.

But "the little worm" of politics did not leave him alone. And why should it? He was the son of the precursor of the Revolution in Veracruz, and his destiny was to continue making the Revolution—by other means. His father's old friends, like Cándido Aguilar, encouraged him. "The lawyer Alemán came to see him, with his portfolio, with his cheap little suits. He was still very poor."[11] He had no doubts about what party to follow: In 1929 he had joined the PNR. In the year of 1932, when his first son, Miguel, was born, the lawyer launched his candidacy for Alternate Deputy in the Coatzacoalcos assembly. The Franyutti family, who were rich landowners, proposed one of their relatives, and the PNR chose him. Alemán had to yield:

> My forces were neutralized, depriving me, perhaps, of a first victory in electoral contests; I had mortgaged my house, sold some of my furniture and gone so far as to pawn a fine watch given me as an honor, all to gather funds with which to finance my campaign. Furthermore I had the bitter experience of being threatened by Felipe Fernández, who was known as *El Tigre*—an assassin for hire well known in the region, who bragged of having killed 28 people. . . . The party had made its decision.[12]

The experience left him with a lesson: To do politics one first had to make money.

Parallel to its legal labors, and without any ideological contradiction, the law office dedicated itself to business. The exportation of fruit to Holland, the exploitation of a silver mine in Taxco, the making of cider or the sale of wood were the first ventures. They ended badly, but soon the heavens opened, thanks to an idea of Ramos Millán: urbanization. Fraccionamientos México ("Land Lots of Mexico"), the new company created by the group of friends, was devoted to buying up large areas of land bordering the city and located on old haciendas—ruined, abandoned, or in danger of expropriation. The remaining members of the Porfirian aristocracy were motivated to sell, even at low prices, in order to shift their money into urban businesses. The young lawyers saw their opportunity, and with the help of powerful local military men they obtained soft loans.

Their first real estate ventures were in the rugged environs of Cuernavaca, which the young lawyers urbanized with the support of

General Ávila Camacho, to whom they donated a fine piece of land. Later they advanced into the extensive Anzures plains—near where Calles, the *Jefe Máximo,* lived. They converted the property of the Polanco Ranch into a luxurious residential colony. And at one end of the Chapultepec Forest, with a clear view of the Castle, Alemán and his friends built and sold elegant residences, in California styles. Within six years, they also acquired, at very low prices, Los Pirules Ranch north of the capital, which in time became the heavily populated Satellite City. Alemán would note in his memoirs: "That professional activity . . . [provided me with] the necessary security to pursue my political career free of pressure."[13]

Don Eugenio Méndez commented at the time to a fellow Veracruzan that Miguel Alemán "is the best politician in the history of Veracruz." And he was proving it. In 1934 he directed the Cárdenas campaign in Veracruz. A year later, as a reward, Cárdenas tried to name him Minister of the Court, but because of his age Alemán had to settle for Magistrate of the Superior Tribunal of Justice for the Federal District and Territories.

He would spend little time in that "work hall." On December 1, 1936, at the age of thirty-three, Miguel Alemán became governor of the state of Veracruz. His cabinet was formed of "fellow students." He would later write, "It is obvious the friendship uniting us since our time in the Faculty of Jurisprudence counted decisively in these choices."[14]

His fundamental objective was to do away with the old revolutionary passions that had destroyed his father and to assemble a constructive series of achievements. So he issued a new Rental Law (the issue had been a source of agitation in the port city for twenty years). And on the occasion of a visit from President Cárdenas, he ended twenty years of religious tension in his state by allowing the churches to reopen, with only a single request to Bishop Guízar y Valencia: "The one thing I beg of you is that you do it not with scandal but with decency and decorum." Another of his triumphs was the unification of the peasants. The era of General Adalberto Tejada had been marked by a constant confrontation of the "red" agrarian leagues (radical, encouraged by the government, some connected with the Communists), the "white" leagues (defenders of large and small property), and sometimes even "yellow" groups supporting moderate reforms. But in 1936 Cárdenas's Agrarian Reform in itself diminished the impetus for agitation. There were still peasants without land, but there was enough land to distribute to the peasants. Alemán moved swiftly. On March 27, 1937, he called a huge meeting of peasants in the baseball stadium of Jalapa and secured agreement among the various groups for the election of a new guiding committee for the Peasant League.[15]

After the oil expropriation, Cárdenas received very strong support from Governor Alemán, who not only immediately made positive statements but traveled throughout Mexico and headed a block of governors who supported the presidential decision and were firm in demanding that the companies hand over their installations. Few were surprised when Ávila Camacho appointed Alemán coordinator of his campaign and then took him into his cabinet as Minister of the Interior.

Encouragement of tourism, support for the Mexican cinema, and prison reform were other benign facets of his term as governor. The repression of independent journalists was its dark side. When Miguel Palacios Macedo published, in *El Universal*, a series of critical articles on the inflationary policies of the regime, the director of the newspaper received a polite phone call recommending that his contributing writer cease contributing. It was a Porfirian method and only the beginning of a long list to follow.

Thanks to the suspicious but in any case opportune death of Maximino Ávila Camacho, the road lay open for Alemán to become the first civilian President of the Mexican Revolution. But there were some obstacles. The United States embassy, always alert, favored the Minister of Foreign Relations, Ezequiel Padilla, and distrusted Alemán because of his connections through the unions with the left, particularly with Vicente Lombardo Toledano. *Time* magazine published an article titled "Aided by Death," maliciously referring to the deaths of Manlio Fabio Altamirano (a precandidate for governor of Veracruz whose murder by political enemies had led to Alemán's becoming the candidate) and of course Maximino Ávila Camacho's last banquet. A few months would pass until—in a confidential meeting in March 1946—Alemán made his intentions clear to Guy Ray, first secretary of the United States embassy in Mexico. Lombardo would have no influence on his cabinet. An Alemán government would request technical aid from the United States to rehabilitate the railroads and to develop Pemex, the state oil company. The aid Mexico needed for industrialization would come from the United States, not Great Britain and least of all Russia.[16]

Alemán would display a new kind of campaign—modern, loud, and lavish. Madero had toured the country preaching his democratic creed. Cárdenas had ridden on horseback through thousands of villages, listening to the complaints of the peasants. But the cub was something else entirely:

> Alemán swept along, on flat-bottomed trucks, surrounded by girls; his best "campaign slogan" was his own youthful smile, natural, contagious, optimistic, and never absent. The Revolution had dis-

> mounted from its horse to board, in public, the flat-bottomed trucks overflowing with lovely ladies, the new Adelitas [the beloved in a Revolutionary ballad] of the first generation of sons of the Revolution; and in private the Cadillacs with their long tails, and the Packards . . . while Alemán reviews the country that he has already come to know well since his poor and provincial beginnings, in a joyful campaign to the notes of *La Bamba* [the popular song that meant Veracruz to all Mexicans], which would become the second national anthem during his six years.[17]

La cargada ("the charge")—a key term in the Mexican political vocabulary—was total. Onto Alemán's chariot climbed the workers (truckers, electricians, streetcar drivers, teachers, lithographers), women's groups, bureaucrats, deputies, a Socialist Lawyers' Front, the governors, legislatures, Communist intellectuals, and, of course, the CTM, in whose presence the candidate had given his initial speech: "Workers must be conscious of the fact that excessive demands will be harmful to them. The country demands industrialization."

Present at this speech in the Iris Theater was the socialist intellectual Lombardo Toledano, who perhaps thought that Alemán was in harmony with his own personal, primarily Marxist conception of the history of Mexico. In Lombardo's view, the Revolution had ended the semifeudal stage but had not produced a thriving capitalism, the stage that precedes socialism. One had to support the national bourgeoisie in order to reach that stage and from there leap to the happy reality that only existed in the USSR, that "land of the future" Lombardo had visited for the first time in 1936 and to which he would frequently return, always enthralled by its "achievements." But the university graduate Alemán was about as socialist as Lombardo was North American. He had other plans in mind. He would apply his capitalist convictions to the country, not as a means but as an end.

His friends, who were now more his friends than ever, turned their thoughts to business enterprise for the country—and for themselves. And didn't they deserve it? Like the eponymous leader of their generation, the founder of Grupo H-1920, they had reached the heights after many years of faithful collaboration with the generals. They had acted willingly and for the best—not like the intellectuals Vasconcelos and Gómez Morín, who tried to form an electoral base independent of the Party of the Mexican Revolution and therefore failed. Willingly and for the best—not like Lombardo Toledano, who was constantly exerting ideological pressure from the left, or the heterodox Cosío Villegas, solitary and resentful. Willingly and for the best—organizing "round tables" to plan solutions for national problems: tourism in Acapulco, sugar in Cuernavaca, oil in Ciudad Madero. Willingly and for the best, señores, willingly and for the best, in a spirit that the President's friend Fernando Casas Alemán concisely expressed to a journalist after the triumph of his candidate in the elections of July 1946: "On the first of December, we will break a piñata. Let's see what we can get."

What they got was in fact everything that fell out of the piñata. The Sonoran generals and Cárdenas had included a high proportion of men from the lower classes in their cabinets. With Alemán, middle- and upper-class people completely displaced the more humble. The recruitment of his friends and teachers was truly impressive; at least eleven of his old school friends came to hold high public positions.[18] Other friends secured official contracts and every sort of opportunity, licit and illicit, for economic prosperity. The long-ago agreement of mutual aid signed in 1927 was honored with interest. The "friendification" that Molina Enríquez had so valued in Porfirio Díaz appeared again, refurbished and up to date, during Alemán's six years of office.

Once the presidential sash had crossed his body, Alemán began to act decisively. He had promised the modernization of Mexico through the twofold path of industrial growth and an increase in agricultural production; and he went to work at once. He amended Cárdenas's agrarian legislation, introducing—as the PAN had suggested in a proposal initially rejected by the Congress—protection for agricultural and cattle-raising properties that already held, or would be in a position to receive, certificates of immunity. In essence, the amendment protected private property from any threat of expropriation and determined the maximum size of small holdings in accordance with variations in crops and conditions: one hundred hectares in the case of irrigated land, three hundred for commercial crops like sugar or bananas, five hundred for cattle-raising areas. The changes gave small landowners security and confidence again but met with harsh criticism from the official left (Cardenista and Lombardista) and of course from the independent left.

But Alemán did not feel that he was sinning; he was prepared to prove that his "theory of abundance" worked. He ignored criticism, assured himself of the support of the National Peasant Confederation (from then on entirely subservient to the presidential power), and set in motion the most ambitious project of agricultural growth in Mexican history.

Government investment in agriculture rose from 12 percent of gross national investment under Ávila Camacho to 20 percent under Alemán.[19] A good part of it was assigned to building numerous dams in

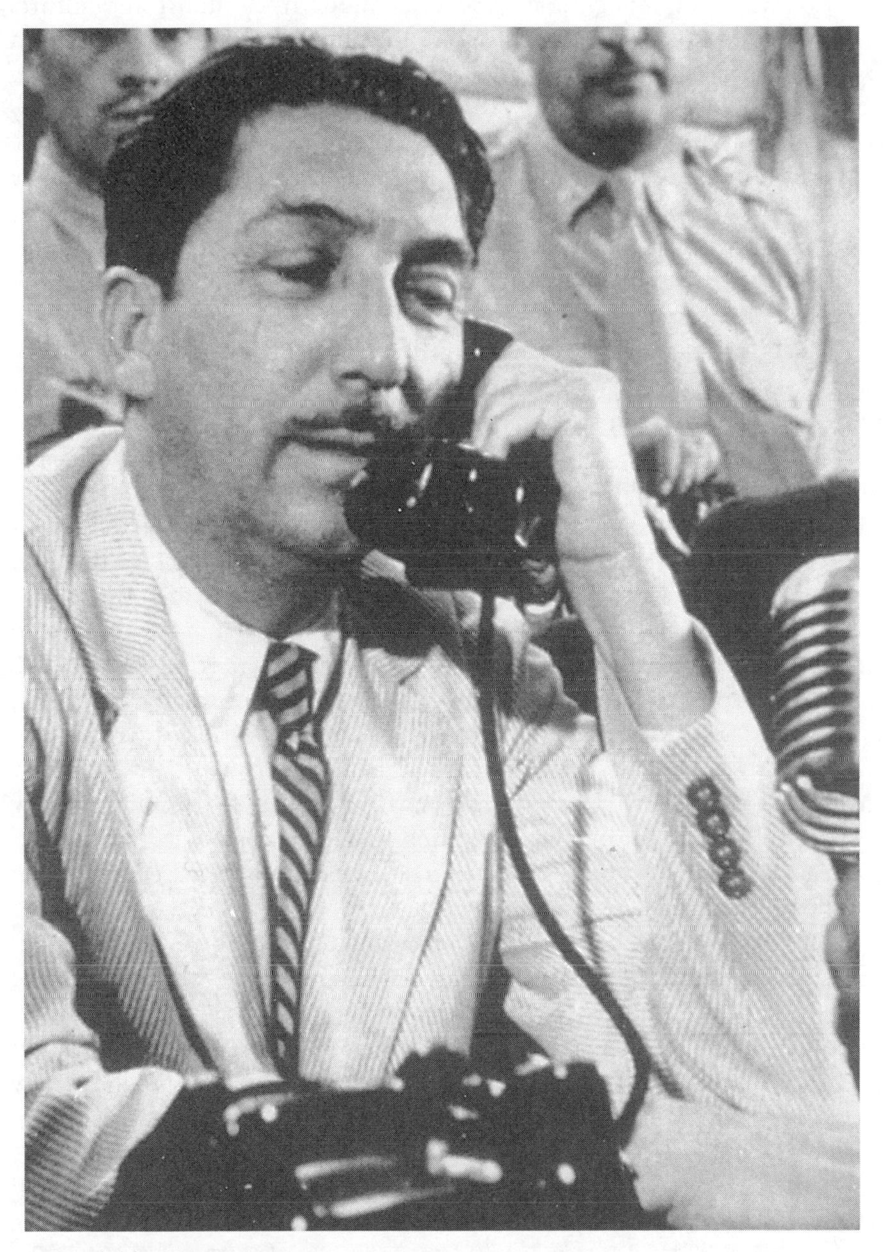

Miguel Alemán, 1948
(Archivo Fotofija)

the potentially productive areas of the country, those that already practiced or were potentially able to practice modern, capitalist agriculture: the northern states of Sonora, Sinaloa, Chihuahua, Tamaulipas, and Baja California. The engineer Adolfo Orive Alba continued to be in charge of irrigation, but his Committee was upgraded into a Ministry for Hydraulic Resources.

Irrigation was the most important but not the only chapter. Thirty thousand tractors were imported. With the help of the Rockefeller Foundation, new varieties of wheat and corn were developed. A Committee for Corn was established (headed by Alemán's old buddy Gabriel Ramos Millán, who came to be christened "the Apostle of Corn"), and it efficiently distributed the improved seeds. The management of the Banco de Crédito Agrícola was professionalized, and its funds were primarily directed toward small landholdings.

These and other programs showed real results. Agriculture grew at a rate of 8 percent during Alemán's six years, 2.8 percent higher than the gross national product. The production of various commercial crops increased impressively: between 1946 and 1952 (a drought year), the value of cotton, the country's principal export product, rose from 80 million to 199 million pesos; coffee went from 78 million to 283 million; tomatoes from 78 to 179 million; sugar from 159 to 292 million; wheat from 139 to 375 million (it had reached a peak of 442 million in 1951). The two products of greatest importance to the people since pre-Hispanic times (corn and beans) performed in a similar fashion, from 680 to 1.6 million and from 94 to 180 million pesos, respectively. Agriculture fed the cities and left a surplus for export. And agricultural products rose from 44.3 percent to 55.7 percent of total exports.[20]

But ex-President Cárdenas criticized the changes in agricultural legislation. He believed they would favor the renewed concentration of land into a few hands. He was not wrong. Many politicans—using their own names or those of family members, friends, or surrogates—bought up land and then exploited the opportunities for loans and the newly created infrastructure. People pinned a name on the new agricultural entrepreneurs: "nylon farmers."

One day, passing by the great estate of the Pasquel family in Ciudad Valles, Cárdenas saw a sign that read, THIS IS PRIVATE PROPERTY. And the general noted in his diary: "Should be expropriated." But yet, always the institutional man, Cárdenas did not carry his disgust to the point of an open break. Alemán entrusted him with planning the development of the Tepalcatepec river basin. That "hot country" of Michoacán attracted the general: "I spent my first years as a revolutionary in that area. The unhealthiness of the area and the precarious conditions in which the peasants live oblige me to participate in their life."[21] Shortly afterward, Cárdenas submitted his project to Alemán, and the President named him a member of the Committee for Tepalcatepec.

Certainly the best argument for Alemán's program was its economic success. Of the 934 collective agricultural experiments (the "collective *ejidos*") registered in 1940, only 688 were still operating in 1950. Many had shut down or adopted patterns of individual production. There was a clear tendency within the *ejidos* toward renting out plots of land and carrying on other disguised commercial activities. The Businessman President was realizing the dream of the Sonorans: the creation of a modern Mexican agriculture. What those generals could not have imagined—and perhaps would have condemned—was how agriculture now came to subsidize the new paradigm of Mexican wealth: the industrialization centralized in Mexico City.

Alemán once said—perhaps seriously—that he wanted "all Mexicans to have a Cadillac, a cigar and a ticket to the bullfights." During his term many cigars were smoked, the bullrings were filled every Sunday, and though 25 million Cadillacs were not sold, some hundreds were. It was said that Alemán "taught Mexico to think in millions," and that in his time "there was peace, there was tranquility, there were jobs." Both claims were accurate up to a point.

Alemán did change the scale of the economy, and many Mexicans did benefit from it. He treated the industries of the infrastructure with indulgent affection: electricity, energy, communication, and transport. And he immediately began to institute policies to protect and foster those industries that had begun to prosper during the war. Industry grew at an average rate of 7.2 percent annually. In 1940 there were 13,000 industrial establishments in Mexico, 65 percent of which were devoted to food and textile production. In 1950 there were 73,000, with food and textiles constituting only 48 percent.[22] The most dynamic areas were chemicals, paper and cellulose, and steel. The policy of replacing imported with Mexican goods became part of the Mexican landscape. A soft-drink factory was even opened to compete with Coca-Cola. Many of the country's important businesses were founded in those years: Condumex (electric wire), ICA (the country's most important construction company), Grupo Chihuahua (cellulose), Telesistema Mexicano (the leading Mexican television network), Tubos de Acero de Mexico (steel pipes), Industrias Ruiz Galindo, Industrias Nacobre. Foreign investment (especially from the United States) flowed into various areas, where plants were opened that manufactured textiles, rubber, and chemicals. IEM Westinghouse opened the first factory for domestic electrical appliances in Mexico.[23]

The fact that this increase was based on deficit financing, with the new industries importing more than they exported, was seen as a natural aspect of the economic takeoff. Few then imagined that it would become chronic. The great majority of these investments also had one feature in common: They were localized in Mexico City. The planners of the time must have been aware of this centralization, but they would have certainly regarded it as just another natural process. What basically happened was that the new paradigm, rather than industrial, was urban, and urban meant the City of Mexico. The modern dream reinforced an ancient residue of Aztec and Spanish imperialism. Mexico City again became (as it was for the Aztecs) "the navel of the moon." From its industrious heights (temporarily subsidized by the rest of the country, especially through agricultural production but also through mining and oil), progress was expected to spread throughout the whole country. There was something monstrous in the idea of sending raw materials seventy-three hundred feet up into the air at the center of the country in order to transform them and supply the entire nation. And almost no one imagined that this subsidy would become not only chronic but exponential.

Under Alemán, medium-size businesses grew large, small ones became middle-level enterprises. "The Alemán people took over Mexico, but they made it grow," recalled a young industrialist who developed his business in those years. The story of this young man is symbolic of the era. He had studied chemical sciences at the University of Mexico but wanted to earn his living on his own rather than becoming a bureaucrat or a salaried technician. With an old freestanding Chandler printing press, he opened a street-level printshop. Suddenly the policy of replacing foreign with Mexican goods knocked at his door. His particular niche in the graphic arts was the manufacture of posters, tickets, labels for medicines, and especially folding boxes for cosmetics. The Spanish Civil War and then the Second World War had favored the development of Mexican industries, largely with Spanish-refugee capital. At the end of the war, French products began to arrive (and in the 1950s American brands), but the tariff barriers kept the prices of imported goods extremely high. Foreign companies decided to manufacture their products in Mexico and feed the growing internal market. The industrialist traveled to the United States and bought used machinery. At the end of the Alemán period, he already owned five offset printing presses and a plant of his own.

In his new prosperity, he came up with a "Revolutionary" idea for a wall at the entrance to his factory—he would hire a disciple of Diego Rivera to paint a mural. The subject is the printing plant as an allegory of wider change. In the country, in the open air, a rural schoolteacher teaches a group of humble peasants to read. On one side of the mural, the young businessman, his partner, and their employees are working at the Chandler press. Two boys sell newspapers with the headlines ALL MEXICO MUST KNOW HOW TO READ and THE PRINTING PLANT AT THE SERVICE OF CULTURE. In the distance, almost imperceptible, smoke is rising from a city.

But this mural as well, though intended otherwise, was a sham. It was all very beautiful but all a lie. The factory was in Mexico City, not in the country. The schoolteacher showing the alphabet to peasants was in reality a very urban secretary. The factory produced folding cardboard boxes for cosmetics, not primers to fight illiteracy. The printshop was at the service not of culture but of Yardley or Jean Patou perfumes. And yet the young businessman thought of himself as a Revolutionary.

WE ARE BUILDING THE PATRIA announced the signs on many public works constructed during the Alemán era. It was the epoch of "great accomplishments." The construction industry showed unprecedented development. Highways became what railroads had been during the Porfirian age, the symbol of progress—more than eleven thousand new kilometers of progress: The Cuernavaca road from Mexico City, which was the first superhighway in Mexico, with separate roads for traffic south and north; the Pan-American Highway, the Cristóbal Colón, which crossed Mexico from the Rio Grande to Suchiate in Chiapas; and the Acapulco route, with a modest bridge (which seemed very impressive at the time) over the Mezcala River. Alemán also built important new railways, like the one that crossed the desert of Altar in Sonora to connect Baja California with the rest of the country. When he inaugurated that route, he did not cut a ribbon but burned through a chain with a blowtorch, to symbolize the effort that had been required.

The fever for construction went beyond highways and dams. There was the titanic project of leading the waters from the Lerma River to Mexico City. (Diego Rivera, to mark its completion, created a work of painting and sculpture representing Tlaloc, the Aztec god of rain.) And a modern network of airports was inaugurated—including Mexico City, Acapulco, Tijuana, and Ciudad Juárez.

Tourism was a matter of first priority. Before Alemán (who began his work on the tourist industry while Minister of the Interior under Ávila Camacho), Acapulco was little more than a pleasant, tranquil port noted for the seventeenth-century fort of San Diego. The old Acapulco had a few hotels in traditional Spanish or colonial Californian style (known as "Coca-colonial"), beaches with wooden chairs, and fishing boats on the ocean. Alemán "revolutionized" Acapulco. He constructed the airport, urbanized the nearby bay of Puerto Marqués (and is said to have abused the local peasants in the process), enlarged the great panoramic highway that circled the bay, and without too much protest allowed it to be named the Miguel Alemán Coastal Highway. From then on, Acapulco began to fill up with modern hotels and a growing international ambience that foreshadowed the later "Jet Set."

More than three centuries after the poet Bernardo de Balbuena had written his *Grandeza Mexicana* (The Grandeur of Mexico) in 1604, another poet, Salvador Novo, produced a literary description of Mexico City titled "The New Grandeur of Mexico."

The parallel was justified. If a metaphor exists for Alemán's project, for its unquestionable grandeur but also its megalomania, it was the development of Mexico City as a faithful mirror image of the man and his work.

The modern urban growth of the city bore the distinctive mark of each political period. The Sonorans built public buildings and schools and favored the creation of new residential neighborhoods (*colonias*). The Maximato and Cardenismo passionately encouraged working-class life. In 1940 Mexico City was still a peaceful city of 1.76 million inhabitants in a country of somewhat more than 20 million. A celebrated line of the poet Alfonso Reyes described it well: "Traveler, you have reached the clearest region of the air."

Following a centrifugal model first initiated by wealthy families during Porfirismo, "respectable" middle-class people, during the twenties and thirties, left their homes in the center of the city (which gradually filled with slums and working-class neighborhoods) to move to the new *colonias* on the periphery. The wealthy whose money dated back to the Porfirian age usually lived in the Colonia Juárez or Las Lomas. The new rich created by the Revolution (including the Sonorans Obregón and Calles) preferred Roma, Condesa, or Lindavista. During the forties the middle class went to Roma Sur, Polanco, Condesa, Hipódromo, or sometimes a little farther south, to the up-and-coming Colonia Del Valle or the old Tlacopac. In the colonial towns of the Valley of Mexico like Tacubaya, Coyoacán, San Ángel, Tlalpan, Mixcoac (each one characterized by its own culture and traditions), one could still breathe the provincial atmosphere that had been preserved through the centuries.

Then came Alemán. The look of the city changed precipitously and for good. Its three million inhabitants were in for new experiences:

automobile traffic; the first skyscrapers—prodigies of hydraulic engineering in a zone of substantial and severe seismic instability and soft subsoil; but above all, the new avenues—long arms that reached from the inhabited city toward the outer reaches of the valley, absorbing the old small towns and crossing uninhabited plains. Only Alemán could have had the vision to turn an entire small river—the Río de la Piedad—into a pipeline bringing water to the city or in 1950 to build Mexico City's first speedway (the Miguel Alemán Speedway), constructed in a series of graded descents, or to extend the already extremely long Avenida de los Insurgentes all the way to the Ciudad Universitaria (University City), then being feverishly erected, in huge surges, across the zone of volcanic rock (*pedregales*) to the south.

Mexico City, since pre-Columbian times, had nourished a solid urbanist tradition. The Aztecs, the Spaniards, and the Mexicans had all been great architects and engineers. Alemán channeled this tradition, communicating his own audacity to his substantial and gifted generation of architects. In 1949 the first multistory apartment building in the country was opened for residence—the Multifamiliar Miguel Alemán. It was the work of the architect Mario Pani. The original project had called for the construction of two hundred living units. Pani proposed to put 1,000 apartments within the same boundaries and as his contractor for the project chose the young engineer Bernardo Quintana, recent founder of the most important construction business in Mexico, the ICA. The "Multifamiliar" would ultimately include 1,080 apartments. It would be a landmark building for having implemented the ideas of Le Corbusier before that architect finished building his own Marseilles apartment complex, completed in 1952.

Often guaranteed against loss with public funds, private architecture caught the constructive frenzy. Great American chain stores like Sears and Woolworth built their Mexican headquarters on the commercial Avenida de los Insurgentes. The first parking structures were erected. But perhaps the new burst of private building most characteristic of the time was the opening of the new residential *colonia*: Jardines del Pedregal de San Ángel. This *pedregal* was a vast expanse of lava, produced by the eruption of the Xitle volcano some two thousand years ago, and had seemed destined to remain a rare natural feature of the Valley of Mexico, a lunar landscape with exuberant vegetation (prickly pear, dwarf thornbushes, many kinds of cactus) inhabited only by iguanas and rattlesnakes. The creator of this *colonia*, the architect Luis Barragán, had the idea—inspired by the painter of Mexican volcanoes, the eccentric Dr. Atl—that the rock formations could be turned into steep, flowering islands surrounded by gardens. In the natural course of

The new urbanization of Alemán
(Private Collection)

things, the young businessman with the printing factory came to build a house of moderate size in the Pedregal. Nearby, the politicans of the Alemán era were raising mansions. Significantly, two PRI senators who would later become Presidents settled in that area: Adolfo López Mateos and Gustavo Díaz Ordaz. The Pedregal became—and there was no doubt about it—the *colonia* of the "Institutional Revolution."

But if the new city of Mexico mirrored the era of Alemán, the Ciudad Universitaria mirrored that mirror. On another area of volcanic rock within the same Pedregal de San Ángel, architects conceived and constructed an immense complex. It would resemble a pre-Hispanic city, with tall pyramidlike buildings and long expanses of gardens and grass. The Central Library designed by Juan O'Gorman was the most striking construction. It was made up of two shapes, one low and translucent (the reading room) and the other (the stacks) tall and enclosed and proclaiming the knowledge that it contained through a brilliantly colored mural of mosaic stones resistant to all weather. The library is an exceptional but insistently assertive building. It is impossible not to look at it. It shouts and waves, demanding your attention.

And all the other impressive buildings as well are monuments to the skills of a great school of Mexican architects, synthesizing their expression of Mexican history with emblems of the sense of grandeur characteristic of the Alemán era. The stadium, for instance (designed by Augusto Pérez Palacios), utilized a natural depression in the ground and then surrounded it with an embankment of earth—a technique used in the construction of dams—compacted through a technical process. Concrete was saved in this way and used to raise the balanced double corbel of the press box and inner balcony. The *tepetate* stone—excavated in order to set the foundations of the structure into the solid rock—was applied to the sloping walls of the stadium, which then came to resemble a volcanic crater like those that can be seen from its steps.

At a cost of approximately $25 million, the Ciudad Universitaria was the President's monument to his regime. And so, to mark its completion, a statue was erected at the center of an immense esplanade. It was of course a statue of President Miguel Alemán.

The "political miracle of Mexico"—until 1968—aroused the wonder of Mexicans and foreigners. It is said that De Gaulle considered it remarkable, that some African countries sent specialists to study the Mexican government with a view to adopting the system, and that it was the envy of the military "gorillas" of Latin America. There were professors in the United States who wound up viewing it as the politicians wanted them to see it: not as a sham but as reality—an eccentric, revolutionary, one-party democracy. When the euphoria was over, when the miracle had shown its huge limitations, Gabriel Zaid, a young Mexican engineer (who is a business consultant and also a noted poet and essayist), defined it precisely: "The Mexican political system is the greatest modern business that the Mexican genius has created."[24] It was a system with both Spanish and Aztec roots, with Liberal ancestors (Juárez and Porfirio), a Sonoran father (Calles), a godfather from Michoacán (Cárdenas), but as a full-fledged business it was the production of the greatest modern Mexican businessman: Miguel Alemán Valdés.

The Mexican political system in its developed form differs from the ordinary dictatorships of Latin America. It is an institutional regime and thereby more modern. The system centers on the investiture of the presidency (in the Presidential Chair), not on the person of a tyrant who may have erupted onto the political scene and installed a regime outside any law, based only on his own charisma or armed strength. The cardinal principle of no reelection is another modern characteristic. It is valid for the presidency, the governorships, the municipal presidencies (equivalent to U.S. mayors), and it also holds for deputies and senators, federal and state. It is the untouchable dogma that effectively protects the country from a single man remaining in power and even against the possibility of his prolonged and dominating influence.

An essential difference between the Mexican system and totalitarian regimes shows in another modern feature—its respect, not total but broad, for civil liberties. Widespread terror, intolerance, and forms of massive repression based on the ideological hegemony of a race, creed, or doctrine do not form part of the Mexican mentality, substantially inoculated against intolerance since the formation of the mestizo Mexican people (though it cannot be said that the culture is totally free of discrimination based on "racial" factors such as color). The political tradition of Mexico has also had its positive effect. The ruler was always supposed to inspire respect and even fear, but at the same time he had to be or at least appear to be patriarchal, as in the days before independence, and later to seem tolerant and liberal, in the nineteenth-century tradition. Although the nation was subordinate to the President and the President was subordinate to no one, there has been considerable space for autonomy in economic, social, religious, and cultural life. Politics was the exclusive territory of the so-called Revolutionary Family, but as the decades passed, various episodes and circumstances would slowly enforce its coexistence with independent and opposition groups.

It is neither a dictatorship nor a totalitarian regime. But neither is it a representative, democratic, and federal republic as prescribed in the

Constitution. There were those who noticed the similarity between the present Mexican political system and that of the Porfirian era: a "monarchy in republican clothing," as Justo Sierra had said. With the great reservation of the "no reelection" principle, Porfirismo was, in effect, its true antecedent. Cosío Villegas would coin another long expression: "six-year absolute monarchy hereditary through collateral transfer," an accurate phrase that nevertheless discounts the formal respect for constitutional forms and the pretensions to Revolutionary faith. It is an authoritarian regime that is difficult to define, and perhaps José Vasconcelos was the one who best described it as "Collective Porfirismo."

If the key to the Porfirian social contract was in the personal linkage of each social group with Don Porfirio (what Molina Enríquez described as the "friendification" [*amificación*] of Mexico), the key to the Revolutionary social contract was an expanded and improved re-editing of the Porfirian method of "bread or the bludgeon." The assumption, which came to be widely shared, was that all individuals or groups could rise—or at least not lose hope of rising—in the social and economic scale, provided they did it amicably *within the system*, not independently outside of it. Gabriel Zaid noted that, as with great American corporations like General Motors, the business of Mexico offered a wide range of services (security, political stability, peace, economic growth, public works, the infrastructure, education, health and assistance, social security, etc.). But its principal objective has been to coordinate power within the country.

There exists in Mexico "a dynamic market in buying and selling obedience and goodwill." From on high and like a waterfall, the power at the center auctions or concedes contracts, lucrative positions, and public employment to bidders who can offer the best assemblage of obedient clients. "The essence of this social contract, the balm that calms souls, reconciles minds, and resolves contradictions is state money. . . . Politics does not consist of winning public elections but of rising within the system." The voters, who were supposed to be the owners of the corporation, lose control to the bureaucrats, who for all practical purposes are the real owners. And they, in their turn, depend on the will of "El Señor Presidente," who is at once both chairman of the board and executive director, for six years, of the enterprise.[25]

Through the creation of the PNR, Calles had brought the generals and regional caciques under control, offering them the bread of power every six years but threatening them with the bludgeon for the slightest dissident action. Calles had made one major mistake. He had attempted a hostile takeover of his only competitor—the Church—by applying the bludgeon. Many of the peasants of Mexico had responded with their own bludgeons. Cárdenas, on his part, had abandoned the erroneous objective of eliminating the Church and instead gave the "instrument" (as he called the ruling party) economic domination over the country and ownership of the nation's oil. And then he strengthened the PRM by directly absorbing—as political armies in reserve—both the peasants and the workers, both the bureaucrats and the soldiers. Ávila Camacho completed the cycle, offering the Church the bread of reconciliation and definitively excluding the military and the regional caciques from the center of power. But a young entrepreneur was needed, someone with fresh energy who could get the "instrument" moving. The static structure oriented toward the Aztec, Hapsburg, and Bourbon past—constructed by caudillos, jefes, and generals—had to be converted into a dynamic corporation looking toward the future—and organized by university graduates. Miguel Alemán was that businessman president who would succeed in modernizing and energizing the "instrument."

The business of the State put a most ancient device to modern use—what Octavio Paz would describe as "the transmission of the Aztec archetype of political power." In many ways Porfirio Díaz—and Juárez before him—had behaved like a *tlatoani*. He was seen, treated, and feared as such. But by retaining power, he would in the end identify that power with his own person, not with his position. Despite his respect for republican forms, he continued being the caudillo, the strongman at the margins of the law, the one who made and imposed his own law: "Don Porfirio" first and then "El Señor Presidente." But now the system incorporated "the secret supremacy of the Aztec model." Like the *tlatoani*, the modern President always relies on legality. His power is absolute, almost sacred, but it resides in the investiture, not in the man. "The *Tlatoani*," wrote Paz, "is impersonal, sacerdotal and institutional."[26]

The culminating moment, the acid test for the system, comes around every six years. The Aztecs had developed a complex procedure for choosing their *tlatoani*, and through a highly mysterious process of transmission, the pattern has appeared almost intact in twentieth-century Mexico. This is the method of *tapadismo* ("secret deliberation"). A conclave of nobles and military chieftains, meeting in complete privacy, would discuss the selection of the heir to the throne. The group of prominent men would remain sequestered until they "hit upon" the successor. (According to other sources, the former *tlatoani*, just before his death, had already designated his heir.) In any case, the unveiling took place, the "Revelation" *(destape)*, and the chosen one could finally appear "before the people. . . . All of them gazed at him."[27]

Porfirio Díaz had tried out the method successfully in the selection of his compadre Manuel González (who was president for four years, between 1880 and 1884), but then he had no need to use it. Every four years Díaz was both "secret deliberator" and the "candidate as yet unrevealed" (*tapado)* who chose himself. Well into the revolutionary era, Carranza died for the error of "revealing" Bonillas in 1920, and the assassination of Obregón was partially a consequence of his attempt to follow in the footsteps of Don Porfirio (1928). Calles, with no trouble at all, was able to launch Cárdenas in 1934, but for Cárdenas to do the same with Ávila Camacho in 1940 cost him considerable effort and almost triggered a new revolt. As late as 1946, the "Revelation" of Alemán met with some problems. But in 1952 the rules were finally clarified and adjusted to arrive at an Aztec level of refinement.

Once "revealed," the heir travels across the country in a long presidential campaign. He flagrantly pretends to be seeking votes, but what he is doing in reality is symbolically validating the legitimacy of the Revelation, behaving *as if* the vote of the people is what counts. The democratic sham is necessary to keep the wheel of power turning. His campaign is a patriotic pilgrimage. He allows the people to see him and recognize him. And he is accompanied by a caravan of candidates for public positions. His young fellow speakers are *jilgueros* ("songbirds") who chant the beauties (*primores*) of the Mexican Revolution. Everywhere he is received "with brass bands," with "ecstatic applause." Banquets are organized, also discussion groups and public addresses. The candidate hears requests and complaints, receives letters, makes promises: schools, drainage, roads. He orders his secretaries to write everything down. Although he says that he needs the votes of the people, in reality he does not need them at all because thanks to the vote of the Great Elector—the outgoing President—the election has already been won. Nobody has any illusions about the result.

And besides, apathy and ignorance are the best allies of the system. By 1952 there were ten million possible electors but only two groups voted (essentially the captive votes of peasants and workers and a small group of thinking voters—often in opposition—from the middle class). The rest consciously abstained or simply ignored the elections. The true heart of the system rests on the traditional, premodern political culture of the majority of Mexicans, for whom *politicians are the legitimate owners not only of power but of the nation.* And so politicians, most especially the President of the Republic, have behaved with almost total impunity.

As a whole, then, the Mexican political system has been a contemporary, functional version of a very ancient mode of organizing power.

Pouring "old wine into new bottles," the modern corporation has operated to produce a centuries-old product: the patriarchal control of sovereigns, the filial obedience of the governed. And nevertheless—despite the fact that "the market of goodwill" paid much better than that of freedom—submission was not universal. The system ruled over the country, but the country was not the system.

Supported and legitimized by the electoral system of the PRI, the President directly ruled over three great subordinate power groups—beyond his own proper executive sphere and that of the city of Mexico, whose Regent formed part of his cabinet. These were the formal, corporate, and real power centers. The formal powers were the Legislature, the judiciary, the state and municipal governments (and the government bureaucracy). The corporate powers were mostly to be found within the PRI, in its peasant, worker, and popular sectors. The real powers had guns and armed men: the caciques and the army.

Inhabiting a gray area (enjoying relative independence but disunited among themselves and always under at least the potential control of the President and his tentacles) were other powers: the press and the media, the Church, the businessmen, the cultural enterprises and the intellectuals, the State universities and other institutes of higher education including the teaching and student bodies.

Within the narrow dark zone of the opposition there continued to exist the intransigent, the courageous, the visionaries, and sometimes the crazies—an opposition of the right and the left, of liberals and of independents.

For six years the President enjoys absolute power. The man who enters office owes his post to the man who is leaving it. And the man who leaves office leaves free and clear, knowing that the new man will cover his back. The President-Elect can amply practice nepotism, but not to the extreme of passing on the Presidential Chair to his biological sons or brothers (the example of Maximino was an object lesson). Those who are chosen come from a definite clan, not physical but political. They have to be members of "the Revolutionary Family."

In its origins, the "Revolutionary Family" had been a closed confraternity of pistol-toting generals surrounded by "scribblers" (*tinterillos*) who wrote speeches for the generals and hymns of praise, which they themselves delivered. Beginning with the "institutional" stage and the formation of the corporation, the "Revolutionary Family" cleaned up its manners and broadened its membership. In effect, any young man from the city or the provinces with intelligence, shrewdness, ambition, and the proper contacts could become a member of the family through various routes—the party, a union, a school, a city government—and from there he could begin to climb, toward the apex of the corporation. To the question "What would you like to be when you grow up?" nine out of ten Mexican boys would answer: "President of the Republic." To sit in the Presidential Chair and draw the tricolor sash across your chest has long been the Mexican dream of glory. When asked once—in his old age—if he had ever dreamed of being President, Daniel Cosío Villegas, the most critical of Mexican intellectuals, answered directly, "Never, ever, at any moment . . . have I stopped wanting to be President."[28]

Through mere formalism or rhetoric, the President of Mexico was called "the highest holder of the mandate of the nation." That was his legal character, but in reality the Presidents obeyed no mandate but their own. They were not "holders of a mandate" but sovereigns. The Constitution of 1917—promulgated under Carranza—encouraged this unlimited concentration of power. It rooted sovereignty in the soil, the subsoil, the waters and skies of the Nation, in the people as a collective entity. From there it passed to the State, which transmitted it to a specific government, and from there, finally, it was entrusted to the President. The only control that could be exercised on a President in office (aside from foreign pressure) was whatever that President—through temperament or conviction or "republican austerity"—consented to exercise upon himself.[29]

Miguel Alemán decided to test the limits of presidential power and discovered that beyond those external limits (the power of the United States, for example) and a fixed internal limit (the sacred principle of "no reelection"), he could exert, as they used to say in the times of the viceroys, "his royal desire." Like their distant predecessors, the Presidents of Mexico could treat public assets as private property. They could distribute money, privileges, favors, positions, recommendations, sinecures, lands, concessions, contracts. The Sonorans—including Obregón and Calles—had become great agricultural businessmen and landowners. To develop their agricultural enterprises, they could count on the generous assistance of the new official banks through what were euphemistically called "preferred loans." "The Revolution," some of the generals used to say, "did us justice." The Agrarian Reform of Cárdenas (who also owned a number of small ranches) slowed down but did not eradicate the process.

Stories of the corruption current under Alemán would fill a long series of volumes. Many of his friends, inside and outside government, accepted the offer of presidential "bread" with enthusiasm and became wealthy through official concessions, not always illegal but often immoral. A mentality developed that encouraged and developed monopolies and special concessions. An automobile salesman would secure the right, with no competition, to sell his cars for government use. Bureaucrats who had previously owned businesses sold their goods in great quantities to the State, and others established new businesses to supply or serve their own ministries—with prices and conditions of sale set by themselves. A medical Subdirector of Social Security founded an ad hoc drug business. One of Alemán's cronies even lifted $40 million out of the Treasury through manipulating the Foreign Trade Bank.

If the government announced a construction project, the bureaucrats—in their own names or those of others—would buy up the land adjoining the project and develop it later, when it already had its new inflated value. The toleration of smuggling and traffic in merchandise and property confiscated during the war from Italian and German proprietors were other varieties of that peculiar process of collective capitalization that—with all its colonial antecedents—was a creation of the Institutional Revolution.[30]

Of course the Revolutionary generals had also enriched themselves through their positions. Obregón, Calles, Cárdenas, and Ávila Camacho owned ranches they would not have been able to acquire solely through their military salaries. Many of the other generals of the Revolution—famous and obscure—paid themselves back for their services by seizing the haciendas of the Porfirian elite. But the scale of corruption attained by Grupo H-1920 was something that had never been seen before. The French essayist Jean François Revel, commissioned by the Parisian review *Esprit* to prepare an article on "Mexican democracy" would write in horror (under the pseudonym of Jacques Severin): "You can do any kind of business you want in Mexico provided that you first 'come to an agreement' with the governor of the state or some important federal personality. As long as you can 'interest' the politicians, Mexico is a paradise for businessmen.[31]

The new rich, whom the people called "*tanprontistas*" ("so-quickists" because they had "so quickly" made their fortunes), behaved in a way that matched the name: They raised mansions like Hollywood film sets, held bacchanalian parties, poured out rivers of money. Their conspicuous display, no longer restrained, filled the pages of the newspapers. Jorge Pasquel, one of Alemán's childhood friends, a powerful customs official and great smuggler, the overlord of vast landed estates, was a lover of the famous actress María Félix. She recalls in her memoirs:

> When he courted me, he was at the height of his power, because his friendship with the President opened all doors, in Mexico and abroad.

> While I was making the movie *Maclovia* on the lakes of Michoacán, he showered me with attentions. Once I told him on the phone that the hotel at Pátzcuaro—where I was staying with the whole film crew—had run out of ice; and the next morning he sent me a seaplane with a refrigerator. I was impressed and thanked him and then he wanted to send the plane every day with special food dishes and delicacies. It was an excessive display of luxury in sharp contrast with the poverty of the place. So then I asked him—instead of sending me caviar and lobsters—if he would fill the plane with sacks of maize, rice and beans to be distributed among the Indians of Janitzio.[32]

"We live in muck," Lombardo Toledano would declare in 1952, having repented a thousand times that he had helped launch the Cub of the Revolution on his way to the Presidential Chair. "The bite, the stickup, the final ploughing, the stuffing, the little drink, a string of names that have been invented to describe this immoral process. Justice has to be bought, first from the policeman, then from the prosecuting attorney, then from the judge, then from the mayor, then from the deputy, then from the governor, then from the minister . . . "[33]

Lombardo did not say "and then from the President of the Republic," but there was no need. It was well known that President Alemán continued doing business from his political position and that he acquired or expanded his shares in a whole host of businesses, including of course his favorite area: hotels and tourism in general. And if the President himself could use his power for business deals, then a similar permissiveness easily pervaded the whole corporation.

The ordinary citizen tolerated the universality of bribes either because he believed that politicians were the *owners* of power and could do their "royal will" (*regalada gana*—a common idiom for "whatever you want"), or because he knew that there was no effective recourse against the abuse of power by the privileged. The humble people of the city got their vicarious revenge by going to the Follies Theater to laugh at the political sketches of the comedian Jesús Martínez, known as "Palillo" ("the Toothpick"), "the scourge of phonies."[34] Or word of mouth passed on to the people what could not be said in writing or in public, like these verses attacking Alemán and his group—composed as soon as he was no longer President:

> *Ali Baba with his forty rats*
> *has left the people wearing only sandals*
> *but the sultan feels very much at ease*
> *spending away his millions in Paris.*

* * *

Alemán was a man who "exuded charm," his friends exclaimed and his enemies muttered. Certainly, that "man with the eternal smile" was irresistible. He would give his courteous attention to whoever needed it, when it was needed and for as long as it was needed. And he especially charmed women. Rumors had him involved with every well-known actress, even with María Félix herself, who was rumored to have constructed an underground tunnel from her home in Polanco to the official residence of Los Pinos. "He didn't drink, he didn't smoke but he did everything else,"[35] asserted one of his doctors. But it was power that really attracted him. Power was the essential business, the great national industry. Around 1950, he began exploring ways to retain it.

Some of them were symbolic. He wanted to become a *doctor honoris causa* of the University, and of course he achieved it (though some said his doctorate was *ignoramus causa*). He had a mania for putting his name on every public work that he inaugurated (plazas, schools, avenues, hospitals), but near the end of his term he did something the country had not seen since the days of another President from Veracruz. He himself unveiled his own statue in the Ciudad Universitaria as Santa Anna had done with his equestrian statue that once stood in the old Plaza del Volador.

Within the country there were no further barriers left to him except a limit on how long he could remain in power. There was the inconvenient rule of "no reelection" to prevent him from returning in full to the good old Porfirian days. But in 1951, the fifth year of his government, anything seemed possible. A national campaign was mounted to glorify Alemán. The CTM named him "Worker-in-Chief of the Fatherland"; in the assembly of the PRI, Gustavo Díaz Ordaz, a senator from Puebla, proposed (and his proposal was accepted) to enrich the party's official philosophy with "the thoughts of President Alemán." In this context of exaltation, Alemán began to seriously explore the possibility of being reelected or of extending his term.

Perhaps he would have accomplished it, had it not been for the firm opposition of the old Revolutionary generals, especially Cárdenas as well as Ávila Camacho, who strongly opposed what he described, with his customary courtesy, as the "clumsy attempts of the false friends of Licenciado Alemán." But none of the generals would react with greater vehemence than Alemán's old political godfather, who had extended his protection over the Alemán family ever since that far-off year of 1915—his father's friend Cándido Aguilar.

At the beginning of his term, Alemán had named Aguilar an executive member of the National Peasant Commission and had assigned him an office next to his own in the National Palace. The general, a passionate agrarianist, began to take his job seriously, which naturally led to confrontations with some of Alemán's friends, particularly with Senator and Colonel Carlos I. Serrano, who had taken away the lands of some peasants near Michoacán. Shortly after these disagreements, Alemán would ask for the resignation of his beloved "Jefe." Months later, according to the testimony of Dr. Justo Manzur Ocaña (with whose family Alemán had lived in his youth, shortly after coming to Mexico City from Veracruz), Aguilar defused an attempt by Cárdenas to overthrow Alemán through a military coup. If true, this would have been the umpteenth service rendered by the old general to his cub Alemán. But in August of 1950, they again fell out:

> There was propaganda going around that called for the reelection of Alemán. And among members of *ejidos* in Yucatán, there were people who made speeches in favor of reelecting Alemán. Then General Aguilar went to speak with him and said to him, "It's going to cost you your life." Alemán answered him, "No, general, I'm not thinking of having myself reelected." And General Aguilar approved of those words and went to the newspaper and said, "I have just spoken with Alemán and he says he is not going to reelect himself." And, poof! the reelection collapsed.[36]

With the option of reelection (or extending his term) blocked off, he was left with the possibility of a hidden reelection, something he had already considered. He would yield power to one of his friends, a man who owed everything to him, the Regent of Mexico City, Fernando Casas Alemán, famous for the oriental splendor of his house. (Some called him "the idiot prince.") So certain was Casas Alemán of being "the right man" that he ordered reams of publicity printed. Even the American magazine *Newsweek* reported his impending "Revelation" as a fact. It is not clear why the attempt failed. It seems there were divisions within Alemán's inner circle, where there were some who wanted to veto Casas. But the attitude of Lázaro Cárdenas was decisive. By playing a careful political game, floating the idea that he was willing to support an opposition candidate (General Miguel Henríquez Guzmán), Cárdenas forced a "Revelation" other than the man Alemán would have ideally preferred. The President was forced to realize that not even reelection through a surrogate would be allowed him, and he chose his Minister of the Interior, a sixty-year-old civilian named Adolfo Ruiz Cortines.

The internal limits had been clarified once again. The President could not reelect himself or reelect himself through relatives or through friends so close to him that they could become more than relatives—accomplices. The first commandment of power in Mexico had been reinscribed: "You will yield power after six years."

"That was another time," Alemán explained years later, to an animated group dining with him, and he then proceeded to recount, in exuberant detail, his contribution to the political maneuvers of 1940. He had been the director of Ávila Camacho's campaign. He knew that Mexico City was heavily pro-Almazán. "These elections cannot be lost," he had said to his candidate. "Do what you have to do," the gentleman general is supposed to have answered. And then Alemán initiated a military operation for the seizure of polling places that resembled the standard work of Gonzalo N. Santos. He set up his headquarters in an establishment for renting automobiles, where he concentrated soldiers and arms. When reports arrived that a polling place had been "seized" by Almazán supporters, Alemán would send brigades to openly steal the ballot boxes, firing their machine guns if necessary, into the air or with deadly effect. "Mission accomplished," he might have then reported to his chief. "That's how they won the elections. . . . my reward was the Ministry of the Interior . . . that was another time."[37]

Before the birth of the PNR in 1929, presidential elections were settled with guns. With the PNR already in existence, but before the institutionalization of the Mexican political system, elections ran the risk of rivalry between Revolutionaries, a situation that generally brought back the guns. Vasconcelos had as many or more revolutionary credentials than his rival, Ortiz Rubio; Almazán had taken part in more battles than his opponent, Ávila Camacho. In both of these elections, the difference between winners and losers was a matter of membership in the Party of the Revolution, but the victories still were more than doubtful, and they cost a lot of blood. It was not until 1946, with the transformation of the PNR (which was the current party name) into the PRI, that the resort to bullets considerably decreased. The secret of success was the development by the PRI of a complex, original, and Machiavellian technology for electoral control.

The old electoral law issued by Carranza in 1916 stipulated that the polling places "belonged" to the citizens who first arrived to take control of them. It was a law that converted the polling station into a "space" of power. Many of the terrible incidents of 1940 were a result of this situation. One had to "seize" the polling places before the enemy did, drive him out of the voting spaces he had already occupied, and then gain control of the ballot boxes by force or at the very least intimidate opposition voters. The new law of 1946 assigned the management of elections and polling places to District Vigilance Committees controlled by the government. It was then, in the elections that brought Alemán his victory, that the modern era of the PRI truly began.

The methods instituted boggle the imagination, and they included every step of the electoral process, from its preliminaries to the last minute of the count. Months before the first Sunday in July (the day on which every six years an election is held for "the Big One," the presidency), a skillfully assembled and selective voting list was compiled, everywhere in the country. Those suspected of sympathizing with the opposition were listed separately, and the members of the PRI were marked for special status. Independent voters were denied credentials and replaced by imaginary voters supplied with "provisional credentials." All bureaucrats were ordered and the corporate organizations of workers and peasants were heavily pressured to vote overwhelmingly for the official candidate, at the risk of losing their positions, jobs, or land (the bludgeon method) or else with the promise that their perquisites would increase or improve (the bread method). Often these votes were used as "stuffing" days after or before the election, kept apart in separate ballot boxes that were added to the final count. Huge public buses transported peasants from remote areas, clutching their ballots already marked for the PRI, to vote—at a randomly selected polling place or one that had already closed.

"Valid Voting," the first part of Madero's slogan, was adulterated in a thousand ways. If the vote is going badly at a polling place, shock troops assault and vandalize it, steal the ballot boxes, threaten the voters, expel (frequently with police help) the representatives of other parties, and, occasionally, use their guns. A record is lost, a photographer taking pictures of fraud in operation is beaten up by gunmen, the ballot boxes are filled before voting even begins. There are endless cases of voters registered two or three times; others bring fifty or a hundred credentials with them; some people are paid to enter and vote as many times as they can. Days later, while the count is still going on, the media are ordered to announce, beforehand, the victory of the official candidate.

In the elections of 1952, the official candidate, Ruiz Cortines, faced three principal contenders: General Miguel Henríquez Guzmán, the lawyer Efraín González Luna (for PAN), and Vicente Lombardo Toledano, the candidate of the *Partido Popular* he had founded in 1948. According to *El Popular*, Lombardo's organ, the electoral violations of 1946 were repeated in new and the usual ways. On several farms in Tlalpan (near Mexico City) peasants transported from the state of

Morelos were assembled to form "shock troops" for the elections. Voters arrived wearing yellow buttons, a sign that they should be allowed to vote as many times as they wanted. Bricklayers from the Ciudad Universitaria and workers from the District Department of Transport and Sanitation traveled in "flying squads" to vote without credentials or any registration on a voting list. Officials in charge of polling places directly took on the job of stealing ballot boxes. Secret polling stations were opened. In San Luis Potosí, according to Lombardo, "even primary school children were required to vote, and in the capital of that state, soldiers voted over and over again."[38]

Antonio Mena Brito, a former PRI militant, director of the Youth Action wing of the party in the fifties, once admitted that "the PRI is an organ to legitimate elections." It had always been that, but Cárdenas gave the party real power and organically integrated the worker and peasant masses. Alemán's PRI retained the corporative pattern but put it all at the service of the electoral machinery. "In the months preceding 'the Big One,'" recalls Mena Brito, "the PRI was pampered." It received every possible assistance from the government in services and sometimes in cash. "Any minister would see you if you flashed your PRI card. They would ask me 'what do you want? paper?' And they gave it to me. They would invite me to eat, to dinner. But when the elections were finished, the PRI became a kind of beggar before public power." Years later, President López Mateos (1958–1964) would say to Mena Brito that—between one Big One and the next—"the PRI has to find work."[39]

General Sánchez Taboada, at the head of a group of young lawyers, did make an attempt to institute primary elections within the party. Toward this purpose they supported the National Confederation of Popular Organizations (the CNOP), which was supposed to represent the middle classes. They tried to hold such elections in a few states, but the opposition of the peasant and worker sectors of the party forced them to call off their efforts. After that, the sharing of the electoral pie for the legislative power at municipal, state, and federal levels was always firmly entrusted to corporative representatives and handled behind closed doors. The sectors of the PRI, especially the workers, knew that they could count on a portion of the pie, but the last word on this and all electoral questions came from the owner of both the pie and the knife: the President of the Republic.[40]

And once a decision was made on the presidential successor, any dissent or disagreement would simply fall away. All that counted then was the campaign, the speeches, the leaflets, the demonstrations, the "charge." The electoral machinery would grind into motion. The hour for the Big One had come.

* * *

As the modern Mexican political system became firmly established, the President would eventually come to choose ministers, assistant ministers, chief administrative officials, senators, deputies, judges, governors, the Regent of the Federal District, ambassadors, and some municipal presidents. All of them had a common objective: "to stay on the good side of the boss," beginning with the boss immediately above them and, at the highest level, the Boss of all Bosses: El Señor Presidente. As was natural enough, "job mania"—that endemic illness of the country that José María Luis Mora had censured in the nineteenth century—continued to be a national passion. Many Mexicans dreamed of having a *chamba* ("a soft job") in one of the many bureaucratic offshoots of the executive power, earning money and rising through the levels and coming to feel that their own personal progress was also the public progress of the *patria*. It was fitting that an old friend of Alemán, Cesar Garizurieta ("the Opossum"), cofounder with Alemán of the group Socialists of Veracruz and a man who had lived his life safely sheltered in public positions, coined one of the most celebrated statements of the Mexican political dictionary: "To live outside the budget is to live in error." When years later he found himself without work, the Opossum, faithful to his perception, felt left out in the cold and killed himself.

A courtier culture was being re-created in the country. Instead of settling national problems in the public square, the "Revolutionary Family" grew accustomed to "washing its dirty linen in private" and turned to palace intrigue, the shadows, whispers, the verbal "stab in the back." The newspapers would supply only meager information poorly presented, and so rumor became the favored medium for political "disclosures." A national pastime—along with soccer, bullfighting, wrestling, and boxing—was trying to guess what might be going on in the mind of the man who went about his business in the old palace of the viceroys.

There was only one Presidential Chair, but the person who sat in it filled two functions: presiding over the government and acting as the chief of state. The first role involved an immense range of duties and the use, at his discretion, of enormous financial resources. The second carried with it a whole panoply of ceremonial display. As had been true in the Porfirian age, the President was both the supreme power of the nation and the high priest of the fatherland. September, "the month of the *patria*," was the most active time for the sacramental aspects of the presidency. While peddlers of little national flags appeared everywhere in the streets, the President was performing his own patriotic activities. On September 1 he gave the *Informe*, his State of the Nation Address to the Legislature. (In 1951 Alemán arrived to deliver the speech after passing under a series of triumphal arches, as had the viceroys of New Spain.) No one interrupted or questioned the President's address. Each year, a PRI deputy merely responded with a panegyric to the "goals achieved by the present administration." On September 13, the President paid honor to the "Child Heroes" of Chapultepec. On the fifteenth he shouted the *grito* in the Zócalo.

Around 1954, the young poet Jorge Hernández Campos, fascinated, intrigued, and disgusted by the display of Alemanista power, wrote a poem that registered the bewilderment of a number of Mexicans when they contemplated the great "totem," perpetually reborn every six years, who governed them:

I'm the most excellent Mr. President Don So and So of Something
and when the earth trembles
and the masses heaped together
all of them in the Zócalo roar
and I shout Viva Mexico!
what I really mean is Viva me!

"In the opinion of the public," wrote Cosío Villegas, "there is nothing so despicable as a deputy or a senator . . . they have come to be the standard for the filthiest human debasement."[41] He was only slightly exaggerating. Few now remembered the great parliamentary tradition of Liberal Mexico. And the deputies "now" had little in common with those courageous representatives of the nation at the chaotic Convention of Aguascalientes in 1914 or with the utopian radicals of the Constitutional Convention of 1917. Nor even with the deputies who in the wild twenties settled their differences with bullets. When Cárdenas forced the resignation of deputies loyal to Calles and imposed a Legislature more to his liking, he gave the coup de grace to Congress. By 1940, the deputies and senators had returned to Porfirian orthodoxy: They were a club of the President's friends.

The passage of time further increased the docility of the Congress. Very rarely did one encounter the strange case of a "representative" actually representing his district or the senators their states. The deputies campaigned every three years and the senators every six. They gave speeches, they accepted petitions, they made promises. But everyone knew it was only a ritual. Once elected, they would rarely drop by their districts, since they owed their positions not to election by the voters but to the will of the Great Elector who had previously "zeroed in" on their names printed in the lists of "contenders." When in 1943, after the legislative elections had already been held, the executive authority changed his mind about supporting the winning candidate from Oaxaca and swung, at the very last moment, toward another man, the slighted candidate mounted the rostrum of the Lower House and publicly committed suicide.

Part of the problem stemmed from the constitutional requirement that extended the principle of "no reelection" to legislators, automatically limiting the pressure toward relations between citizens and the legislator who was supposed to represent them. In any case, the passage of deputies or senators through the Lower or Upper House was just that, a passage, a step toward the heights of public administration, an entrance-level job in the business of Mexican government, where one had to look good and obey in order to continue climbing.

According to the Constitution, the deputies had two primary obligations. One was to review the previous year's public expenditures without "limiting themselves to investigating whether the amounts were spent or not in accordance with the respective portions of the budget but to pass on to an examination of the accuracy and justification for the sums spent and the responsibilities for what may have taken place." The other was "to examine, discuss and approve the budget for the following year, and determine the taxes necessary to cover it."[42] Not a single Legislature took these requirements seriously. In general, the deputies approved the budgets at once or with unimportant modifications (usually increases), and they would close their eyes whenever the President spent more than had been allotted.

The makeup of the Legislature reflected the distribution of positions within the PRI. A system of quotas had been established in the days of the PRM, and the PRI continued them: so many deputies for the various sectors—workers, peasants, middle-class professionals. The future President chose the senators. Alemán, for example, decided that he would allow Gustavo Díaz Ordaz to be the senator from Puebla for 1946–1952 even though another candidate had already been "revealed" and despite the fact that the political group supporting Díaz Ordaz had worked for the benefit of the late Maximino Ávila Camacho. Alemán was familiar with the efforts of Díaz Ordaz as a deputy in the Legislature from 1943 to 1946, and therefore he agreed. Díaz Ordaz paid him back handsomely. Especially important among his innumerable services were his defense of the regime against attacks on its supposed immorality and his proposal that the Nobel Peace Prize be awarded to the President.

Although there were incidents of internal discord between sectors of the PRI represented in the Houses, the general rule for deputies and senators was to abstain from proposing legal initiatives and to vote "as a block," "rapidly" "dropping the formalities" for whatever the executive proposed. The public began to identify the deputies—insofar as they identified them at all—with the image of an arm raised to vote "yes!"

Jean François Ravel was struck by the strange "republican" ceremony of September 1, 1951, when a joint session of the Upper and Lower Houses listened to the *Informe*. Except for the special feature that year of the triumphal arches, the process was an annual one:

> After passing below the triumphal arches erected for the occasion—at the expense of various states and unions—between his residence and the Congress, the President reads his report to the deputies. It is a long list of the benefits he has spread across the nation during the past year. All the radio stations in the country broadcast his speech and the day is a holiday so that the people can listen to him. After he finishes, the deputies applaud, thank the President and assure him of their total support.[43]

The observer had missed a few details. Besides the deputies and senators, representatives of the "living forces" of Mexico come to hear the report: the judges of the Supreme Court, the governors, the army, union leaders, peasants, representatives of youth and business groups, and a good number of people specially invited (the Church was always excluded). At the end of the speech, a PRI deputy responds in behalf of Congress, with unremitting praise. (When in 1945 the deputy Herminio Ahumada had the audacity to mention the word *God* in his response, and then went on to castigate "false Revolutionaries," to point to "the political backwardness of Mexico" and to admit "the tragedy of Mexican democracy," he was thrown to the wolves. They stripped him of his seat.) The President leaves the building in an open car (preferably a Cadillac), acclaimed by the crowd (often transported there by the PRI but including people curious to see El Señor Presidente close up), in the midst of a hail of confetti (courtesy of the PRI). Then comes the "hand-kissing": all the invited guests lining up to salute the President at the National Palace.

Cosío Villegas's judgment on the behavior of the official deputies and senators was utterly accurate: "The revolutionary congresses have been as servile as those of Porfirismo."[44] The legislative power was a sham power. And nevertheless, the entrance of the Panista opposition into the Chamber of Deputies in 1946 created a small difference. At least 4 of the nearly 150 deputies would raise their hands to vote "no!"[45]

* * *

Before the long list of abuses committed by the executive power in the Revolutionary era, only once did a member of the Supreme Court of Justice resign: Alberto Vásquez del Mercado, in May 1931. Ortiz Rubio, who was then President, had forced exile on the famous intellectual Luis Cabrera for criticizing the new government. The same judges of the Court who kept silent when, in protest, Vásquez del Mercado read his resignation, thanked him afterward in private.[46]

Here as well, the text of the Constitution worked against independence. The Constitution of 1917 completely dispensed with the system of popular election of judges established in 1857 and entrusted the selection of the Supreme Court to the Legislature meeting in joint session. Beginning in 1923, the principle of appointment for life was also approved. Unfortunately, at the suggestion of Calles, in 1928 the designation of the judges was entrusted to the President, subject to the approval of the Senate (which served the President and would always approve his selections). In 1934 Cárdenas eliminated life appointments and replaced them with six-year terms to run parallel with the presidential term. The purpose was to give every President his own court (in both senses of the phrase).

The sensible reforms of Ávila Camacho and his spirit of self-criticism did not, in the case of the Supreme Court, renew the legendary prestige it had enjoyed during the epoch of the Restored Republic (1867–1876), the only time (aside from the brief dawn of Madero) when the country attempted an authentic division of powers. Although the judges were well paid and now once again appointed for life, the Supreme Court never really took on any case of national importance, nor did it alter its subservient relation to the executive power, and it never formulated any new constitutional proposals to strengthen its own power or efficiently defend individual rights.

There were various opportunities to do so. The electoral abuses against the PAN candidates to the Lower House in 1946 and 1949, amply documented, came to the attention of the Supreme Court. Applying the third paragraph of Article 97 of the Constitution would have allowed the Court to carry out an investigation, if it determined that individual rights had been violated. The PRI proposed to nullify the paragraph, arguing that it would "endanger the autonomy and dignity" of the Court "for it to become involved in political disputes." PAN, on the contrary, suggested that the article should be applied, since action had not been taken with due speed and efficiency and even some lower judges had been guilty of interference, thus justifying a full investigation. One Supreme Court judge, Fernando de la Fuente, had the courage to declare, "If the Court eliminates the political power delegated to it by Article 97 of the Constitution, it is reduced to a minor court of appeals and will no longer be the High Tribunal, the guardian of the Constitution. Its decisions will not be respected by any authority . . . and caciques will remain completely free to commit excesses and oppress the people."[47] In the end the wording of the article was not modified. But neither were its provisions put into effect.

The appointment of a woman (María Lavalle Urbina) to the Supreme Court was a sign of symbolic progress. But it meant very little in the face of the Court's silences and its failures to act. When Judge Luis G. Corona made a vitriolic statement concerning "the garden Alemán and his government have made out of the Pedregal de San Ángel" and alluded to the constructions of the Ciudad Universitaria as "a mausoleum under which the dignity of Mexico has been buried," the independent magazine *Siempre!* criticized him in an article titled, "You Speak Up a Little Late, Señor Corona." The magazine was right, of course, because the judge was daring to make his critical statement six months after Alemán had left office. The Supreme Court had in fact become only "a minor court of appeals," and the highest court in the land was only a sham power.[48]

Federalism and the free town council (the *municipio*) were also institutions written into the Constitution, but they too were a sham. Alemán continued the process of undermining the power of the states, especially when he was dealing with military men. In the first eight months of his term, ten governors were replaced for one reason or another. The dismissal of General Marcelino García Barragán, governor of Jalisco, was perhaps his most disgraceful action. The general, in 1946, had committed the sin of supporting the presidential candidacy of Henríquez (he would sin again in 1951–1952), and in February 1947 he had only two more weeks to serve, before passing on power to his successor. Alemán, apparently to demonstrate with absolute clarity just who was in charge, engineered his impeachment and named an interim governor to replace him. García Barragán tried to block the move through an appeal to the Supreme Court, but the justices, as was to be expected, put off considering the case until the general had already lost his office.[49]

The pretext for removing a governor was never important. There were more than enough legal methods available: cancellation of powers, impeachment, resignation, unlimited leave, replacement, interim appointments. What mattered for any new President was to replace problematic governors inherited from his predecessor, not only for reasons of loyalty and political control but also to facilitate cutting up the pie.

The *municipio* was a long-standing institution in Spain and Mexico. The first decision taken by Hernán Cortés after landing his men on the shores of Mexico was to create the municipal council of La Villa Rica de Veracruz. This germ of local democracy had operated with reasonable efficiency during the colonial period, even in indigenous areas. The liberal and federalist legislation of the nineteenth century willingly accepted the institution of the *municipio,* and it retained its prestige and efficacy even under Porfirismo. In fact the great revolutionary chieftains Obregón and Carranza had been municipal presidents (equivalent to mayors). No other executive would match Carranza in defending the freedom of the town councils. For him, governing the city of Cuatro Ciénegas had been his school for learning to govern Mexico. "The free town council," proclaimed the Carrancista newspaper *El Pueblo*, "will become the seedbed for citizens of a great, free, strong and cultured country."

But dreams were not reality. On the one hand the ancient practice persisted of local domination by caciques. The strongman of a region would select and remove municipal presidents at will or simply install himself in the office. And eventually the budget allotments for municipalities became so meager that honest people often declined to be chosen or when chosen refused to take office even if threatened with imprisonment. The salary of the national Finance Minister could be thousands of times higher than the wages assigned to the overseer of a municipal treasury and hundreds of times greater than the entire budget of a poor municipality.[50]

In the era of the Institutional Revolution, the PRI would "be elected" to nearly all municipal presidencies. The candidates that PAN began to offer had to run a true gauntlet, not even to win but to reach home alive and healthy on the day of an election. In 1947, for example, the PRI deputy Enrique Bravo Valencia warned his constituents in Zamora, Michoacán: "We cannot allow PAN to win again in the next municipal elections. The PRI is the party of the government, the official party, and therefore the only Party. . . . The government gave you land and created unions for you and therefore you must support the party of the government. If we find out that you stand with the National Action Party, the government will take away your lands and expel you from the unions."[51]

The first case of municipal elections won by Panistas had occurred in that area and in that year, in Quiroga, Michoacán. Beginning with Alemán up until 1970, PAN would win additional victories until they held 40 municipal presidencies—out of 27,000. The system could allow the people to make the mistake here and there of electing a municipal president from the opposition; or on rare occasions a governor had no choice but to accept a local deputy from outside the PRI; at the very most it could even tolerate the presence of four PAN deputies in the Lower House. But that any of the 64 senators or 31 governors appointed by the President should come from outside the PRI? Never.

Ávila Camacho had succeeded, for good, in turning two of the major real powers into dependents of the system: the caciques and the army. If Gonzalo N. Santos had dreamed of a cabinet position in 1940, he understood, in 1946, that when Alemán asked him to wait a while for a place in the cabinet, the wait was only a polite fiction. "In politics, what you have to wait for does not happen," Santos would write with the wisdom of age.[52] Even though he, like many other caciques, kept full control of his feudal domain for many years, the armed rebellion of a cacique against the government had now become unthinkable.

As for the army, Alemán merely finished the work of his predecessor. He founded the Bank of the Army and Navy, created the first suburban military *colonias*, put younger men into the top commands, and—a masterstroke of politics—allowed himself the luxury of choosing Gilberto R. Limón as his Secretary of Defense, a man who had served with the forces that encircled General Miguel Alemán González until they drove him to his death in 1929. One of Don Porfirio's famous sayings had been: "In politics, I have no loves and no hates." As far as hates went, President Alemán seemed to agree fully.

From his two predecessors, Alemán inherited the obedience of the peasants who had been incorporated into the PRI. Despite the protests, toward the end of his term, by the old Revolutionary generals who became active in the formation of a new party (the Authentic Party of the Mexican Revolution, the PARM), Alemán would not budge on his revisions of agrarian policy and could rely on the support—only tacit at times but nonetheless effective—of the National Confederation of Peasants. And also, behind the closed doors of the PRI, new generations of politicians were putting their effort into wrapping up new packages that could counterbalance the workers and the peasants—obedient middle-class groups as presents for the men in power. The corporate variety of the colonial age reappeared, under the most varied designations, within the structure of the PRI. Someone would assemble a package of women professionals, or of revolutionary architects, or of music composers.

Turning the workers into an integrated, dependent sector of the PRI was a far more complicated task, and one that was never completely accomplished. Calles, Morones, Cárdenas, and Lombardo had used the workers as armies on reserve for them within civil society. The

President Alemán visits a factory, 1950
(Archivo Fotofija)

workers had become conscious of their power, and some of their leaders thought that Mexico could become a union or corporate state, headed by the workers, that is to say by the leaders of the workers. If Alemán wanted to create a stable basis for a market economy regulated by the State, he had to establish new precedents, and he had to do it in confrontation with the unions of the two great state enterprises: Mexican Oil and the National Railways.

"There must be no illegal work stoppages," Alemán had very clearly warned on December 1, 1946, in his inauguration address. Some days later he received the first threat of such an action from the Oil Workers' Union. His response was rapid. On the morning of December 19, soldiers took over the country's gas stations and all the installations of the Atzcapotzalco refinery. The distribution and consumption of gasoline never skipped a beat. The union leaders called it an assault. Alemán's people called it a necessary measure. Marco Antonio Muñoz, governor of Veracruz between 1950 and 1956, recalls:

> Unfortunately . . . the leaders had become bullies . . . they had gotten used to calling illegal work stoppages any time of the day or night. . . . How could we develop and strengthen the economy of a country that was subject to the whims of these leaders? They had to be controlled. . . . When a group wants to disturb order, the government imposes order and President Alemán did it, making people understand that public peace took precedence over the interests of a single group.[53]

During that meal among friends when Alemán was feeling lively and speaking clearly, he described what had happened behind the scenes. Before the scheduled work stoppage, he had ordered Antonio Bermúdez, the agile and capable director of the Mexican oil industry, to give the union a salary increase of 10 percent, with the option of going as high as 15 percent. The union leaders rejected the offer: They wanted a few more percentage points. After the action by the army, the parties returned to the bargaining table. It was the turn of the union leaders to accept an increase of 15 percent, but management lowered its offer to the original 10 percent. The leaders had no option but to agree. A peace-making dinner was then held, and the President attended. One of the leaders, who was slightly liquored up, said to him, "But we were just trying things out, Sr. Presidente!" To which Alemán answered, only half joking: "Well, you tried me out all right, you sons of bitches!"

But Alemán's action had already deeply tried the soul of the most powerful unions in the country, and they began to think things over carefully. In March 1946, most of them (the telephone workers, the miners, the oil workers, the telegraph operators, the streetcar drivers, the electricians, the cement workers) left the CTM and joined the always militant railroad workers in forming a new general union independent of the regime, the United Confederation of Workers (CUT). It had a membership of 200,000. The caudillo of the newborn organization was the leader of the railroad workers' union, Luis Gómez Z., who had formerly been a candidate for the leadership of the CTM but had been thwarted by an alliance between Lombardo Toledano and Fidel Velázquez.[54]

Reduced but not defeated, relying on a multitude of small unions and the rapid recruitment of bureaucrats and peasant organizations, the CTM weathered the storm and changed its slogan, replacing the socialist "For a Classless Society" with the nationalistic "For the Emancipation of Mexico." Lombardo was the inspiration behind the change, but by the middle of 1946 he had announced his intention to form a political party independent of the government, the Partido Popular. Fidel Velázquez had promised to support him, but after a few months the alliance between the worker and the intellectual was broken. Velázquez produced a historic announcement: No worker who was a member of the CTM could be part of any party but the PRI. All the worse if it were a party—as Fidel Velázquez said—of "communists" such as Lombardo was trying to form. The rupture between Lombardo and the CTM in November 1947 would begin to cement the firm compact between the government and the most important union organization in Mexico.

The worker's movement would have a caudillo for more than just six years. Fidel Velázquez would become a Porfirio Díaz of the working class. Born in 1900 in San Pedro Atzcapotzaltongo in the state of Mexico, he had worked in the fields as a young man. When he came to Mexico City, he became a carpenter's apprentice, then an employee of the milk company El Rosario, where in 1924 he organized a union of the milk workers of the Federal District, affiliated with the powerful CROM under the union caudillo Morones. From then on he functioned as the primary representative of the group of leaders known as "the Five Little Wolves": Amilpa, Yurén, Sánchez Madariaga, Quintero, and Velázquez himself. During the 1930s he gradually distanced himself from Morones's CROM (which began to fall apart after the assassination of Obregón) and moved closer to the CGOCM of Lombardo Toledano, with whom he founded the CTM in 1936. The next year, he succeeded in isolating the Communists within the CTM (they would never move back into positions of power). Under Ávila Camacho—taking advantage of Lombardo's frequent absence as he spent time traveling through other countries as Secretary of the Workers' Center for Latin America—Velázquez strengthened his leadership position in the General Secretariat of the CTM. And when Lombardo, in 1947, tried to split the CTM from the PRI and draw it to his new *Partido Popular,* the other "little wolves" and certainly the big wolf Fidel declined to join his adventure.

From that time on, the key to the success of Fidel Velázquez lay in his maintaining a precise idea of the limits of his own power. Unlike Morones, who had been Minister of Industry and Labor for Calles and had dreamed of becoming President of all the Mexicans, Velázquez was content with being the president for life of the Mexican workers. Unlike Lombardo Toledano, who often subordinated practical union work to Marxist theory (and who also wanted to be President), Don Fidel was a rare bird, an ideological chameleon: He would step to the right and step to the left according to which better suited his specific mission: *mediating* between the government and the workers. Many intellectuals and left-wing leaders considered him a manipulator of the working class, "a lackey for the interests of the government in collusion with big business." But the fact that he has held on to his union power for more than half a century (in 1997 he is still going strong) suggests that he has managed impressively. Within his union, from the top of the pyramid, Fidel Velázquez was as faithful as Porfirio to the method of "bread or the bludgeon." Everything was allowed within the CTM except the independence of its member unions, whose slightest movements had to be approved by "Comrade Fidel." His famous smoked glasses, his laconic way of talking between his teeth, his piercing irony are part of Mexican political legend. Many would consider Fidel the most important Mexican politician of the last sixty years.

Midway through 1948, the peso was devalued by almost 50 percent. The unions took up sides. With the President first and foremost was the CTM, the spinal column of the PRI, moderating its demands in the name of "nationalist unity." Also standing with the regime, although not as solidly as the CTM, were various older workers' organizations like the CROM, the CGT, or the textile unions. The progovernment groups encompassed about 500,000 workers. Opposed to the President, or at least fighting for some independent space, stood the CUT and a new union grouping organized by Lombardo (all in all some 330,000 workers). Alemán did not wait for the unionized workers of the other great state enterprise—the railways—to "try him out"; it was he who decided to test them first.

Since the beginning of 1948, in the central secretariat of the Association of Railroad Workers, a picturesque individual had succeeded

Gómez Z. His name was Jesús Díaz de León, whose nickname was *"El Charro"* because he used to come to union assemblies dressed like a cowboy out of Mexican folklore. Once he was up in the saddle at the forefront of the union, the *Charro* executed a dangerous trick. With government and police support, he carried out a coup d'état against his own executive committee, expelling everyone who had been close to Gómez Z. or to the Communist faction. Gómez Z. himself was accused of fraud, harassed, then jailed. He was released after six months when no proof could be found that he had diverted funds from the railroad workers' union into the foundation of the CUT, but the union was now totally out of his hands. Discreetly, Alemán telephoned him and advised him to stop being an agitator. As for Díaz de León, his "destiny" or at least his place in union history was decided. Since then, in Mexico, any union leader who sells himself to power has been called "a *charro* leader."[55]

There were only a few strikes during Alemán's six years (Altos Hornos, the Electric Company, the Ford Motor Company). After the defeat of the CUT, an attempt was made to create a new General Union of the Workers and Peasants of Mexico (UGOCM), but the Ministry of Labor refused to register it as a legal union. The President—designated by the CTM as "Worker-in-Chief of the Fatherland" and its "Honorary Secretary General"—finished his term as he had begun it, with the same inflexible resolution: "There must be no illegal work stoppages." And so when the miners laid off by the Santa Rosita Mine of Coahuila headed toward the capital on their "march of hunger," it was not Alemán but Cárdenas who intervened to calm their anger. But the mine was not required to make any concessions. And the miners were not rehired.

The rules of the game had been set for the middle of the century. "Bread" if you stood with the government; "the bludgeon" if you were against it. Significantly, in December 1951, the oil workers rejoined the CTM. Behind them would come a long procession of unions returning to the fold.

If all the leaders of the CTM had been *charros*, the CTM could not have survived, because the workers, for all their submissiveness, would not have stood for it. The successive bludgeonings applied by Alemán had established the terms of the compact and traced the limits for maneuvering, but without a wide and varied offering of bread on the part of the government to the leaders, and through them to the workers, any arrangement would have crumbled.

The workers in that printing enterprise founded by the young industrialist were a good example. They belonged to an old union organized in 1920 and part of the CTM: the Union of Graphic Arts. Factory elections were held to select delegates and subdelegates, and they were real elections. The union, in its turn, also periodically renewed its executive committee or, when appropriate, reelected it. Within the unions, in contrast to national political life, the dogma of "no reelection" did not operate, a fact that could lead to problems but also allowed the leaders and representatives to get to know one another much better. The union's internal elections were genuine, so authentic that sometimes they ended with gunfire or with one of the two contenders behind bars.

At the union local, in the evenings, the executive committee attended to the many problems brought before it by the workers: firings, leaves of absence, requests for promotion, complaints of bad treatment, cash loans, recommendations for relatives or friends. In a large waiting room decorated with portraits of the President and heroes of the Reform and the Revolution, many members would wait to see the union leader. They might be there for hours, because the leader puts no time limits on his agenda. Overall, thousands may come to see him, but he personally listens to and speaks with each one, and he knows each worker by name. The leader is the patriarch of the organization, the president of the miniature republic, a minor Fidel Velázquez.

The negotiations with the business are not a sham. It is true that labor lawyers or the bosses hold secret discussions with the leaders, it is true that the leaders sometimes receive bribes, but there is a limit to what is done behind the backs of the workers, because opposition forces exist within the union. If an opposition faction discovers a devious arrangement, it will not hesitate to denounce it before the leaders of the CTM, bringing it to the attention, more precisely, of the judge of judges, the man himself, "comrade Don Fidel," whose door is open to all.

In exchange for mediating with the bosses, the union requires total loyalty. The "penalty of expulsion" (*cláusula de exclusión*) can always be applied by the union and the worker can then be fired (something the factory cannot do, unless it is willing to provide three months pay plus additional days for each year of seniority). The business cannot employ nonunion workers or hire them on its own. All hiring passes through the union. If a worker does not fulfill his union responsibilities (participating in the parade on May 1, voting for the PRI every six years, attending a meeting), he receives a severe reprimand. In the case of union bureaucrats, they may lose their posts.

Many of the workers in this factory have come from the country and feel that the move to an industry and the city was an unquestionable improvement over the life of their parents. They never lose their connection to their villages; they receive letters and they send money. There are some who ask for leave every year to go back for a few months and plant their small bit of land; they do not want to lose their rights in the *ejido*. But the majority are now definitely established in the city. Some have already bought their "little lot" in the city or a used car or an apartment not far from work. In their "humble house," as they call it, they have nearly all the basic appliances of the time: a washing machine, a refrigerator, an iron, a radio.

Life was hard for these workers in the age of Alemán, but without any need for statistics, they knew that their situation was better than the great majority of Mexicans.

In this political solar system, three forces circled around the sun of the President and his electoral machinery: the dependent powers (formal, real, and corporate); those not so dependent (the press, the Church, the entrepreneurs, the university, the intellectuals) and the almost invisible planets that surrounded him at a distance and in darkness and were barely or not at all dependent upon him—the opposition on the left and the right. Naturally enough, a visitor from a democratic galaxy (like Revel—the essayist "Severin" from *Esprit*) would judge the display with the eye of an astronomer or astronaut—and sometimes like a bewildered astrologer.

For the rather few Mexicans who regularly read the newspapers at the beginning of the fifties, it was a given that the press did not honestly, thoughtfully, or independently report on political events or politicians. The press was at best a sound system, at worst a full member of the governmental chorus. The reader could absorb the details of last night's crime of passion, the American Major League baseball scores, all the important moments of the bullfights, the guide for a good Catholic to masses and saints' days, news about films and shows and the fiestas and dances of high society's "300 and a Few More" (the title of a gossip column of the time). He could even keep up with the most recent intellectual currents of the modern world, thanks to the excellent literary supplements, which had been a weekly tradition for decades and which complemented the literary reviews that circulated among the educated minorities of the country. Whether he was interested in art, radio, cinema, the sciences, or the humanities, he could count on dozens of regular publications, some of a quality equal to their counterparts in Europe or the United States (*Trimestre Económico*, *Cuadernos Americanos*, *Historia Mexicana*, *Problemas Agrícolas e Industriales de México*.) But if he was looking for honest information or disinterested opinion on the real national situation, he had to turn to the unwritten press: gossip, stories, and rumors.

Accustomed to *Le Monde* or *L'Humanité*, "Severin," understandably, had harsh things to say about the Mexican press:

> You can read the Mexican press for months without meeting with the least little article that really criticizes the government. All the newspapers show a fawning respect for the world of political power, completely accept its most insignificant statements, and never exhibit any independent investigation, any honest reporting on the real situation of the country. And their circulation is insignificant. Mexico City has a population of 2,000,000 and the daily papers that offer political news, six or seven of them, print about 300,000 copies. The provincial newspapers, all of them together, print no more than 300,000. Which means only 600,000 people in a country of 25,000,000 read newspapers with any political news.
>
> And yet the daily newspapers are big business and the publishers are major financial powers. A newspaper lives on its advertising, on paid articles, on blackmail (all the major businesses have to pay up in one way or another or else they might read "Coca-Cola is bad for your health" or some brand of cigarettes "can cause harm to your eyes"). But above all the newspapers feed on politics. It is a matter of playing the game with those who are in power and earning a share of the sure business opportunities that are the perquisites of the politicians.
>
> The owners of the newspapers reap these benefits. As for those who do the writing, their salaries are miserable. Some of them are forced to find other means of access to the company safe and it can be said that, in this respect, honesty is all the more admirable in that it certainly does not pay. In Mexico, a daily newspaper consists of forty to fifty pages. Twenty are entirely filled with advertising; ten carry social news, entertainment and sports; six to ten pages are devoted to enormous displays in commercial type: they are messages from various organizations, states, unions, chambers of commerce or private groups and addressed to the authorities of the government, whether to thank them, ask them for something, wish the president a happy birthday, etc.
>
> All the above brings in money. The rest, five or six pages, reports the official line and emphasizes if possible the event the government wants to stress. In general an inauguration or an official trip earn an eight-column headline on the first page. For international news, the agency dispatches are reproduced verbatim, faithfully echoing the line from Washington.
>
> The press then is part of the governmental system, among

> whose primary beneficiaries are the owners of the papers. And the government can also exercise, as a last resort, a radical means of control. It has a monopoly on the importation and distribution of newsprint. As a result, the newspapers are in constant contact with a government credit institution, the *Nacional Financiera*, to which they are all more or less in debt.
>
> Thus the press is shackled in a hundred ways and besides it cheerfully puts up with its chains.[56]

The article incidentally drew angry letters from Mexican governmental authorities to the Mexican embassy in Paris. They urged their representative to do everything they could to have the offending issue of *Esprit* recalled or removed from the newsstands. Apparently, some members of the Mexican governmental bureaucracy believed they might be able to operate in Paris as they did in Mexico City.

And yet, despite the semiofficial character of the principal Mexico City newspapers (*Excélsior*, *El Universal*, *Novedades*) and even in the case of the official government organ, *El Nacional*, the Mexican press never fell into the dogmatic extremes of newspapers in the fascist or Communist countries. What kind of press was it, then? How can it be explained?

Always vigilantly aware, the intellectual Daniel Cosío Villegas, a troublesome comet in that planetary system, arrived at a precise definition: "It is a free press that does not make use of its freedom."[57] The phenomenon seemed especially painful to him because during those years, while he was beginning his new role as a historian of modern and contemporary Mexico, Cosío Villegas was immersed in an intellectual and political world much closer to his temperament and convictions, the environment of nineteenth-century liberalism. With nostalgia and fascination for those times when Mexico, as at no other moment, had almost seemed to be a Western democracy, he was looking through hundreds of political newspapers of the era (argumentative, satirical, analytical, literary, Catholic, Conservative, Liberal) that had been issued across the years of the Reform and the French Intervention and in some cases survived till the year 1896 when Porfirio Díaz, as one of his projects, established the first "mass-market" newspaper in Mexico, *El Imparcial*.

Before that time, newspapers in Mexico had been not a business but "an extension of the activities of a man of letters." Someone would start a newspaper to express and defend a system of ideas and would find in freedom of expression reason enough for its existence and the condition for its success. But with the creation of *El Imparcial*, things changed because of three factors that were basically still operative in the age of Alemán: the regime of Porfirio Díaz was dictatorial, the publishers of the newspaper were his friends, and it was a genuinely mass-market enterprise, with freedom of expression (or the defense of it) no longer a requirement for success.

If the legislative power was, at best, only "a quiet and potential censor" of the executive power, if the judicial power was reduced to administering routine justice in conflicts between individuals but "would hesitate to maintain an opinion contrary to the executive, should he be in conflict with another state power and even with a single person," then Mexico urgently needed the so-called fourth power to exert its historical role as it had done in the nineteenth century and, at times, in the era of Calles and Cárdenas, when even *Excélsior* and *El Universal* could function as part of a critical, independent, and even oppositional press. In the age of Alemán, these remains of independence tended to vanish. He had already enforced a covert censorship of the press when he was Minister of the Interior. As President, he reinstated the Porfirian model of press relations with some new security devices. Cosío Villegas would write:

> The press knows that it cannot really oppose the government, which has a thousand ways of controlling and even destroying it. It also knows that many of these modes of pressure can have a legal and even elegant appearance. Consider, for example, a restriction on newsprint based on a scarcity of foreign currency; or a huge increase of tariffs on the importation of paper or machinery; or instigating a strike which is then legitimized by the labor courts with the vote of the governmental representative deciding the case, etc.[58]

In the middle of 1951, "et cetera" descended with full force on an ephemeral opposition magazine. It was called *Presente* and was published by Jorge Piñó Sandoval and a group of independent journalists. For thirty-six weeks, in editorials and cartoons, *Presente* was committed to criticizing the corruption of the Alemán regime. It specifically named the *tanprontistas* who had "so quickly" filled public posts and bought or built their dream mansions, which the magazine would describe in detail. Of course *Presente* had no advertising. It depended on being accepted by the reading public, and it was so successful that every issue would sell out in a few hours. The government tried various means of indirect pressure against Piñó: It suspended his newsprint supplies and delivered much more expensive Finnish newsprint, forcing him to raise the price of the magazine; it punished contributors to the magazine by excluding them from any professional relationship with the government (banning, for example, the literary work of the writer Margarita Michelena from the official radio program *La Hora Nacional*). When these cautionary methods had no effect, the goverment turned to the bludgeon. Just like the old days when Porfirio Díaz's police raided the press of the Flores Magón brothers, the police of the institutional-revolutionary regime broke into the offices of *Presente* and destroyed the printing plant. The magazine still continued to publish, but was finally forced out of existence. Piñó went into hiding and then left for Argentina, afraid that he might meet the fate of a journalist who had recently been shot down or another who had been imprisoned. Not long afterward, his return trip paid by the President, Piñó Sandoval came back to Mexico. As he had done with the oil workers, Alemán succeeded in getting the hardened journalist to accept the "bread" (and avoid the "bludgeon"). In the course of time, Piñó would write newsreel narration for the government.[59]

Shortly before these events, while the sword was still hanging on its thread over *Presente*, Alemán came up with the idea for another of his great inaugurations. Not a building or a highway but a ceremony that, given the immediate circumstances, took on a certain cynical resonance. It was "the Day of Freedom of the Press." Theoretically, the government was paying homage to the successors of the great journalists of the nineteenth century. In practice the government was demanding then, and would demand year after year, that the press put the stamp of acceptance on its dependent status, glamorized as independence. José Pagés Llergo, the newspaperman who had enthusiastically praised Hitler, reflected its effect a few days later. He had changed his opinions and was now a democrat. (His later experience with Alemán was destined to push him further in that direction.) But now he wrote in the magazine *Hoy*: "What *Hoy* has said about the present regime would have caused a great headache for the magazine under any other government. And I would say more. I think no government could have shown itself not only more respectful but more amicable and helpful to the press than this present government."

Pagés would regret his positive sentiments. He would later be quietly fired from his magazine for publishing a photograph that showed the son-in-law of the (by then) former President Alemán with his eyes glued to the half-exposed breasts of a French model while his wife—Beatriz, the daughter of the ex-President of Mexico—glared at him with a look of profound irritation. The press had received its instructions: "that they could write what they feel like writing but stay clear of the President of the Republic or the Virgin of Guadalupe."

It was better to go after the bread than provoke the bludgeon, as the great novelist of the Revolution, Martín Luis Guzmán, implicitly admitted. He was then director of the magazine *Tiempo*, which he had founded in 1942. When the government, with considerable brutality, broke up a meeting (at the Juárez Hemicycle in the Alameda) of supporters of Henríquez for President, Guzmán refused to publish any report that disagreed with the official version of events. "I have the obligation," he is supposed to have told his collaborators, "to mutilate and deform the truth if it contributes to the political objectives supported by *Tiempo*." His principal editors submitted their collective resignation.

It was "a free press that did not make use of its freedom," because it was profitable to make no use of it. The sale of newspapers to the public was a secondary interest. The true client was the public advertiser—the government—with its full-page advertisements or the news stories that were also purchased, though under the table, or the private advertiser—the business entrepeneur—whose prosperity was often due to governmental concessions and who had absolutely no interest in having his newspaper attack the government. Or "it was a free press that did not make use of its freedom" because for a journalist to use it could cost him dearly, much more than money. It could cost him his life.[60]

The Mexican press was economically prosperous with up-to-date technology, but its strength was not devoted to any cause at all. It could boldly attack communism (which in fact was one of its favorite targets during the fifties), but it never mentioned—or even hinted—at the paradox of its own situation, which was that in Mexico, freedom of the press was more theoretical than real, and the press was a stunted "fourth power," not as dependent on the "first power" as deputies, senators, generals, caciques, union workers, or the peasants on the *ejidos*, but dependent nonetheless. If the new instructions were "nothing against the President, ever," the same concentration of power in the hands of the President also meant, in practice, "with the President, always."

The politics of reconciliation between the Church and the State was a reality that would not be reversed, and the Church was using it to reconstitute its ancient wealth (though on a lesser scale and through the use of intermediaries) and especially to support, without serious restrictions, its specific areas of interest: pastoral activities and education. Only the repression directed against the fascistic Sinarquistas and the democratic PAN of León (in Guanajuato) in 1945—both had connections, differently, with elements of the Church—would cast any shadows over a peaceful coexistence destined to be not only definitive but cordial.

At the other end of the world from those raging radicals of the twenties and thirties, the suave university men of Alemán's generation had no problem with their families going to mass, their wives giving money to the Church, their children being educated in the prestigious private Catholic schools of the Marist brothers or the Sisters of the Sacred Heart.

And during the era of Alemán, the Church also began a campaign for moral values in the Distrito Federal of Mexico City and its environs. Bishop Hernández Hurtado recalls, "The censorship by the Church of films, shows, and magazines was done through leaflets, functioned on a national level, and was very successful."[61] And with reason. In the census of 1950, 90 percent of the country declared itself Catholic. The Church had so much power over public opinion that a campaign in the press against a popular brand of soap, stigmatizing it as "protestant," was enough to make the sales drop catastrophically. In this case, the company found a solution by wrapping each bar in a portrait of the Pope, but when the anathema was pronounced not against a "protestant" business but against Protestants themselves, the results could lead to blood. There were incidents—usually at the instigation of a village priest—where Protestants were attacked and beaten and their houses burned. And even occasionally (on November 5, 1949, in the village of Tlacochahuaya in Nuevo León, for instance, or in San Pedro Ixtlahuaca, Oaxaca, in 1950—where the victim was a pastor) a Protestant would be murdered.[62]

But despite such moments, the country in general enjoyed a climate of religious tolerance. In the Distrito Federal alone, there were fifty-five explicitly Catholic publications, thirty-five Protestant, and seven of other religions, including two in Yiddish. The political system had turned a good part of the Mexican people into its dependents, but there was no remnant of those vague totalitarian tendencies of the twenties and early thirties, when Calles preached the need to "take control of the consciousness of our children." Freedom of thought continued to be effective and complete, like many other civil liberties the Liberals of 1857 had imprinted upon the Constitution.

The most powerful businessmen in Monterrey owed an old outstanding debt to Manuel Gómez Morín. It was not only an economic but a moral debt. As a corporate lawyer he had saved them from possible bankruptcy during the economic collapse of 1929, by coming up with the idea of private bond issues—for the first time ever in Mexico. But during the forties, when Gómez Morín tried to collect from his friends and associates by involving them with the PAN, they refused. They had resolutely tried to resist the centralization of the labor movement into the CTM during the forties and had serious disagreements with Cárdenas in 1936. But when this chance came to convert their well-earned economic independence from the center into open political militancy, almost all of them decided against it. President Alemán knew that Monterrey had been, ever since the end of the nineteenth century, the leading city in the process of industrializing the country, and of course he was willing to appoint a state governor sympathetic to big business.

And these men of Monterrey were the most independent business magnates in the country, who in 1943 had just founded their own Technological Institute, entrepreneurs who owed their success to their own efforts. (One hundred fifty new industrial enterprises per year were established in Monterrey during the forties.) Insubordination was even less to be expected from the textile manufacturers of Puebla, dependent on the exorbitant tariffs against foreign imports, or the older businesses of Mexico City supported in multiple ways by the government. When it did come, it was narrowly restricted to financial, industrial, commercial, or banking issues. And as little as it was, it was more than the nouveaux riches—businessmen who had risen to affluence through infinite means of direct and indirect access to government funds—ever dared.

Under Alemán, many businesses were started on the basis of government concessions, with the natural concomitant of an "officialist" mentality. Their founders may have come from old Porfirian wealth that had managed to save part of its rural patrimony by converting it into urban investment or else from the business class born during the Revolution or, more likely, from among the politicians to whom the Revolution "had done justice." In any case they knew their major customer, sometimes their only customer, was the government. For them political opposition was meaningless, against a regime that had brought the Revolution into the business world or more precisely had turned the Revolution into a gigantic enterprise for industrial promotion and political control. To be rich and to be part of the opposition would have seemed an insanity proper only to the idealistic lawyers of PAN or to another strange and romantic breed—the handful of "Marxist millionaires" whose priest was a man of rigid and impeccable morality, Narciso Bassols, the purest Jacobin intellectual of the Mexican twentieth century.[63]

Under the Hapsburgs and even later under the Bourbons, the University had been totally integrated into the purposes and activities of the State. An absolute consensus existed on the reasons for the legitimacy of the State, the extent of its power, and its responsibility for guaranteeing justice and equality. From the halls of the universities came the intellectual cadres who served the Monarch in matters human and divine. In Mexico, the Pontifical University—founded in the sixteenth century by Viceroy Mendoza—had closed its doors in 1833, after the first rush of Liberal legislation, but during the Centenary Fiestas Justo Sierra had reopened the University in its secular incarnation. It was one of the many ways through which the viceregal past would reappear in the modern world. For that high priest of the "Religion of the Fatherland," the National University was meant to replace its colonial and Catholic counterpart within a modern and patriotic environment, and to be a fountain of intelligence at the service of the State—to become, in Sierra's words, "the brain of the nation."

In 1917, the University was on the edge of extinction. It was seen as an elitist leftover from the Porfirian past. Its defenders at that difficult moment were the "Seven Sages," a group that deserved its name. A myth began to develop around the University, a premise that later became an axiom: Only by passing through the University (and not any university but *the* National University—the UNAM in Mexico City) could one earn the credentials needed for any aspirations to power.

Like Calles, many revolutionary military leaders had been teachers. Others resented not having been able to acquire enough education. Most of them respected the academics and students of the University, not enough to hand over power to them but still they were willing to give them technical responsibilities and, when appropriate, positions as assistant ministers and sometimes even ministers. Cárdenas was the exception. He distrusted the University, and though he did not exclude its graduates from the government, he made them share his cabinet and middle-level positions with his friends, unassuming people from Michoacán with much less formal education.

During the thirties, in the heyday of the politics of the masses and socialist ideology, the University had gone through a period of friction with the government. It was its punishment for supporting the candidacy of Vasconcelos. The labors of Gómez Morín as Rector had saved the financial and academic autonomy of the University, which for some years became a small island of critical independence in relation to the government.

The estrangement between the University and the government did not destroy the myth. Instead the opposite happened. The relative rebelliousness of the University raised its value on the stock exhange of Mexican politics. Ávila Camacho saw it that way and took a higher percentage of University graduates into his cabinet. In 1945 he gave the University the equivalent of a constitution, its Organic Law, which among other articles assigned the choice of a Rector to a government committee composed of University men themselves.

If Cárdenas had become the owner of the myth of the Revolution, Alemán made the myth of the University his own. From the early days of his electoral campaign, he placed the University at the center of his plans. The great national problems were to be discussed, decided, and later resolved by University men. What hope could an almost illiterate general have to hold power in a modern country? Of what use were hours of battle in comparison to hours of study? Engineers, architects, doctors would finally fulfill the "sophocracy" dream of the Seven Sages. They would be headed by the new holy caste, the elite of the elite, the lawyers who had degrees from the National University.

Miguel Alemán's lucky star also crossed the calendar. In September 1951, a few months after the inauguration of the Ciudad Universitaria (home of the University myth and the sacred space of modern Mexico), an anniversary dawned. It would be exactly four hundred years since the original founding of the Pontifical University by Viceroy Mendoza. Alemán invited foreign academics to commemorate the event with great pomp and circumstance. "Intellectuals from all over the world come to pay their homage to the University of Mexico," the newspapers proclaimed. "When the buffalo covered the plains of Missouri, in Mexico they founded the University," Rector Luis Garrido reminded his audience in his address. Now the University men of the Alemán era, who thought of themselves as Revolutionaries, could lay claim to being heirs of a four-hundred-year-old tradition.

From then on, the career of Alemán became a paradigm for Mexican University graduates. Instead of Vasconcelos's "The spirit will speak for my race," a more accurate motto would have been, "From the halls of the University to the halls of power." (In the Faculty of Law at the time, they used to say, "Here you study to be President.") What University man in his right mind would think of opposing the government?[64]

When the moment came to move their studies into the Ciudad Universitaria, they sang a good-bye serenade to their old schools in the center of town. And they left their student riots behind as well, a Mexican tradition since the age of Don Porfirio. Alemán himself, in 1929, had been a witness to the violent energy of his pro-Vasconcelos friends, but now the only politics the students pursued was within the University itself.

There were some student disturbances during the Alemán era, especially in the provincial universities, those that were not *the* University. At that same intimate meal with his friends, Alemán exchanged memories with one of the student leaders who had given him problems but who had, with the passing of the years, become governor of a state. "Do you remember?" asked Alemán. President Alemán had called him in and asked him what he would like: money, a scholarship, foreign travel. Without compromising himself on any one selection, the young man naturally chose all three. He spent a number of months in Europe, courtesy of the national budget. When he would run out of money in any remote European capital, he would go to the Mexican Consulate, write out a request to the President, and immediately receive his funds, cash on the table.[65]

It was a sign of the era of Alemán that student fervor was now devoted to sports. In 1952, a screaming crowd filled the stadium of the University City to see the game—of North American football—between the two great colossi of Mexican college sports. They had been mortal enemies since the end of the thirties: the White Donkeys of the Polytechnic Institute (wearing cherry red and white) and the Pumas of the university (wearing blue and gold). The Pumas won. The "Guélum," the Polytechnic cheer, was stilled. Over and over, the roar of the "Goya," the University cheer, resounded. In the twilight, the seats on the University side of the stadium filled with celebratory torches. The sky was clear. As the students left the stadium, they could see the statue of the patron saint of their personal city: Miguel Alemán, dressed in his toga of *Doctor Honoris Causa*.

All the university men were Mexicans, but only about 25,000 men and women (of 25 million Mexicans) formed the body of university graduates during the era of Alemán. Almost all the intellectuals held university degrees, but no more than a few dozen university men were intellectuals. The doctors, engineers, and especially the lawyers in the government were all university graduates, but they were not intellectuals. They were part of the social "estate" of "university men," a group destined to gain growing access to power until they became the dominant force in Mexican public life. "Intellectuals" in Mexico were and are men of letters with a public reputation, who express their opinions in print on issues of general interest.

Almost all the men of the Reform had been intellectuals. During the ephemeral decade of the Restored Republic, they had actually held power, with excellent results for the country. They were "fiercely, proudly, arrogantly, absurdly and irrationally independent." Very few intellectuals in the later history of Mexico would merit such descriptions.

After taking power in 1876, Porfirio Díaz had reconstituted the viceregal arrangements. Once again the intellectual was organically incorporated into the State. Even men who were viscerally critical, perhaps acidly so, like Francisco Bulnes, surrendered their specific intellectual weapons to the regime, in order to work "from within" and "build the nation." Entire generations of real or potential writers on politics used their pens to write poetry, history, or novels, to move freely among all genres except the forbidden zone of political criticism. Among intellectuals, bit by bit, a feeling of dependence on the power of the government began to grow. In his personal diary, where he dared to write down severe criticisms of the Díaz regime, the novelist Federico Gamboa noted, around the year 1895:

> Why do I necessarily want to live in the employ of the government? Why don't I learn how to do something else? . . . It is the old tacit agreement. For our living we count entirely on the government, and every government—from the viceroyalty to the present day—counts on the fact that we count on them.[66]

Don Porfirio had the intellectuals, in one concise characterization, "tied to him by their guts."

During the Revolution, intellectuals played a role that was far from admirable. Porfirio Díaz had given them minor jobs, scholarships, and grants, but the new Madero regime refused to continue these practices and left many of them without work. The foreseeable result was support, by the majority of intellectuals, for Victoriano Huerta. There is no greater paradox than the fact that the cabinet with the greatest intellectual weight in Mexican history (after that of the Restored Republic) was formed by Huerta. When his regime fell in July 1914, a caravan of intellectuals set out on the road to exile. The Porfiristas (Bulnes, Pereyra, Rabasa) and the Huertistas (among them some of the later members of the "Atheneum of Youth") took refuge in the United States or in Europe. Some of them, like Alfonso Reyes, became professional exiles.

During the years of battle, almost every revolutionary general had his group of *licenciaditos* ("favorite college graduates"), who wrote speeches and drew up programs, laws, and proclamations. Strictly speaking, most of them were not intellectuals but university men with political ambitions. Some of them had come to the Revolution because they were moved by an authentic passion for the redemption of the Mexican people. But except for José Vasconcelos under Obregón, no Mexican intellectual was ever completely trusted by a revolutionary jefe or gained direct political power.

By the time Alemán became president, the major intellectuals of the Atheneum had either died (Caso, Henríquez Ureña), were living in epicurean (Reyes) or mystical (Vasconcelos) retirement, or were still active, like Martín Luis Guzmán practicing modern journalism but at the service of the government. Some of their old disciples—the Generation of 1915—lived "mute and immobile" at the margins of power, or confronted it in more or less unsuccessful fashion. Lombardo founded the *Partido Popular* in 1948, supported by another intellectual of the left, Narciso Bassols. On the other ideological pole, Gómez Morín continued being the president of PAN. Cosío Villegas had asked for an indefinite leave from the Ministry of Foreign Relations and the Bank of Mexico (two institutions where he had been working) and a temporary leave from the Fondo de Cultura Económica (the publishing house he had founded). With the help of a Rockefeller Fellowship, he devoted himself full-time to historical research and only made occasional public appearances. One of them was in September 1949: a commemoration of the fifteen years of the Fondo de Cultura Económica. President Alemán attended. Cosío Villegas—who always dressed like an English gentleman—would appear in his usual correct dress but with a striped shirt (highly unusual for him), as if to subtly underline his disagreements with the triumphalism of the Alemán era. His new political essays—published in *Cuadernos Americanos* and forming a sequel to his *Crisis of Mexico*—earned him the esteem of an intellectual minority in Mexico and Latin America, but unlike the essays of Vasconcelos during the twenties (which had attacked financial corruption among the Revolutionaries and also had been strongly anti-American), those of Cosío Villegas never reached a wider public.

With a few such exceptions, the relationship between the government and the intellectuals had returned to the old, palmy days of Don Porfirio. "Jobs and more jobs" was the watchword. Competing with themselves, emulated by an entourage of imitators, converted into national icons, the mural painters Rivera and Siqueiros (Orozco had died) continued to rely on generous state patronage while they painted the walls of public and private buildings, hotels, theaters, the homes of artists and high society. The poets held minor jobs in various secretariats but following in the steps of Jaime Torres Bodet tended more and more to concentrate into the Ministry of Foreign Relations. In 1950, aside from Torres Bodet (then secretary-general of UNESCO), Foreign Relations provided work for the poet José Gorostiza (who until 1949 had been director-general of the diplomatic service) and Rodolfo Usigli himself, author of *The Imposter* (*El Gesticulador*), who was Second Secretary at the Mexican Embassy in Paris, one step above the Third Secretary, a thirty-six-year-old poet named Octavio Paz, who in those days was in the process of publishing—under the aegis of *Cuadernos Americanos*—perhaps the most important book of meditation on national self-knowledge in the Mexican twentieth century: *The Labyrinth of Solitude* (*El laberinto de la soledad*).

In the days of Don Porfirio, Justo Sierra had given fellowships, jobs, subsidies, and every sort of support to intellectuals and artists through the University, the Department of Fine Arts, and the National Museum. Alemán followed the same prescription. The architects had their six gilded years with a wave of public works, especially the Ciudad Universitaria. At the new Institute of Fine Arts, the visual artists, the musicians, the poets, and others found their home. At the Instituto Nacional Indigenista and the Institute of Anthropology, under the watchful eyes of their founder, the archaeologist Alfonso Caso, the anthropologists, archaeologists, and historians did their work. The salaries paid by the University were low but not insignificant. The same could be said of the research fellowships awarded by the Colegio de México, which also financed pleasant postgraduate trips to Paris, where the government of Alemán had promoted the first international Mexican artistic exposition and opened a Casa de México. If an intellectual happened to be a Communist and had spent some years in prison on the Marías Islands, but had talent and was named José Revueltas, the doors of the government were closed to him but he could still write scripts for the Mexican cinema. And for all the intellectuals, as a last resort, there was the modest income available through cultural journalism.

The regime of Alemán, in only six years, had achieved what it took Don Porfirio thirty years to do: turning the intellectuals into indirect dependents of the regime, "gripped by their guts." As they had done during Porfirismo, the intellectuals consented to the sacrifice of their political freedom, a sacrifice that could have disastrous moral consequences. The writer César Garizurieta had said, cynically, "to live outside the budget is to live in error." Yet many intellectuals could legitimately echo the words of Cosío Villegas when he said that modern Mexico would not have come as far as it had without the contributions of its intellectuals. And besides, support from the State permitted them to continue doing what was most important to them—their personal creative work.

Light-years from the magic circle of dependency (where the real, formal, and corporate powers resided), beyond the concentric ring of those not so dependent (the circling planets of the press, the Church, big business, the University, the students, and the intellectuals) could be found the strangest creatures in the Mexican political system. They were not insubordinate individuals or political groups. They were just not dependent.

To the ideological right of the system, the government repressed the homegrown fascist movement of Sinarquismo, canceling its registration as a party and in effect making its political activities illegal. But it felt obliged to tolerate the persevering and peaceful activities of the PAN. In 1949, ten years after he had founded the party, Gómez Morín stepped down as its president. He was still convinced that "the political issue is at the core of our nation's problems," but he thought that the first phase of the party's existence was completed. Alemán's regime had incorporated many of the PAN's initiatives (especially in the area of economics) and would go on adopting more of them through the years (and so would later Presidents). Gómez Morín now thought that new times called for new strategies. It was then he coined a formula that would become famous: "We have to move souls." He meant—to conquer the political commitment of the people in the states and the cities.

If the PRI kept on winning "absolutely everything," the PAN had the advantage of an almost religious patience. "They can defeat them again and again," observed a newsman of the time, "they can expel its most respectable leaders from the Legislature accusing them of being reactionaries. It doesn't matter. National Action stays on its feet and keeps moving ahead." And in effect, for PAN, electoral abuses against them were "their daily bread": meetings drowned out by taxi drivers blowing their horns, sympathizers sent to jail, death threats to their candidates. Four or five seats in the Lower House every three years, one or two municipal presidencies every six years seemed a meager harvest for all these sacrifices and humiliations. But the Panistas did not retreat. They were convinced that they were sowing for the future.

To the left of the political system, there was no planet but instead a belt of small asteroids: the various factions of the left. The oldest was the Communist Party founded by the adventurous Comintern agent, Manabendranath "Indú" Roy (also instrumental in creating the first Communist Party of his native India) during the age of Venustiano Carranza. A small party, combative, honest, disciplined, fiercely committed to its ideology, it had over thirty years gone through many ups and downs. Powerful during the presidencies of Obregón and Calles, it had been violently banned during the Maximato, influential in union and agrarian circles under Cárdenas, and then seen its influence decline under Ávila Camacho. Although in 1946 it had supported the candidacy of Alemán, the Businessman President—once he was sitting in the Presidential Chair—treated the Communist Party the same way he did the Sinarquistas at the opposite end of the ideological spectrum. He canceled their electoral registration. During the six years of Alemán, many Communists went to jail.

The government was not responsible for all the misfortunes of the left in Mexico. Almost from the time it was founded, the Communist

Party of Mexico began to reproduce the intolerant attitudes of its original model. In Mexico, as in the Soviet Union, there were expulsions, excommunications, persecutions, and denunciations, all in the name of the truth revealed by Marx. The assassination of Trotsky (carried out in Mexico City by a Stalinist agent) and the Hitler-Stalin pact were stones against which many Communists stumbled, incapable of morally or doctrinally justifying these actions. Nevertheless, despite external persecution and internal purges, the Communist Party survived clandestinely or in prison, waiting for better times.

And what is more, during the age of Alemán, communism was gaining prestige in a number of student, university, and union circles. In Mexico, as elswhere in Latin America, the USSR retained a halo of attractive mystery. When in 1951 the poet Octavio Paz—then living in Paris—wanted to publish information on the Soviet prison camps, nobody in Mexico dared to do it because nobody believed him. He had to publish his report in the Argentine magazine *Sur*. The facts were convincing, but Paz himself—who since his youth had put his faith in the socialist revolution and had confronted the Civil War in Spain—made a point of concluding with the words: "It is inaccurate . . . to say that the Soviet experience is a condemnation of socialism. The crimes of the bureaucratic regime are its own and completely its own, and not of socialism."[67]

Mexican high culture was beginning to be overwhelmingly a culture of the left. And perhaps the man with the greatest prestige on the Mexican left of the time was the leader, orator, philosopher, editor, and teacher Vicente Lombardo Toledano, whose ideology was a synthesis of Marxism and nationalism. He taught that Mexico would advance to socialism but its progress had to be gradual, supported by progressive groups of the national bourgeoisie, and as a matter of first importance, it had to be strongly anti-imperialist.

During Alemán's electoral campaign and perhaps influenced by the example of his old fellow student Manuel Gómez Morín, Lombardo hatched the idea of founding "a new organ of combat," the "true party of the Revolution." In 1948, the *Partido Popular* (Party of the People) would be born, a "party of the masses for the purpose of defending national independence, for improving the life of the people and for promoting and energizing the true industrialization of the country. . . . A party that will be democratic, national, revolutionary, anti-imperialist, and will include workers, peasants, progressive intellectuals and other nuclei of the petty bourgeoisie of town and country."[68]

Lombardo had originally staked his hopes on attracting the CTM to his new party. An idealist and an ideologue rather than a practical leader, he underestimated the political pragmatism of the "Five Little Wolves." Although his old disciple, Fidel Velázquez, continued to admire him and call him *maestro*, the CTM refused to allow its members to join any party but the PRI. Lombardo had the same experience with the workers that Gómez Morín had with the businessmen. Neither group was willing to accompany the intellectuals on the adventure of creating a political base independent of the system. They were both left alone, small and inoffensive asteroids at a distance from the sun.

In 1949, midway through Alemán's term, the PP participated in its first legislative elections. That same year, the party split in two. It had run a truly popular candidate for the governorship of Sonora, but the government refused to recognize his victory. (The only concession the government granted the PP was one seat in the national Legislature.) Indignant at the fraud in Sonora, Narciso Bassols, the leader of the more radical faction, declared that the PP could not behave like the PAN and accept "the crumbs of three or four seats" sent their way by the government. But Lombardo took a more conciliatory attitude, which led to Bassols and his group immediately resigning from the party. In a later lecture on "the political problem of the Mexican Revolution," Bassols would deliver one of the most severe condemnations ever formulated against the system that Alemán had perfected:

> The most serious . . . consequence of the prolongation of the present electoral system is not only the process of falsifying the results of an election . . . but the fact that the entire political life of the country is being poisoned, paralyzed, suffocated. . . . The apparatus of mystification that characterizes the elections leads nowhere but to lack of action, has no stronger political life than the impetus which leads it to avoid any true political life.[69]

Lombardo Toledano would run for President in 1952. Even though electoral machinations were used against him, he did not hesitate to hold a secret meeting with President-Elect Ruiz Cortines and congratulate him on assuming the responsibility for correcting the course of the Mexican Revolution. From then on, it was no secret that the People's Party received money from the government budget. The later fate of the PP had been cast. It would selectively oppose actions by the PRI that it considered retrogade or support actions that it considered progressive. It became the loyal opposition on the left. The choice of what to support or oppose would, as with so many things in Mexico, be the responsibility of one man, in this case Lombardo Toledano.

* * *

Among the many generals not dependent on the system—and who in 1951 supported the presidential candidacy of Miguel Henríquez Guzmán—two old partisans still had considerable moral prestige: Francisco J. Múgica and Cándido Aguilar. After the government's mounted police charged, swinging their truncheons, to break up a meeting of Henríquez supporters on the Alameda and left dead men on the ground behind them, both of these generals were persecuted by the authorities. The Mexican system was not, even remotely, a regime of direct terror, but the old Porfirian pattern—"bread or the bludgeon"—would be applied as usual. If you were insubordinate, you paid the price with prison and sometimes with your life. Múgica had to hide in the trunk of a car to escape an attempt by the National Security Police to arrest him. Aguilar was imprisoned for forty days in Veracruz. Months before, he had written a letter to his lifelong protégé, Miguel Alemán:

> Your government has had many successes but also made various mistakes; perhaps the worst of them was allowing many of your collaborators to enrich themselves. These errors or mistakes would be forgiven by future generations if you, transcending the squalid interests of a group or of your friends, should rise to the height of the great patriots and statesmen and establish democracy in Mexico, which we have not been able to achieve after forty years of struggle.[70]

The man whom Aguilar wanted to convert into the "Father of Mexican Democracy" had now sanctioned an electoral and physical assault on the members of a legal political grouping. For Aguilar, the result was a feeling of deep resentment against Alemán. He never wanted to see him again. Only on his deathbed—according to Dr. Justo Manzur Ocaña—did the resolve of the old Carrancista weaken:

> I was at the door of the room when Miguel Alemán arrived. The general was already very weak. I entered and I said, "Señor, Licenciado Alemán is here and wishes to see you. . . . Should I let him come in or not?" "Tell him to come in." And he did. . . . I saw how Alemán opened his arms to him. He said to him, "Jefe, Jefe," and he moved to embrace him. They were reconciled. And he died a few days later, on March 20, 1960.[71]

And then there was the world of popular humor.

Ever since the twenties, in modest neighborhood tents or review theaters, the Mexican people would be entertained by political jokes with double meanings or parodies by famous comedians. Calles had endured and tolerated them. Cárdenas had given these performers their own theater. Ávila Camacho had winced at the gibes directed against his brother Maximino, but he had not moved against the political humorists. In 1946 the theater of satire was healthier than ever.

But when Alemán's University graduates stepped onto the political stage, they imposed censorship on the theatrical stage. Scripts had to be closely examined by officials of the Ministry of the Interior, who would also attend and scrutinize every performance. "Once," Jesús Martínez *Palillo* ("the Toothpick"), remembered, "I was brought before the Chief of Police. I didn't answer him and he slapped me twice in the face. There was nothing I could do but spit at him. They threw me bleeding into a cell." It would not be the last time. For the rest of the age of Alemán, the Toothpick would go onstage with a permit in his pocket to avoid summary arrest.[72]

On July 20, 1948, a day on which the Lyric Theater was presenting the premiere of the play *The Fourth Power*, the director of the Office of Performances ordered the theater shut down because the play contained "political allusions" against President Alemán. In protest, all the theaters of the city closed their doors. They would eventually reopen, and *Palillo* would continue flailing away at "the immoral politicians, octopuses sucking up the national budget," but the precedent had been set—political humor was fine, just so long as it did not mention the First Officer of the Nation. From that time on, the political joke gradually began to abandon the tents and the theaters to take refuge in private spheres: the cafés, the streets, conversations on the telephone and in the bedrooms.

Outside the system as well was the formidable exterior power of the United States of America. And if the U.S. Department of State had any doubts about the authenticity of Alemán's attitude toward the United States, his stance of cooperation and reconciliation, they would have been dispelled through some of the first actions of his government. Under Alemán, the relationship between the two countries took on the hues of a honeymoon.

In March 1947, for the first time in history, a president of the United States visited Mexico City. And it was not on just any date at all. A century before, in March 1847, the American forces under Winfield Scott had shelled Veracruz and a few months later were fighting Mexican soldiers in what is now the Park of Chapultepec. And now President Truman was visiting Chapultepec to pay homage and lay a floral wreath at the monument to the Niños Héroes.

Alemán returned the visit by going to Washington in May 1947. A

Mexican President had not been to Washington since 1836, when former President Santa Anna, taken prisoner after his Texan debacle, had to listen to the stern reprimands of Andrew Jackson. Since then there had been no official visits at all by a President to the United States. The meeting between Franklin Roosevelt and Ávila Camacho, in the Mexican city of Monterrey, was already an omen of cordiality, but the taboo had continued in force.

Alemán's visit broke it. He was received with great honor. In his speech to Congress, he mentioned "the absolute faith in democracy and essential love for liberty" that united the two countries. In a gesture unknown since the age of Porfirio, Alemán invited U.S. interests to invest in Mexico and succeeded in getting the American government to entertain the prospect of supporting projects—for the Mexican infrastructure—to be proposed by the Mexican government. On the thorny problem of a possible return of the American oil companies to Mexico, Alemán was very precise. The general participation of foreign enterprises in the ownership and exploitation of Mexican oil was out of the question because it was against the law, but his government was prepared to issue permits for drilling and exploration.

Truman was committed to supporting the economic development of Mexico. Any difficulties encountered in the process were smoothed over by the outbreak of the Korean War. For the first time since Porfirismo, the United States opened a line of credit to Mexico (for $150 million, which the country could use to develop its oil facilities), lent an additional $56 million for the reorganization of the railways, and signed an agreement on emigration, all without Mexico's being required to agree to a military pact (which other Latin American countries had to do) or become part of the anti-Communist bloc that the United States was creating in Latin America. In some American newspapers, President Alemán was crowned with the epithet "Mr. Amigo."[73]

But there was another American, a true and very long-term "amigo" of Mexico, a single but intensely knowledgeable voice, who made a very different judgment on the work of Miguel Alemán.

When the regime was at the height of its glory, when new public works named after the President were being inaugurated day after day and there was talk (in lowered voices) of his possible reelection, when the Mexican government was turning its back on the rural community and resolutely adopting the paradigm of an urban and industrial nation, when almost the entire society had been subordinated to the government through the method of "bread or the bludgeon" and Mexico was enjoying its first honeymoon with the United States after a century of bitter disagreements, the Spanish translation of a book was published (in 1951) that would anger all the politicians and intellectuals of Mexico—except for Daniel Cosío Villegas. Its author was perhaps the best American friend Mexico has ever had: Frank Tannenbaum. The book was titled *Mexico: The Struggle for Peace and Bread.*

No American had traveled, studied, and understood Mexico as Tannenbaum had. And he had done it for more than twenty-five years. He was a social philosopher whose political progress, intellectual formation, and moral stance had given him a privileged position from which to observe and evaluate the Mexican Revolution.

On the fourth of March 1914, on the day that he turned twenty-one, the young Tannenbaum had shared the front page of *The New York Times* with Pancho Villa. While Villa was leading his *dorados* to the capture of Torreón, Tannenbaum was leading a brigade of 190 unemployed men in the peaceful but peremptory seizure of the Church of Saint Alphonsus on West Broadway in New York, in order to demand their rights: bread and a place to sleep. He was then a young anarchist, who frequented the circle of Emma Goldman and Alexander Berkman and was a fervent member of the IWW. Later he would go to Columbia University and become a leading American sociologist, but he would never lose his youthful rage at injustice or his hopes for a better world.

Tannenbaum first visited Mexico in 1922. It was a very bright time for the country. Vasconcelos had begun his educational crusade; the missionary teachers were moving across the nation; Diego Rivera was painting his murals at the Ministry of Education; thousands of copies of the classics were being printed and thousands of them were being given away; indigenous culture was again beginning to nourish a sense of Mexican identity; the Constitution guaranteed the working class a legal status equal to that of the most advanced countries; the land that the haciendas had seized from the villages was now being returned to the people. Tannenbaum was one of the first to bear witness to this new dawn. From then on—as a Latin American author would write—"he felt the bliss of Mexico down to his bones."[74]

He became a great admirer of the peasant revolution, crystallized for him in the figure of Emiliano Zapata. Later he would become a personal friend and partisan of Lázaro Cárdenas and his program of Agrarian Reform. But now, in 1951, he was writing that Mexico had lost its way. What was the social cost, he asked, of industrializing the country? Protected behind a wall of high tariffs, the three groups who benefited from Alemán's policies were building a monopolistic alliance against the rural majority of the population, whom he considered Mexico's real "treasure": "We have here what is almost a conspiracy by a few owners, one hundred thousand workers, and a few governmental officials to provide the mass of the population with desperately insufficient clothing at excessively high costs."

The situation was not confined to textiles. A small privileged working class that included no more than 20 percent of the population had been developing without any connection to the great mass of Mexicans. The situation, Tannenbaum pointed out, contained "all the elements of tragedy." The support of the government for industry did not greatly impress him. An industrial policy that raised the cost of living for an entire nation in order to benefit a minority seemed to him doubtful at best.[75] Far better would be the widespread encouragement and diffusion of small-scale rural projects.

Suddenly, the friend was now "a former friend." For the Mexican economists, his name became taboo. His only public defender was Daniel Cosío Villegas. Five years earlier, Cosío himself had suffered universal condemnation for supporting not very different ideas. Since 1940 he had been thinking that the "legendary wealth" of Mexico was just that, a legend, and that the country should opt for a modest and balanced future, living in equal part on its agriculture, its industry, and its mining.

In his defense of Tannenbaum, Cosío Villegas presented his own pendulum theory of Mexican history. Since Independence, the country had been zealously seeking the two great goals of the Western world: individual political freedom and general material well-being. Incapable of gaining both at the same time, it had alternated periods of liberty without progress and periods of progress without liberty. In either case, the undue emphasis on one goal over the other had led to revolution. Without progress but with liberty the Restored Republic had fallen into the hands of Díaz the dictator. With progress but without liberty, Porfirian Mexico had shattered to pieces. For him, the lesson was obvious: The country had to achieve a moderate advance in both directions, without permitting one to prevail over the other. But where did the country stand, in 1951?

In the worst imaginable condition. Mexico was not choosing the way of progress, because Alemán's program would not lead to *general material well-being* but to a restricted variety, of benefit to specific groups. Industrial growth favored the three castes of government officials, industrial workers, and entrepreneurs. This narrowing focus of the economic program carried with it serious dangers, but Cosío Villegas saw something even worse, which seemed to have escaped even Tannenbaum's attention:

Daniel Cosío Villegas, 1953
(Private Collection)

The so greatly praised material progress (and not only the minuscule industrial progress) is used like a stream of light to blind the eyes of the people with its glow and prevent them from seeing their own open wounds . . . their political wounds! What is more, the man in charge . . . asks for . . . the recognition that all this material progress is his personal achievement though it has been accomplished, not with the personal fortune of the man in charge but with the funds of the country; not with the hands of the man in charge but those of the Mexican worker; and at a cost that—were you to know its full and exact extent—would almost shock you to death. And so he demands that each achievement carry his own name so that future generations may see it everywhere, as if he were a god.[76]

19

ADOLFO RUIZ CORTINES

The Administrator

In 1952, on the first of December, the day on which power changes hands every six years in Mexico, the perennial smile was wiped right off Miguel Alemán's face. It is a custom for the new President to receive the presidential sash from the hands of the outgoing President, take the required oath, and then give his inaugural address. When Calles passed on the office to Obregón, Cárdenas to Ávila Camacho, Ávila Camacho to Alemán, the words had been cordial—praise for the outgoing President and then a summary of the new program. But this time the incoming President did something unexpected. As his speech developed, the audience began to hear notes of censure directed toward the triumphant self-assertion of the Alemán era. And the conclusion left nothing in doubt. Repeatedly wagging a critical finger in Alemán's direction, Adolfo Ruiz Cortines used serious words: "I will not permit the principles of the Revolution nor the laws that guide us to be broken. . . . I will be inflexible with public servants who stray from honesty and decency."[1] According to many who knew him, Alemán "hated the old son of a bitch" from that moment on.[2]

There had been many jokes about Ruiz Cortines's age. He had "escaped from the tombs of the pharoahs." And when the near-legendary Sergeant De la Rosa—last survivor of the War of the French Intervention and a picturesque character 112 years old—came up to

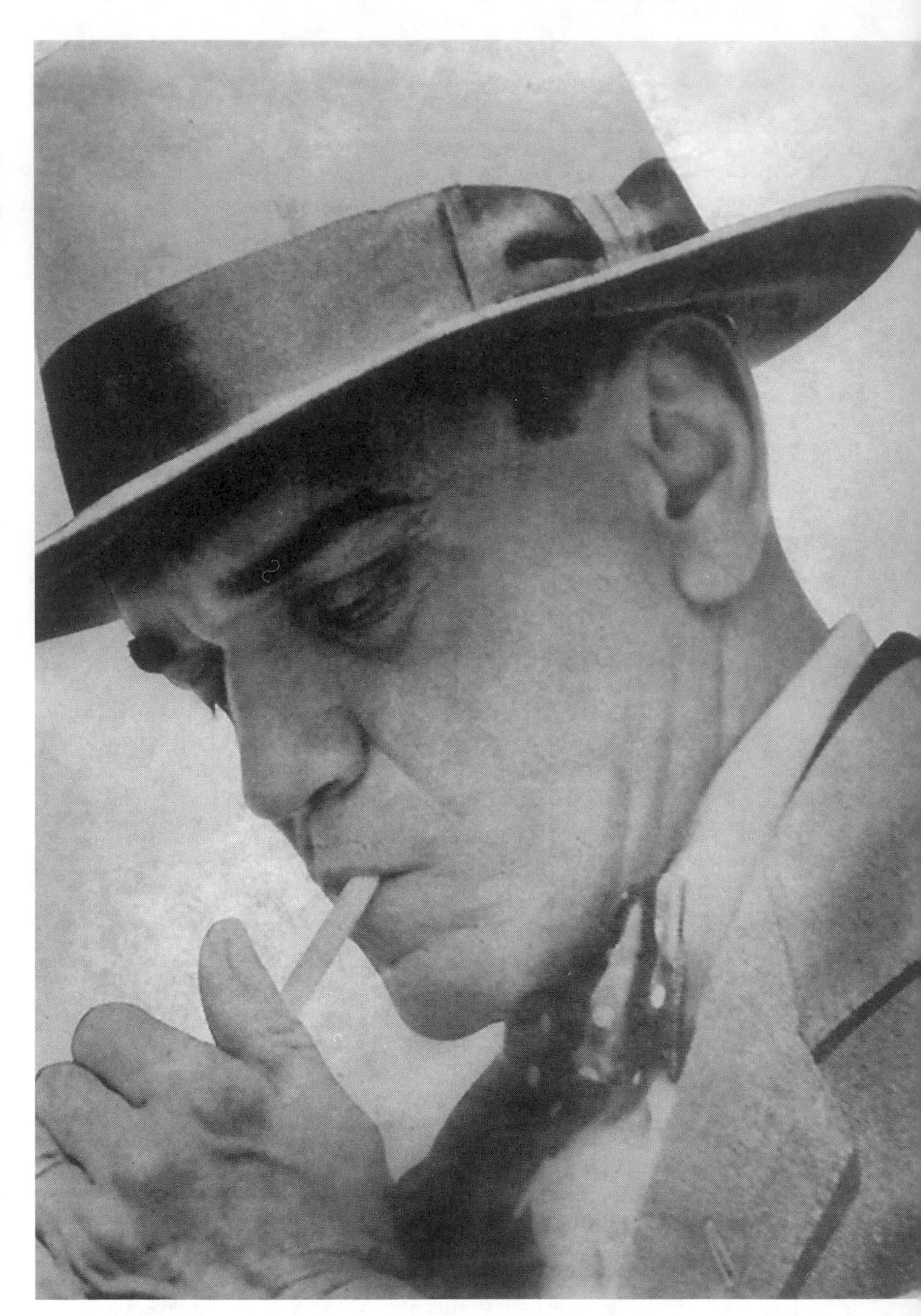

Adolfo Ruiz Cortines, 1952
(Archivo Fotofija)

congratulate him on the presidency, it was said that "one of his grandsons is visiting the President." A native of Veracruz, where humor is second nature, Ruiz Cortines took it all in stride: "They elected me to be the President, not a seed bull."[3] He was not terribly old—only sixty-two—but compared with the "cub" Alemán, who was not yet fifty, he seemed an elder statesmen. And no previous Mexican President—not even the prototypical "Old Man," Venustiano Carranza—had come to the Chair past the age of sixty.

His acceptance address had been the first surprising signal of a swing in the pendulum of power. Many more would follow. When he formed his cabinet, Ruiz Cortines chose not only younger university graduates (as Alemán had) but also more experienced people somewhere between his and Alemán's age. He refused to give jobs to his friends—and he had many good friends, especially among his domino-playing buddies from Veracruz. He continued to play his favorite game with them but would not give them preference for political favors, again in clear contrast with Alemán.

The day after he announced his cabinet, he published a complete and detailed list of his property and assets: a house in Mexico City, a farm (jointly owned with a friend) in Veracruz, some modest savings, a 1948 Lincoln, his wife's car, and their furniture. It came to about $34,000.[4]

He then required all 250,000 public employees to do the same, with the clear warning that their statements would be immediately checked and then examined again at the end of his presidency. When the Finance Minister sent him a check worth about four thousand dollars for his "special expenses," Ruiz Cortines returned it, contending that his salary was enough for him.

He rejected the usual offer—from the automobile dealers—of a latest-model car to begin the presidential term, and though his wife, Doña María Izaguirre, had great influence over him, she could not persuade him to let her keep the three hundred gifts she received for her birthday. The First Lady had to be satisfied with those that came from old friends, not one single gift more.[5]

There was a theatrical element in all this display of honesty. Who had ever heard of the President's chauffeur getting a ticket for an illegal U-turn from a traffic cop—known as *mordelones* ("big biters") in Mexico because they have become the essential symbol of petty corruption? But this along with his less eccentric actions sent a clear message to bureaucrats as well as the public: Ruiz Cortines was now the President, and he would not tolerate dishonesty and waste.

He soon moved from symbols to action. He ordered a temporary suspension of all government contracts so that each project could be reconsidered. In one case, his Minister of Communications reported the receipt of a bill for a highway, 120 kilometers in length, that existed only in that very request for payment. Ruiz Cortines fined the contractor three times the amount he was trying to collect. With the "so-quick-ists" of the Alemán era, he showed no mercy. The stroke of a pen ended Jorge Pasquel's total control of oil distribution or the transportation monopoly that had earned its overlord, the former director of the nation's social security system, some millions of dollars.[6]

"We are still a poor country," he said in his first *Informe* in September 1953, and offered a flood of discomforting facts: 42 percent of Mexicans were illiterate; 19 million peasants were living in extreme poverty; 60 percent of the population received the benefits of barely a fifth of the national revenue; in the last ten years the population had grown by 6 million, many of whom had no other options but to cross the northern border as "wetbacks."[7] To deal with "the crisis of Mexico," Daniel Cosío Villegas in 1946 had called for a purging of the men in power and a renewal of the principles of the Revolution that he thought had been abandoned. Frank Tannenbaum had proposed rethinking the emphasis on industrialization and returning to the agrarian emphasis that had been represented by Cárdenas. President Ruiz Cortines could agree with neither of these prophets. Although he belonged to their generation (actually—born in December of 1889—he was older than either), Ruiz Cortines was politically part of the Institutional Revolution. In no way did he consider the Revolution to be either a closed cycle or in crisis, still less a program exhausted or dead. The Revolution was just as alive and present as it had been in 1910, but it did need a vast cleansing of men. There was nothing in his proposed six-year program to indicate a correction toward the social models of the twenties and thirties or to contradict the objectives of Alemán. Only men and forms were at fault.

The man first and most at fault had been Alemán himself. Perhaps Ruiz Cortines could have moved against him in court or applied greater legal pressure, with the threat or reality of prison terms, against Alemán's friends. He chose not to do it and with the choice firmly fixed a new rule for the Mexican political system: Former Presidents are legally untouchable. But there was no way to shield the image of the ex-President from public opinion or the press. With Ruiz Cortines the press felt itself comparatively free to serve as a safety valve for the moral complaints of the ordinary citizen. One PRI general, Adolfo León Osorio, dared to accuse Alemán and his friends of having drained the country of $800 million.

While he traveled through Europe with the Brazilian actress Leonora Amar or crossed the Bay of Acapulco in his huge yacht, the smile gradually began to return to Miguel Alemán's face. He knew that all former Presidents passed through a kind of Purgatory and if he were strongly criticized by one group, in other spheres he could still count on great popularity. Perhaps he already understood that he would never hold the only power specifically reserved for ex-Presidents, the kind of moral ascendancy that Cárdenas would exercise until the day of his death. But—in exchange for that—Alemán could dedicate himself to commercial activities and to the development of the Mexican tourism industry that was of such great interest to him.

Fundamentally, little had changed. The political system organized by the business genius of Miguel Alemán was entering a second stage, in which its dependent powers—and those not so dependent as well—would steadily become more and more dependent on the center. There had only been a shift at the top of the organization. The founder of the business had given way to the manager. But because power was so concentrated into the hands of any President, a new style would have inevitable practical consequences. What kind of man the President was, as always, would be of great importance.

In the presidential Olive Train that in May of 1920 was carrying the wandering forces of Venustiano Carranza toward Veracruz, there was one car that seemed to be full of high explosives, marked with the usual warnings on danger and the necessary precautions. Inside was something other, dangerous in its own way: 150 million pesos in gold—the National Treasury. When the long train of the former *Primer Jefe* reached Aljibes station, Carranza, with his small escort, had to abandon it and take the road to the Sierra of Puebla that would lead him to his death. A young civilian attaché, assigned with the rank of major to the troops of his relative by marriage, General Jacinto B. Treviño (in rebellion against Carranza), was put in charge of the Treasury and ordered to escort it back to the capital, there to sign it over in person to the new President of Mexico, Don Adolfo de la Huerta. The young man was of course far from the center of power and had nothing in common with the President except his first name. He was Adolfo Ruiz Cortines.

The painter David Alfaro Siqueiros knew him in his youth and described him as "very much the young man about town," with his white summer suit and straw fishing hat, with the deep dark eyes and thick eyebrows of his Andalusian ancestors, "a quintessential Veracruzan 'dandy.'"[8] In the brothels of the port, he was known as *El Faquir* ("the Fakir") (perhaps because of his slender good looks and his "oriental" liking for women).[9] In the Villa del Mar dancing hall, they called him *Cintura Brava* ("the man with the agile waist").[10] A good drinker but not a big drinker, a good baseball player but not a star, he had a particular passion for sitting down with three friends at a table in the Portal de Diligencias, the most "typical" café of Veracruz, and spending hours absorbed in that curious game (very popular in Mexico) of prediction, deception, statistics, and chance called dominoes.

He had been a man of numbers almost since childhood. Born in 1889, shortly after the death of his father, Adolfo Ruiz Tejeda, he left school very early to learn how to keep account books conscientiously. Nowhere in Mexico could the skill be as useful as in the commercial port of Veracruz. At the age of fifteen he was working as a junior accountant.

He joined the Revolution, and though he saw gunfire close up in 1915 at the battle of El Ébano in his native state, his contributions were restricted to the account books, as an army paymaster. His moment of glory came in May 1920—the return from Aljibes to Mexico City as guardian of the National Treasury in the Olive Train.[11]

When he left the army, he entered an environment so congenial to him that he would remain there from 1926 till 1935: the Department of National Statistics. Daniel Cosío Villegas was then a young teacher—recently returned from studies at Harvard: "For three solid months, from Monday to Friday, I taught the course in statistics. Seated in the first row taking copious notes, always attending with religious punctuality, would be Don Adolfo Ruiz Cortines."[12] He put those classes to good use, very quickly becoming one of the foremost Mexican authorities on statistics.

Early in the thirties, he solidified his reputation with a group of technical articles published in the magazine *Crisol*. Written without the least literary flair but completely free of rhetoric, the articles proposed ideas that were unusual in their day but would be amply justified in the future: the worth of a politically autonomous bureau of statistics, the need to deal with overcrowding in the big cities (at a time when Mexico City only had a population of 1.23 million), the importance of making the country conscious of the population problem. In a Catholic country where this last issue has almost always been taboo, Ruiz Cortines expressed his opinion that reckless population growth could lead to "serious social disturbances."[13]

At the age of forty-five (late to begin a political career) he became Chief Administrative Officer (*Official Mayor*) of the government of the Distrito Federal, where, beyond his administrative, technical, and political duties, he had to deal and dance with seventy-three bureaucratic organizations, to be "the torero to those tigers."[14] (It was during this period that he first met Miguel Alemán.) He moved through various posts—which included managing the finances, efficiently and honestly, for the Ávila Camacho presidential campaign. In 1944 he became the governor of his native state. It would be a prudent and efficient administration. He kept corruption at bay through investigatory commissions and doubled the state treasury. His personal honesty became as famous as his continuing devotion to the game of dominoes, which he continued to play at his old haunt, the Portal de Diligencias. Ávila Camacho (who had great esteem for him) recommended Ruiz Cortines to Alemán, and—at the sudden death of the previous minister—he was chosen to be the new Minister of the Interior.[15]

His moral qualities, unusual in themselves, stood out even more amid the rush to riches of the booming Alemán years. Surrounded by corruption, disorder, lavish and wasteful spending, he projected an immaculate image. But there was nothing simplistic about the man. "You have to swallow a lot of toads in this business of politics," he said once. His way was to wait, half in the shadows, with humility and skepticism for whatever unexpected event might happen. And suddenly it did. When the dominoes of politics offered him the pieces for a perfect game, when he knew that he was the *tapado* ("the unannounced presidential successor"), he reacted in (for him) a natural way: He and a couple of friends bought tacos at a stand and then went to see a movie at the Cine Metropolitan.[16]

The *viejito* ("nice old man") would become President. He was not there to innovate but to strengthen, to consolidate, and above all to watch over a historical legacy that he felt had been entrusted to him. With the methods of a statistician, with spartan morality and the shrewdness of a master domino player, he would go on repeating his moment of glory within the Revolution, he would continue guarding the Treasury of the Nation.

Significantly, he would not make his presidential tours by car or plane but by rail, and not in just any train—he would use the same Olive Train for which he had taken responsibility in May 1920. Twenty kilometers an hour was slow progress, but Ruiz Cortines was in no hurry. He had waited half a century to climb to the presidency, his mandate required this of him, and besides—in the private compartment of his train—you could really play some good games of dominoes.[17]

"Today we're going to do a little inspecting," the President said to Antonio Ortiz Mena—Director of Social Security—asking him to accompany him early in the morning to the Calle del Factor, a traditional center for bakeries and tortilla stands. At the hour when the distributors were just beginning to sell, the President "fell upon them" by surprise, checked their prices, and weighed their products with a little scale he had brought along in his car. He was looking for a fair ratio between expenses, product, and prices.[18]

Cárdenas had given out land like bread, although the practical effect of the distribution had lowered agricultural productivity and so reduced the supply of bread. Alemán was more interested in bulls, cigars, and Cadillacs than bread. With neither the messianic temperament of Cárdenas nor the megalomania of Alemán, Ruiz Cortines saw his work as simply that of a good administrator—to put things in order, count up and measure needs, establish priorities, and verify results. And he would use careful methods to seek sensible results. With his characteristic public solemnity, he described himself as "not a caudillo, nor a unique man, neither a savior nor an executioner but a servant."[19]

In 1954 he had to pay the bill for Alemán's economic acceleration. During Holy Week he devalued the peso from 8.50 to 12.50 to the dollar. It was clear that the new exchange rate undervalued the peso, but Ruiz Cortines said that he "was not going to spend his six years devaluating."[20] He had taken a step he thought would be long-lasting, and time proved him right. The next devaluation of the peso would not come until 1976.

He always, inevitably, knew the statistics. He was aware there was not much land remaining that could be usefully distributed. And he was convinced, as much or more than Alemán, that land distribution of itself could not resolve the problem of rural poverty. During his six years, *ejido* grants shrank to only 3.2 million hectares. Nevertheless, when the opportunity came to expropriate a real landed estate like that belonging to the heirs of William Randolph Hearst in Sonora, he did it without commotion, in a rapid but orderly manner.[21]

He initiated a number of national projects, including a highly successful National Campaign for the Eradication of Malaria, and he was always very conscious of costs. None of the projects involved a significant increase of the public sector over the private. Like the good man of commerce that he was, Ruiz Cortines had an elemental faith in the market. Many of his efforts would be shelved by later Presidents with different ideas about financing. But generally, until 1970, the economic ethic of Ruiz Cortines prevailed in Mexican public administration over those of Cárdenas and Alemán, with their concentration (though very differently directed) on state investment. When Ruiz Cortines left office in 1958, the national debt was negligible, amounting to only $64 million.

In his role as inspector of the nation, he would come up against zones of corruption or established interests. He did not jail the corrupt, but he was willing to act against them. And he also came up with ingenious methods for controlling smaller expenses. One of them was the "colored pencil" approach, a practice initiated during his time as Minister of the Interior. When functionaries had come to him asking for additions to their budgets, he would amiably receive them and show his concern with a request for payment directed to the proper office and written with a colored pencil. The color was a message. Red, for instance, meant "Pay all of what this says to pay." Green meant "Only give him half of what I'm asking you to give him."[22]

Educated in the days of Don Porfirio, Ruiz Cortines shared in and promoted—with anachronistic and singular fervor—what Justo Sierra had called "the Religion of the *Patria*." In those days, many children would listen to the *Hour of the Nation* (*Hora Nacional*) every Sunday night at ten o'clock on the radio. (It can still be heard now, at the same time.) The program had gone on the air in 1942, but during the presidency of Ruiz Cortines and under his influence, it was given a dramatic format. The children would hear the drama of Guillermo Prieto throwing himself before the Conservative soldiers about to kill Juárez and declaiming, "Stop! Brave men do not murder!" or the old professor of literature Erasmo Castellanos Quinto in the midst of the random military slaughter of the Tragic Ten Days, walking across the dangerous Zócalo on his way to teach his regular course at the National Preparatory School. When the children turned off the radio at eleven o'clock, they could dream of being Cuauhtémoc reborn. Ruiz Cortines loved to hear young orators deliver patriotic sermons, and in his own person, profoundly ceremonious, he fused the liberal patriotism of the nineteenth century and the revolutionary fervor of the twentieth into a single cult. "He was the greatest officiating priest of power that we have ever had," said a man who had been a high official in the PRI of those days.[23]

Ceremony for Ruiz Cortines always involved giving proper precedence to history and to the institutions of the *patria*. Once he visited the state of Chihuahua for a political meeting and was received by the governor. At once he asked, "Where is the mausoleum of our father Hidalgo?" (Hidalgo had been executed in Chihuahua.) They went together to the Church of San Francisco. The President then requested that representatives of the state Legislature, courts, and the municipality of Chihuahua be asked to come there as well. A photograph was taken. "Now, all right," said Ruiz Cortines. "We can get down to politics."[24]

In a different way but equally impelled by a ceremonious respect for the institutions of the *patria*, he would insist on a particular mode of precedence whenever he visited San Luis Potosí. The province of course had a governor, but the man who gave the orders was the cacique *ad aeternum,* Gonzalo N. Santos. He had been a close friend of Ruiz Cortines since they had been through the battle of El Ébano together back in 1915. When the Olive Train would arrive at San Luis Potosí, his "brother Gonzalo" would always have to wait his turn to embrace his compadre. Before opening the doors, Ruiz Cortines would tell his staff, "First the governor."[25]

The position of the President, the "mandate," was to be respected down to the smallest details. When he occasionally used an obscenity—the foulmouthed Veracruzan springing up for a moment—he would quickly correct himself with the phrase "a word said with my excuses to the mandate."[26] His usual dress—a sober suit and his famous bow tie—was also an expression of his "mandate." But the most important feature of his behavior was the general sense of gravity that he impressed upon the presidential position. A surprising testimony to its impressiveness comes from one of his fiercest enemies, Othón Salazar, the head of the Teachers' Union, who suffered at the hands of the regime after a celebrated strike near the end of Ruiz Cortines's term: "He impressed me from the moment he left his office to talk with us. He moved with solemnity, and it was there as well when he spoke with us. I was impressed with his sense of personal grandeur." [27]

Ruiz Cortines made exacting demands on language and expression in every area of his administration except for its most important usage in government—to tell political truth. Or more accurately, he was very rigorous and even elegant in formulating the lie that had, since 1946, come to control the political life of Mexico. During his long inaugural speech, he had said that the reciprocal respect between the three powers of the Union and those of the states and a strengthening of municipal government "would invigorate institutional life," but a few paragraphs later, he was talking about the politics of Mexico as if it were the Garden of Eden: "Freedom, democracy and the Revolution are the fundamental bases of our development."[28]

Under Ruiz Cortines, the Mexican political system reached the height of its development and stability. In this respect, the aims of Alemán and the austere bureaucrat who succeeded him were identical. Ruiz Cortines—through his honesty and demanding personal style—in fact contributed a moral weight that even further sacralized the already lavishly sacred institution of the presidency.

The system operated with greater efficiency than under Alemán. In line with the style of Don Adolfo, the PRI further refined the tool of official rhetoric to grease its electoral machinery. General Gabriel Leyva Velázquez, the president of the party, affirmed at a major PRI convention:

> Honorable delegates! Very near here, in the National Palace, an exemplary citizen is governing the nation, a man filled with patriotism and virtue, the honorable President of the Republic Don Adolfo Ruiz Cortines . . . the Party of the Revolution proclaims with pride that the people is its guide, the Constitution its slogan and Adolfo Ruiz Cortines its flag.[29]

This President further tightened the screws of control on the real armed powers within the country: the army and the caciques. Like Alemán, he conceded only two cabinet posts to military men, and he lowered the military portion of the budget from 9.7 percent to 8 percent.[30] A greater problem for him was the "spent cartridges" (*cartuchos quemados*), the old generals who insisted that Alemán had betrayed the principles of the Revolution. Some of them were still surly, especially those connected with the Federation of the Parties of the People, who had seen their candidate, General Henríquez Guzmán, steamrollered by the PRI machine (and themselves physically assaulted) during the 1952 elections. They were still playing with the idea of an uprising. In January 1954, when a group allegedly loyal to Henríquez attacked a military barracks in Ciudad Delicias, Chihuahua, the government used the opportunity to dissolve the Federation. Some "spent cartridges" were expelled from the PRI or forcibly retired from the army. With others, the government simply waited patiently for them to die. The movement that had coalesced around Henríquez gradually faded away, but some of its younger militants moved on to more radical positions, like the agrarian leader Rubén Jaramillo in the old Zapatista area of Morelos.

Other *cartuchos quemados* chose to form their own political party (the Authentic Party of the Mexican Revolution), one of whose leaders described its curious origin: "Don Adolfo Ruiz Cortines . . . conceived the idea that all of us Revolutionaries still alive should join together. . . . We had the honor of having the President himself accord us our official registration," which is to say that the party was an invention of the President.[31] He was perhaps trying to counterbalance the influence of the only military man with genuine and long-standing moral authority: Lázaro Cárdenas. The former President continued to be a dissident figure along the margins of the system, but without ever leaving it.

Ruiz Cortines dealt with the surviving caciques (like Bonifacio León in Nuevo León, Lázaro's brother Dámaso Cárdenas in Michoacán,

Cándido Aguilar in a portion of the state of Veracruz, or, of course, Gonzalo N. Santos in San Luis Potosí) through a policy of temporizing with them and avoiding confrontation. These caciques had been of great help to the government during the presidential campaign of 1951 in often violent action against the "Henriquista" generals. The government saw no reason to fight with them. In fact, "he quieted them all, he attracted them all to him, he kept them calm, and he made them collaborators of the government. . . . He fought them without a rifle, you could say by slapping them with his hat (*con sombrerazos*)."[32] Once Gonzalo N. Santos registered his displeasure at only being offered the minor and stressful post of ambassador to Guatemala as a reward for his long-standing loyalty. ("You're going to send me to Guatemala! I'm going to send you to go f— your mother! I'm not Don Dumbo Nobody that you can just send me out of the country!" and when the screaming brought the commander of the Presidential General Staff, General Hernández Bermúdez, striding into the room, Gonzalo said, "Get this m—f— cop out of here or I'll do it with the barrel of my pistol!") But Ruiz Cortines would keep his temper and finally end the conversation by directing Santos to "Go back and rule over your caliphate!"[33]

As for the formal powers—the courts and the Legislature and the governors—the Supreme Court showed some independence and dignity in finding for 41 percent of the cases brought against the government, even though they were all minor matters. (It should be remembered that the Supreme Court of Mexico—in total contrast with the United States—virtually never hears a case of national importance.) But with the national Legislature and governors, the system of control and subordination grew even stronger. Toward the governors, Ruiz Cortines acted as had Alemán. He removed them at his own discretion. His term began with seven "of his own" governors and twenty-two whose allegiance had been to Alemán. Through the usual civil pressures, he got rid of them one by one until—by the end of his term—he was in firm control of twenty-eight out of twenty-nine governorships. In recalcitrant instances, the army would rattle bayonets to induce the change or he would send a high official from the Ministry of the Interior to impress some governor with the virtues of resigning his office. (The Chief Administrative Official of the Ministry of the Interior, Gustavo Díaz Ordaz, a firm and energetic man, executed that task of persuasion in Tabasco and Guerrero.) But it was in the Legislature that the increasing ascendency of the President was most evident. In Ruiz Cortines's first Congress, an average of only 3 percent of the deputies would ever cast a vote against a government proposal. (In his second Legislature the rebellious figure rose to all of 5 percent.) The procedures for selecting deputies and senators began to take on a more deeply Porfirian coloring.[34] A new leader of the Association of Railroad Workers—David Vargas Bravo—had succeeded "*el Charro*" Díaz de León. One day the President said to him, "You are going to be the senator from San Luis Potosí." But Vargas answered that, though he had been born in San Luis, he had left the state when he was five years old and therefore nobody knew him there. Besides, he wasn't even a member of the PRI. "No problem," answered the President. "Gonzalo will take care of things. I'll speak to him." And Gonzalo N. Santos cooked up the necessary documentation. "I did nothing," said Vargas years later, "not even campaign. I was the senator because the President named me and that was it."

Vargas would have a bitter experience when he came to take his position as a senator too seriously. During a discussion on broadening the punishments for railway accidents caused by worker carelessness, he defended his own union with such vehemence that the leader of the Senate—José Rodríguez Clavería—called him aside and said: "We've had it up to here with you . . . either you get the goddamn hell out of here or you'll face the consequences." Vargas had to leave the country and not return for some years. "*Charrismo*" returned to the Association of Railway Workers in the person of Samuel Ortega.[35]

While Ruiz Cortines was governor of Veracruz, President Alemán had given women the right to vote in municipal elections. Now, in his own first week as President, Ruiz Cortines extended that right to all Mexican elections. His proposal was—as so often—unanimously approved by the Legislature. But the President who had implemented this reform showed little respect for women as voters (or voters in general). When a woman from Chiapas scolded him for his choice as governor of that state and asked him how he could have selected such a man, Ruiz Cortines responded, "If I let you do the choosing, you would tear each other's hair out."[36]

The peasants organically incorporated into the PRI through the National Confederation of Peasants (CNC) had no greater room for maneuver. They continued to be the "political herd" of the PRI. When there was a presidential tour, they would be driven to the proper village or city in truckloads. If they refused to come, they could lose their land. If they did come, they would get a sandwich, a soft drink, a few pesos. They would be issued banners and placards, carefully instructed what they should shout when a particular political personality mounted the dais. (The same would happen when they voted; the ballots would be passed out collectively.) The orators they heard would praise the participation of the peasants in the Revolution and there would be numerous mentions of Zapata, whose ideals were still "alive" within the PRI.

The CNC supported, "with mystical fervor," the rural initiatives of President Ruiz Cortines and obtained certain concessions in exchange. Owners of cattle who had been issued "certificates of immunity" against expropriations had to supply a certain percentage of seed bulls to the *ejidos*, the debts of some peasants to the Banco Ejidal were forgiven, electricity came to a number of *ejidos*. It was the old technique of giving out "bread," which worked for the PRI in all the political spheres of the country. But the peasants came cheap. Their very geographical dispersion made greater investment unnecessary. Sometimes an authentic peasant leader would surface and the system would turn to the other pole of the Porfirian formula: "the bludgeon." There was the honorable peasant leader Jacinto López, who led the occupation of various landed estates that had been legally disguised as small private properties in the states of Sonora, Coahuila, Nayarit, Colima, and Baja California. López went to prison more than once, but his closeness to the People's Socialist Party of Vicente Lombardo Toledano favored the success of some of his initiatives. In contrast, from April 1954, the leader of sugarcane workers in Morelos, Rubén Jaramillo, was on the run in the mountains, pursued by the state police.[37]

The other great corporate sphere, that of the workers, also continued its dependence on the center. But Ruiz Cortines did act immediately to heal some of the wounds inflicted by Alemán. The miners of Nueva Rosita in Coahuila, whom Alemán had driven from their jobs without a second thought in 1951, were allowed to go to work again on the Lázaro Cárdenas Dam under the supervision of that general himself. And after the devaluation of 1954, when the CTM asked for a 24 percent increase in the salaries of its members and set the date for a general strike, a firm no by the government and industry would have led to a general crisis. The system recognized its limits and put the job of arranging a settlement into the hands of Adolfo López Mateos, the skillful and charming Minister of Labor who actually settled five thousand threatened strikes in only eleven days, with wage increases ranging between 10 and 16 percent. It seemed that everybody was happy with the results, especially the President, who then coined one of his high-sounding phrases: "Mexico off to fecund and creative labor!"

At the end of Alemán's term, an organization had been created—the CROC (Revolutionary Confederation of Workers and Peasants)—which tried to unite groupings independent of the CTM. It was led by the fighting union leader, Luis Gómez Z., but it only slightly dented the power of the CTM. And in 1955 the CTM would create an unexpected chorus by forming a new bloc (the BUO—Bloque de Unidad Obrera) which included nothing less than its two former ferocious rivals, both of whom had long seen their best days go by: the anarchist CGT and the CROM of the old labor leader Luis N. Morones.[38]

To judge by the incidence of strikes (there were fewer than two hundred a year between 1953 and 1957—a much lower figure than normal—and most of them would be settled in a very short time by the smooth mediation of López Mateos), the labor picture was the best of all possible worlds (until 1958) for the system as administered by Ruiz Cortines. But within the corporate sphere of labor, a few dissonances sounded, single notes that would later build into a threatening symphony. The most resonant of them began to be heard in July 1956, when the teachers' union mobilized throughout the country to demand a salary increase. Headed by Othón Salazar and Encarnación Pérez, Section IX of the National Union of Educational Workers took a step that would have great repercussions in the labor earthquakes that would shake the system during the electoral year of 1958. They withdrew from their national union, which was part of the CTM.

The greatest press event of the period was the founding of the magazine *Siempre!* ("Always!") by José Pagés Llergo, former idolator of Hitler and the editor who had lost his job at the magazine *Hoy* for publishing the picture of Alemán's son-in-law ogling the breasts of a French model. *Siempre!* became a rare and important publication in the spectrum of the Mexican press. It was open to a whole range of views, from the left through the center to the right, and carried off the small miracle of creating a kind of public plaza where each writer could express his point of view on the political events of the moment. It printed ferocious and well-documented attacks on the waste and corruption of the Alemán regime, but toward the reigning President it adopted an attitude of moderate support.[39] And when the labor crisis came in 1958, *Siempre* took the side of the institutional enterprise and the President, criticizing "the insolent, radical attitude of the workers." Even in *Siempre!*, important to many young minds of the time for its comparative independence, the fundamental rule of the Mexican press was still in force: "Don't touch the President or the Virgin of Guadalupe!"

Meanwhile, processions in honor of the Virgin of Guadalupe had become more frequent and less troubled by any shadows of interference from the State. The Church was continuing its long march toward making up for lost time. It built new churches and absorbed steadily larger groups of middle- and upper-class students into its schools. Ruiz Cortines, once questioned by someone who felt that the clergy was exerting a scandalous influence on public life, answered with words that would have smoothly filled the mouth of Don Porfirio: "In the next elections, the clergy is going to be of help to us . . . and besides, they will not be offering a candidate for President."[40] (As an incidental anecdote, on the very day of those 1958 elections, a procession of the faithful was heading toward the shrine of the Virgin of Guadalupe when they were commandeered to vote in a nearby zone where there was considerable opposition to the PRI. It looked as if, in the days of Ruiz Cortines, even the Virgin was a partisan of the PRI.)[41]

Elsewhere, there was an air of domestication during that general political lull of the 1950s. The University graduates continued to climb the bureaucratic pyramid. Most of the intellectuals, as usual, took their places within the official world. Even the old warriors Diego Rivera and David Alfaro Siqueiros had calmed down. Rivera was painting portraits of the richest women in Mexico, and Siqueiros was doing a mural for the government in the Castillo de Chapultepec. As for the students, their attitude during the years of Ruiz Cortines was mostly festive, still without any mass concern for issues of national politics. And the doors of the PRI were always open to young men who wanted "to serve the *patria*." There were occasional student disturbances. The Politécnico in 1956 witnessed a confrontation between students and the administration over dormitory issues. Ruiz Cortines got tired of it and sent the army in—but without any significant violence—to sweep away this "seed of disintegration." In 1956, when the students at the University seized a number of buses to protest a rise in fares, the demonstrations remained very good-natured and ended in a way appropriate to that peaceful era. A delegation met with the President, who admonished them with his proverbial seriousness. In response, the students did not react like "rebels without a cause." Instead they gave the President an enthusiastic old college cheer.[42]

During the fifties, the PAN made a striking turn to positions that brought the party closer to the Church. Gómez Morín remained the moral leader, but he was no longer party president. In 1956, Alfonso Ituarte, a former member of the Catholic Association of Mexican Youth, won that position, and his views prevailed over the secular wing of the party, men (such as Adolfo Christlieb Ibarrola) more like Gómez Morín. Unable to attract the support of the Mexican business community, who had few problems with upholding the system, PAN moved toward the right, and many of its party positions were filled by men who did not distinguish religious from political issues or, more precisely, were concerned with the salvation of their souls and replaced politics with religion. Christlieb called them "the group of the pious" and "pissers of holy water," cruel epithets, but yet the fact that he could express them proved that democracy continued to exist within the party.[43]

In the congressional elections of 1955, PAN won only 8 percent of the national vote. But in Mexico City its proportion was far greater, close to 43 percent. One year later, there was a controversial election in Ciudad Juárez. The results of one polling station—where PAN had a margin of twelve hundred votes—were annulled, giving the election to the PRI candidate, René Mascareñas, over PAN's Alfonso Arronte. To the accusations of fraud, Ruiz Cortines answered, "I will never permit a city with that name to fall into the hands of the reaction."[44]

PAN's presidential candidate in 1958 was a businessman who *had* chosen to oppose the system: Luís H. Álvarez, a liberal textile manufacturer from Gómez Morín's home state of Chihuahua. During his campaign, he received death threats and there were some aborted attempts on his life. At the end of the usual rigged elections, PAN was allowed six seats in the Legislature. The party assembly decided that the six should refuse to serve. Christlieb wrote, "National Action will not sell its worldly and moral birthright in Mexican politics for the miserable plate of lentils of a few congressional paychecks."[45] But four of the six deputies refused to resign their seats and were expelled from the party. Briefly, the PAN was discouraged. It was hard to fight against the system, especially with a president like Ruiz Cortines, impeccably honest in his personal affairs but as dogged in defense of the system as if it were another Olive Train, this time carrying the politics of the nation. And, besides, the *viejito* felt a contempt for the PAN that came through as humor. He called them "mystics of the vote."

On the night of July 28, 1957, a strong earthquake shook Mexico City and overthrew the very symbol of the *patria:* the gilded Angel of Independence. In disbelief, the people of the city walked along the Paseo de la Reforma to see the pieces scattered on the asphalt. One year later would come another earthquake, not from nature but from the labor unions, the first real test of strength for that structure built for eternity, the modern Mexican political system.

In April 1958, the government sent in its police to break up a demonstration by the MRM, the Revolutionary Movement of Teachers. Within a short time (along with some "protest demonstrations" by the students), they were confronting other union movements: the electricians, the telegraphers, the oil workers, and the most powerful of them all, the railroad workers. The pattern was always the same. Before the union marchers could cross the streets of the city and reach the Zócalo, the government would use the police, the "grenadiers" (*granaderos*)—a tactical force specialized in repressing civil demonstrations—and disperse them with swinging clubs.

There was a clear moral and social justification for the labor unrest of 1958. Salaries, or their buying power, were seriously in decline. But the conjunction and intensity of labor agitation was also built on a political flaw or fault line within the system. It was the sixth and final year of a presidency. Once the successor has been "revealed," during the high tide of the presidential campaign and before the first Sunday in July when the elections are held, the outgoing President has to deal with the inevitable weakening of his power before the growing—but as yet not absolute—power of the man who will succeed him. Within this gray zone of the Mexican political calendar, even if the old and new Presidents are the best of friends, the situation is confused, a kind of dyarchy exists that can favor the jockeying of forces eager to put themselves in a favorable position for the coming six years. A number of labor leaders understood this process and acted upon it. They knew that Ruiz Cortines would not want to risk his image in history, washing an almost bloodless presidential term down the drain with a major repression in its final hour. They also knew that the next President—the Minister of Labor, Adolfo López Mateos—still legally lacked the executive power to make decisions. Within this gap they acted, and they acted efficiently. Their various movements did not end with college cheers for the President.

The leader of the dissident teachers, Othón Salazar, would describe, long afterward, another essential background to the wave of strikes in the last years of the fifties: the rank growth, throughout the Ruiz Cortines years, of *charrismo*, a union leadership at the service of power: "The *charrismo* created by Miguel Alemán in time became part of the structure of the State. . . . The unions, their leaders . . . no longer defended the interests of their members." [46]

On the first day of September 1958, when he gave his last State of the Nation Address, Ruiz Cortines confronted the labor insurgency, especially the agitation by the teachers and the railroad workers. With only three months left to his presidency, he indicated that the government would use its "maximum energy" against those workers who were disturbing public order with their protests. Five days later, plainclothes police came to Othón Salazar's door and, without identifying themselves, tried to talk their way into the house. When no one would let them in, they broke open the door, knocked his children aside, and took Salazar away in his shirtsleeves:

> They held me for four days. Nobody knew where I was. I didn't even know where I was, because they took me there blindfolded. . . .

> I was treated with arrogance, humiliated by the police . . . it was then that I felt what it must be like when you know you're going to die. But I had a lot of inner reserves that . . . helped me to remember the great men of Mexico, who had not trembled when their lives were in danger.[47]

Salazar was released on that occasion, but he would be picked up again, six times in all: "They never beat me. It was all psychological torture." In earlier years, his union local had been, as he said, "a democratic island, in a sea of *charro* unionism," but in 1958 his movement had become something more, the radiating center for a series of actions by independent unions.

The political and directly repressive moves against the teachers had from the first been taken out of the hands of the Minister of Education. Direct responsibility had passed to the Ministry of the Interior, and in particular to a man who had shown outstanding gifts for acting with determination and "without legal quibbles," ever since he had joined that ministry. The official who took charge was not the rather weak Minister himself, Luis Angel Carvajal, nor his Assistant Minister, Fernando Román Lugo, but the Chief Administrative Official of the Ministry, Gustavo Díaz Ordaz. He was a politician who had been formed under the harsh hand of Maximino Ávila Camacho, and for him Ruiz Cortines was a "*maestro*."[48] But there was an essential difference between Díaz Ordaz and Ruiz Cortines. The old man used to say that "in politics, the shortest line between two points is a curve." For his disciple, "in politics, the best line is the hard line."[49]

In comparison to the major labor groups in the country, the 110,000 workers on the national railroads had to endure low salaries and slender benefits. But they also had to support the shame of their union being—ever since the internal coup of 1948—the prototype of the "*charro* union." Although the "*Charro*" himself, Díaz de León, was no longer union chief, his successors were tarnished, to a certain degree, with the stigma of illegitimacy. In 1958, the two groups against whom Díaz de León had carried out his purge—the independents (in the Grupo Hidalgo 96, headed by Gómez Z.) and the leftists—united, recognizing an opportunity during those months when power was not entirely at ease in the Presidential Chair. They hoped to win better salaries and benefits for the railroad workers and reverse the insult of 1948.

Ruiz Cortines had put a man in charge of the railways who was irascible and inflexible, disciplined and a disciplinarian, and determined at all costs to modernize the railways. Roberto Amorós was a "Prussian," all work and no play, who often slept in his office.[50] At the extreme opposite end of the political spectrum, a labor leader rose to face him—Demetrio Vallejo, a man in the mold of the Flores Magón brothers (and from the same state of Oaxaca). He had belonged to the railroad workers' union since 1934. Vallejo was the leading member of a committee of dissidents in 1958 who decided to ask for improved benefits and a salary increase much higher than the 200 pesos the *charro* union was requesting or the 180 that Amorós was willing to concede. They wanted a general increase of 350 pesos per month per man.

At the beginning of June, Vallejo put pressure on management by calling work stoppages of two hours in the national railway system. Generally the workers responded to his appeal, repudiating their "company-union" representatives. An orator at a union assembly on July 26 dared to reveal that the aims of the movement exceeded issues of salary and benefits: "We don't need leaders! Our leaders are useless because they're not fighting to improve the lot of the workers but for their own bastard interests, to be deputies and senators . . . the leaders are traitors to the proletarian workers' movement. . . . Fidel Velázquez . . . is the worst sellout . . ." That same day, information began to arrive that ten sections of the union across the country were throwing their *charro* executive committees out of office.[51]

It was only a matter of days until the Big One, the presidential election of 1958. It was absolutely essential that the country be at peace. The Committee—which had effectively replaced the Union—had lowered the magic figure to 250 pesos, but the work stoppages continued. Now they were eight hours long. On June 30, President Ruiz Cortines intervened. He split the figures in half—between Amorós's offer and the workers' demands—and came up with a decision of 215 pesos. The labor situation was calm for the elections.

Once the Big One was over, the Committee had another demand—new union elections to remove the *charros* and "achieve a real cleansing of the union." The administration resisted their request, and work stoppages began again on the railroads. Receiving the vocal support of business interests and the press, the government decided to repeat Alemán's direct action with the oil workers in December 1946. The army seized the railroads and arrested some two hundred workers (killing three in clashes). It was announced that a supposedly subversive document had been discovered in one of the local offices, calling for a great labor covenant between the railway workers and the teachers, telegraph operators, and oil workers. But the government did not arrest Vallejo, leaving the way open for him to negotiate under pressure, and he immediately sat down at a table with Amorós.

Supported by the groups that had been badly treated since the *charro* takeover of 1948 (including the faction headed by Gómez Z.), Vallejo negotiated more forcefully than the government had imagined and won an agreement from the enterprise to permit union elections in which he himself could run as a candidate for Secretary-General of the union. The elections were honestly conducted, and Vallejo won with a huge majority.

The new Executive Committee took power on August 27. "This union will be apolitical, absolutely apolitical," declared Vallejo, as he explained that the workers would be free to belong to whatever party they wanted. It seemed to be a promise of genuine union democracy, but the immediate action he took toward his former allies of the Hidalgo 96 Group promised something very different. He excluded them from power. Gómez Z. accused Vallejo and his partisans of being "absolutely political" and linked to the three left-wing parties—the Communists, the People's Socialist Party of Lombardo Toledano, and the Worker and Peasant Party (a splinter group from the Communists). And he said that the Communists had "the absurd intention of converting the Union into a communist party."

One of the first things Vallejo did was to pull the railway workers out of the BUO, the Bloque de Unidad Obrera. It was a very dangerous precedent for Fidel Velázquez, a step that could lead toward a new stampede of unions out of the CTM, as had happened in 1947. During all this period of conflict, the unions within private industry remained under the control of the system, perhaps because they enjoyed salaries and benefits that were far superior to the public sector, which was the stage for the union insurgency. It was clear that the government of the "stingy" (*agarrado*) Ruiz Cortines had carried his fiscal economizing to an extreme.

In any case, Fidel Velázquez waited no longer and with his usual cunning moved to the attack. "More papist than the Pope," he requested a general increase of 25 percent for the unions of the CTM, and made an ideological swerve toward the left. All of a sudden, he declared that social justice was "a myth in Mexico"; in no other country in the world was the distribution of wealth "as unjust as in ours"; the profits of twenty-five big businesses were "greater than the entire federal budget"; all price controls were "absolutely ineffective." The government, in turn, conceded a salary increase. There had to be peace for the first of December, when the new *tlatoani* would take control of the State.

Hounded by internal dissension and by a hostile press, which saw him as the peon of a "red conspiracy," Vallejo agreed to suspend his tactic of work stoppages and accept the most recent proposals of management. He would wait to "try out" the new President and to make his move toward a unionism different from what the system had imposed. It remained to be seen whether it would be an independent union movement or an arm of the parties of the left.[52]

Perhaps it was the doubtful historical balance of the Agrarian Reform that contributed, deep down and in compensation, to the growing radical shift of former President Lázaro Cárdenas toward the political left. When his cherished plans for the great *ejido* distribution of the Nueva Italia Hacienda finally ended, over the passage of years, with the entrance of multinational companies buying up land for new and vastly profitable crops, someone claims to have seen him weep at the failure of his dream. In 1939, near the end of Cárdenas's term of office, Vicente Lombardo Toledano had advised a friend not to use ideological terms when speaking to Cárdenas: "He is a man of the left, but no Marxist-Leninist."[53] Ten years later, in the midst of the Alemán years and their financial euphoria (of which he disappoved), Cárdenas wrote in his diary, "Only socialist culture can make Mexico and other 'colonial' nations truly free and fair."[54] It was the height of the cold war, and Mao had just come to power. As time went on, Cárdenas spoke out against America's role in the Korean War and, much nearer home, was angered by the CIA-sponsored coup—in 1954—that overthrew Jacobo Arbenz, the president of Guatemala whose program had been inspired by Cárdenas's actions as President of Mexico. That same year he caused a small scandal (to official minds) by attending the funeral of the painter Frida Kahlo, whose coffin—honored within the official precincts of the Palace of Fine Arts—was covered with a red flag bearing the hammer and sickle.

In 1956 he would receive the Stalin Peace Prize, but far more significant historically was a small step he took that year, when he interceded with President Ruiz Cortines to prevent the impending expulsion from Mexico of a small group of Cuban radicals. He had only met their leader two days earlier and found him "a young intellectual with a vehement temperament and the blood of a fighter."[55] The President yielded to the request of the elder statesman. Of course none of the parties involved could have known the historical road that would open—after a chance to plan and organize in Mexico—for Fidel Castro and his followers.

Cárdenas was an extreme but also representative example of the strong shift toward the left in the Mexican political and intellectual classes during these years. It was natural that a leader already on the left—like Vicente Lombardo Toledano—would begin to make frequent visits to Moscow and China, add the adjective "Socialist" to the name of his People's Party, and publicly assert his commitment to Marxism-Leninism. But the trend—especially in the mode of anti-Americanism—had its influence on circles completely alien to the politics and ideas of communism.

The Ministry of Foreign Relations, for instance, turned contrary to American initiatives and clearly suspicious of U.S. motives. When President Eisenhower visited Mexico for the inauguration of the Falcon Dam, Ruiz Cortines read a speech in which he indirectly criticized the United States by mentioning Juárez's injunction to respect the rights of others. And the Ministry of Foreign Relations—at the Tenth Inter-American Conference in Caracas in 1954—opposed American pressure on Latin American governments to fight communism, arguing that it could best be opposed through economic progress. Ruiz Cortines gave the same answer to Vice President Nixon's obsessive insistence on the threat of communism, when he took him to visit a shantytown in Mexico City and said, "Look, Mr. Vice President, this is the most widespread 'ism' in Mexico, 'hungerism,' and this is what I am interested in eradicating so that the 'isms' you worry about will not arise."[56]

But the principal impulse for the generalized turn toward the left came from the labor insurgency. Cárdenas—in his diary written for himself and for history—registered his surprise at the "frigidity of the government" in the face of labor demonstrations. The teachers asked him to intercede in their conflict, and he went to the President on their behalf, but it did not prevent the onslaught of the police or the arrest of the union leaders, charged with the crime of "social dissolution," a legal tool codified under Ávila Camacho. The director of Petróleos Mexicanos, the Mexican national oil industry, confided to Cárdenas in private that he thought the workers had good reason for their demands, but in public the grenadiers responded to those same workers with teargas and clubs. Cárdenas was convinced that "the labor movements do not represent direct attacks on the government but they are requests dealing with concrete problems—the high cost of living and the struggle to renovate union leadership. They are not to be wondered at."[57]

In October 1958, in the offices of the Ministry of Foreign Affairs, opposite the old statue of the "Little Horseman" (the Bourbon monarch Charles IV) on the Paseo de la Reforma, two Mexican writers looked out the window and saw a union demonstration marching by. Suddenly, the police charged, breaking heads with their clubs. A few days later, another march passed along the same route. There were students in it, and intellectuals. Some of them were chanting slogans doubting the virtue of the President's mother. Spontaneously, the two writers decided to join the march. They were Octavio Paz and Carlos Fuentes.

The act was a small prediction of the future. But for the moment, the presidency of Don Adolfo Ruiz Cortinez embodied the system at its calmest and most successful point. In power was a man who had chosen, of his own free will, to be responsible (true—on the whole—of his chief administrators as well). The chances of selection had set an honest and reasonable emperor on the throne, an exception in many ways to the heirs who would succeed him. But there was nothing within the system—no checks and balances—to prevent less competent, less sane, less merciful men rising to the presidential power. For now, the skillful and clever *viejito*, past master of the domino board of Mexican politics, was presiding over general social stability and moderate economic growth. And the rings of the wheel of power were still basically stable, firmly in orbit around *El Señor Presidente*.

20

ADOLFO LÓPEZ MATEOS

The Orator

The day before Adolfo López Mateos would officially accept his nomination as the PRI candidate for the presidency of Mexico, his home was besieged by governors, political and union leaders, newsmen, photographers, cameramen, childhood friends, people who had known him in college—every one of them eager to congratulate, to embrace, stroke, touch, or even just to see the man who would be the future President of Mexico. Outside, in the front yard, a trio was singing the new corrido "Hurray for the Rooster" ("*Viva el Gallo*"), with its insistent chorus:

> *Viva beloved Mexico*
> *and Adolfo López Mateos!*
> *May their names remain united!*
> *That's what the people want!*[1]

On the following day, November 17, 1957, a Sunday, the day of the week when Mexicans fill the stadiums for soccer games and the sun and shade seats of the bullrings for the *corridas de toros*, in the morning precisely in a soccer stadium, before the thousands bused into the city at government expense, the candidate accepted the nomination and the stadium exploded with cheers and applause. He had a smile on his face—a softer and more honest smile than that of Miguel Alemán—but it hid feelings, preoccupations, and sufferings that few people were aware of.

A few days later, López Mateos would organize an almost military defense of his house to ward off the swarming herd of politicians looking for jobs, lucrative favors, or at least promises, an avalanche of desire someone aptly described as "the charge of the buffalo."

Soon he was on the first flight of his electoral campaign, and Antonio Mena Brito, one of his aides, noticed that he was tired and worn out and never took off his dark glasses. He made no attempt to hide his fear, his inner tensions, still less his physical pain, those persistent migraines that would suddenly assault him and throw a hard cast across his face, so different from the pleasant and distinguished smile that normally marked his expression. It was to hide these bursts of pain that he wore those glasses.[2]

He would be the first genuine orator in the history of the modern Mexican presidency. None of the Revolutionary caudillos—not to speak of the sphinx that had been Porfirio Díaz—were good public speakers. They had their "songbirds" (*jilgueros*) for that job, their "golden beaks" (*picos de oro*) in charge of dropping well-turned phrases. Neither Ávila Camacho, nor Alemán, nor even Ruiz Cortines (so firmly sculptured in his convictions) had the gift of words. For that job, they used the young graduates from the Faculty of Law, men who emulated the style of Lombardo Toledano, ready to move mountains with a "burning verb." Now the Institutional Revolution could rely on the luxury of a President who was a real orator, in his youth the best of them all. Yet López Mateos, strangely enough, did not seem to be enjoying his new position. He preserved, intact, his nineteenth-century oratorical style, as showy as it was old-fashioned: "Here! Before the infinite sea, perpetual invitation to adventure . . . !", but he would ask the young orators who shared the platforms with him to "be brief, the heat wipes me out."[3] When he was already President, flying on a work trip, and one of the motors on the plane suddenly stopped, López Mateos remembered and slightly modifed a statement attributed to Porfirio Díaz during a moment of danger when a train seemed about to derail: "Think of what the nation would gain, if this plane should fall!"[4] Did he want to be President? In a sense, of course he did, like almost every Mexican. But perhaps not at a deeper level. He was formed less for the world of power than for bohemia, for art, for love, and, unfortunately, for illness.

The date and place of his birth are uncertain, but it is most probable that the fourth son of Mariano López, a dentist, and Elena Mateos, a

Adolfo López Mateos
(Archivo Fotofija)

teacher, was born in Atizapán de Zaragoza in the state of Mexico on May 26, 1909. Following the strange common denominator of many Presidents of Mexico (Juárez, Díaz, Obregón, Cárdenas, Ávila Camacho, Ruiz Cortines) López Mateos as well lost his father when he was young. Doña Elena had published a book of poetry in her youth, but economic need (in the midst of the Revolution) brought her to Mexico City, where she became the director of an orphanage and limited her literary efforts to writing gentle letters of admonishment to her children, so that she could avoid criticizing them in person. She was a great lover of the opera and frequently took her children with her to performances.

Adolfo attended primary school on a scholarship, irregularly frequented the Escuela Nacional Preparatoria between 1923 and 1925; and in 1926, a resident by then of Toluca, he entered the famous Instituto Científico y Literario of the State of Mexico (founded by Ignacio Ramírez). He was not a dedicated student. In many courses he just squeezed by through taking special exams. There were clear reasons for his lack of application—his love for oratory, for trekking and camping in the country, and for love.

"In love, one drowns, as in the sea; I will be like a drowning man in the immense sea of your love," he declaimed and then wrote down in a letter to his girlfriend Celestina Vargas, who melted when she heard him. Seventy years later, she would remember how "handsome" he was, the charming modulations of his voice, that "perfectly lovely smile . . . that honest laugh."[5] In 1925 he met Eva Sámano, a young teacher, and finally married her (at the insistence of his mother) after a courtship of twelve years. Before, during, and after that courtship, he had many affairs.

His words were as appreciated in public as in private. He showed a phenomenal gift for formal oratory. In 1929, in the annual competitions for public speakers sponsored by the newspaper *El Universal* (a devoutly desired forum for many young men in a country that admires eloquence), López Mateos became the oratorical champion of the Distrito Federal. But he would become even better known—within modest limits—for his speech later in the same year, pulsing with fervor about the Spanish language: "It is an idiom of bronze, a language of bells and of cannon but also a language of gold and silver, that has translated the mystical ecstasies and amorous raptures of a race."[6]

By some arbitrary decision of the judges, he only won second place on that occasion, but his schoolmates carried him out on their shoulders.

Soon afterward, he put his speechmaking gifts to work in the service of a higher and more dangerous cause, the presidential campaign, in 1929, of José Vasconcelos against Ortiz Rubio, the candidate designated by the *Jefe Máximo,* Plutarco Elías Calles. Like other young men of the time, he had read the novel *Sascha Yeguilev* by the Russian author Leonid Andreyev and could see himself in the character of the hero, the pure and handsome young man who tragically gives his life for the transformation of his country. He was part of the Student Directorate of the Pro-Vasconcelos Convention, and his comrades remembered that he was "valiant and combative." He would get up on a stack of soft-drink boxes and harangue the workers: "The country is in danger! Only Vasconcelos can save it!"[7] At the most heated point in the Vasconcelos campaign, when his friend Germán del Campo was murdered by PNR *pistoleros*, López Mateos himself was attacked and might well have met the fate of Sascha Yeguilev. Vasconcelos himself would recognize the fact, years later: "I clearly remember from that time and can never forget the sight of the bandaged head, the noble wounded head of a young man who at that moment symbolized the entire Patria: it was Adolfo López Mateos."[8]

After the elections and subsequent persecution of the student followers of Vasconcelos, Adolfo crossed into Guatemala for some months. He returned in 1930 to finish his studies in economics at the University. Some of his friends from the Vasconcelos movement, seared by resentment, had converted to Marxism. Others had joined the government of General Benigno Serrato, who had succeeded Cárdenas as governor of Michoacán and then become an opponent of Cardenismo. López Mateos made his decision to quietly mount the chariot of the Mexican Revolution. Through the political contacts of his fiancée's father, he went to work for the governor of the state of Mexico—General Filiberto Gómez—and by 1931 had become the private secretary to his successor, Carlos Riva Palacio, a loyal follower of the *Jefe Máximo*. He would again be using his oratorical gifts, serving the same regime that had given him his head wound and killed his friend Germán del Campo. (But he would remember that baptism of blood. When already the presidential candidate, he would speak of himself as belonging "to a frustrated, embittered generation, the Vasconcelistas.")[9]

He was to spend a decade in the limbo of bureaucracy, earning little, politically unknown. In 1941 he would give a speech at a banquet that deeply moved the strongman of the state of Mexico, Isidro Fabela. Fabela had been a close collaborator of Carranza and had created the phrase "the man with the flowering beard" to describe Don Venustiano. His own speeches were sometimes so sentimental that some of his friends coined the verb *fabeliar* to mean affected, syrupy speech. But he was also a man of considerable achievement, a prolific writer and a notable internationalist who represented Mexico skillfully and gallantly before the United Nations. He was now very close to President Ávila Camacho, during a period of unrest in the state of Mexico after the tumultuous 1940 elections. (Ávila Camacho himself had experienced the effects of López Mateos's oratory. He had heard him speak in memory of Morelos, and the oration had brought tears to his eyes.)[10]

Fabela became Adolfo's political father. He taught him many of the secrets of politics as he understood it, a mixture of oratory, flattery, suaveness, and courtesy—politics as an offshoot of diplomacy. Fabela first made him Director of the Literary and Scientific Institute of the state of Mexico, where he spent two years but was uncomfortable with the political haggling he had to confront. Hard and manipulative politics, even in a university context, were unnatural to him and made him ill at ease. Affably treating his students to a drink or taking up the challenge of a lovely young waitress at his favorite café who said to him, "You have such a wonderful mouth that I would like to kiss it," by giving her a five-minute kiss in public before sitting down again to finish his coffee—these were actions with which he felt completely comfortable.[11]

And then Fabela pointed his finger once more (the *dedazo*—"pointing with the finger"—is the literal Mexican political term), making him an Alternate Federal Senator (able to vote only when the incumbent is absent) for the post held by Fabela himself. López Mateos resisted: "I have no vocation for politics. I prefer the quiet life of a scholar at the Institute to the hazards of political activities for which I am not suited." But he accepted, and it decided his fate. When Fabela was named the Mexican ambassador to the World Court at The Hague (and the presidential candidate Miguel Alemán agreed to his replacement), Adolfo López Mateos became a senator. The same day that he took the oath, only moments afterward, in the House itself, word was brought to him that his mother had just died. It was not something he could ever forget, as if his hesitancy about entering politics had somehow been mysteriously confirmed.[12]

But in his new position, he put his own special abilities and preferences to good use. More diplomat than politician, he attended conferences, took part in international committees, and displayed his speechmaking ability at a meeting of UNESCO in 1947. It seemed like the perfect life for a roving orator, but to his close friends he confessed: "I'm proud to be a functionary and to be able to serve Mexico, but how I miss those days when I was nothing and had nothing. . . . I miss them with pain in my soul . . . but anyhow . . . "[13]

In 1951, his good friend Adolfo Ruiz Cortines became President and asked him to become Minister of Labor. From the first, Ruiz Cortines recognized his value. He was an attractive man, a diplomat, a conciliator, all the qualities most needed to deal with the labor issue—clearly taking shape as the most intractable problem on the political agenda. And of course, Ruiz Cortines loved to hear him talk. On February 5, 1957, on the centenary of the Liberal Constitution of 1857, López Mateos delivered an address that was really an exaltation of the socially oriented Constitution of 1917, but which the political class, always attracted to empty and categorical phrases, acclaimed above all for one line: "The Constitution is not our law but our shield and our flag." Nine months later, the orator would be "revealed" and on his way to the Presidential Chair.[14]

José Vasconcelos—then seventy-six years old—had lived to see one of his former followers rise to the presidency of the republic. He said that he was "delighted with the choice of an intelligent and good man," and some of López Mateos's comrades on that civic crusade of 1929 would come to fill important positions in the national Legislature. But there would be none in the cabinet or among the President's advisors. López Mateos was not about to risk a new defeat by taking the romantic opposition into power. One former political romantic in power—he himself—would do. He wanted a strong cabinet, one that could compensate for his own limitations, his own aversion to maneuverings and great tensions. Especially at the end of 1958, with the railroad workers, the teachers, the oil workers, and the students taking their resentments into the street.

He would choose tried, loyal, and determined men, like Uruchurtu, "the Regent of Iron," to preside politically over the Distrito Federal or the fearsome former Callista and opponent of Cárdenas, General Gómez Huerta, to become commander of the Presidential General Staff—the personal army of the President. But the most important pillars of the new political structure would be the Ministries of Finance and the Interior. For his Finance Minister, López Mateos was to choose Antonio Ortiz Mena, the director of the Social Security Institute, a famous and highly regarded financial expert. For the Ministry of the Interior, with its vast police powers and control over elections and electoral procedures, he chose his friend and comrade from the Senate, Gustavo Díaz Ordaz, the man who could act "without legal quibbles" and was considered both intelligent and tough. It was a very able cabinet, one of the best group of administrators Mexico has ever had (though not as thoroughly honest, in some cases, as the cabinet of his predecessor, Ruiz Cortines).

Some days before he was to take the presidential oath, López

Mateos had a migraine attack and was carried out of a public function on a stretcher. And on the very day he was being sworn in, he suddenly stopped speaking and sat down for a moment. Someone heard him say, "How my arm hurts!" When he stood up again to take the oath (which is done with a fully extended arm), he could only half reach out as he spoke the words, trying to swing the arm straight but not succeeding. It was a chance event, but in retrospect it would almost seem to symbolize the ambiguity he felt about his high responsibilities.

And in fact, on the following day, in the old mansion on Avenida Bucareli where he had his office, Díaz Ordaz understood that it was he, rather than his friend Adolfo, who would execute the day-to-day business of power in the country for the next six years. He was far more suited to hard work, detail, manipulation, and severity than his bohemian former fellow senator. The problems and the people "should come to him and not go beyond him," said Luís M. Farías, Díaz Ordaz's Director of Information. "This is what it's about, that they shouldn't bother the President of the Republic. Simply put, this is the issue."[15]

The Mexican political system and the railway workers' movement led by Demetrio Vallejo were like two trains destined inexorably to collide head-on. The first axiom of the system was dependency, above all that of the workers for whose benefit, supposedly, the advances in labor legislation had been instituted. To permit the independence of a union as powerful and pugnacious as the railroad workers' must have seemed, from the point of view of the system, like derailing the train of the Revolution. Once that car jumped the track, it could be followed, in a wild progression, by the telegraphers, the teachers, the streetcar drivers, the telephone operators, the electricians, the pilots and stewardesses, the students, the peasants—all the "living forces" that had been so patiently taking their places within the system, accepting the bread and fearing the bludgeon. Vallejo and his group were confronting power with an open refusal to be content with "bread" and an olympian disdain, almost recklessness, before the threat of "the bludgeon."

Nor were they a group likely, on their side, to be flexible if the need arose. The new union leadership committee was dominated by the Communist Party and the dissident communists of the Worker and Peasant Party together with some representatives of Lombardo's People's Party. Luís Gómez Z., the leader of the independent group in the Association of Railroad Workers, said, perceptively, some years later, "The Communists have always been friends of the road and not the lunchbox (*itacate*.)" (The *itacate*—a Náhuatl word—is literally the package of food an Indian carries with him when he travels "on a road.") Gómez Z. respected the ideological commitment and the honesty of the Communists. In the era of Ávila Camacho he had granted them leadership positions, and in the time of Alemán had shared prison with them as well as the humiliation of being subject to *charrismo*, a leadership who served management and not the workers. But now the independents and the Communists had become irreconcilable enemies. They were separated not only by a different approach to labor militancy but also by distinct conceptions of politics and, ultimately, of history. Gómez Z.—like the other independent leaders including those opposed to him such as David Vargas Bravo—were sons of the Mexican Revolution and moved within its parameters. They were not trying to achieve the dictatorship of the proletariat, nor did they themselves want to dominate the Mexican working class (something Fidel Velázquez was ably doing at the time). They wanted to lead the railroad workers, to enrich their *itacates* with better salaries and benefits (and also at times to fatten their own personal *itacates*). The Communists—willing to reduce their own "lunchboxes" to the point of starvation and with no interest in personal enrichment—were completely devoted to ideological objectives. "They wanted to eliminate the CTM and afterward perhaps the government," said Gómez Z.[16] And David Vargas Bravo—destined, with Gómez Z., to play an important role in later events—would also insist, "It's the truth! They wanted to bring down the government," and express his anger at the Vallejo forces for having persecuted men who had not supported the work stoppages in 1958: "They beat up . . . dissident workers every day . . . they had proscription lists . . . like the French Revolution."[17]

No other group of workers had been as much a part of the Revolution as the railway men. Who could forget that the Revolution had run on the rails? Some of Villa's *dorados* had been railway men, including his messenger of death, Rodolfo Fierro. Railway men had helped Madero cross the border, after his escape from Porfirio's prison. A railroad man, Margarito Ramírez, had contributed to rescuing Obregón from pursuit by the soldiers of Carranza. All the fighting tradition of these workers was concentrated into this moment, at the beginning of 1959, when they confronted the goverment. They were known for having a surplus of "balls, lots of balls" (*huevos, muchos huevos*). They had shown it in 1958, when the great majority of them supported their leader, Demetrio Vallejo, a worker like themselves. It was not ideology that motivated them. "The workers were not Communists," Vargas Bravo remembered. What moved them was the strength of an honest and militant leader, the will to reject *charro* leaders and government "puppets" (*peleles*), the promise of better salaries and benefits, and the heady rush of having already "made the government yield."[18]

But there was no lack of *huevos* on the other side. President López Mateos would live through the confrontation with a degree of moral suffering that fused with his intermittent spells of physical pain. He would lose more than forty pounds during those months. It was not courage that he lacked, it was the stomach to withstand tension. His bohemian sensibility would have much preferred diplomacy to confrontation. But his Minister of the Interior (and Díaz Ordaz's fellow Oaxacan, Gilberto Suárez Torres, Assistant Attorney General for the Distrito Federal) was making his plans to "grab those strikers by the balls." Gustavo Díaz Ordaz would say and believe that "our destiny and our history as a country are at stake." Finance Minister Ortiz Mena would later point to the fact that this "was the beginning of his moment" for Díaz Ordaz. And he remembered the Minister of the Interior discussing possible plans, "to provoke an incident with the army, to force an outbreak of violence." Those were Díaz Ordaz's solutions.[19]

Under Vallejo's leadership, the Association of Railroad Workers began to behave with unusual confidence. They not only set the date for a forthcoming strike but commented publicly on the policies of management, proposed the creation of a council with broad worker participation, and issued communiqués about the oil industry. While the administration of the railroads declared that they were "totally incapable of resolving the demands" put forward by the union, a wave of labor solidarity began to grow, threatening the hegemony of the CTM. The verbal assaults of Fidel Velázquez about the "Communist red" railway workers did not prevent the creation of dissident groups of union purifiers among the telephone operators, streetcar drivers, telegraphers, and oil workers. A show of force in the streets of the city that would have brought together these groups and more was banned by the government. But Vallejo was not intimidated into postponing the strike, which began at midday on February 25, 1959.

The leader of the railway electricians, David Vargas Bravo, had been excluded (along with his followers) from the central committee of the union. López Mateos called him in and asked his opinion on the actions of his fellow (and rival) union leaders. "If you yield, you lose. It's a pretext," he answered. But nevertheless, the government and the union did come to an agreement: a 16.66 percent increase on their 215 pesos per month, 62 million pesos for medical and pharmaceutical services, 30 million for constructing new residences for workers who lived far from the railways or in small villages, the renewal of thirty-five clauses that had been abrogated during the times of *charrismo,* and a promise by the company to increase its charges for foreign mining companies—"a patriotic victory that must be recognized!" Lombardo Toledano noted in *Siempre!*[20] The news of the victory triggered enormous enthusiasm among dissident union groups and university students. The writer Carlos Monsiváis, who was then a student, remembers: "When the Vallejo people won the first strike, we went to the Buenavista railroad station. Othón Salazar arrived with a group of teachers. Vallejo spoke, a band played *La Rielera*—'The Railroad Woman' ('I am a railroad woman/I have my Juan/He is my life/I am his love'). There were tears in everybody's eyes."[21]

But a few days later, things suddenly changed. Two railroads—the Ferrocarril Mexicano and the Ferrocarril del Pacífico, which operated semi-independently, as did their unions—had not been included in the contract. A new strike date was set. The management of these railroads dragged their feet on conceding the same terms as had been concluded with the rest of the railroads. But a mediation committee (*Junta de Reconciliación*)—composed of government, labor, and management representatives under the jurisidiction of the Ministry of Labor—refused to sanction the impending strike. The date was finally fixed for Holy Week, when it would affect millions of vacationers. And when no agreement was reached, the two railroads were struck on March 25.

The next day, Holy Thursday, the government arrested some of the leaders and dismissed thirteen thousand workers from the Mexicano and Pacífico railroads. But the door was not closed to negotiations. On Easter Saturday Vallejo and the Minister of Labor, Salomón González Blanco, held a conciliatory meeting. The government agreed to sanction the same 16.66 percent pay increase it had accorded to the rest of the railway system. Vallejo only had to do one thing. He was to visit President López Mateos and reach a direct agreement with him, so that the gesture could be perceived as coming directly from the President. But the stubborn Vallejo offended López Mateos by insisting that he had to bring a tape recorder along, because he could not take any action without keeping the Executive Committee fully informed. To make things worse, the Executive Committee insisted that there must be a total strike on all the railroads, in solidarity with the Mexicano and Pacífico. Events were moving toward the crisis.

A year later, behind bars in the old Lecumberri Prison, Vallejo wrote that if the union had yielded on March 28—without the release of those arrested and the rehiring of those who had been fired—the Committee would have lost its position and *charrismo* would have taken control of the union. Four years later, still in prison, he wrote down another opinion: "It would have been logical to call a truce, to reorient and reorganize our forces, to seek the militant solidarity of other labor organizations." He went on to say that he would have taken that step, but the representatives of the PC, the POCM, and the PP—who dominated the Committee—had insisted on the nationwide strike.

He would later learn—still during his time behind bars—that the prudent Lombardo had advised accepting the government proposals but that the Communist Party and the Party of Mexican Workers and Peasants had refused and carried the movement toward its final consequences. On Easter Saturday, when no agreement had yet been signed and Vallejo's desire to bring a tape recorder into the President's office still rankled within the sensitive soul of López Mateos (and was offensive to the whole pattern of secrecy on which all the men of the system had been raised), the Executive Committee called the railway men out on strike.[22]

"I was perfectly aware," wrote Vallejo, "or at least I had a sense of the danger we were running in voting for the full strike, because only those blinded by the excitement of our victories or die-hard labor theorists would not have been conscious of it." In a lightning attack throughout the entire country on that very same Easter Saturday, March 28, the police, the army, and squads of special agents arrested ten thousand railway workers. There were no polite words. They took men down with their clubs and their bayonets. Orders had clearly been given to make an example of the strikers. Those arrested were thrown into prison, and some were sent to army headquarters, the notorious Campo Militar Número 1. In Monterrey, a leader died under torture—Román Guerra Montamayor—and they threw his body across a railroad track, first putting lipstick on him and painting his nails red as a mark of his political affiliation and presumably as an attempt to suggest that the murder had been a crime of passion among homosexuals.

At one o'clock in the morning on Easter Sunday, March 29, David Vargas Bravo was roused out of bed by a call from Díaz Ordaz himself. He was to report "immediately" to the Ministry of the Interior on Avenida Bucareli. There he was told what had happened to his fellow railway workers. Vallejo, the entire Executive Committee, and all the local union secretaries throughout the country were in jail. Vargas Bravo was ordered to collaborate with Gómez Z.—who had already spoken previously with the President himself and was waiting there in the office—in order to reestablish railroad service. "You choose your Committee," said Díaz Ordaz. "We'll take care of the rest. Decide. The Committee will meet today. And you'll be given authority by the Ministry of Labor." Their orders were to get the railroads running "using people who will make things look as if they've gone back to normal." On that same day, as part of the blitzkrieg, the enterprise announced the firing of at least ten thousand workers. (Some estimates put the figure as high as twenty-five thousand.)

Immmediately afterward, as Vargas Bravo himself reports, "Gómez Z. and I split up the responsibility for the whole republic in my office and the Committee was recognized." In the two days that followed, they broke the strike. Men hired by Gómez Z. and Vargas Bravo ran the famous "phantom trains," empty of any passengers or cargo but escorted by the army. It was a noisy simulation of reestablished service. Without dispatchers or telegraph operators, the electricians who were Vargas Bravo's men "funneled the trains through the stations" and succeeded in confusing many strikers, a considerable number of whom returned to work.

The prisons were bulging. There was no room for all those who had been arrested. "They had taken everybody in," Vargas Bravo remembered. Díaz Ordaz asked him to go to the Lecumberri Prison and indicate who could be released. He did it and freed about half the prisoners. On Tuesday, March 31, the Mexican press announced that, among the 800 prisoners in Lecumberri, "150 have been identified as Communist agents, including some who were directors of the union, which is simply intolerable and cannot be allowed in Mexico." On the following day, the accusation was even more precise. It was the followers of Vallejo who were seeking "the overthrow of the government of the Republic and afterwards they would dictate a new Constitution. . . . Vallejo has committed the crime of treason to the Patria."

Months later, thirty-five leaders of the movement, including Vallejo and the real brain behind the action—Valentín Campa—received sentences of between four and fourteen years in prison. They were accused, among other things, of violating Article 145 of the Penal Code, the crime of "social dissolution," which Ávila Camacho had created to combat the fascists during the Second World War and which now in practice allowed the government to imprison whomever it decided to consider an enemy of Mexico. A new and illustrious prisoner accompanied them, an alleged "traitor to the fatherland" like them, the last living member of Mexico's great triumvirate of muralists—David Alfaro Siqueiros.[23] After traveling through Latin America calling López Mateos an "impostor" and a "sellout," Siqueiros had been arrested by General Gómez Huerta, the commander of the President's General Staff, and brusquely taken off to the Lecumberri Prison. For the crime of "social dissolution," he would spend almost the entire six years of the López Mateos era behind bars, until freed by presidential decree in July 1964.

From his prison cell that year, Vallejo would be writing: "I hope that this painful experience may serve as a lesson to the revolutionary parties, that they should not directly intervene in union struggles, but be satisfied with the role of modest advisors when that is required of them and that they never become arbitors to decide what must be done on a given problem."[24]

Two incompatible systems coming from opposite directions had crashed head-on. The Mexican political system would not tolerate a Trojan horse within its rigid corporate organization. The inflexible ideological system of the Communist Party and its dissident counterpart (the POCM) staked everything on yet another attempt to take the Winter Palace or at least to create a new level of class consciousness. Could an agreement have been reached? With difficulty, because the opponents were too much alike in critical ways—both showed a similar intolerance, and a total democratic illiteracy. Both shared responsibility for the crash, but the brutality would be the historical burden of the government.

And caught in the middle were the workers, men with *huevos* but above all people who wanted both a freer "road" and a "lunchbox" with something extra in it to take home that night to their families.

The pattern had been established. Nothing could or should alter the magic circle of dependency. The system had responded ferociously against the real or supposed beneficiaries of the Revolution—against the workers. From that moment, any rebellious unions who wanted to "try out" the government knew what they could expect to face. For each insurgent action (or threat of one) there would be an even stronger reaction on the part of the authorities. The political system, in chorus, lined up behind the actions of the government. In the Lower House the deputies repudiated "the foreign elements which the union leadership supported with the intention of strengthening the unjust movement that they were sponsoring."[25] The judicial power maintained a prudent silence. Big business enthusiastically approved. The Church did not open its mouth but breathed easier after the blow had been delivered to the enemies of Christianity. Even *Siempre!*, the least dependent magazine among the "not so dependent" elements of the press, criticized the "unbelievable blindness" of Vallejo, who had "put the government into an extreme situation, which would have become an alley with no way out, if it had not been resolved as it was." Only the younger intellectuals, the socialists and the Communists, the most politicized students, and of course the weak parties of the opposition (the PAN to the right and the PPS to the left) attacked the measures.

At the end of 1959, the Legislature approved a fundamental amendment to Article 123 of the Constitution, the official locus for labor relations in Mexico. A "Section B" was added to the article. It contained a good portion of "bread": in salaries, holidays, vacations, overtime, and bonuses for workers in the service of the government. On the "bludgeon" side of the ledger, so many conditions, ambiguities, and restraints were placed on the right to strike that permission to do so would in fact always now be left to the discretion of the State. In addition, government employees were forbidden to join any union except the FSTSE (Federation of Union Workers in the Service of the State).

What is surprising is that—even given this context—workers who had been raising the flag of union independence since 1958 did not completely lower it. The telegraphers struck, and the government fired their leaders. Othón Salazar's independent teacher's movement continued its public protests, and when the teachers marched out of the Escuela Normal Superior on August 4, 1960, the government hurled their grenadiers, mounted police, and plainclothes *judiciales* against them. Oil workers who mounted a protest in front of the Monument to Independence were driven off with fire hoses. It was the law of the jungle; protests met with tooth and claw: the truncheon and the canister of teargas.

The airline pilots were allowed to form a union, but the first time they went on strike, the government seized the planes and the entire operation. "Seizure" (*requisa*) was a method without any constitutional basis, mentioned only in secondary laws and there sanctioned only when services of the utmost necessity were in danger. But the government, as always, had the right to interpret a situation. It made broad and lax interpretations and now freely began to use the tool of *requisa*. When the telephone operators struck in 1962 and their work sites were seized, the authorities declared, in the best tradition of the Bourbon emperors, "The practice of seizure is not subject to any deliberation. It is a matter of a presidential decree, and a decree is not discussed, it is obeyed."

Meanwhile, in overall union leadership, the government was both adjusting the safety valve by using new men and relying on a very long-term and popular intermediary in Fidel Velázquez. Though it was the strikes and retaliations that filled the front pages, the quiet labors of Velázquez at the head of the CTM were perhaps more decisive. Millions of workers in the private sector willingly accepted the pragmatic methods of the man who represented them before management and government and was winning them a steady increase in their *itacate*s. Since the days of Alemán, the CTM conventions had continually reelected him.

He would soon break Porfirio Díaz's record of thirty years in power. (Eventually he would double it.)

But the union unrest, primarily in the public sector, was real, and the judicious permission for the rise of new leaders was an important method of control. Some, like the electrician Rafael Galván, were old friends of López Mateos but independent of the CTM and with ideological convictions moderately to the left. On December 4, 1961, the President himself attended the inauguration of a new collective labor organization, the CNT, uniting a number of public sector unions and headed by his friend Galván. Others among these new leaders had their own base of support but were closely connected to the CTM, like the charismatic leader of the oil workers, Joaquín Hernández Galicia (*La Quina*), who took control of that restless union by carrying "bread or the bludgeon" to extremes. When he took power, *La Quina* announced, very directly, that anyone who protested his policies would be expelled from the union. And he also rained down so many benefits on his workers that the costs became a scandal. But the payback to the system in terms of obedience seemed to justify the mounting bill. And with the most revolutionary union of them all, the Association of Railroad Workers, the President chose none other than Luis Gómez Z. to run the organization. López Mateos so appreciated his services of 1959 that he referred to him as his "relative" and would give him the honor of officially "revealing" the new president-to-be in 1963.[26]

Meanwhile, the employees of that printing plant founded in the age of Alemán would proudly hang a huge picture of their own leader, Antonio Vera Jiménez, on the walls of the union office. It showed him with President López Mateos and President John Kennedy, who had visited Mexico with his wife in 1962 and prayed at the Basilica of the Virgin of Guadalupe. The life of the thousands of workers affiliated to that union had gone on without any major upsets. Their salaries and benefits had risen year by year. They punctually paid their dues, marched past the National Palace every year in the May Day Parade, played soccer on Sunday mornings, went to the bullfights, the wrestling matches, the boxing bouts where they could enjoy seeing, at a ringside seat with his friends, the President of the Republic (and great boxing fan) Adolfo López Mateos. The printshop employees liked the President. When they shouted *"Viva López Mateos!"* they meant it. Many of them, perhaps most of them, were completely unaware of what had happened to the railway workers or else attributed it to "political matters," which were not their concern. If Campa and Vallejo, with their sacrifice, had hoped to raise the class consciousness of the average Mexican worker, they had failed.

* * *

In 1961 the President sent an initiative to Congress, a new social security law for the benefit of members of the armed forces. It was of course, as usual, approved unanimously. The army appreciated this gesture of support, but it never placed any conditions on its own unlimited support of the government. The military had fully absorbed the function Ávila Camacho had assigned to it—it was now "the immaculate defense of our institutions."

Frequently put to use against labor unrest, the army was also called upon in these years to "impose order" in three other areas of conflict—a small faction within its own military hierarchy, some university campuses, and a certain number of peasant leaders who resisted the passive servility of the CNC (National Confederation of Peasants) and continued to wave the old, legendary banner of Emiliano Zapata.

General Celestino Gasca—an old Revolutionary who had once been governor of the Distrito Federal under Obregón—announced well ahead of time that he would lead a rebellion "to overthrow the government" on September 15, 1961. Properly appreciative of his courtesy—as Mexicans generally are—the army arrested Gasca and his partisans with time to spare. In the two traditionally Catholic cities of Puebla and Morelia, the army patrolled the streets to calm rioting between two contending political factions: left-wing students and teachers versus the middle and upper classes connected with Church and business interests. One student was killed in Morelia. Army battalions also intervened—early in the sixties—to put down protests against two successive governors in Guerrero, the most violent of Mexico's states. There were dozens of dead. But for the opposition to the government, the deepest and most long-lasting memory of military action during this period would be the murder of Rubén Jaramillo, the peasant leader of the Zapatista region of Morelos.[27]

Besides being an heir to Zapata's drive for agrarian justice, Jaramillo was a Methodist preacher. Under Cárdenas, he had supported the creation of the Zacatepec Sugar Mill—which was organized as a cooperative—in the hope that it would bring prosperity to the historical descendants of Zapata. Inefficient and corrupt administrations had soon canceled that hope. Jaramillo, who preached sermons against alcohol and other vices among the peasants, felt that he could do no less than advise them legally as well and protect them against the outrages they had to endure. He formed his Committee for the Defense of the Sugarcane Workers (*Comité de Defensa Cañera*), and the government began to persecute him, until he had to flee with a group of men to the mountains.

Ávila Camacho had tried to reach Jaramillo through the method of persuasion that had served him so well with the Cristeros. He spoke with him; he offered him amnesty, fields, money, but ultimately he failed because Jaramillo's grievance was not about religion but about the land, a loyalty he felt with religious force. He kept on fighting for years, as an independent candidate for governor of Morelos (in 1946), as a militant in the presidential campaign of Henríquez Guzmán (in 1951), and through the following decade in struggles against various local caciques, constantly putting pressure on the authorities, calling for the distribution of land to thousands of peasants who had their right to it. He himself lived with incredible modesty: farming his little bit of land and wearing clothes sewn by his wife.[28]

On March 23, 1962, he and his family were murdered—apparently at the instigation or with the foreknowledge of General Gómez Huerta, chief of the Presidential General Staff (which consists of regiments under the President of Mexico's personal command), and certainly with the agreement of the President. The writer Carlos Fuentes investigated the crime and described the life of Jaramillo to a mass audience. He visited the region shortly after the events and then published a moving and dramatic report in *Siempre!*—to the surprise and anger of López Mateos, who arranged for *Siempre!* to lose all government advertising. But that magazine was daily becoming more independent, more worthy of respect, under its editor-in-chief, José Pagés Llergo. Fuentes wrote:

> They pushed him down. Jaramillo could not hold himself back, he was a lion of the field, that man. . . . He threw himself at the party of murderers; he was defending his wife and his children, and especially the unborn child; they brought him down with their rifle butts, they knocked out an eye. Epifania flung herself on the murderers; they tore her rebozo, they tore her dress, they threw her on the stones. Filemón cursed at them; they opened fire and he doubled over and fell beside his pregnant mother, on the stones. While he was still alive, they opened his mouth, picked up fistfuls of earth, pulled open his mouth and laughing filled it with earth. After that it went fast; Ricardo and Enrique fell riddled with bullets; the submachine guns spat on the five fallen bodies. The squad waited for them to stop breathing. But they went on living. They put their pistols to the foreheads of the woman and the four men. They fired the finishing shots.[29]

Five thousand came to the Protestant burial in Jojutla, the town from which Emiliano Zapata had ridden out half a century before to make the Revolution of the South. Fuentes would preserve the voices of dignity and grief of the peasants of Morelos: "They've killed for us those who were the feet and arms of the forsaken."

Though it was not yet widely recognized, a qualitative change was taking place in a crucial sector of society: the more politicized students from the middle class, the kind of people who read *Siempre!* or the more radical magazine *Política.* The passionate sixties had begun in Mexico on January 1, 1959, with the victory, as unexpected as it was inspiring, of Fidel Castro, whose army entered Havana to begin the last year of the fifties. The railway workers' strike had sent a wave of enthusiasm across college campuses, as if revolutionary persistence would be enough to move mountains and displace dictators and dictatorships. "To live was to go to demonstrations," wrote Carlos Monsiváis: "People, awake!—Let the consciousness of the people awake!—Grenadiers! Murderers! In the demonstrations, orgasmically, all the intensity poured out that could not be expressed since Calles except in support of the government."[30]

After the movements of the railway workers and the teachers had been broken, there was no more room for mere "protest demonstrations," or any college cheers for the President. Student politics left the campus of the Politécnico and the University to raise its voice about the politics of the nation:

> There was a big [student] demonstration due to begin at the teachers' college. . . . By eight in the morning the building was surrounded by the mounted police and all kinds of secret agents. . . . The mounted police charged the demonstration head on. Some of us tried to sing the National Anthem but there was no time. I ran into a building and from the flat roof, we watched the relentless, interminable pursuit. Forty students who were trapped . . . were the first to fall. The kids from the vocational schools had brought planks with nails in them and chains but any attempt at resistance was impossible.[31]

A fierce antagonism began to develop between two antithetical forces: the student and the grenadier. Although not all the students were ideologically to the left (during this period there was also a growth in the old and powerful student militancy of the right that dated back to the thirties), a new interpretation—from the left—was taking root in the consciousness of Mexican students: the so-called revolutionary Mexican state was in reality an ignoble front man "for the bourgeoisie and for imperialism." Opposed to it were the classes of the workers and the peasants and their faithful spokesmen: the students, the artists, and the intellectuals.

In contrast to Ruiz Cortines, who had changed almost all of his predecessor's governors, the era of López Mateos was marked by an almost seamless stability at the state level. Only in Baja California, in San Luis Potosí, and in untamed Guerrero were there problems that required the resignation or removal of a governor. The merits of the governors he favored were often similar to those of Colonel José Ortiz Ávila, governor of Campeche between 1961 and 1967, who dated the beginning of his march toward the governorship to a favor he had done for López Mateos during the President's first State of the Nation address to the Legislature. He was assigned to sit next to a vocal Panista deputy named Molina, with a pistol pointed at him (hidden under his coat but obvious to Molina). The deputy was not to interrupt the President. Nor did he. After which, López Mateos explained to the colonel that politics had to be done with considerable brains but if you added *huevos* to the brains, it made a much better dish.[32]

Many governors of the period must have demonstrated similar qualifications. On the other hand, what happened in the state of San Luis Potosí, the "caliphate" of Gonzalo N. Santos, was a world apart from the normal process of state government under López Mateos. One of Santos's many memorable sayings was "Morality is a tree which either feeds you blackberries or isn't worth a f—." San Luis Potosí had been his private hunting preserve, where he imposed what he called his "law of the three *ierros*: "prison (*encierro*), exile (*destierro*), and burial (*entierro*)." He was said to have no need in his state "for administrators, only gravediggers." In 1958, in his old age, he still had much power and wealth, living at his ranch El Gargaleote—of 87,000 hectares—with his family and his bodyguards, among them the famous *Mano Negra,* who prided himself on never leaving "the dead for which he was responsible" fewer than five hundred meters away from the highway.[33]

The situation in the state of San Luis Potosí was all the sadder contrasted with its great Liberal history. The earliest opponents of Porfirio Díaz had all been Liberal leaders from San Luis Potosí. It had also been home to the courageous newspaperman Filomeno Mata, who had published the anti-Díaz *El Diario del Hogar* for thirty years in between jail terms. And Madero had titled his program *El Plan de San Luis* (written while he was imprisoned there). But the triumphant Revolution had badly repaid this state that had bred so many of its precursors. It had been the domain of two consecutive caciques, Saturnino Cedillo (who had tried to lead the last modern military rebellion—against Cárdenas—and died in the attempt) and then the "Dark Chestnut Horse" himself, Gonzalo N. Santos, originally from the Huasteca region (as a result of which, he mentioned in his memoirs, should he by chance have to go to hell, he was "used to hot country").

At the end of the forties, the Rector of the Autonomous University of San Luis Potosí decided to give Santos a medal of recognition for having granted the university autonomy during his governorship. Dr. Salvador Nava, an ophthalmologist who taught at the school of medicine and was a member of the University Council, challenged the decision and refused to sign his name to the official document. The university, he said, had received its autonomy in 1923 from Governor Rafael Nieto. Santos would not forget the insult.

In 1958, Salvador Nava would form a political front, for the first time uniting a spectrum of organizations in opposition to Santos. He was proposing to remove the cacique through democratic means. Though his group did not sympathize with any party, they recognized the practical utility of fighting "within the PRI against one of its members." And so his Federation of Professionals and Intellectuals (FPEI) of San Luis Potosí joined the National Confederation of People's Organizations (the CNOP), a corporate arm of the PRI for all its middle-class sectors.

It looked like a good time for action. Everyone knew that López Mateos had been a Vasconcelista and that Santos, as a leader of *pistoleros* for the government candidate, had participated in violence against them. He was regularly accused of having fired—or at least orchestrated—the gunshots that had killed young Germán del Campo. Santos would repeatedly deny his responsibility for that particular death and may have been innocent of it, since he was hardly in the habit of covering up his many murders: "Look, I'm not going to lose any sleep over one damn dead man more or less!"

On November 1, 1958, Gonzalo N. Santos accomplished a miracle. Nava's party became the Civic Union of San Luis Potosí, uniting PAN, the Communists, the Sinarquistas, professional people, and even some formal sectors of the PRI—all as one (for the time) against Santos. And the idea arose of nominating Dr. Nava as their candidate for municipal president.

Within a few days, Nava could count on the support of 80 percent of the railway men and various other workers' groups. On November 19 a local newspaper announced that there were "more than a thousand armed men ready to finally flush out Santos." On November 20, Santos did not appear for the traditional parade commemorating the beginning of Madero's Revolution. His man the governor, Manuel Álvarez, was pelted with rotten eggs and fled the city. That same day, on the bandstand in the central square, a mock scaffold was erected for the governor and the cacique, with a banner that read SANTOS, ASSASSIN OF STUDENTS! The army moved in.

Indignant at the intervention of the troops, a body of citizens went to the capital, trying to get the attention of Ruiz Cortines's Minister of the Interior, Ángel Carvajal, who could not see them because "he was just going out." His Chief Administrative Official, Gustavo Díaz Ordaz (who would become the Minister within a few days), took the bull by the horns and met with the citizens. He told them he could not help Nava's group because "they were Sinarquistas," which the delegation indignantly and vehemently denied, insisting that they were members of the PRI.

On the twenty-sixth, the citizens issued an ultimatum. Unless the governor was removed, there would be a total strike. The next morning arrived and with it the strike. On December 2, the day after López Mateos became President, the new Minister of the Interior, Gustavo Díaz Ordaz, told Nava that the results of the coming elections (on December 7) would be honored. The strike continued and there was some violence, but then a new military commander arrived and announced: "The people must not confront the people." He meant that the army was no longer available to back Santos, and it was the end of the *cacicazgo* ("cacique rule"). The vote on December 7 went 26,319 for Nava and 1,683 for Santos's candidate. The sign THIS HOUSE VOTED FOR NAVA appeared in many windows. Businesses announced, "Buy here, we are Navistas." Fidel Velázquez himself presided over a change of local union leadership in the city, removing Santos's men from power. On December 23 Congress recognized Nava's victory. The people of San Luis Potosí (the *potosinos*) had the sweet experience of Christmas in a free city. Democracy had won a victory without precedent in the contemporary history of Mexico.

Nava would run a model government. Every week he made a radio address, informing the public on the progress of his administration. Every day he hung a list on the walls of the town hall, detailing the use of public money. At the end of his first *informe* he called a meeting in the central square and before fifteen thousand people rang the new "Liberty Bell," cast from the melted-down bronze of plaques that had borne the name of Gonzalo N. Santos, which the people had torn from schools, markets, plazas, streets, and dams.

With Santos's governor removed, the star journalist of the magazine *Siempre!*, Francisco Martínez de la Vega, took over the governor's chair. He was an old friend of López Mateos and a moderate leftist, suspicious for that reason of the popular mayor of the state capital, whom he identified with "currents traditionally opposed to the principles of the Mexican Revolution." Though Nava was far from being a man of the right, he did not hide his Christian sympathies. But his government was a rainbow of parties. When Díaz Ordaz reproached him for including the Communists, he answered that he had to—they had supported him—although their responsibility (the administration of city cemeteries) would be a difficult base from which to agitate. Beyond local issues, however, the case of San Luis Potosí was disturbing to the central headquarters of the political system: the PRI and the Ministry of the Interior. That unease turned into alarm when Nava, at the beginning of January 1961, resigned as mayor to contend for the governorship of his state.

He actually thought that he could be nominated within the PRI. He had never been a member of any other party. He had been a fine municipal president, and he had led the fight against Santos. But he was seen as a newcomer, who had not paid his dues through steadily rising within the PRI. So he was told, at party headquarters in Mexico City by the president of the party, General (and lawyer) Corona del Rosal, that he simply could not be the candidate.

When he returned to San Luis Potosí and reported the interview, his people insisted that they should continue "as independents" opposing the official PRI candidate. The "Navista movement" spread throughout the state, and the aggressions against them began. In May 1961, Jesús Acosta, the campaign coordinator in the important Huasteca region of Tamuzunchale, was shot dead. PRI representatives arrived in San Luis Potosí with the mission of explaining to Nava that he must withdraw. But they could not persuade him. "Do you understand the consequences of all this?" they asked the Navistas. "Do you understand what could happen to you?" They were warned, but they did not foresee what really would happen, after widespread electoral fraud gave the victory to the government candidate at the beginning of July.

Between July and September, the *potosinos* carried out an extensive campaign of civil disobedience. On August 20 the military commander of the zone warned Nava that he would be held responsible for any "agitation," and that it would be met with "the necessary energy." All public demonstrations were prohibited and all attempts to demonstrate were halted by the army. And then, on September 15, the central government closed the books. Government provocateurs, firing randomly, appeared on the roofs; at midnight all the lights in the city went out; a few citizens connected with the civil disobedience movement were killed in the streets. The army took over the headquarters of the Navista Committee. The next day they arrested Nava and his closest collaborators.

They were taken to Campo Militar Número 1 and then to Lecumberri Prison. The opposition newspaper, *Tribunal*, was raided and its presses smashed. Nava was charged with the usual crime of "social dissolution," but other crimes were thrown in—stockpiling weapons, incitation to rebellion, and more. He was let out on bail a month later. The major demonstrations were over, but for two years Nava continued to offer public and political resistance. By February 1963 the government lost its patience. He was to be persuaded more directly. Dr. Nava was arrested and tortured, not out of pure sadism but to *make him yield.* His body covered with bruises, he was released, and for over two decades—until a new political situation had begun to develop—he never made another public speech.

Nava was neither a Communist nor a Sinarquista, nor a member of PAN. He had functioned within the PRI, but he was a maverick, a citizen who wanted to exercise his political liberty within his own city and within his own state. He found that the attempt—consecrated since 1857 in the Constitution—was equivalent to the crime of "social dissolution." Any aspirant to real power had to operate not merely inside the ruling party but according to the formal structures of command and the accepted forms—within that party—for distributing power.[34]

At the end of 1958, General Lázaro Cárdenas, sixty-two years old, left Mexico for the first time in his life (except for a brief visit in his youth to an American border city). He spent some days in New York. In that Empire City of urban intensity, it was Central Park that impressed him most: "an agreeable contrast to the blocks that human technology has constructed to concentrate economic power, the product of international financial organization and the sweat of peoples of this and other continents."[35]

Once he began traveling, the mileage he logged—as usual with him—was impressive: eight countries of Western Europe, Poland and Czechoslovakia in the East, then the USSR, China, and Japan. Western Europe was of little concern to him. He was not interested in art or museums; he had eyes only for the landscapes—the olives of Trevi or soil erosion in Tuscany. Russia and China filled him with enthusiasm. And on January 1, 1959 (while he was visiting Nice), news came that nourished his sense of hope. The young intellectual "with the blood of a fighter" whom he had saved from deportation in 1956, Fidel Castro, had overthrown Batista in Cuba.

On July 26, 1959, Castro spoke before hundreds of thousands of euphoric Cubans in the main square of Havana, announcing his decision to reassume the position of premier of Cuba. Beside him, waving to the crowd, was General Lázaro Cárdenas. The young Mexican writer Elena Poniatowska described the enthusiasm centering on the leader of the Cuban Revolution and the former President of Mexico. Looking at Cárdenas, she noted how "erect . . . strong and healthy" he looked, "the most gallant tree of Mexico." On the flight back, she spoke with the general. Showing the renewed ardor of his youth, Cárdenas praised Castro, foresaw a luminous future for Cuba, and also spoke of the new Cuba as a mirror of his hopes for Mexico. In his opinion, the Mexican Agrarian Reform had not failed. It could be revived by increasing available agricultural loans from a billion to five billion pesos annually and by providing seed, machinery, education, health services—all that Cuba would now initiate for the benefit of the Cuban peasants, the *guajiros*. He had been deeply moved to hear the chant from the Cuban crowd: "The Agrarian Reform goes on!" He had faith that the Cuban Revolution would also move Mexico in a better direction, and he felt himself strong enough to take the lead, not from a position of official power now but through a tenacious and intelligent opposition from the left. "And communism?" asked Poniatowska. "Everybody says that communism is slowly invading the Caribbean—" The general cut her off with a firm answer: "There is no communism in Cuba, only poverty."[36]

Back in Mexico, Cárdenas would confront President López Mateos in the first round of their antagonistic relationship. Cárdenas had taken the position of a moral advocate for the political prisoners of the regime—the railway men and the teachers. Throughout the six-year term, he would repeatedly ask for their freedom. López Mateos would always refuse, contending that Cárdenas was badly informed on the true intentions of those who had been arrested. In 1960, on the occasion of the showy celebration of the Fiftieth Anniversary of the Revolution, Cárdenas would write in his diary: "What a contradiction and what sarcasm, to hear the Minister of the Interior Díaz Ordaz . . . say that civil rights are guaranteed."[37]

In the following year came the attempted invasion of Cuba at the Bay of Pigs. Cárdenas, in his sixty-fifth year, decided to leave for Cuba and fight alongside the government. His plane was grounded, by order of the President of Mexico. He was taken off the plane by force, but he then went to the Zócalo, mounted the hood of a car, and insisted on speaking to the gathered crowd. For thirty minutes, the police prevented him from talking, while he was given, in the words of Carlos Fuentes, "the longest ovation I have ever heard . . . for that man who twenty-three years earlier had proclaimed the nationalization of the oil from the central balcony of the National Palace and now, at street level, was defending the independence of a small, threatened nation."[38]

He was finally allowed to speak. He thundered against the Bay of Pigs invasion, against the United States for supporting "dictatorial regimes," and in favor—speaking especially to the young people in the audience—"of a campaign against the country which thinks it can control everything with gold." López Mateos could barely contain his fury. Cárdenas was behaving with a political boldness that verged on insubordination. For much less, Cárdenas himself had sent Calles into exile. In one of their many uneasy conversations, López Mateos told him, "They say the Communists are weaving a dangerous web around you," to which again Cárdenas would respond that misery, not communism, was the problem in Latin America.[39] He was speaking from his own experience. During his term of office he had fought against poverty and he had kept the Communists at the margins of power. Without becoming its President ("I cannot: I am an ex-President"), he now presided over the creation, in August 1961, of the Movement for National Liberation (MLN), a new attempt to unite the movements and personalities of the left: followers of Lombardo, former members of the PRI, Communists, Trotskyists, and intellectuals.

According to Daniel Cosío Villegas—more aware than almost anyone else of what went on between the lines of Mexican politics—López Mateos was hoping that Cárdenas would lead an armed rebellion, so as to be finished with him once and for all. The former President was applying so much pressure that many believed he would split the trunk of the system. It was a possibility that hung in the air as Cárdenas enjoyed his alliance with the members of the young intellectual left who, in love with the Cuban Revolution and disenchanted with the Mexico forged by Alemán, took shelter in the beneficent shade of "the most gallant tree of Mexico," the symbol of the Revolution that had come to life again.[40]

The paradigm of generations formulated by the Spanish thinker Ortega y Gasset was (like all formulas) far more limited than its creator imagined. But applied to the intellectuals of the Revolution and their successors in Mexico, Ortega's formula seems to describe a reality. The Generation of 1915—with leaders like Gómez Morín, Lombardo, and Cosío Villegas—had created the Revolutionary institutions. The next generation—"the Generation of 1929"—had consolidated that order of things behind the leadership of Alemán. According to the scheme of Ortega y Gasset, the following generation should play a critical role in judging its predecessors.

While Alemán was President, a magazine had been founded that gave its name—*Medio Siglo*—to this new generation of intellectuals who came into their own in the middle of the century. They followed the lead of a somewhat older man, the essayist and poet Octavio Paz (who was to become Mexico's first Nobel Prize Laureate in Literature). Paz was the son and grandson of revolutionary intellectuals (his father a Zapatista, his grandfather a Porfirian journalist who became a Maderista). In a much more profound sense than Miguel Alemán, he was a true "cub of the Revolution." As a young man Paz had become a Marxist. He had gone to Spain—as a writer in support of the Republic—during the Spanish Civil War. But he was disillusioned early by Stalinism, had passed through a Troskyist period, and then turned back, along the trail of his own roots, to a profound and beautifully written meditation on the Revolution in his prose masterwork, *The Labyrinth of Solitude*.

During the fifties, the generation of *Medio Siglo* paid its final tribute to revolutionary nationalism, following Paz's lead in the search for *mexicánidad* (the deepest expression of which Paz located in the Revolution itself). They looked for "Mexicanness" in colonial and pre-Hispanic history and in the phenomenology of everyday life. And then the clearest thinkers made a judgment on that long, drawn-out process of looking inward that had begun in the twenties. They decided that it could easily run aground on self-absorption and solipsism. To the second edition (in 1959) of *El laberinto de la soledad*, Paz himself added a postscript that included the idea that Mexicans were now, for the first time in their history, "contemporaries of all men."

Responding to this assertion, Mexican intellectuals began to do something substantially new for them. As a conscious choice and not as exiles, they began to travel. A number of them went to France and were drawn to Jean-Paul Sartre, whose personal existentialism was moving in those years toward social criticism and political commitment to working-class and anticolonial struggles. Nevertheless, back in Mexico and into the late fifties, the hero of the *Revista Mexicana de Literatura* (Carlos Fuentes and Emmanuel Carballo were its editors) remained Albert Camus, the defender of radical philosophical liberty. These intellectuals were still looking for "a third way,"—"neither Eisenhower nor Khrushchev: new forms of life and human community."

Cuba changed all that. The whole intellectual spectrum of Mexico celebrated the triumph of Fidel Castro, from the old rightist and anti-Yankee Vasconcelos (who died in June of 1959) to the rising constellation of currents on the left. For the "Generation of the Midcentury," Cuba was not a historical event. It was a religious revelation. Articles, reportage, and dazzling pictures of the young Cuban *barbudos* enraptured intellectuals and students and most especially made those of the same age as the Cuban warriors feel uncomfortable with their old and moldering Revolution.

The rise of Cuba across the waters was accompanied, in Mexico, by the government repression of the railroad workers. The comparison seemed to further underline the differences between the two revolutions. In 1960, at the age of fifty, the Mexican Revolution seemed more frozen than ever. How could it justify what it had done and was doing to the heirs of the *rieleros* of the Revolution? The young intellectuals—who had already criticized the turn away from Cárdenas's legacy—found few significant differences between Alemán, Ruiz Cortines, and López Mateos. Not without reason, they saw the succession as a generational hegemony—"Alemanismo" in different shades—and denounced the ugly currents that ran through all three regimes, pointing to the ostentatious display of the new bourgeoisie, the administrative corruption, and all the lies in the press. They condemned the leadership of the CTM, which for them was no more than "*charro* unionism." The economic model in vogue earned no respect from them, because they felt that it ignored social justice. And they detested the farce of official, "Revolutionary" language.[41]

Following in Cárdenas's footsteps, they came to feel that the best way to defend the authentic Mexican Revolution lay in defending the Cuban Revolution. Along with Cárdenas, they founded the MLN. As their duty, they took on a responsibility to express the needs of the people with clarity and precision. The values of the Constitution of 1857 for them were secondary: federalism, individual liberties, the division of powers, democracy. Their credo was a combination of Cardenismo (completing the Agrarian Reform, defending natural resources) and Marxist socialism (a strengthening of the role of the State, subordination of the bourgeoisie to the interests of the nation, militant anti-Americanism). Their general preoccupation, in theory, was to remedy social and economic injustice in Mexico, to reduce inequality and poverty, but unlike Cosío Villegas (who had spent years studying agricultural economy) or Frank Tannenbaum (who had come to know the country close to the soil), the new intellectual generation was entirely urban, lacked any practical experience in peasant or working-class life, and showed a tendency, constantly increasing, to conceptualize the political issues of the country and the world in ideological terms.

Carlos Fuentes was, without doubt, the Mexican prototype of the committed intellectual—a man who most represented that moment in time. In 1958 (when he was thirty years old), he had published his powerful first novel, *La región mas transparente* ("Where the Air Is Clear"—the words drawn from a famous Alfonso Reyes poem describing a Mexico City long vanished: "Traveler, you have reached the clearest region of the air"). The novel was a critical kaleidoscope of the City of Mexico and its different social strata. From then on, he was continually active and prominent, as a writer, an editor, and—after his visit to Havana in 1960—an attentive observer and partisan of the Cuban Revolution, interpreting the problems of Mexico in the light of that new experience. His celebrated second novel—*The Death of Artemio Cruz*—would depict (through its onetime revolutionary protagonist) the corruption and decline of the Mexican Revolution. Fuentes saw Cuba as not only the political but the historical nemesis of the United States, a living argument for socialism as the wave of the future in contrast with "the Lockean individualism . . . the protestant spirit" represented by the United States.

After the failed invasion at the Bay of Pigs, he would write a brief, intense piece celebrating the construction of socialism in Cuba and presenting the Cuban Revolution in terms more Hegelian than Marxist, as a poetic moment in the ascent of the peoples of the world toward freedom:

> The march of the Roman slaves to Calabria, the Way of the Cross, the winter camp at Valley Forge, the march of the men of France along all the roads of France to form their revolutionary federations, the adventure of the Army of the Andes, the assault on the Winter Palace in Petrograd, the charge of the Zapatista guerrillas, the defense of the Bay of Pigs by the Cuban militias, they are not the history of a group of men in one fixed place and time: they are the history of all men in an embrace beyond centuries and frontiers.[12]

Seen at a distance, the commitment of Carlos Fuentes and his generation to the Cuban Revolution seems excessive. But it was not so at that moment. Almost the entire intellectual world of Latin America shared a fervor—or at least a sympathy—for the Cuban Revolution. Solidarity with Cuba was felt everywhere, Yankee imperialism was repudiated, the search was on for the right road to socialism or, at least, a mixed economy totally directed by the State.

As time passed and it became clear that the Caribbean David had fallen into the hands of the Russian Goliath, the intellectual fields of the Mexican left began to divide behind new fences. The most radical, linked to the Communist Party or other sects of the revolutionary left, saw the Mexican Revolution as a cadaver. It had been the last bourgeois revolution in history, a sort of eccentric and very late French Revolution with nothing to offer the socialist future except its record of sacrifice. The economic backwardness of Mexico was no longer a valid pretext for delaying the achievement of socialism. Gradualist solutions were consigned to the garbage heap. Had not Mao Tse-tung proved that a peasant society could leap past the stages prescribed by Marx and arrive at socialism via the rapid road?

The intellectuals of *Medio Siglo* followed a different pattern. The sympathies they felt for Fidel Castro would take years, and sometimes decades, to lessen or disappear. But their political positions would shift toward a moderate left not very different from what Vicente Lombardo Toledano had advocated. They believed that the Mexican Revolution was close to death but was not yet a corpse. It could be renewed. To reinvigorate its memory, Fuentes had written his impressive report on the death of Rubén Jaramillo, heir to the heritage of Zapata. And for the same reason he and his friends moved closer to Lázaro Cárdenas.

Fuentes accompanied Cárdenas on one of the tours the general was in the habit of making through the Mexican countryside. He testified to the "tide of peasants" who venerated Cárdenas and enthusiastically accepted the credo of this man already bathed in legend: the need for a total agrarian reform, a demand for union democracy, the rejection of private enterprise, the belief in a corporate state that would integrate the masses as component social armies, invariably a defense of Cuba and a resentment, bordering on hatred, toward the United States, including a rejection of Kennedy's "Alliance for Progress."

Without reading Marx or Sartre, Lázaro Cárdenas understood the need to form an alliance with the young intellectuals. Together with the general, they would pressure the Mexican system toward the left. In the sixties, these younger intellectuals had more personal space within which to act. The university, the press, and the growing middle-class public of readers allowed them to live with some degree of independence from the system.

The famous statement attributed to Porfirio Díaz: "Poor Mexico, so far from God, so near the United States," deserved the addition, in 1959, of "and so near to Cuba." The government had to find a reasonable balance between the American Goliath and the Cuban David supported by the Russian Goliath. There were no easy answers. To unreservedly favor Cuba would result in unbearable tension with the United States and with Mexican business interests and the Church; to unreservedly favor the United States would result in unbearable tension with the revolutionary government of Cuba, and with the intellectuals and other sectors of the Mexican left who might become more radical. The system entrusted this delicate mission to its Minister of Foreign Relations, Manuel Tello, his Assistant Minister, the poet José Gorostiza, and, along with them, an impressive group of career diplomats formed in a tradition that reached back to Díaz and Juárez (and even earlier)—to those years when the proverb was minted, "A Texan can win when he fights with a Mexican, but he is lost if he parleys with him."

In the case of Cuba, the government of Mexico spiritedly defended the traditional principles of "nonintervention in the internal affairs of countries" and "respect for the self-determination of nations." If it had to swallow—at a 1960 meeting of the Organization of American States (OAS)—the *Declaración de San José,* which condemned "the intervention or threat of intervention from outside the continent in the affairs of the American republics" (the USSR had just announced that it would support the Cuban Revolution), the Mexican delegation added a reservation to its vote, emphasizing that the declaration "should not be interpreted as a condemnation of the Cuban people whose aspirations for social and economic betterment can count on the fullest support of the Mexican people." When President Oswaldo Dorticós of Cuba visited Mexico in June 1960, López Mateos welcomed him with words comparing (and praising) the Mexican and Cuban revolutions. But when a senator used the occasion to attack the United States, Manuel Tello calmed the U.S. State Department by explaining that foreign policy in Mexico "in accordance with our Constitution" was the province of the President.[43]

Meanwhile, during the Eisenhower era, except for its objections to the CIA-sponsored invasion of Guatemala, Mexico kept its bridges open to the United States through mutual presidential visits and almost complete support for U.S. foreign policy—with the exception of Cuba. (Mexico, for instance, voted in the UN against recognition of China.) When Kennedy became president, the relations grew even closer, not only because of the Alliance for Progress but because Cuba had now defined itself as communist, a fact that, far from aggravating the problem for Mexico, actually opened the door to a solution.

After the failed invasion at the Bay of Pigs, the Mexican government began to harass political demonstrations of solidarity with Cuba, impose severe police controls on travel to and from the island, confiscate pro-Cuban propaganda, and display a tolerant attitude toward business and church groups that were waging an intense campaign to discredit the Cuban Revolution. At a 1962 meeting of the OAS at Punta del Este, Uruguay, Manuel Tello declared, "There undoubtedly appears to be an incompatibility between belonging to the Organization of

American States and professing a Marxist-Leninist government, as there would be with a profession of absolute monarchy,"[44] but Mexico nevertheless refused to vote for the United States–sponsored resolution to expel Cuba and invoked a procedural subtlety: The OAS charter did not provide for the possibility of expelling a member nation. And at the time of the October missile crisis, Mexico joined the unanimous resolution of the OAS council, calling for the Soviet missiles to be withdrawn from Cuba, but once again added a reservation—the resolution should not be used to justify an invasion.

Kennedy's government rapidly recognized the agile balancing act involved in Mexico's foreign policy. Mexico could not join the pan-American condemnation of Cuba for two reasons. One of them was the weight of Mexican history. Juárez had said, "Peace is respect for the rights of others," and the idea of national sovereignty—based on the need to protect the nation against the menace from the north—had long been a national obsession. The other limitation was more practical and immediate. Cuba was beginning to export its revolution. Venezuela had already complained of Cuban aid to rebel groups. The idea was to create "foci" of revolution, "one, two, a thousand Vietnams" in Latin America, and the dream would carry Che Guevara to Bolivia and his death. The political environment in Mexico seemed potentially fertile for such a "Vietnam." There was still the open wound of the railway workers' strike, and the state universities were full of impatient young men who talked of "going to the mountains." In the rugged, impoverished sierra of Guerrero, guerrilla "foci" had already begun to appear. The government of Mexico had to buy itself security, and the price after all was not excessive. Toward Cuba, Mexico had to be the exception to the rule. And it could build its political choice on a genuine moral issue—the sincere initial sympathy for a revolution that had appeared to be the daughter of liberty and justice and not of communism, born of Martí, not of Marx.

On his visit to Mexico in June 1962, Kennedy asserted that the fundamental aims of the Mexican Revolution coincided with his Alliance for Progress. A year later, in a ceremony deliriously celebrated by the Mexican press, the United States—garnering international praise—returned to Mexico a tiny stretch of frontier territory, El Chamizal, which had been a bone of contention since the War of 1847 (and had cropped up again as an issue in 1911). Although 177 square kilometers of desert did not add a great deal to the national wealth, it was a recovery torn from the belly of the giant. Perhaps the Mexican Revolution was a corpse or else on its deathbed, but it was still winning battles.

As for Fidel Castro, he was always grateful to the Mexican government for its stance on Cuba. In 1964 he declared: "To Mexico, to the government of Mexico that has maintained the strongest position, we can say that you inspire us with respect . . . and that we are prepared to commit ourselves to maintaining a policy . . . of not meddling with the internal affairs of any country."[45]

The old proverb could well be amplified. Not only the Texans (supply "the Americans") but also the Cubans were losers "if they parleyed with the Mexicans."

To repel the threat of Cárdenas and his MLN, López Mateos would choose to imitate Cárdenas and revive the memory of earlier moments in the Revolution. He too could expropriate a great industry now in the hands of foreigners—electric power, controlled by the American Bond and Share Company and the Belgian-owned Compañia de Luz y Fuerza del Centro. The difference between this situation and 1938 was that—now in 1960—both foreign companies wanted to sell. After the union disturbances of 1959, the government had forbidden them to raise their prices, which they had been expecting to do as usual. The President of the World Bank, Eugene Black, protested in person to López Mateos and only succeeded in being literally thrown out of the President's office. (For some years thereafter, the World Bank shunned Mexico.)

On September 27, 1960, the government—having arranged the terms of purchase—peacefully took over the electricity industry with all its personnel and installations. Economically, the operation was much cleaner and cheaper than had been the case with the oil. Politically it lacked the drama of 1938, but it paid the government substantial dividends—they had recovered natural resources and so could wave that most popular of Cardenista banners.[46]

Another memory to be revived was one close to the President's heart—the educational impetus of Vasconcelos. He himself had edited books of popular culture during the thirties. In February 1959 he created the National Commission for Free Textbooks, under the direction of the great narrator of the Revolution, Martín Luis Guzmán. It would distribute millions of required texts in Mexican primary schools. And then López Mateos launched an "eleven-year plan" to raise the level of education in the country as well as following the path of his master Vasconcelos by restoring the practice of free student breakfasts for primary-school pupils. As another echo of the broad cultural enthusiasm of the twenties, he planned and inaugurated a series of museums honoring the history of Mexico: the Museum of Natural History, of the City of Mexico, of the Viceroyalty in the former monastery of Tepotzotlán, and the magnificent Museum of Anthropology, presided over by the immense bulk of a statue of Tlaloc, the Aztec god of rain, and bearing on its wall an appeal to Mexicans "to look with pride into the mirror of your past."

To compete with Cárdenas in agrarian reform, López Mateos distributed 16 million hectares, more than what Ávila Camacho, Alemán, and Ruiz Cortines all together had done, and surpassed only by the 17.9 million distributed by Cárdenas himself. By 1964, 25 percent of the national territory had been expropriated and shared out, But there was a problem. If the country consisted of 200 million hectares and only 15 percent were really suitable for agriculture, what kind of land had López Mateos distributed? Mountains and crags, said his critics.[47]

But no one, not even the recalcitrant General Cárdenas, could deny many of the public achievements of the regime. Typhus, smallpox, and yellow fever were eradicated, malaria greatly reduced. And in 1961 a new Medical Center, of international quality, was inaugurated. The name of Mexico continued to gain more luster on the international economic (and political) markets. The years of "order, peace, and progress" since 1940 were beginning to pay off, but the process of reestablishing Mexico's credit was accelerated by the active diplomacy of Adolfo López Mateos.

His old passion for wandering the Mexican countryside found a creative outlet during his presidency. He would be Mexico's ambassador to the world. If Cárdenas had not traveled until he was sixty-two, López Mateos on the contrary would carry his cordial smile and traditional Mexican courtesy along all the roads of the planet. He went to the United States and Canada, Latin America, Europe, and Asia. The Cuban Missile Crisis caught him crossing the Pacific, but he had faith that his country was safe, thanks to the efficient services of his Minister of the Interior, "Gustavito" Díaz Ordaz.

He would travel accompanied by much of his cabinet and invited representatives of the Mexican business community. The public grew accustomed to his absences and began to call him López Paseos ("López Promenades") or "the Gulf [*golfo* also means "lazy man"] of Mexico," but the appellations were all good-natured. After one of his long trips, as if he were returning from the battle of Thermopylae, he announced—with his oratorical voice and gestures from the balcony of the National Palace—to a crowd that had gathered out of interest as well as those who had been bused in, "Mexicans! I take my place again in the Fatherland, with the same flag . . . !"

He had regained El Chamizal for Mexico, he had graced the nation's honor with his support before international organizations for nuclear disarmament, he had won—for Mexico—designation as the site of the Olympic Games to be held in October 1968. He represented the new enthusiasm among the Mexican middle classes—to go out and see the world. And also to receive great statesmen as guests of the nation. The apotheosis of all this was the visit to Mexico, in 1964, of Charles de Gaulle, whose words proved that the country had regained a modest but visible and respected place among nations: "France admires Mexico for its political stability, economic development, and social progress." There had to be something mysterious and extraordinary to this "Institutional Revolution," De Gaulle would say to the frantic ovations of a public that adored great leaders.[48]

"Mexico is not a neutral country. It is an independent country," proclaimed López Mateos, and the moderate left could not dismiss these words as mere populist rhetoric. The regime had shown its relative independence in its political stance toward Cuba and its closeness to the emerging countries of the Third World. (López Mateos admired Tito.) Perhaps for this reason, and in view of the tenuous renewal of positive action that the regime seemed to be promising, General Cárdenas made a decision based on his acute political instincts and sounded the retreat. Forced to choose between the "institutionality" of a "Revolution" that he himself had constructed and the opposition inspired by a socialist revolution that he admired but was foreign, Cárdenas chose the institution.

Much to the disappointment of many members of the MLN, he accepted the post—which López Mateos had been insistently offering him—of chief executive for the Balsas River Committee, a project somewhat similar to the American Tennessee Valley Authority. When his comrade Heriberto Jara reproved him for accepting the position while all the political prisoners were still behind bars, Cárdenas answered that he could use the post to successfully protect the iron deposits of Las Truchas in Michoacán from the greed of foreigners. And besides, he felt that the MLN had served the purpose of displacing the Revolution toward the left but that it had not solidified as a group that could unite the separate "authentically revolutionary" forces of the country. It was time to leave it to its fate.

And yet his cooperation with the regime was always conditional and ambiguous. He would again request the release of political prisoners—in a letter to the President toward the end of the six years. Some prisoners including Siqueiros were freed to mark the conclusion of the presidential term, but Cárdenas never received an answer to his letter. By then he had committed himself to supporting Gustavo Díaz Ordaz for the succession. He knew that backing the official candidate was a golden rule in the regime of "institutionality." This time he did not haggle. Besides, the campaign declarations of the candidate had reassured him: "The agricultural problem is the most pressingly serious on the horizon of Mexico. . . . The mere distribution of the large landholdings that still remain— openly or concealed—will not lighten the anguish of many millions of Mexicans."[49]

Cárdenas's support for the candidacy of Díaz Ordaz was the kiss of death for the MLN (which vanished within a few years) and caused serious problems for the Central Campesina Independiente (Independent Peasant Center), which had also been founded in the shade of "the most gallant tree of Mexico." And so, while the PAN was taking advantage of a new political reform that created "party deputies" and enabled it to increase its representation in the Lower House to twenty seats, the political left dissolved again into its separate components. Only one powerful sector—the cultural or intellectual left—remained strongly afloat. And in the universities, the students, in all their variety, still attended to the politics of Mexico and the world with an alertness that had not been theirs for decades.

Though the regime (at the very beginning of López Mateos's term) had made free and fierce use of the bludgeon on insurgent unions, it was somehow not enough to convince the most powerful private business groups that the system would never permit the Communists to capture the labor movement. Deceived by the Revolutionary rhetoric of the government and its praises for the Cuban Revolution, big business was suspicious of statist, socialist, and (their ultimate fear) Communist proclivities in the President of the Republic. López Mateos, on his side, was prone to using verbal formulas that seemed designed to set off alarms throughout the business community. In one speech, he spoke of his regime as a government "of the far left within the Constitution," a phrase that was no more than rhetorical sophistry floated against the pressure from the left to discredit the Mexican Revolution and also perhaps a bit of windy compensation for the recent jailings of union leaders. But the business community read the words as a clear sign of a change in the air: Yesterday Havana, tomorrow Mexico City.

Economic consequences followed. Investments went down, capital left the country, tourism declined. It was not a divorce but a distancing. For a time the newspapers were filled with paid advertisements by various business groups warning the public on the dangers of "the state taking control of the economy," which would be "totally harmful and unsettling for Mexico."

Finally, on November 24, 1960, a paid announcement appeared in the newspapers that summed up the attitude of the business community. It was signed by all the great organizations of Mexican private initiative—

industrialists, bankers, large wholesale and retail businesses. It was titled, "Which Way Are We Going, Señor Presidente?" The statement registered "positive unease" about the direction of the government. The businessmen conceded that recent nationalizations by the government were "fully justified from a political point of view," but would the process stop with the electrical industry or would it move into new areas like steel, sugar, transportation? The successful experience of various industrialized countries—they argued—and the recent disasters suffered by Peronism in Argentina argued that the government should not buy but sell enterprises. The businessmen reiterated their commitment to "the traditional thesis of collaboration with the government" but at the same time (courteously of course) asked for a clear response to the question formulated in their title.[50]

And the government answered at once—and with unusual amiability—through its talented Minister of Finance, Antonio Ortiz Mena. The objective of the government, he explained, was "to favor the economic development of the country without competing against private initiative." The government had taken over some enterprises, but the purpose had been to avoid bankruptcies and the loss of sources of employment and, of course, business. The words and the tone of the answer calmed the minds of the private business community. But there still remained the issue of Cuba. Working together with representatives of the Church, big business pressured the government to unequivocally condemn communism and the "tyrannical and interventionist" government of Fidel Castro at the OAS meeting in Punta del Este. And there, of course, Minister of Foreign Relations Manuel Tello pulled an ace out of his sleeve, condemning communism but declining to support the expulsion of Cuba.

The right signals had now been given. It had been a matter, more than anything, of changing the way the business community *perceived* the words and acts of the government. On a golf course, Antonio Carrillo Flores, López Mateos's ambassador to the United States (who had been Finance Minister under Ruiz Cortines), had this exchange with his friend, Juan Sánchez Navarro, one of the most important business leaders in the country and the man who had actually composed the "Which Way" statement:

> CARRILLO FLORES: But don't you people have enough freedom?
> SÁNCHEZ NAVARRO: Yes, but the government threatens it.
> CARRILLO FLORES: You should learn to distinguish form from content.[51]

The businessmen began to distinguish the backdrop of Adam Smith economics from the foreground of socialist phrases. Relations between the government and private business would become as cordial as they had been in the age of Alemán. And businessmen came to appreciate the economic measures taken by the government and applied by the brilliant Ortiz Mena—the development of the borders, of tourism, of a merchant marine, the industrialization of the sisal plantations in Yucatán, the construction of new petrochemical plants, the development of the automobile industry. Good public administration maintained the external debt at reasonable levels; the autonomy of the Bank of Mexico was scrupulously respected. The thesis that only private initiative could productively manage enterprises was challenged by the facts and by the numbers, achieved by a particularly efficient and capable group of government bureaucrats whose success depended, above all, on the old Porfirian tradition, respected by all the Revolutionary governments: autonomy accorded to the finance and Treasury areas of government, which allowed independent and expert action by the succession of heirs to Porfirio Díaz's master of finances, José Ives Limantour.

The entrepreneurs would come not only to reaffirm their political dependence on President López Mateos (and his successors to come) but to see it as a blessing. The politicians would take care of politics. The businessmen would attend to business. After all, said one prominent banker, "We're all in the same boat. We are all Revolutionaries."

The Church as well had reacted to López Mateos's rhetoric on socialism and Cuba and been pleased to discover that it would not extend beyond the bounds of speechmaking. A nationwide campaign had been initiated—with signs sprouting in the windows of houses and cars signaling CRISTIANISMO SÍ! COMUNISMO NO! Steadily rebuilding since the era of Ávila Camacho, the Church—along with its expanding charity and undisturbed pastoral activities—had become the leading proprietor of private education in the country. And now that the government—with enough indirect words and direct actions—had made it quite clear that Mexico City would *not* become Havana, the Church could, more and more, feel itself at ease with the system.

But inside the Church itself, something new was building. After the anticlerical measures of Calles and the consequent Cristiada, institutions had been set up abroad to protect and educate the future Mexican priesthood. The most important seminaries were Montezuma College in the United States and the Pío Latino Seminary in Rome (which also taught candidates for the priesthood from other Latin American countries). And while studying in Rome during the early sixties, a number of these young Mexican priests would be caught up in the winds of change initiated by Pope John XXIII and the Second Vatican Council. They would take the *opción preferencial por los pobres* ("the decision to commit themselves to the poor") and would return to Mexico as proponents of the new Liberation Theology. The Jesuits would soon give up educating the wealthy (leaving that enterprise to groups like the Marian Fathers) and turn their attention to "the Church of the poor." Within the hierarchy of the Church, Sergio Méndez Arceo, the bishop of Cuernavaca, became the leader of the new left-wing theology. And at an activist, grass-roots level of priests who would join or lead movements among the peasants, a bold liberalism began to develop in parts of the north and a more socialist and radical perspective among young priests in the impoverished and far more Indian south, especially in the state of Chiapas.

At the close of his term, the balance for López Mateos was overwhelmingly positive. The system had passed through its first great test and had maintained the subordination—complete or partial—of virtually all groups within the society. Only in the new zones of the left (in culture, the universities, and those within the Church committed to "the cause of the poor") was a movement of opposition under way. It would grow to be as radical as the Cristeros or the Sinarquistas but inspired this time not by the slogan of *Viva Cristo Rey!* but by the triumph of the Cuban Revolution.

During the following three decades, the workers would not rise against the system again. They would instead defend it at moments of authentic social insubordination from the middle class. The CTM would continue to strengthen its hegemony over labor. The Communists would not regain any real influence over the unions. Demetrio Vallejo and Valentín Campa would spend ten years in prison, years they would never see again, Campa thinking of "the road" and Vallejo of "the lunchbox." And the peasants would not find another leader with the moral dimensions of Rubén Jaramillo.

But the economy was growing and the personality of the President was radiant. Outside of the left in general—who made their negative judgments about him—and the worker or peasant groups directly harmed by his regime, López Mateos was a very popular man. Abroad they spoke of the "Mexican miracle." Within the country, a joke spread far and wide: "When they take orders from López Mateos on the day's agenda, they ask, 'What do we do today, a trip or a woman?'" (The last two nouns take on alliterative life in Spanish: *viaje o vieja?*) The joke was often very close to the truth. Even when he did not happen to be traveling the world, he could be seen speeding around Mexico City in his brand-new Maserati and, on parade routes, knocking over a barrier to kiss a pretty girl on the cheek. As for his many love affairs, López

Mateos knew that the people appreciated them. They showed that "our President is macho"; and when he finally decided to take a new wife, it would be the young and beautiful Angelina Gutiérrez Sadurní, a daughter of the Mexican upper class.

But behind the glittering facade, he had been a sick man for years, plagued with ferocious migraines, feeling himself ever closer to total physical collapse. And the agony of the body had been compounded for him by the tension in his mind, the inner loneliness, the remoteness that he really felt from the responsibilities of power. Shortly before he left office, he called in an old and close friend, Víctor Manuel Villegas. He had arranged a scholarship for him, with the pay of an ambassador, so that Villegas could go to Europe and earn the doctorate in fine arts that he had always desired. López Mateos received him in his dressing gown, showed him his recent and sudden baldness and the way the robe hung on him, and said to his friend, "I'm finished." Villegas, overjoyed at the great favor, tried to cheer him up, telling him that he would soon be leaving the presidency and would be able to share with his friends the pleasures of life that were his due. But the President answered him: "I no longer have any friends, compadre, and you are no longer my friend. . . . You are not because I am not the man you used to know. I'm a man deformed. For six years I've almost never had a dialogue with anyone."[52]

He had always felt himself alien to power, alien to himself when in power. And for that reason, he had lived in flight. Now that the doors of freedom were opening, there was no way he could again become what he had been. And he also knew that he was a sick man. Villegas went off without saying another word, and he never saw his old friend again.

A year after López Mateos left office—acclaimed by enthusiastic crowds—he had another attack of terrible, crippling pains in his head. This time it was diagnosed as a cerebral aneurysm. People from all social classes came to the hospital, stunned and worried about the fate of the popular ex-President. An operation revealed that he had seven aneurysms—swollen and tumorous blood vessels in the brain. The doctors saw no hope for anything but a long and painful death.

He would gradually lose control of his body. An emergency tracheotomy would cost the great orator his power of speech. His consciousness leaked away, and for two years he lived in a vegetative state until, submerged in silence and pain, he died on September 22, 1969, ignorant of the silence and pain of his country, whose destiny he himself had placed in the stern hands of Gustavo Díaz Ordaz.

21

GUSTAVO DÍAZ ORDAZ

The Advocate of Order

The father of Gustavo Díaz Ordaz was a man whose work, for the best years of his life, had been to command. During the last decade of the Porfirian era, Ramón Díaz Ordaz was the appointed Political Chief (*Jefe Político*) over various towns in his native state of Oaxaca, and at the beginning of 1911, in the dying moments of the Díaz regime, he was the political and police administrator of San Andrés Chalchicomula in the neighboring state of Puebla. Madero's triumph swept away his position along with most of the Porfirian bureaucracy. He returned to the state of Oaxaca with his family. Gustavo, the second of his four children, had just been born, on March 12, 1911, to Ramón's wife, Sabina Bolaños Cacho, who was a teacher and a very stern and pious woman.

Don Ramón then held various jobs. He was a bookkeeper at one point, and at another the administrator of a hacienda. Money was always a problem, and the family had to uproot itself more than once. They were living in the state capital of Oaxaca in 1924 when an earthquake struck and ruined their house. For a month, the family had to camp out in the garden. In 1926 Gustavo began to attend the Institute of Arts and Sciences of Oaxaca, where Benito Juárez and Porfirio Díaz had studied. His fellow students would remember him as a serious, methodical boy with a good memory and a certain defensive arrogance.

He had a dry, cutting politeness; and though he did not look for trouble, he would not run away from it either. His teachers were almost all disciplinarians. One of them—his *maestro* of zoology and botany, Agustín Reyes—was blind, and he would put Gustavo in charge of keeping unruly students in order.[1]

It was the heyday of Vasconcelos, the philosopher from Oaxaca, the "learned Madero," but Gustavo did not seem to be oriented toward the politics of opposition. One of the women who went to school with him would remember that "he did not like things to be loose," by which she meant not only politicians but his attitude toward the fiestas that were traditionally celebrated at the Institute. His duty was to study. His relaxation was basketball, where he was a good shooter and played on a team that made it to the national tournament in Mexico City.[2]

The family went through some hard times. A generation back there had been money and property, but it had all gone with the Revolution. At one point, they had to endure the shame of being thrown out onto the street for nonpayment of rent. A maternal uncle—a prominent state official—grudgingly took them in. When the uncle had important visitors, Don Ramón and his family had to huddle into a corner of the kitchen, so as not to be seen or heard.[3]

Gustavo dressed in coarse twilled linen—cheap material but always clean and pressed. He was poor but respectable. In his studies lay his future, and if he could not buy books, he borrowed them and read them at night, sometimes by the light of the streetlamps on the main square.

His elder brother, Ramón junior, returned from Spain—where, for a time, he had studied in a Paulist seminary—in order to help the family, and he got a job teaching Latin at the Institute. The students made fun of his measured mode of speech, his Castilian pronunciation, but especially of his looks. He had a big mouth and huge teeth, and they nicknamed him *Diente Frío* ("Tooth Out in the Cold"). One day a misbehaving student was expelled from the classroom but lingered in the doorway, thrusting out his jaw to mock Ramón Díaz Ordaz. Gustavo stood up and rushed the boy, hammering him with punches.[4]

It would have been a normal enough action in any circumstances—a brother enraged over an insult to a brother. But in the Díaz Ordaz family, it was something more. Gustavo was not only defending his brother, he was defending himself. "My grandmother," said Díaz Ordaz's eldest son, "discriminated against my father and favored his brother Ernesto." Gustavo too had huge protruding teeth and was skinny and bony. Ever since he was a little boy, the sense of being ugly had tortured him. His mother would freely say to anyone, "But what an ugly son I have!" It would take its toll on his life.[5]

While Gustavo was compensating for the instability of his family fortunes and his own lack of good looks with determined attention to his studies, a killer earthquake struck Oaxaca, on January 14, 1931. As if nature were trying to persuade him that its only law was disorder, chaos, and chance, his house was once again demolished. Oaxaca was strewn with dead. People fled the city. (In one year the population shrank from 29,000 to 12,000.) People sold their property at pitiful prices. Among them were the Díaz Ordaz family who moved, in two stages, to the city of Puebla. Behind them they left Oaxaca for good, and the memory of their ancestor, José María Díaz Ordaz, who had been a Liberal governor of the state and "martyr of the *patria*," killed in battle toward the end of the War of the Reform. Before them lay a deeply uncertain future.[6]

The city of Puebla, in the early thirties, was still profoundly marked by its history of religious devotion. In the nineteenth century, the rebellion against the Liberal President Comonfort—which lit the fuse for the War of the Reform—had been organized in that city. It was out of Puebla that the Conservatives had marched toward the capital. From that moment, the tension between liberals and conservatives in Puebla could only be compared to the situation in the other great clerical center of the Mexican provinces: Morelia (once Valladolid). Both cities were breeding grounds for extremists.

The state reflected its capital. It was a place where it had been difficult to introduce revolutionary reforms. In 1924 Vicente Lombardo Toledano himself had been its young governor, and he had tried to set precedents for the rights of workers in an area of high labor density. (Puebla had been the center for Mexico's textile factories since the middle of the nineteenth century.) The manufacturers had protested to Obregón, and the general had immediately removed the intellectual governor. General Leónides Andrew Almazán, when he was himself governor, had run into the same problem with local hacendados because of his attempts at agrarian reform. He was forced to resign by the government of Calles. When Cárdenas became President, the progressive forces in the state thought their opportunity had come, but General Maximino Ávila Camacho could not let the opportunity pass to become absolute master of the state of Puebla. In one of those acts of political balancing that Cárdenas performed (and which still remain difficult to comprehend), Maximino's violence-racked electoral "victory" over the progressive candidate, Profesor Gilberto Bosques, was allowed to stand. Bosques and the leaders of the teachers who had supported him were forced to leave the state. While the Agrarian Reform and new modes of labor relations were taking root elsewhere in the country, none of it happened in Puebla. Maximino would construct roads and buildings but permitted no bridges to the workers. Puebla "fell into one of the worst dictatorships imaginable . . . it seemed suspended in time."[7]

For the Díaz Ordaz family now installed in the state capital, just making a living continued to be the principal problem. Don Ramón found work as a bookkeeper, then as an inspector of markets. Gustavo had to take jobs to pay for his studies. In 1932 he became an office boy in the state Ministry of the Interior. He would later fill other positions connected with courts and law enforcement. For a few years he was an "agent of the court" (*diligenciario*). His function was to execute judicial orders, from simple ones like summonses to more complicated actions like evictions for failure to pay rent, which might involve scenes of high drama—court agents (supported by the police) breaking locks, throwing furniture and belongings of the now homeless people into the street, and installing new locks everywhere to prevent them from returning.

He would later serve as a prosecuting attorney, work that he began while he was still a *pasante* in law (a college student who has fulfilled all his requirements but still has to write and deliver a thesis to earn a degree) and then as a First-Degree Penal Judge, dispensing justice on state-level offenses ranging from theft to homicide with mitigating circumstances.[8]

Already equipped with this respectable practical record in peremptory execution of the law and prosecution of crimes and delinquents, Gustavo became a lawyer at the age of twenty-five. It was customary to celebrate the degree with a party, but he was too poor to do so and instead took his friends out for a drink. There were guitars, and the guest of honor—with his fine speaking and singing voice—stood up and sang. He had just enough money to pay the bill when suddenly one of his professors ordered another drink—a *sangría*. To pay up, Díaz Ordaz had to borrow two pesos from a friend.[9]

But in the meantime, that voice perhaps—"grave, well modulated, cultivated with a certain amount of painstaking care"—had helped the ugly duckling win a beautiful bride, Lupita Borja, the daughter of a rich and respected lawyer. Now married (after a five-year courtship), Díaz Ordaz, when renting their first apartment, insisted on paying for one month in advance. He was told it was unnecessary but responded, "Let me do it. It puts me more at ease." After all, he knew what it was like to throw people out on the street (and to be dumped on the street himself). He had no intention of taking any risks.[10]

He would work briefly in a lawyer's office, but he gravitated, like his father, toward public service and the practice of power. And then the cacique of the state of Puebla, Maximino Ávila Camacho himself, noticed him in 1938 and made him president of the Central Council of Conciliation and Arbitration. "Maximino saw that he wore spurs" (*espolones*—not riding spurs but the weapons of a fighting cock), said Gonzalo Bautista O'Farrill, Maximino's chief tactical advisor, a professional politician who was not the cacique's servile employee but the man who calmed, oriented, and—only when nothing more could be done—threw up his hands and stepped aside while Maximino reached for his guns.[11] There was an explosive labor situation at the time in the state of Puebla, expecially in Atlixco, which had become a battlefield between the left-wing unionists of Lombardo's FROC and the gunmen of Morones's CROM. Díaz Ordaz hesitated to accept the position because of his lack of experience in labor issues, but no one said no to Maximino. "Learn fast," said the cacique, and his disciple—always the good student—did just that.[12]

Díaz Ordaz left his mark on the Council. He put order into the chaotic proceedings and showed an even hand, not leaning toward one group or the other and willing to hand down judgments against the employers as well. "It is not pressure that will support your case but the law," he would lecture to the workers of Atlixco, where the army often had to intervene between the warring labor factions.[13]

His reputation for competence and loyalty carried him up the rungs of power. His next job would be Vice-Rector of the University of Puebla. He really became the official in charge, since the Rector, a prominent doctor, spent most of his time attending to his practice in Mexico City. As Vice-Rector, Díaz Ordaz's slogan was "To Help Anyone Who Wants to Study." He ruled the university with a firm hand especially when, within the usual run of student conflicts and disturbances, he believed that he had detected "external forces."[14]

At the age of twenty-eight, in 1939, he became the highest judicial official in the state, the President of the Superior Court, and he would later move on to a state administrative post, as Secretary of Government. He had become the third-most powerful politician in the state of Puebla, two rungs below the cacique Maximino, one rung below the governor, who was then Gonzalo Bautista. Gustavo functioned as a channel between the two. And he kept them both happy. Maximino liked him for his firm and heavy hand; Bautista admired his intelligence, his memory, his punctuality, and his capacity for work. Díaz Ordaz learned from both of them. He performed so well over a two-year period that Bautista predicted he would someday reach that place where everyone longed to be—the Presidential Chair. When Mexico entered the Second World War, Bautista began to spend much of his time inspecting military bases in Texas. Once again, Gustavo Díaz Ordaz filled a top position while, in name, being only the assistant. One day, signs with a patriotic slogan supporting the war effort appeared on the walls of the city:

FOR YOUR SONS, YOUR LIFE.
FOR THE PATRIA, YOUR SONS.

The author was the Secretary of Government (and de facto governor of Puebla), Díaz Ordaz.[15]

In 1942 Don Maximino sent his fighting cock to compete in the national pit. He chose him to join the Lower House of the Legislature, as representative from the first district of the state of Puebla. The chief of the Puebla CTM, which controlled the district, spent election day—when they had nothing to do but wait for Díaz Ordaz's election to be declared—in the office of the Minister of Government, along with another politician and Díaz Ordaz himself. "It was a Sunday," he recalled, "and . . . we were just waiting. How to pass the time? Gustavo noticed . . . a calendar on the wall from the National Lottery, with the picture of a flag on its staff, some clouds, the sky, nothing else. So we set ourselves to cutting it up, shuffling up all the pieces and then putting them together again. That's how we spent the whole afternoon."[16]

This was not a single, chance event. It was Díaz Ordaz's favorite pastime: assembling jigsaw puzzles. He had done it since he was a child. He would pick out pretty pictures (or have them bought for him) of horses running across fields, of old houses in the country. Then, in his own private woodworking shop, he would glue them to a sheet of plywood, turn the wood over and cut the board up into different shapes—little animals, fishes, whatever—using a saw and a small hammer. He would polish the pieces. Then he would mix them all up thoroughly and spend hours making order again out of the chaos. It was a natural thing for him to improvise a puzzle when he had nothing more to do than await the announcement of his election.[17]

He moved on to the Senate, though the new governor of Puebla, Carlos I. Betancourt, would have preferred one of his own followers. The presidential candidate—Miguel Alemán—stepped in and arranged an agreement. Díaz Ordaz would take the seat and Alfonso Moreira—a Betancourt man—would become the Alternate Senator (*Senador Suplente*). But once in the Senate, Moreira became a bit of a problem for Díaz Ordaz, arguing with him and trying to assert his influence in public. Until, one day, the senator confronted Moreira and warned him that if he kept this up, "he should understand the consequences." From then on, Moreira behaved impeccably, because it was well known that his opponent was a man who carried a gun.[18]

Nor was he afraid to draw it. Once he took his wife to a movie—*A Farewell to Arms*—and noticed that a man had followed them to the theater and was still behind them afterward, keeping at a distance, while they were walking back home. After dropping off his wife, he slipped out of his house again by a side door and circled around to come up on the man, who was standing at a corner and lighting up a cigarette. Before the stranger knew he was there, Díaz Ordaz had a gun on him and was relieving him of his pistol. The man had no choice but to confess that he was from the Puebla State Security Police (commanded at the time by the murderous Alfonso Vallejo Aillón). Díaz Ordaz took his identification papers as well and sent him off running. The next morning he appeared at Governor Betancourt's office. The governor tried to give him a traditional embrace ("My dear friend! How are you?"), but Díaz Ordaz laid the identification and the pistol on his desk and said, "Next time it happens I'll show the ID and the pistol on the floor of the Senate. . . . So, many thanks, and that's all I have to say."

Betancourt kept him from storming immediately out of his office by blaming the affair entirely on Vallejo Aillón. Díaz Ordaz stayed to drink a shot from the governor's special stock of the best imported whiskies but did not believe his excuses. He had, however, won something important: Betancourt's respect—and fear.[19]

During Díaz Ordaz's time in the Upper House, he became good friends with a man who, to all appearances, was his exact opposite—Adolfo López Mateos. They were an odd couple. The one—physically unattractive, unsociable, withdrawn, studious, meticulous, bristly, and sedentary, a man with few friends. While the other was a handsome and likeable person, open, bohemian, with pleasant manners and a superficial sensibility—to whom many men were friends, and many women. They obviously complemented each other, and in the Senate they both became conduits for the wishes of President Alemán. López Mateos did the traveling and the speechmaking; he was the legislative ambassador of the regime. Díaz Ordaz did the donkey work; he was the legislative lawyer of the system, but he also frequently mounted the podium to defend his point of view, always with energy, sometimes harshly.

And like all self-respecting senators, Díaz Ordaz was a soldier of the system, and specifically of the President. He called for Alemán to be given the Nobel Peace Prize and described the bitter labor conflict of the Santa Rosita mines as "an internal problem between two groups of workers." One of his strongest speeches was an attack on the Henriquista partisan Antonio Espinosa de los Monteros, who had dared to criticize—in Washington, D.C.—the corruption and monopoly of the PRI: "a Mexican not worthy of the name who plunges a traitorous knife into the very heart of the Fatherland . . . If there have ever been human rights in Mexico . . . it is now!"[20]

The natural next step would have been to move up to a governor's position, which—unlike the Senate—could involve real power. But in that direction he had been blocked. Even though ex-President Manuel Ávila Camacho liked Díaz Ordaz, Betancourt, when he finished his term as governor of Puebla, would pass on the job to a candidate even more in Ávila Camacho's good graces—one of his own brothers, General Rafael Ávila Camacho. It was a disappointment for the senator. It was clear that his ascent could now only take place on the federal level. With his usual dedicated labor, he threw himself into the campaign of Ruiz Cortines, during which he once addressed a crowd in terms unusual even for the authoritarian tone of Mexican political language. It was necessary to "support a political axis which will exercise power in an absolute manner, because we cannot believe that everyone can hold or obtain his maximum universe of power."[21] He was on his way now, on the road of national politics, ready to confront the jigsaw puzzle of the future.

When López Mateos went to Puebla—on behalf of President Ruiz Cortines—to offer "Gustavito" a job, he had the pleasure of trying out a new and flashy car, a Ford Crown Victoria that belonged to the man he had asked to accompany him, Agustín Arriaga Rivera, a young PRI leader, whose merits as company very much included the new car that López Mateos wanted to drive.

But Díaz Ordaz was not easily persuaded. He saw the job—Legal Director for the Ministry of the Interior—as "charity . . . a meager position," in which he was not willing to bury himself. Nor did he feel any affection for the Minister himself, Ángel Carvajal, with whom he had some old problems. But after all, López Mateos insisted to his friend, one does not say no to the President of the Republic. In the end, given the additional promise of a rapid ascent, Díaz Ordaz reluctantly accepted.[22] The ascent came faster than he had expected. On February 5, 1953, the Chief Administrative Officer of the Ministry was appointed to the Supreme Court and Díaz Ordaz took over. It was his entrance ticket to national politics.

One of his first assignments was to remove two governors from power. As he had once carried out domestic evictions, it was now his turn to perform political evictions, which he did with total efficiency.

When he returned from managing the removal of Governor Manuel Bartlett in Tabasco—where a poisonous spider bit him and kept him ill and surly for days—he is supposed to have submitted his resignation to Minister Ángel Carvajal. But even though personal relations between them were not good, Carvajal refused to accept it. And for good reason. He saw before him a man who could really work.[23]

In the mid-fifties, at the zenith of the system, day-to-day political activities could be boring. And besides, Ruiz Cortines did not easily delegate power. The situation favored a lax atmosphere at the Ministry of the Interior. Carvajal was said to spend much of his time at the movies, though always leaving a light on in his office to make it seem that he was busy late into the night. He was also said to have a poor relationship with his Assistant Minister, Fernando Román Lugo. Given all this, the business of the Ministry began to flow toward the office of the Chief Administrative Official.

In 1956, when labor trouble started with the dissident Teachers' Union, the Minister of Education threw the hot potato to the Interior Ministry, where it immediately landed and was handled on the desk of Gustavo Díaz Ordaz. In the crucial year of 1958, his friend "Adolfito" already "revealed" (*destapado*) as the next president, Díaz Ordaz began to act, more and more, as the man in charge of business in the Ministry of the Interior. Antonio Mena Brito remembered being called to his office and told, without even being asked to sit down, by Díaz Ordaz as he chewed on a pencil: "You're going to be a senator!"[24]

Once Adolfito was seated in the Chair, Díaz Ordaz officially took over the Ministry of the Interior with all the grace of a tidal wave. Immediately and "without legal quibbles," he fired the long-standing office manager, Acosta Ralú. He charged into all the issues sent his way—removing a governor in San Luis Potosí, installing a governor in Chiapas, confronting the strikes of a new wave of labor militancy, and most especially articulating his strategy to "grab the railroad workers by their *huevos*."

His project of erecting a barrier so that "no one should bother the President of the Republic" was virtually an act of psychological complicity with his friend Adolfito. The President had no interest in difficult matters or in details. He did not refuse to make decisions or to concern himself with problems in critical moments, but his preference was for tinsel and oratory, for the *viajes* and the *viejas* (the "trips" and the "women"). And so he needed somebody who would, as he said, "take care of the store." There was no one better at this than Gustavito. He assumed his responsibility with the same commitment as his "maestro" Ruiz Cortines, but he was completely devoid of the ex-President's sense of humor. He was not playing dominoes; he was firmly fitting jigsaw puzzles together.[25]

Throughout the years of López Mateos, in every situation of conflict, Díaz Ordaz was directly involved. If the President lost forty pounds, the "Thin Man" or "Goofy" (*Tribilín* in Mexico)—as many of his critics called him because of his facial, not psychological or sartorial, resemblance to Walt Disney's amiable and casually dressed dog—grew in stature. When there was union, peasant, student, or electoral repression, it was ultimately directed from Díaz Ordaz's office. No fly flew across the political map of Mexico without its direction being noted in the old Porfirian building on Avenida Bucareli.

But one of Díaz Ordaz's most important moments of power—with the greatest impact on his future career—was a foreign issue, the question of Mexico's reaction to the Cuban Missile Crisis. López Mateos was traveling in the Pacific, and United States ambassador Thomas Mann came to the Minister of the Interior with a question from the president of the United States. Where did Mexico stand on the crisis? Díaz Ordaz answered that Mexico had always supported Cuba's right to have defensive weapons but that these missiles were clearly offensive weapons, controlled by the Russians, which could threaten the United States or Mexico. Immediately afterward he called the President, who had just landed in Hawaii, and told him what he had said to the ambassador. President López Mateos then made the official announcement. It would not have looked right for Díaz Ordaz to do it. (Perhaps the Minister for Foreign Affairs might have taken the responsibility. But he was on the same junket, traveling through the Orient with López Mateos.)

Far from being bothered by Díaz Ordaz's decision, the President was relieved. He had suffered terrible migraines on this trip but now felt even more certain that a man was in charge back home who could protect the country from headaches. He would later confess that it was at that moment he decided that Gustavo Díaz Ordaz would be his successor. He had not only answered the U.S. ambassador admirably; he had also carried out López Mateos's instructions to prevent any disturbances and keep the country orderly and tranquil. In fewer than twenty-four hours, all the leaders of the left had been picked up and stored for the time being in prison. This was the man for the next "Revelation."[26]

Luis M. Farías, Information Director for the Ministry of the Interior, was a radio professional. He had been an announcer, and it was second nature for him to think about projecting a good image of his chief. On his very first day of work, he asked that an official photograph be taken. As he was instructing the photographer to try this or that angle,

Díaz Ordaz grumbled at him, saying that he "was not Coca-Cola" and adding: "I have no angle. I'm ugly, that's how I am. People should not only respect the Minister of the Interior; they should be a little afraid of him. . . . It's healthy for the country. I'm ugly enough so that people can be afraid of me."[27]

Farías reminded him of what Cromwell had said when a court painter was about to do his portrait: "Paint me as I am, warts and all." Díaz Ordaz liked the comparison. "Yes, I would rather be Cromwell than some pretty boy."

He did not want to be a pretty boy, but he would have preferred to be or feel himself to be less ugly. His mother had hammered it into him, and the resentment was strong enough for him to tell his children about the "discrimination" he had suffered at her hands. Farías himself, tired of looking for the right angle, had to admit that his chief looked "insignificant, ugly, with jutting teeth, skinny and rather short."[28] The star reporter of *Excélsior*, Julio Scherer, would decribe him as "two little spheres for eyes, sparse eyelashes, yellow skin with scattered brown moles, thick lips and a nose with a wide base."[29] His secretary, Urbano de Loya, would later note, "His ugliness complex! By God, what a problem that was!"[30]

To struggle against it, he marshaled a whole battery of defensive attitudes and compensations. The most obvious was the cultivation of his few physical virtues, most especially his voice. Farías attested that when he had his first meeting with him, "this ugly, skinny, awkward man, without any humor to him, became an attractive person because of how well he spoke."[31] He had worked on that quality for many years. To improve it, he had learned to play the guitar and could sing old Mexican songs, without any real musical cultivation but with impressive resonance and a good ear.

Another external tactic was how elegantly he dressed. He would have the finest suits made for him, and he would order his initialed shirts from a specialty store in London. He was obsessed with dressing well or, rather, tortured by the possibility that he might be dressing badly. Once, as he was in the process of matching and rematching suits and ties, he said to his secretary, "How complicated life can get! Remember when I only had one suit! I had no choice but to wear that coat with that tie! Now I have so many that I'm never free of worries."[32]

His greatest gift, of course, was his mental agility and especially the power of his memory. "His intelligence," someone said, "stripped away his ugliness." It was as if rejected by matter—by his own material being—Gustavo Díaz Ordaz felt the need to retreat to a life of the mind, there to gather his strength and then reach out to establish control over matter. Throughout his years as a student and all through his political career, he had displayed an incredible capacity to retain information; he had become a living archive of people with all their identifying qualities—names, professions, problems, and weaknesses. And perhaps that will to dominate from the vantage point of the mind explains his passion for jigsaw puzzles, a diversion he never abandoned and kept on practicing, with the same devotion, all the while he was President. Whatever puzzles he ordered from abroad, he could solve in no time. The more pieces the better. But none of them had enough pieces to satisfy him, so he would shut himself in his woodworking room at home, put his saw to use, and create his own puzzles. To *impose* chaos on matter and then play at putting it together again, over and over through the mastery of his mind.

His meager physical virtues and his powerful mental faculties could not completely obscure the reflection he still saw in the mirror. An additional compensation, another one of his subtle tactics, was to neutralize the gibes of the world. He liked to be told the jokes that circulated about him in the streets. He would make up others by himself, even sharper. People used to say that he was all "beak and ears," that "López Mateos should get the Nobel Prize because finally, thanks to him, they've discovered the 'missing link,'" or "Díaz Ordaz should move the capital of Mexico to Yucatán because only there could somebody ever say to him 'Good-bye, handsome'" (*Adiós lindo*—a common farewell in Yucatán). The President seemed to accept all this punishment with good humor. And he would add his own witty remarks. At a state banquet in honor of American president Lyndon Johnson, he made an unexpected reference to his "personal ugliness" and repeated one of his own most celebrated jokes about himself: "Somebody told me that people from Puebla have a reputation for being two-faced. I answered him, 'Do you think if I had another face that I would go out on the street with this one?'" When he was in Tehuacán (also the name of a mineral water) during his presidential "campaign," he woke up to find the walls of the city plastered one morning with the usual sort of slogan that appears punctually every six years: "Tehuacán with Díaz Ordaz!" On seeing the exhortations, the candidate said, "It must taste horrible!"[33]

And so what he called his "personal ugliness" pursued him not only in the humor from the streets but in his own conscious and public attempt to impose himself on the process, to master the caustic humor of others with his own self-lacerating words.

Since his bodily world had no harmony or beauty, it was not worth the trouble to cultivate it or its possibilities. Better to retreat inward, toward the spaces of solitude where no friends were necessary. "Our father was not a man for friends," his sons would later say.[34] A family man, a good protector and provider for his wife and his three sons, he pulled away from any excessive closeness to people with whom he was acquainted, even his relatives, especially after he became President. And worse, day by day, he would see (in truth and in his imagination) the ranks of his enemies increase. His closest friend, Herminio Vásquez, realized suddenly that, even in private, he had to change from "Gustavo" to the formal second person and be addressed, instead of by the first name he had heard from him for years, as "Sr. Vásquez."[35]

In Díaz Ordaz's case, even his personal honesty and degree of detachment from money seemed part of his disdain for the world of the senses. As President, he would move away from another friend, one much attached to the senses all his life long—López Mateos. The new President showed an implicit disapproval of his predecessor's traveling habits. ("This is called the 'seat' of the executive power . . . this is where the President should be.")[36] And he seemed to develop a judgmental coldness toward "Adolfito." The pleasures of the roads of the world were not for him.

The enjoyment of food was a pleasure he surely did relish: He was a man from Puebla, a city much praised since colonial times for the best and most varied cuisine in Mexico. To be a Poblano and not able to indulge yourself at the table was like a foretaste of hell. But Díaz Ordaz paid a physical price for his sullen retentiveness, for his "difficulty in opening up to anyone." Or at least he expressed his character in his guts as well as his mind. From his youth on, he was plagued with severe gastrointestinal problems—stomach pains, nausea, constipation. To eat the soups, meats, sauces, and legendary sweets of Puebla was to invite long hours or days of discomfort and pain. And of course the same was true anywhere in Mexico—at the banquets of his campaign or the many feasts offered to a President of the Republic. Sometimes he would yield to temptation and pay heavily. Most of the time he accustomed himself to pretending that he was eating. And special food would be cooked for him afterward—without grease, without this, without that.[37]

The many fine (and meticulously catalogued) wines in his large wine cellar; the many cases of champagne he received as gifts; the shellfish, the green turtle, the lobsters—all things that he loved—had to be tasted or drunk in tiny rations. "It will take me six years to digest my discomforts," he said to a friend at the beginning of his *sexenio*. As if his stomach were speaking for him.[38]

He had something like a hundred pairs of shoes because he had trouble with his feet. When he tried to play the presidential sport par excellence—golf—he would usually develop painful calluses. "He kept all these shoes in boxes, each one marked with a label: 'Moccasins, black,' 'Sport shoes that I bought in Los Angeles,' 'Blue, for the light blue pants.'" And for him almost everything else, including his life, was ticketed and marked, everything laid out meticulously, everything planned in detail on little cards—"such and such a day, this community, the time, the meeting, the proper suit, shirt, underwear, shoes."

If somehow something was out of place—a bottle on his desk, one little card exchanged for another, an impertinent statement—Don Gustavo would exclaim, "This can't be! What a stupid thing!" and he would begin to stride around his office in irritation, brooding on the flaw. When other forms of compensation failed him, when reality resisted his imposition of order, when the pieces did not fit, when he could not swallow or digest what he had stored up within him—the final, infallible recourse always remained: violence. At those times, building up toward action, he would shut himself up and no one would dare to disturb him, for any reason. Even his wife would be afraid to knock on that door.[39]

When he became President of Mexico, Díaz Ordaz projected his personal situation onto the political life of the country. It was a world marked by pressure, constriction, retention, narrowness, strict order. "The Presidency of the Republic," he used to say, "is like an orange-squeezer. Only the pulp remains." Surely he had "Adolfito" López Mateos in mind, who left office "an old man" and who had originally approached the gates of power as if they were steps to the gallows. Yet he himself would take his seat in the Presidential Chair overcome with intense and long-lasting anguish. The very word obsessed him: *angustia*. He said to his fellow Poblanos in a campaign speech: "I want the Poblanos to give me, more than their votes, their hearts, so that when I come to suffer the greatest anguish that a man can endure—and if I must be President, I must suffer it—I will know that the hearts of the people of Puebla are with me."[40] His son Gustavo Junior said that he used to talk about the "anguish" that swept over him when he said good-bye to López Mateos and sat down alone for the first time in the office of the President within the National Palace: "He said it was awesome . . . he lived with that anguish for six years."[41]

To contain the overwhelming pressure, he would assert the sanctity of the presidential position more than any other twentieth-century President. He would speak not only, like Ruiz Cortines, of the "dignity of the presidential investiture," but of the "majesty of the office." His concept of the position was almost explicitly theocratic: "From here it is possible to feel when the people are satisfied, angry, demanding, united, indifferent or cold." At the same time he felt himself "dramatically

Gustavo Díaz Ordaz
(Archivo Fotofija)

alone." He would say that "one of the tasks most difficult for a man in power . . . is to find the truth within the dense forest."[42] Only Calles, obsessed with transforming his illegitimate origins, resembled Díaz Ordaz. But unlike Calles, Díaz Ordaz was no visionary, no reformer, but a man with a single fixed idea—to conserve at any cost and "without any change . . . the directions that Mexico has taken" as well as his own rigidly maintained attitudes and opinions: "I am not going to alter my behavior now that I have taken on the highest responsibility."[43]

He had already asserted in 1960 that the political leaders of his time should be "a solid bridge across which the new generations must pass." As President, his mind saw only two poles. On the one hand, strength and authority and the majesty that only he embodied and represented; on the other, obscure threatening forces, foreign to Mexico, with a will to sow disorder and anarchy across the national jigsaw puzzle.

This was the man to whom "Adolfito" had passed on the presidency so that he might govern Mexico during the most rebellious and libertarian of decades.

The new style of power took effect immediately: no forms or ceremonies, no privileges for relatives, no personal influence—only the pure and naked application of power. Those who worked closely with Díaz Ordaz had to learn five rules by heart: 1) Tell me the truth; 2) Don't give me any excuses; 3) If you break the law, then break it, but don't let me know; 4) Be careful about the information you give me; and 5) I will not change my cabinet, because "you don't change horses in the middle of the river."[44]

The six years of Díaz Ordaz's presidential term would be a period of growing economic strength for Mexico, as was true in much of the world during that prosperous decade. But the credit for this achievement has to be assigned to Díaz Ordaz and the work of his brilliant Finance Minister, Antonio Ortiz Mena—whom the president reappointed, after Ortiz Mena's six years of service to the López Mateos government. (Ortiz Mena—as the other most influential minister—had been a possible choice for President. He had met with Díaz Ordaz and told him that the presidency did not interest him, but that he wanted to continue being a strong Finance Minister and would support the candidacy of the Minister of the Interior. It was an offer of loyalty—a value of supreme importance to Díaz Ordaz—and it won his respect and his loyalty in return.)[45]

In economic matters, Díaz Ordaz had always shown great responsibility. His meticulous, retentive character (and his personal honesty) favored a cautious attitude toward the monies entrusted to his responsibility. Under López Mateos, he had opposed Mexico's bid for the Olympic Games, expressing doubts about the cost to Mexico or the benefits it would supposedly bring to the country. With Ortiz Mena managing the economy, Mexico's gross national product grew by 6 percent a year (8 percent in 1968) while inflation was a minor factor, at 2.6 percent per annum, lower than in the United States. Through the sixties (from 1958 to 1970, during Ortiz Mena's twelve years in charge of the Ministry of Finance) the real buying power of wages in Mexico increased by 6.4 percent per year. The exchange rate of the peso vis-à-vis the dollar never wavered, at 12.5 pesos. The International Monetary Fund treated (and sometimes used) the peso as one of the world's stable currencies. And in 1970 Mexico's external debt would be only $4 billion, with all of the money carefully invested in development projects. That year—when Díaz Ordaz and Ortiz Mena left office—the country had a balanced budget.

Under Díaz Ordaz, the country experienced the "Mexican Miracle," a high point of economic progress. Ortiz Mena once defined the process as *El Desarrollo Estabilizador*, "The Stabilizing Development."[46] Mexico even invested in new petrochemical projects in Central America. And Díaz Ordaz—though he explicitly avoided the traveling music that had marked the rhythms of his predecessor—made at least two other foreign gestures that deserve respect. He visited and sought to improve relations with Guatemala—a country long suspicious of its "colossus of the north." (The Guatemalans felt that the state of Chiapas rightfully belonged to them.) And before a joint session of Congress in Washington, he urged the Americans to remove their customs restrictions against Mexican products and permit Americans to freely buy and import from Mexico, while—more broadly—calling on the American legislators to recognize the economic needs not only of Mexico but of all the other nations of Latin America.[47]

In the case of individuals, it can often happen that traits favoring them in one area or period of their lives will do damage to them elsewhere. When the life of a country depends too narrowly on the personal qualities of an individual, a nation can then suffer the same paradox. The obsessive need to restrict and manage was of help to Díaz Ordaz in preserving financial responsibility. Dealing with human beings was something else.

When Miguel Alemán sat down in the Presidential Chair, the oil workers had "tried him out" and encountered the army. The railroad workers had tested López Mateos and run smack into Díaz Ordaz (and the army). There were no dissident unions to "try out" Díaz Ordaz in December 1964 because all their leaders were in Lecumberri Prison, accused of "social dissolution." For him, the probe came from an unexpected direction, a sector of the professional middle class—the doctors.

The recently created ISSSTE (Institute of Social Security at the Service of Workers of the State) and other government agencies like the Ministry for Health and Assistance offered its doctors salaries and benefits that left much to be desired in comparison with the Mexican Institute of Social Security (also a governmental institution but in far better shape because of the capable direction of Ortiz Mena between 1952 and 1958). The interns and residents were particularly exploited, with meager salaries for the long consecutive hours of work required of them. In November 1964 there was a rumor that they would not even get their usual Christmas bonuses. Two hundred residents and interns of the Hospital 20 de Noviembre decided that they would threaten to strike unless conditions were improved. The outgoing López Mateos administration dismissed them. On the night of the day that the news of the firing arrived, they formed a new union, the Mexican Association of Resident Doctors and Interns (AMMRI) and issued a list of requests for the improvement of their conditions and other general issues relating to the hospitals.

It was a movement with nothing red about it, "white" down to the color of the robes that the 1,500 members of the AMMRI wore as they filed peacefully toward the Zócalo, though the Ministry of the Interior had denied them a permit. They wanted to speak with the President, and one of their spokesmen was granted the right to address him "for three minutes." Díaz Ordaz tolerated the quick discourse and then responded in a brusque and irritated burst. "He scolded us as if we were school children." As President of the Republic he was not going to be the court of first appeal, he had to respect his investiture, he had grave responsibilities to handle, not all the minor business of the country.[48]

The strike spread, to forty-six hospitals across Mexico. The new director of the ISSSTE ordered the Christmas bonus paid immediately to all residents and interns in all the hospitals under his jurisdiction. But it was too late to buy the doctors off. They wanted answers to their other complaints and voted for a new strike. But then they decided not to carry out their threat, because they believed the government was seriously "studying" their demands.

In January 1965 a new organization was formed, of senior doctors, the AMM (Alliance of Mexican Doctors.) They took a less radical stance but identified with the same list of grievances. Díaz Ordaz granted them an interview and was rather more polite than he had been to the younger doctors: "If someone puts a gun to your stomach and asks you for your money, you would answer, 'Please lower the pistol so that I can see how much money I have in my pocket.'" But of course, being Díaz Ordaz, he did know just how much money he had. His real concern was to avoid any appearance of being coerced, for fear of setting off a new wave of strikes. But—at an even deeper level—his resistance had a more personal motivation: "I am not a man who lets himself be pressured."

The events then took a number of turns. The residents and interns insisted on continuing the strike, but the senior doctors urged a return to work in order to let the President demonstrate his good faith. With the votes of a bare majority, the strike continued in the forty-six hospitals for ten days after the interview with the President, on January 21. No news came from the government. The doctors decided to show their good faith and return to work on the thirtieth. Still only silence from the President's office. The more radical AMMRI voted for a national strike, to begin on the twentieth of February.

Without a previous word to the union representatives, two days before the strike date, Díaz Ordaz gave the press a copy of a decree he had delivered to the Ministry of Health and Assistance. It was his unilateral and definitive reply to the medical movement. He ordered some increases and established additional benefits for the student doctors. The union made a "counterproposal" that included a few technical modifications, little more than a matter of form, which the government considered reasonable. But there was a semantic objection from the President's office. The doctors should not have called their statement a "counterproposal," but an "interpretation of the presidential decree." If the President was not a man who yielded to pressure, neither was he a figure who could receive counterproposals. Like the Pope, he had to both be and appear infallible.

But still things did not seem to have gone badly. A committee of doctors visited the President in March to thank him. It looked as if the situation had reached a happy ending, worthy of the era of Ruiz Cortines—a little bread and no bludgeon. But Díaz Ordaz was not Ruiz Cortines. The medical pieces of the system had gotten out of order. They had to be aligned, and he would do it.

The President began to wield two weapons against the doctors: union pressure and a propaganda campaign to discredit them. The doctors of the ISSSTE were pressured to make their unions part of the FSTSE (Federated Union of Workers in the Service of the State), an organ of the CTM and therefore of the PRI. The organizations bravely refused to give up their independence, but many doctors could not resist and joined a government union. At the same time, the government enlisted the press, which under Díaz Ordaz was thoroughly dependent on the system (except for the magazines *Siempre!* and *Política* and the newspaper of the moderate left, *El Día*). The newspapers began to publish a series of full-page advertisements signed by the government-linked unions or offered anonymously, all of them painting the doctors Communist red. Nonexistent associations like "The Committee for the Defense of the Dignifying of the Medical Class" or "The National Association of Professionals and Intellectuals" offered descriptions of the dissidents as "a privileged class apart from the common people," men who "were stabbing the nation in the back." False rumors were spread about patients who had died for lack of medical attention. The entire objective of the campaign was to goad and harass the doctors and to break their independence. And then Díaz Ordaz got what he wanted—a confrontation. In the climate of growing tension directed against them, the militant AMMRI and the reticent AMM fixed the date for a new strike on April 20, 1965.

For Díaz Ordaz, "interests orchestrated from a great distance" were clearly at work, inciting chaos. If the doctors should appear to have won their case, any other group of employees or citizens might use the same "strange and suspicious" tactics (demonstrating in the streets), fueled by who knows what "shameful and dark" interests (he generally meant the KGB, though sometimes he might include the CIA), imagining that they could make "inappropriate and impossible" demands (increases in wages and benefits) upon the government.

When the doctors went on strike again, he soon set a date for them to return or be fired and proceeded to "orchestrate" one of the most effective pressure techniques the system had at its disposal: large-scale workers' meetings condemning and threatening the doctors. There were immense demonstrations in which street sweepers and garbage collectors—supplied with free sandwiches and soft drinks—shouted antidoctor slogans. And it worked. The physicians gave in. Díaz Ordaz then issued another decree that offered some general increases, with absolutely no mention of any of the other points that had been at issue. The message was clear. Whatever was to be gained would come as a grace from on high, never as something won from below.

Unrest was still widespread among the doctors—friction (and even a threat of violence) between those who accepted the end of the struggle and those who wanted to go back on strike. A few days before the first presidential *Informe*, a squad of grenadiers entered the Hospital 20 de Noviembre—the original flashpoint of the doctors' movement. The order was given to clear the building in ten minutes. The army brought in military medical personnel to provide basic services and moved those who were dangerously ill to a military hospital on the outskirts of the city. It was a calculated act of terror. That was its only real purpose.[49]

Some doctors were arrested, others were subsequently fired. Over fewer than twelve months, the medical profession had now received an intensive lesson in contemporary Mexican politics. The longest, most intense applause during the President's State of the Nation Address was a response to his description of the doctors as delinquent and criminal.

In political life more narrowly defined—the sphere of the parties—Díaz Ordaz's accession marked the end of an attempt by Carlos A. Madrazo (who had been president of the PRI for a year) to introduce democratic reform within the party. Madrazo—who as a young man in the thirties had been to the far left of the political spectrum—wanted to create the competition characteristic of a party in a modern democracy: internal elections and criticism as well as self-criticism among different options and platforms. Though he knew that a previous attempt to democratize the party had failed during the mid-forties, he had now managed to institute party elections, "from the bottom up," in a majority of the 2,328 municipalities throughout the country.[50] He had traveled everywhere giving fiery speeches in favor of the "new tonic" and against "undesirables" within the PRI who opposed the reforms. The sympathies of young reformers within the PRI were with him, but not those of the people who really counted—the President and his men (and corporate sectors of the PRI, primarily the unions). Díaz Ordaz forced him out of his position, ending the experiment, and when Madrazo died in an airplane accident in 1969, it became one of those sudden deaths of controversial politicians that would linger as a question in the air of Mexican politics.

The PAN—the most important and, for all practical purposes, the only opposition party—made some advances during the midterm election of Díaz Ordaz's *sexenio*. For the first time, PAN put up candidates in 99 percent of the electoral districts. They won nineteen seats in the Lower House, took slightly more than 11 percent of the overall votes, and won municipal elections in eight cities of the state of Sonora, including Hermosillo, the capital. It was the first time in history that they had been allowed to win a state capital, and a few months later, in Yucatán, they won another with the city of Mérida.[51]

But PAN had been denied other municipal victories through fraud. And to the chiefs of the PRI, Mérida seemed to be perhaps an unacceptable "additional loss." The Minister of the Interior, Luis Echeverría, let Finance Minister Antonio Ortiz Mena know that the President wanted tax auditors sent to "the rich people" of Mérida so that they might be moved to "think about their own interests" and stop supporting PAN. In the end, PAN would be allowed to govern the city, which they did for three years, but visitors in 1970 noted that "the white city" was unusually dirty. There were hardly any city services because state and federal budgets had been closed to the PAN municipal president.[52]

Since the days of Porfirio, there had never been so complete a climate of subordination in the country. There was almost no suggestion of independence in that first circle of the system's true dependents—neither in the army, nor in the thousands of official unions, nor in the peasant organizations. The 32 governors, the slightly more than 2,300 municipal presidents, the deputies (minus a handful of non-PRI people), the senators, and the judges all snapped to attention before *El Señor Presidente*. And some of the forces that normally occupied the second circle—of the not entirely dependent—moved into complete subordination during the reign of Díaz Ordaz.

Big business was extremely satisfied with a President who had publicly declared himself to be anti-Communist and also with a management of the economy (by Antonio Ortiz Mena) that was not only responsible and honest but remarkably accomplished. It was very clear to them now "which way" the President was going, and they followed along. The Church, at least at its highest levels, strongly approved of the President's disapproval of Marxism and appreciated the private manifestations of devotion evidenced by the President and his family: the gift of a top-of the-line automobile to the Bishop of Puebla, the papal blessing on the marriage of Gustavo Junior, the public piety of Doña Guadalupe. If Díaz Ordaz exercised power in an almost theocratic fashion inspired by his Oaxacan heroes and predecessors (especially Juárez, but also Díaz), he behaved toward the Church much like a traditional Puebla conservative. And the Church paid him back with interest. In 1968, for the first time in history, the episcopate would issue a declaration in favor of the Mexican Revolution.

The press in general had turned slavish. The practice of the *embute* (the "stuffed envelope") had its golden age under Díaz Ordaz and became firmly rooted in the traditions of Mexican journalism. Money—in cash or check, often in substantial figures—came directly from the President's office to writers on the presidential beat. "Take it, if it doesn't corrupt you," a journalist once said to another, demonstrating that the very concept of corruption had been emptied of meaning.

There was an incident in 1966—merely funny on the surface but meaningful in substance and in effect fatal to one newspaper. *El Diario de México* published two photographs on its front page: one of President Díaz Ordaz, the other of two mandrils. Through a proofreader's error the captions were interchanged, and the Mexican public would read, under the picture of the President: "The zoo has been enriched. In the photographic section, there are pictures of some of the new specimens acquired by the authorities for the diversion of the citizens of our capital. . . . These monkeys were put into their individual cages yesterday."[53]

A few days later, the newspaper was forced to announce—on its front page—that it was going out of business: "President Díaz Ordaz orders the death of *Diario de México*. A typographical error causes the severe decision." Not a word from the other newspapers, though presumably they strengthened the numbers and training of their proofreaders.[54]

The President with the hard hand had a very thin skin for political criticism. He once jailed the eighty-year-old Dr. Bernardo Gastélum—who had been Calles's Minister of Health—for writing critically about his policies. And the extremely slender roster of opposition journals was finally reduced even more by the disappearance of *Política*, forced out of existence by a selective process—affecting only them—of increases in the price of newsprint. They paid, among other things, for a mistaken prophecy in 1963 when the journal had printed a picture of Díaz Ordaz, then Minister of the Interior, above the caption "He will not be President."

In 1965 the most prestigious publishing house in Mexico—the Fondo de Cultura Económica—also ran into the President's displeasure, when it published a translation of *The Children of Sánchez*, the painstaking and powerful presentation of the life of a poor Mexican family by the American anthropologist Oscar Lewis. Never before had the President of Mexico considered it necessary to descend into the kitchen of cultural life and directly censure the internal activities of the intellectuals, let alone on the issue of an already published book. But Díaz Ordaz did not want to be reminded of the ugly face of Mexico (nor for Mexicans to focus on that image) while the country—in the full tide of economic development—was preparing to become the Olympic capital of the world in October 1968. A careful reference to poverty in the normal Presidential *Informe* on the first of September would always draw sympathetic and pious applause for the compassion exhibited by the leader of the nation, but it was not desirable to have the subject presented all through a long book published by a house that was financially dependent on the state—though resolutely independent in its internal decisions. Moreover, this publishing house was headed by a foreigner. Wasn't there by chance some Mexican who could take over?

Arnaldo Orfila Reynal had been born in Argentina but had directed the Fondo de Cultura Económica since 1948. He was an editor of immense dynamism, and Mexican literature owed him an incalculable debt. He had published Juan Rulfo, Octavio Paz, Carlos Fuentes, Juan José Arreola, produced an edition of the complete work of Alfonso Reyes, discovered many new novelists, poets, and essayists as well as giving Mexico its first translations of classic modern authors like Heidegger and Lévi-Strauss. He was now close to seventy years old, but Díaz Ordaz hounded "the foreigner" out of the position he had filled with such splendor. Orfila would submit his resignation in the glow of universal sympathy from the entire intellectual and cultural world of Mexico and then found, with his own money and the support of others, an independent publishing house truly opposed to the government: Siglo XXI Editores.[55]

The famous cardiologist Ignacio Chávez, Rector of the University, seems to have suffered a similar fate, though it is not clear as to how much pressure Díaz Ordaz actually exerted upon him. (The President would later deny forcing him out.) His successor—Javier Barros Sierra—was an erudite and sensitive man who was not destined to have very cordial relations with Díaz Ordaz. But before any future developments, there would be the meeting between them in which the new Rector discussed building new schools for the University on land that belonged to an *ejido* and lay on the border of the Distrito Federal and the state of Mexico. Díaz Ordaz's comment was, "If there are any problems, Javier, which police force do you want brought in?"[56]

Díaz Ordaz in a speech once said, "In every young man, there is a substance which is generally pure, generous, idealistic."[57] But he knew that the young were restless and unruly; he felt they were by nature lawless and easily manipulated by "outside agitators." In fact, to think of them was to think of conflicts and, consequently, of police.

During one of the great decades of change in the history of the Western world, the demand for order and authority by one of the most rigid governments in modern Mexican history (fueled by the neurotic need for control of the man who sat in its seat of power) would confront, with mounting intensity, the claim by young people to freedom and independent judgment—during a period when it was good though often confusing, exhilarating but also dangerous and sometimes fatal to be young. The first encounter of importance came in the provinces, at the University of Michoacán—the famous Colegio de San Nicolás (founded in 1540) in the city of Morelia.

Like the student disturbances during the fifties in Mexico City, the direct motive seemed trivial—an increase in bus fares—but the federal government believed it was seeing the dark hand of Communist activism. Suddenly a student from Guerrero (who had political connections with the PRI) was shot to death. With statements prominently displayed in the press, the government blamed "professional agitators involved with foreigners." But the boy had actually been killed by police dressed as students. The governor of the state was Agustín Arriaga Rivera, whose flashy new car had earned him the company of López Mateos on the drive to Puebla that first drew Díaz Ordaz into the work of imposing order on the nation as a whole. As he tells it, the shot was accidental and the students had then carried the body through the streets of Morelia, in a huge demonstration, "presenting him as a victim of the government."

The events took place on October 2, 1966. More demonstrations followed, with the utopian objective (in those days) of forcing the removal of the governor. A representative of the President arrived to explain that the government could not be coerced. The student Strike Committee maintained its demand. "Then Díaz Ordaz," Arriaga remembers, "acted as he always did, violently."[58] The President ordered his Interior Minister, Luis Echeverría, to work with the Assistant Minister of Defense to coordinate the occupation of the university. He would end the student uprising and also, himself—on his own initiative, not that of the students—get rid of Arriaga, who had not shown a hard-enough hand. The governor had in fact by then taken control of the situation, but he knew that his political career was over when he saw the paratroopers arrive on October 6.

They marched through the streets of Morelia by night, to the beat of a drum. And then they moved into the university, a military occupation like some "action of a Latin American gorilla government" but it was happening in Mexico, where the children of the middle and upper classes at their studies were not in the habit of seeing men in helmets with rifles on their shoulders, patrolling the precincts of a university as if it were a village in Vietnam. This was the college, as Jaime Labastida, President of the Federation of Professors, would say, "founded by Father Vasco de Quiroga, the oldest university still functioning in the Americas, where Hidalgo had been Rector and Morelos had been a student . . . now raided for the first time."[59] The army found nothing to suggest the "terrible conspiracy" supposedly being fashioned there, not even stones, let alone guns or Molotov cocktails.

The students maintained a careful and exemplary composure. They would walk by the soldiers as if they were not there. The governor was then informed on the telephone—by Díaz Ordaz's Minister of Defense, Marcelino García Barragán—that private student residences were to be entered and thoroughly searched. The governor was to secure a list of addresses and hand them over to the local military commander, an old Michoacán general and politician named Ireta, who was shocked at the order—and told the governor he was, but followed instructions and took the list. It was another series of evictions for Díaz Ordaz, at bayonet point, through coordinated movements of troops. By some miracle no one was killed.[60]

The movement had been broken. The left-wing School of Higher Studies—which had taught a mixture of science and Marxism—was permanently closed. Some student leaders were sent to jail, others went into hiding or fled the city. A year later, a similar operation was carried out at the University of Sonora.

In his *Informe* of 1966, delivered a little more than a month before the army occupation of the University of Michoacán, Díaz Ordaz had indicated (and foreshadowed) his reaction to the threat against the principle of authority that—as in so many other countries during those years—unrest among the young represented for the guardians of vested power and custom: "Neither claims of social and intellectual rank, nor economic position, nor age, nor profession nor occupation grant anyone immunity. I must repeat: No one has rights against Mexico!"[61]

It was an obvious threat against the universities and especially against the students. Years before, such warnings had become reality against opposition workers and peasants or independent political candidates like Dr. Salvador Nava. Under Díaz Ordaz, the bludgeon had gone into action against doctors, attempts at democratic reform within the PRI, the politicians of the PAN, critical journalists, opposition magazines, an internationally respected editor, and soon against thousands of students in two provincial universities.

Exactly a year after his warning in the *Informe* of 1966, Díaz Ordaz stood up before the Legislature—on September 1, 1967—for his next State of the Nation Address and said: "For a person in authority not to exercise the power conferred upon him by the law is as harmful as for him to abuse that power." There would be "no rights against Mexico," and the President was more convinced than ever that he knew what "Mexico" meant. All he had to do was look into his own mind.

The young man was about as old as the modern Mexican political system, born in the forties, at the beginning of the era of Alemán. He was not only a product and a beneficiary of the system. As a University student of the middle classes, he was in a sense its favorite son.

He had studied in the venerable National Preparatory School, founded by Benito Juárez. There—not in his family—he first became aware of politics. Someone gave him a copy of *Listen Yankee!* by the American sociologist C. Wright Mills, who felt that Marxist politics made more sense in Latin America than anywhere else. A friend told him to read Lombardo Toledano in *Siempre!* or lent him an explosive issue of *Política*. In his philosophy class, a teacher introduced him to the work of Jean-Paul Sartre, who was then shifting—fortunately, according to the teacher—from an asphyxiating, solipsistic existentialism toward a position imbued with Marxist hope. And he began to feel an underlying sense of shame because of his class. What had he done, what would he do in the future, for the disinherited of his country?

From preparatory school he went on to the University, to the Faculty of Engineering of the UNAM. In contrast to the faculties in the humanities, Engineering prided itself on being "apolitical." Actually, many of its professors expressed opinions to the right of the Mexican political spectrum, and some students were members of a militant anti-Communist organization: MURO. But the spirit of the student body around him was still, in 1965, more playful than ideological. The young man went through the traditional, cruel, but relatively harmless process of freshman hazing and inflicted it on others in 1966. It was the tail end of the age of innocence.

A small group in Engineering maintained contacts with the faculties that were truly political: Economy and Philosophy, Letters and Political Sciences. Becoming friendly with those students within his own discipline, the young man gained access to a new intellectual universe. Marxism was in fashion, but—as opposed to the thirties—its locus was the academy, not unions or even political parties. Its influence spread to Mexico especially from the universities, the intellectual magazines, the publishing houses, and the cafés of Paris. The professors in the faculties of the humanities were largely members of the "Generation of the Mid-century." They were followers of Sartre and of Marx, they had studied in Paris, they had gone to Cuba, they had written about Cuba, they believed firmly in a socialist future for Mexico. The young man could be found in a more moderate band of student opinion, not carried away by the intellectual and political passions of those classroom revolutionaries, more interested in Camus than Sartre, more wary of the "socialist paradise" of Cuba.

In 1967 he began to read translations of a philosopher whom Hitler had driven out of Germany, a member of the Frankfurt School of social thinkers, who preached a kind of social messianism through a synthesis of Freud and Marx. The young man literally devoured *One-Dimensional Man* and other books by Herbert Marcuse. He filled them with notes and markings, copied out quotes, found in those pages a prophecy whose accomplishment was just around the corner and could be summed up in the epigram Marcuse had drawn from his old friend and colleague Walter Benjamin: "It is only for the sake of those without hope that hope is given to us." If Mexico and Latin America seemed to have no hope, that fact alone could transform them into the promised land of a future liberation.

To build the future, to rediscover the Indian, to educate the worker. These were the youthful slogans and illusions of the era, a vague and sentimental idealism that only at times became practical action. The world was seen as the struggle of light and darkness. On the side of the blessed were those who suffered, the "wretched of the earth" in the words of Franz Fanon, the intellectual prophet of the Algerian Revolution (and herald of Black Liberation). The forces of darkness were the exploiters, the bosses, all the lackeys of Yankee imperialism. When Che Guevara died in Bolivia and the publishing house of Siglo XXI printed his diary, young men all over Mexico wanted to be like him. He was the icon on the walls of the cities, the symbol of the new man, the pure and incorruptible hero. Latin American protest songs came into fashion. The campuses opposed the war in Vietnam and all the right-wing "gorillas" of the Central American dictatorships.

The young man was somewhere at the margins of all this fervor, finding perhaps his greatest pleasure in Mexican literature and the "boom" of Latin American writing—the novelists of "magic realism" and social criticism. He read *The Labyrinth of Solitude*, Octavio Paz's meditations on the Mexican conciousness, as if it were Scripture, and responded to Carlos Fuentes's criticisms of the course of the Revolution in *The Death of Artemio Cruz*. He was dazzled by the mysterious and magical pages of Juan Rulfo, where the power and weight of authority in Mexico was carved into images of massive but also crumbling stone. A friend said to him, "Cárdenas was the last President who governed for the poor." He began to believe that since 1940 the Mexican Revolution had been betrayed.

This young man had more in common with those of the same age in Paris, Prague, Berlin, or California than with his parents. In every song by the Beatles, he saw some of himself. Everywhere in the Western world, the rebelliousness of the sixties was a reaction to the conservatism of the fifties, which had been a time for much of the world to "settle down" after the horrors of the Second World War. The new mood was based on disdain for conservative values, a drive for sexual liberation (permitted by the Pill), the search for pleasure, and an anger at social injustice, but it was also based on affluence.

There was money around, or money to be earned, and the young of the middle classes could afford to be disdainful toward it. The public face and image of the times were set in the United States, in the realm of behavior but also at a whole new level of publicity. The limited beatnik movement of the late fifties and early sixties—to be a member of which you essentially had to be an artist or pretend to be one—became a literary trend and then vastly expanded into the far more colorful hippie movement—which anyone could join by wearing the right clothes or growing long hair—brandishing a hedonism that was profoundly threatening to old-fashioned conservatives.

In Mexico too—a country with strong traditions of parental authority—many young people came to think of their parents as *la momiza* ("the mummies"), men and women far too easily scandalized by the "tangled locks" (*greñas*) of their sons and the miniskirts of their daughters. Not to speak of premarital sex or experimentation with drugs, or for that matter even the "existentialist" cafés that filled the Zona Rosa—the nightlife center of the city—with all their dangerous air of new freedoms and innovations on a direct line from Paris.

The young man would never join the Communist Youth groups or seriously consider "going to the mountains" or truly "making the Revolution." He located himself among the three definitions that Octavio Paz proposed in his 1967 book, *Corriente Alterna* (*Alternating Current*): the mutineer, the rebel, and the revolutionary. By the mutineer (*revoltoso*) Paz meant the person who rises against power spontaneously, instinctively, with uncontrollable violence, like the Zapatistas in Mexico. By the revolutionary (*revolucionario*), he meant an individual committed—with some apparatus of rational concepts and measure of decision—to a party or movement in revolt. The rebel (*rebelde*) was "individualist, solitary," disobedient and unruly toward authority. It was within that category that the young man felt himself described.

And rebellion was the distinctive mark of his generation. Perhaps somewhat more bookish than most of his friends and acquaintances, the young man would share, with them, the same patricidal vocation. For him it was more important to negate than affirm, to criticize rather than propose. And youthful rebellion in general had one favorite target: the Mexican political system. Even though most students had at best poor and sketchy information (if any) on what had happened to the railway men and the teachers and the doctors and their fellow students in Morelia and Sonora, the young knew that the government of Mexico—and above all President Díaz Ordaz—was a repressive force. The PRI for them was nothing but its omnipresent propaganda, which filled them with a greater nausea than Sartre's classic existential malaise. And to each of the protagonists of the system, the youth of the sixties would apply a single defining term: the army meant fear; the union leaders were sellouts (*charros*); the peasants were exploited; the deputies were "puppets" (*peleles*); the governors were imposed; the municipal presidents were nonexistent; the press was bought; the business entrepreneurs were exploiters; the Church was fanatic; Lombardo's PPS was "government;" the PAN, reactionary; but the University was "autonomous." About themselves the students had, if not a messianic, at least a lofty opinion. There was a Latin American protest song popular at the time: "Long live the students because they are the yeast/of the bread that will leave the oven with all its full flavor."

Capitalism, imperialism, colonialism were so many terms for the devil. The sacred words were *socialism* and *revolution. Freedom* and *democracy* retained their nobility and purity, but they lacked any palpable weight. When the news of student rebellions in Europe and the United States reached Mexico, hundreds of thousands of young Mexicans felt that the hour was coming for them. A small minority, mostly Communist militants, dreamed of a revolution based on clear ideas, what Octavio Paz had termed "lucid violence." In a passionate pamphlet by Carlos Fuentes titled *Paris, 1968*, some foresaw a coming birth of freedom for them and for Mexico.

A specter of rebellion was rising everywhere. Students of the world, unite! You have nothing to lose but your boredom! The young man was about to become an anonymous participant in a luminous and terrible experience, the Mexican '68.

Everything began on July 22 of that year with a minor conflict between two groups of preparatory (high school–age) students. A pickup game of touch football at the square of La Ciudadela in Mexico City ended in a fight between players from a vocational school (part of the scholastic structure connected with the Politécnico) and students from a preparatory affiliated with the University. The grenadiers arrived, waded in to break up the disturbance, and began swinging their clubs. What followed was an out-of-control police riot. The vocational-school students fled to their nearby building, and the grenadiers pursued them into the school, where they struck out indiscriminately at students, without asking who might or might not have been in the square.

There were injuries, and a wave of protest began to rise against police brutality. On July 26, two parallel demonstrations moved along streets near the center of the city. The Estudiantes Democráticos—a group with Communist connections—was commemorating Fidel Castro's assault on the Moncada Barracks (the first spark of the Cuban Revolution). The other march, convoked by the National Federation of Technical Students, was simply protesting against the recent police violence. At one point, the two demonstrations met, and someone proposed that they march together, into sacred territory. "To the Zócalo!" They were intercepted by a charge of the mounted police.

During the following days, the government arrested much of the Central Committee of the Communist Party. Some preparatory students were also taken into custody. Their schoolmates, in reaction, occupied some of the preparatory schools and declared indefinite strikes. At the Politécnico itself, the Higher School of Economics declared itself to be on strike. The demands were "Stop the repression!" "Punish those responsible!" On the first block (radiating out from the Zócalo) of the center of the city, the students halted traffic. Stores pulled down their shutters. The students set fire to buses emptied of their passengers. The police launched their teargas canisters. Fighting went on for a week, and there were estimates of four hundred wounded and a thousand in jail.[62]

Díaz Ordaz was traveling on the Pacific coast, accompanied by some of his cabinet including his Minister of Defense, Marcelino García Barragán, when he received a call from Interior Minister Echeverría. It was a warning about the dangers of an impending "slaughter." Díaz Ordaz then either gave a command or let Echeverría issue an order.

It would be the army again, into the streets, this time carrying out a military attack against the formidable target of a preparatory school. Soldiers in Jeeps and light tanks rolled through the center of the city, on their way to capture the San Ildefonso Preparatory School. By night, they blew away the wooden gate (of carved wood dating back to the eighteenth century) with a blast from a bazooka and poured onto the school grounds and into the building.

Through the military occupation of the San Ildefonso School, the government elevated the situation from a local, primarily police matter to an issue of national security. The Regent of the Distrito Federal, General (and lawyer) Alfonso Corona del Rosal—Ávila Camacho's old polo-playing friend—gave a press conference in which he announced that there existed "a carefully planned action of agitation and subversion. . . . In my opinion, it is caused by elements of the Communist Party." Interior Minister Echeverría condemned the "vile and naive interests, extremely naive, trying to divert the upward ascent of the Mexican Revolution."[63]

In a meeting on the esplanade of the University, Rector Javier Barros Sierra lowered the national flag to half-staff. He then made a speech of public protest against the army occupation of University buildings—the San Ildefonso Preparatory School, which was a formal part of the University. On the same day, the Rector directly appealed to the students, calling on them to "contribute fundamentally to the cause of freedom" and made a bold and unheard-of decision to himself lead a march of more than fifty thousand people for some kilometers into the city, turning then to march back to the University. On that march, for the first time, the slogan was heard that would be chanted all through 1968: "People! Join us!" (*¡Únete pueblo!*)

From the apartment buildings constructed by Miguel Alemán, families applauded the students and the courageous Rector who was leading them: "Good for you, boys!" "Go! Go!" "We're with you!" Only a few blocks away, the grenadiers were lined up waiting. They would have attacked the march, if it had turned toward the Zócalo. The traditional University cheers (the *goyas*) were now shouted outside the sports stadiums; they had become inspirational cries, *gritos* for a civic battle.[64]

It was certainly not the first mass student demonstration in the history of Mexico City. Since the Vasconcelos campaign of 1929, the students—at times—had taken to the streets. But a demonstration had never been led by a man of the system—the Rector himself—asserting the literal meaning of the autonomy of the University and leading an orderly and peaceful march of protest against the assault on a preparatory school. The young man from the Faculty of Engineering saw the Rector at the head of his fellow students—the rebels of his time and a man in power daring to lead them. To oppose injustice, to protest the hand of the police raised against the rights of the young. He was with his girlfriend. They felt the presence of each other and the presence of all their fellow students—the University taking on an entirely new significance of community and strength. It was the moment to do something more than just read or talk. They joined the lines of the marchers.

The demonstrators felt an ineffable sense of freedom, the heady rush of treading on forbidden ground. An immense NO swept them all up, surged from their voices, from their steps, swelled up toward the heavens, a NO to authority. The air was lighter and clearer. You could breathe in a different way. Soon the chant was raised for what was truly going on: *¡Ganar la calle!* ("Win the streets!") When they returned to the University, the marchers sang the national anthem.

In Guadalajara, President Díaz Ordaz delivered his first public comment on the events. He deplored the "shouting matches . . . the deplorable and disgraceful occurrences," and—including himself in the picture—asked the parties concerned to set aside "the self-love that distorts perspectives so that we may resolve these problems. . . . A hand is extended: It is the hand of a man who across the brief history of his life has known how to be loyal. Mexicans will say whether this hand was left outstretched in the air."[65] Raised in the protest culture of the sixties, the students were contemptuous of the President's words. A graffito was painted on walls across the city: "We want a paraffin test on that outstretched hand!" The students insisted that there had been deaths in the skirmishes. The government totally denied it. The hand was left hanging in the air. And so ended the formative stage of the Student Movement.

During the first days of August, the National Strike Committee (the CNH) was formed, with representatives from almost all the schools of higher education in the capital and much of the rest of the country. Shortly afterward, the Coalición de Maestros (Coalition of Teachers) was organized in support of the students. All groups united behind the *Pliego Petitorio* ("the List of Requests"), issued on August 4 and consisting of six points: 1) removal of the two most important police chiefs; 2) dissolution of the repressive tactical forces, like the grenadiers; 3) respect for the autonomy of the university; 4) payments to anyone who might have been injured; 5) repeal of Articles 145 and 145A of the Penal Code—the laws on "social dissolution"; and 6) the release of all political prisoners.

The students had to measure the strength of their movement. On August 13 a hundred thousand students marched from the Casco de Santo Tomás, site of the Politécnico, to the Zócalo. "We won the streets when we entered the Zócalo on that Tuesday because it broke a taboo," recalled Salvador Martínez della Roca, "*El Pino*" ("the Pine Tree"), leader of the Committee for the Struggle of the Faculty of Sciences. The students carried no weapons, but their words were weapons: "Mexico! Freedom! Mexico! Freedom!" The slogans were innocent ones, meant to instruct: "Books, yes! Bayonets, no! Books, yes! Grenadiers, no!" On the banners and signs, the citizens of the city could read, YOU DO NOT DOMINATE MAN, YOU EDUCATE HIM! or THESE ARE THE AGITATORS: IGNORANCE, HUNGER AND POVERTY. They passed out flyers, and one of them said, "People of Mexico! You can see that we are not vandals nor are we rebels . . ." In the Zócalo, there were protest speeches. In one of them, Eduardo Valle, a student leader from the Faculty of Economy of the UNAM, called the political stability of Mexico a false facade, described the regime as senile, and characterized the Student Movement as a struggle of the people.[66]

In the same way that they invented slogans as they went along, so the students were discovering, entering, and conquering a territory almost unexplored in Mexico—that of direct democracy. Each faculty, school (a "school" is a less important division within a Mexican university), or other educational center had at least one leader. About 250 leaders made up the National Strike Committee, most of them between twenty-two and twenty-six years old. The leaders held regular meetings in their faculties or schools and would communicate the wishes and decisions of their "bases" at the general meetings. Deliberation in both forums was completely open, and all resolutions were submitted to a vote. Official propaganda portrayed these meetings as secret revolutionary councils. The truth was quite different.

The students did not have weapons or make any plans to arm themselves. At one of the general meetings, someone proposed the creation of defense units to be called "security columns," but the committee as a whole indignantly rejected the idea. It is true that many of the visible and influential leaders were or had been members of one or another left-wing party: *pescados* (of the Communist Party), *mamelucos* (the Maoists), *troskos* (members of the Third Trotskyist International), *anarcos* (anarchists). But only 10 percent of the leaders who formed the CNH had any party membership at all, and even among that 10 percent, some broke off their party connections during the heat of the conflict because they could not accept their respective bureaucratic "lines." There were Catholics in the movement, PAN members, even priests. The great mass of young rebels were inspired by a direct, libertarian enthusiasm. Their actions were motivated more by emotions than ideas, or when ideas were at issue, specifics were more important than overall ideologies.

In the meetings, the debates would have a practical content; they had to be—the constantly repeated magical invocation of the time—"to the point" (*concretito*). As *El Pino* remembers, "There were those who would go off for an hour on Althusser, Marx, and Lenin, but what interested the people was: *What are we going to do tomorrow?*"[67]

Not even a great veteran of the left like the writer José "Pepe" Revueltas could generate his kind of ideological revolutionary passion among the students. The influence of Revueltas on the Movement was little more than symbolic. He was a venerable "doddering old man" (*ruco*) but a *ruco* nonetheless:

> One day Revueltas tried to read a description of his socialist project to the CNH. He called it "Cognitive Democracy." Nobody there had a very clear idea of what democracy was, but *cognitive* democracy, nothing, nil! Pepe had a folder of sixty, eighty, a hundred pages. . . . They whistled him off the stage after a minute and a half. "This old man with a goatee, what is he talking about?". . . Just a

> minute and a half of Marx plus who knows how many other names out of nowhere and the murmurs began and then Pepe had to step down.[68]

Not even in the (more conservative) technical schools could "doddering old men"—even if they were pure revolutionaries—command this generation. In the Faculty of Sciences, Dr. Elí de Gortari, who had been a central figure in the Student Movement at the University of Michoacán in 1962, was much admired. But the leader would be a charismatic young man in a wheelchair named Marcelino Perelló. In the overflowing auditorium of the Faculty of Engineering, among calls for a return to classes and others demanding that the strike go on forever, Heberto Castillo's voice commanded attention. He was a prestigious teacher in civil engineering—who had invented and patented an important formula for structural calculations—and he was an impassioned leftist. He had been a member of the MLN and had attended the "Tricontinental" meeting of liberation movements in 1967 in Cuba. The students respected him, but the man they followed was a young and fiery orator: Salvador Ruiz Villegas.

In the air there was something of the old Vasconcelos enthusiasm that had reigned among their student ancestors forty years before. It was from that period that they drew their system of organizing for action in "brigades." Luis González de Alba—a student leader—would later say:

> After the meetings we would meet in three rooms and decide where each brigade should go. We would divide the cans for collecting money and the leaflets that we needed to distribute. It's worth noting that in those days we used to pass out six hundred leaflets a day and collect one or two thousand pesos. Then we would hold impromptu meetings in the streets, and it was not only us, the students, who spoke, but we would invite the public to step up. And they would say—clearly, decisively, frankly—that they were with us. It was the golden age of the Movement—from the twelfth to the twenty-seventh of August.[69]

Though the students would leave their schools to visit working-class neighborhoods, factories, and other towns, the real stages for the Movement were the campuses of the University and the Politécnico. During that brief "golden age," celebration and deliberation were the most obvious qualities of life. The students lit bonfires at night and sang songs from the Spanish Civil War, but there were also pickup games of soccer on the University Esplanade. The two campuses had become immense agoras—public spaces for the practice of democracy. Against a system that enforced their obedience through the benefits of "bread" or the punishments of "the bludgeon," the students offered every possible synonym for the word *dialogue*: *discussion, deliberation, dissent.* In total contrast to the usual methods of the PRI, everything was discussed between equals within the Movement, everything was subject to a vote. "They could accuse us of anything, but never of acting without open consultation; there was so much consultation that it made us move too slowly." An almost ridiculous exaggeration of the parliamentary spirit was a natural response to the political asphyxiation imposed by the system, the total absence of true political discussion in Mexico.

Its *public* character was the very heart of the Movement. The students went out to the public, they sought the public, they publicly discussed their ideas, they made their proposals matters of public record. Their fundamental weapon for all this was the mimeograph machine. González de Alba would remember: "All night long stretched out to sleep in a corridor in the Ciudad Universitaria, I would hear the sound of the mimeograph machine printing flyers."[70] From the profusion of those leaflets arose one of the dark rumors circulated about the Movement. Where did all that money come from? But it was an easy charge to defuse. Paper and ink were the only expenses, and they came from the stocks of the schools themselves, through soliciting money from the public, and through student contributions, which were enough and more for the few statements that the CNH placed in newspapers and even to pay for the strikers' meals.

In the middle of August the National Strike Committee pressed the government to "publicly respond" to its list of requests. The students wanted to initiate a "public dialogue" with the regime: "We would emphasize that to establish dialogue in a public form has advantages because it involves the massive and democratic participation of everyone."[71]

There were two reasons for the student insistence on the public character of the movement. If the leaders were willing to agree to private arrangements or anything "under the table," they would have immediately lost the support of their base. But a deeper motive also had its effect. The confidential nature of the Mexican political system—private, secret, almost mafia—was an essential key to its operation. If they could submit official truth to public scrutiny, the students would be "stripping the PRI naked" or at least striking at its Achilles' heel. The government was well aware of this and refused to send representatives to a public debate called at the Ciudad Universitaria on Tuesday, August

20. Besides student leaders from the CNH and representatives from the Coalition of Teachers, speakers included a member of the Association of Fathers of Families of the IPN (the Politécnico), a representative of the University, a peasant from Zapata's state of Morelos, and the leader of the youth groups of PAN, Diego Fernández de Ceballos, who pointed out that "the Movement has finally touched on one of Mexico's taboos: the Presidential Position."[72]

That day, Professor Heberto Castillo proposed doing a television program in which various university instructors would explain—to the public—the reasons for the Movement. To everyone's surprise, the program was telecast the very next day as part of a series called *Anatomías*, produced by Jorge Saldaña. Castillo appeared on it and spoke with complete freedom and great restraint: "The Student Movement has no intentions of subverting the institutional order. The student leaders are prepared to initiate a dialogue with the highest authorities."[73] That night the students could dream of marching with their hands raised in the V for Victory.

On August 22, the newspapers were filled with the news of the Soviet tank columns rolling into Czechoslovakia and the crushing of the "Prague Spring." It was pitiful to see Dubček's dejected face on the nightly news, as he read the retractions forced upon him by the invaders, or the faces of those young Czech students confronting the tanks. It was a relief to turn back to Mexico. Here there were no tanks in the streets, but signs of respect and conciliation. The Minister of the Interior contacted the CNH by phone and communicated an official willingness to begin the dialogue. The youthful obsession with the public nature of the dialogue was so extreme that the CNH assembly conducted an interminable discussion on the Byzantine issue of whether telephone calls from the government to the CNH were or were not *public*. The final decision (and resolution) was negative. A telephone call was not public; the authorities had to make a genuinely public declaration of their willingness to negotiate. Despite all this, communication was not broken off. And in fact there always had been contacts between some individual leaders and officials of the government. The CNH now requested a list of such contacts and again insisted that the debate must take place publicly.

During those days, the press carried a swelling flood of full-page paid government advertisements attacking the Student Movement, but the Minister of the Interior maintained contacts by letter and by phone. It was "the CNH who set the conditions, not the government." It became known that the official participants in the dialogue would be Echeverría himself, the official government attorneys of the republic and the Distrito Federal, plus the Minister of Education, the writer Agustín

Yáñez. Heberto Castillo demanded more: Radio, television, and the newspapers must be present.[74]

On the twenty-seventh of August, the Student Movement reached its zenith and of itself provoked the beginning of its fall. An imposing demonstration marched to the Zócalo. The cheers had a different quality from those of a few weeks earlier: "We don't want the Olympics! We want the Revolution!" "We don't want the *Informe*! We only want the truth!" Many chants and signs attacked the grenadiers. There were groups merely asking for "Respect for the Constitution," but others, more hotheaded, broadcast their defiance and their anarchic spirit of celebration: "We will rise in revolt when we feel like it!" The target for the greatest amount of verbal abuse was the President of the Republic, Díaz Ordaz:

Tell me why, tell me, Gustavo!
Tell me why you are a coward!
Tell me why you have no mother!
Tell me, Gustavo! Why?

On the fences, on the posts, on the very walls of the National Palace, some students pasted pictures of Che Guevara alongside elaborate insults directed at Díaz Ordaz. Someone painted an equation: GDO = DOG. One sign showed the profile of the President on the wider but similar profile of an ape.

The meeting began at eight o'clock at night. Roberto Escudero from the Faculty of Philosophy thundered against "the tyranny, the dictatorship and the injustice" and demanded that "the Government appear before the people." While some students entered the cathedral and with the permission of the priest began to ring the bells, orators called for the release of political prisoners, assailed the deputies, the *charro* union leaders, the press bought and paid for. They read a message from the railway man Demetrio Vallejo written from his prison cell. The meeting had become an Assembly of the People. To each question raised by the speakers, the crowd would respond with a resounding YESSSS! The site of the impending dialogue had to be determined. Shouts went up of "The Zócalo! The Zócalo!" At one point, the crowd decided to demand that the President come out to meet the crowd. In a great chorus, over and over they chanted the words: "COME OUT ON THE BALCONY, MONKEY WITH A BIG SNOUT! (*¡SAL AL BALCÓN, CHANGO HOCICON!*)"

The defiant Assembly decided that the government would have to conduct the dialogue on September 1 at ten o'clock in the morning, one hour before the State of the Nation Address. During the meeting, some students draped a red and black flag on one of the lampposts of the Zócalo (and later hauled it down). The Assembly made a final decision to leave a contingent on guard in the Zócalo, who could camp out there until the next Assembly, scheduled for September 1. When the meeting ended, it was almost ten o'clock. "The country was ours," said the charismatic leader from Chapingo, Tomás Cervantes Cabeza de Vaca, many years later.[75]

It may have seemed so, but it was not. Because for that night of August 27 when the students were drunk on jubilant drafts of protest, for that explosion of parricidal energy, that caricature of the instantaneous revolution, for the collective desacralization of patriotic and national symbols—the flag supplanted, the bells of the cathedral rung in rebellion, the threat to burn the gates of the National Palace, the painted slogans and the face of Che Guevara imprinted on the walls of the National Palace itself—for all this they would not be forgiven.

The comments on the origins of his mother, the foul and venomous words, struck a sensitive chord in Díaz Ordaz partly because they reflected his own behavior within his everyday political environment. One of his paid writers who knew him well spoke of his "barroom language" and the "angry coarseness of the words" that Díaz Ordaz used to humiliate his subordinates, including of course his ministers.[76] But the worst thing—the inadmissible sin—was the fact that their crude humor was directed at the presidency. And something more. The cruel sentences and images multiplied, as if in infinite mirrors, the greatest, original wound—the "personal ugliness" of *El Señor Presidente.*

Within the personal vision of the President, the Student Movement was no more than the final and most complex puzzle in a long series that had begun with the union movements at the end of the fifties and continued through the successive conflicts of his own *sexenio*: doctors, students in the provincial universities, guerrilla activity in Chihuahua and Guerrero. In 1965 a group of young militants (connected with the Communist Party) had tried to attack a military barracks in Chihuahua, emulating Fidel Castro's attack on the Moncada Barracks, which—though unsuccessful—began the chain of events that would lead to the Cuban Revolution. But the young Mexicans met only death and a burial without coffins at the explicit order of the governor of Chihuahua: "They fought for soil, no? Then give them soil until they're stuffed with it." Had they lived, they would have taken to the mountains of Guerrero—Mexico's wildest and most violent state (75 percent rural, with a minimal literacy rate and after Chiapas the second-poorest zone in Mexico). There, in 1968, two small guerrilla groups were operating, led independently by teachers who had been born in Guerrero—Genaro Vázquez Rojas and Lucio Cabañas.

These guerrilla leaders had long histories of leftist activism and were dedicated Marxists. In the state of Guerrero, where the strongest law was often "the law of the machete," both men had witnessed the military slaughter of civilians. These rebels—each with only a few hundred followers—held out for years (Cabañas until the mid-seventies) in the rugged sierra of Guerrero. For Díaz Ordaz, all these happenings had a single common denominator: They were products of a Communist conspiracy. "He was convinced," remembers the business leader Juan Sánchez Navarro, "that there was an international plot internally supported by groups of the extreme left with the intention of changing all of Mexican society to Marxist doctrines and he was horrified at the thought that this could happen to the country."[77]

In his memoirs, Díaz Ordaz left a detailed listing of other pieces of the conspiratorial puzzle, which he considered irrefutable and felt certain he could fit together into a coherent whole. There was, for instance, the presence of Mexican representatives—among them Heberto Castillo—at the "Tricontinental" meeting in Cuba in 1967. Somehow, the President believed that plans were laid there to prevent Mexico from successfully hosting the Olympic Games.

Certainly the "Tricontinental" had other, more ambitious objectives: the active exportation of the Cuban Revolution to underdeveloped countries, especially in Latin America. But toward Mexico (and despite some impassioned speeches by Mexican delegates to the meeting), Cuba maintained a special policy, clearly based on Castro's appreciation of Mexico's refusal to vote for the expulsion of Cuba from the OAS and its continued maintenance of full official relations. Everywhere else in Latin America, yes, but the Revolution was not packaged for export to Mexico.

Yet Díaz Ordaz went on assembling his pieces. There had been an international Communist Party meeting in Sofia in 1967. The head of the Mexican Communist Party, Arnoldo Martínez Verdugo, had gotten a specific order (the President was convinced of it) to stir up trouble *between* student groups so as to stimulate student discontent—all for the purpose of obstructing the Mexico Olympics. Further confirmation of the plot (worldwide, of course, against Mexico) was the Paris student uprising of May 1968. The President carefully examined photos, slogans, and pamphlets (the words presumably translated for him), though he found the slogans as disgusting as the revolt against authority itself.

On the students equating love with the revolution, he commented, "Will they make love as badly as they make the revolution? They have my sympathies." And the President began to discuss building up more military technology suitable for urban warfare, "defensive arms . . . small tanks (*tanquecitos*)" and armored Jeeps.

When the student demonstrations began, Díaz Ordaz felt that they were a natural consequence of the French precedent—the same international forces of the left were zeroing in their sights on Mexico. De Gaulle had used the army—that was an excellent move, crying out to be emulated. But Díaz Ordaz also insisted on his own interpretation of what De Gaulle had done.[78] When Daniel Cosío Villegas—who had begun to publish a weekly column in *Excélsior*—pointed out that De Gaulle had given explicit orders that the troops never fire their weapons (a fact reported by the entire French press), Díaz Ordaz refused to believe it, and wrote a private response to Cosío Villegas. His purpose was to justify his own use of the army (against a preparatory school) to perhaps the most independent and prestigious intellectual in the country at that time. And he argued that, after all, Cosío Villegas had not been there in Paris during the deliberations of the French government. De Gaulle surely could not have imposed any limitations on the use of arms in the face of an imminent "civil war."[79]

Every day Díaz Ordaz seemed to acquire new pieces to fit into his puzzle. Where did he get his information? From the very beginning, he had refused to personally approach the students, because he would not "put the presidential investiture at risk." Given his surly and withdrawn nature, he naturally came to rely on his own official government sources, in particular on the agencies that dealt with police or security matters: his Ministry of the Interior and the Federal Bureau of Security. The foreign press would have given him additional insights, but the President did not read foreign languages nor was he interested in acquainting himself with stories that criticized his actions. For the national press he had little respect, even though it was almost entirely obedient to the government. For him, "The difference between politicians and journalists is that they see with one eye while we use both."

He used to divide the letters sent to him into two piles, marking them in his own hand with either the words *adherence* or *solidarity* or *Interior* (because the critical letters were then dispatched directly to the Ministry of the Interior). He treated the published opinions of the intellectuals by the same criterion: for him or against him. The issue was never an attempt to understand the phenomenon (other than through his preset schemes) but only to calculate the range and degree of support.

His police spies, attending the meetings of the Movement, would surely tell him what was really going on.

How many students went to the demonstrations (some of them with clearly more than 100,000 participants)? His experts used maps, assigned three people to each meter, and came up with 15,000 or 16,000, tops.

Who promoted, directed, and benefited from the demonstrations? "International communist groups," of course, but there was also "a nauseating complicity" because, beyond the Cubans and the Soviets, the American FBI seemed to be involved, as well as Catholic groups like the right-wing student militants of MURO or the international organization Opus Dei or the Jesuits. The PAN too played its part. But the major responsibility lay with "the pro-Soviets and pro-Maoists."[80]

As for the mass of students, incomprehensibly available to be used "against one of the most beautiful events in the world . . . the Olympic Games," Díaz Ordaz would not grant them any capacity for discernment, any merit, any sentiments of generosity. "Youth!" he said to the paid writer who knew him well. "Those sons of bitches are not youth! They're nothing! Blood-sucking parasites! Beggars, cynics, illiterates! . . . Stinking filth! And they don't even have the balls to really stand up and fight, to start what they call their battle! Their battle! Those sons of . . ." and so on.[81]

Where did the students get their "enormous sums" of money? Well, he believed that there had been the order (from Sofia, Bulgaria) to give them money through the Cuban embassy, but that most of it came out of the government funds assigned to the University. There must be other sources too. Former officials of the López Mateos government, resentful of Díaz Ordaz, must be contributing to the student coffers. And naturally there were students "who got money in the mornings at the Cuban and Russian embassies and then showed up at night to get their subsidies from the United States."[82]

Many of these "facts" were clearly false, even ludicrous; others could never be proved. Only a few corresponded to reality. PAN and the Jesuits had expressed their support but through a free and genuine adherence to the democratic aims of the Movement, not because of any tawdry political calculations. The subservience to the Communists, the "huge sums" of money, the "collections from the embassies" were all absurd, but the President made no independent attempts to verify anything. The pieces he had been given by his agents (and Echeverría's Ministry of the Interior) fit smoothly into the jigsaw puzzle he was in the process of solving—the vast, illimitable conspiracy against Mexico and against him. And it was a "conspiracy" he felt in his guts, filling him with "anguish," as more than one report testifies. He saw himself "living in the midst of an immense ocean of problems."[83] Among other fears (and this one was shared by his brilliant Finance Minister, Ortiz Mena), he thought that sinister international forces—in the same way "they" had forced the devaluation of the French franc after May 1968—had a similar objective in Mexico: to create an economic conflict and loss of confidence in the Mexican economy. The next grim step would be devaluation of the peso.

On the night of August 27, when the students with shouts and insults were calling for him to come out on the balcony, the President was not in the National Palace. None of the information conveyed to him "was a delight to my ears." But an egregious bit of misinformation prompted much thought on his part. The red and black anarchist flag that had been hung from a lamppost (and then removed) was transformed into an elaborate anecdote. Student pistoleros were supposed to have forced their way into the cathedral at gunpoint and then hung their banner on the flagpole of the cathedral itself. A plot, to involve the clergy! The event had never happened, but for President Díaz Ordaz, it was another massive (and damning) piece of the puzzle.[84]

He was also convinced that Heberto Castillo, at that meeting, had established a claim to "maximum, indiscriminate, almost untouchable authority." Weren't they beginning to call him "the Little President" (*el Presidentito*)? Díaz Ordaz blamed Heberto Castillo for the seizure of the Zócalo, for turning it into an encampment, "with the intention of molesting the President on his entrances and exits [to and from the National Palace]." He imagined what the scene would be like, if he were to actually allow a dialogue to occur in the Zócalo: "the President of the Republic, seated on the bench of the accused, answering questions and enduring insult and ridicule. Then would come physical pressure to sign some document."

To insult the National Palace by plastering it "with emblems of a Latin American guerrilla" was to desecrate "the symbol of our nationality." And many citizens agreed that the students, on the night of August 27, had crossed a line of blasphemy. For the President, the final piece of the jigsaw puzzle had now locked into place. These rebels had "insulted Mexico." There was no doubt that this required an immediate response, and one that would have "neither limit nor end."[85]

Among the students it was said that "López Mateos would have come out on the balcony." Of course Díaz Ordaz was not in the palace to do so, but a gesture of some kind could have meant a great deal, especially in Mexico, a country enamored of symbols. A woman who had been a student would later say: "I think that if the President had come out on the balcony and confronted the crowd despite . . . the insults . . . he would have won over many of the comrades."[86] But Díaz Ordaz was incapable of making any gesture. He was not, unlike López Mateos, a man of gestures. He only believed in actions that could end a situation. And he had a crystalline conviction: "There exists an imperious need to fully maintain the principle of authority."[87]

THEY WERE EVICTED was the headline in *El Universal* on August 28. Striking at dawn, in a lightning operation with armored trucks, the police had cleared away three thousand demonstrators who were camped out, on guard, in the Zócalo. Police and students fought one another throughout the center of the city. The army made its appearance as well. Soldiers pursued students all the way to the statue of Charles IV (*El Caballito*) near the Alameda, where they attacked a small van with University identification, breaking the windows with their rifle butts and beating the passengers, who were mostly women. Some units of police and soldiers harassed students on their trek or drive back to the Ciudad Universitaria, leaping out of vehicles to cut groups off and making free use of their truncheons.

As the morning light grew stronger, new units of police arrived to patrol the area, protecting a demonstration organized by the government and made up of bureaucrats calling for "an apology to the flag." Fighting broke out again between police and students. The army returned, with fourteen tanks that advanced against the crowd. Fighting spread and intensified and—from the buildings—people dropped garbage and bottles and flowerpots down on the soldiers. Young people confronting tanks—the images of Prague had come alive on the streets of Mexico City.

Later in the day, the Minister of Defense spoke: "We made a Revolution to overthrow a dictatorship and then to destroy a usurpation. We do not want nor will we accept another dictatorship." The man delivering this statement was not just another general. It was Marcelino García Barragán, one of the last veterans of the Revolution still active in political life. He had been a Henriquista, had been removed from office by Miguel Alemán in a humiliating manner when he was governor of Jalisco—two weeks before his term expired, merely as a demonstration of power—but nevertheless, of all human qualities, loyalty counted most for him, in this case loyalty to the institutions of the government and above all to the highest among them—the authority of the President. "We have the impression," he added, "that the people support us, because they are tired of the rioters."[88]

In the following days, the signs of a hardening government line multiplied. Heberto Castillo was intercepted while driving his car and beaten by police clubs. Tanks were stationed at the outskirts of the University and the Politécnico. Government provocateurs appeared on the campuses and elsewhere—sometimes wearing masks—to attack students with clubs and blackjacks, metal pipes and rods. On August 31, Vocational School 7 (a preparatory school linked to the Politécnico) was attacked by two hundred armed men in plainclothes who beat up students and threw them into cattle trucks. The police had been given a clear order. "Disorder" was to be punished "with arrest" and then bringing the culprits up on charges "before the authorities." The instructions were systematically executed—with zeal—at various points in the capital.

The President, from his office at the National Palace, had a different view of events. According to him, riots and bus burnings were still continuing and the morale of the police had declined, because they were forced to confront much larger numbers who were "better armed." So the order had been given to arm the police more strongly, and—as was to be expected—"the police were retaliating." Díaz Ordaz also claimed that there were continual demonstrations in the Zócalo but that they were "in general orderly."[89]

In fact they did not even exist. From the morning of August 28, the CNH had understood the price they were paying for the verbal and symbolic excesses of the evening of the twenty-seventh. They sent no demonstrations into the Zócalo, and they rapidly repeated their willingness to enter into a dialogue, while offering explanations for what had occurred. The CNH had never ordered the painting of slogans or the raising of the red and black flag. The bells had been rung with the consent of the cathedral priest, Father Jesús Pérez. Heberto Castillo, who was recovering from two wounds to his forehead, softened his position on the nature of the dialogue. It was required "more than ever," but it did not have to be transmitted by radio and television, "as if it were a Roman circus." On August 31, a day before the presidential *Informe*, the CNH repeated its desire for a peaceful dialogue, without pressure from the army or the police, and extended an olive branch to the President:

> Even during the celebration of the 19th Olympic Games, we pledge to sweep the streets every day . . . to serve as ushers, as interpreters. . . . Before being students in a struggle to achieve greater freedom and democracy, we are Mexicans. . . . We are not against the Olympics. We want Mexico to fulfill its Olympic commitment with dignity.[90]

It was the students who were now "extending their hand." The President would show, in his *Informe*, whether it would be left hanging in the air.

"Insult does not offend me. Slander does not touch me. Hate has not arisen within me." Thirty-six seconds of applause greeted these words of Díaz Ordaz. He went on to suggest a national debate on the laws against "social dissolution" while refusing to repeal them; to deny that there were any political prisoners while hinting that, if there were no "illegitimate pressures," he might consider freeing some of these people who did not exist. But the events going on within the country had to be clearly understood "in the context of international information on similar bitter experiences." Only force could work against such opponents. He went on to give a long list of the damages, robberies, annoyances, and economic upheavals supposedly suffered by society at the hands of the students. And he added a claim to outrage on behalf of Mexican womanhood, asserting that "so many women" had been "abused in a filthy manner," who could after all have been the wives, mothers, sisters, or daughters "of any Mexican." He followed with praise for the "modest, heroic, ordinary men (*juanes*) of the army" and concluded with a threat that—in the context of what was to come—deserves to be long remembered: "We will go as far as we need to go."[91]

On September 3, the CNH met to discuss the *Informe*. They did not concern themselves with the list of requests, even less with the absurd charges of the President. (No one except Díaz Ordaz had suggested that the students were guilty of rapes. Had there been the slightest suggestion of proof, it would have been all over the front pages of the official press.) The CNH was, instead, broadening its critical propositions to include the entire political system:

> In Mexico, the system of political oppression and centralization in the exercise of power has reached such an all-embracing extreme—from the level of the policeman on the street all the way up to the President—that a simple struggle for minimal political liberties (like a demonstration in the streets and a request that political prisoners be released) confronts the most ordinary citizen with the crushing apparatus of the State and its nature of despotic, inexorable domination against which there is no appeal.[92]

The students were speaking with one another from a whole new perspective: They were moving from emotion to intellect and from there to the realm of politics. Their criticisms of the "old and obsolete system" aroused the attention of one of the oldest critics of that system,

Cosío Villegas, who began to move closer to the students. "Mexico is living on old ideas in a new world," he wrote on September 13, "and as there is no public life in Mexico, as the greatest political wisdom is silence, men of politics have become small and enigmatic."[93] The whole apparatus now bored the students. They had stopped "diverting themselves with that political circus." There was only one cure, and it was what the students were proposing—"to make the public life of the country public."

The last few days had been a time of tension and of hope. The CNH was beginning to gauge the seriousness of its position. Now it was tenaciously insisting on the dialogue, giving assurances that there would be no insults, not even cheering. They would even consider conducting the dialogue on paper. The Rector had called on the students to return to their classes. And then the CNH announced a new demonstration for Friday, September 13, and ceremonies to celebrate Independence Day, on the fifteenth, at the campuses of the University and the Politécnico. Weakened by the assaults upon it, the Student Movement, on that day of the thirteenth, would conduct a new patriotic battle, one that many would remember. It was the day of the "Silent Demonstration."

Overhead, along the line of march, a helicopter flew almost at the level of the treetops. Below, 200,000 young people marched with handkerchiefs tied over their mouths. Luis González de Alba would describe it:

> We had barely left the forest of Chapultepec, we had walked only a few blocks when the lines began to grow. All along the Paseo de la Reforma, the sidewalks, the greens, the monuments, and even the trees were full of a multitude who within a hundred meters doubled our numbers. And from those dozens of people and then hundreds of thousands, you could hear only the footsteps. . . . Steps, steps on the asphalt, steps, the sound of all the marching feet, the sound of thousands of feet that were advancing. The silence was more striking than the multitude. It seemed like we were trampling all the verbosity of the politicians, all the speeches that were always the same, all the demagoguery, the rhetoric, the piles of words never backed up by actions, the stream of lies, we were sweeping it all away under our feet.[94]

On September 15, a festive crowd filled the esplanade of the Ciudad Universitaria. It was a joyful celebration of Independence Day. At a big outdoor party, they played the traditional game of "mock marriages" and Heberto Castillo was asked to "unite the couples." There came a moment when students began to call on Castillo to "give the

Demonstration of students and teachers at the University, 1968
(The banner reads "Don't shoot!")
(Archivo Fotofija)

grito!" Which he did, in a lighthearted way, shouting out *Viva Mexico*! And the party went on, for some hours, a fiesta of freedom. Among the laughing boys and girls hugging each other on the esplanade, the young man from the Faculty of Engineering and his girlfriend kissed and celebrated the Independence of their country, the independence of their generation, their still vivid hopes for a better and broadly shared future.

In Díaz Ordaz's memoirs the whole scene is transformed. It becomes a new act in his long psychodrama of conspiracy, a new piece in the menacing puzzle. "Heberto Castillo succeeds in establishing himself as the supreme authority within the CNH." *El Presidentito*, according to Díaz Ordaz, "had them wave the national flag, played the national anthem and spoke the ritual words. And from that moment he thought of himself, and some groups in the CNH began to consider him as the President of the Republic. I realized, on the next day, that if we have 'little presidents' . . . then they could form a 'little State' with a 'little President' within a State."[95]

It was now not only a matter of "insulting Mexico" in its patriotic symbols of the flag, the National Palace, the Zócalo, and the cathedral. This was an attempt to replace the President in one of his most sacred ritual functions: delivering the *grito*. They had to be given "a little demonstration."

On September 18 the army occupied the University. There was no resistance offered by the students, but five hundred people were nevertheless arrested, including university officials. For two weeks the students roamed around the outskirts of their violated campus. Nonresistance did not prevent clubbings of students and other random military violence like the destruction of files. One terrified young student would lock herself in a bathroom, without food, for fifteen days. And not only the UNAM was the object of this "little demonstration." The most prestigious academic institution in the country, the Colegio de México, was sprayed with machine-gun fire from a moving car.

"An ugly scandal!" wrote the President, referring to the criticisms of his actions, which he completely discounted. They were evidently unaware that "the University was part of Mexico." Nor did they have his kind of information. Within the University, according to him, they had already formed no fewer than two cabinets, one around Rector Javier Barros Sierra, the other of course headed by the *Presidentito* Heberto Castillo. The "cabinet" of Barros Sierra was even violating unwritten rules of sexual power, admitting Mexican womanhood (in this case apparently unthreatened by rampant student lust) to positions on the cabinet—though the right order of things was not entirely overturned, since almost all of them were, for some odd reason, "standing in for their husbands" (presumably busy at conspiracies in dark corners of the city). And yet there was "a humorous note"—one of Barros Sierra's putative "appointees" would actually be the first instance in Mexico of a woman minister. The other "cabinet," headed by the *Presidentito*, was made up of intellectuals anxiously awaiting real power. These older rebels worked during the day at the University and "relaxed in the evening by going to demonstrations."[96]

On the nineteenth, the day after the invasion of the University, Rector Barros Sierra protested the assault and said that "young people need understanding, not violence." García Barragán, the Minister of Defense, implicitly responded when—in answer to a question about a clash between civilians and soldiers—he said, "The troops are not about to take this kind of thing any longer."[97] Each day, more people were arrested. That the opposition PAN should censure the government violence within the forum of the Legislature was no surprise, but even a member of the PPS, "the loyal opposition of the left," rose to defend the students. Why did Madrazo—who had been the liberalizing president of the PRI—not mount the podium and make a similar statement? "I don't know what happened to me," he would say months later, with true grief. He was sure that he could have prevented some terrible things. "The young would have listened to me," he said.[98] If not the young, at least some of the members of the PRI, because there were those among them who dared to rise and criticize the army's blitzkrieg against the University. But those who applauded from the benches thought, mistakenly, that these Priístas had spoken, as usual, under orders. The disobedient party members would receive a harsh reprimand and then be compelled to retract their words publicly.

This small problem of a few honorable men resolved, the PRI leaders in both houses of the Legislature launched a campaign of slander against the Rector. He would respond by resigning on September 23, giving as his reason (and his only response to the attacks): "We all know whose orders they obey in Mexico."[99]

There had not been a single proven instance—and the government press would have displayed such proofs ad infinitum—of a student carrying any real weapon other than words. But for Díaz Ordaz (as for the students) words *were* weapons. He used them as such, they reflected the natural flow of his atavistic feelings, linked to the physical discomfort that he felt with himself, with his appearance and his inner health ("the acidity of his life," said the paid writer who knew him). He had not the slightest trace of grace to him, as if the image he had seen all his life in the mirror made him see other people, other things with incredible coarseness. "I have never met another man," said the same writer, "with such a persistent and deeply rooted inability to love other people."[100]

During those days, the President called two meetings, one after the other: with representatives of the parties and then with the editors in chief of the newspapers. Both meetings followed the same pattern—a stream of generalized warnings: I will not hesitate to do anything, the *patria* is in danger, we have to save Mexico. Some witnesses remember that the President seemed out of control. He would lose track of his words, break off his sentences. No one of course interrupted him or even remotely dared to raise an objection. And some of them thought they were being made accomplices to a grave governmental decision.

Mexico City was living under a barely camouflaged state of siege. On September 24 the army seized the Casco de Santo Tomás. The students of the Politécnico fought against them and their police cohorts for hours. Pseudo-students appeared on the scene. They were armed, and they would shout things like "Long live the students!" while breaking store windows and destroying property. The army carried high-powered M1 rifles. Some student leaders were seized and sent to the Lecumberri Prison. At the Ciudad Universitaria—and in defiance of the tanks surrounding the UNAM—students massed outside the home of Barros Sierra, calling for him to withdraw his resignation. "If Barros Sierra resigns, we will go to the mountains!" The former Rector came out on his balcony. At first he just stood there, without a word. And then he announced that he was accepting their request. He would withdraw his resignation. Díaz Ordaz noted in his memoirs, "What a shame! But after all the University was full of garbage and filth!"[101] Rector Barros Sierra resumed his official activities. And on September 30, the army withdrew from the campuses of the University and the Politécnico.

On Wednesday, October 2, Daniel Cosío Villegas delivered an article to *Excélsior* that would be printed two days later. That elderly man of the mind—whose journey of action and intellect had been a long one, all the way back to the Revolution itself—knew that at this moment Mexico was playing out its destiny. He wrote that he was certain the students were in the right and that it could all work toward the greater health of the country:

> For the first time in a quarter of a century, authority, accustomed to official applause (insincere but very loud), has been obliged to recognize the existence of dissident public opinion. . . . The very life of that which with pride we call our revolution, the Mexican Revolution, is at stake. Will the country honorably emerge from this difficult test? . . . The entire nation awaits the outcome, this time to make an irrevocable judgment.[102]

Tanks in the Plaza de las Tres Culturas, Tlatelolco, October 2, 1968
(Archivo Fotofija)

* * *

The Olympics were due to open in a week. On the morning of October 2, at the home of Barros Sierra, three student leaders had begun the first official discussions with two representatives of the President: Andrés Caso and Jorge de la Vega Domínguez. The Student Movement had obviously been weakened. Its leadership had been battered and depleted. Those leaders still at liberty called a public meeting for that evening. The site was chosen naturally and spontaneously—a place that had already witnessed eight such meetings. It was the Plaza de las Tres Culturas (the Square of the Three Cultures) in the Tlatelolco neighborhood of Mexico City. It drew its name from its location at the center of a complex that joined a series of modern apartment buildings (for middle- and working-class people) with the ruins of an ancient pre-Hispanic temple and a colonial church rich in history. The meeting would begin at five in the afternoon.

By morning, tanks had surrounded the square. Soldiers were sitting on the tanks, or elsewhere in the sun, cleaning their bayonets. Despite the mounting violence in recent days, the leaders did not consider the military presence particularly unusual or dangerous. Tanks and soldiers had been on the streets ever since that first march led by the Rector. But they decided that the meeting should be brief. The Movement was just beginning to recover from the occupation of the two campuses. There was no need to run any risks. The CNH had even issued an explicit order. Only those of the leaders scheduled to speak should attend the meeting. But devotion to the cause and the hope of continuing the struggle brought out many people. The largest numbers gathered near the loudspeakers and the microphones, which had been placed on the third floor of one of the modern buildings, the Edificio Chihuahua.

There were between five and ten thousand people in the square. It was a young crowd, and they were in a good mood. Many of them were seated, or lying out, near the pre-Hispanic ruins, not far from the ancient Great Temple and the charnel house (*osario*) where the skulls of sacrificial victims had been stored. (Tlatelolco had been a great and separate city, the second-most important in the Aztec Empire.) Flanking them was the Church of Santiago Tlaltelolco (the Náhuatl spelling), where the Franciscan Father Bernardino de Sahagún had—in the sixteenth century—established a school for the Indian nobility.

Some of the students noticed an unusual presence. Young men "with very short hair," wearing a white glove or a knotted white handkerchief on their left hands, were circling the Edificio Chihuahua. From a distance, the army stood watch. Their commander was General José Hernández Toledo, who had led the invasion of the University of

Morelia in October 1966, the assault on the preparatory school that had led to the birth of the Movement, and the occupation of the National University on September 18. The leader who was functioning as master of ceremonies was urged to hurry. The meeting should be over, the people should disperse, before anything unexpected could happen. Suddenly men wearing the white gloves or handkerchiefs moved together, up the two access stairs of the Edificio Chihuahua to the third floor, where they ordered the leaders to drop down flat on the floor of their balcony. It was about six-twenty in the afternoon. Two helicopters came flying in low over the plaza, in steadily narrowing and threatening circles. They were attack helicopters, clearly armed for war. Suddenly they dropped flares: "one green and one red." (Other versions say that one flare was thrown from the church.) People in the crowd shouted, "Don't move! Don't run!" And the firing began, from the upper floors of the buildings and from the helicopters.[103]

Jean François Held, correspondent for the French weekly magazine *Le Nouvel Obervateur*, who had been in the Middle East and in Vietnam, would write, "Never have I seen a crowd fired on like that." Some of the young men with the white gloves, whom he had thought were students, forced him to take shelter in an apartment that was full of police. The men with the white gloves were constantly coming and going. He overheard one of them say, "Twenty-four hours ago we got orders to come here, wearing something white on our hands, without papers but with our pistols."[104] They were the Battalion Olympia, specially trained as a security force for the Olympic Games. A man who served in the battalion, Ernesto Morales Soto, would later state, "We had orders, when a flare was dropped . . . to station ourselves at both entrances [to the Edificio Chihuahua] and prevent anybody from entering or leaving."[105] One of their assignments was obviously to arrest the leaders. We may never know whether a direct order was given—in so many words—to pour heavy weapons fire down from a height on a trapped, unarmed crowd. Or whether it was simply that no limits were placed on how much shooting they were free to do.

The student leader Gilberto Guevara Niebla saw the members of the Battalion Olympia shooting directly at the crowd with "pistols, machine guns, rifles of different calibers." Then the army began to advance toward the Chihuahua Building, in combat formation. The flares had been their signal too.[106]

Some squads of soldiers entered the square from the other side, coming in behind the Chihuahua Building, which was raised on pylons. There was a moment when the bullets of the Battalion Olympia—who were firing pistols and heavy-caliber weapons from the third-floor balcony they had occupied—began to land among (and hit) the soldiers who were emerging from under the building. "It was a disastrous operation from the military point of view, absurdly conceived."[107] The soldiers swung their guns toward what they thought was an attack and opened up directly at the forces that were their partners in the operation. They had no radio communication to tell them otherwise. The men on the balcony—some of them now wounded—threw themselves down on the ground among their student prisoners and shouted in chorus, "Battalion Olympia!"

Down on the expanse of the plaza, people were fleeing in terror. A desperate human wave swept from the Chihuahua Building toward the far end of the square only to meet the advancing soldiers and turn and race away again until death might come from a bullet, a bayonet, or through falling and being crushed by the crowd that was running—with absolutely nowhere to turn—for their lives. Bullets were descending like rain on the plaza. And soon—because it was October in Mexico City—rain began to fall from the sky and run with the blood. The army and the police and the paramilitary gunmen had created a closed circle of hell.

The army responded to the sound of gunfire coming from positions they could not identify. Soldiers would confirm this reaction: "As I was advancing, I heard shots coming from up in some buildings, against me and against the rest of the soldiers in my unit."[108] There were some soldiers wounded (and killed). The entrance trajectories that doctors later had a chance to examine—in many wounded students and a much smaller number of soldiers—had come overwhelmingly from above and from military-issue weapons. It was a terror operation, perhaps partly out of control (but who can say how much?), with not only the Battalion Olympia but other snipers stationed in the buildings, pouring down waves of "stray shots" that the government could claim had been fired by students. The most mysterious of these "stray shots"—at the very beginning of the action—struck General Hernández Toledo, who was immediately taken to a hospital.

It was an odd bullet, a .22-caliber. During the massacre, one of the members of the Battalion Olympia, cursing at a wounded student, referred to the ".22 bullets . . . that your little friends have shot."[109] It was as if he knew beforehand that the students were to be connected with that specific grade of low-caliber bullet (which soldiers at least would not be using). It had to appear that students had fired first. What better evidence than the wounding of the commander of the army by a bullet that could not be normal military issue? Was there a police sharpshooter, stationed at some convenient point, with the assignment of putting a .22 bullet—very unlikely to kill—somewhere into the body of the general? An investigation might have determined something—pro or con the hypothesis—but there would never be any investigation.

The shooting went on, intensely, for sixty-two minutes, according to one witness. Then it began to diminish, almost coming to a stop. Suddenly new volleys began, and behind them an abnormally loud burst of fire. One of the small tanks had opened up, shooting into the square. It was a quarter after seven. As if the tank had been yet another signal, a new round of firing began. And it continued till eleven at night. After that, all through the darkness and through the following day, stray shots could still be heard around the Plaza de las Tres Culturas. Not only pistols and rifles and bayonets were used in the massacre but tanks, bazookas, machine guns. About five thousand soldiers fired fifteen thousand rifle rounds. By nightfall, two thousand people were in custody. Many of them were stripped, beaten, taunted, and abused.

It is certain that orders had been given not to leave any evidence behind. There were to be no obvious traces. Lights were extinguished. Telephone service to the entire area was cut. Photographers were told they would lose their cameras if they tried to take pictures. A foreign journalist who refused to surrender his film was stabbed a couple of times with a bayonet to make him understand the need for "order."[110] Even ambulances were forbidden to directly enter the killing ground. "They didn't let them in! They arrived, wailing like women gone crazy! They were stopped! They told them to turn off their sirens, to put out their lights!"[111] The embargo on information extended even to the hospitals. The police would prevent people from getting in to see their wounded relatives. There seems little doubt that most of the bodies were trucked away and burned. (Luis Echeverría—when President after his years of service as Díaz Ordaz's top cop—is presumed to have given an order to burn bodies after a later minimassacre of his own. If so, the order was given effortlessly, as if it had been given before.)

Only the voices remain, of the murderers, of the relatives, of the witnesses, of the arrested leaders, all of them collected by the writer Elena Poniatowska in homage to the fallen:

> —The blood of my daughter would have splashed on the shoes of all those boys running back and forth through the plaza.
>
> —"Brother, are you hit? My brother, answer me!" . . . Julio was fifteen years old. He was studying at the Vocational School Number 1, near Tlatelolco. It was the second time in his life that he went to a political meeting. He had invited me to go that day. The first time was when we both went together to the silent demonstration. My

father died shortly after Julio. He had a heart attack at the shock. He was his youngest child, his only son. He used to repeat over and over, "But why my son?". . . My mother went on living, I don't know how.
—Most of the corpses were lying on their stomachs, swelling up under the rain but their faces were turned up and back like trampled flowers, like the flowers spattered with mud and crushed in the gardens of the Chihuahua Building.
—I saw people fall on the stairs and when I was already on the roof I saw a kid of fourteen or fifteen running ahead of us . . . and they ran him through with a bayonet.
—There were bodies lying on the concrete, waiting to be carried away! I could count many from the window, at least 68! They kept on piling them up under the rain. . . . I remember that my son, Carlitos, was wearing a green jacket and I looked for him in every body . . .
—The helicopter opened fire and you began to hear the sound of shells from the sky. They were firing like savages. The Edificio Chihuahua began burning, because of the gunfire from the sky.
—Don't make the body of my child disappear, like you've done with the others! Don't do it to me! Even if he's dead, even if he's among the dead, I want to see him![112]

There are a few testimonies to soldiers protecting women, guiding an elderly *señora* out of the plaza, one telling another to "shoot in the air! They're not criminals! They're only kids! Shoot in the air!" There are no witnesses to any acts of mercy on the part of the government battalions and special police target-shooting at the crowd and dragging students off to the trucks.

When he received news of the massacre of Tlatelolco, Octavio Paz—then the Mexican ambassador to India—resigned his post in protest. Lázaro Cárdenas went into a profound depression. He did not want to believe that "the soldiers of the Revolution" had used their arms against the students. For a time, he would protect Heberto Castillo (and others) against arrest, though the government would find and seize Castillo later, in 1969.

A witness who was present the morning after the massacre said that the plaza was red with blood that workers were trying to cover up with sawdust. The great temple of Tlatelolco had stood here, across its many years of human sacrifices. Octavio Paz, some weeks later, would compare the massacre to the Aztec rituals, "trying to terrify the population, using the same methods . . . as the Aztecs."[113] It was a frightful thing to feel that this site seemed to attract death, that centuries before the students had come here, rivers of Indian blood had flowed from the pyramid, and that in Mexico the ancient gods could well laugh at the passage of time, at the stubborn return of the blood.

But there is perhaps a more precise comparison than the Aztec sacrifices—which were after all meant to calm and reassure the people of Tenochtitlan, convincing them of their own continuity through the ritual offering of captives. The massacre was more like some acts of the conquistadors—Cortés at Cholula closing off all exits to the great square and, while the unarmed Indian nobles danced in his honor, ordering his gunmen to massive slaughter as a mode of terror and control.

In 1915, President Eulalio Gutiérrez had said to José Vasconcelos, "The Mexican landscape smells of blood," without knowing that the voice of an ancient Aztec poet was speaking through him:

And the smell of the blood moistened the air.
And the smell of the blood stained the air.[114]

In his memoirs, the President recorded his personal version of what had happened. Since the students had not been able to take control of the National Palace, they were trying to take control of the Plaza de las Tres Culturas, with the objective of eventually seizing the Ministry of Foreign Relations, which overlooked the plaza.

They had already managed to "infatuate" many correspondents who had come to Mexico for the Olympic Games. One of the student leaders arrived in the first of a cortege of four cars—all the passengers armed with submachine guns, as was the leader himself, whom Díaz Ordaz even identifies as Sócrates Amado Campos Lemus. "In the middle of it all, it seems that he was the most manly (*hombrecito*) of them all." The army was then "just waiting." Their assignment was to prevent the seizure of the Ministry of Foreign Relations. The students (suddenly depicted as masters of some mysterious technology) managed to interfere with the President's radio communications. But their leaders gave the order to capture the Ministry.

The army moved to stop "the rioters" from fulfilling their intentions. Yes, there was a flare, "nothing but a simple semaphore." Díaz Ordaz then had General Hernández Toledo—that specialist in assaults on universities and preparatory schools—walking into the square "with a megaphone," where he makes a request for "good sense." He takes a bullet through the shoulder, shot with a descending trajectory. More shots—claims Díaz Ordaz—came down from above, "where there were no soldiers, there were no police, only 'them,' who were doing the shooting." He then adds, like a literary master of reticence (given what had happened in reality during the time span covered by his phrase), "Their shooting did not last long."

There was a "roundup" of agitators, who were sent to Campo Militar Número 1, to be "questioned." Of course they had to "confiscate the submachine guns." And then he writes about the students, at last, gaining "their own sweet dead"—*sus muertitos*. (Mexican Spanish is a language addicted to diminutives, with various overtones of meaning. Adding the diminutive "*-ito*" to the word for "dead" both reduces the victims and conveys a polite sense of affection. A waiter in a restaurant is being courteously deprecatory toward the subject at issue—such a small thing that he can offer you!—when he talks about bringing you a *cafecito* rather than a simple coffee. Referring to the many dead he had strewn behind him as *muertitos* was something the cacique Gonzalo N. Santos was prone to do.)

The President deplored the cost the students had paid to win their *muertitos*. They had shot down their own people. This was demonstrated by the "fact" that the greater part of the dead and wounded, "both rioters and soldiers" showed "clearly vertical trajectories" in their bullet wounds. It had all been the young "idealists" firing their submachine guns from the buildings.

Then the President's mind turned to the wounded and the dead and especially the doctors and the ambulances going about their "sad duties." Almost immediately he felt impelled to repeat as if he were trying to convince himself: "They finally have their dead. And at what cost! And possibly shot by their own comrades!" Within a few lines, Díaz Ordaz had used three different formulas to describe those "responsible." First it was "they" (like that, between quotes), then it was the young "idealists" firing down at their own comrades, then it was "possibly" them who had killed their fellow students. Was he trying to moderate something he knew was a lie? Quickly the narration swerves away to thoughts of the coming morning, when the city and the country and millions of Mexicans will go about their normal lives. And yet it had been a bloody and painful day, "certainly for some Mexican homes. Lacerating and painful for all Mexicans."

But he offers not the slightest sentiment of remorse. Whatever he may have believed (or told himself or was setting down for the world to be told), he would then add something with the absolute ring of truth and coherence, at least as coming from the mind of Gustavo Díaz Ordaz. He would express his sense of having fulfilled his duty: "They

Bullet holes in window overlooking the Plaza of Tlatelolco, October 3, 1968
(Photo Hemanos Mayo. Archive general de la Nación)

want to change this Mexico of ours. They want to change it for another which we do not like. If we want to preserve it and we remain united, they will not change what is ours."[115]

The President's account is of course riddled with fantasies or lies. There was never any attempt (nor does anyone else claim there was) to seize the Ministry of Foreign Relations. Neither Sócrates Amado Campos Lemus nor his friends came to the meeting armed with submachine guns or even pistols. No student had any firepower (a few of them admitted years later that they had brought small handguns for protection but that they had quickly thrown them away before being seized by the Battalion Olympia). The trajectories of the bullets were mostly downward, but nobody in Mexico believed it was the students who had killed their own comrades. Not even "possibly." Díaz Ordaz never mentions the Battalion Olympia.

By the seventies—when he composed his memoirs—the participation of this group was public knowledge, but many details would remain hidden. The Battalion Olympia had originally been formed from elements of the army. Their specific assignment was "rapid assaults," which in the case of Tlatelolco meant arresting students, especially the leaders. Why, then, did they fire on the crowd? Were they the only ones who fired on the crowd? Were they "them"? One student saw an older man, in his thirties, with light brown hair of ordinary length, firing point-blank at the leaders on the balcony and then "with a chilling tranquility" emptying his last two shots at all the people fleeing back and forth in their trap. The man was wearing a white glove on his left hand, but given his age and his haircut, he could not have been a member of the Battalion Olympia, though he was pretending to be. Who was he? Who had placed him there with orders to fire on the crowd?[116]

"It's not true that we tortured people in Campo Militar No. 1. The police had a special installation," General Felix Galván López would say. He had been Chief of Staff for Defense Minister Marcelino García Barragán. "They took students there, professors, whoever they wanted. Imagine what may have happened in those rooms, what the plainclothes police (*judiciales*) may have done."[117]

General Marcelino García Barragán says that the army had no orders to attack the students. They were there to clear the plaza. They had fired against sharpshooters they believed were students but were really members of special military groups like the Battalion Olympia or else plainclothes police who had mingled with them, posing as members of the groups. "The events of Tlatelolco," according to García Barragán, "were a real trap laid for the army." But if so, who laid the trap? Who were the provocateurs?[118]

Luís Echeverría "certainly had a role in it." At his job, he was known to be "a working machine": accomplished, efficient, responsible. He was always the first to arrive, the last to leave. When the President invited him to play golf, he would approach it as if it were a work assignment. He would arrive on the course not merely early but at five-thirty in the morning. Many people testify to his servility toward Díaz Ordaz (which he had earlier shown to other superiors). And he had gotten quite close to Díaz Ordaz since the days of Ruiz Cortines. He knew him like the lines of his own hand.

On the evening of October 2, Echeverría was having coffee with a distinguished pair of guests, the painter David Álfaro Siqueiros and his wife, Angélica Arenal. Siqueiros was of course the most celebrated former political prisoner in Mexico, an icon of the left. Was he invited by chance that night, to hear the telephone ring and Echeverría answer and behave in a way that was "possibly" surprised, at the news that there had been deaths at the Plaza de las Tres Culturas in Tlatelolco?[119]

After almost three decades, the central questions are still unanswered. By whom, when, and how was the strategy for October 2 planned? Who directed the Battalion Olympia? Who were the men who pretended to belong to it? Who shot at General Hernández Toledo? Who piloted the armed helicopters? Who threw the flares, and from where? Who and how many died at Tlatelolco? No one knows except their relatives. Many of them will not speak. They only set out candles on the Plaza de las Tres Culturas every October 2. Within the culture of secrecy and private decision—often secrecy for life—that marks the modern Mexican political system, will we ever have a list of the men who were responsible? Will they ever be directly and exhaustively judged, if not before the courts, then before history?

During the inaugural ceremonies of the Olympic Games on October 10, it was announced over the public-address system that the President of Mexico, Gustavo Díaz Ordaz, was present in the stadium. Many people had no idea of the magnitude of what had happened. Others preferred to block it out, to forget it. Abroad, the images of Tlatelolco appeared on the nightly news, but in Mexico the mass media omitted or distorted the truth. Only one radio station—connected with CBS and broadcasting in English in the capital—gave any details on that night of October 2. Almost nobody knew that the students in custody were being beaten and tortured, because the press was not permitted into the prisons or Campo Militar Número 1 (nor would they ever have been permitted to publish any information they might have discovered).

There was a moment of spectacle in the stadium, just before the

President officially opened the games. An immense kite in the form of the dove of peace, the symbol of the Mexican Olympics, was floated into the sky above the presidential box. Days before, on the morning of October 3, on the Peripheric Beltway that circles the city, someone had splashed that Olympic symbol—the white dove of peace—with red paint. Soon afterward, the young man from the Faculty of Engineering had driven by it. He remembered a professor who had often talked about economic progress in the Mexican sixties. It was true, the professor had said, that the distribution of wealth was still unequal, but nevertheless—and to be celebrated—was the fact that Mexico was at an economic "takeoff point," poised to become a world-class, competitive economy. And it was all due to good government and to wise economic policies. The young man was not devoid of a capacity for practical judgments. The arguments—in the lecture hall—had seemed of interest, but none of that mattered a damn to him now. What filled his mind was an image of the dove of peace covered with blood.

One year later, in the fifth *Informe* of his presidential term, Díaz Ordaz would spend considerable time on the events of October 2, 1968. He would use all his usual formulas. The Student Movement had been an ambush, a provocation, terrorism, subversion, illegality, an ignoble exploitation [of the occasion of the Olympic Games], rancor, violence, adventurous willfulness, absolute and irrational refusal of every formula for compromise, and, above all, anarchy, over and over again, "anarchy!" His government had been the embodiment of responsibility, peace, tranquility, institutions, freedom, progress, sovereignty, flawless patriotism, strict discipline, indomitable loyalty, and, above all, order. Over and over again, "order!" At the most dramatic moment of his speech, he raised his voice and said: "I totally assume the responsibility: personal, ethical, juridical, political and historical, for the decisions of the government in relation to the events of the past year."[120]

While leaving the Legislature after the speech, one experienced politician said to another, "Echeverría is the *tapado*"—the President's as-yet-undisclosed choice to succeed him. By taking the responsibility completely upon himself, Díaz Ordaz was opening the way for his successor, the man who had won him over through his loyalty, his Minister of the Interior during the massacre of Tlatelolco.

On October 11, 1968, Cosío Villegas wrote, "The government will fall into a disrepute that nothing and nobody will ever wash away."[121] But President Díaz Ordaz seemed to feel certain that he had solved the jigsaw puzzle of the conspiracy against Mexico.

And yet he lived with great unease through what remained of his term. "He knew that he was going to be left alone, isolated . . . and he would grotesquely refuse any gift . . . he wouldn't let anyone praise him . . . he was dry, peremptory, he viewed life with the greatest harshness."[122] His gastric problems worsened. He had to have operations for the loosening of a retina. His wife was broken by the tension, lived apart from him in a separate residence, and would die within four years of cerebral arteriosclerosis. And he became enraged at Luis Echeverría, now the "revealed" candidate, who through a series of symbolic actions was trying to reconcile himself with the students and disclaim any responsibility for 1968. Echeverría had suddenly become "disloyal," and Díaz Ordaz almost decided to cancel his nomination. The decision would have been supported by the Minister of Defense, but the President finally decided against it. His son, Gustavo Junior, remembers that he was immersed in his specific mental affliction, he was "much more anguished" than ever. He slept very badly, and he began to count the days remaining till the end of his term.[123]

In June 1970, President Díaz Ordaz arrived by helicopter at the Estadio Azteca. Mexico was hosting the second-most important athletic event in the world (for many, even more important than the Olympics), the World Cup soccer tournament. The President of the nation would of course inaugurate the games. By appearing on such a significant occasion for the Mexican public, he was submitting himself to their verdict. The judgment was general and broad—repudiation, a democratic plebiscite of whistles and insults. He would have liked to put order into this anarchy, but not even he, an expert at the job, could have tried to solve the jigsaw puzzle of 110,000 people.

In 1977, ex-President Díaz Ordaz was appointed ambassador to Spain. He was infinitely bothered by the assignment. He had neither the will, the taste, nor even the information to fill the function. After leaving power on December 1, 1970, he had stopped reading newspapers or attending to the news on televison or radio, as he told the press on one of the rare occasions when he spoke to them from his retirement. Now, to the usual questions raised about 1968, he responded by banging on the table and punctuating his words with his index finger. His greatest source of pride, he said, was 1968, because it had given him the chance to serve the country, to rescue it from disorder. "I do not have my hands stained with blood."[124]

He stayed in Spain a week and resigned. Sometime after his return to Mexico, he was diagnosed with the colon cancer that would kill him. Before his death, he gave his memoirs—in manuscript—to Gustavo Junior, "to do with them as you wish."

The unpublished memoirs present him as the only repository of truth and reason, the victim of a universal rejection. It is unlikely that they are all lies. They testify as well to his lack of information, his erroneous assertions, his significant omissions. They portray the mental landscape within which he made his decisions. The Mexican political system and the personal psychology of Díaz Ordaz had converged in a presidency of absolute power furnished with poor and dishonest information. In Mexico, the king was not only naked, he was blind.

Luis Tomás Cervantes Cabeza de Vaca was one of the most popular and charismatic student leaders of the Movement of 1968. He was a strong, powerfully built, and impetuous young man. Like most of his fellow leaders, he would go to jail, but he was arrested five days before the massacre. He had been in hiding, and one of his comrades broke under police pressure and informed on him. The man—Leobardo López Arreche—would later commit suicide.

On October 1, Cabeza de Vaca was confined in a cell seven feet long and three feet wide. They took him out the next day, kept him elsewhere for a few hours, and then put him back in. On the wall, while he was gone, a slogan had been painted, "F___ your mother, Díaz Ordaz murderer." It was signed, "your father, Cabeza de Vaca." They ordered him to "wipe it off with your head!" That same day, at ten o'clock at night, they handed him over to the military.

And men from the Dirección Federal de Seguridad (the Mexican equivalent of the FBI) as well as from the army (contrary to the claims for military purity made by Galván López) put a hood over his head and tied his arms behind his shoulders. A major demanded information on nonexistent stockpiles of arms, on payments to the Student Movement (including an attempt to force him to make a statement implicating Madrazo, former president of the PRI and a critic of authoritarian rule). The questions were punctuated with powerful blows to the stomach from a sergeant. The fist then descended, to his testicles. When he fell to the ground in agony, they used their booted feet on him. They wanted him to give them the names of other students. Who had led the "columns" at Tlatelolco? It was the first time he had heard of anything having happened on the second of October. He kept telling them he had been arrested days earlier. They still wanted names. And then they began to use electricity—probably cattle prods—on the hooded man with his arms painfully lashed behind him. Electricity at his testicles, his rectum, his mouth.

Sometimes they would stop, briefly. They would tell him that they were calling for the firing squad. Then the electric torture would resume, and the heavy fists. He was writhing on the ground "like a snake in a bonfire," weeping and screaming and cursing. Again, for a moment, they stopped. He could hardly breathe, his flesh was trembling from the electricity, his mouth was as dry as death. A soldier said, "Don't get your hopes up! You Communist pigs! If we get tired, there are gringos nearby to take over," and another one said, "Sir, the firing squad is ready!"

They heaved him up by his armpits and dragged him somewhere and tied him to a post. They said they were going to let him talk to his comrades before he died. He asked if he could see them, but during this entire torture session they would never take off his hood. He heard the voice of Sócrates Amado Campos Lemus—who had either been broken early and without much labor or was, more likely, a paid police agent—trying to get him to acknowledge fantastic charges, on the grounds that they were both "about to be executed," or else feeding him lies about one of the women leaders of the movement. Tita (he was told) had been a government spy and was off free and easy in the streets with fifty thousand pesos in her pocket.

He would not yield. He would not give in to their demands for false information. They told him he was going to be shot, they told him he was going to be castrated. In each case they carried out a mock execution and then a mock castration. On October 3, at seven in the morning, they threw him into a cell at the Lecumberri Prison, where he was kept for twenty-eight days, without a cot or a blanket, without being allowed to go out to a bathroom, and fed only twice a day. They would hand him a glass of cornmeal gruel through a small opening in the door of his cell. And during that month, he saw no one, not even his jailers.[125]

During the days in which the prisoners were tortured and "judged," a "Political Index" was widely circulated. The Ministry of the Interior was the source of the document. Its purpose was to prop up the judgments by detailing the supposed crimes of the leaders. Raúl Álvarez Garín had "given the order to fire on anything that moved in the Plaza de las Tres Culturas." Roberto Escudero had "directed the squadrons dedicated to seizing buses as well as the brigades that painted insulting slogans." González de Alba had been in charge of "actions of sympathy" for Cuba and Che Guevara. Many of the accusations were based on the "confessions" of Sócrates Amado Campos Lemus, the leader whom Díaz Ordaz had described as "the most manly of them all" for his supposed captaincy of a gang of machine-gunners.

The verdicts against the leaders of the Student Movement of 1968 form one of the most shameful chapters in the sad and servile history of the judicial power in Mexico. The judges in the service of the President were guilty of multiple violations of the law. One of them was the printing—prior to the trials—of the prison sentences. The 113 defendants finally "tried" were indiscriminately accused of robbery, homicide, assault, incitement to rebellion. Among the more incredible cases was that of Gilberto Rincón Gallardo, a member of the Communist Party, arrested during the early days of the Movement. He had been born with two shortened and almost useless arms. The crime for which he would go to prison was "burning twelve buses with grenades thrown from a great distance."

They would emerge from prison in 1971, to oppose the government from various watchtowers: the press, the university, and—years later—the nascent parties of the left, no longer clandestine or persecuted but duly recognized by the government. Time had not tarnished their convictions, but it had dimmed their illusions. Roberta Avendaño, Tita, would say: "You felt that you were really going to resolve the problem, without recognizing that there were so many interests behind the scenes who would not allow it. I admire the comrades who still go out to demonstrations and shout their slogans. I have never been able to go to another one. I stand along the side and watch and I feel like I want to cry."[126]

Another leader, Gilberto Guevara Niebla, would say, twenty-seven years later, "We are all broken people." But today the great majority of the leaders of the Student Movement of 1968—now in their fifties—are seeking some way to change the life of Mexico in the direction of democracy, so as to give meaning to the sacrifice that "broke" them. Many of them—and their generation—will always bear profound scars, but they have mended themselves, and they are acting "for Mexico."

V

PAST AND FUTURE: THE DECLINE OF THE SYSTEM

The Student Movement of 1968 opened a crack in the Mexican political system where it was least expected: among its greatest beneficiaries, the sons of the middle class. On their own account, they rediscovered that "man does not live by bread alone." Their protest was not in behalf of revolution, it was for the broader cause of political freedom. As had been the case with the doctors, the government did not know how to handle middle-class dissidence except through the same violent methods (loaded threats or loaded guns) that had given them effective results with the workers and the peasants. Here, their action had the opposite effect. The system was not broken, but it was gravely wounded.

And yet, for a time, it seemed that nothing had happened. Neither Mexicans nor foreigners pulled their savings or investments out of Mexico. In 1968 the peso remained stable. The economy grew by nearly 10 percent with almost no inflation. It was business as usual. And the forces circling around the sun of presidential power had reached an almost total state of acquiescence and obedience. The PRI had abandoned its experiment at democratization. The army had never been more faithful, under the command of a former Henriquista dissident who had been given the opportunity to demonstrate its loyalty. The old-style caciques—once almost untouchable in their private domains—had faded out of the national picture. (Only

Gonzalo N. Santos wanted to revive his "glory days" and offered to continue shooting down students, but Díaz Ordaz calmed him by reappointing him to the army, as a brand-new general.)[1] The Congress, the courts, the state, and local administrations as well as the corporate sectors of the PRI—worker, campesino, and other organizations—firmly supported the government. None of the ex-Presidents had any weight, except for Cárdenas, who retained his moral influence but was a very tired man, directing what would be his final efforts to the practice of state charity in the miserably poor Mixtec area of Oaxaca.

The press remained almost totally servile; the Church was for the first time publicly and steadily in favor of a Mexican Revolution that under Díaz Ordaz gave their educational, charitable, and pastoral work considerable room to grow; and the businessmen confidently rode in "the same boat" with their explicitly "anti-Communist" president. As for the intellectuals of the Generation of the Midcentury, they continued to teach classes on Marxism at the University and publish their books or translations, but they often collected their salaries at the Ministry of the Presidencia (the office—now abolished—that handled expenditures directly authorized by the President) and crossed their fingers with hope for the ascent to power of their employer, the Minister of the Presidencia, Emilio Martínez Manatou.

The outermost circle—those who were not dependent on the government—had become impotent. The PPS was clearly a party that supported the PRI. The PAN lived on the crumbs it received, and the President felt that he never even had to speak with its current president, Adolfo Christlieb Ibarrola. It was a universal pattern of subordination. Truly outside were only those primarily urban elements not directly integrated into the budget: the Communists and now a substantial number of university students—dissident sons of the middle classes.

Díaz Ordaz had not come to power by accident in 1964, nor had it been the result of some capricious act of bad judgment on the part of his predecessor. He was so natural and genuine a product of the system that he was *in fact* President of Mexico for twelve years: from 1958 to 1970, first as the real man in charge under López Mateos, and then, officially, during his own six years. In confrontation with the "testings" of the system by dissident unions (some of them connected with the Communist Party), the system had attracted, favored, and created the politician it required. From the final years of the fifties, the governmental business—to avoid giving up any of its power—required an agent of the court, a prosecuting attorney, a man with the temperament of a police chief or authoritarian general. This was Díaz Ordaz's historical role.

And he believed, religiously, that the system could not yield a particle of power without losing its very existence, its *being*. Absolute concentration of power is in itself a tragedy and a constant peril. But in the case of Díaz Ordaz, the *tlatoani* was not Hadrian but Commodus or Nero. He was a man who saw enemies or detractors everywhere, and when he did not see them he invented them. The immense jigsaw puzzle of the country had to be thoroughly solved. There could be no leftover pieces, nothing incomplete. The students wanted something totally different—to open up the public forum, to create an agora of living opinions and forces and ideas. It was a fearful image for Díaz Ordaz, a synonym for chaos, for puzzle pieces turned into lives with no firm interlocking edges.

When Díaz Ordaz spoke of Mexico as "our democracy," he almost certainly believed that the word had meaning—democracy Mexican-style. But the students were convinced that the system had adulterated language itself and that Mexico was no "democracy." Sadly, they never had time to give greater substance and organization to their criticism. They spoke of dialogue but they were poorly prepared as political tacticians (or in political knowledge). Many of them were isolated within their own monologues of protest and their passionate, amorphous celebration of "freedom." They did not consider (though of course they never had the time) the formation of a political party or at least a permanent and autonomous organization. And the leaders (and participants) from the far left were a small minority, but they had their effect on radicalizing certain actions or positions that moved events inexorably toward fiercer government action.

But what more could Mexican students in their early twenties have known and done in that time of infinite hope and delusion? They shared—with their fellow students in other countries—the myth of an ultimate transfiguration, of the achievement of personal freedom in itself altering the world. And if a number of them also believed in the transfiguring myth of the Revolution, so did many people everywhere. In May 1968, when the students rose in Paris, Octavio Paz theorized that perhaps the revolution prophesied by Marx would be accomplished not by workers but by the vanguard of the students. In a few months, he would change his judgments, but if revolutionary hope could sweep away—for the moment—a man of ideas at the age of fifty-four, who could blame the students of Mexico for their hopes and their dreams?

The chapter on Tlatelolco in Díaz Ordaz's memoirs ends with this sentence: "Mexico will be the same before and after Tlatelolco and perhaps will go on being the same—in what is most important—because of Tlatelolco."[2] In these words—as with so many of his others, more specific and less far-reaching—this *tlatoani* could not have been more mistaken. Daniel Cosío Villegas would describe the Student Movement as the greatest example of civic devotion he had witnessed in almost thirty years. It would not be forgotten, and it would be imitated "tomorrow."[3] The student leader Eduardo Valle, *Búho* ("the Owl"), would offer a more specific prophecy:

> I think the Movement will have its effect on the children.... In generations that lived it ... seeing their elder brothers move ... into action, hearing stories of the days of terror, feeling them in their blood.... The government of this country will have to be very wary of those who were ten or fifteen in 1968.... They will always remember the assaults upon, the murders of their brothers.[4]

After Tlatelolco, Mexico would never be the same as it was before and it would not be the same—in what is most important—because of Tlatelolco. The men of the system—threatened (and insulted) by rejection on the part of their "prodigal sons"—would not draw the lesson that in fact their system was not eternal and had to be altered in favor of democracy. The Presidents who followed Díaz Ordaz would instead, desperately or vigorously and at immense cost to the country, put their energies into patching up the tarnished architecture.

The Mexican business of power, in 1968, went through a brief season of totalitarian temptation. Its armed employees had carried out a mass murder to strengthen the authority of the chairman of the board. To a degree and for a while, the act had its effect. The hand of the Presidents would be strengthened. Tlatelolco left a legacy of defiance but also a legacy of fear. The next few executives of Mexico would have, in many ways, a freer range and greater power than their immediate predecessors—an ominous echo of the past. But Mexico was not—and would not become—a Latin American gorilla state. There had been a profound loss of legitimacy on that dark night of Tlatelolco. The next President would be a man who had been the right hand of Díaz Ordaz's hard line, and yet he was to become immediately obsessed with making the people forget that he had ever done it. Though the deep-seated cult of authoritarian government in Mexico would not recognize the fact, 1968 was both its highest point of authoritarian power and the real beginning of its collapse.

That decline has lasted twenty-nine years, with multiple twists and turns. Though the direction now is clear—toward the bankruptcy of the system and its inevitable transformation—the process is still an open one, in a sense a free fall. The history of this last phase—the decline of the enterprise established by Miguel Alemán—cannot truly yet be written. What is possible is a sketch. There is no way to know how it will all come out, and many important facts are still hidden within the minds of the living, locked away in silence. But the constants of modern Mexico have retained their force and insistence—the pull between the future and the past, and the enormous, swollen significance of who sits in the Presidential Chair.

22

THE PREACHER

Once again—for the fourth time out of five civilian presidencies—a Mexican executive would pass on power to the man who had been his Minister of the Interior, his top policeman and executor of force and control. But this exchange was between a President and a Minister who had shared twelve years of particularly intense police labors. For six years, under López Mateos, Luis Echeverría had served Díaz Ordaz as Assistant Minister of the Interior until he moved up to Minister when Díaz Ordaz shifted to the Presidential Chair. As he was traveling across Mexico during that process of self-display called the "campaign tour," Echeverría would show a phenomenal capacity to identify people by name and surname and past history. It had been one of his functions at the Ministry.

But those traveling with him knew virtually nothing about him. There were big businessmen, big labor leaders (including Fidel Velázquez himself), the usual journalists and politicians but also university teachers, intellectuals, and even a few students—the victims of 1968. Nobody really knew Echeverría. That anonymity had been another one of his functions at the Ministry.

He was a man who had come up entirely through the party. His first job in government—beginning in 1946 and lasting eight years—had been as private secretary, always showing servile obedience, to the man who for six of those years was President of the PRI, General Rodolfo Sánchez Taboada.[1] Between 1954 and 1958 he had been Chief Administrative Official in the conflict-ridden Ministry of Education, where he

Luis Echeverría
(Archivo Fotofija)

had shown a certain gift for negotiation. From there, Díaz Ordaz had taken him into his service. He was much less widely experienced than any of his predecessors, except in one respect. He had been well-schooled—from 1946 on—in political maneuvering.

Who was he? During his campaign, he projected an image of force. In every little town, he promised something, thousands of promises, as if he were in charge of all the economic resources of the planet, as if votes really counted for anything and he was determined to win them. Hospitals, tourist centers, schools. A highway, right here! if the people should choose to make him President.

"I thought I knew him, but I was mistaken," Díaz Ordaz would say. He had been so quiet, so loyal, so ready to take responsibility, that no one had either the chance or the means to get to know him. He seemed like a perfect Mexican patriarch, with nine children and an accomplished wife (the daughter of a powerful Jalisco cacique and director of a folkloric dance troupe). He was an athletic man, and he was especially devoted to the memory of Juárez, memorabilia and representations of whom took up a decent amount of space in his large colonial house. He offered himself as the model Mexican, stepped down into life from a mural by Diego Rivera.

Once President, his proposed objective became to introduce a radical change in the historical direction of the country. He would become a new Cárdenas. He would return to the nationalist, peasant origins of the Revolution and its concern with social justice. But he would infuse these ideals with the ideological content that his intellectual peers of the left had been formulating since the 1960s—the university teachers and writers of the Generation of the Midcentury, trained in academic French Marxism. Echeverría would insist that he belonged to that group, speaking of "this generation in whose name we have reached the presidency."

To achieve his goals, he was willing to redesign the entire economic stance of Mexico. It was time to jettison "Stabilizing Development." Ortiz Mena—seeing the economic structure he had painstakingly built up over twelve years being destroyed—would recall trying to teach Echeverría economics (in private sessions) during the presidency of Díaz Ordaz. It was no wonder, Ortiz Mena would say, that Echeverría as a University student had miserably failed mercantile law.[2]

And yet, Echeverría's program was not an anachronism, nor were his social goals mistaken. The abandonment of the peasants and the unequal distribution of income were another and deplorable face of the "Mexican miracle." They had to be corrected, and the fashionable approach involved the methods recommended by the CEPAL (Economic

Commission for Latin America). The CEPAL favored state intervention in economic affairs, high tariffs and the replacement of imports with locally manufactured products, and, as the necessary consequence, a closed and protected economy. Echeverría would religiously assimilate these ideas. As he came to power, a much more radical experiment was being initiated by Allende in Chile. Mexico would try to move along the same road.

But beyond his declared intentions—which Echeverría assumed without cynicism, with true conviction—his design was essentially in the tradition of Alemán (at the beginning of whose term, Echeverría—as a young man—had first entered public life). He was the child of a political system that he wanted to preserve. And to do that, it was necessary to neutralize the democratic impulses of 1968. He would have to draw those people from the Student Movement (battered and worse in 1968) back to the "chariot of the Revolution." The policy designed to reach this goal would be called—with Orwellian flair—a "democratic opening."

Echeverría brought University professors into his cabinet, made them his advisers, or else offered them important posts in the public sector. One of the intellectual caudillos of the Generation of the Mid-century, Pablo González Casanova (author of a fundamental critical analysis—*La democracia en México,* published in 1965), would become Rector of the University, after Barros Sierra's death in 1971. The most famous member of the generation, the writer Carlos Fuentes, would accept the position of ambassador to Paris and actively support Echeverría's government.

It was harder—and sometimes impossible—for him to win over the young men and women of the "Generation of '68." A number of them had chosen to take up arms against the government. There were urban guerrillas kidnapping and killing businessmen and politicians in Monterrey, Guadalajara, and the Distrito Federal. Some belonged to the "23rd of September League," named after the failed assault on the Madera Barracks in Chihuahua, led by the Gamiz Brothers in 1965. There were various other guerrilla "foci," especially in the mountains of Guerrero. To those who had made such choices, no "opening" was offered, only the customary "bludgeon." Between 1970 and 1976, Mexico would go through a largely secret and poorly documented version of the same "dirty war" that unfolded in other Latin American countries, fought between a part of the Generation of '68 and the powers of the government and army.

But to those willing to walk the way of peace, the Echeverría government made many-sided offers that were difficult to resist, a glittering display of practical and rhetorical bread. The first substantive action—

in 1971, during the first months of the term—was an amnesty for the imprisoned leaders of 1968, both students and teachers, as well as the other long-standing political prisoners including the railway men Campa and Vallejo. And as the six years progressed, the government would steadily and substantially increase its subsidies to the universities and technical institutes of the capital and the provinces, where many of the Generation of '68 were now employed.

For the UNAM itself, the budget would grow by 1,688 percent between 1968 and 1978. If an academic job was not what a young person wanted, there was the public sector, a tree of plenty growing ever more leafy, its dependents rising from 600,000 in 1970 to 2.2 million in 1976 (an annual growth rate of 28 percent, twenty times higher than any other form of employment in the country). Although the great majority of the 1968 leaders tried to reenter active life in areas of relative independence (the embryonic political organizations of the left led by the indomitable Heberto Castillo, the universities, or the left-wing journalism that was growing stronger as the "democratic opening" heated up), there were few who could afford to "survive outside the budget."

The highest percentage (78 percent) of UNAM graduates ever on a President's cabinet, sympathetic support for the government in left-wing intellectual circles, even a '68 student leader whom Díaz Ordaz had identified as a major agitator in Guadalajara—Francisco Javier Alejo—elevated to director of the Fondo de Cultura Económica publishing house and also to the status of economic guru for the Echeverría regime—these moves and more seemed to be clear evidence that Echeverría was no Díaz Ordaz. He was seen as representing the progressive ideology that the committed intellectuals had formulated for Mexico after the triumph of Fidel Castro: more power for the State, a steadily reduced range of action for private enterprise, the end of *charro* union leaders, investment in agriculture, a healthy distancing from the United States. Not to support Echeverría would be "an historical crime."[3]

Then Gabriel Zaid submitted an article to the cultural supplement of *Siempre!* in which he wrote: "The only historic criminal in Mexico is Luís Echeverría." José Pagés Llergo—director of the magazine—supported a subeditor's decision not to print it (perhaps remembering his own, hasty long-ago judgment on the breast-ogling photograph of Miguel Alemán's son-in-law). Pagés Llergo invoked the old prohibition: "Nothing against the President or the Virgin of Guadalupe."[4]

In his article, Zaid refused to forget Echeverría's active complicity in the events that were remembered as the Massacre of Tlatelolco. Due to the rejection of the article, he stopped writing for *Siempre!* and took his work and intellect to the new magazine *Plural*, directed by Octavio Paz, who had returned to Mexico after ten years abroad to become a leading, independent critic of the regime. The magazine was connected with the newspaper *Excélsior,* which under the leadership of Julio Scherer García was displaying a degree of independent journalism not seen in Mexico since the presidencies of Benito Juárez or Francisco Madero.

It was not only because of Tlatelolco that Zaid branded Echeverría a criminal. There were also the still mysterious but brutal and highly suspicious events of Corpus Christi Thursday—June 10, 1971.[5]

The leaders of the '68 student movement had been released from prison. To demonstrate that their cause was still alive, they had called for a demonstration. The march would begin at the Politécnico, in the Casco de Santo Tomás.

No one expected trouble, let alone an organized ambush. But groups of young men suddenly charged the march. They were carrying long rods (kendo-style martial arts batons) and they began battering the students, then seizing some and dragging them off to a side street where police vehicles sat disguised as private cars and even ambulances. The bloody and beaten captives were thrown into them. There were also women agents involved, who lured students toward the side street where the thugs would attack. (The use of "Mexican womanhood" for political repression had come a long way since Díaz Ordaz's merely verbal invocation of their supposedly outraged honor.)

Two young friends, veterans of the '68 Student Movement, ran into a building and from their vantage point on the roof saw a group of the stick-wielding hoodlums assemble and set off in a collective military trot, cheering "*Arriba el Ché Guevara!*" ("Hurray for Che Guevara!") in chorus, each with a yellow stick in one hand and a rock in the other. They passed untouched and unbothered before some small antiriot tanks lined up on the road. Then they began to break store windows, continually shouting, "Ché Guevara! Ché Guevara!" Photographs exist, of these men on the run. They were, very clearly, *not* students but a well-drilled—though completely implausible—attempt to pose as students. It seemed like the Battalion Olympia all over again.

One of the young men had seen his girlfriend die on the Plaza de las Tres Culturas of Tlatelolco. He sat down in the stairway of the building and began to weep, fiercely. Among the residents and among those who had taken refuge, a rumor was running that they would come into the houses, as they had at Tlatelolco. A thirteen-year-old-boy came running up and said that there was a doctor nearby totally distraught, talking about a couple of hundred dead, of people bayoneted.

Los Halcones, government provocateurs dressed as students, Corpus Christi Thursday, 1971
(Archivo Fotofija)

The two young men made a decision to try to get to a bus stop. They reached it safely, and the bus they mounted was full of young people with swollen faces and bloodstained clothes. One of the hoodlums smashed his stick into the bus, at the same time waving it on. They drove away through speeding ambulances and the sound of gunfire. On the following day the young men were to learn that the paramilitary thugs had entered two hospitals, wielding their sticks, to finish off wounded students in their beds.[6]

No one knew—or would know—how many died on that Thursday of the Body of Christ. It would be years till some of the real facts came out. President Echeverría appeared on television the same night to announce that he was ordering an immediate investigation of what had happened, "no matter where the blame may fall." On June 11, the city newspapers—all of them—experienced a fleeting moment of absolute freedom. They reported the facts accurately (so far as they were known) and with righteous anger. The mysterious paramilitary group was given a name, *Los Halcones* ("The Falcons"). And within a few days, two high officials submitted their resignations—the Regent of the Distrito Federal, Alfonso Martínez Domínguez, and the Chief of Police, Rogelio Flores Curiel. But then the promised investigation never happened. The official word was that "emissaries of the past" had laid a trap for the progressive President within his own regime. But now they were out of office. He had freed himself of them. And of the stain of 1968.

The story was accepted by much of public opinion and by the intellectuals who had become part of the regime. Others openly maintained their doubts and pushed for an investigation. It would come in time but unofficially, from public-minded journalists.

And Echeverría proceeded on his course. An eminent and aged intellectual, whom the President assiduously courted, observed him at first with cautious respect and then with growing astonishment. The President awarded Daniel Cosío Villegas the Premio Nacional de Letras in November of 1971. The old Liberal, who had just completed the final volume of his monumental *Historia Moderna de México*, accepted the award because he felt that "we are beginning to breathe a climate of political freedom in Mexico." But Echeverría puzzled him. In his articles for *Excélsior*, he at first admired the President's dedication to the problems of the campesinos and the energy he constantly displayed: "He thinks his six years are a semester." But then, observing the President more closely with scientific curiosity and publishing a continual stream of observation in a way no one had really dared to do for a hundred years, he noticed that Echeverría suffered from a kind of verbal incontinence, an "almost physiological" need to address Mexico and the world. And he would dub Echeverría, "the Preacher" (*el Predicador*).

The President's values—his urge to renew the country—seemed real enough. But increasingly Cosío Villegas began to feel that his methods might lead to disaster. After a series of statements on his lack of desire for traveling anywhere except in Mexico, Echeverría became a new López Mateos and never stopped traveling until his term ran out. He moved through the world, with an immense retinue, and he talked and talked. It was as if he were seeking "international consecration." By the end of 1973, Cosío had diagnosed Echeverría as an incurable case of loquacity, monomania, and genuine mental disturbance.[7] (Interestingly enough, another old man—with no moral authority or intellectual weight but a shrewd practical sense of character—agreed with Cosío Villegas. The aging gunman Gonzalo N. Santos, who knew Echeverría well, confided to his son that he had always thought the man—though skilled at hiding it—was *loco*.)[8]

Echeverría would dispatch his Minister of Foreign Relations to "settle" the conflict between the Arabs and the Israelis. He would put himself forward to lead all the countries of the Third World. He would—like earlier Presidents—fish for the Nobel Peace Prize, but only Echeverría could think of trying to neutralize the person whom he saw as his greatest rival—Mother Teresa—by requesting her support for his candidacy. (The Mother showed her vast capacity for Christian charity by actually offering it.)[9] And he would make the announcement that—on the day his term expired—he was perfectly prepared to be elected secretary-general of the UN.

Internally, the now despised "Stabilizing Development" had provided the country with a substantial monetary surplus. It had to be spent, and if more was needed, it would have to be printed or borrowed. Spending, for Echeverría, was investing; and both activities seemed to him good, productive actions in and of themselves, abstracted from any specific realities. Huge scale was the distinguishing mark of all his projects. Once his Minister of Agriculture suggested that sunflowers were more productive than corn. Without knowing anything at all about sunflowers (or bothering to find out), Echeverría called for a substantial investment—a complete failure because the peasants knew nothing about sunflowers either (nor did anyone try to teach them) and the planting, done at the wrong time, yielded a zero harvest.[10] When his Minister of the Treasury, Hugo Margáin, explained to him that such things existed as an internal and an external debt—and that the country had reached its limit in both—Echeverría accepted his resignation and made him ambassador to Great Britain.[11] He gave the job to his friend, José López Portillo, a man with absolutely no economic or significant political experience, and then he appointed the young former '68 student leader, Francisco Javier Alejo (a neo-Keynesian) as Assistant Minister—therefore, by default, dumping the country's economy into the young man's hands. After which Echeverría made a declaration at odds with the entire history of modern Mexico: "Starting from this moment, the economy will be managed from Los Pinos [the President's residence]."

Nothing like this had ever happened before. Benito Juárez and Porfirio Díaz had religiously respected their Finance Ministers Dublán and Limantour. All the Revolutionary governments, without exception, had followed the same rule, not only with the Ministry of Finance but also, after its foundation in 1925, with the Bank of Mexico. But Echeverría had to cover the costs of his enormous spray of promises and expectations, scattered the length and breadth of the country from his campaign days on. It was the way he would show himself as more revolutionary, more radical than the young. With the money (printed or borrowed) raining down from his hands, he would wash away the responsibility for 1968.[12]

Cosío Villegas, from the pages of *Excélsior*, warned that the President had lost all sense of degree and direction in his management of the economy. In every way it seemed—no matter what his intentions—that he was constantly doing the wrong thing. When he inexplicably decided to attack Fidel Velázquez as a *charro* labor leader, he only strengthened the power of the tough old man, who felt called upon to demonstrate the limits of his subordination to the government by demanding and receiving substantial pay increases for his unions. Velázquez then gained the renewed and continuing support of the workers and helped to nourish the worst inflation Mexico had seen in decades.[13]

Then Echeverría loudly decided to take on big business ("all these little rich types"—*riquillos*—he said in an *Informe*), but the businessmen immediately formed a new centralized organization for exerting pressure (the Consejo Coordinador Empresarial), shipped money out of the country, delayed investments, raised prices, and added their own fuel to the fire of inflation. "Our President," wrote Cosío, "does not understand that by consenting to and even encouraging the inflation, he is digging a grave for the highest of his intentions—the equitable distribution of government revenue."[14]

Inconsistency, contradiction, sheer unconsciousness reigned supreme throughout his policy. The government announced the end of protectionism for Mexican enterprises—and then the internal market was promptly protected to a degree never seen before. Foreign investment was not only discouraged, it was repudiated. At the level of pure farce,

Echeverría would do things like threaten the board of directors of Coca-Cola in an attempt to force them to give him their secret formula.[15] If Kafka were reborn a Mexican—someone said at the time—he would be a chronicler of everyday life.

At the base of it all was the original sin of the Mexican political system. Cosío Villegas summed it up yet again: "the character of absolute monarchy that marks our government." Echeverría did, in a sense, respond to the old man's incisive criticisms. In 1974 the government financed the publication and circulation of a pamphlet, *Danny the Disciple of Uncle Sam,* which explained that Cosío was "a parasite," an exploiter of agricultural workers, a convinced Communist, and—somehow also—in the pay of the United States government.[16] The old man responded with another brief book that immediately became a best-seller: *El estilo personal de gobernar*. His thesis was that "the personal style of governing" staked the national destiny on the life and character of a President. And as for Echeverría, he described him as a man incapable of dialogue. He was "monologue" come to life, incapable of conversation, able only "to preach."[17]

The unregenerate old liberal would die on March 10, 1976. "Have you seen the size of our external debt?" he said, not long before, to a disciple who frequently visited him. "It's risen to almost $26 billion. We are screwed."[18]

He was not alone in his critical views among the great old men. Back in 1970, shortly before his death on October 19, Lázaro Cárdenas had left a moving political testament, in which he listed the ways in which Mexico had strayed from the original program of the Revolution and then reassumed the legacy of Madero, for himself and as his message to the future. "No reelection" had to be complemented "by the valid worth of the vote." It is said that, after his death, magic rites were performed in the Tarascan countryside, in an attempt to bring him back to life. And many Indian homes in Michoacán would keep a picture of him on an altar, with candles lit as offerings, so that he might watch over them from the other world. There are villages in Michoacán where the day of his death is still honored like the day of a saint. He had written of the Indians of Mexico, "I will not forget them." Year after year, many of them remember him, with offerings on their altars and in their hearts.[19]

Ruiz Cortines would die in 1973, in his beloved port of Veracruz, still talking about politics as "the need to swallow toads."[20] Lombardo Toledano had died in 1968 itself, convinced that the CIA was responsible for the Massacre of Tlatelolco. The friend of his youth, Manuel Gómez Morín, would die in 1972, lamenting the misfortunes of Mexico, complaining that his party, the PAN, had not become "what he hoped for."[21] In the seventies, after contending in thirty-five "endless fights," the PAN was so tired, the political reforms introduced by Echeverría in 1973 were so meager, that the party decided not to present a presidential candidate for 1976.

By the middle of that year, the failure of Echeverría's populist experiment was undeniable. It had been an attempt at a kind of white-collar (rather than working-class) Peronism, a vast expansion of bureaucratic employment, replete with left-wing rhetoric but based only on an "innocent neo-Keynesian" manipulation of the economy, "worthy only of some economics student just about to get his degree." The "new Cárdenas" would finish as another Alemán, with the far more impossible dream of bringing not only a number of school friends but the whole class of university graduates to comfort and power.[22] And like Alemán, he himself would leave office with immense wealth acquired during his six years. But during the same *sexenio*, the peso had fallen from 12.5 to 22 per dollar. The foreign debt had sextupled. Real salaries—the buying power of the average income—was half what it had been during the much maligned "Stabilizing Development."

But there had still been one real, tangible achievement: "the resurgence of a more democratic and open public life." The newspaper *Excélsior* was at the forefront of this process. Then, in July of 1976—four months before the end of his term—the President orchestrated a coup against the director of the best newspaper in Mexico. For years he had resented the independence of Julio Scherer García. He had managed, secretly, to deprive the paper of most of its advertising from private business sources, with the object of driving it into the arms of the government, which would then heroically save the newspaper, naturally at the cost of its freedom of speech. The form of pressure was not an unusual one, but the projected libretto—Echeverría, the progressive savior—bore the mark of his almost schizophrenic character, seeking not only submission but the cachet for himself of "the good man." The pressure, however, had not worked.

In 1976 the government successfully used another old method developed to perfection by the State postrevolutionary enterprise. It was the technique of manipulating the workforce. The newspaper was a cooperative, and Scherer could be removed through a vote of the members. First the government sent peasants to occupy (within the *Distrito Federal*) lands that belonged to the *Excélsior* cooperative. (The occupiers claimed that the land was theirs, but they peacefully vanished once the coup was accomplished. They had served their purpose of demonstrating that the government meant business and that worse things might happen if enough members of the cooperative failed to understand what had to be done.) With a climate of anxiety created and a vote about to be taken, paid hoodlums arrived on the premises and shouted down any attempt by Scherer to address the assembly. Fear did the rest, and a government man—Regino Díaz Redondo—was voted in to replace Scherer. It was a disgraceful blow against the best of Mexican journalism, but, like virtually everything Echeverría did, the knife would turn in his hands and cut his own flesh.[23]

The coup against *Excélsior* liberated—for good and all—an important sector of Mexican journalists and intellectuals. Four important periodicals would be founded and nourished, directly or indirectly, by the numerous intellectuals and journalists who broke off their connections with *Excélsior* in the wake of the coup. There would be the first daily newspaper (young, leftist, and independent) to genuinely reflect the sensibility of '68, *Unomásuno*. (This group would later split, with the most progressive faction initiating *La Jornada*, a strong critic of the government.) A monthly magazine—*Nexos*—also came into being, affiliated with veterans of '68 and circulating in more academic environments. They had been preceded—in November 1976—by the foundation of *Vuelta*, directed by Octavio Paz, a new, independent literary and critical journal that would soon be respected throughout the Spanish-speaking world. And in that same month, Julio Scherer García himself started the magazine *Proceso*, which would become the most widely read weekly in Mexico, a forum where the reader could encounter what had formerly been unpublishable.

It would be in *Proceso*, eight years after the event, that the darkness around Corpus Christi Thursday, the minimassacre that echoed Tlatelolco, would truly begin to clear.

Echeverría, with his "democratic opening," had tried to take control of the Student Movement, to become the most fervent of its advocates, to be the man who "preached" its essential values. And his lavish spending had won loyalties, among the young and those older and more experienced. But in 1975, when he appeared before the students themselves at the UNAM, boldly venturing into their space for what was supposed to be the culminating moment of his regime—the definitive washing-away of the blood—he received a response he did not expect.

He had traveled the world (constantly) in some measure to gain the respect of these students. He had spent billions of dollars. He had modeled himself after Cárdenas, and supported Allende before General Pinochet had killed him and tortured and executed thousands of other Chileans. He had visited Castro. But the students shouted "Murderer!" at him, and threw things. A stone drew blood on his forehead. As his aides rushed him out of the area, he was cursing furiously, "Young fascists! . . . manipulated by the CIA! That's how Hitler Youth, Mussolini's Youth used to scream!" until they pushed him into the trunk of a car to maneuver him safely out of danger.

Four years later, on June 11, 1979, *Proceso* published an interview conducted by Heberto Castillo, the *presidentito* whom Díaz Ordaz had imagined at the head of a vast conspiracy against Mexico and against him. The horrifying testimony was that of Alfonso Martínez Domínguez—the Regent of the Distrito Federal, who had been forced to resign in June 1971. He said the whole action of Corpus Christi Thursday had been planned by Echeverría. Martínez Domínguez (who had the paramilitary "Falcons" on his payroll) had been called in and told that he was to take the blame for the events and resign. He was instructed to call his family together and inform them that "he was doing a service for the President of Mexico." "Echeverría," Martínez Domínguez said to Castillo with tears in his eyes, "wouldn't let me go, held my face like this, squeezing my jaw." It had all been a plan to intimidate the released student leaders of the left—and at the same time offer the President a chance to be a hero against "the forces of reaction." He would be able to throw the blame on others, as far away as he could from himself, for Tlatelolco and the rest of the actions he had supervised, in 1968, on behalf of the paranoid President Gustavo Díaz Ordaz. It was a maneuver of elaborate political conspiracy and also, in some real way, insane. And on Corpus Christi Thursday as at Tlatelolco, no one would ever know how many had died. According to Martínez Dóminguez, Echeverría had snapped out an order by telephone to burn the bodies.

23

THE GAMBLER

Echeverría would pass the presidency on to a friend of his youth, a man of pure creole descent who would write in *My Times*, his voluminous memoirs: "I was raised in the nobility." José López Portillo y Pacheco was very much aware of being a direct descendant of the first conquistadors. Some of his ancestors had discovered the Philippines or led the settlement of Sonora and Sinaloa. Later they would serve the Emperor Maximilian, Porfirio Díaz, and Victoriano Huerta. His mother, in her old age, would say on television, "We lost everything when the revolutionaries came."[1]

The presidents of Mexico—even Miguel Alemán or Luis Echeverría—had always been forced to operate within at least one constraint: the relative poverty of the country in respect to the industrialized world. All that would change with the discovery of large, new oil fields on the southeast coast. Suddenly it seemed that the imperial dreams of the viceroyalty could finally be realized, after only a delay of some centuries. And then chance (and Echeverría) entrusted the helm of the country to a classic nineteenth-century creole—Santa Anna reborn.

He had been a lawyer in private practice till he was forty (and taught as well, as a professor of law at the UNAM, for a decade of that time between 1947 and 1958). He had shown the typical disinterest in politics that had so marked the nineteenth-century creole ruling class. His father, an engineer and a historian of the colonial period, once had written: "Sometimes Mexican history has twelve Judases and no Jesus Christ." It had not seemed a sphere worth close attention (or dangerous

José López Portillo
(Archivo Fotofija)

risks) for the successful lawyer, José López Portillo III. But he did like to dream about history and let his mind spin fantasies on "myths, dogmas and mysteries." His moderate wealth gave him enough space to build a family (with an attractive wife), spend some time painting canvases, and improvise Hegelian theories on the history of Mexico. At his own expense, he would publish an indulgent and bizarre pamphlet called *Don Q.*, in which, half joking (but only half), he saw himself as the latest incarnation of Quetzalcóatl, a myth that obsessed him, "a lord greater than Prometheus" who had sought to civilize his people and persuade them to give up their bloody sacrifices, and then been defeated through deception by his brother Tezcatlipoca but who would eventually return—incarnate, triumphant, and beneficent—in the Spanish Conquest. López Portillo even had a polychromed plumed serpent sculpted on the outside of his home, in the plush Pedregal de San Ángel, as a homage to his Quetzalcóatl.[2]

But in 1958, when López Mateos became President, López Portillo was drawn to "look close up" for the first time at a President and began to feel within himself "a process of sacralizing the man who holds power and a necessity to analyze it." He closed his office and entered public service, in minor positions, helping to formulate plans for development. But his real interest continued to be "theory," such as his lucubrations on "*chromatic* democracy in Mexico" (which explained why the PRI had the historic right to monopolize the colors of the national flag).[3]

He would move up, through three administrations, but never into a truly important post until the friend with whom he had once taken a youthful trip to Chile and Argentina, Luis Echeverría, made him his figurehead Minister of Finance. And three years later came the "Revelation": He would be the next President.

It must have been a moment of mystical rapture for him. Destiny, history, mystery, the cosmos, Providence had all tranformed him into the great *tlatoani* of Mexico. Perhaps he felt himself more Quetzalcóatl than ever. But there was a different play—in four acts—about to be staged in the theater of the nation.

First there was his initial presentation. He gave an inaugural address that was one of the best in the history of Mexico. His theme was not the "Mexican miracle" but the crisis of Mexico. With a tone of authentic moral strength, he announced that the country had to unite, heal its wounds, and move on as one. There would be two years of recuperation, two of consolidation, then two of growth. What a relief it was to hear a sensible message after six years of demagogy! Words seemed to take on value and meaning again. He would establish an

"Alliance for Production." To the poor, the disinherited of Mexico, he offered a message of hope and a request for pardon. And he won the instant support of much of the country, for his obvious desire to do what was right, for his goodwill. It was a plebiscite of feeling, far more important than the election, in which López Portillo did not even have to be bothered with an opposition candidate.

So many times before, the people of Mexico had put their trust in a man of destiny. The fate of the nation would once again ride on the goodwill of one man, but this was a man with *charisma*! More of *that* than any President for a century and a half! "A gentlemanly, good-looking . . . rather melancholy-looking person, fine dark eyes, soft and penetrating, and an interesting facial expression. . . . one would have said that this is a philosopher, living in dignified retirement," the Marquesa Calderón de la Barca had written of Antonio López de Santa Anna. It could have been a prophetic sketch of how the new López would photograph.

Both Lópezes were gallant, both rode fiery horses, both wore long sideburns, both were natural orators who could speak spontaneously and emotively into the ears of the *patria*. López Portillo loved to improvise speeches, looking at the eyes of his audience. And both Lópezes were definitely *macho*! "I was very macho," wrote the modern López in his memoirs, "never a coward ready to retreat. . . . I accepted the prestige of being macho and I lived it intensely: responding to every challenge . . . and with the stubbornness of a child, the arrogance of a young man and the foolishness of an old man, I will never back down. The word of a macho!"[4]

In the background, you can hear the singing of another archetypal creole, the screen star Jorge Negrete: "Jalisco! Don't back down!" And López Portillo recounted the stories of how, in his youth, in school and in the streets, flanked by his sidekick, "Blacky" Durazo, he would take on anyone who looked at him cross-eyed or didn't look at him when he should have.[5] (He would later make "Blacky" the Chief of Police of Mexico City, where Durazo would become fabulously wealthy on a salary of $350 a month.)

The inaugural speech over with, his term begun, another characteristic was revealed, in common between the two Lópezes, another recall of nineteenth-century creole attitudes. Neither of them had much taste for sitting in the Presidential Chair. López Portillo would wander through the corridors of the National Palace, looking at pictures of old Presidents, joking—with a note of wonder and disbelief—about his rise to power, as if he had been the winner in a cosmic lottery.[6] He found someone to take over the hard work of government, a well-known historian of Mexican liberalism, Jesús Reyes Heroles from Veracruz. Reyes Heroles was a man who harked back to the old Spanish tradition of the intellectual in power, like the writers Gracián or the great Quevedo, who had played important roles in the courts of the Hapsburgs. For him, intellectual life and political life were an integrated whole. He was more like a Chinese mandarin than a Machiavelli.

Reyes Heroles's overriding slogan was "Change in order to conserve; conserve in order to change." He had been president of the PRI under Echeverría and had more than once succeeded in bringing him under temporary control. Serving López Portillo with care and intelligence—for the first two years of the term—he accomplished an even more difficult task. He broke through the mutual rigidity between the government and the political left, which had lasted for half a century. The Political Reform of 1978, created and put into effect by Reyes Heroles, legalized the Communist Party and other organizations of the left. It was a genuine historical amnesty, a true, not feigned "democratic opening." The government recognized the left as a political force in its own right; and the left renounced the option of violence. During the same two years, López Portillo—who had learned something about economics through on-the-job training as Echeverría's Finance Minister—opted for continual small adjustments in economic policy (between Keynesianism and tighter money practices), describing himself as "the pointer on the scale."[7]

Reyes Heroles was keeping political life on an even keel. And for two years, the economic adjustments seemed to be working. The economy had revived and appeared to be recovering its rhythm. The oil deposits (first discovered in 1972) were now showing signs of being far richer than had been suspected. Surely the economic problems of the country would prove manageable.

But as López Portillo's eyes grew wider at the prospect of a black ocean of oil, a new trait (shared like the others with the long-ago López) began to surface in him. If Santa Anna had thought of himself as "the Napoleon of the West," his successor began to feel that he actually could fill the role of Quetzalcóatl and lead Mexico to the "administration of abundance" down a road of black gold.

"The occupation and activity of politics have little to do with mythomania and the overestimation of self that lead to delusions of grandeur," Reyes Heroles would say, as he was forced to resign by this new version of José López Portillo y Pacheco—the uncrowned king. The wild ride of Act II was about to begin.

The moderate plan for three two-year periods of growth went out the window. Its substitute was a program so beyond measure or limit that in retrospect Echeverría seemed almost restrained. "Not to make use of the opportunity," the President later explained, "would have been cowardice."

And now he wanted power, all the power, not only for himself but for all his relatives. His sister and his cousin already held high government positions. But his son—the economist José Ramón López Portillo—was still "outside the budget." He made him Assistant Minister of Programming and the Budget and then would refer to him as "the pride of my nepotism." But the government was still lacking a closer, more intimate presence. Not his wife—who was squandering fortunes on her trips to Europe accompanied by a grand piano at which to display her gifts as a concert performer. The government seemed in need of his mistress, a beautiful young woman from the Generation of '68, dark as La Malinche, with a doctorate in physics, the ex-wife of Luís Echeverría's oldest son: Rosa Luz Alegría. If the other López, in his middle age, had married a fifteen-year-old beauty, this López was not going to be left far behind. He was already married but he still could name the young woman his Minister of Education. Horrified at the destiny awaiting the Ministry once filled by Vasconcelos, Reyes Heroles—who had not yet resigned—persuaded "Pepe" to change his mind. The President made Rosa Luz Alegría Minister of Tourism instead.

On his birthday, before an appreciative public, López Portillo performed a number of tricks on horseback. Someone would say at the time that it was lucky for Mexico the President hadn't named a horse to his cabinet. And besides being a horseman, he was a boxer, a tennis player, a gymnast, a painter. He had no limits.

And with all that oil out there, neither did the finances of the State. As had been done under Echeverría—but with much greater resources that entitled Mexico to far larger loans—vast sums were wasted in bureaucratic salaries and current expenditures but also as investments that produced low yields (or none, or considerable losses). Every area of the economy was included: railroads, nuclear energy, dozens of speedways for Mexico City, a huge expansion of the steel industry (for which there was absolutely no demand). Everything was to be modernized by the end of the *sexenio*.

PEMEX, the state-owned oil company and the largest public enterprise in Mexico, was a typical example, at the very center of the black ocean of wealth. Investments were made with no sense of order or coordination. At the cost of $1.5 billion, a pipeline (for natural gas) 750 miles long was constructed to the border before signing any purchase agreement with the United States. It would later be used locally but not to send gas across the border. The immense PEMEX tower—the biggest white elephant in the history of Mexico—was constructed as if to show that yes, in Mexico, we too have skyscrapers. In Tabasco, at the center of the boom, a whole new city was built. The result was predictable. In 1981, 87 percent of every dollar of PEMEX's assets was owed to foreign banks, forming a fifth of the total foreign debt of the country.[8]

The pharaonic atmosphere emanating from presumably eternal fields of oil also contaminated private businesses, like the powerful ALFA group of Monterrey. The organization broke with its traditional (and beneficial) policy—dating back to Porfirismo—of distrust toward the central government. They began to buy up businesses in wholesale lots, with no discrimination and paying generously. Why bother to bargain? Then they hired young men, fresh out of school with marvelous degrees and no experience at all, to run their new factories. They paid them millions of pesos. All the money to erect this vast new pyramid came from foreign banks. The sky was the limit. And there are no barriers in the empty air.[9]

The President—and many Mexicans who identified with him in 1980—were drunk less on power than on that favorite virtue of nineteenth-century creole dreams: glory. Like Santa Anna, López Portillo would visually assert his own right to that immaterial radiance by mounting, in bronze effigy, a horse. The equestrian statue of the President was erected, appropriately, not in Mexico City (where the other López had placed his, to be later destroyed) but in the money city of Mexico, Monterrey.

Two engineers, in print, were among the few dissenting voices. Gabriel Zaid had spent fifteen years studying the needs of the Mexican countryside. In a series of articles, he attacked the uses to which all this oil wealth was being put, calling—as Frank Tannenbaum had decades earlier—for a total change in the economic orientation of the country, a rejection of all the showy projects in favor of this great opportunity to develop programs that could meet the humble but vital needs of the poor: expenditures that could lead to a well-designed sewing machine, a packet of improved seeds, a better and inexpensive system for collecting rainwater. And Heberto Castillo—calling attention to all the haste and waste of the oil orgy—reminded the public that "the oil will not last forever."[10]

The economic prop for it all was even more fragile than the oil supply itself. It was the per-barrel price of oil. A decline in that one economic indicator could turn a dream into a nightmare.

It was time for Act III, a highly melodramatic part of the play that—were it fiction instead of sad reality—had much to offer for those with cruder tastes in entertainment. It was the devil-may-care macho staking it all. López de Santa Anna was "capable of throwing his medals—if he had no money left—before the spurs of the fighting cocks and wagering his glory on the chance of a match." López Portillo's cockpit was the country. His stakes were the economic future of Mexico. For four years he had been throwing his bets on the sand, and he was sure that he was winning. Suddenly, in 1981, some of Mexico's customers were offered cheaper oil by other producers. Jorge Díaz Serrano, the director of PEMEX, made the natural decision to lower Mexican prices, by four dollars a barrel.

When the President heard of it, he fired him. (The consolation ambassadorship this time was to the USSR.) López Portillo would not allow reality to trifle with "his oil." His economic guru at the time—José Andrés de Oteyza—fully agreed with the President's action. López Portillo decided not to lower but to *raise* the price of a barrel by two dollars, warning Mexico's customers that if they failed to buy now, they would get none of "his oil" in the future. But the buyers were perverse enough to pay him no attention. He had lost his biggest bet.

Would he back off? Never. Cowards back down, machos attack. He would support the value of "his peso." But it was obvious that the dollar was worth more and more, and the business public (and those of the ordinary public who could) rushed to exchange their artificially supported pesos for cheap—and far more valuable—dollars. The game was not over, it could not be over for López Portillo. He kept on borrowing. Between July and August of 1981, $9 billion left the country.

He was terrified of having to devalue the peso. With almost suicidal insistence, he felt it as a threatened devaluation of himself as a President and as a man. He addressed the nation, calling for the defense of the peso, the maintenance of his whole economic "structure," which he was willing to defend *como un perro*, "like a dog." The pathetic comparison did him no good. The public kept buying dollars. His plea to the nation was another lost bet. In February 1982, he had to devalue the peso and himself.

It was the worst devaluation in Mexican history. The peso plunged from 22 to 70 per dollar and kept on going down. (The incoming President, Miguel de la Madrid, would have to set the exchange rate at 150 per dollar as the first action of his government.) The new López, like the old one, veered from euphoria to depression. And in August 1982, a financial earthquake shook the international markets. Mexico was claiming "a temporary cash-flow problem," but the country, in reality, was bankrupt.[11]

López Portillo faced the nation on September 1, 1982, the day of his last *Informe*. "I was responsible for the helm but not for the storm," he declared. And then he blamed it all on the bankers and the "exporters of dollars" (*sacadólares*), and with the stroke of a pen—he was no coward that backs down—he nationalized the banks.

But he did know that this was the final act and that the play was over. As Santa Anna had done—when he attributed the loss of Texas to everyone and everything but himself—López Portillo then wept before the joint session of Congress (at the very moment that he mentioned the poor, the disinherited, whose pardon he had asked—and promised to earn—six years earlier).

The old nineteenth-century pattern of the creole elite had renewed itself in a minicycle that encompassed the ruin of the country and the ruined reputation of José López Portillo III. In the nineteenth century, led by Santa Anna, they had lost half the territory of the nation. In the eighth decade of the twentieth century, López Portillo and his train of advisors had mortaged all of Mexico.

24

LOST OPPORTUNITIES

Miguel de la Madrid, born in 1934, was fourteen years younger than López Portillo and had been a former student of his in the Faculty of Law at the UNAM. He was a man who also recalled the nineteenth century but a later, much more sober and admirable phase than the false glitter of Santa Anna. De la Madrid brought an element of the Liberal era back to the political discourse of Mexico. Among the Revolutionary Presidents of Mexico, only Carranza and Madero had similar affinities, though they had been born to the era while De la Madrid had chosen to make it his subject of study. His attitude—the old liberal perception—involved a view of the State as an imperfect human creation, steering clear of any fascination with either vast Hegelian necessity or divinely ordained Thomistic structure. Perhaps such an attitude—and the character of the man who held it—could offer the country a chance to confront its true dimensions and its genuine political needs.

Mexico was still suffering in 1983 from the open wounds of Tlatelolco and the Thursday of Corpus Christi, from the demagoguery of the Echeverría era and the unbelievable irresponsibility with which the government—between 1977 and 1982—had let enormous wealth flow away forever through its careless and uncaring hands. And the public pain was even fiercer because some understood—and many partly perceived—that the collapse had not been inevitable. To err may be human, but not to that degree. The sensation of having been victim of a huge deception, the evidence of almost hallucinatory corruption, the daily shifting sacrifices imposed by the economic crisis, all this tied a knot that was difficult to loosen, a knot of bewilderment and anguish.

In choosing Miguel de la Madrid, his almost polar opposite, López Portillo seemed not only to be selecting a trusted subordinate—in the by now traditional mode of institutional succession—but admitting the bankruptcy of his own personality (not only of his reputation and—by the way—of Mexico). If so, it was his only admission. Even the humblest peasant had been a witness to the overwhelming publicity campaign, promising "black gold for all" followed—in so short a time—by the message: "We are experiencing a wartime economy." And all this without any public explanation about the reasons for the disaster or even a minimal admission of responsibility.

The new President promised from the first not to promise the impossible. Inflation had to be controlled and the medicine would be very bitter. Neither he nor many other Mexicans saw any alternative. But the relevance of the diagnosis and the quality of the surgeon were not in themselves enough to mend an injury of that magnitude. The lack of limits on the presidential power—worse than ever after the applications of terror by Díaz Ordaz and Echeverría—had led to the prostration of the Mexican economy. The value of democracy—of the right of the people to choose and judge and control—had never seemed clearer. A whole range of societies had been able to discover that value. Even the unnatural child of Hitler and Mussolini, the former Franco Spain was now a democracy after the death of its caudillo in 1975.

The new President had the intellectual and moral sensibility to choose a road of political progress. But like his predecessor—though he was otherwise so unlike him—he had no political experience. His work had been in banking and government finance. One man had pointed his finger (the *dedazo*) and made him President of all Mexicans. In his campaign, he had spoken of democracy, and a litany of old Liberal values: representation, the republic, federalism, the strengthening of the legislative and judicial powers. And the people had responded, they had actually come out to vote on July 6, 1982, in numbers not seen since the days of Almazán or Vasconcelos, who had of course both been opposition candidates. There was something in the air, in this willingness to vote (68 percent for De la Madrid). And it was not devotion to the PRI, which received far smaller votes for its legislative candidates. Of course a strong opposition candidate might have reduced the proportion, and there was no alternative to De la Madrid. But yet the people did choose to indicate their support for his "moral renovation of society," which they saw as a declaration of war against corruption. In this election, the act of voting seemed to be taking on a new weight, poised perhaps at the verge of becoming something truly meaningful.[1]

Miguel de la Madrid Hurtado *(far left)*
(Archivo Fotofija)

And the cumulative experiences of the last decade and a half—with the devastating climax of economic collapse—had begun to grow a new harvest of opposition. Many Mexicans had had enough. "The tiger" was stirring in a number of forms. If you went into many towns and villages through the south and center of the country—the heart of "Old Mexico"—you would find radical slogans on the walls and equally intense opinions. In the southeast, some committed bishops were quietly spreading the left-wing gospel of Liberation Theology. A challenge to authority was growing in the north, expressed in the renaissance of the PAN, elections fought—with increasing intensity and popular support—against the PRI, a growing independence in the regional press, and a critical attitude on the part of the Church and even of elements in the business community. Once again, as in 1908, things were changing—generations and ideas and the political geography of the country.

The strongest renewal of the PAN came in the state of Chihuahua—where its founder, Gómez Morín, had been born. In 1983, they won some municipal presidencies, most especially in the two principal cities—Chihuahua itself and Ciudad Juárez. In the city of Chihuahua, the victory had been conceded—an unprecedented act—by the PRI candidate before the famous "electoral alchemy" of the ruling party had a chance to dissolve and recombine the voting results. The winner was Luís H. Álvarez, who had been the PAN candidate for President of the Republic in 1958. The victor in Ciudad Juárez was Francisco Barrio, a vigorous and outspoken young businessman. Oscar Ornelas, the PRI governor of the state—a man very similar to De la Madrid in his background and nineteenth-century liberal orientation—had made a decision not to use strong-arm methods in the elections. After the PAN municipal victories, he announced that he could work with a PAN mayor. But De la Madrid visited Chihuahua to publicly reprove him. Before huge assemblies orchestrated in the purest PRI style, he declared that "those who think that the party of the Revolution is in crisis are deceived." Now he was speaking not as a liberal but as a member of the "Revolutionary Family." No one was surprised that electoral fraud returned to the picture—against the PAN—in the 1985 gubernatorial elections for the states of Nuevo León and Sonora.

On September 19, 1985, the worst earthquake in the history of the country struck Mexico City. The government, stunned, unprepared, reacted slowly and clumsily. As another sign—if one were needed—of how the system had frozen down to its very heart, the Ministry of Foreign Affairs proudly announced that "under absolutely no conditions" would they request aid, least of all from the United States. The public was not only willing to accept aid; they were begging for it. The disaster was in every way immense. Close to twenty thousand people were killed.

In total contrast to the official rigidity, the civil population—and most especially young people and women—showed courageous strength and resolution. From the first moments, the streets were filled with students from the preparatory schools and the University and the Politécnico, spontaneously organizing teams to save victims and help those who had lost everything. Thousands of boys—from every social class—risked their lives to enter the ruins and "get people out" (*sacar gente*). Hundreds of private automobiles pasted on a red cross or flew a red flag and crossed the city like ants looking for the wounded. At one site in the neighborhood of Tlatelolco—where the damage was very severe—a boy of fifteen or sixteen took charge of rescue operations. Everybody obeyed him—the police, the soldiers, the student volunteers. The city was living through the same phenomenon of affirmation and solidarity as it had in 1968, but the direction was reversed. The students were not shouting "People! Join us!" Instead, they had joined with the people.

In an improvised crusade, the young volunteers obtained and distributed supplies, services, and information. Contributions of all kinds came in to the private and public universities; the students organized an information system that matched up needs and resources; the university kitchens prepared food and the warehouses stockpiled supplies for continual distribution. And the thousands of volunteers went out everywhere—to all the neighborhoods and shelters and parks, to the sidewalks and the buildings in ruins—filling the considerable governmental vacuum, trying to supply the immediate needs of the deeply wounded population of Mexico City.

Virtually all of this was run by the students themselves. And beyond the young with the legs and energy and commitment to take on the hardest work, the people of the city—with really few exceptions—helped and did not harm one another. It was an outpouring of the virtues of the Mexican people, a shared civic baptism by fire and human sympathy.[2]

And in its terrible way, it was also an opportunity for the government—despite the profound inadequacy of its responses to the disaster. Given the extent of the destruction, it could have been the time to rebuild by decentralizing the organization of the country. A fifth of the population lived in the capital, consuming half of the country's imports. The national debt was in large part the debt of the capital. Food, housing, transportation, services—all of them were subsidized in the Distrito Federal. It could have been the moment to begin the process of dismantling the pyramid and returning economic resources and political autonomy to the states and cities. But it was an opportunity that the government let slide.[3]

And in July of 1986 would come another moment when historic change was possible—the gubernatorial elections in Chihuahua. The support in that state for the PAN was a function of the general resentment throughout the country but also of the urge for autonomy—a view of the center as the fountain of evils—that was a constant in the history of Chihuahua as of other northern states. The PAN candidate for governor was the popular municipal president of Ciudad Juárez, Francisco Barrio, charismatic as a person and spiritually fueled by his Charismatic Catholic faith—an activist through and through. Eighty percent of the electorate lived in the cities of Ciudad Juárez and Chihuahua (also governed by a PAN mayor). A poll showed that the state favored the PAN by three to one.

PAN was poised to win the election, and with an overwhelming majority. Barrio would have become the first non-PRI governor since the institutional structure had locked into place. Behind him were not only the vast majority of the people of Chihuahua but much of the press and diverse currents in the Church. There were the "ecclesiastical base communities" in the northeast of the state led by Father Camilo Daniel, who preached a "balanced form" of Liberation Theology, in his case (and generally in the north) inspired not by Marxism but by the Christian anarchism of the Gospels. He had led peasants in a number of successful acts of civil resistance. And from the mainstream Church, from the archbishop of Chihuahua himself, came statements no cleric had dared to make since the days of the Cristiada: "The people are tired of deceptions. There will be violence if the vote is not respected."

There was even a likeable and independent-minded left-wing candidate—Antonio Becerra Gaytán of the Unified Socialist Party of Mexico—running a plainspoken and forthright campaign. And he said that he was perfectly prepared to accept the judgment of the masses. It looked as if it was going to be a real election.[4]

The PRI, at that historical moment, could have raised the flag of democracy. Only within the party itself would there have been any protest and then only from a part of the political bureaucracy and the corporate unions. The various branches of government would have unanimously applauded. The army would not have resisted any action of the President. The old-fashioned, regionally all-powerful caciques (who could veto the federal government within their own domains) were part of history. (As if to put a final point and closure on the breed, Gonzalo N. Santos died that very year.)

In the circle of relative independence, there were now more journals and newspapers willing to take stands, more intellectuals living "outside the budget" and producing criticism of the system for their growing number of readers. The Church in general was adopting more liberal values and attitudes. The students had shown their valor and civic devotion during the earthquake. Even some young businessmen, after the economic disaster and the nationalization of the banks, were beginning to admit what was now an open secret: The Mexican political system had to be reformed through an opening toward democracy.

Among the political parties and intellectuals of the left—true independents—the atmosphere was equally promising. To the right of center, PAN was finally gathering the fruit of its long, democratic journey and was offering the prospect of a peaceful change of power in some of the more economically developed states and cities of the country, especially in the north. The parties on the left independent of the government (notably the old Communist Party, now the Mexican Party of Socialist Unity, and Heberto Castillo's Mexican Workers' Party) were—like the left-wing intellectuals—much more receptive to a democratic commitment. They supported the Sandinistas and the guerrillas in El Salvador but also would acknowledge that Solidarity in Poland had ended the dogma of the Communist workers' paradise. Gorbachev's perestroika was seen as a great hope. In private some of them even criticized Fidel Castro.

It was a formidable opportunity. And the international reaction would have been enthusiastic. But the De la Madrid government would not move, would not disturb things, would not accept the offer history was putting in its lap. The PRI "electoral alchemy" began mixing its beakers, creating its own results. Business as usual—a great victory for the PRI.

PAN organized a vast movement of civil resistance. Three prominent PAN leaders—including Luís H. Álvarez—went on a hunger strike. Twenty-one prominent intellectuals signed a manifesto of protest that appeared in newspapers across the world. Manuel Bartlett, De la Madrid's Minister of the Interior, invited them to a dinner and explained that a PAN victory in Chihuahua would have opened the door to three historical enemies of Mexico: the Church, the United States, and the business class. The writers refuted his arguments, denied the existence of this triple conspiracy or that a single governor from a party other than the PRI would somehow be an assault on the sovereignty of the nation. Bartlett would not budge. At the end of the meal, he suggested that if there had been fraud, it was patriotic fraud. And any disturbances that might occur—he warned them—would encounter the forces of public order.[5]

The government of "moral renovation" had lost its last opportunity to lead Mexico toward democratic change. In 1987, a few months before the "Revelation" of the next PRI candidate (which had always meant the next President), De la Madrid, in private, expressed his satisfaction about the course of his government. Inflation (which had risen drastically during the *sexenio*) had been brought under control through the innovative idea of a "pact" between government, workers, and business. The government had begun to sell off some of the virtually bankrupt enterprises—there were more than a thousand of them—acquired during the wild spending years of Echeverría and López Portillo. The economy was being liberalized, and it was clear that the next President should be someone who could consolidate the economic change of direction.[6]

Public opinion would not turn against De la Madrid after he finished his term. Díaz Ordaz would be haunted till his death by the bloody ghost of Tlatelolco, Echeverría by his lies and lunacies. When López Portillo entered a restaurant, people would sometimes bark, to remind him of his immortal sentiment: "I will defend the peso like a dog." De la Madrid could walk in the street and, though he might not be applauded, he was certainly not insulted. The people may very well have appreciated the breathing spell of those six years. But they also knew that De la Madrid had confused prudence with passivity and sometimes with immobility.

Within the PRI itself, a "critical current" had developed, demanding democracy. It was headed by the former governor of Michoacán, Cuauhtémoc Cárdenas, the son of "Tata Lázaro." His name was a symbolic fusion of the populist nationalism of Mexican history. Along with other independent Priístas (like Porfirio Muñoz Ledo, former Minister of Work and Education, who had once been president of the party), Cárdenas saw a historical opportunity and picked up the banner of democracy, where De la Madrid had dropped it with disdain. He would carry it with him outside the PRI.

That was where those "populists," belonged, said De la Madrid, they belonged on the left, among people who were "dogmatic and sometimes crazy." Of Cuauhtémoc Cárdenas and Muñoz Ledo—"As far as I'm concerned, let them go! Let them form another party!"[7] It was too late to create another party for the coming election, but Cárdenas agreed to run as the candidate of a coalition of small parties. The other candidate of the left—Heberto Castillo of the PMT—had the wisdom to withdraw in favor of Cuauhtémoc Cárdenas. The PRI candidate—the official opposition—would be the brilliant young economist and former Minister of Programming and the Budget, Carlos Salinas de Gortari.

25

THE MAN WHO WOULD BE KING

Not in their worst nightmares could the lords of the PRI have imagined what would happen to them on the sixth of July, 1988. As they had done six years before, the electorate came out to vote but not in support of the official candidate. They came to the voting booths to punish him, to insist on the harm that had been done them for which no compensation had been offered, to issue a radical demand for change. When the first results came into the Ministry of the Interior on Avenida Bucareli, the proportions in favor of Cárdenas were so alarming that an unexplained (and inexplicable) computer glitch was conjured out of the air, so that time could be gained to manipulate the results electronically. The government called it—with involuntary humor, with poetic justice—"a breakdown of the system." For the ordinary citizen, it was not the computer network but the Mexican political system that had crashed.

The public would never know the real results of those elections. Carlos Salinas de Gortari was declared President. It had been a whole new method for dealing with the ballot boxes. Gonzalo N. Santos, after all, was dead. The bullets were electronic and lodged not in human bodies but in the body of democracy. At the height of his presidency, Salinas de Gortari ordered the records—hidden away in the cellars of the Legislature—to be burned. It is hard to believe they did not contain the evidence of his defeat. Cuauhtémoc Cárdenas struggled tenaciously—in the public forum—against what he considered to be the theft of the country. An order from him would have sent Mexico up in flames. But perhaps in memory of his father, the missionary general, a man of strong convic-

Carlos Salinas de Gortari
(Archivo Fotofija)

tions but not a man of violence, he did the country a great service by sparing it a possible civil war. Instead he formed a new political party of the left. Its name was also a commitment: the PRD, the Party of Democratic Revolution.

Once in office, Salinas had to gain some measure of credibility and he had to do it fast. (Legitimacy was beyond his grasp.) In a smoothly executed raid carried out by a military squadron, he arrested the tough and powerful leader of the Oil Workers Union, Joaquín Hernández, *La Quina*, who for years now had been behaving like a regional cacique—dispensing favors and sometimes fatal displeasure—in the northeast. But it was not for his real crimes that Salinas moved against him. Hernández was said to have supported Cárdenas's campaign and financed a libelous pamphlet against Salinas titled "An Assassin in the Presidency" (dealing with the accidental killing—a true event—of a servant by Salinas when he was four years old). *La Quina* (charged with some of his real crimes as well) would spend the *sexenio* in prison. Salinas had made a significant point. No one within the political system was going to "try him out." He was the one who would see how much others could take. The reaction among the Mexican public was a mixture of admiration and relief. Salinas—"the little guy"—grew in stature. He was a man with *huevos*.[1]

And also political and economic talent and the talent to surround himself with other talented men. His administration was the heyday of the economists—young graduates of prestigious American universities. Salinas himself had a degree from Harvard and was only forty years old. After the disasters of the last decade, the bright young men were concerned above all with reinvigorating the economic growth of Mexico. Their mission was to send the country striding out—this time to stay—into the First World.

Miguel de la Madrid, with varying success, had fought inflation, but otherwise he had been at best a caretaker rather than a creative gardener of the economy. He had begun the process of privatizing unproductive state enterprises—which Salinas would put into high gear—but he had left a foreign debt of $102 billion, zero economic growth, and an 8.6 percent decline in real salaries. The peso had continued to lose its value in relation to foreign currencies. It was 925 to the dollar when Salinas took office.

The whiz-kid economists attacked at full steam. Pedro Aspe Armella (Finance Minister, MIT) balanced the budget, and by 1991 they had achieved a surplus. Inflation was down to a reasonable rate—at least in comparison to the recent past—of 20 percent. In 1990 Salinas renegotiated the foreign debt and reduced both interest and principal by

35 percent. Privatization proceeded rapidly (and the "Ivy-Leaguers" were encouraged by the similar moves of the Spanish socialists under Felipe González). It would later become clear that considerable corruption had accompanied the process of privatization. New millionaires and billionaires appeared across the map of Mexico. By the end of the *sexenio*, 85 percent of public enterprises had been closed, sold, or allowed to go bankrupt. As a result, the treasury had gained $22.5 million and ended an annual hemorrhage of $4.5 million.[2] In Mexico, businessmen and investors gave their hearts to Salinas. And abroad, Mexico was being transformed—in the eyes of the business world—from the ugly duckling of 1982 to the new "model child" for economic development. The President's fame soared. Worldwide awareness of his name would not surpass that of Emiliano Zapata, but it was prominent enough in newspapers and magazines throughout the world.

It was "perestroika a la Salinas," and its crowning achievement would be the NAFTA agreement with the United States (and Canada). Even coming up with the idea of the North American Free Trade Agreement was a violation of the Eleventh Commandment of official Mexican mythology: Thou Shalt Not Trust Americans. Salinas understood that it had become a myth, with strong historical foundation but inappropriate to the present day. He proposed and boldly pushed the idea. The best international salesman ever born in Mexico, committed (with his "Ivy Leaguers") to a lobbying effort unusual for the Mexican political class, Salinas offered effective social and economic arguments to the Americans. The agreement—he asserted—would reduce Mexican emigration (especially the illegal variety) to the United States. It would take advantage of the complementary nature of the two economies, and it would improve the competitiveness of the North American area as a whole in the face of the European and Asian trading blocs. The Americans were persuaded.

Among the Presidents of Mexico, there have been some who were totally committed to the call of the future—which for them meant the economic development of the country. This emphasis was always at the expense of a concern with political freedom. In a very modernized form (and with polished methods of presentation), Salinas de Gortari and his generation shared that quality in particular with Porfirio Díaz. They were an imperious and impatient group, the new *científicos*, the "enlightened despots" of the computer age. Like the mode in which they had come to power, their weapons would be quick decisions, the careful selection and manipulation of information, not violence or floods of rhetoric.[3]

The political system—the business of the nation—would be modernized and streamlined, but there would be no careless jettisoning of effective procedures, even if they happened to be premodern. In dealing with labor, for instance, Salinas may have imprisoned *La Quina,* but he had excellent relations with the lifetime president of the workers, Fidel Velázquez, eight-eight years old and still the great pragmatist, a partisan of the "lunchbox" and thoroughly disdainful of any ideological "road." With him, Salinas reaffirmed—in 1988—"the indestructible historic pact" between "the Revolutionary government and the working class." In almost any other Latin American country, a drastic readjustment of salaries (which was a feature of the new *pacto*) would have brought the workers down into the streets in huge demonstrations. But not in Mexico. "It is one of the advantages of corporativism," commented Salinas.

The PRI would be gradually (but not abruptly or dangerously) reformed. That job would be supervised by his political son and fellow economist, Luis Donaldo Colosio (Northwestern—not the Ivy League but still prestigious), as the new president of the party. Colosio's intention—across a decent and safe expanse of time—was to convert the PRI into a party of citizens, not of sectors. He would talk about his goal with enthusiasm, as if it were right out there within the reach of his hand. A first test—which the Salinas government passed with flying colors—came in 1989 with a clear electoral victory for the PAN in the state of Baja California Norte. Salinas instructed Colosio to accept, without hesitation, the election of the first non-PRI governor the party had ever permitted. The decision gave the government a moral reinforcement, which underscored, in retrospect, De la Madrid's error in Chihuahua. The United States did not annex the state, nor did it fall, lock, stock, and barrel, into the control of the Church or big business. In this case, democracy had triumphed, but the PRI powers in Baja California would not forgive the regime—nor Colosio—for their humiliation.[4]

The PRD, grievously wronged by the fraud of 1988, was born with its back turned to the government of Salinas, which it saw as a mortal enemy. Salinas was not interested in arranging any agreements with Cárdenas (who called him "the usurper," "Señor Salinas," or simply "Salinas" but never "the President"). As he reached a legal and political modus vivendi with the PAN, he felt he had no need of the PRD. They were to be isolated, with the rest of the left. Only the profound Political Reform perpetually postponed—democracy—would have gained them a voice in the affairs of the country. The call for that *Reforma Política* was growing in volume and breadth, now articulated in print (to one degree or another) by the majority of the intellectuals who in Mexico—as in the rest of Latin America and other Catholic societies like Poland and Italy—have often assumed the role of secular priests.

But the truth was that Salinas de Gortari never took that call seriously. None of the arguments convinced him—least of all the obvious moral ones such as the capacity of democracy to form responsible citizens, who would have the maturity to discuss their disagreements without resorting to weapons or brute pressure. The pretext for putting off democracy was, as always, an assumed danger from the left—even in 1989, when communism was collapsing, the cold war was ending, and (on the other side of dictatorship) gorilla governments like Pinochet in Chile and Stroessner in Paraguay were being replaced by the vote. Only three governments in Latin America would continue to shut their doors to democracy: two geographical islands (Haiti and Cuba) and the historical island of Mexico.

The economy first, the Political Reform second, and in its own good time. Meanwhile, alongside the bright young men who could negotiate with the Americans in their own language using the same economic buzzwords, Salinas slipped some of the "hard-liners" of the PRI (their opponents—and the public—called them "dinosaurs") into vitally important posts of power: Fernando Gutiérrez Barrios as Minister of the Interior, Manuel Bartlett (who had managed the Chihuahua situation for De la Madrid) in Education, Carlos Hank González in Agriculture—men for whom democracy within the party (let alone for Mexico) was not a far-off prospect but an absurdity. And from on high, with the sense that he could do no wrong, and that power—which he had not earned—was his by right of intellectual stature and almost of inheritance (his economist father had been a minister in López Mateos's cabinet), Salinas continued his reforms—of the economy and of the public face of authority.

Toward the peasants, he undertook the most ambitious economic and social changes of any government in the last fifty years. They were reforms of historic importance, and in part they were valid reforms. But they were imposed from above, with no consultation. They involved no less than a modification of the "untouchable" Article 27 of the Constitution. The peasant was given the right to set his own terms for his own land. He could retain communal membership in the *ejido* or convert his land into private property. It was a measure that gave peasants the opportunity to free themselves from the protective but frequently oppressive hand of the "faceless boss"—the local, state, or federal government. To maintain its ties with the peasants (and forestall attacks—that could benefit the PRD—on its abandonment of the legacy of Lázaro Cárdenas), the government formulated a program that would offer funds and projects directly to the peasants, without a cumbersome bureaucracy. The program, called "Solidarity" (*Solidaridad*), was very successful in most of the country. Television carried a whole slew of advertisements praising its achievements. Yet it was also a typically corporate manipulation of the campesinos. There were criticisms—never proved—that the program was really the seed for a new party, which Salinas would lead once his presidency was over. But for most of the peasants, the program was apparently welcome. The government was helping them—with cash on the table.

Another constitutional change—to Article 130—made a final peace with the Catholic Church, which was given juridical personality and full internal autonomy. All restrictions on the performance of religious rites were removed. Priests were free to give their public opinions and to vote. For the *religiosísimo* Mexican public, it seemed only a confirmation of conditions that had long been taken for granted, but the Church courteously and astutely accepted the offer. It was a seal on the reconciliation between "the Two Majesties" initiated by the pious Ávila Camacho.

Salinas saw the nationalization of the banks by López Portillo for what it was—the gesture of a distraught and defensive latter-day Santa Anna, which should and would be reversed. Public employment continued to be a central source of "bread," maintained at the level reached in the previous *sexenio*—the highest in history—of almost four and half million employees, much of it unfortunately useless to the country and essentially parasitic. The administration was committed to "bread," and honestly reluctant—in the age of instant communication—to resort to "the bludgeon."[5]

But in political life, the approach continued to be piecemeal. There were symbolic changes in the Legislature, where the presence of PAN and PRD deputies introduced real debate if not real influence. Perhaps the strongest piece of symbolism had happened before Salinas actually took office, when the PRD leader, Porfirio Muñoz Ledo, rose to question President De la Madrid during his *Informe* of September 1, 1988. It was a challenge—in the wake of the great computer glitch—to the "sacred aura" of the "presidential investiture," the first time such a thing had ever happened. The press and the public appreciated the importance of the gesture. The lords of the PRI screamed "traitor!" at him from their seats in the Legislature. But across the nation at large, the PRI had to face the fact—especially in the last three years of Salinas's term—that an important sector of modern Mexico no longer trusted the presidential system. Especially, in various states, they chose to vote massively for the PAN (and, to a lesser degree, for the PRD). In the state of Guanajuato, at the center of the country, the PRI refused to accept the gubernatorial victory of the PAN. Both PAN and the PRD strongly protested, as did a good deal of the nation.

Ever since the scandal in Chihuahua, the spotlights of the international press (their intensity greatly heightened by the gross electoral fraud of 1988) would swing toward elections in even minor Mexican towns. (Electoral problems in the tiny village of Tejupilco had made the front page of *The New York Times.*) *The Wall Street Journal* called the Guanajuato election results fraudulent. Under broad pressure, the PRI would pull its governor out, to be replaced by an interim Panista. In San Luis Potosí, Dr. Salvador Nava, who in the late fifties had overthrown the power of Gonzalo N. Santos and had then tried to run for governor and been crushed for his will to independence (he was now seventy-five years of age and ill with cancer), returned to the public cause of democracy. He ran for governor and the PRI unleashed all its electoral devices against him—short of active violence. One hundred thousand phantom voters, seizure of ballot boxes, falsified vote counts. Salinas gave the victory to the candidate of the PRI. Nava called for a huge protest march from San Luis Potosí to Mexico City. "The usurper will not have a moment of peace, neither by day nor by night," he announced. Crowds followed him, especially the young. Before he reached the Zócalo, Fausto Zapata, the PRI governor, resigned. But it was only a temporary victory. The PRI would decide to retain the governorship. And in Michoacán, where the PRD was strong, Salinas—as part of his policy of isolating the left—refused to allow the PRD to assume an interim governorship.[6]

The centralization of power in Mexico could have allowed a daring and reformist statesman (which Salinas claimed he was) to change the system with the same boldness that Gorbachev had shown in Russia. But in 1993, as the next elections approached, Salinas was full of self-exaltation. There were strong critics of his economic programs. On the left, his measures were totally rejected, but even more liberal opinion had its doubts. The overly rigorous financial reform had squeezed small and medium businesses. Large, productive foreign investments had not materialized, except for short-term volatile commitments to the Mexican stock market. And most important, if the First World dream was to be realized, where was the great increase in exports, destined to make Mexico a rival of the "Asian Tigers"? It was not happening because the government was systematically overvaluing the peso, avoiding devaluation to maintain its own popularity.[7] Critics felt that the vaunted economic miracle was no more than an adjustment for the benefit of the system created by Alemán—the nation as a business and the business of power.

But there was plenty of national and international acclaim for Salinas's achievements (and very little notice paid to the continuing political backwardness of the country). People were seduced by Salinas's intelligence, by the renewed and welcome international respect for Mexico and its leader. Democracy could wait. All the bitterness of that election night in July 1988 had turned to honey for Salinas. The voters had rejected him then. Now it was he who would reject them—their aspirations, their claims to control their own destiny. And it would be his fatal error.

On November 17, 1993, the Congress of the United States approved the North American Free Trade Agreement. For Salinas de Gortari it was not merely a triumph but the very essence of glory. A few days later he "revealed" Luis Donaldo Colosio, his political son, as the next PRI candidate. He had chosen his most loyal follower—instead of other possibilities seen publicly as more independent. Many felt that Salinas was following in the path of Calles, if not of Porfirio Díaz. It seemed likely that he was in fact reelecting himself under another guise. And more miraculous triumphs might be in store for him. The presidency of the World Trade Organization? Perhaps an elevation again to chief executive of Mexico, a symbolic coronation in the year 2000? He could be the king who would inaugurate the new millennium, lord and master of the Mexican political system.[8]

But as had happened to Porfirio Díaz in 1910, after the Centenary Fiestas, history was preparing an unexpected surprise for Carlos Salinas de Gortari. At the blithest point of his surge toward the future, all the forces of the Mexican past would rise up to block his way.

26

THE THEATER OF HISTORY

"I want there to be democracy, that there not be inequality. I want a life with dignity—deliverance as God speaks of it." These are the words of the Indian José Pérez Méndez, a member of the Zapatista Army of National Liberation (EZLN), a clandestine guerrilla organization of around twelve thousand men. On the morning of January 1, 1994, they had apparently come out of nowhere to take control of three cities in the state of Chiapas: two small towns, Margarita and Ocosingo, and the old colonial city of San Cristóbal de las Casas. In San Cristóbal they burned recent state records. There were limited encounters with the Mexican Army, 145 listed deaths, and hundreds wounded.[1] After a week and a half, the Zapatistas withdrew and—their forces still intact—retreated into the dense Chiapas jungle.[2]

Although the weekly magazine *Proceso*—in the middle of 1993—had revealed the existence of a guerrilla movement in Chiapas, the immense majority of Mexicans knew nothing at all about them, neither their origin nor their intentions. Though the government did have earlier information about the Zapatistas, they had been afraid that taking military action would hurt their chances for negotiating the NAFTA agreement. They had held back, betting on the possibility that the attractions of *Solidaridad* would draw the Indians back to their fields.

Mexican public opinion received the news with immense surprise and confusion. An Indian rebellion in Mexico, in the age of NAFTA, near the end of the twentieth century? Inconceivable. A guerrilla war in Mexico, when the long, ferocious guerrilla war in El Salvador had just ended with the signing of a peace pact in Mexico City? Not possible.

Suddenly, with the uprising in Chiapas, many Mexicans felt that history was crashing down on top of them. History as a synonym for ancient arrears, old habits of mind, even of myths deeply rooted among the people. Chiapas seemed to embody the past—latent, unresolved, still vigorously alive. The outbreak looked like an eruption of the lava of history.

Chiapas and Yucatán include much of the historical territories of the ancient Mayas. In contrast to the pattern of racial integration across most of the country, *mestizaje* was uncommon in both states. Since the age of the Conquest, the Mayans and the Spaniards had been separated by thick walls of distrust, which had led them into the only true ethnic wars in Mexican history: the rebellion of the Tzeltales in Chiapas in 1712; the terrible War of the Castes in Yucatán between 1847 and 1850 (which trailed on into the beginning of the twentieth century); and the war of the Tzotziles against the creoles of Chiapas (the *coletos*) in 1867 during the presidency of Juárez. These were wars of extermination in which the Indian population tried to somehow reverse the Conquest. The Indians wanted to put an end to hacendados, caciques, governors, and even priests—all of them mostly white men, with some mestizos, who were exploiting and humiliating them. In Yucatán, most of this ethnic tension gradually disappeared during the twentieth century, in the wake of modernization and increased *mestizaje*. But Chiapas would remain the worst island of shame within Mexico. It was a place that the Agrarian Reform had ignored, the most backward state in the country, where tyrannical conditions prevailed on the coffee plantations and the cattle ranches. Until quite recently, Indians were often not allowed to walk on the same sidewalks as whites. The legacy of this ethnically rooted despotism—felt as such by the various Mayan-speaking communities—was a feature at all levels of the treatment of Indians by authority, right up to the moment of the EZLN uprising.

When Pérez Méndez said, "I want . . . deliverance as God speaks of it," it is not likely he knew that he was emulating his own Tzotzil ancestors who had rebelled in 1712, following a prophet who had taken the name of Don Sebastián de la Gloria. After announcing "the death of the King and of God" and the coming of a Virgin who had "appeared in the jungle," De la Gloria and his fevered militants attacked villages—among them Ocosingo—where they killed the whites and sacked haciendas and sugar mills. As was happening in 1994, many of the local Indian villages refused to accept the new faith or follow that theocratic experiment, which would end by being drowned in blood. A century and a half later, when Juárez was President, the mestizo caudillo Díaz

Cuzcat had led Chiapas Indians in another holy war, against the government. It had lasted three years and—in its final phase—included a leader from Mexico City, an engineer named Fernández Galindo, who exchanged his suit for Indian dress and led the final fighting. But the result was the same as in 1712. What would be the fate of this third appearance of ethnic and religious war on the soil of Chiapas?[3]

José Pérez Méndez spoke as a prisoner, captured not by the Mexican Army but by peasants like himself, the inhabitants of the village of Oxchuc, where the EZLN had fought the army and left dozens of dead. It was evident that not all peasants or Indians sympathized with the Zapatistas. In Ocosingo and Altamirano, the press reported, "the rebels were forced to retreat in the midst of displays of rejection by the population." In time it came to be known that the bulk of the Zapatista army came from Las Cañadas in the mountains (*los Altos*) of Chiapas, an area without water, postal service, schools, hospitals, or electricity, where a defiant Indian community—pressured by growing numbers (7 percent a year in some zones)—had been concentrated for decades.

They were a group seeking redemption—they were in need of it—and a message of redemption was what they received from their religious leaders. As if Mexican history were truly a sacred scripture that continually rewrites and re-forms the people of Mexico, the scene of the drama was the land of Fray Bartolomé de las Casas, the most celebrated defender of the Indians, whose call to dignity—"Humanity is one"—had moved Charles V to issue *Las Nuevas Leyes* in 1542, officially abolishing Indian servitude and slavery in the conquered territories. It was in the city of San Cristóbal—then called Ciudad Real—that Las Casas, in 1541, had confronted the feudal landowners and denied them the sacraments unless they freed their Indians. His spirit seemed to have slipped across the centuries to take form again in the figure of Don Samuel Ruiz, bishop of San Cristóbal de las Casas.

The Indians revered him. They called him "Tatik," "our father" in the Tzeltal Mayan dialect. Born in 1924 in Zamora, Michoacán, in the heart of Cristero country, he had been a student at the Pío Latino College in Rome for Latin American priests, Rector of the Seminary of Irapuato, and an active and enthusiastic participant in the Second Vatican Council, where John XXIII had opened the gates of the Church to a deeper social commitment. He had come to Chiapas at the beginning of the sixties and—shocked by the social reality he encountered—decided to head a Church that would be "less sacral, more social." Like Las Casas, Ruiz was a prophet. He preached "the charity of the truth," trying to make the Indians conscious of the conditions of injustice under which they were living. He wanted to save them not in another life but in this one, within History. To the Indians, he was "the stream that never fails," "the ceiba tree that protects us." And Ruiz would say, "Certainly we have something to do with the rebellion, because as a result of Christian reflection, we urge the Indians to recover their dignity."[4]

Bishop Samuel Ruiz's struggle was linked with the "ecclesiastical base communities," lay organizations connected with that fraction of the Catholic Church that had "chosen action in behalf of the poor." Their ideology was drawn from the activist (and sometimes violent) wing of Liberation Theology. "Those who taught us the catechism, those who lived in the same village, they asked us to join the guerrillas," explained one of Pérez Méndez's comrades. Nevertheless the six thousand "catechists" of the area (lay Catholics, usually married, entitled to teach doctrine) were not, properly speaking, men under arms. They were soldiers of the faith, instructed by other prophets close to Don Samuel, like the former priest of Ocosingo, the Dominican Gonzalo Ituarte, who had been a pastor to 250 Indian communities and was now Ruiz's assistant. Ituarte would recall with pride that his Dominican order had contested the legitimacy of the Conquest and always maintained a perspective different from the other orders. In contrast to the Jesuits (concerned with learning and also with ritual, partisans of social change but partisans of peace) the Dominicans also preached social justice but, when the time came, supported the guerrillas. In Ituarte's office, in the episcopal palace of San Cristóbal, there is an immense map of Chiapas, carefully studded with pins of all colors: Each pin is a "catechist." "I am *Tijwanej*," Ituarte would say. "The Tzeltal word means 'he who stings, he who stimulates.'"[5]

But those who "sting" or preach did not command the uprising. It was a relatively young man, an elegant talker, a successor to Sebastián de la Gloria and Fernández Galindo, who had stepped up to the microphone since that first of January. He was, according to a witness who saw him during the days when the EZLN occupied San Cristóbal de las Casas, "a white man, robust but not heavy. He looks agile, probably he wears a mustache, he is not pockmarked. He is likeable, courteous, and educated."[6] The photograph or televised image of this guerrilla leader wearing a ski mask, smoking a pipe, or writing a text—with cartridge belts crisscrossing his chest and a Casio watch on his wrist—would be flashed throughout the world. He would become the kind of legend Mexico had not witnessed since Emiliano Zapata. He was Insurgent Subcommander Marcos.

His likely identity would remain unknown for more than a year. (He himself wears his mask and will not say who he is.) He would later

"Subcommander Marcos," Rafael Sebastián Guillén Vicente
(Archivo Fotofija)

be identified as Rafael Sebastián Guillén Vicente, born in Tampico probably in 1957. He had been a student of philosophy, the energizing spirit behind literary groups, a Marxist of the French school (he had written his thesis on Althusser), a teacher of graphic design and communications, an imaginative actor, humorist, short-story writer and poet, a volunteer worker in Sandinista Nicaragua, and, since 1983, a member of the Front for National Liberation (FLN). This guerrilla group had originated—like others operating in the country during the seventies—from the trauma of 1968.[7] Marcos would have been been about eleven years old in 1968, one of those about whom the student leader Eduardo Valle had prophesied: "The government of this country will have to be very wary of those who were ten or fifteen in 1968. . . . They will always remember the assaults upon, the murders of their brothers."

At the beginning of the eighties, the leadership of the FLN had published its *Statutes*. Its aim then was to set up a "dictatorship of the proletariat." The strategy would be to "initiate the struggle in those places where 'irredentist' masses are ready to take up arms, making use of the geographic evaluations duly assessed by our commanders." The state of Chiapas, in particular the diocese of Samuel Ruiz and more narrowly the zone of Las Cañadas, offered an ideal location for "irredentism," for a defiance of the center by a group with a strong sense of its own ethnicity and grievances.[8]

The initial proclamations of the Zapatistas in 1994 mentioned no other goals but the removal of "the dictator" (Salinas), the defeat of the Mexican federal army, social justice, and the formation of a "free and democratic government." One of the Indian *commandantes*, suddenly, remembered the remains of another faith: "We want socialism . . . in our case it will be different. Here it is going to work."[9] It was the last time the word *socialism* would appear in their assertions. They would present their war as a specifically Mexican, indigenous rebellion, one that was not inspired by Marxism like the recent guerrilla uprisings in Central America. With time, information would come to light about the crisis within the clandestine organization, provoked by the fall of communism. In 1992 Marcos had written with irony: "Socialism is dead. Long live conformism and reform and modernity and capitalism." The militants had begun to abandon the Zapatista ranks. "Whole villages left us," Marcos would later say. "We were about to lose it all."[10]

There had also been a time, before the rebellion, of friction between Bishop Ruiz and the guerrilla leadership. Don Samuel wanted to awaken and mobilize the civil society of the country, and bring opinion to a point where it would induce a miracle of change in the direction of justice and democracy. But he stopped short of violence. "These people came to mount a saddled horse," he said about the guerrillas. There were priests who strongly disagreed with him, and others who called for reaching a balance between the weapons of war and peace. "This is not the place for the Word of God," said Marcos, after (in 1993) imposing his arguments for an armed uprising on some of his comrades who were opposed, "This is not the place for the government of the Republic. Here we are going to have the Zapatista Army of National Liberation." It was the voice of a guerrilla of the seventies and eighties who had added indigenism to his ideological arsenal.[11]

Socialism was dead, but the Salinista amendments to Article 27 of the Constitution favored a resort to arms. There had been no democratic sounding of the will of the peasants, no national political reform that could have encouraged local response and local accountability. In Chiapas, where the Agrarian Reform had barely arrived, it was now officially declared over. Landless peasants now had no hope of changing their situation. Most of the land was concentrated in only a few hands. Immense landed estates held certificates of immunity against any confiscation of their holdings. And the Salinista version of agrarian reform—which gave the peasants the right to become direct owners of their land—also potentially stripped them of the protection that the *ejido* system had afforded them against the land being taken away to pay off debts. Many peasants in Chiapas felt they would now surely lose their lands to creditors or large landowners. All that was missing—and both things came to pass—was a drop in the price of coffee and the North American Free Trade Agreement. How—the Indians asked themselves—were they going to compete with U.S. farmers? Many of them made a quick calculation and decided not to wait for a slow death. It was better to follow Marcos and rebel. They sold the little they had—some cows, some tools—and used the loans from the Banco Rural and the funds that came to them from Solidarity for buying weapons.[12]

Marcos would create an Indigenous Revolutionary Committee, in order to give the impression that the movement was directed by the oldest inhabitants of the continent. Anchoring his struggle in Mexican history was one of his first strokes of genius. Right from the beginning, he said that the Zapatistas were "heirs to the true forgers of our nationality." All they wanted was that there should be "democracy, that there should not be inequality," deliverance as God speaks of it.

If Salinas had chosen to change the political structure of Mexico, if true democratic procedures had been in place, the Zapatista uprising might not have happened. Or at least it would have seemed far less legitimate. Like Porfirio Díaz in 1910, Salinas had "woken the tiger." And yet the fratricidal violence feared by many Mexicans did not develop.

Apparently Marcos believed that the original action would touch off a nationwide rebellion. He talked about moving on to Mexico City. Some of the Zapatistas were not even armed, only carrying carved wooden guns (though at least some of the military poverty was for public image and consumption, since Marcos would later say that the Zapatistas also had rather well-armed units—with AK–47s, Uzis, grenade launchers, and night-vision scopes—stationed on the access roads to resist a possible frontal attack by the Mexican army on San Cristóbal).[13] But essentially the leadership must have trusted in the likelihood of a chain reaction, a rising against the government that would spread across Mexico. Immediately after the occupation of the towns, Salinas was under considerable pressure to release the full force of the army. The president of the workers, Fidel Velázquez, said to him, "Let's exterminate them!" Velázquez had lost his father in the Revolution and had himself been seriously wounded. He and other powerful men close to the President called for action that would have been very bloody but surely devastating and final for the EZLN. With the precedent of Tlatelolco in the country's past, and strongly affected by a concern for international opinion but also by some degree of moral resolution that let him decide against turning Mexico into El Salvador or Guatemala, Salinas made a decision for which he merits great respect. He ordered the army to dislodge the Zapatistas from the cities, drive them into the jungle and then set up positions of containment, watchfully, without unleashing the bombers and the modern weapons against which the Zapatistas could have deployed little but their courage. And the Mexican Army—as it has done since the days of Cárdenas—obeyed the orders of the President.

The guerrillas had believed that they could set the country on fire, but (except for a few incidents) the country did not begin to burn. The EZLN returned to the jungle, where Marcos, in practice, gave the famous saying of Clausewitz ("Politics is war by other means") a twist more than modern—in fact authentically postmodern. Politics became guerrilla warfare by other means. He had now become the romantic Hero of the Mountains of the Southeast, the poet who wrote in a Spanish genuinely infused with Tzeltal meanings and overtones.

After what Marcos defined as "five hundred years of poverty and exploitation," the Indians of Mexico were taking the initiative: "Now they will have to share our fate, for better or worse. They had their chance to turn their eyes toward us and do something to fight the gigantic historical injustice the nation has done to its original inhabitants."[14]

Few people managed to remember that Mexico was an exception to the rule of ethnic discrimination in the Americas; nor did many point to the very special character of Chiapas in Mexican history. *Mestizaje*, not ethnic hatred, has been the overriding reality in most of Mexico and modern Mexican history. The social oppression of the poor has been an immense problem. But racism is not a major feature of the Mexican consciousness (a point that too often perhaps can become clear to Mexicans when they cross the northern frontier, though it is true that, outside Chiapas as well, in certain pockets of poverty—like the Tarahumara region of Chihuahua—racial consciousness is a contributing factor to social oppression.) However, the mestizos of Mexico (which means—except for the pure Indians themselves—nearly everybody, including Marcos) maintained an incomprehensible silence about this feature of Mexican history. The rising was supposed to be a revenge for the Conquest, and that was enough.

But in reality it was a rebellion that was and remains localized both geographically and historically;[15] it did not spread violence—though it did spread political and social awareness to the rest of the country. And the elections of 1994 were cleaner, the leaders of the PRI somewhat more receptive to the attractions of democracy, due to the shock waves they felt from Chiapas. The real impact of the uprising came from its power as gesture and then from the conversion of Marcos into perhaps the first "postmodern guerrilla."[16] It was through his faxes from the jungle, his interviews for the newspapers and television that the image of Marcos acquired a virtually impregnable position. On the Internet, he moved across the world, appealing to the romantic idealism of an age that had lost many of its myths. The military phase of the war had been brief and totally unsuccessful, but in a way the guerrillas had triumphed by other means, through the media.

The eruption of Zapatismo outside the system provoked immediate problems within it. Luis Donaldo Colosio—the PRI candidate, the expected future President—seemed like a broken man since the outbreak in Chiapas on January 1. His friends knew that Colosio—a very gentle, courteous, and prudent man—was made for concord, not for conflict. Since his *destape* (the "Revelation" that he was the PRI candidate), he had seemed insecure, hesitant about his impending ascent to power. After the Zapatista uprising, he became deeply depressed. He spoke to almost no one. In the cold silence of his library, he sat listening to Bach sonatas. His eyes—which always had a melancholy look to them—now seemed those of a defenseless child, looking out at a landscape of fear. Should he or should he not go to Chiapas? Should he make a statement about the situation or not? He was tormented by another personal tragedy. His wife had terminal cancer. They had both known it for some time and even so had jointly decided that Colosio should accept the nomination.

Luis Donaldo Colosio
(Archivo Fotofija)

He would keep repeating that he "wanted to be President" as if trying to convince himself in the face of his own uncertainty. "I swear by my children that I prefer not to win than to win through fraud,"[17] he said, but he knew that fraud was second nature to the political system that was proposing him for President. And his campaign seemed to be in trouble. Public enthusiasm was lacking, despite the usual PRI artifices for conjuring up a crowd. The other man who had been most mentioned as a possible presidential candidate, Manuel Camacho (public administration, Princeton), Salinas's Regent of the Federal District, was drawing a great deal of attention. Many felt Salinas had not chosen Camacho because as President he would be independent of Salinas. But now Camacho had been put in charge of peace negotiations with the EZLN in Chiapas and appeared to be making significant progress. There was a rumor that Colosio felt the President had abandoned him, that he had become almost an orphan. It was even said that he was thinking of renouncing his candidacy, but that his wife had given him the strength to continue.

On March 6, 1994, he made a brave and startling speech in which he distanced himself from Salinas. He described a Mexico that was still impoverished, beaten down, still a Third World country. He promised that he would carry out the Political Reform (implementation of democracy) and that he would separate the PRI and the government. Two weeks later, in the crowded aftermath of a political rally in Tijuana, Baja California, as Colosio was moving through the crowd, a man got close enough to shoot him dead.

Had he been murdered or executed? Was Salinas the hand behind the crime? Almost impossible, despite Colosio's apparent rupture with him before even reaching the Presidential Chair. It was certainly clear, after the outbreak in Chiapas, that so grave a crime as the murder of the presumed future President of Mexico would mire the government in disrepute, drive away investors, and destroy Salinas's achievement. Did Aburto, the man with the gun, act on his own? He would behave as if he were mentally unbalanced, even under harsh police interrogation. It is not entirely impossible that he was a solitary assassin. Or was he somehow enlisted or manipulated into the act by the capos of the "Revolutionary Family" who had seen their power diminish under Salinas, and who wanted to pay him back for his sin of blocking the normal free movement of the political elite—and wanting to take control of the system (through Colosio as his stand-in) and turn himself into a king. Of all the scenarios, this one seems the most likely. Later, with the revelation of the alleged crimes of Salinas's brother Raúl, yet another possibility surfaced. Could Raúl have ordered Colosio killed, fearing that Colosio as President would break with Carlos

Salinas and reveal Raúl's immense financial corruption and connections with major drug cartels? Whatever the truth of the case, and even though the military power of the Chiapas guerrillas was much weaker than had at first appeared, their success at international communication, the mere fact of their persistence, now merged with the assassination of a future President. The bullet that killed Colosio finished the work begun by Marcos. It fragmented the system that now appeared to be morally bankrupt.

In August 1994, Mexicans elected the PRI candidate Ernesto Zedillo (who had been Colosio's campaign manager) to the presidency. This time even foreign observers felt that it had been a substantially fair election though it was a vote, most obviously, against violence, not in favor of the system; and the media, especially television, certainly favored Zedillo. (The events in Chiapas probably cost the PRD and its candidate, Cuauhtémoc Cárdenas—who had no involvement with the Zapatista uprising—the votes of many Mexicans uneasy with this return of the past.) Shortly afterward, another political assassination, of José Francisco Ruiz Massieu, secretary general of the PRI—shot down by a gunman in the street—sent tremors through the nation. On December 1 Zedillo took office in a climate of uncertainty and fear.

The loss of public confidence by Mexicans in their own country had to eventually influence the confidence in Mexico of the outside world and especially of foreign investors. Mexico had attracted considerable foreign savings through offering high interest rates (as well as drawing investors to the Mexican stock market). A financial misstep (or the appearance of one) could send those funds rushing back across the border. It happened—at the very beginning of Zedillo's term.

The peso had long been overvalued by the Salinas government, and political considerations—the desire to maintain public loyalty to the PRI—had prevented the series of gradual, small devaluations that were called for. In the first days of the Zedillo government, Jaime Serra Puche, the new Minister of Finance, clumsily announced a sudden devaluation, with the public—national and international—poorly prepared for the shock. It was a public relations disaster and far worse. The money began to flee, and "the Crisis" (*la Crisis*) of Mexico's apparently solid new economy began. It had been a house of cards after all, and now it was collapsing.

In the modern climate of worldwide financial interconnection, there was considerable fear that the Crisis would spread to other Latin American countries and beyond. The Clinton government stepped in to defend Mexico's financial situation, but the economy would remain sensitive to assaults on the peso. And then it was discovered that the murder of Ruiz Massieu apparently could be traced directly back to Raúl

Ernesto Zedillo Ponce de León, 1995
(Photo by Aaron Sánchez)

Salinas (whose sister—as a further sordid complication—had once been Ruiz Massieu's wife till the marriage ended in a bitter divorce). The motive seemed likely to have been a fear of revelations about Raul's illicit gains (hundreds of millions of dollars at the least) during his brother's *sexenio*. The latest "man of destiny," Carlos Salinas de Gortari, fell into total disgrace. Popular opinion began to blame him for *la Crisis* and speculate about his possible connection with the death of Colosio. In 1995, his older brother, Raúl, accused of ordering the assassination of Ruiz Massieu, would enter the maximum security prison of Almoloya de Juárez, in the state of Mexico. Rejected by the Mexican people, Carlos Salinas de Gortari would disappear into exile. For a long time, no one even knew where he was. He has recently surfaced in Ireland, seems to spend much time in Cuba, and sometimes visits the United States, waiting for and working toward a virtually impossible public rehabilitation.

During his campaign, a few months before he was shot down, Luis Donaldo Colosio was asked (by a prominent journalist) to give his opinion of the Salinas family. As if already half aware of the future, he answered with a question, "Have you seen *The Godfather*?"

Every national history has its own script, but few have turned so consistently toward the past as has Mexico. Scenes have been reactivated and replayed over and over again. There has been a constant theatricality, at times deliberate, often unconscious, and evident above all in a Mexican aesthetic of death. Sometimes they have been scenes of pure gesture, without roots in any contemporary Mexican reality. Marcos and the Zapatistas are quite another kind of performance. In the social situation of Chiapas, death itself is not a matter of theater. It is a reality as vivid as life. The Indian dead, "*their* dead," as Marcos writes, do not even earn a place in a national accounting, "so mortally dead of a 'natural' death, I mean of measles, whooping cough, dengue, cholera, typhoid, mononucleosis, tetanus, pneumonia, malaria . . ." According to Marcos, these "natural" dead in Chiapas come to 14,500 a year. "The old men say that the wind, the rain and the sun are speaking in a different way, that with so much poverty, one cannot continue harvesting death, that it is the hour to harvest rebellion."[20]

The drama of Zapatismo is a real one. The movement has intense roots and reasons that genuinely exist in Mexico. But there are also actors within it who consciously and unconsciously have taken on roles. And it is a play that is directed, after all, by a man who was in his youth (as those who knew him assert) a highly accomplished actor. The Indians of Las Cañadas are real Indians, but they "represent" all the Indians of Mexico, supposedly "exploited for five hundred years." Samuel Ruiz is a dedicated man with faith in the possibility of justice, but he is also "the new Las Casas." The Dominicans and their "catechists" are genuine soldiers of that same faith, but onstage they are the new missionaries who "sting" with their militant consciousness. And there is Subcomandante Insurgente Marcos. His commitment to altering the social reality of Chiapas and Mexico is authentic (his democratic convictions less certain). He has been in the jungle with the Indians since 1983. It is a real and some would say admirable life choice. But his specific role as leading man is less clear.

Marcos has been trying to make history—to compose it—*through a deliberate theatrical use* of the sacred scripture of Mexican history. He portrays himself and his followers as the legitimate heirs of *all* the just wars in the history of Mexico, "the struggle . . . against slavery, the War of Independence against Spain led by the Insurgents," the resistance against the American invaders, against the French, against "the Porfirian dictatorship that denied us the fair application of the Laws of the Reform" and led to the people rising in rebellion behind "their own leaders . . . Villa and Zapata, poor men like us." And against his Zapatistas, trailing back through history, he arrays all the standard villains and fuses them together: "the same men who fought against Hidalgo and Morelos, the same men who betrayed Vicente Guerrero . . . who sold half our soil to the foreign invader . . . who brought a European prince here to govern us . . . who formed the dictatorship of the *científicos* . . . who resisted the expropriation of the oil . . . who massacred the railway workers in 1958 and the students in 1968 . . . the same men who strip us today of everything, absolutely everything."[21] The centuries are summoned up to compose a mythical drama, a morality play of heroism and redemption, no longer an internationalist Marxist tableau but a huge mural of Mexican history.

It is of course true that without their act of violence, without the gesture of rebellion, Marcos and the Zapatistas could not have taken their place onstage. But even Fidel Castro, hardly a pacifist, has said, negatively, of Marcos: "He talks a lot about war and death."[22] And the traditional Mexican reverence for death, the exaltation of martyrdom, comes out in words like "Brothers, do not abandon us, take our blood as your nourishment, do not leave us alone, let it all not have been in vain!"[23] Marcos may possibly gain his martydom. But Mexico no longer requires the blood of men as nourishment. There has already been too much blood in Mexican history.

The Zapatistas have said that they have chosen to "give death back its meaning." But in the midst of so much death, (including their own) it is preferable to give life back its meaning. In a Mexico isolated, torn up by war, or in a Mexico turned truly tyrannical, everyone—except perhaps a few at the top of the heap—would lose.

The civic energy that the Zapatistas have released could be channeled toward the goals of justice, freedom, and democracy that the majority of Mexicans sincerely desire. Marcos could set an example, by seeking the leadership of the Mexican political left or by heading a Mexican Zapatista party. Unfortunately he has not taken off his mask, he has not publicly revealed his face and his name, he has not renounced the historical work he has been composing. He has not put an end to his play within the larger play of the modern Mexican system.

But we should not lose perspective. The theater of the Zapatistas, with all its drama and flair (and international acclaim) is a limited and isolated phenomenon. Even though, in 1996, another guerrilla group, the People's Revolutionary Army (EPR), carried out small actions in a few southern states and briefly took over the front pages (with photos of its leaders wearing ski masks like Marcos), the EPR soon essentially became a clandestine forum for political assertions, and it seems clear that they are—at least up to now—far less numerous than the EZLN and without any broad popular (or intellectual) following. As these lines are being written, in early 1997, the government is still carrying on its protracted negotiations with the EZLN who are still encamped in their zone of toleration in the Chiapas jungle. But it is in the *real* theater—the one that most counts, the theater of the PRI on the stage of the nation—that the future of Mexico will be decided.

The Zedillo government has completely repaid (well ahead of time) the loan that it received from the United States. On the economic front, Mexico does seem to be emerging from *la Crisis* earlier than expected, though many problems still remain. In 1996 the economy grew at a rate of 4 percent, and the major economic indicators are now positive. On the stage of politics, the signs are far more ambiguous and highly dependent, as they have always been in Mexico, on the personality of this particular President.

Ernesto Zedillo himself, a Yale Ph.D. in economics and a strong advocate of the market economy, is a rather unusual man to be President of Mexico. He grew up in Mexicali in Baja California (a city on the American border) and is more at ease with American ways than any other President of Mexico has ever been. He has no connection with the established "Revolutionary Family" and is a man of scrupulous personal honesty, recalling in this respect Adolfo Ruiz Cortines. He comes from a relatively poor background and he, his wife, and his children (before moving to Los Pinos) lived in a modest middle-class home and (highly unusual in Mexico for a family of any standing) had no servants. He is intelligent but has shown that he can be intolerant of criticism. Fate and tragedy have given him the opportunity perhaps to preside over great changes in the history and political system of his country.

There were hopeful signs of a political opening near the beginning of his administration. Early in 1995, he convoked an important meeting between the three principal parties (his own PRI, the PAN, and the PRD) to formulate a National Agreement for Democracy, designed to expedite the Mexican transition to a true democratic system. And he would later speak of the need to establish "a healthy distance between the President and the PRI," floating the prospect of a presidency far more linked to the needs of the people than to the corporate interests of the PRI.

Unfortunately, two years after the National Agreement for Democracy, the situation is much more uncertain. There have certainly been progressive moves of a kind unimaginable under Salinas. In almost all cases, state and municipal elections have been honest and the PRI has accepted its losses. PAN now holds four governorships and 255 municipal presidencies, so that the party governs (at the local level) 38 percent of the Mexican people. The PRD holds 221 municipal presidencies, involving about 9,000,000 people, 10 percent of the population. An important achievement has been the new status of the Federal Electoral Institute, now directed by independent citizens rather than government functionaries. The press and radio have never been freer. Television has begun to move in the same direction. The businessmen—given the weakened financial condition of the state—are much more on their own than they have ever been before. Intellectuals in general earn their living not in government ministries but from a growing public of readers. The Church is in the process of taking back some of its old influence and power (and there is a certain danger in the attempt by some of its representatives and partisans to legislate morality.)

On the other hand, much still seems the same. Fidel Velázquez—96 years old—remains the labor president of Mexico. Congress and the majority of the state and municipal governments are under control of the PRI. And President Zedillo, at the end of 1996, had hardened his positions, probably in response to recent (and serious) PRI electoral losses and a growing number of defections from the party. The PRI has broken off its ongoing discussions (with the PAN and PRD) on Electoral Reform. The key issue was that of government-subsidized campaign financing, with the PRI wanting to maintain a large official purse, presumably for its traditional purposes of influencing and buying votes. And Zedillo in general has now adopted a hard-line manner, talking tough, presenting an image of "strength." It seems to be a campaign decision, taken in view of the impending elections of 1997, an attempt to project the aura of the presidential office, to sell the traditional "strong" President as a value, inducement, and reason for voting PRI.

There will be important mid-term elections in 1997 throughout Mexico for the Chamber of Deputies, some key governorships, as well as (for the first time during the era of the PRI) the mayor of Mexico City (the Regent of the Distrito Federal), who has up to now been appointed. The results will be a vital indication as to whether Mexico can move into the next century as a true multi-party democracy, with the major parties sharing and alternating power. The prospect that President Zedillo can lead the way toward that transition appears more doubtful now than it seemed two years ago. Without a reforming leader from within the system, it is likely that the process of change will be far more hazardous and uncertain.

And if the political system does not progress, there are possible dangers on the horizon. The army—as it has for half a century—has continued to be the loyal servant of the government, the "immaculate defence" of the national institutions. But its practical role in the country is growing, not only through its confrontations with minor guerrilla actions or its entrance into the war against the drug cartels. More important, many people now see the military as a possible and viable (and far more respected) replacement for the police. The "top cop" of Mexico City for instance is now a highly regarded general. Faced with the entrenched tradition of police corruption throughout the country, ordinary citizens as well as more influential sectors of the population tend to trust the army far more than they do the police. Violent street crime—and, even more, the fear of it—has increased in the wake of *la Crisis* as has—at the level of the wealthy—the threat of kidnappings for ransom. Over the last decade, Mexican drug cartels have become more closely linked with the *narcotraficantes* of Colombia and much more involved in the highly lucrative cocaine trade. These gangsters have immense financial resources and in a culture as porous as the Mexican political system—with its long tradition of openness to corruption—those resources are a grave danger to the integrity of the government and the State. (The army itself has recently been compromised by these temptations.) Without a true transition to responsible democracy, all these factors could wear away the national patience, worsen the environment of daily life and even—as remote as it now seems—lead the military for the first time in half a century to claim a justification and right to power. The result of the elections of 1997—and the reactions of the government to those elections—will reveal a great deal about the possible future of Mexico.

It is time "to make public life public," as Cosío Villegas wrote, to go out into the street where ninety million Mexicans carry on their lives, most of whom want above all a decent existence, in which a democratically elected government assures its citizens a reasonable standard of living. And security and justice and stability and liberty.

The ordinary Mexican is no longer obsessed by the gravitational pull of the past. Intoxication with history is now more an issue for political and intellectual elites. In the midst of the Crisis, in a national mood of confusion and unease, today's Mexican is turning toward the future. And the man and woman in the street have begun to understand that, even if the lack of democracy is not Mexico's foremost problem, the country's other problems cannot be resolved without democracy. These are issues of the past and the present and the future, including the ancient social and economic problems that Mexico has endured as "the land of inequality." Without a legitimate division of powers, the President, if he wishes, can reign as an absolute dictator for six years. Without a solidly based and independent system of justice, the corrupted "Revolutionary Family" will continue exploiting "public posts as private property,"[24] sacking the country as it has from the days of Alemán to Salinas de Gortari. Without a truly efficient and honest civil service, neither a just system of taxation nor a way of delivering benefits directly to the poor are possible, as modes for reducing the enormous inequalities between great wealth and great poverty. Without a reliable and honest police system, the streets will be insecure and the financial influence of drug cartels will grow geometrically. Without true and effective federalism, the capital will continue to exercise a form of imperialism over the provinces and the cities. Without democracy—the ideal of Madero (and less completely of Juárez)—any economic reforms, even if they move in the right direction, will always be fragile and endangered.

To continue the theater of history is to be condemned and condemn the country, dramatically or grotesquely, to endless repetition. There has been and there is a different possibility. Mexican society could finally reconcile itself with its origins. Not to forget them—but to bury at once and forever Cuauhtémoc with Cortés, Hidalgo and Iturbide, Morelos and Santa Anna, Juárez with Maximilian, Porfirio with Madero, Zapata and Carranza, Villa and Obregón, Calles with Cárdenas, all of them reconciled within the same tomb. But Mexico would have to be less pious toward its modern actors. There can be no reconciliation with Tlatelolco. A democratic Mexico would go after the truth about what happened there, in all its details. It would answer those questions that have gotten no response. It would give faces and names to the dead. But there should not be vengeance either—only a mature contemplation of the entire recent history of Mexico and all the wrong directions it has taken. Then Mexicans could begin to compose a new history for themselves, free of that part of the past that is only weight and sickness. The history of Mexico could then begin to be the story of all Mexican lives. Democracy would print a final period, a closure for good and all to the biography of power.

SOURCE NOTES

Introduction

1. Genaro García, ed., *Crónica oficial de las fiestas del primer centenario de la independencia de México* (Mexico, 1990).
2. Ibid., p. 76.
3. Ibid., p. 82.
4. Charles A. Hale, *El liberalismo mexicano en la época de Mora* (Mexico, 1987), p. 199.
5. Daniel Cosío Villegas, *El Porfiriato, Vida política exterior,* vol. 1, *Historia moderna de México,* 2 vols. (Mexico, 1985), p. 253.
6. Ralph Roeder, *Hacia el México moderno: Porfirio Díaz,* 2 vols. (Mexico, 1985), vol. 1, p. 69.
7. For the economy during the Porfiriato see Fernando Rosenzweig et al., in Cosío Villegas, *El Porfiriato, Vida económica, Historia Moderna de México,* 2 vols. (Mexico, 1985).
8. Cosío Villegas, in *El Porfiriato, Vida política exterior,* vol. 1, pp. 692–732.
9. García, "Discursos," in *Crónica oficial,* Appendix 100, p. 53.
10. James Creelman, "President Díaz, Hero of the Americas," *Pearson's Magazine* (New York), vol. 19, no. 3, Mar. 1908, p. 244.
11. See Carleton Beals, *Porfirio Díaz* (Mexico, 1982); Cosío Villegas, *El Porfiriato: Vida política interior*; vol. 1, 2, *Historia Moderna de México*, 2 vols. (Mexico, 1985). Roeder, *Hacia el México moderno: Porfirio Díaz*; José López Portillo y Rojas, *Elevación y caída de Porfirio Díaz* (Mexico, 1927); Francois Xavier Guerra, *México, del antiguo régimen a la revolución* (Mexico, 1988).
12. See the Bernardo Reyes Papers, Fondo DLI, at Centro de Estudios de Historia de México, CONDUMEX.
13. See Roeder, *Hacia el México: Díaz,* vol. 1, p. 294.
14. *El Monitor Republicano* (Mexico), Oct. 7, 1886, Año 24, Epoca Quinta, p. 1.
15. Cosío Villegas, *El Porfiriato, Vida política interior,* vol. 2, pp. 526–27.
16. Andrés Molina Enríquez, *Los grandes problemas nacionales* (Mexico, 1978), pp. 133–35.
17. Cosío Villegas, "La hacienda pública," in *El Porfiriato: Vida económica,* vol. 2, pp. 887–972.

18. Federico Gamboa, *Diario de Federico Gamboa (1892–1939)* (Mexico, 1977), p. 78.
19. Jorge Fernando Iturribarría, *Porfirio Díaz ante la historia* (Mexico, 1967), p. 12.
20. Creelman, "President Díaz," p. 242.
21. Quoted in Jean Meyer, *La Revolución Mexicana* (Mexico, 1991), p. 35.
22. García, *Crónica oficial,* pp. 159–60.
23. Ignacio Ramírez, *Obras,* 2 vols. (Mexico, 1960), vol. 1, p. 317.
24. Justo Sierra, *Evolución política del pueblo mexicano,* vol. 12, *Obras completas,* 15 vols. (Mexico, 1991), p. 173.
25. Ibid., pp. 102–3, 251–52.
26. Ibid., pp. 251–52.
27. Ibid., pp. 396–97.
28. Ibid., pp. 396, 399.
29. Lucas Alamán, *Semblanzas e ideario* (Mexico, 1978), p. 169.
30. Arturo Arnaiz y Freg, "Alamán en la historia y en la política," in *Historia Mexicana* (Mexico), vol. 3, no. 1, Oct.–Dec. 1953, p. 254.
31. Alamán, *Semblanzas,* p. 151.
32. Lucas Alamán, *Historia de México,* 5 vols. (Mexico, 1985), vol. 5, p. 947.
33. Justo Sierra, *La educación nacional,* vol. 8, *Obras completas,* p. 190.
34. Justino Fernández, *El arte del siglo XIX en México* (Mexico, 1960), p. 210.
35. See Guillermo Tovar y de Teresa, *La ciudad de los palacios: Crónica de un patrimonio perdido,* vol. 1, (Mexico, 1992), 2 vols., pp. 35–38.
36. Bernal Díaz del Castillo, *Historia verdadera de la conquista de la Nueva España* (Mexico, 1964), pp. 160–61.

Part I

Chapter 1

1. See Justino Fernández, *El arte en el siglo XIX* (Mexico, 1960), pp. 133–40; Francisco Sosa, *Apuntamientos para la historia del monumento a Cuauhtémoc* (Mexico, 1887).
2. Miguel León Portilla, *Toltecáyotl: Aspectos de la cultura náhuatl* (Mexico, 1991), p. 21.
3. Fray Diego Durán, *Historia de las Indias y de Nueva España e islas de la tierra firme,* 2 vols. (Mexico, 1967), vol. 2, p. 246; H. B. Nicholson, "The Chapultepec Sculpture of Motecuhzoma Xocoyotzin," in *El Mexico Antiguo* (Mexico, 1961), vol. 11, pp. 379–423.
4. Gabriel Zaid, "Problemas de una cultura matriotera," in *Cómo leer en bicicleta* (Mexico, 1986), pp. 163–64.
5. José Miranda, *Las ideas y las instituciones políticas mexicanas* (Mexico, 1978), pp. 81–82.
6. Gabriel Méndez Plancarte, "Francisco Xavier Clavijero," in *Humanistas del siglo XVIII* (Mexico, 1941), pp. 35–37.
7. See David Brading, *The Origins of Mexican Nationalism* (Cambridge, 1985), pp. 3–23; also quoted in Brading *Orbe Indiano, de la monarquía católica a la república criolla, 1492–1867* (Mexico, 1991), pp. 620–22.
8. Josefina Z. Vázquez, *Nacionalismo y educación en México* (Mexico, 1979), pp. 87–91.

9. Justo Sierra, "Catecismo de Historia Patria," in *Textos elementales de historia,* vol. 9, *Obras completas,* 17 vols. (Mexico, 1984), pp. 396–98.
10. See José Emilio Pacheco, ed., *La poesía mexicana en el siglo XIX* (Mexico, 1965), pp. 165–79.
11. Justo Sierra, "Discurso inaugural . . . ," vol. 5, *Obras completas,* pp. 432–36.
12. Justo Sierra, "Exportación arqueológica," Oct. 28, 1880, vol. 5, *Obras Completas,* pp. 25–28.
13. Ibid.
14. Moisés González Navarro, *El Porfiriato, Vida social,* in Daniel Cosías Villegas, ed., *Historia moderna de México* (Mexico, 1985), pp. 3–41.
15. John Leddy Phelan, *The Millenial Kingdom of the Franciscans in the New World* (Berkeley and Los Angeles, 1970), pp. 86–91; see Brading, *Orbe Indiano,* pp. 122–48.
16. Brading, *Orbe Indiano,* p. 99.
17. Quoted in José M. Gallegos Rocafull, *El pensamiento mexicano en los siglos XVI y XVII* (Mexico, 1951), p. 26.
18. See Brading, *Orbe Indiano,* pp. 108–13; for the "Apologética historia sumaria"; see also Edmundo O'Gorman, *Cuatro historiadores de Indias* (Mexico, 1972), pp. 120–24.
19. See Henry Raup Wagner, *The Life and Writings of Bartolomé de las Casas* (Albuquerque, N.M., 1967), pp. 213–16, 222–24.
20. See Robert Ricard, *La conquista espiritual de México* (Mexico, 1986), pp. 148–49; James Lockhart, *The Nahuas after the Conquest* (Stanford, 1992), p. 205.
21. Zumárraga to the bishop of Mexico, Aug. 27, 1529, in Joaquín García Icazbalceta, *Don Fray Juan de Zumárraga, primer obispo y arzobispo de México,* 4 vols. (Mexico, 1988), vol. 2, pp. 169–245.
22. Quoted in Luis Villoro, *Los grandes momentos del indigenismo en México* (Mexico, 1987), pp. 39, 56, 57.
23. Fray Bernardino de Sahagún, *Historia general de las cosas de la Nueva España,* 4 vols. (Mexico, 1956), vol. 1, p. 29.
24. Vasco de Quiroga, "Información en derecho," in Rafael Aguayo Spencer, ed., *Documentos* (Mexico, 1939), pp. 61, 64, 68, 134.
25. Elena Isabel Estrada de Gerlero, "Las utopías educativas de Gante y Quiroga," in *El otro occidente, los orígenes de Hispanoamérica* (Mexico, 1992), p. 128.
26. Ibid., p. 132; see also Felipe Tena Ramírez, *Vasco de Quiroga y sus pueblos de Santa Fe en los siglos XVIII y XIX* (Mexico, 1977).
27. See Vasco de Quiroga, "Reglas y ordenanzas para el gobierno de los hospitales de Santa Fe y Michoacán," in *Documentos,* pp. 247–67; see also Francisco Miranda Godínez, *Don Vasco de Quiroga y su colegio de San Nicolás* (Morelia, 1972).
28. See Ramón Xirau, *Idea y querella de la Nueva España* (Madrid, 1973), pp. 88, 92.
29. See Rocafull, *Pensamientos mexicano,* pp. 159–64; Silvio Zavala, *Filosofía de la Conquista* (Mexico, 1977), pp. 94–98.
30. See Navarro, *El Porfiriato, La vida social,* pp. 150–53, in the series *Historia Moderna de Mexico,* ed. Daniel Cosío Villegas.
31. Emilio Rabasa, "El problema del indio," in *Evolución histórica de México* (Mexico, 1972), p. 198.

32. Francisco Bulnes, *El porvenir de las naciones hispanoamericanas* (Mexico, 1992), pp. 26–27.
33. Justo Sierra, "*Evolución política del pueblo mexicano,*" vol. 12, *Obras completas* (Mexico, 1991), p. 369.
34. See Charles A. Hale, *The Transformation of Liberalism in Late Nineteenth Century Mexico* (Princeton, 1989), pp. 228–31.

Chapter 2

1. See Fray Bartolomé de las Casas, *Brevísima relación de la destrucción de las Indias* (Buenos Aires, 1966), pp. 56–65.
2. José María Luis Mora, *México y sus revoluciones,* 3 vols. (Mexico, 1950), vol. 2, p. 171.
3. Pérez Verdía quoted in Josefina Z. Vázquez, *Nacionalismo y educación en Mexico* (Mexico, 1979) p. 70.
4. See Ignacio M. Altamirano's prologue in E. del Valle, *Cuauhtémoc: Poema* (Mexico, 1886), pp. xxxiii–xxxiv.
5. See Luis Villoro, *Los grandes momentos del indigenismo en México* (Mexico, 1987), pp. 149–68.
6. Justo Sierra, "Cortés no es el padre de la patria," in *Obras completas,* 15 vols. (Mexico, 1991), vol. 9, pp. 191–94.
7. Genaro García, ed., *Crónica oficial de las fiestas del primer centenario de la independencia de México* (Mexico, 1990), pp. 138–42; see also *El Imparcial, The Mexican Herald,* and *El País,* Sept. 16, 1910.
8. See Fray Diego Durán, *Historia de las Indias y de Nueva España e islas de la tierra firme,* 2 vols. (Mexico, 1967), vol. 2, p. 459.
9. Ibid., p. 490.
10. Ibid., pp. 495–96.
11. See Hugh Thomas, *Conquest of Mexico* (New York, 1994), p. 185.
12. See José Luis Martínez, *Hernán Cortés* (Mexico, 1992); Miguel Alonso Baquer, *Generación de la conquista* (Madrid, 1992); Francisco Castrillo, *El soldado de la conquista* (Madrid, 1992).
13. Thomas, *Conquest,* p. 308.
14. Martínez, *Hernán Cortés,* p. 497.
15. See Lucas Alamán, *Disertaciones sobre la historia de la Republica Mejicana,* 3 vols. (Mexico, 1969), vol. 1, pp. 112–51; see also Martínez, *Hernán Cortés*, pp. 385–468.
16. Alamán, *Disertaciones,* vol. 1, p. 112.
17. Martínez, *Hernán Cortés*, pp. 767–95; Thomas, *Conquest,* pp. 600–601; Francisco de la Maza, *Los restos de Hernán Cortés* (Mexico, 1947), pp. 153–74.
18. See David Brading, *Orbe Indiano, de la monarquía católica a la república criolla, 1492–1867* (Mexico, 1991), pp. 158–60, 165–67.
19. Emilio Rabasa, "El problema del indio," in *Evolución histórica de México* (Mexico, 1972), p. 225.
20. Nancy M. Farris, *Maya Society under Colonial Rule: The Collective Enterprise of Survival* (Princeton, 1984), pp. 186–90.
21. Jerónimo de Alcalá, *Relación de Michoacán* (Morelia, 1980), pp. 347–56.
22. Charles Gibson, *Tlaxcala in the Sixteenth Century* (Stanford, 1967), pp. 181–89.

23. Charles Wright, *Conquistadores otomíes en la guerra chichimeca* (Querétaro, 1988), p. 79.
24. See Bartolomé de las Casas, *Historia de las Indias,* 3 vols. (Mexico, 1951), vol. 3, pp. 177, 178, 274, 275.
25. Justo Sierra, *Evolución política del pueblo mexicano,* vol. 12, *Obras completas,* 15 vols. (Mexico, 1991), p. 56.

Chapter 3

1. "Crónica oficial de las fiestas del primer centenario de la independencia de México," in *El Imparcial,* Sept. 16, 1910.
2. Hugh Thomas, *Conquest* (New York, 1994), p. 59.
3. Alexander von Humboldt, *Ensayo político sobre el reino de la Nueva España* (Mexico, 1991), pp. 51–98.
4. Jonathan I. Israel, *Race, Class and Politics in Colonial Mexico 1610–1670* (Oxford, 1975), pp. 9, 38.
5. Diego Garcés, "Relación de Ajuchitlán," in René Acuña, ed., *Relaciones geográficas del siglo XVI,* 10 vols. (Mexico, 1987), vol. 9, p. 38.
6. Juan Martínez, "La relación de la ciudad de Pátzcuaro," in *Acuña,* vol. 9, pp. 200–201.
7. See Israel, *Race, Class,* pp. 69–70.
8. Juan de Solórzano y Pereyra, *Política Indiana,* 2 vols. (Madrid, 1774), vol. 1, p. 222.
9. Humboldt, *Ensayo político,* p. 60.
10. Solórzano, *Política Indiana,* vol. 1, p. 221.
11. Israel, *Race, Class,* p. 71.
12. Fray Bernardino de Sahagún, *Historia general de las cosas de la Nueva España* (Mexico, 1982), p. 514.
13. Solange Alberro, *Del gachupín al criollo o de cómo los españoles dejaron de serlo* (Mexico, 1992), pp. 89, 122, 214; James Lockhart, *The Nahuas after the Conquest* (Stanford, 1992), pp. 235–38; see also Serge Gruzinski, *La conquista de lo imaginario* (Mexico, 1989), pp. 180, 194.
14. Sahagún, *Historia general,* p. 138.

Chapter 4

1. See Cristóbal de la Plaza, *Crónica de la Real y Pontificia Universidad de México* (Mexico, 1931); "Traslación de la estatua ecuestre de Carlos IV," in *El Monitor Republicano,* Sept. 8, 1851, Mar. 25–Oct. 2, 1852; Enrique Salazar Hijar y Haro, "Los trotes del caballito," in *Costo de modernidad, México en el tiempo,* no. 3, (Mexico, Oct.–Nov. 1994), pp. 17–23; *Fuera Carlos IV de la plaza de Agustín Primero* (Mexico, 1822), Lafragua collection no. 217.
2. Guillermo Tovar y de Teresa, *La ciudad de los palacios, crónica de un patrimonio perdido,* 2 vols. (Mexico, 1992), vol. 1, p. 19.
3. See José R. Benítez, *Historia gráfica de la Nueva España* (Mexico, 1929); Lucas Alamán, *Disertaciones sobre la historia de la República Mejicana,* 3 vols. (Mexico, 1969), vol. 1, pp. 147–51; Ignacio Rubio Mañé, "Obras públicas y educación universitaria," in *El Virreinato,* 4 vols. (Mexico, 1983), pp. 12–214.
4. Alexander von Humboldt, *Ensayo político sobre el reino de la Nueva España* (Mexico, 1991), p. 119.

5. Bernardo de Balbuena, *Grandeza Mexicana* (Mexico, 1974), p. 13.
6. Richard M. Morse, "Toward a Theory of Spanish American Government," in *Latin American History* (Austin, Tex.), pp. 509–10; Richard M. Morse, *El espejo de Próspero, un estudio de la dialéctica del Nuevo Mundo* (Mexico, 1982), pp. 46–50, 64, 66.
7. Morse, *Espejo de Próspero,* p. 70.
8. José Miranda, *Las ideas y las instituciones políticas mexicanas* (Mexico, 1978), pp. 103–8.
9. Ibid., p. 109.
10. See J. H. Parry, *The Sale of Public Office in the Spanish Indies under the Hapsburgs* (Berkeley and Los Angeles, 1953), pp. 6–20.
11. Miranda, *Las ideas,* pp. 110–11.
12. See Jonathan I. Israel, *Race, Class and Politics in Colonial Mexico* (Oxford, 1975), p. 66.
13. Morse, "Toward a Theory," p. 510.
14. John Leddy Phelan, "Authority and Flexibility in the Spanish Imperial Bureaucracy," in *Latin American History,* vol. 11 (Austin, Tex., n.d.), pp. 739–49.
15. David Brading, *Mineros y comerciantes en el México borbónico (1763–1810)* (Mexico, 1995), pp. 50–51.
16. John Lynch, *Las revoluciones hispanoamericanas 1808–1826* (Barcelona, 1976), p. 12.
17. Herbert J. Priestley, *José de Gálvez, visitador-general of New Spain, 1765–1771* (Philadelphia, 1980), pp. 140–41, 346–47.
18. Luis Navarro García, "El virrey marqués de Croix," in *Virreyes de Nueva España (1759–1779),* 2 vols. (Seville, 1967), vol. 1, p. 265.

Chapter 5

1. James Creelman, "President Díaz, Hero of the Americas," in *Pearson's Magazine* (New York), vol. 19, no. 3, Mar. 1908, p. 245.
2. Luis González y González, "El linaje de la cultura mexicana," in *Vuelta* (Mexico), vol. 6, no. 72, Nov. 1982, pp. 14–23.
3. Zumárraga to the emperor, Apr. 15, 1540, in Mariano Cuevas, *Documentos inéditos del siglo XVI* (Mexico, 1975), p. 99.
4. Lockhart, pp. 235–37.
5. See Moisés González Navarro, *El Porfiriato, Vida social,* in Daniel Cosío Villegas, ed., *Historia moderna de México* (Mexico, 1957), vol. 4, p. 469.
6. Guillermo Prieto, "A la Virgen de Guadalupe," in Gabriel Zaid, ed., *Ómnibus de poesía mexicana* (Mexico, 1979), p. 428.
7. Ernesto de la Torre Villar and Ramiro de Anda, *Testimonios históricos guadalupanos* (Mexico, 1982); see Edmundo O'Gorman, *Destierro de sombras, luz de origen y culto de la Nuestra Señora de Guadalupe del Tepeyac* (Mexico, 1991); Francisco de la Maza, *El Guadalupanismo mexicano* (Mexico, 1981).
8. See David Brading, *Orbe Indiano, de la monarquía católica a la república criolla, 1492–1867* (Mexico, 1991), pp. 375–80, 382–87; see also Fausto Zerón-Medina, *Felicidad de México* (Mexico, 1995).
9. Justo Sierra, *Evolución política del pueblo mexicano,* vol. 12, *Obras completas,* 17 vols. (Mexico, 1984), p. 47.
10. See León Lopetegui and Félix Zubillaga, S.I., *Historia de la Iglesia en*

América española, 2 vols. (Madrid, 1965); Mariano Cuevas, *Historia de la Iglesia en México,* 5 vols. (El Paso, Tex., 1928); Alberto de la Hera, *Iglesia y corona en la América española* (Madrid, 1992).

11. François Chevalier, *Land and Society in Colonial Mexico, the Great Hacienda* (Berkeley and Los Angeles, 1963), pp. 229–62.
12. Juan de Ortega Montañés, *Instrucción reservada al conde de Moctezuma* (Mexico, 1965), p. 89.
13. Sierra, *Evolución política*, pp. 113–15.
14. See José M. Gallegos Rocafull, *El pensamiento mexicano en los siglos XVI y XVII* (Mexico, 1951); Mauricio Beuchot, *El tratado de Francisco de Naranjo para la enseñanza de la Teología, siglo XVII* (Mexico, 1994); José M. Gallegos Rocafull, *Estudios de historia y de filosofía en el México colonial* (Mexico, 1991).
15. Juana Inés de la Cruz, "Respuesta a sor Filotea de la Cruz," in *Obras completas* (Mexico, 1981), p. 836.
16. See Octavio Paz, *Sor Juana Inés de la Cruz, o las trampas de la fe* (Mexico, 1983); Elías Trabulse, *Sor Juana Inés de la Cruz: Florilegio, poesía, teatro, prosa* (Mexico, 1979); José Pascual Buxó, *El enamorado de sor Juana* (Mexico, 1993); *Obras completas de sor Juana Inés de la Cruz*, 4 vols. (Mexico, 1951–1957).
17. Jonathan I. Israel, *Race, Class and Politics in Colonial Mexico 1610–1670* (Oxford, 1975), pp. 130–36;
18. See Genaro García, "Don Juan de Palafox y Mendoza, su virreinato en la Nueva España, sus contiendas con los P.P. Jesuítas, sus partidarios en Puebla, sus apariciones, sus escritos escogidos, etc.," in *Documentos inéditos o muy raros para la historia de México* (Mexico, 1974), pp. 519–663.
19. "Relación que de orden del rey dio el virrey de México don Antonio Sebastián de Toledo, marqués de Mancera, a su sucesor, el excmo. señor don Pedro Nuño Colón, duque de Veragua, en 22 de octubre de 1673," in *Instrucciones y memorias de los virreyes novohispanos,* 2 vols. (Mexico, 1991), vol. 1, pp. 620–21.
20. "Francisco Xavier Clavijero," in Gabriel Méndez Plancarte, *Humanistas del siglo XVIII* (Mexico, 1941), p. 28.
21. González y González, "Linaje," p. 17; See Gregorio M. de Guijo, *Diario 1648–1664,* 2 vols. (Mexico, 1953).
22. Quoted in Antonio Gómez Robledo, "La conciencia mexicana en la obra de Francisco Xavier Clavijero," in *Historia Mexicana* (Mexico), vol. 19, no. 75, Jan.–Mar. 1970, p. 363.
23. See David Brading, *Church and State in Bourbon Mexico: The Diocese of Michoacán 1749–1810* (Cambridge, 1994), pp. 16–19, 32–38, 40–46, 82–97, 97–100.
24. Ibid., pp. 72–76.
25. Ibid., pp. 62–81; Nancy M. Farris, *Crown and Clergy in Colonial Mexico* (Oxford, 1968), pp. 91–92.
26. Farris, *Crown and Clergy,* pp. 87–108; Brading, *Church and State,* pp. 9–16.
27. Luis González y González, *Once ensayos de tema insurgente* (Zamora, 1985), pp. 14, 71–90; Plancarte, *Humanistas*; Miguel Cascón, S.I., *Los jesuítas en Menéndez Pelayo* (Santander, 1939), pp. 215–26; Angel Santos, *Los jesuítas en América* (Madrid, 1992).

28. See David Brading, *The Origins of Mexican Nationalism* (Cambridge, 1985), pp. 24–38; Brading, *Orbe Indiano,* pp. 627, 633–40; Edmundo O'Gorman, ed., "El Heterodoxo guadalupano," in *Obras completas de Fray Servando Teresa de Mier,* 4 vols. (Mexico, 1981).
29. O'Gorman, "Heterodoxo," vol. 1, pp. 241–44.
30. Manuel Abad y Queipo, *Representación sobre la inmunidad personal del clero, reducida por las leyes del nuevo código, en la cual se propuso al rey el asunto de diferentes leyes, que establecidas, harían la base principal de un gobierno liberal y benéfico para las Américas y su metrópoli* (Valladolid, Dec. 11, 1799), in José María Luis Mora, *Obras completas, política,* 8 vols. (Mexico, 1987), vol. 3, pp. 15–75.
31. Ibid., p. 163.
32. See Romeo Flores Caballero, "La consolidación de vales reales en la economía, la sociedad y la política novohispanas," *in Historia Mexicana,* vol. 18, no. 71, Jan.–Mar. 1969, pp. 334–78.
33. Manuel Abad y Queipo, "Escrito presentado a don Manuel Sixto Espinosa, del Consejo de Estado y director único del príncipe de la paz en asuntos de Real Hacienda, dirigido a fin de que suspendiese en las Américas la Real Cédula del 26 de diciembre de 1804 sobre enajenación de bienes raíces y cobro de capitales píos para la consolidación de vales reales," in *Mora,* vol. 3, pp. 100–15.
34. Manuel Abad y Queipo, "Representación a la primera regencia en que se describe compendiosamente el estado de fermentación que anunciaba un próximo rompimiento y se proponían los medios con que tal vez se hubiera podido evitar" (Valladolid, May 30, 1810), in *Mora,* vol. 3, p. 140.
35. Ibid., p. 141.
36. Farris, *Crown and Clergy,* pp. 242–53.
37. González y González, "Linaje," p. 17.

Part II

Chapter 6

1. Juan E. Hernández y Dávalos, *Colección de documentos para la historia de la guerra de Independencia,* 6 vols. (Mexico, 1877–1882), vol. 1, p. 126.
2. Carlos Herrejón Peredo, *Hidalgo antes del grito de Dolores* (Morelia, 1992), p. 23.
3. Hernández y Dávalos, *Colección,* vol. 1, p. 722; see Luis González y González, "De cómo llevarse con los próceres," in *Once ensayos de tema insurgente* (Morelia, 1985), pp. 91–97.
4. Lucas Alamán, *Historia de México,* 5 vols. (Mexico, 1990), vol. 1, p. 226.
5. José M. de la Fuente, *Hidalgo íntimo* (Mexico, 1910), pp. 126–28.
6. Antonio Pompa y Pompa, *Proceso inquisitorial y militar seguidos a D. Miguel Hidalgo y Costilla* (Morelia, 1984), pp. 14, 54, 74, 83, 96; Carlos Herrejón Peredo, *Hidalgo: Razones de la insurgencia y biografía documental* (Mexico, 1987), pp. 109, 131, 135, 136, 149; See "Don Miguel Hidalgo y Costilla," in *Boletín bibliográfico de la secretaría de hacienda y crédito público* (Mexico, 1967), p. 375.
7. Pompa y Pompa, *Proceso inquisitorial*, pp. 116–17; Herrejón, *Hidalgo: Razones,* pp. 107, 108, 112, 170; see Nicolás Rangel, "Estudios literarios

de Hidalgo," in *Boletín del Archivo General de la Nación* (Mexico), vol. 1, no. 1, Sept.–Oct. 1930.

8. Enrique Arreguín, *Hidalgo en San Nicolás: Documentos inéditos* (Morelia, 1956), pp. 131–32.
9. Alamán, *Historia,* vol. 1, p. 227.
10. Francisco Morales, *Clero y política en México (1767–1834)* (Mexico, 1974), pp. 150–51; See Juan Hernández Luna, "El mundo intelectual de Hidalgo," in *Historia Mexicana* (Mexico), vol. 3, no. 10, Oct.–Dec. 1953, pp. 157–77.
11. Hernández y Dávalos, *Colección,* vol. 1, pp. 119–20. Herrejón, *Hidalgo: Razones,* pp. 208, 245; Luis Castillo Ledón, *Hidalgo: Vida del héroe,* 2 vols. (Mexico, 1972), vol. 2, pp. 31–32; see also Agustín Rivera, *El joven teólogo Miguel Hidalgo y Costilla, anales de su vida y de su revolución de independencia* (Mexico, 1954), pp. 50–51.
12. Alamán, *Historia,* vol. 1, p. 246.
13. José María Luis Mora, *México y sus revoluciones,* 3 vols. (Mexico, 1965), vol. 3, p. 221; Francisco Bulnes, *La guerra de Independencia Hidalgo-Iturbide* (Mexico, 1969), pp. 127–42.
14. Alamán, *Historia,* vol. 1, p. 298.
15. Ibid., p. 227; see Vicente Fuentes Díaz, *El obispo Abad y Queipo frente a la Independencia* (Mexico, 1985).
16. Hernández y Dávalos, *Colección,* vol. 2, p. 404; Herrejón, *Hidalgo antes del grito,* p. 78.
17. Mora, *México y sus revoluciones,* vol. 3, p. 17; Alamán, *Historia,* vol. 1, p. 243.
18. Mora, *México y sus revoluciones,* vol. 3, p. 17.
19. Alamán, *Historia,* vol. 1, p. 244.
20. Herrejón, *Hidalgo: Razones,* pp. 242–43.
21. Castillo Ledón, *Vida del héroe,* vol. 2, pp. 140–41.
22. Hernández y Dávalos, *Colección,* vol. 1, p. 126.
23. Mora, *México y sus revoluciones,* vol. 3, p. 118.
24. Alamán, *Historia,* vol. 1, pp. 316, 387; vol. 2, p. 75; see Manuel Carrera Stampa, "Hidalgo y su plan de operaciones," in *Historia Mexicana,* vol. 3, no. 10, Oct.–Dec. 1953.
25. Mora, *México y sus revoluciones,* vol. 3, p. 118.
26. Pompa y Pompa, *Proceso inquisitorial,* p. 234; Herrejón, *Hidalgo: Razones,* p. 312; Jean Meyer, *Los tambores de Calderón* (Mexico, 1993), pp. 70–73.
27. Pedro García, *Con el cura Hidalgo en la guerra de Independencia* (Mexico, 1948), p. 161.
28. Ibid., pp. 161–62.
29. Hernández y Dávalos, *Colección,* vol. 1, p. 185; Pompa y Pompa, *Proceso inquisitorial,* pp. 234, 237; Herrejón, *Hidalgo: Razones,* pp. 315, 321, 329.
30. Pompa y Pompa, *Proceso inquisitorial,* pp. 239, 241, 245–46; Herrejón, *Hidalgo: Razones,* pp. 245, 317, 319, 323.
31. Pompa y Pompa, *Proceso inquisitorial,* pp. 239, 241, 245–46; Herrejón, *Hidalgo: Razones,* 245, 317, 319, 323.
32. Pompa y Pompa, *Proceso inquisitorial,* 239, 241, 245–46; Herrejón, *Hidalgo: Razones,* pp. 245, 317, 319, 323.

33. De la Fuente, *Hidalgo íntimo,* pp. 126–28.
34. Hernández y Dávalos, *Colección*, vol. 1, p. 190.
35. Herrejón, *Hidalgo: Razones,* pp. 339–40; Luis Villoro, *El proceso ideológico de la revolución de independencia* (Mexico, 1967), pp. 87–93.
36. De la Fuente, *Hidalgo íntimo,* p. 529.
37. Alamán, *Historia,* vol. 5, p. 221.
38. Ibid., p. 204.
39. Manuel Abad y Queipo, *Colección de escritos más importantes que en diferentes épocas dirigió al Gobierno* (Mexico, 1813), p. 59.
40. Carlos Herrejón Peredo, *Morelos: Vida preinsurgente y lecturas* (Zamora, 1984), pp. 162, 170–71.
41. Herrejón, *Morelos,* p. 217.
42. Ibid., pp. 161–67.
43. Ernesto Lemoine, *Morelos su vida revolucionaria a través de sus escritos y de otros testimonios de la época* (Mexico, 1991), p. 287.
44. Alamán, *Historia,* vol. 2, p. 71.
45. Herrejón, *Morelos,* p. 214.
46. "Causa instruída contra el Generalísimo D. Ignacio Allende," in Genaro García, *Documentos históricos Mexicanos,* 8 vols. (Mexico, 1985), vol. 6, p. 45; see Luis Villoro, "Hidalgo: violencia y libertad," in *Historia Mexicana,* vol. 2, no. 6, Oct.–Dec. 1952.
47. Carlos Herrejón Peredo, *Morelos: Documentos inéditos de vida revolucionaria* (Zamora, 1987), p. 178.
48. Herrejón, *Documentos inéditos,* p. 308.
49. Lemoine, *Morelos su vida,* p. 45; see Wilbert H. Timmons, *Morelos, sacerdote, soldado, estadista* (Mexico, 1983), pp. 57–58.
50. Herrejón, *Documentos inéditos,* p. 237.
51. Lemooine, *Morelos su vida,* p. 222
52. Museo Nacional de Arqueología, Historia y Etnografía, *Morelos: Documentos inéditos y poco conocidos,* vol. 1 (Mexico, 1927), pp. 357–58.
53. Lucas Alamán, *Semblanzas e ideario* (Mexico, 1978), p. 83.
54. Lemoine, *Morelos su vida,* p. 190.
55. Alamán, *Semblanzas,* p. 87.
56. Herrejón, *Documentos inéditos,* p. 210.
57. Lemoine, *Morelos su vida,* p. 328.
58. Ibid., pp. 211–12.
59. Ibid., p. 234.
60. Ibid., p. 287.
61. Timmons, *Morelos, sacerdote,* p. 88.
62. Lemoine, *Morelos su vida,* p. 288.
63. Alamán, *Historia,* vol. 2, p. 220.
64. Gabriel Zaid, *Ómnibus de poesía mexicana* (Mexico, 1973), p. 164.
65. Lemoine, *Morelos su vida,* p. 279–80.
66. Hernández y Dávalos, *Colección,* vol. 2, pp. 227–28.
67. Lemoine, *Morelos su vida,* p. 181.
68. Ibid., p. 264.
69. González y González, et al., *El Congreso de Anáhuac* (Mexico, 1963), p. 14.
70. Lemoine, *Morelos su vida,* p. 182; Villoro, *Proceso ideológico,* pp. 98–105.

71. Ibid., p. 246.
72. Simon Bolívar, *Doctrina del Libertador* (Venezuela, 1976), pp. 73, 74.
73. Alamán, *Historia*, vol. 3, p. 209.
74. Lemoine, *Morelos su vida,* p. 255.
75. Ibid., p. 384.
76. *Correo Americano del Sur*, in García, *Documentos*, vol. 4, p. 193.
77. Lemoine, *Morelos su vida,* p. 474.
78. Alamán, *Historia*, vol. 4, p. 118.
79. Ibid., pp. 215, 217; Carlos Herrejón Peredo, *Los procesos de Morelos* (Zamora, 1985), pp. 63–64.
80. Herrejón, *Procesos,* p. 64.
81. Alamán, *Historia,* vol. 4, p. 216.
82. Ibid., pp. 221–22.
83. Ibid., pp. 219–20.
84. Herrejón, *Procesos,* pp. 162–64.
85. Alamán, *Semblanzas*, pp. 76–90.
86. Niceto de Zamacois, *Historia de México,* 20 vols. (Mexico, n.d.), vol. 18, pp. 172–73.
87. "Discurso pronunciado por el señor General don Porfirio Díaz, Presidente de la República, al recibir del Excelentisimo señor Embajador de España las reliquias de José María Morelos, el 17 de septiembre de 1910," in Genaro García, ed., *Crónica oficial de las fiestas del primer centenario de la Independencia de México* (Mexico, 1990), Appendix no. 57, pp. 23–24.
88. Justo Sierra, *Evolución política del pueblo mexicano,* vol. 12, *Obras completas,* 15 vols. (Mexico, 1991), p. 158.

Chapter 7

1. Alexander von Humboldt, *Ensayo político sobre el reino de la Nueva España* (Mexico, 1973), p. 30.
2. See Carlos Navarro y Rodrigo, "Vida de Agustín de Iturbide," in *Memorias de Agustín de Iturbide* (Madrid, 1919); Mario Mena, *El Dragón de fierro: Biografía de Agustín de Iturbide* (Mexico, 1969); Silvio Zavala and José Bravo Ugarte, "Un nuevo Iturbide," in *Historia Mexicana* (Mexico), vol. 2, no. 6, Oct.–Dec. 1952, pp. 267–76.
3. Navarro y Rodrigo, *Memorias de Agustín de Iturbide*, p. 239.
4. Lorenzo de Zavala, "El Historiador y el representante popular Ensayo critico de las revoluciones de México desde 1808 hasta 1830," in *Obras históricas* (Mexico, 1969), p. 56.
5. Lucas Alamán, *Historia de México: Desde los primeros movimientos que prepararon su independencia en el año de 1808 hasta la época presente,* 5 vols. (Mexico, 1985), vol. 5, p. 55.
6. "Expediente relativo a la solicitud de libertad hecha por varias mujeres parientes de insurgentes, aprisionadas por orden del coronel Iturbide, 8 July 1816–11 Feb 1819," in Genaro García, *Documentos históricos mexicanos: Obra conmemorativa del primer centenario de la independencia de México,* 8 vols. (Mexico, 1985), vol. 5, p. 387.
7. Navarro y Rodrigo, "Agustín de Iturbide," pp. 237–38; see Edmundo O'Gorman, *Hidalgo en la historia* (Mexico, 1964).
8. Quoted in Alamán, *Historia*, vol. 4, p. 14.
9. Ibid., vol. 5, pp. 56–57.

10. Ibid.
11. Ibid., vol. 4, p. 451.
12. Ibid., vol. 5, p. 56.
13. "Proclama con que D. Agustín de Iturbide anunció el plan llamado de Iguala," Feb. 24, 1821, in Alamán, *Historia,* vol. 5, appendix 6, p. 8.
14. Humboldt, *Ensayo político,* pp. 118–19.
15. Alamán, *Historia,* vol. 5, p. 333.
16. Iturbide to the bishop of Guatemala, Oct. 10, 1821, in Alfonso Trueba, *Iturbide: Un trágico destino* (Mexico, 1959), p. 114.
17. Agustín Iturbide to Miguel Negrete, Oct. 23, 1821, in Trueba, *Iturbide,* p. 123.
18. Javier Ocampo, *Las ideas de un El peublo Mexicano en la consumación de su independencia* (Mexico, 1969), pp. 263–78.
19. Alamán, *Historia,* vol. 5, p. 609.
20. Zavala, *El historiador,* p. 131.
21. "Discurso dirigido al Congreso después de haber jurado como emperador de México," May 21, 1821, in Trueba, *Iturbide,* p. 150.
22. Agustín Iturbide to Simón Bolívar, May 29, 1822, in Trueba, *Iturbide,* p. 151.
23. Gabriel Zaid, ed., *Ómnibus de poesía mexicana* (Mexico, 1989), p. 165.
24. Henry George Ward, *México en 1827* (Mexico, 1981), p. 111.
25. "Discurso," Feb. 11, 1823, quoted in Timothy E. Anna, *El Imperio de Iturbide* (Mexico, 1991), p. 188.
26. Ibid.
27. See Magnus Mörner, "Una carta de Agustín de Iturbide en 1824," in *Historia Mexicana,* vol. 18, no. 52, Apr.–June 1964, pp. 593–99.
28. Quoted in Trueba, *Iturbide,* p. 212.
29. See José Gutierrez Casillas, S.J., *Papeles de don Agustín de Iturbide* (Mexico, 1977), pp. 359–60.
30. Simón Bolívar to Francisco de Paula Santander, Jan. 6, 1825, in *Simón Bolívar, Doctrina del Libertador* (Sucre, Venezuela, 1976), pp. 185–86.
31. Lucas Alamán, *Semblanzas e ideario* (Mexico, 1978), p. 170; see Rafael Aguayo Spencer, "Alamán estadista," in *Historia Mexicana*, vol. 3, no. 10, Apr.–June 1953, pp. 279–90.
32. See Michael P. Costeloe, *La primera república federal de México 1824–1835* (Mexico, 1983); Carlos Herrejón Peredo, *Guadalupe Victoria, Documentos I* (Mexico, 1986).
33. See José Fuentes Mares, *Poinsett: Historia de una gran intriga* (Mexico, 1964), p. 99; Joel R. Poinsett, *Notas sobre México* (Mexico, 1973).
34. "Una mirada sobre la América Española," Apr.–June 1829, in Bolívar, *Doctrina,* pp. 283–84.
35. Ibid., p. 286.
36. Zavala, "El Historiador," p. 500.
37. See Enrique Olavarría y Ferrari and Juan de Dios Arias, "México independiente 1821–1855," in *México a través de los siglos*, 5 vols. (Mexico, 1970), vol. 4, pp. iii–viii.
38. Quoted in Fernando Díaz Díaz, *Caudillos y caciques* (Mexico, 1972), p. 134.
39. Bolívar, *Doctrina,* p. 251.
40. Ibid., p. 323.

41. José María Luis Mora, *Obras sueltas* (Mexico, 1963), p. 498; see Charles A. Hale, *Mexican Liberalism in the Age of Mora 1821–1853* (New Haven, Conn., 1968).
42. Alamán, *Semblanzas*, p. 153.
43. Alamán, *Historia,* vol. 5, pp. 686–87.
44. Antonio López Santa Anna to Pedro de Landero, Jan. 17, 1830, in José C. Valadés, *México, Santa Anna y la guerra de Texas* (Mexico, 1982), p. 97.
45. Díaz, *Caudillos,* p. 84; see Ohland Morton, "Life of General Don Manuel de Mier y Terán, As It Affected Texas-Mexican Relations," in *Southwestern Historical Quarterly,* vol. 46, no. 1, July 1942.
46. *El Sol*, Sept. 26, Oct. 1, 1829, quoted in Díaz, *Caudillos,* p. 85.
47. "¿Quien vive?" Sept. 26, 1829, quoted in Díaz, *Caudillos,* p. 86.
48. Valadés, *Mexico, Santa Anna.*
49. Zavala, "El Historiador," p. 113.
50. Valadés, *Mexico, Santa Anna,* p. 113.
51. "Manifiesto del presidente de los Estados Unidos Mexicanos a sus conciudadanos," Mexico, June 18, 1833, Archivo General de la Nación de México, Serie: Archivo de Guerra, v. 1015, 2a. pte., fs. 4 (Archivo General de la Nación, México).
52. Ibid.
53. Manuel Mier y Terán to José María Luis Mora, Nov. 1831, in Mora, *Obras sueltas*, p. 35.
54. "Manifiesto."
55. Ibid.
56. Mora, *Obras sueltas*, p. 11.
57. "Manifiesto que de sus operaciones en la campaña de Tejas y en su cautivero dirige a sus conciudadanos el general Antonio López de Santa Anna," May 1837, in Genaro García, *Documentos inéditos o muy raros para la historia de México* (Mexico, 1991), p. 125.
58. Lucas Alamán to Duque de Terranova y Monteleone, Apr. 19, 1834, in Alamán, *Documentos diversas* (Mexico, 1947), vol. 4, p. 337.
59. Lucas Alamán to Duque de Terranova y Monteleone, June 28, 1836, in ibid., p. 343.
60. Quoted in Díaz, *Caudillos*, p. 138.
61. Ibid., p. 144.
62. Santa Anna to Ministro de Guerra, in *El Cosmopolita*, Dec. 8, 1838, quoted in Díaz, *Caudillos,* p. 145.
63. Justo Sierra, *Evolución política del pueblo mexicano,* vol. 12, *Obras completas*, 14 vols. (Mexico, 1977), p. 220.
64. Carlos María de Bustamante, "Apuntes para la historia del gobierno del general Antonio López de Santa Anna," quoted in Díaz, *Caudillos,* p. 166.
65. Fanny Calderón de la Barca, *La vida en México durante una residencia de dos años en ese país* (Mexico, 1959), p. 440; similar accounts can be found in Frank Sanders, "México visto por los diplomáticos del siglo XIX," in *Historia Mexicana* (Mexico), vol. 20, no. 79, Jan.–Mar. 1971, pp. 368–411.
66. José María Bocanegra, "Memoria del Secretario de Estado y del Despacho de Relaciones Exteriores y Gobernacion de la Republica Mexicana correspondiente a la administracion provisional, en los años 1841, 1842 y 1843," Jan. 1844, quoted in Díaz, *Caudillos,* p. 165.

67. Calderón de la Barca, *Vida en México,* p. 462.
68. Guillermo Prieto, *Memorias de mis tiempos* (Mexico, 1976), p. 361; see also Juan Suárez y Navarro, *Historia de México y del general Antonio López de Santa Anna* (Mexico, 1987).
69. Prieto, *Memorias,* p. 369.
70. Alamán, *Semblanzas,* pp. 140, 142.
71. Agustín Yáñez, *Antonio López de Santa Anna: Espectro de una sociedad,* Conference at El Colegio de Mexico, Sept. 21, 1971, quoted in Díaz, *Caudillos,* p. 148.
72. Calderón de la Barca, *Vida en México,* p. 408.
73. Ibid., pp. 87–88, 98.
74. Cited in Moisés González Navarro, *Anatomía del poder en México (1848–1853)* (Mexico, 1977), p. 390.
75. Lucas Alamán to Antonio López Santa Anna, July 23, 1834, quoted in Valadés, *Alamán,* p. 340.
76. Quoted in Valadés, *Alamán,* p. 432; Alamán, *Historia,* vol. 5, p. 688.
77. Alamán, *Historia,* vol. 5, p. 688.
78. Lucas Alamán to Duque de Terranova y Monteleone, May 28, 1847, in Alamán, *Documentos,* vol. 4, p. 446.
79. Alamán to Terranova y Monteleone, June 28, 1847, in Alamán, *Documentos,* vol. 4, p. 448.
80. Prieto, *Memorias,* pp. 420–21.
81. Lucas Alamán, "Nuestra profesión de fe," in Luis González y González, *Galeria de la Reforma* (Mexico, 1983), p. 130.
82. Alamán to Terranova y Monteleone, Oct. 28, 1847, in Alamán, *Documentos,* vol. 4, p. 455.
83. Alamán to Terranova y Monteleone, Nov. 28, 1847, in Alamán, *Documentos,* vol. 4, p. 457.
84. Ibid.
85. "Antonio López de Santa Anna a sus compatriotas," Sept. 16, 1847, quoted in Díaz, *Caudillos,* p. 215.
86. Ibid.
87. Alamán, *Historia,* vol. 5, p. 688.
88. Mora to the Mexican government, June 1848, in José María Luis Mora, *Diplomática,* vol. 7, *Obras completas,* 8 vols. (Mexico, 1988), p. 194.
89. Arturo Arnáiz y Freg, "El Dr. Mora, teórico de la reforma liberal," in *Historia mexicana,* vol. 5, no. 20, Apr.–June 1956, p. 569.
90. González Navarro, *Anatomía del poder,* p. 35.
91. José María Luis Mora to the Mexican government, Oct. 30, 1849, in Mora, *Obras completas,* p. 286.
92. Alamán to Terranova y Monteleone, May 13, 1848, in Alamán, *Documentos,* vol. 4, p. 471.
93. Alamán, *Semblanzas,* p. 169.
94. Alamán, *Historia,* vol. 5, p. 903.
95. Ibid., p. 927.
96. "Revista política de las diversas administraciones que la República Mexicana ha tenido hasta 1837," in Mora, *Obras sueltas* (Mexico, 1963), p. 129.
97. Jesús C. Romero, *La verdadera historia del himno nacional* (Mexico, 1961), p. 72; see also Luis Reyes de la Maza, *El teatro en México en la época de Santa Anna,* 2 vols. (Mexico, 1979), vol. 2, pp. 256–57.

98. Ibid., p. 57.
99. Ibid.
100. Ibid.

Chapter 8

1. Melchor Ocampo, "Viaje a Veracruz, Puebla y sur de México," in Melchor Ocampo, *Obras completas,* 3 vols. (Mexico, 1900), vol. 3, p. 592.
2. See Raúl Aurreola Cortés, *Ocampo* (Michoacán, 1992); Francisco de la Maza, "Melchor Ocampo, literato y bibliófilo," in *Historia Mexicana* (Mexico), vol. 11, no. 41, July–Sept. 1961, pp. 104–18.
3. Melchor Ocampo, "Representación sobre reforma del arancel de obvenciones parroquiales," in *Obras completas,* vol. 1, pp. 1–17.
4. Ocampo, "Viaje," p. 592.
5. Quoted by Angel Pola in Ocampo, *Obras completas,* vol. 1 , pp. 3–4.
6. "Respuesta primera a la impugnación de la representación sobre Reforma de obvenciones parroquiales," in Ocampo, *Obras completas,* vol. 1, p. 75; see José Bravo Ugarte, *Munguía, Obispo y Arzobispo de Michoacán 1810–1868* (Mexico, 1967).
7. "Impugnación a la representación sobre Reforma de obvenciones parroquiales," in Ocampo, *Obras completas,* vol. 1, pp. 34–35.
8. Ibid., pp. 41–42. See Robert J. Knowlton, "La Iglesia mexicana y la Reforma: Respuesta y resultados," in *Historia Mexicana*, vol. 18, no. 72, Apr.–June 1969, pp. 516–34.
9. Ocampo, ibid., verse from Matthew, 10:9, pp. 7–10.
10. Ibid., p. 166.
11. Ibid., p. 167.
12. Ibid., p. 53.
13. Ibid., p. 54.
14. Ibid., p. 355.
15. See Daniel Moreno, *Los hombres de la Reforma* (Mexico, 1961); Porfirio Parra, *Estudio histórico-sociológico sobre la Reforma en México* (Guadalajara, 1906); José María Vigil, "La Reforma," in *México a través de los siglos*, 5 vols. (Mexico, 1970), vol. 5.
16. Justo Sierra, *Juárez, su obra y su tiempo* (Mexico, 1980), p. 91, 136.
17. Ibid., pp. 104, 178.
18. Francisco Bulnes, *Juárez y las revoluciones de Ayutla y Reforma* (Mexico, 1905), p. 284.
19. Daniel Cosío Villegas, *La Constitución de 1857 y sus críticos* (Mexico, 1980), pp. 166–67.
20. Ibid., p. 120.
21. Melchor Ocampo to A. García, Mar. 8, 1853, in Ocampo, *Obras completas,* vol. 2, p. 291.
22. "Discurso de protesta como presidente electo de los Estados Unidos Mexicanos, el 1. de diciembre de 1871," in Benito Juárez, *Documentos, discursos y correspondencia,* 15 vols. (Mexico, 1964–1970), vol. 15, p. 572.
23. Benito Juárez, "Manifiesto al pueblo del estado de Oaxaca con motivo de haber sido reelecto como gobernador," Aug. 12, 1849, in *Documentos,* vol. 1, p. 654.
24. Francisco Vasconcelos, *Apuntes históricos de la vida en Oaxaca en el siglo XIX*, s.p.i.

25. Sierra, *Juárez*, p. 438.
26. Ibid., p. 19; see also Rafael Zayas Enríquez, *Benito Juárez: Su vida, su obra* (Mexico, 1971); Andrés Henestrosa, *Los caminos de Juárez* (Mexico, 1985).
27. Benito Juárez, "Apuntes para mis hijos," in *Documentos,* vol. 1, p. 91.
28. Sierra, *Juárez*, p. 19.
29. See Jorge Fernando Iturribarría, *La generación oaxaqueña del 57* (Oaxaca, 1956); *Historia de Oaxaca*, 5 vols. (Oaxaca, 1982).
30. "Discurso patriótico pronunciado por el Lic. Benito Juárez en la ciudad de Oaxaca, 16 de septiembre de 1840," in Juárez, *Documentos,* vol. 1, pp. 480, 482–83.
31. Ibid., p. 483.
32. Ibid., pp. 480, 483–84.
33. See Ronald Spores, et al., *Benito Juárez: Gobernador de Oaxaca, documentos de su mandato y servicio público* (Oaxaca, 1987).
34. Benito Juárez to Bishop of Oaxaca, Jan. 25, 1848, in *Documentos,* vol. 1, p. 519.
35. "Circular a los señores gobernadores de departamentos de Oaxaca," Jan. 24, 1849, in *Documentos,* vol. 1, pp. 612–13.
36. Benito Juárez, "Exposición al soberano Congreso de Oaxaca al abrir sus sesiones," July 2, 1849, in *Documentos,* vol. 1, p. 625.
37. "La Crónica, periódico del gobierno del estado de Oaxaca," July 3, 1850, in Benito Juárez, *Miscelánea,* 3 vols. (Mexico, 1987), vol. 3, p. 37.
38. Bulnes, *Juárez y las revoluciones,* p. 177.
39. "Juárez desde una nueva faz," in Benito Juárez, *Discursos y manifiestos*, 3 vols. (Mexico, 1987), vol. 2, pp. xxii–xxiv.
40. Benito Juárez, "Exposición al soberano Congreso de Oaxaca al cerrar sus sesiones extraordinarias," Feb. 29, 1848, in *Miscelánea*, vol. 3, p. 18.
41. Juárez, "Apuntes," pp. 269–71.
42. Benito Juárez, "Manifiesto al pueblo del estado de Oaxaca con motivo de haber sido reelecto como gobernador," Aug. 12, 1849, in *Documentos,* vol. 1, p. 654.
43. See Robert J. Knowlton, *Los bienes del clero y la Reforma mexicana, 1856–1910* (Mexico, 1985); Walter V. Scholes, *Política mexicana durante el régimen de Juárez, 1855–1872* (Mexico, 1976).
44. Constitución de 1857, quoted in Felipe Tena Ramírez, *Leyes fundamentales de México 1800–1976* (Mexico, 1976), p. 601.
45. Anselmo de la Portilla, quoted in Daniel Cosío Villegas, *Constitución de 1857,* pp. 98–99.
46. Clemente de Jesús Munguía, "Circular que el obispo de Michoacán dirige al muy ilustre y venerable cabildo," in *En defensa de la Soberanía, Derechos y Libertades de la Iglesia* (Mexico, 1973), p. 99; see Jan Bazant, "La desamortización de los bienes corporativos de 1856," in *Historia Mexicana*, vol. 16, no. 62, Oct.–Dec. 1966.
47. Ignacio Manuel Altamirano, *Obras completas,* 6 vols. (Mexico, 1986), vol. 6, p. 142. See Moisés Ochoa Campos, *Ignacio Manuel Altamirano: Discursos cívicos* (Mexico 1984).
48. See Jacqueline Covo, *Las ideas de la Reforma en México: 1855–1861* (Mexico, 1983); Anselmo de la Portilla, *México en 1856 y 1857: Gobierno del General Comonfort* (Mexico, 1987); Mario Guzmán Galarza, *Documentos básicos de la Reforma* (Mexico, 1982).

49. Luis González y González, *La Ronda de las Generaciones* (Mexico, 1984), p. 17.
50. José Fuentes Mares, *Juárez, los Estados Unidos y Europa* (Mexico, 1991), p. 194.
51. "Testamento de Melchor Ocampo," June 3, 1861, in Ocampo, *Obras completas,* vol. 2, pp. cvii–cviii.
52. Quoted in Egon Caesar Conte Corti, *Maximiliano y Carlota* (Mexico, 1971), p. 39.
53. Ibid., p. 35. See José Luis Blasio, *Maximiliano íntimo: Memorias de un secretario particular* (Mexico, 1966).
54. Maximilian to the archduchess Sofía, 1858, in ibid., p. 64.
55. Maximilian to the emperor Francisco José, Apr. 17, 1859, ibid., p. 67.
56. Maximilian to King Leopold of Belgium, 1859, ibid., p. 69.
57. Ibid., p. 57.
58. Ibid., p. 69.
59. See José Fuentes Mares, *La emperatriz Eugenia y su aventura mexicana* (Mexico, 1976); Manuel Rivera Cambas, *Historia de la Intervención Europea y Norteamericana en México y del Imperio de Maximiliano de Habsburgo*, 3 vols. (Mexico, 1987).
60. Conte Corti, *Maximiliano y Carlota,* pp. 222, 245.
61. Ibid., pp. 219–23. See Alfred Jackson Hanna and Kathryn Abbey Hanna, *Napoleón III y México* (Mexico, 1981).
62. José C. Valadés, *Alamán: Estadista e historiador* (Mexico, 1987), p. 432. See José Manuel Hidalgo y Esnaurrízar, *Proyectos de monarquía en México* (Mexico, 1962).
63. Maximilian to the Count Rechberg, Feb. 18, 1863, in Conte Corti, *Maximiliano y Carlota,* p. 152.
64. Ibid., p. 261. See E. Lefevre, *Historia de la Intervención Francesa en México, Documentos oficiales recogidos en la secretaría privada de Maximiliano* (Brussels and London, 1869).
65. José María Iglesias, *Revistas históricas sobre la Intervención Francesa en México* (Mexico, 1987), p. 411.
66. Paula de Kolonitz, *Un viaje a México en 1864* (Mexico, 1984), p. 62.
67. "Discurso pronunciado por Maximiliano al desembarcar en Veracruz," quoted in José Fuentes Mares, *Juárez, el imperio y la república* (Mexico, 1982), p. 50.
68. Iglesias, *Revistas históricas,* p. 444.
69. Maximilian to his brother Carlos Luis, July 10, 1864, in Conte Corti, *Maximiliano y Carlota,* p. 288.
70. King Leopold to Maximilian, Dec. 25, 1863, ibid., p. 222.
71. Maximilian to his brother Carlos Luis, Oct. 4, 1864, ibid., p. 291.
72. Iglesias, *Revistas históricas,* p. 476.
73. Maximilian to José María Gutiérrez de Estrada, Apr. 30, 1864, in Conte Corti, *Maximiliano y Carlota, p.* 293.
74. Maximilian to Dr. Jilek, Feb. 10, 1865, ibid., p. 316.
75. Maximilian to José María Gutiérrez de Estrada, Apr. 30, 1864, ibid., pp. 294, 305.
76. Iglesias, *Revistas históricas,* pp. 465, 551–54.
77. Ibid., p. 566.
78. Maximilian to Dr. Jilek, Feb. 10, 1865, ibid., p. 316.

79. Iglesias, *Revistas históricas,* p. 518.
80. Carlota to the Empress Eugenia, Apr. 27, 1865, ibid., p. 346.
81. Maximilian to King Leopold, May 12, 1865, ibid.
82. Francisco de Paula de Arrangoiz, *México desde 1808 hasta 1867* (Mexico, 1985), p. 588.
83. Quoted in Conte Corti, *Maximiliano y Carlota,* p. 371.
84. Iglesias, *Revistas históricas,* p. 698.
85. "Proclama de Maximiliano para justificar el decreto del 3 de octubre de 1865," in Daniel Moreno, comp., *El sitio de Querétaro* (Mexico, 1989), p. 289.
86. See Juárez, *Documentos,* vol. 9, pp. 119–30.
87. King Leopold to Maximilian, Nov. 12, 1865, in Conte Corti, *Maximiliano y Carlota,* p. 386.
88. Maximilian to Countess Binzer, Feb. 3, 1866, ibid., p. 402.
89. Ibid.
90. Maximilian to the Count of Hadik, Feb. 3, 1866, ibid., p. 403.
91. Carlota to Maximilian, July 1866, ibid., p. 454.
92. Vicente Riva Palacio, "Adios, Mamá Carlota," quoted in Gabriel Zaid, *Ómnibus de poesía mexicana* (Mexico, 1973), pp. 172–73.
93. Benito Juárez to Andrés S. Viesca, Aug. 7, 1866, in Juárez, *Epistolario* (Mexico, 1957), p. 367.
94. Conte Corti, *Maximiliano y Carlota,* p. 485.
95. Carlota to Maximilian, Sept. 9, 1866, ibid., p. 499.
96. Stefan Herzfeld to Agustín Fisher, Sept. 5, 1866, ibid., p. 529.
97. Archduchess Sofia to Maximilian, Jan. 9, 1867, ibid., p. 553.
98. Samuel Basch, quoted in Daniel Moreno, *Sitio de Querétaro,* p. 79.
99. See José Fuentes Mares, *Proceso de Fernando Maximiliano de Habsburgo, Miguel Miramón y Tomás Mejía* (Mexico, 1966); Manuel Ramírez Arellano, *Últimas horas del Imperio* (Mexico, 1869).
100. Samuel Basch, quoted in *Sitio de Querétaro,* p. 127.
101. Concepción Lombardo de Miramón, *Memorias* (Mexico, 1989), p. 588.
102. Cited by Conte Corti, *Maximiliano y Carlota,* p. 587.
103. Samuel Basch, quoted in *Sitio de Querétaro,* p. 104.
104. Quoted in Conte Corti, *Maximiliano y Carlota,* p. 598.
105. Benito Juárez to Pedro Santacilia, Jan. 1, 1867, in Juárez, *Correspondencia Juárez-Santacilia 1858–1867* (Mexico, 1972), p. 343.
106. "Manifiesto al pueblo de México," July 15, 1867, in *Historia documental de México,* 2 vols. (Mexico, 1984), vol. 2, p. 350.
107. Juárez to Pedro Santacilia, Dec. 1, 1865, in Juárez, *Epistolario,* p. 337.
108. Juárez to Santacilia, Jan. 12, 1865, ibid., p. 293.
109. Juárez to Santacilia, Jan. 12, 1865, ibid.
110. Juárez to Santacilia, Jan. 6, 1865, ibid., p. 294.
111. Juárez to Santacilia, Feb. 9, 1865, ibid., p. 296.
112. Juárez to Santacilia, Sept. 21, 1865, ibid., p. 324.
113. Juárez to Santacilia, Feb. 23, 1865, ibid., p. 297.
114. Juárez to Santacilia, Mar. 23, 1865, ibid., pp. 303–4.
115. Quoted in Miguel Galindo y Galindo, *La Gran década nacional,* 3 vols. (Mexico, 1987), vol. 1, pp. 113, 120.
116. Quoted in José Fuentes Mares, *Y México se refugió en el desierto* (Mexico, 1979), p. 105.

117. Emilio Rabasa, *La organización política de México. La Constitución y la dictadura* (Madrid, n.d.), p. 145.
118. Ibid., p. 143.
119. Benito Juárez to Guillermo Prieto, Oct. 1, 1865, in Juárez, *Epistolario*, p. 329.
120. Ralph Roeder, *Juárez y su México*, 2 vols. (Mexico, 1958), vol. 2, pp. 426–27.
121. Juárez to Santacilia, Dec. 21, 1865, in Juárez, *Epistolario*, pp. 338–39.
122. Bulnes, *Juárez y las revoluciones*, p. 608.
123. Guillermo Prieto, "Discurso del 16 de septiembre de 1861," quoted in Bulnes, *Juárez y las revoluciones*, p. 611.
124. See Mario Treviño Villarreal, *Rebelión contra Juárez, 1869–1870* (Monterrey, 1991); Ireneo Paz, *Algunas campañas: Memorias* (Mexico, 1885).
125. Carleton Beals, *Porfirio Diaz* (Mexico, 1982), p. 332.
126. Quoted in Luis González y González, "El Subsuelo Indígena," in *La República Restaurada, Vida social*, Daniel Cosío Villegas, ed., *Historia moderna de México* (Mexico, 1974), p. 273; see Vicente Pineda, *Sublevaciones indígenas en Chiapas* (Mexico, 1986).
127. González y González, "Subsuelo indígena," pp. 278, 279.
128. Ibid., pp. 280, 281.
129. Roeder, *Juárez*, pp. 471, 473, 479.
130. "Decreto del Congreso de Colombia en honor del presidente Juárez," May 2, 1865, in Juárez, *Epistolario*, p. 318.
131. Sierra, *Juárez*, p. 438–49.

Chapter 9

1. James Creelman, "President Díaz, Hero of the Americas," in *Pearson's Magazine* (New York), vol. 19, no. 3, Mar. 1908, p. 232.
2. Ibid., pp. 231, 234, 238.
3. Ibid.
4. Ibid.
5. See Rafael de Zayas Enríquez, *Porfirio Díaz* (New York, 1908); Angel Taracena, *Porfirio Díaz* (Mexico, 1983).
6. Creelman, "President Díaz," pp. 249–50.
7. Porfirio Diaz, *Memorias*, 2 vols. (Mexico, 1983), vol. 1, p. 84.
8. Luis González y González, "El subsuelo indígena," in *La República Restau rada, Vida social*, Daniel Cosío Villegas, ed., *Historia moderna de México* (Mexico, 1974), p. 269.
9. Charles Etienne Brasseur, *Viaje por el istmo de Tehuantepec* (Mexico, 1981), p. 152. See Miguel Covarrubias, *Sur de México* (Mexico, n.d.).
10. Juan B. Carriedo, *Estudios históricos y estadísticos del estado Oaxaqueño*, 2 vols. (Mexico, 1949), vol. 1, p. 173.
11. Brasseur, *Viaje por el istmo*, p. 159.
12. Ibid.
13. Ibid., p. 160.
14. Juan Carlos Díaz interview with Lila Díaz, unpublished (Mexico City, 1985).
15. Creelman, "President Díaz," p. 241.
16. Thelma D. Sullivan, "Tlatoani and Tlatocayotl in the Sahagún

Manuscripts," in *Estudios de Cultura Náhuatl* (Mexico, 1980), pp. 226–27.
17. Fray Diego Durán, *Historia de las Indias de Nueva España e Islas de Tierra Firme*, 2 vols. (Mexico, 1984), vol. 2, p. 401.
18. Bernardino de Sahagún, *Historia de las Cosas de la Nueva España*, 4 vols. (Mexico, 1956), vol. 2, pp. 90–97.
19. Sullivan, "Tlatoani and Tlatocayotl," p. 229.
20. Durán, *Historia de las Indias*, vol. 2, p. 407.
21. Creelman, "President Díaz," p. 237.
22. Cited in Nigel Davis, *The Aztec Empire: The Toltec Resurgence* (Oklahoma, 1987), p. 102.
23. Andrés Molina Enríquez, *Los grandes problemas nacionales* (Mexico, 1978), pp. 146, 145.
24. Francisco Bulnes, *El verdadero Díaz y la Revolución* (Mexico, 1992), p. 34.
25. Quoted in Charles R. Berry, *La Reforma en Oaxaca: Una microhistoria de la revolución liberal, 1856–1876* (Mexico, 1989), pp. 27–28.
26. Benito Juárez, "Apuntes para mis hijos," in *Documentos, Discursos y correspondencia,* 3 vols. (Mexico, 1964), vol. 1, p. 95.
27. Díaz, *Memorias,* vol. 1, p. 84.
28. Benito Juárez to Pedro Santacilia, Dec. 7, 1866, May 15, 1867, in Benito Juárez, *Epistolario de Benito Juárez* (Mexico, 1957), pp. 392–93.
29. Federico Gamboa, *Diario* (Mexico, 1971), pp. 88–89.
30. Molina Enríquez, *Grandes problemas,* pp. 132–33.
31. Ibid., pp. 136–40.
32. Ibid., p. 147.
33. Creelman, "President Díaz," p. 237.
34. Molina Enríquez, *Grandes problemas,* p. 134.
35. Ibid., p. 135.
36. Bernardo de Balbuena, *Grandeza mexicana* (Mexico, 1974), p. 91.
37. Gamboa, *Diario*, p. 77.
38. Luis Navarro García, "El virrey Marqués de Croix," in *Virreyes de Nueva España (1759–1779)*, 2 vols. (Seville, 1967), vol. 1, p. 265.
39. See *Estadísticas económicas del Porfiriato, fuerza de trabajo y actividad económica por sectores* (Mexico, n.d.); Chester C. Kaiser, "J. W. Foster y el desarrollo económico de México," in *Historia Mexicana* (Mexico), vol. 14, no. 53, July–Sept. 1964, pp. 81–101.
40. *Revolucíon Mexicana: Crónica Ilustrada* (Mexico, 1976), vol. 1, pp. 1–18.
41. Justo Sierra, *Evolución política del pueblo mexicano,* vol. 12, *Obras completas*, 15 vols. (Mexico, 1977), p. 280.
42. Creelman, "President Díaz," p. 245.
43. Daniel Cosío Villegas, *El Porfiriato, Vida social, Historia moderna de México* (Mexico, 1985), p. 254.
44. Quoted in Carleton Beals, *Porfirio Díaz* (Mexico, 1982), p. 332.
45. See Nelson Reed, *The Caste War of Yucatan* (Stanford, 1993).
46. González y González, "Subsuelo indígena," p. 302.
47. "Memorial de la fundacion del Ajusco," quoted in Gabriel Zaid, *Ómnibus de poesía mexicana* (Mexico, 1989), pp. 48–50.
48. Sierra, *Evolución política,* p. 119.

49. Quoted in Leticia Reina, *Las rebeliones campesinas en México 1819–1906* (Mexico, 1988), p. 165.
50. Ibid., see also Jean Meyer, *Problemas campesinos y revueltas agrarias (1821–1910)* (Mexico, 1973).
51. González y González, "Subsuelo indígena," p. 316.
52. Ibid., p. 154.
53. Enrique Krauze, *Emiliano Zapata: El amor a la tierra* (Mexico, 1982), p. 27.
54. Emilio Rabasa, *Evolución histórica de México* (Mexico, 1972), p. 101.
55. Enrique Krauze and Fausto Zerón-Medina, *Porfirio. La ambición,* 6 vols. (Mexico, 1993), vol. 3, p. 58. Jorge Fernando Iturribarría, "La política de conciliación del general Díaz y el arzobispo Gillow," in *Historia Mexicana,* vol. 14, no. 53, July–Sept. 1974, p. 92.
56. Ibid., p. 88.
57. José López Portillo y Rojas, *Elevación y caída de Porfirio Díaz* (Mexico, n.d.), p. 348.
58. Bulnes, *Verdadero Díaz,* p. 29.
59. Iturribarría, "Política de conciliación," p. 96.
60. Justo Sierra, "El día de la Patria," in *La Libertad* (Mexico), Sept. 16, 1883, in Sierra, *Textos elementales de historia,* vol. 9, *Obras completas,* p. 110.
61. Sierra, "Catecismo de historia patria," vol. 9, *Obras completas,* p. 395.
62. Cited in Jean Meyer, *La Revolución Mexicana* (Mexico, 1991), p. 35.
63. Daniel Cosío Villegas, *El Porfiriato, Vida política interior,* 2 vols., *Historia moderna de México* (Mexico, 1985), vol 2., pp. 292, 317–55.
64. Justo Sierra to Porfirio Díaz, Nov. 1899, in Sierra, *Epistolario y papeles privados,* vol. 14, *Obras completas,* pp. 96–97.
65. Creelman, "President Diaz," pp. 231, 237, 324.
66. See John M. Hart, *El anarquismo y la clase obrera mexicana, 1860–1931* (Mexico, 1980); Charles A. Hale, *La transformación del liberalismo en México a fines del siglo XIX* (Mexico, 1991).
67. "Informe de Rafael Zayas Enriquez a Porfirio Diaz," Aug. 6, 1906, quoted in Ralph Roeder, *Hacia el México moderno,* 2 vols. (Mexico, 1983), vol. 2, pp. 273–76.
68. Ibid., p. 37.
69. Francisco I. Madero, *La sucesión presidencial en 1910: El Portido Nacional Democrático* (Mexico, 1908), p. 237–38.
70. Ibid., p. 237, 284.
71. Cited in Stanley R. Ross, *Madero* (Mexico, 1977), p. 89.
72. Beals, *Porfirio Díaz,* p. 378.
73. "Brindis pronunciado por el señor general Porfirio Díaz, presidente de la República en el banquete que ofreció a una parte del cuerpo diplomático especial de Estados Unidos, el 11 de septiembre de 1910," in Genaro García, ed., *Crónica oficial de las fiestas del primer centenario de la Independencia de México* (Mexico, 1991), Appendix 102, p. 54.

Part III

Chapter 10

1. José Vasconcelos, *Don Evaristo Madero: Biografía de un patricio* (Mexico, 1958).

2. Interview with Alida Viuda de Madero by the author, Mexico, 1985; see also Carlos B. Madero, *Relación de la familia Madero* (Parras, Coahuila, 1973), pp.1–27.
3. Francisco I. Madero, "Memorias," in *Epistolario (1900–1909), Archivo de Don Francisco I. Madero,* 2 vols. (Mexico, 1985), vol. 1, p. 4.
4. Ibid.
5. Ibid., Armando de María y Campos, "Experiencias espiritistas," in *ABC* (Mexico), Feb. 28, 1953; "Las memorias de Francisco I. Madero," in *ABC*, no. 31, Jan. 31, 1950; José Natividad Rosales, *Madero y el espiritismo* (Mexico, 1973).
6. Madero, "Memorias," p. 8; Gabriel Ferrer de Mendiolea, *Vida de Francisco I. Madero* (Mexico, 1945).
7. Alfonso Taracena, *Madero, vida del hombre y del político* (Mexico, 1937), pp. 19–24; see also José Natividad Rosales, "Madero boxeador y bailarían," in *El Dictamen* (Veracruz), Feb. 21, 1932; Madero to his father and grandfather in "Francisco I. Madero Papers," Fondo LXIV–3, Centro de Estudios de Historia de México, CONDUMEX.
8. Pedro Lamicq, *Madero* (Mexico, 1958), p. 56.
9. Manuel Márquez Sterling, *Los últimos días del presidente Madero* (Havana, 1917), p. 174.
10. "Cuaderno manuscrito de Madero con sus comunicaciones espiritistas," 1907–1908, 1901–1907, Archivo particular de la Señora Renée González.
11. Quoted in Daniel Cosío Villegas, *El Porfiriato, Vida política interior,* vol. 2, *Historia moderna de México* (Mexico, 1985), p. 472.
12. Taracena, *Madero,* p. 295.
13. Francisco Madero to his brother Evaristo, Aug. 24, 1906, in *Epistolario,* vol. 1, p. 166.
14. Francisco Madero to Espiridión Calderón, June 9, 1905, in "Manuscritos de Francisco I. Madero-León de la Barra," Fondo X–1, Centro de Estudios de Historia de México, CONDUMEX.
15. Comunicaciones de Francisco I. Madero con "José," in "Cuaderno manuscrito."
16. Ibid.
17. Interview, Alida Viuda de Madero; Francisco Madero to his father, Francisco, May 16, 1906, *Epistolario,* vol. 1, p. 154; José C. Valadés, *Imaginación y realidad de Francisco I. Madero*, 2 vols. (Mexico, 1960), vol. 1, p. 140.
18. Francisco Madero to his father, Francisco, Jan. 20, 1909, *Epistolario,* vol. 1, pp. 297–98.
19. Francisco I. Madero, *La sucesión presidencial en 1910: El Partido Nacional Democrático* (Mexico, 1908), pp. 179–85, 230–41.
20. Ibid., p. 310.
21. Quoted in Gabriel Ferrer de Mendiolea, *Vida de Francisco I. Madero* (Mexico, 1945), pp. 58–59.
22. Quoted in Stanley R. Ross, *Madero* (Mexico, 1977), p. 59.
23. Francisco Madero to his father, Jan. 8, 1909, in *Epistolario*, vol. 1, p. 293; Ross, *Madero,* pp. 84–90; on the Antireelectionist party, see Santiago Portilla, *Una sociedad en armas* (Mexico, 1995), pp. 53–66.
24. Francisco Madero to Aguirre Benavides, Apr. 20, 1910, in *Epistolario*, vol. 2, p. 126.

25. "Discurso del señor don Francisco I. Madero, candidato a la presidencia de la república, pronunciado en los balcones del hotel de France, el domingo 22 de mayo de 1910, en la ciudad de Orizaba, Veracruz," quoted in Federico González Garza, *La Revolución mexicana: Mi contribución político-literaria* (Mexico, 1936), p. 420.
26. Ross, *Madero,* p. 92.
27. "Plan de San Luis," Oct. 5, 1910, quoted in González Garza, *Revolución mexicana,* pp. 203–11.
28. Francisco Madero to Eduardo Maurer, Nov. 8, 1910, in *Epistolario*, vol. 2, p. 311.
29. Ramón Prida, *De la dictadura a la anarquía* (Mexico, 1958), pp. 275–88; Alfonso Taracena, *La verdadera revolución mexicana, 1900–1911* (Mexico, 1991); Hans Werner Tobler, *La revolución mexicana: Transformación social y cambio político, 1876–1940* (Mexico, 1994); José C. Valadés, *Historia general de la Revolución Mexicana*, 9 vols. (Mexico, 1985), vol. 1.
30. Rosales, "Madero boxeador," p. 10; Ferrer, *Vida de Francisco Madero,* pp. 13–14.
31. Charles C. Cumberland, *Madero y la Revolución Mexicana* (Mexico, 1981), pp. 140–76; Jorge Vera Estañol, *Historia de la Revolución Mexicana: Orígenes y resultados* (Mexico, 1967). Francisco Vázquez Gómez, *Memorias políticas, 1909–1913* (Mexico, 1982), pp. 171–84.
32. Vázquez Gómez, *Memorias políticas,* p. 183.
33. Vicente T. Mendoza, *El corrido de la Revolución Mexicana* (Mexico, 1990), p. 45.
34. Quoted in Ross, *Madero,* p. 170; see Edith O'Shaughnessy, *Huerta y la revolución* (Mexico, 1971).
35. Friedrich Katz, *La guerra secreta en México*, 2 vols. (Mexico, 1982), vol. 1, p. 61.
36. Cumberland, *Madero y la revolución,* pp. 199–211; Ross, *Madero,* 186–95.
37. Mercedes González Treviño to her son Francisco I. Madero, Aug. 18, 1911, quoted in Taracena, *Madero,* pp. 486–87.
38. León de la Barra to Francisco I. Madero, Aug. 20–21, 1911, in Manuscritos de Francisco I. Madero–León de la Barra, Fondo x–1, Centro de Estudios de Historia de Mexico, CONDUMEX.
39. Francisco Madero to León de la Barra, Aug. 25, 1911, in Manuscritos de Francisco I. Madero–León de la Barra, Fondo X–1, Centro de Estudios de Historia de Mexico, CONDUMEX.
40. Márquez Sterling, *Últimos días,* pp. 286–87.
41. Manuel Bonilla, Jr., *El régimen maderista* (Mexico, 1962), p. 11.
42. See Cumberland, *Madero y la Revolución,* pp. 213–38; Prida, *De la Dictadura a la anarquía,* pp. 368–437; Ross, *Madero,* pp. 239–62; Taracena, *Madero,* pp. 555–81; Valadés, *Historia general,* vol. 2, pp. 64–115; Manuel González Ramírez, *Fuentes para la historia de la Revolución mexicana: Planes políticos y otros documentos* (Mexico, 1974); Luis Liceaga, *Félix Díaz* (México, 1958).
43. Márquez Sterling, *Últimos días,* pp. 324–25.
44. See Cumberland, *Madero y la Revolución,* p. 230; Francisco Bulnes, "Los tremendos idealistas trágicos," in *Los grandes problemas de México*

(Mexico, 1970), pp. 62–68; Katz, *Guerra secreta,* vol. 1, pp. 116–18; Márquez Sterling, *Últimos días,* pp. 343–77; Ross, *Madero,* pp. 226–30.

45. See Ferrer, *Vida de Francisco Madero,* pp. 124–33; Andrés Molina Enríquez, *La Revolución agraria de México de 1910 a 1920,* 5 vols. (Mexico, 1933–1937), vol. 5, pp. 11, 122–23; Ross, *Madero,* pp. 226–38; José Vasconcelos, *Breve historia de México* (Mexico, 1937), pp. 534–35.
46. William H. Beezley, "Francisco I. Madero," in *Essays on the Mexican Revolution: Revisionist Views of the Leaders* (Austin, n.d.), pp. 3–24; Cumberland, *Madero y la Revolución,* pp. 239–62; "Discurso de Francisco I. Madero, el 16 de septiembre de 1912, en la apertura de sesiones del Congreso," in *Los presidentes de México ante la nación 1821–1966,* 4 vols. (Mexico, 1966), vol. 3, pp. 22–43.
47. On the "Tragic Ten Days" see Lamicq, *Madero,* pp. 101–33; Michael C. Meyer, *Huerta: A Political Portrait* (Lincoln, Neb., 1972), pp. 45–63; John P. Harrison, "Henry Lane Wilson, El trágico de la Decena," in *Historia Mexicana* (Mexico), vol. 6, Jan.–Mar. 1957, pp. 374–405; Francisco L. Urquizo, *Viva Madero* (Mexico, 1969); Henry Lane Wilson, *Diplomatic Episodes in Mexico, Belgium and Chile* (New York, 1927), pp. 273–88.
48. Katz, *Guerra secreta,* p. 122.
49. Márquez Sterling, *Últimos días,* pp. 415, 416.
50. Ibid., p. 457; Taracena, *Madero,* pp. 588–89.
51. José Vasconcelos, "Ulises Criollo" in *Memorias,* 2 vols. (Mexico, 1982), vol. 1, p. 440.
52. Ibid., p. 441.
53. Guadalupe Appendini, "Ante la acusación de Madero, Aureliano Blanquet admitió: 'Sí, soy un traidor,'" in *Excélsior* (Mexico), Jan. 24, 1985.
54. Márquez Sterling, *Últimos días,* pp. 466–68.
55. Ibid., pp. 503, 507.
56. Katz, *Guerra secreta,* p. 132.
57. Márquez Sterling, *Últimos días,* pp. 570–71.
58. José C. Valadés, et al., *Francisco I. Madero: Semblanzas y opiniones* (Mexico, 1973), pp. 243–60; Katz, *Guerra secreta,* pp. 134–35; Ross, *Madero,* pp. 310–11.

Chapter 11

1. See Jesús Sotelo Inclán, *Raíz y razón de Zapata* (Mexico, 1943).
2. Ibid., Alicia Hernández, *Haciendas y pueblos en el estado de Morelos, 1535–1810* (Mexico, 1973); "Querellas de Anenecuilco contra haciendas," Ramo Tierras, expediente 23; vols. 26–63, Archivo General de la Nación (Mexico).
3. José Zapata to Porfirio Díaz, June 17, 1874, quoted in *Memorias y documentos,* vol. 11, *Archivo del general Porfirio Díaz,* 24 vols. (Mexico, 1947–1958), pp. 142–43.
4. Ibid., vol. 2, pp. 300–301.
5. Ibid., vol. 2.
6. John Womack, Jr., *Zapata y la revolución mexicana,* (Mexico, 1969), p. 62.

7. Sotelo, *Raíz y razón,* p. 526.
8. Ibid., p. 531.
9. Ibid., pp. 63–75.
10. Valentín López González, "Los plateados de Morelos," manuscript; interview with Claudio Lomnitz, Nov. 5, 1985; Sotelo, *Raíz y razón*, pp. 191–94.
11. Antonio Díaz Soto y Gama, *La revolución agraria del sur y Emiliano Zapata su caudillo* (Mexico, 1976), p. 245; Antonio Díaz Soto y Gama, "Cómo era Zapata," in *El Universal* (Mexico), Apr. 7, 1943; Serafín M. Robles, "Zapata, agricultor," in *La Prensa* (Mexico), June 17, 1936; interview with Valentín López González by the author, Mexico, Apr. 1984; Octavio Paz Solórzano: "Emiliano Zapata," in José T. Meléndez, *Historia de la Revolución Mexicana* (Mexico, 1936), p. 320.
12. Interview with Gustavo Baz, Mexico, Jan. 1983; Fernando Alpuche y Silva: "El General Emiliano Zapata," *El Nacional* (Mexico), July 1941.
13. Serafín M. Robles, "Emiliano Zapata era todo un charro," *La Prensa*, May 1, 1936; Díaz Soto y Gama, "Cómo era"; Octavio Paz Solórzano, "El caballo del cura," *Hoguera que fue* (Mexico, 1986), p. 184.
14. Sotelo, *Raíz y razón*, pp. 426–46; Luis Gutiérrez y González, "¡La viuda de Zapata!" *Hoy* (Mexico), Mar. 28, 1953.
15. General Héctor F. López, "Cuándo fue consignado Zapata," *El hombre libre* (Mexico), Apr. 5, 1937.
16. Sotelo, *Raíz y razón*, p. 498.
17. Ibid., p. 203.
18. Womack, *Zapata y la revolución,* pp. 70–78.
19. Paz Solórzano, Hoguera . . . , p. 184.
20. Translated from the Spanish text in Robert Redfield, *Tepoztlán: A Mexican Village. A Study of Folk Life* (Chicago, 1973), pp. 200–201.
21. *Emiliano Zapata y el movimiento zapatista: Cinco ensayos* (Mexico, 1980), pp. 128, 140, 142.
22. Vicente T. Mendoza, *El corrido de la Revolución mexicana* (Mexico, 1990), p. 50.
23. Womack, *Zapata y la revolución,* pp. 93–94; Paz Solórzano, Hoguera . . . , pp. 259–61.
24. Emiliano Zapata to Francisco León de la Barra, Aug. 17, 1911, "Archivo del General Jenaro Amezcua," Fondo VIII–2, carpeta 1, legajo 33; Francisco I. Madero to Emiliano Zapata, Aug. 22, 1911, Fondo VIII–2, carpeta 1, legajo 55; Centro de Estudios de Historia de México, CONDUMEX; Victoriano Huerta to Francisco León de la Barra, Aug. 21, 1911, quoted in Womack, *Zapata y la revolución,* p. 117.
25. Ramón López Velarde, "Zapata," in *La Nación* (Mexico), July 1911; Sotelo, *Raíz y razón,* p. 314; Porfirio Palacios, *Emiliano Zapata, datos biográfico-históricos* (Mexico, 1960), pp. 58–62; Salvador Sánchez Septién, *José María Lozano en la Tribuna Parlamentaria, 1910–1913* (Mexico, 1956), pp. 34–35.
26. Quoted in Womack, *Zapata y la revolución,* p. 124.
27. Alfonso Taracena, "Madero quiso refugiarse con Zapata," *Revista de Revistas* (Mexico), Sept. 22, 1957.
28. Díaz Soto y Gama, *La revolución agraria,* pp. 263–65; *Documentos inéditos sobre Emiliano Zapata y el cuartel general* (Mexico, 1979), p. 98.

29. Díaz Soto y Gama, *La revolución agraria,* p. 252.
30. "Plan de Ayala," Nov. 25, 1911, in Manuel González Ramírez, *Fuentes para la historia de la Revolución mexicana: Planes políticos y otros documentos* (Mexico, 1974), pp. 73–83.
31. Womack, *Zapata y la revolución,* p. 194.
32. Ibid, pp. 134–38, 147–48, 159–67, 170–74; see also Domingo Díez, *Bosquejo histórico-geográfico de Morelos* (Morelos, 1967); José Ángel Aguilar, *Zapata: Selección de textos*, (Mexico, 1980).
33. Marte R. Gómez, *Las comisiones agrarias del sur* (Mexico, 1961), p. 37.
34. Felipe Ángeles, "Genovevo de la O," *Cuadernos Mexicanos* (Mexico), no. 11, 1984.
35. *Zapata y el movimiento . . .*, p. 134.
36. Mendoza, *Corrido,* p. 55.
37. Díaz Soto y Gama, *La revolución agraria,* pp. 181–204; see Vito Alessio Robles, "La Convención Revolucionaria de Aguascalientes," in *Todo*, June 15, 1950; Florencio Barrera Fuentes, *Crónicas y debates de las sesiones de la Soberana Convención Revolucionaria* (Mexico, 1964); Basilio Rojas, *La Soberana Convención de Aguascalientes* (Mexico, 1961), pp. 105–20.
38. See Eric R. Wolf, *Las luchas campesinas del siglo XX* (Mexico, 1972).
39. José Vasconcelos, *La tormenta* (Mexico, 1936), pp. 203–26.
40. Recuerdos de Don Jacinto Díaz Zulueta, trasmitidos a su hija Carmen Díaz; see also Womack, *Zapata y la revolución,* pp. 215–16; Fernando Benítez, *Lázaro Cárdenas y la Revolución mexicana, El caudillismo,* 3 vols. (Mexico, 1977), vol. 2, pp. 53–54.
41. Quoted in John Womack, *Zapata and the Mexican Revolution* (original edition in English, New York, 1968), p. 220.
42. Ibid.
43. "Pacto de Xochimilco," González, *Fuentes para la historia,* pp. 113–21; Berta Ulloa, "La revolución escindida," *Historia de la Revolución Mexicana*, 23 vols. (Mexico, 1979), vol. 4, pp. 43–46.
44. Paz Solórzano, *Emiliano Zapata,* p. 268; Womack, *Zapata and the Mexican Revolution,* pp. 201–2; Martín Luis Guzmán, *El águila y la serpiente* (Mexico, 1966), pp. 385–87.
45. Octavio Paz Lozano, *Posdata* (Mexico, 1970), p. 89.
46. François Chevalier, "Un factor decisivo de la revolución agraria de México: 'El levantamiento de Zapata' (1911–1919)" *Cuadernos Americanos* (Mexico), Nov. 6, 1960, pp. 165–87; *Zapata y el movimiento;* Paz Solórzano, *Hoguera . . .*, p. 185.
47. Fernando Horcasitas, *De Porfirio Díaz a Zapata: Memoria Náhuatl de Milpa Alta* (Mexico, 1968).
48. Díaz Soto y Gama, *La revolución agraria,* pp. 259–60; "Zapata y Villa creyentes," *El Universal*, Apr. 6, 1949; see Chevalier, "Factor decisivo"; Paz Solórzano, "Emiliano Zapata," pp. 178–84, 340–41; Horcasitas, *Porfirio Díaz a Zapata.*
49. Redfield, *Tepoztlán*, pp. 197–204; SEP-INAH, *Zapata y el movimiento;* Paz Solórzano, "Emiliano Zapata," p. 321.
50. *Documentos inéditos,* pp. 10, 39, 43, 127, 161, 163, 177.
51. Womack, *Zapata and the Mexican Revolution*, pp. 237–38; Gómez, *Comisiones agrarias,* pp. 62–77.

52. Ibid.
53. Díaz Soto y Gama, *La revolución agraria,* pp. 272–74; Bernardino Mena Brito, "El verdadero zapatismo," *El Universal,* July 8, 1941.
54. Miguel León Portilla, *Los manifiestos en Náhuatl de Emiliano Zapata* (Mexico, 1978).
55. Womack, *Zapata and the Mexican Revolution,* p. 247.
56. Ibid., p. 264; Palacios, *Emiliano Zapata . . .*, pp. 58–62; Gómez, *Comisiones agrarias,* p. 91; Baltasar Dromundo, *Vida de Emiliano Zapata* (Mexico, 1961); Adolfo Gilly, *La revolución interrumpida, México, 1910–1920: Una guerra campesina por la tierra y el poder* (Mexico, 1971).
57. Zapata to Octavio Paz Solórzano, Mar. 1, 1917, Paz Solórzano, "Emiliano Zapata," p. 374.
58. Arturo Warman, "El proyecto político del zapatismo," Conference II on Comparative Peasants Rebellions in Mexico (1982); see also "Leyes zapatistas," in Archivo del General Jenaro Amezcua, Fondo VIII–2, legajos 106, 199, 214, 216, 221, 235, Centro de Estudios de Historia de Mexico, CONDUMEX.
59. Gómez, *Comisiones agrarias,* p. 92; Paz Solórzano, "Emiliano Zapata," pp. 369–73.
60. Miguel R. Delgado, "El testamento político de Otilio Montaño," *Excélsior* (Mexico), Jan. 21, 1919; Juan Salazar Pérez, "Otilio Montaño," *Cuadernos Morelenses* (Morelos, 1982).
61. Díaz Soto y Gama, *La revolución agraria,* pp. 256–59, 263.
62. José González Ortega, "Cómo murió Eufemio Zapata," *Todo,* Jan. 13, 1944.
63. Valentín López González, *Los compañeros de Zapata* (Morelos, 1980), pp. 31–33; see also León Portilla, *Los manifiestos.*
64. Womack, *Emiliano Zapata and the Revolution,* pp. 279–83; Germán Lizt Arzubide, *Zapata* (Mexico, 1973); "La encrucijada de 1915," *Historia de la Revolución Mexicana,* vol. 5; Armando de Maria y Campos, "Reto de Zapata al Constitucionalismo," *Revista ABC* (Mexico), no. 10, Jan. 17, 1953.
65. Valentín López González, "La muerte del general Emiliano Zapata, *Cuadernos Zapatistas* (Morelos, 1979); Elías L. Torres, "No te descuides Zapata," *Jueves de Excélsior* (Mexico), Apr. 1937.
66. Salvador Reyes Avilés, "Parte oficial sobre la muerte de Zapata," Apr. 10, 1919, quoted in Paz Solórzano, "Emiliano Zapata," pp. 377–78.
67. *Zapata y el movimiento . . .*, p. 136; Redfield, *Tepoztlán,* pp. 201–4; Ettore Pierri, *Vida, pasión y muerte de Emiliano Zapata* (Mexico, 1979), pp. 33–43, 225–48.
68. Salvador Martínez Mancera, "Perdura en el sur la leyenda de que Emiliano Zapata no ha muerto," *El Universal Gráfico* (Mexico), Apr. 13, 1938.
69. Ibid.; also *Zapata y el movimiento,* p. 143.
70. Octavio Paz Lozano, *El laberinto de la soledad* (Mexico, 1959), p. 128.
71. Ibid.
72. Armano de María y Campos, *La Revolución Mexicana a través de los corridos populares,* 2 vols. (Mexico, 1962), vol. 1, pp. 225–28.

Chapter 12

1. Ramón Puente, "Francisco Villa," in *Historia de la Revolución Mexicana* (Mexico, 1936), pp. 239–40; see John Reed, "Villa asesino, bandido y consumado hombre malo," in *The World*, May 8, 1914; Elías Torres, *Vida y hechos de Francisco Villa* (Mexico, 1975), pp. 9–11.
2. Puente, "Francisco Villa," pp. 240–41.
3. Roberto Blanco Moheno, *Pancho Villa que es su padre* (Mexico, 1969), pp. 175–78.
4. See Reed, "Villa asesino"; Ramón Puente, "La verdadera historia de Pancho Villa, por su médico y secretario," *Excélsior* (Mexico), Mar. 23, 1931.
5. Friedrich Katz, "Pancho Villa: Reform Governor of Chihuahua," in Georges Wolfskill and Douglas Richmond, eds., *Essays on the Mexican Revolution: Revisionist Views of the Leaders* (Austin, Tex., 1979), pp. 26–31.
6. John Reed, *Insurgent Mexico* (Reprint: Berlin, 1969), p. 122.
7. Ibid., p. 123.
8. Mario Gill, "Heraclio Bernal, caudillo frustrado," *Historia Mexicana* (Mexico), vol. 4, no. 13, July–Sept. 1954, pp. 138–58.
9. Puente, "Francisco Villa," pp. 249–51.
10. Ibid., pp. 251–53; Patrick O'Hea, *Reminiscences of the Mexican Revolution* (London, 1981), pp. 171–72; Federico Cervantes, *Francisco Villa y la revolución* (Mexico, 1960), pp. 216–22.
11. See "Juvenal," "¿Quién es Francisco Villa?" *Verdades históricas* (Dallas, 1916), and also Martín Luis Guzmán, *Memorias de Pancho Villa* (Mexico, 1968), pp. 216–22.
12. Guzmán, *Memorias,* pp. 230–37; Puente, "Francisco Villa," p. 257.
13. See Luis Aguirre Benavides, *Las grandes batallas de la División del Norte al mando del general Francisco Villa* (Mexico, 1964).
14. Silvestre Terrazas, "El verdadero Pancho Villa," *Boletín de la Sociedad Chihuahuense de Estudios Históricos* (Chihuahua), vol. 6, no. 10, Sept. 1949, pp. 290–95; vol. 6, no. 11, Oct.–Nov. 1949, pp. 307–10; vol. 7, no. 6, Nov. 1950, pp. 453–55.
15. Quoted in Margarita de Orellana, "Pancho Villa, Movie Star," *La mirada circular* (Mexico, 1991), p. 90.
16. Ibid.
17. John Reed, "Las batallas desesperadas de Villa preparatorias de la captura de Torreón," *The World*, Apr. 12, 1914, p. 2.
18. Vicente T. Mendoza, *El Corrido de la Revolución mexicana* (Mexico, 1990), p. 77.
19. Cervantes, *Villa y la revolución,* pp. 140–43.
20. Vito Alessio Robles, "Narración del Combate de Paredón," *Todo* (Mexico), Jan. 5, 1956, art. 331.
21. Rafael F. Muñoz, *Vámonos con Pancho Villa* (Mexico, 1950), pp. 109–10.
22. Cervantes, *Villa y la revolución,* p. 114–15; Terrazas, "El verdadero," pp. 453–55; see also Reed, "Villa, asesino."
23. Felipe Ángeles, "La batalla de Zacatecas," *Historia de la Revolución Mexicana* (Mexico, 1936), p. 284.
24. Puente, "Francisco Villa," p. 259; see also Felipe Ángeles, "Batalla de Zacatecas," p. 284.

25. Reed, *Insurgent Mexico*, pp. 126–30; Friedrich Katz, "Agrarian Changes in Northern México in the Period of Villista Rule," *Contemporary Mexico* (Los Angeles, 1976); Katz, "Pancho Villa," pp. 34–43; Jesús J. Lozano, "Francisco Villa, gobernador del Estado de Chihuahua," *Boletín de la Sociedad Chihuahuense de Estudios históricos* (Chihuahua), vol. 6, no. 9, Aug. 1949, pp. 176–80; Cervantes, *Villa y la revolución,* pp. 73–88; Terrazas, "El verdadero," pp. 662–65.
26. Reed, *Insurgent Mexico*, p. 134.
27. John Reed, "El socialismo de Villa se funda en la necesidad," *The Sun*, Jan. 2, 1914, p. 4.
28. John Reed, "Con Villa en México," *Metropolitan Magazine* (New York, Feb. 1914); "Juvenal," "¿Quién es," pp. 8–9.
29. Reed, "Con Villa."
30. Quoted in Edmundo Valadés, "Los jefes de la lucha de 1910 vistos por novelistas y cronistas de la Revolución," *Excélsior*, Nov. 21, 1982; see also Muñoz, *Vámonos*, p. 18, José Vasconcelos, *La tormenta* (Mexico, 1983), p. 543; Martín Luis Guzmán, *El águila y la serpiente* (Mexico, 1975), p. 50.
31. Guzmán, *Águila,* p. 49.
32. Reed, "Villa is brutal . . . " *The World* (New York), Mar. 1, 1914.
33. Guzmán, *Águila,* p. 256.
34. O'Hea, *Reminiscences,* p. 160.
35. Puente, "Francisco Villa," pp. 242–44; Ramón Puente, *La dictadura, la revolución y sus hombres* (author's edition), pp. 305–11; see also Rafael F. Muñoz, *Pancho Villa, rayo y azote* (Mexico, 1955).
36. See Fernando de Ita, "Soledad de Villa recuerda su vida," *Unomasuno* (Mexico), Mar. 18, 1983; Luz Corral, *Pancho Villa en la intimidad* (Chihuahua, 1976); Juan Durán Casahonda, "Las mujeres de Pancho Villa," *Todo*, Jan. 2, 1934.
37. Guzmán, *Memorias,* p. 280.
38. Reed, *Insurgent Mexico*, p. 123.
39. Ibid., p. 179; Terrazas, "El verdadero," pp. 307–10.
40. Guzmán, *El Águila . . . ,* pp. 199–214.
41. Reed, *Insurgent Mexico*, p. 143.
42. O'Hea, *Reminiscences*, pp. 162–63.
42. "Juvenal," "¿Quién es," pp. 11–19; see Puente, "La verdadera"; Guzmán, *El Águila . . . ,* pp. 215–57, and Luis Aguirre Benavides, *De Francisco I. Madero a Francisco Villa* (Mexico, 1966), pp. 101–5.
44. Reed, *Insurgent Mexico*, p. 127.
45. Ibid., p. 139.
46. Felipe Ángeles, "Genovevo de la O," *Cuadernos Mexicano* (Mexico, 1984), no. 11.
47. See Federico Cervantes, "Felipe Ángeles: Revolucionario, idealista y desinteresado," *El Universal* (Mexico), Nov. 26, 1947; José C. Valadés, "Los compadres de Pancho Villa," *Todo*, Sept. 26, 1933; "Juvenal," "¿Quién es," p. 16; Guzmán, pp. 253–54. Alberto Calzadíaz, *General Felipe Ángeles,* vol. 8, *Hechos reales de la Revolución*, 8 vols. (Mexico, 1982).
48. Álvaro Obregón, *Ocho mil kilómetros en campaña* (Mexico, 1973), p. 169; and Enrique Beltrán, "Fantasía y realidad de Pancho Villa," in *Historia Mexicana*, vol. 16, no. 61, July–Sept. 1966, pp. 71–84.

49. Ibid., pp. 203–5.
50. Berta Ulloa, "La revolución escindida," *Historia de la Revolución Mexicana*, 23 vols. (Mexico, 1979), vol. 4, pp. 59–99, 161–68.
51. "Pacto de Xochimilco," Dec. 4, 1914, in Manuel González Ramírez, *Fuentes para la historia de la Revolución: Planes políticos y otros documentos* (Mexico, 1974), vol. 1, pp. 113–21.
52. Ulloa, "Revolución escindida," pp. 59–61.
53. Guzmán, *El Águila . . .*, pp. 407–36.
54. Obregón, *Ocho mil kilómetros,* pp. 299–427.
55. Reed, *Insurgent Mexico*, p. 49.
56. "14 Members of Staff Die with the General Urbina," *Evening Star* (El Paso, Tex.), Sept. 16, 1915; "Juvenal," "Quién es," pp. 26–28.
57. Rafael F. Muñoz, *Relatos de la Revolución* (Mexico, 1981), pp. 181–88.
58. Quoted in Berta Ulloa, "La lucha revolucionaria," in *México y el mundo,* vol. 5, *Historia de sus relaciones exteriores*, 8 vols. (Mexico, 1991), p. 280; Friedrich Katz, *La guerra secreta en México,* 2 vols. (Mexico, 1982), vol. 1, pp. 349–50; Charles Cumberland, *La revolución mexicana* (Mexico, 1975), pp. 285–88.
59. Alberto Calzadíaz, *El fin de la División del Norte,* vol. 3, *Hechos reales,* pp. 98–107.
60. Celia Herrera, *Francisco Villa ante la historia* (Mexico, 1964), pp. 156–58, and "C. Herrera responde a Blanco Moheno," *Excélsior*, July 22, 1955; Marte R. Gómez, *Pancho Villa: Un intento de semblanza* (Mexico, 1972), p. 63; Enrique Beltrán, "Fantasía y realidad," pp. 67–87; Torres, *Vida y hechos,* pp. 67–87.
61. Muñoz, *Vámonos . . .*, pp. 132–40; Rafael Trujillo Herrera, *Cuando Villa entró a Columbus* (Mexico, 1973), pp. 313–19; Víctor Ceja Reyes, *Yo, Francisco Villa y Columbus* (Chihuahua, 1987), pp. 44–57; Torres, *Vida y hechos,* pp. 81–83.
62. Puente, "Francisco Villa," p. 266.
63. Muñoz, *Vámonos . . .*, p. 148.
64. Calzadíaz, "General Felipe Ángeles," p. 131.
65. Ibid., p. 140.
66. Puente, "Francisco Villa," p. 268; Adolfo Gilly, "Felipe Ángeles camina hacia la muerte," in Odile Guilpain Peuliard, *Felipe Ángeles y los destinos de la Revolución mexicana* (Mexico, 1991), pp. 7–39; *Documentos relativos al general Felipe Ángeles* (Mexico, 1982).
67. Mendoza, *Corrido*, p. 104.
68. Puente, "Francisco Villa," pp. 269–70; see John W. F. Dulles, *Ayer en México: Una crónica de la Revolución (1919–1936)* (Mexico, 1977) pp. 64–70.
69. Víctor Ceja Reyes, *Yo maté a Villa* (Mexico, 1960), pp. 145–87; Alberto Calzadíaz, *Muerte del Centauro,* vol. 7, *Hechos reales.*

Chapter 13

1. Ildefonso Villarello, *Historia de la revolución mexicana en Coahuila* (Mexico, 1970) pp. 11–81.
2. Jesús Castro Carranza, *Origen, destino y legado de Carranza* (Mexico, 1964–1970), pp. 21–46; Mariano Escobedo to Benito Juárez, Mar. 27, 1865, and Apr. 27, 1865, *Benito Juárez, Documentos, discursos y corre-*

spondencia, 15 vols. (Mexico, 1966), vol. 9, pp. 728–32, 795–97; Benito Juárez to Jesús Carranza, Mar. 3, 1868, in ibid., vol. 13, pp. 191–92.

3. Castro Carranza, *Origen*, pp. 106–36; interview with Rafael Carranza H. by the author, Mexico, Mar. 1985; see also Julia Carranza to Torreblanca, Nov. 20, 1924, in Archivo Plutarco Elías Calles.
4. Daniel Cosío Villegas, *El porfiriato, Vida política interior, Historia moderna de México* (Mexico, 1985), vol. 2, pp. 466–70; Rafael Carranza, "Conference Held at Museo-Casa de Carranza" (Mexico), Mar. 12, 1985.
5. Porfirio Díaz to Bernardo Reyes, Feb. 3, 1894, in "Archivo del General Bernardo Reyes," Fondo-DLI, carpeta 20, leg. 3861, Centro de Estudios de Historia de Mexico, CONDUMEX.
6. Cosío Villegas, *El porfiriato, Vida política,* vol. 2, p. 423.
7. Bernardino Mena Brito, *Carranza, sus amigos y enemigos* (Mexico, 1953), pp. 249–50, 267, 344–45.
8. Quoted in Alfonso Taracena, *Venustiano Carranza* (Mexico, 1963), pp. 33–34; William Weber Johnson, *Heroic Mexico: The Violent Emergence of a Modern Nation* (New York, 1968), p. 146.
9. William Beezley, "Governor Carranza and the Revolution in Coahuila," *The Americas*, vol. 33, no. 1, Oct. 1976; see also "Venustiano Carranza al pueblo de Coahuila," Aug. 1, 1911, quoted in Taracena, *Venustiano Carranza,* pp. 48–52.
10. Miguel Alessio Robles, "El viejo pachorrudo y conspirador," *El Universal* (Mexico), Feb. 26, 1940; see also Ramiro Beltrán, "Así veía Madero a Carranza en 1909," *Hoy* (Mexico), Aug. 18, 1951.
11. Alfonso Junco, "Madero y Carranza," Mena, *Carranza,* pp. 245–52; Taracena, *Venustiano Carranza,* pp. 52–79.
12. Blas Urrea (Luis Cabrera), *La herencia de Carranza* (Mexico, 1920), p. 10; José Vasconcelos, *La tormenta* (Mexico, 1983), p. 537–42. Carranza interview.
13. Vicente Blasco Ibañez, *El militarismo mejicano* (Barcelona, 1979), p. 107; see also Alfonso Taracena, *La verdadera revolución mexicana, Sexta etapa (1918 a 1920)*, vol. 6, 18 vols. (Mexico, 1961), pp. 206–7.
14. John Reed, *Insurgent Mexico* (Reprint: Berlin, 1969), p. 249.
15. Martín Luis Guzmán, *El águila y la serpiente* (Mexico, 1966), p. 60.
16. Guzmán, *Águila,* pp. 59–64.
17. Reed, *Insurgent Mexico,* pp. 320–21.
18. "Plan de Guadalupe," Mar. 26, 1913, in Manuel González Ramírez, *Fuentes para la historia de la Revolución: Planes políticos y otros documentos* (Mexico, 1974), pp. 137–44.
19. Taracena, *Venustiano Carranza*, pp. 172–75.
20. Miguel Alessio Robles, "Cómo se conocieron Carranza y Obregón," *El Universal*, Mar. 26, 1927.
21. "Discurso pronunciado el 24 de septiembre de 1913 al pueblo de Hermosillo, Son., por el C. Primer Jefe del Ejército Constitucionalista Venustiano Carranza," in "Archivo del Primer Jefe del Ejército Constitucionalista," 1889–1920, Centro de Estudios de Historia de México, CONDUMEX; see also Arnaldo Córdova, *La ideología de la Revolución mexicana* (Mexico, 1973), p. 97.
22. Berta Ulloa, "La revolución escindida," *Historia de la Revolución Mexi-*

cana, 23 vols., *Periodo 1914–1917,* vol. 4 (Mexico, 1979), pp. 19–38; Martín Luis Guzmán, *Memorias de Pancho Villa* (Mexico, 1965), pp. 379–81, 443; Bernardino Mena Brito, *El lugarteniente gris de Pancho Villa* (Mexico, 1938), pp. 40–41; Isidro Fabela, *Documentos históricos de la Revolución Mexicana,* vol. 1, *Revolución y régimen Constitucionalista* (Mexico, 1960), pp. 271–73.

23. Charles C. Cumberland, *La revolución mexicana: Los años constitucionalistas* (Mexico, 1975), pp. 198–201.
24. Reed, *Insurgent Mexico,* pp. 250–51.
25. Cumberland, "Huerta y Carranza ante la ocupación de Veracruz," *Historia Mexicana* (Mexico), vol. 6, no. 24, Apr.–Jun. 1957, pp. 534–47.
26. Ulloa, "Revolución escindida," pp. 47–58, and "La encrucijada de 1915," *Historia de la Revolución Mexicana,* vol. 5, p. 63; and see also Cumberland, *Revolución,* pp. 138–42, 230.
27. Ulloa, "Revolución escindida," pp. 19–38; see also Cumberland, *Revolución,* pp. 145–74; John Womack, *Zapata y la revolución mexicana* (Mexico, 1969), pp. 210–14.
28. Vasconcelos, *Tormenta,* pp. 592–613.
29. Venustiano Carranza, "Telegrama a los Jefes militares y Gobernadores reunidos en Aguascalientes," Nov. 9, 1914, in Mena, *Carranza. . .,* pp. 189–91; see also Florencio Barrera Fuentes, *Crónicas y debates de las sesiones de la Soberana Convención Revolucionaria,* 3 vols. (Mexico, 1964), vol. 1, pp. 687–708.
30. "Adiciones al Plan de Guadalupe y decretos dictados conforme a las mismas," Dec. 12, 1914, in González, *Fuentes para la historia,* pp. 158–164.
31. Cabrera, *Herencia de Carranza,* pp. 17–32; Cumberland, *La Revolución Mexicana. . .,* pp. 216–17; Córdova, *Ideología de la Revolución,* p. 222; Ulloa, "Revolución escindida," pp. 65–69; González, *Fuentes para la historia,* p. 162.
32. Jean Meyer, "Los batallones rojos de la revolución mexicana," *Historia Mexicana,* vol. 21, no. 81, July–Sept. 1971, pp. 1–37; José Clemente Orozco, *Autobiografía* (Mexico, 1981), pp. 45–47.
33. Orozco, *Autobiografía,* pp. 45–47; see also Ulloa, "La Constitución de 1917," *Historia de la Revolución Mexicana,* vol. 6, pp. 304–5; "Editorial," *El Pueblo* (Mexico), Aug. 1916; Marjorie Ruth Clark, *Organized Labor in Mexico* (New York, 1973), pp. 35–45.
34. "Lo que dijo el 2 de enero en Querétaro el Sr. Carranza," in Félix F. Palavicini (Mexico, n.d.), pp. 257–60.
35. Cumberland, *La Revolución Mexicana. . .,* pp. 224–32; Ulloa, "La encrucijada," pp. 40–42; Lorenzo Meyer, *México y EU en el conflicto petrolero* (Mexico, 1981), pp. 95–99.
36. Jean Meyer, *La cristiada,* 2 vols. (Mexico, 1973), vol. 2, pp. 83–97.
37. Orozco, *Autobiografía,* p. 42.
38. Cumberland, *La Revolución Mexicana . . .,* p. 217.
39. Quoted in Ulloa, "La Constitución de 1917," *Historia de la revolución Mexicana,* vol. 6, pp. 497–98.
40. Frank Tannenbaum, *Peace by Revolution* (New York, 1966), p. 166.
41. Cumberland, *La Revolución Mexicana . . .,* pp. 307–8.
42. Quoted in *Revolución Mexicana: Crónica ilustrada,* 6 vols. (Mexico, 1966–1972), vol. 6, pp. 85–87.

43. Robert Quirk, "Liberales y radicales en la revolución mexicana," *Historia Mexicana*, vol. 2, no. 8, Apr.–Jun. 1953, pp. 503–28.
44. Ulloa, "La encrucijada," pp. 352–57.
45. Robert Freeman, *The United States and Revolutionary Nationalism in Mexico, 1916–1932* (Chicago, 1972), p. 199; Cumberland, *La Revolución Mexicana . . .*, pp. 316–21; Ulloa, "La Constitución," pp. 144–58, 405–12; see also Pastor Rouaix, *Génesis de los Artículos 27 y 123 de la Constitución Política de 1917* (Mexico, 1945).
46. Ulloa, "La Constitución," pp. 325–31; Meyer, *La cristiada,* vol. 2, pp. 91–104; see also *Diario de los Debates del Congreso Constituyente 1916–1917*, 2 vols. (Mexico, 1960), vol. 2, pp. 855–60, 986.
47. Alvaro Matute, *La revolución mexicana* (Mexico, 1993), pp. 139–49; Ulloa, "La Constitución," pp. 417–92.
48. Manuel Gómez Morín, *1915* (Mexico, 1926), p. 20.
49. Orozco, *Autobiografía*, pp. 45–46.
50. Octavio Paz, *El laberinto de la soledad* (Mexico, 1959), p. 181.
51. Ulloa, "La Constitución," pp. 413, 466–67, 512.
52. "Lo que dijo," pp. 257–60.
53. Córdova, *Ideología de la Revolución,* p. 244.
54. Hilario Medina, "Emilio Rabasa y la Constitución de 1917," *Historia Mexicana*, vol. 10, no. 38, Oct.–Dec. 1960, pp. 177–95.
55. Jean Meyer, *La Revolución Mexicana* (Mexico, 1991), pp. 94–95.
56. Javier Garciadiego, "El gobierno constitucional de Carranza" (manuscript).
57. María Eugenia López de Roux, "Relaciones México-Norteamericanas, 1917–1918," *Historia Mexicana*, vol. 14, no. 55, Jan.–Mar. 1965, pp. 445–68; see also Robert Freeman, "EU y las reformas de la Revolución Mexicana, 1915–1928," in *Historia Mexicana*, vol. 19, no. 74, Oct.–Dec. 1969, pp. 189–227.
58. Friedrich Katz, *La guerra secreta en México,* 2 vols. (Mexico, 1984), vol. 2, pp. 37–74; see also Yolanda de la Parra, comp. "México y la primera guerra mundial," *Nuestro México*, no. 8 (Mexico, 1983).
59. Ulloa, "La Constitución," pp. 452–53, 462–65.
60. Blasco Ibañez, *Militarismo*, pp. 36–37.
61. Ibid., pp. 69–86; see also Matute, *Revolución mexicana,* pp. 156, 163.
62. Martín Luis Guzmán, *Muertes históricas* (Mexico, 1959), p. 29.
63. Venustiano Carranza, "Manifiesto a la Nación," *La Prensa* (San Antonio, Tex.), May 18, 1920.
64. Miguel Gil, "Tengo la vida prestada desde 1913, contestó don Venustiano Carranza," *La Prensa* (Mexico), July 27, 1943.
65. Rubén García, "El romano Belisario, figura eterna. Lectura de Don Venustiano Carranza," *El Demócrata Sinaloense* (Sinaloa), Feb. 1950; see Carranza interview.
66. Ramón Beteta, *Camino a Tlaxcalantongo* (Mexico, 1961), pp. 43–72; Francisco L. Urquizo, *Asesinato de Carranza* (Mexico, 1959), pp. 99–105; see also Fernando Benítez, *El rey viejo* (Mexico, 1959), pp. 95–110, Guzmán, *Muertes Historicas*, pp. 83–88.
67. Guzmán, *Muertes Históricas,* pp. 139–40; see also Miguel Márquez, *El verdadero Tlaxcalantongo* (Mexico, 1941), pp. 157–62.
68. Josefina E. de Fabela, "Testimonios sobre los asesinatos de Don Venustiano Carranza y Jesús Carranza," *Documentos históricos,* vols. 18, 19,

28; Legajo sobre la muerte de Venustiano Carranza, Archivo Plutarco Elías Calles.

69. Márquez, pp. 163–65; Fabela, *Documentos históricos,* p. 76; see also Francisco A. Serralde, *Los sucesos de Tlaxcalantongo y la muerte del expresidente de la República, C. Venustiano Carranza, amparo promovido por el defensor de Rodolfo Herrero contra los actos del Presidente de la República y de la Secretaría de Guerra* (Mexico, 1921); Enrique Krauze, "La noche de Tlaxcalantongo," *Vuelta* (Mexico), vol. 10, no. 111, Feb. 1986.

Chapter 14

1. José Rubén Romero, *Álvaro Obregón: Aspectos de su vida* (Mexico, 1976), pp. 3–7; Héctor Aguilar Camín, *Saldos de la Revolución: Cultura y política de México, 1910–1980* (Mexico, 1982), p. 60; interview with Hortensia Calles vda. de Torreblanca by the author, Mexico, 1985.
2. Obregón file, Archivo Plutarco Elías Calles y Fernando Torreblanca (APEC); Hernán Rosales, "La niñez extraordinaria del General Obregón," *El Legionario* (Mexico, 1928), p. 16; see also Camín, *Saldos de la Revolución*; Linda B. Hall, *Álvaro Obregón: Poder y revolución en México, 1911–1920* (Mexico, 1985), p. 28.
3. Djed Bórquez, *Obregón: Apuntes biográficos* (Mexico, 1929), p. 12.
4. Richard H. Dillon, "Del rancho a la presidencia," *Historia Mexicana* (Mexico), vol. 6, no. 22, Oct.–Dec. 1956, pp. 22, 256–69; Juan de Dios Bojórquez, *Forjadores de la Revolución Mexicana* (Mexico, 1960), pp. 3–13.
5. Obregón file, in APEC; see Rosales, "Niñez extraordinaria," pp. 16–17.
6. Álvaro Obregón file, 1909, in "Personajes notables de la Revolución," carpeta 3, no. 499, Centro de Estudios Históricos de México, CONDUMEX (Mexico, D.F.).
7. Bojórquez, *Forjadores de la Revolución,* p. 17.
8. Álvaro Obregón, *Ocho mil kilómetros en campaña* (Mexico, 1973), pp. 1–5.
9. Ramón Puente, *La dictadura, la revolución y sus hombres* (Mexico, 1938), p. 182.
10. Obregón, *Ocho mil kilómetros,* p. 21.
11. Puente, *La dictadura,* pp. 181–83.
12. Quoted in Dillon, "Del rancho," p. 267.
13. Obregón, *Ocho mil kilómetros*, pp. 57–79; Martín Luis Guzmán, *El águila y la serpiente* (Madrid, 1928), pp. 70–71; Francisco J. Grajales, "Las campañas del General Obregón," in Obregón, *Ocho mil kilómetros,* pp. xlviii–lvi.
14. Ibid., pp. 83–154.
15. Ibid., pp. 91–92.
16. Ibid., pp. 122–65.
17. Jorge Aguilar Mora, *Un día en la vida del General Obregón* (Mexico, 1982), pp. 14, 16.
18. Calles vda. de Torreblanca interview. Guzmán, *Águila,* p. 72.
19. Aguilar Mora, *Un día en la vida,* pp. 44–56.
20. Francisco Ramírez Plancarte, *La ciudad de México durante la Revolución Constitucionalista* (Mexico, 1941), pp. 61–63; Berta Ulloa, "La encrucijada de 1915," *Historia de la Revolución Mexicana,* 23 vols. (Mexico,

1979), *Periodo 1914–1917,* vol. 5, pp. 34–35; Aguilar Mora, *Un día en la vida,* p. 24; Hall, *Poder y revolución,* pp. 108–19.
21. Obregón, *Ocho mil kilómetros,* pp. 202–4.
22. Ibid., pp. 261–68.
23. Jean Meyer, "Los obreros en la Revolución Mexicana: Los batallones rojos," *Historia Mexicana* (Mexico), vol. 21, no. 81, July–Sept. 1971, pp. 5, 7–12, 31; Ulloa, "Encrucijada," pp. 107, 114.
24. Obregón, *Ocho mil kilómetros,* pp. 269–70, 299–329; Hall, *Poder y revolución,* pp. 115–30.
25. Obregón, *Ocho mil kilómetros,* pp. 371–72.
26. Guzmán, *Águila,* p. 126.
27. Interview with Miguel Palacios Macedo by the author and Jean Meyer, Mexico, 1973.
28. Hall, *Poder y revolución,* pp. 151–56.
29. Ibid., pp. 163–83.
30. Romero, *Álvaro Obregón,* pp. 40–42.
31. Aguilar Camín, *Saldos de la Revolución,* pp. 23–24; Hall, *Poder y revolución,* p. 31.
32. Oswaldo Burgos, "Los cuentos del General Obregón," *Revista de Revistas* (Mexico), Apr. 11, 1926, p. 17; F. González Guerrero, "Enciclopedia mínima: La memoria del General Obregón," *El Universal Gráfico* (Mexico), Feb. 19, 1929, editorial page; Miguel Alessio Robles, "La figura de Obregón," *Novedades* (Mexico), June 1941.
33. Miguel Alessio Robles, "Obregón: Su vida y su carácter," *El Universal* (Mexico), May 4, 1937.
34. Alvaro Matute, "La carrera del caudillo" *Historia de la Revolución Mexicana,* 23 vols., (Mexico, 1980), vol. 8, pp. 65–77; Hall, *Poder y revolución,* pp. 203–31.
35. Palacios Macedo interview.
36. Matute, *Carrera del caudillo,* pp. 92–102.
37. Narciso Bassols Batalla, *El pensamiento de Álvaro Obregón* (Mexico, 1970), pp. 45–52.
38. Matute, *Carrera del caudillo,* pp. 33–41.
39. Bassols, *El pensamiento,* pp. 45–52; Matute, *Carrera del caudillo,* pp. 30–41.
40. Palacios Macedo interview.
41. Daniel Cosío Villegas, 2 vols. *Ensayos y notas,* 2 vols. (Mexico, 1966), vol. 1, pp. 140–41; Ernest Gruening, *Mexico and Its Heritage* (London, 1928), pp. 518–520.
42. Charles Hall, "The Miracle School," *Hispanic American Historical Review,* vol. 75, p. 2.
43. Cosío Villegas, *Ensayos y notas*, vol. 1, pp. 141–42.
44. Jean Meyer, "Estado y sociedad con Calles," *Historia de la Revolución Mexicana,* 23 vols. (Mexico, 1977), vol. 11, pp. 57, 58, 77–80; Ruth Clark Marjoire, *Organized Labor in Mexico* (Chapel Hill, N.C., 1934).
45. David C. Bailey, "Obregón, Mexico's Accommodating President," in Georges Wolfskill and Douglas Richmond, eds., *Essays on the Mexican Revolution: Revisionist Views of the Leaders* (Austin, Tex., 1979), pp. 87–89; John Foster Dulles, *Yesterday in Mexico: A Chronicle of the Revolution, 1919–1936* (Austin, Tex., 1961), pp. 97–101.

46. Robert E. Quirk, *The Mexican Revolution and the Catholic Church, 1910–1929: An Ideological Study* (Indiana, 1973), pp. 119–37; Jean Meyer, "El conflicto entre la Iglesia y el Estado, 1926–1929," in *La Cristiada*, 3 vols. (Mexico, 1980), vol. 2, pp. 110–40, 304–20.
47. Dulles, *Yesterday*, pp. 158–171; Bassols, *El pensamiento*, pp. 66–73.
48. Antonio Gómez Robledo, *Los convenios de Bucareli ante el derecho internacional* (Mexico, 1933), pp. 204–7.
49. Dulles, *Yesterday*, pp. 108–15, 117, 173–88, 228; Miguel Alessio Robles, "Obregón y de la Huerta," *Senderos* (Mexico, 1930), pp. 167–73.
50. Joaquín Piña, "¡Ya me estoy aburriendo! dijo Obregón" *Así* (Mexico), Sept. 7, 1941.
51. Julio Scherer, "Roberto Cruz en la época de la violencia," *Excélsior* (Mexico), July 1961, pp. 6, 8.
52. Dulles, *Yesterday*, pp. 254–55.
53. Palacios Macedo interview.
54. Obregón-Calles file, 103–HL–22, 102–C50, 42–F–12, in Presidentes, Archivo General de la Nación de México (Mexico, D.F.); file 812.00 Sonora, 812–001055–23, State Department Records (Washington, D.C.), Dec. 1924–July 1928; Bassols, *El pensamiento*, pp. 45–52.
55. Romero, *Álvaro Obregón*, pp. 40–42.
56. Meyer, "El conflicto," pp. 129–31.
57. "Compte rendue d'une conversation de M. Lagarde avec le general Arnulfo Gómez," in Très Secret dossier, Correspondencia Diplomática Francesa (Paris), June 29, 1927, B–25–1, pp. 211–14.
58. Mario A. Mena, *Álvaro Obregón: Historia militar y política* (Mexico, 1960), pp. 118–24.
59. Calles vda. de Torreblanca interview.
60. Aguilar Camín, *Saldos de la Revolución*, pp. 57–59.
61. Ibid., p. 58; Bassols, *El pensamiento*, pp. 99–123; Mena, *Álvaro Obregón*, pp. 118–23; Meyer, "El conflicto," pp. 265–69.

Chapter 15

1. Gerardo Sisniega, ed., *Una página de historia de México* (Mexico, 1934), pp. 6–7; "Semblanza del coronel Juan José Elías," in Elías, genealogía file, pp. 18–40, in Archivo Plutarco Elías Calles (APEC); Héctor Águilar Camín, *La frontera nómada* (Mexico, 1977), pp. 180–82.
2. Carlos Macías Richard, *Vida y temperamento de Plutarco Elías Calles 1877–1920* (Mexico, 1995), p. 37; James Officet, "Los hijos de Pancho: La familia Elías, guerreros sonorenses," in *XIX Simposium de historia de Sonora, Memoria* (Hermosillo, 1984), p. 336; Ramón Puente, *Hombres de la Revolución: Calles* (Los Angeles, 1933), pp. 11–14.
3. Macías, *Vida y temperamento*, pp. 45–47.
4. Ibid., pp. 56–58; Luis Méndez Preciado, *La educación pública en Sonora, 1900–1970* (Mexico, n.d.), p. 40; Gustavo Rivera, *Breve historia de la educación en Sonora* (Hermosillo, n.d.), p. 210.
5. Macías, *Vida y temperamento*, p. 58; see also Juan G. Amaya, *Los gobiernos de Obregón, Calles y regímenes peleles derivados del callismo* (Mexico, 1947), p. 389.
6. Roberto Guzmán Esparza, *Memorias de don Adolfo de la Huerta según su proprio dictado* (Mexico, 1957), p. 389.

7. Macías, *Vida y temperamento,* pp. 71–72.
8. Ibid., pp. 77–94, 103, 109–18; see also Puente, *Hombres.*
9. Ibid., pp. 134–35.
10. Ibid., Guzmán, *Memorias,* pp. 24–26.
11. Ibid., pp. 159–60.
12. Ibid., pp. 162–63; Álvaro Obregón, *Ocho mil kilómetros en campaña* (Mexico, 1973), p. 40.
13. Ibid. (Obregón), pp. 80–82; Águilar Camín, *Frontera nómada,* pp. 54–59; Macías, *Vida y temperamento,* pp. 169, 180.
14. Macías, *Vida y temperamento,* p. 183.
15. Ibid., p. 184; Plutarco Elías Calles, *Informe relativo al sitio de Naco 1914–1915* (Mexico, 1932), pp. 1–28.
16. Alberto Calzadíaz, *El fin de la División del Norte,* vol. 3, *Hechos reales de la Revolución,* 8 vols. (Mexico, 1982), pp. 98–107.
17. Calles, *Informe relativo.*
18. See Angela Encinas viuda de Calles 1917–1924 file, in APEC.
19. See Decretos y Circulares file, vol. 1, pp. 1–24, in APEC; Puente, *Hombres,* pp. 106–8; Macías, *Vida y temperamento,* p. 192.
20. Puente, *Hombres,* p. 106; John Foster Dulles, *Ayer en México: Una crónica de la Revolución, 1919–1936* (Mexico, 1977), p. 121.
21. See Recopilación de materiales históricos de Sonora file, vol. 1, in APEC.
22. Ibid.
23. See Decretos y Circulares file, vol. 2, p. 140, in APEC.
24. Armando Elías Chomina, *Compendio de datos históricos de la familia Elías* (Hermosillo, 1986), p. 158.
25. See Arzobispos file, in APEC; Puente, *Hombres,* pp. 106–8; Claudio Dabdoub, *Historia del valle del Yaqui* (Mexico, 1964), p. 415.
26. See "Escuela industrial Cruz Chávez" file, Aug. 31, 1917, in APEC; Macías, *Vida y temperamento,* pp. 193–95.
27. Francisco Bulnes, *Los grandes problemas de México* (Mexico, 1970), p. 20; Dec. 24, 1925, file no. 812, p. 6,863, (Washington, D.C.), State Department Records (hereafter called SDR).
28. Jean Perier to Poincare, Apr. 29, 1924, no. B–25–1, Correspondencia Diplomática Francesa, Paris.
29. Juan de Dios Bojórquez, *Calles* (Mexico, 1925), pp. 67–77.
30. Puente, *Hombres,* pp. 139–43; Plutarco Elías Calles to Álvaro Obregón, July 25, 1925, Obregón file, in APEC.
31. José C. Valadés, *Historia general de la Revolución Mexicana* (Mexico, 1967), p. 370.
32. See Jewish immigration file, in APEC.
33. Hermann Benzing, *Investigación sobre el viaje del general Calles a Alemania* (Mexico), typescript, p. 12.
34. File no. B–25–1, Correspondencia Diplomática Francesa, Apr. 7, 1926; Alberto J. Pani, *Apuntes autobiográficos*, 2 vols. (Mexico, 1950), vol. 2, pp. 67–68; Interview with Manuel Gómez Morín by the author, Mexico, June 9, 1971; Calles to Obregón, Feb. 2, Apr. 1, 1925, Obregón file, in APEC.
35. Manuel Gómez Morín to José Vasconcelos, Mar. 1, 1926, Archivo Manuel Gómez Morín (hereafter called AMGM).
36. Gómez Morín to Vasconcelos, Sept. 11, 1926, and *El crédito agrícola en*

México, estudio sobre su establecimiento y análisis de su funcionamiento hasta 1931. Bases para su organización de acuerdo con el estado actual del crédito agrícola, pp. 1–3; in AMGM; Manuel Gómez Morín, *El crédito agrícola en México* (Madrid, 1928), p. 11.

37. Gómez Morín, *El crédito agrícola*, pp. 21–41, AMGM.
38. Miguel Palacios Macedo to Manuel Gómez Morín, Jan. 6, 1925, Archivo Miguel Palacios Macedo (AMPM).
39. Correspondencia Diplomática Francesa, Aug. 28, 1926, no. B–25–1.
40. Alexander B. Dye, "Railways and Revolution in Mexico," *Foreign Affairs*, vol. 2, Jan. 1927, pp. 321–24.
41. Daniel Cosío Villegas, "La riqueza de México," *La Antorcha* (Mexico), May 30, 1925; *Saturday Evening Post*, Feb. 26, 1927, pp. 77–78.
42. See Section of official correspondence between the Mexican government and the United States of America's government, respect of the laws on the fracción primera del Artículo 27 Constitucional Mexicano, 1926, Archivo Histórico-Diplomático Genaro Estrada, Secretaría de Relaciones Exteriores de México.
43. "Entrevista con el General Escamilla Garza," in *Excélsior* (Mexico), Mar. 13, 1972.
44. See *Excélsior*; see also *El Universal* (Mexico), Dec. 24, 1926.
45. See Sheffield Papers, Yale University, Manuscript Division, New Haven, Conn., June 28, 1926, Feb. 12, 1927; F. H. Arthur Schoenfeld to James Rockwell Sheffield, Feb. 19, 1927; *Boletín del Archivo General de la Nación* (Mexico, 1979), tercera serie, pp. 43–45; "Declaraciones del General P. E. Calles, Presidente de la República, al New York Times," in *New York Times* (New York), Aug. 6, 1925.
46. Jean Meyer, "Estado y sociedad con Calles," *Historia de la revolución mexicana, periodo 1924–1928*, 23 vols. (Mexico, 1981), vol. 11, p. 25.
47. See Elías Arturo file, Jan. 1927, in APEC.
48. Memorandum of meeting at Mr. Lamont's house at 9:30 A.M., Mar. 31, 1927, in which the Mexicans Mr. Téllez, Mr. Lamont, Mr. Negrete, Mr. Prieto and V. M. (Vernon Mounroe) were present, Dwight Whitney Morrow Papers, Amherst College (Amherst, Mass.); see also Jean Meyer, "Estado y sociedad . . . "
49. Meyer, "Estado y sociedad . . . ," pp. 37–38.
50. Lamont Papers, Amherst College (Amherst, Mass.), Feb.–Apr. 1927; memorandum of the conversation held when Pani and Negrete dined with Morrow on Tuesday evening, Feb. 23, 1927, Mar. 19, 1927, Morrow Papers.
51. *Excélsior*, Feb. 4, 1926.
52. Ambassador Sheffield file, no. 812/404/578, DSR.
53. Ernest Lagarde, Sept. 18, 1926, Correspondencia Diplomática Francesa, carpeta 105, pp. 54, 87.
54. *El Universal*, July 30, Aug. 22–24, 1926; *Excélsior*, Aug. 22–24, 1926.
55. Interview in Arzobispos file, Aug. 21, 1926, APEC; "Morrow to Clark," Aug. 19, 1928, in John J. Burke's Diary, DSR, 812/404/931/G/12.
56. Ibid; *Excélsior*, July 2, 1943.
57. Mons. Mora y del Río to the bishops of Rome, May 27, 1927, Archivo de la Compañía de Jesús; "Algo muy importante debe saberse" manuscript (Mexico, n.d.), Archivo Vita-México, Centro de Estudios sobre la Universidad, AHUNAM.

58. Interview with His Grace Martin Trischler, Havana; Dwight Morrow to Robert E. Olds, Feb. 21, 1928 (Havana), in Morrow Papers; Dwight Morrow to J. Reuben Clark, Oct. 19, 1928, in John J. Burke's Diary, DSR.
59. See *Excélsior,* July 29, 1929.
60. *Diario de Debates de la Cámara de Diputados, XXXIII Legislatura* (Mexico), Oct. 8, 1928, pp. 35–36; Ernest Gruening, *Mexico and Its Heritage* (London, 1928), pp. 372–75.
61. "Reports from reliable source indicate that General Ángel Flores died of arsenic poisoning," Apr. 9, 1926, microfilm no. 812/00, p. 156, DSR.
62. Julio Scherer, "Roberto Cruz en la época de la violencia," *Excélsior,* July 1961.
63. Guzmán, *Excélsior,* July and Oct. 1961.
64. See *El Universal,* July 19, 1928; Jean Meyer, *La Cristiada,* 2 vols. (Mexico, 1971), vol. 2, pp. 144–50; see speech of Sept. 1, 1928, Discursos varios file; and Calles, Informes de gobierno file, in APEC.
65. Discursos . . . ; Informes . . . .
66. Portes Gil, *Autobiografía de la Revolución Mexicana* (Mexico, 1966); see *El Universal,* Aug. 21, Sept. 5 and 7, 1928.
67. *Excélsior,* Dec. 2, 1928, and Jan. 4, 1929; Portes Gil, *Autobiografia . . . ,* pp. 228, 230; *Diario de Debates,* Sept. 1928; José C. Valadés, "A Sáenz se lo llevaron al baño," *Todo* (Mexico), Dec. 9, 1937; *El Universal,* Oct. 14, 1928; Pani, *Apuntes,* vol. 2, p. 111.
68. *Boletín del Archivo General de la Nación de México* (Mexico, 1979), pp. 73–76.
69. Ibid.
70. Gómez Morín to Vasconcelos, Jan. 2, Nov. 3, 1925, June 10, Oct. 8, 1926, Oct. 16, Nov. 3, 1928, in AMGM.
71. Meyer, *La Cristiada,* vol. 2, pp. 339–40.
72. José Vasconcelos, *El proconsulado* (Mexico, 1939), pp. 294–99, 359–64, 412–15; John Skirius, *José Vasconcelos y la cruzada de 1929* (Mexico, 1970), pp. 116–30.
73. Pascual Ortiz Rubio, *Memorias* (Morelia, 1981), pp. 209–11; Daniel Cosío Villegas, *El sistema político mexicano: Las posibilidades de cambio* (Mexico, 1974), pp. 35–50.
74. Lázaro Cárdenas to Plutarco Elías Calles, Oct. 10, 1930, in Cárdenas file, APEC.
75. Abelardo Rodríguez, *Autobiografía* (Mexico, 1962), pp. 145–46; Tzvi Medin, *El Minimato presidencial, historia política del maximato (1928–1935)* (Mexico, 1982), p. 184.
76. Rodríguez, *Autobiografía,* pp. 159–60; Francisco Javier Gaxiola, *El presidente Rodríguez (1932–1934)* (Mexico, 1938), pp. 115, 121; Lorenzo Meyer, Rafael Segovia, Alejandra Lajous, "Los inicios de la institucionalización: 1928–1934" *Historia de la Revolución Mexicana,* vol. 12, p. 168.
77. Meyer et al., p. 167.
78. Ibid., p. 178.
79. Victoria Lerner, "La educación socialista," *Historia de la Revolución,* vol. 6, pp. 58–60.
80. Jorge Cuesta, "El Comunismo en la escuela primaria," *El Universal,* Sept. 29, 1933.

81. Interview with Hortensia Calles viuda de Torreblanca, Mexico, June 26, 1985.
82. José C. Valadés file, in APEC.
83. Calles vda. de Torreblanca interview. José González Ortega, "Calles en la intimidad," *Hoy,* Oct. 19, 1946, p. 26.
84. Comunicaciones del más allá file, in APEC.
85. See "Una ventana al mundo invisible," *Antorcha* (Mexico, 1960), pp. 31, 81; Gutierre Tibón, "Gog y Magog. La conversión del general Calles. El padre Heredia, espiritista," *Excélsior,* Oct. 20, 1958, p. 6.

Chapter 16

1. Centro de Estudios de la Revolución Mexicana, Lázaro Cárdenas, de. *Se llamó Lázaro Cárdenas* (Mexico, 1995), p. xi; Willliam C. Townsend, *Lázaro Cárdenas: Demócrata mexicano* (Mexico, 1959), p. 13.
2. Lázaro Cárdenas, *Apuntes 1913–1940, Obras,* 4 vols. (Mexico, 1972), vol. 1, pp. 5, 6, 9; José Romero, *Lázaro Cárdenas: Su niñez y juventud hasta la época actual a través de mis recuerdos* (Mexico, 1933), p. 13.
3. Cárdenas, *Apuntes,* vol. 1, p. 15.
4. Ibid., pp. 17–18.
5. Ibid., p. 20.
6. Ibid., p. 23.
7. Ibid., p. 51; *Excélsior* (Mexico), Sept. 1935.
8. Ibid., pp. 171–72.
9. Ibid., pp. 82–84.
10. Escuela Cruz Gálvez (1917–1928) file in Archivo Plutarco Elías Calles (hereafter called APEC).
11. Plutarco Elías Calles, *Partes oficiales de la campaña de Sonora rendidas por el General Plutarco Elías Calles Gobernador y Comandante Militar del Estado de Sonora al C. General Álvaro Obregón, Jefe del Cuerpo del Ejército del Noroeste* (Mexico, 1932), pp. 50, 62, 71, 92.
12. Cárdenas, *Apuntes,* vol. 1, p. 142.
13. Ibid., p. 149; Djed Bórquez, *Lázaro Cárdenas (líneas biográficas)* (Mexico, 1933), p. 86.
14. Armando de María y Campos, *Múgica, Crónica biográfica (Aportación a la historia de la Revolución mexicana)* (Mexico, 1939), p. 202.
15. Townsend, *Demócrata mexicano,* p. 38; Bórquez, *Lázaro Cárdenas,* pp. 87, 88; Alfredo Avila y Valencia, "Cuando nos íbamos a quedar sin presidente," *Mujeres y Deportes* (Mexico), Oct. 12, 1935.
16. Victoriano Anguiano Equihua, *Lázaro Cárdenas: Su feudo y la política nacional* (Mexico, 1951), p. 7.
17. De Maria y Campos, *Múgica,* pp. 12, 15–16, 40, 52, 85, 87, 106.
18. *Excélsior*, Sept. 15, 1935, p. 3a, sec. 1.
19. Anguiano, *Lázaro Cárdenas,* p. 47.
20. Cárdenas, *Apuntes,* vol. 1, p. 389.
21. Townsend, *Demócrata mexicano,* pp. 44–45.
22. Luis González y González, *Pueblo en vilo: Microhistoria de San José de Gracia* (Mexico, 1995), pp. 156, 158.
23. File 818–N–12, Section Presidentes: Obregón-Calles, in Archivo General de la Nación (Mexico); O. Ambriz, A. León G., et al., *Historia del agrarismo en Michoacán* (Mexico, 1982), pp. 185–90.

24. Jean Meyer, *La revolución mexicana* (Madrid, 1973), pp. 141–69.
25. Jesús Múgica Martínez, *La Confederación Revolucionaria Michoacana del Trabajo* (Mexico, 1982), pp. 81–82.
26. Múgica Martínez, *La Confederación Revolucionaria,* pp. 35, 162–63; see also Manuel Diego Hernández, *La Confederación Revolucionaria Michoacana del Trabajo* (Mexico, 1982), p. 35.
27. Anguiano, *Lázaro Cárdenas,* pp. 45, 46.
28. Lázaro Cárdenas, "Informe de gobierno," Sept. 1, 1932, mecanuscrito (Morelia).
29. Townsend, *Lázaro Cárdenas,* p. 74; Antonín Piña Soria, *Cárdenas socialista* (Mexico, 1935), p. 92.
30. Nathaniel and Silvia Weyl, "La reconquista de México (Los Días de Lázaro Cárdenas)," *Problemas agrícolas e industriales de México* (Mexico, 1955), p. 310.
31. Ibid., pp. 213–24.
32. Cárdenas, "Informe."
33. Froylán Manjarrez and Gustavo Ortiz Hernán, *Lázaro Cárdenas: 1. Soldado de la Revolución. 2. Gobernante. 3. Político nacional* (Mexico, 1933), p. 77.
34. Anguiano, *Lázaro Cárdenas,* p. 54.
35. Ibid., p. 51.
36. Jesús Tapia Santamaría, *Campo religioso y evolución política en el Bajío zamorano* (Morelia, 1986), p. 214.
37. Ibid., p. 127.
38. Manjarrez and Hernán, *Lázaro Cárdenas: 1. Soldado,* p. 64.
39. González y González, *Pueblo en Vilo,* pp. 244–45.
40. César Moheno, *Las historias y los hombres de San Juan* (Mexico, 1985), p. 145.
41. Ibid., p. 119.
42. Anguiano, *Lázaro Cárdenas,* p. 39.
43. Townsend, *Lázaro Cárdenas,* pp. 69, 70; Gonzalo N. Santos, *Memorias* (Mexico, 1984), pp. 571–98.
44. Cárdenas, *Apuntes,* vol. 1, pp. 287–92, 294, 298.
45. Townsend, *Lázaro Cárdenas,* pp. 97, 101.
46. Alicia Hernández, "La mecánica cardenista," *Historia de la Revolución Mexicana, Período 1934–1940,* 23 vols. (Mexico, 1979), vol. 16, pp. 101, 102.
47. Ibid., pp. 128, 129; Anguiano, *Lázaro Cárdenas,* p. 56–60.
48. González y González, *Pueblo en Vilo,* pp. 143, 154.
49. Cárdenas, *Apuntes,* vol. 1, p. 366.
50. Townsend, *Lázaro Cárdenas,* p. 117.
51. Leonel Durán, *Lázaro Cárdenas: Ideario político* (Mexico, 1972), pp. 130, 131; Marte R. Gómez, "Los problemas de la región lagunera," *El Nacional* (Mexico), Mar. 29, 1941, pp. 24, 29.
52. Cárdenas, *Apuntes,* vol. 1, p. 360.
53. Durán, *Ideario político,* pp. 143–46.
54. Cárdenas, *Apuntes,* vol. 1, p. 360.
55. Gómez, "Problemas de la región," pp. 24, 29.
56. Cárdenas, *Apuntes,* vol. 1, p. 361.
57. Durán, *Ideario político,* p. 154.

58. Susana Glantz, *El ejido colectivo de Nueva Italia* (Mexico, 1974), pp. 101, 102.
59. Ibid., p. 107.
60. Interview with Adolfo Aribe Alva, Sept. 4, 1992.
61. Frank Tannenbaum, "Lázaro Cárdenas," *Historia mexicana* (Mexico), vol. 10, no. 2, Oct.–Dec. 1960, pp. 332–41.
62. "De la alocución del Centro Patronal de Monterrey, Nuevo León, sobre la acción gubernamental y la lucha obrera," *El Nacional,* Feb. 11, 1936.
63. Hernández, "Mecánica cardenista," pp. 157–62.
64. Ibid., pp. 164, 165.
65. Ibid., pp. 182–84.
66. Lorenzo Meyer, *México y los Estados Unidos en el conflicto petrolero (1917–1942)* (Mexico, 1972), p. 310.
67. Weyl and Weyl, "Reconquista," p. 282.
68. Cárdenas, *Apuntes,* vol. 1, p. 371.
69. Meyer, *México y los Estados Unidos,* pp. 319–29.
70. Cárdenas, *Apuntes,* vol. 1, p. 387.
71. Ibid., p. 389.
72. Ibid., p. 390.
73. Vicente T. Mendoza, *El Corrido de la Revolución Mexicana* (Mexico, 1990), p. 147.
74. Tannenbaum, "Lázaro Cárdenas," p. 338.
75. Durán, *Ideario político,* p. 326.
76. Cárdenas, *Apuntes,* vol. 1, pp. 369–70.
77. Townsend, *Lázaro Cárdenas,* pp. 277–86.
78. Cárdenas, *Apuntes,* vol. 1, p. 443.
79. Hernández, "Mecánica cardenista," pp. 198–200.

Part IV

1. Enrique Krauze, *Daniel Cosío Villegas: Una biografía intelectual* (Mexico, 1980), p. 92.
2. "Roberto Blanco Moheno interviews General Lázaro Cárdenas," *Impacto* (Mexico), Mar. 22, 1961, pp. 6–8.
3. Interview with Adolfo Orive Alba, Mexico, Sept. 4, 1992.

Chapter 17

1. José Mendizábal, 7. *Almanaque de efemérides del Estado de Puebla para el año de 1898* (Mexico, 1897), pp. 121–25.
2. Enrique Krauze, *Caudillos culturales en la Revolución Mexicana* (Mexico, 1985), p. 28.
3. Interview with Rafael Ávila Núñez, Teziutlán, Puebla, Oct. 14, 1994.
4. Luis Audirac, *Se hizo de noche* (Mexico, l946), p. 54.
5. Isidro Fabela, *Documentos históricos de la Revolución Mexicana* (Mexico, 1965), vol. 3, pp. 346–47, 496–97; "José C. Valadés interviews General Maximino Ávila Camacho," in *Hoy* (Mexico), May 22, 1943.
6. Gonzalo N. Santos, *Memorias* (Mexico, 1984), p. 678.
7. Alfredo Kawage Ramia, "Partió hace veinte años," *Siempre* (Mexico), Oct. 20, 1975, p. 30; see also Fernando López Portillo, "Memorias," *El Universal* (Mexico), Apr. 27, 1957, p. 2.

8. General Estrada's testimony in Alfonso Taracena, *La revolución desvirtuada* (Mexico, 1971), vol. 8, pp. 42–43.
9. See Gustavo Abel Hernández and Armando Rojas Trujillo, *Manuel Ávila Camacho: Biografía de un revolucionario con historia* (Puebla, 1986), vol. 1, p. 86.
10. Hernández and Rojas, *Manuel Ávila Camacho,* vol. 1, p. 74; see also José Altamirano, *La personalidad del General Manuel Ávila Camacho* (Mexico, 1940), pp. 57–59.
11. Hernández and Rojas, *Manuel Ávila Camacho,* vol. 1, pp. 75–76.
12. Antonio Lomelí Garduño, *Semblanza espiritual de Ávila Camacho* (Mexico, 1957), pp. 67–69.
13. Santos, *Memorias,* p. 647.
14. Interview with Alfonso Corona del Rosal, Mexico, Mar. 6, 1992.
15. Interview with Adolfo Orive Alba, Mexico, Sept. 4, 1992.
16. Santos, *Memorias,* p. 835.
17. The tailor was Saúl Krauze, grandfather of the author.
18. Interview with Justo Fernández, Mexico, Oct. 1992.
19. Orive Alba interview.
20. Editorial in *El Universal* (Mexico), Oct. 1, 1938, p. 3.
21. Ibid., June 25, 1941, p. 3.
22. Editorial in *Excélsior* (Mexico), Oct. 2, 1938, p. 5.
23. Ibid., Mar. 22, 1939, p. 5.
24. Ibid., Aug. 25, 1939, p. 5.
25. "Últimas Noticias," *Excélsior* (Mexico), Apr. 4, 1939.
26. Editorial in *El Popular* (Mexico), Mar. 18, 1939, p. 3.
27. José Pagés Llergo, "Yo hablé con Hitler," *Hoy* (Mexico), Nov. 18, 1939.
28. José Vasconcelos, "La inteligencia se impone," in *Timón,* June 8, 1940, vol. 11, no. 16, p. 9.
29. Lázaro Cárdenas, "Apuntes 1913–1940," *Obras,* 4 vols. (Mexico, 1986), vol. 1, p. 439.
30. Blanca Torres Ramírez, "México en la Segunda Guerra Mundial," *Historia de la Revolución Mexicana,* 23 vols. (Mexico, 1979), vol. 19, pp. 65–80; see also José Luis Ortiz Garza, *México en guerra* (Mexico, 1989).
31. *Tiempo* (Mexico), May 21, 1942, pp. 2–4.
32. Cárdenas, "Apuntes 1941–1956," *Obras,* vol. 2, p. 191.
33. "Valadés interviews Ávila Camacho," pp. 8–10.
34. Interview with Alicia Ávila Camacho de Fernández, Mexico, Oct. 1992.
35. "Interview with Vicente Lombardo Toledano," in James W. Wilkie and Edna Monzón de Wilkie, *México visto en el siglo XX: Entrevistas de historia oral* (Mexico, 1969), pp. 333–36.
36. José C. Valadés, "La Unidad Nacional," *Historia general de la Revolución Mexicana* (Mexico, l985), vol. 10, pp. 74–75.
37. Daniel Cosío Villegas, "La crisis de México," *Extremos de América* (Mexico, 1949), p. 39.
38. Jaime Torres Bodet, *Memorias: Años contra el tiempo* (Mexico, l969), p. 334.
39. Wilkie and Wilkie, *México visto,* p. 336.
40. Luis Gómez Z., *Sucesos y remembranzas,* 2 vols. (Mexico, 1979), vol. 1, pp. 227–41, 298–310.

41. Enrique Krauze, "La reconstrucción económica," *Historia de la Revolución Mexicana*, 23 vols. (Mexico, 1982), vol. 10, pp. 70–82.
42. "Mensaje del presidente Manuel Ávila Camacho el 1 de diciembre de 1940," *El Nacional* (Mexico), Dec. 2, 1940, p. 1.
43. Santos, *Memorias*, pp. 709–10.
44. Ibid., p. 757.
45. Ibid., pp. 768–69.
46. "Gómez Morín interview," in Wilkie and Wilkie, *México visto*, pp. 148–55.
47. Cosío Villegas, "Crisis," pp. 26–27.
48. Adolfo Orive Alba, *La irrigación en México* (Mexico, 1970), pp. 85–92.
49. Marte R. Gómez, "Los problemas de la Región Lagunera," *Vida política contemporánea: Cartas de Marte R. Gómez*, 2 vols. (Mexico, 1978), vol. 1, p. 1,008.
50. Cosío Villegas, "Crisis," pp. 27–28.
51. Wilkie and Wilkie, *México visto*, p. 185.
52. Ibid., p. 204.
53. Ibid., pp. 217–23.
54. Luis Calderón Vega, *Memorias del PAN*, 5 vols. (Mexico, 1992), vol. 1, pp. 176–82.
55. See Franz A. von Sauer, *The Alienated "Loyal" Opposition* (New Mexico, 1974), pp. 98–115; Donald J. Mabry, *Mexico's Acción Nacional: A Catholic Alternative to Revolution* (Syracuse, N.Y., 1973), pp. 30–45.
56. Wilkie and Wilkie, *México visto*, p. 186.
57. Cosío Villegas, "Crisis," p. 41.
58. Torres, "México en la Segunda Guerra," pp. 153–84.
59. Carlos Martínez Assad, "El cine como lo ví y como me lo contaron," in Rafael Loyola, coord., *Entre la guerra y la estabilidad política: El México de los 40* (Mexico, 1990), pp. 339–60.
60. Luis Medina, "Civilismo y modernización del autoritarismo," *Historia de la Revolución Mexicana*, 23 vols. (Mexico, 1982), vol. 20, pp. 81–91.
61. Cosío Villegas, "Crisis," originally published in 1947 in *Cuadernos Americanos* (Mexico), año 6, vol. 32, no. 2, pp. 29–51.
62. Enrique Krauze, *Daniel Cosío Villegas: Una biografía intelectual* (Mexico, 1980), p. 152.
63. Torres Bodet, *Memorias*, p. 442.
64. Interview with Marco Antonio Muñoz, Mexico, Nov. 3, 1992.

Chapter 18

1. Rodolfo Usigli, *El Gesticulador*, in *El hijo pródigo* (Mexico, 1943).
2. See Leonardo Pasquel, *Veracruzanos en la Revolución* (Mexico, 1985); Octaviano R. Corro, *General Miguel Alemán: Su vida revolucionaria* (Jalapa, 1945); Rafael Gallegos Llamas, *Matiz de un revolucionario 1900–1929* (Mexico, 1976).
3. Interview with Pablo Vidaña, Mexico, Oct. 1994.
4. Ibid.
5. Interview with Rita Alafita de González, Mexico, Oct. 1994.
6. Miguel Alemán Valdés, *Remembranzas y testimonios* (Mexico, 1987), p. 34–35.
7. Interview with Rafael Barreiro Gutiérrez, Mexico, July 1994.

8. Interview with Antonio Martínez Baez, Mexico, Nov. 1992.
9. Alemán, *Remembranzas,* p. 56–57.
10. Ibid., p. 82.
11. Interview with Justo Manzur Ocaña, Mexico, June 1994.
12. Alemán, *Remembranzas,* p. 125.
13. Ibid.
14. Ibid., p. 152.
15. Interview with Marco Antonio Muñoz, Mexico, Nov. 1992.
16. Luis Medina, "Civilismo y modernización del autoritarismo," *Historia de la Revolución Mexicana,* 23 vols. (Mexico, 1982), vol. 20, p. 81.
17. Carlos Loret de Mola, "Fue símbolo del sistema," *Excélsior* (Mexico), May 14, 1983.
18. Roderic A. Camp, *Los líderes políticos de México, su educación y reclutamiento* (Mexico, 1983), pp. 164–77.
19. Interview with Antonio Ortiz Mena, Mexico, Nov. 1992.
20. Blanca Torres, "Hacia la utopía industrial," *Historia de la Revolución Mexicana,* vol. 21, pp. 144, 145; see also *Estadísticas históricas de México,* 2 vols. (Mexico, 1985).
21. Lázaro Cárdenas, "Apuntes 1941–1956," *Obras,* 4 vols. (Mexico, 1986), vol. 2, p. 268.
22. Torres, "Utopía industrial," p. 107.
23. See Roderic A. Camp, *Los empresarios y la política en México: Una visión contemporánea* (Mexico, 1990); Elvira Concheiro, et al., *El poder de la gran burguesía* (Mexico, 1979); Robert J. Schafer, *Mexican Business Organizations: History and Analysis* (New York, 1973).
24. Gabriel Zaid, *El progreso improductivo* (Mexico, 1979), p. 217.
25. Gabriel Zaid, *La economía presidencial* (Mexico, 1987), pp. 149–70, *La nueva economía presidencial* (Mexico, 1994), pp. 30–33, 53–56, and *El progreso,* pp. 216–24.
26. Octavio Paz, *Posdata* (Mexico, 1970), pp. 136.
27. Miguel León Portilla, *Toltecayotl: Aspectos de la cultura Náhuatl* (Mexico, 1980), pp. 293–99.
28. Enrique Krauze, *Daniel Cosío Villegas: Una biografía intelectual* (Mexico, 1980), p. 282.
29. Daniel Cosío Villegas, *El estilo personal de gobernar* (Mexico, 1974), pp. 7–14.
30. See Adolfo León Osorio, *El Pantano: Apuntes para la historia. Un libro acusador* (Mexico, 1954), pp. 77–80; *Presente* (Mexico), July 7 and Aug. 11, 1948; *Proceso* (Mexico), May 23, 1983.
31. Jean François Revel ("Jacques Severin"), "Democracie Mexicaine," *Esprit* (Paris), no. 5, May 1952, pp. 783–809.
32. María Félix, *Todas mis guerras,* 3 vols. (Mexico, 1993), vol. 2, p. 60.
33. *El Popular* (Mexico), May 5, 1952.
34. León Osorio, *El Pantano,* pp. 64–66.
35. Interview with Dr. Carlos Soto Maynez, Mexico, July 1994.
36. Manzur Ocaña interview.
37. From a witness, Augusto Elías Paullada, present at the meal.
38. *El Popular,* July 5, 1952; see "Interview with Manuel Gómez Morín," in James W. Wilkie and Edna Monzón de Wilkie, *México visto en el siglo XX*

(Mexico, 1969), pp. 189–92; Luis Calderón Vega, *Memorias del PAN*, 5 vols. (Mexico, 1992), vol. 2, pp. 29, 31, 32, 67, 73–75.
39. Interview with Antonio Mena Brito, Mexico, June 7, 1995.
40. Medina, "Civilismo y modernización," pp. 178–94.
41. Daniel Cosío Villegas, "La crisis de México," *Extremos de América* (Mexico, 1949), p. 23.
42. Zaid, *El progreso,* p. 232.
43. Revel (Severin), "Democracie Mexicaine."
44. Cosío Villegas, "Crisis," p. 22.
45. Luis Calderón Vega, *Memorias del PAN,* vol. 1, pp. 218–19.
46. Krauze, *Caudillos culturales en la Revolución Mexicana* (Mexico, 1985), pp. 285–86.
47. Calderón, *Memorias,* vol. 2, p. 56.
48. *Siempre* (Mexico), July 25, l953.
49. Medina, "Civilismo y modernización," pp. 96–97.
50. Zaid, *El progreso,* p. 331.
51. Calderón, *Memorias,* vol. 2, pp. 71–72.
52. Gonzalo N. Santos, *Memorias* (Mexico, 1984), p. 861.
53. Muñoz interview.
54. Luis Gómez Z., *Sucesos y remembranzas*, 2 vols. (Mexico 1979), vol. 1, pp. 303–11.
55. Interview with Fidel Velázquez, Mexico, 1995.
56. Medina, "Civilismo y modernización," pp. 164–68.
57. Revel (Severin), "Democracie Mexicaine."
58. Cosío Villegas, *Ensayos y notas*, 2 vols., (Mexico, 1966), vol. 1, p. 328.
59. Ibid., pp. 330–33.
60. Rafael Rodríguez Castañeda, *Prensa vendida. Los periodistas y los presidentes: 40 años de relaciones* (Mexico, 1993), p. 22.
61. Zaid, *Nueva economía,* pp. 75–76.
62. Interview with Adolfo Hernández Hurtado, Mexico, Mar. 1993.
63. *Tiempo* (Mexico), vol. 16, no. 403, Jan. 20, 1950, pp. 4–5; vol. 17, no. 435, p. xvi.
64. Zaid, *El progreso,* pp. 207–10.
65. Ibid., pp. 253–78.
66. Witnessed by Augusto Elías Paullada.
67. Quoted in Zaid, *El progreso,* p. 266.
68. Octavio Paz, *Sur* (Mexico), Mar. 1951.
69. Vicente Lombardo Toledano, "La perspectiva de México: Una democracia del pueblo," in *Problemas de Latinoamérica* (Mexico), vol. 2, no. 3, Apr. 15, 1955, p. 73; Medina, "Civilismo y modernización," pp. 120–21; Tzvi Medin, *El sexenio alemanista* (Mexico, 1990), pp. 67–80; Salvador Novo, *La vida en México en el período presidencial de Miguel Alemán* (Mexico, 1967).
70. Narciso Bassols, *Pensamiento y acción* (Mexico, 1967), p. 186.
71. Justo Manzur Ocaña, *La Revolución permanente: Vida y obra de Cándido Aguilar* (Mexico, 1972), p. 276.
72. Manzur Ocaña interview.
73. Matilde Kalfon, "Jesús Martínez 'Palillo': Qué bonita es la nostaliga," *Rino* (Mexico), no. 8, Jan.–Feb. 1992, pp. 19–25.
74. Torres, "Utopía industrial," p. 164.

75. Charles Hale, "Frank Tannenbaum and the Mexican Revolution," *Hispanic American Historical Review*, vol. 75, no. 2, 1995, pp. 215–46; see Frank Tannenbaum, "México: A Promise," *The Survey*, May 1, 1924.
76. Frank Tannenbaum, *Mexico: The Struggle for Peace and Bread* (New York, 1950), pp. 224–25.
77. Daniel Cosío Villegas, "El México de Tannenbaum,"*Problemas Agrícolas e Industriales de México* (Mexico), vol. 3, no. 4, Oct.–Dec. 1951, pp. 158.

Chapter 19

1. *Diario de los debates de la Cámara de Diputados,* Dec. 1, 1952.
2. Interview with Antonio Mena Brito, Mexico, Feb. 9, 1994, and June 7, 1995.
3. Jorge Mejía Prieto, *Anecdotario mexicano: Ingenio y picardía* (Mexico, 1986), pp. 135–40.
4. "The Domino Player," *Time* (New York), Sept. 14, 1953, pp. 28–31.
5. Ibid.
6. Ibid.
7. *Diario de los debates,* Sept. 1, 1953, p. 10.
8. David Alfaro Siqueiros, *Me llamaban el Coronelazo* (Mexico, 1977), pp. 62–63.
9. Gonzalo N. Santos, *Memorias* (Mexico, 1986), p. 886.
10. Interview with José Luis Melgarejo Vivanco, Mexico, Jan. 1993.
11. Ibid.
12. Daniel Cosío Villegas, *Memorias* (Mexico, 1977), p. 104.
13. *Crisol. Revista de Crítica* (Mexico), nos. 15, 16, 24, 26, 29, 33, 41, 44, 71, Mar. 1930–Nov. 1934.
14. Gustavo de Anda, "Adolfo Ruiz Cortines," *Mañana*, Dec. 12, 1973.
15. Interview with Fernando Román Lugo, Mexico, Sept. 1992.
16. Interview with Hesiquio Aguilar, Mexico, Feb. 1993.
17. Ibid.
18. Interview with Antonio Ortiz Mena, Mexico, Nov. 1992.
19. Mena Brito interview, June 7, 1995.
20. Juan José Rodríguez Prats, *Adolfo Ruiz Cortines* (Jalapa, 1990), p. 145.
21. Olga Pellicer de Brody and Esteban L. Mancilla, "El entendimiento con los Estados Unidos y la gestación del desarrollo estabilizador," *Historia de la Revolución Mexicana,* 23 vols. (Mexico, 1988), pp. 226–27; see also "Rescató México el latifundio de Cananea: 261,000 hectáreas," *El Universal* (Mexico), Aug. 22, 1958, p. 1.
22. Luis M. Farías, *Así lo recuerdo* (Mexico, 1992), p. 74–76.
23. Mena Brito interview, Feb. 9, 1994.
24. Ibid.
25. Santos, *Memorias*, pp. 895–97.
26. Julio Scherer, "Entrevista a Gilberto Flores Muñoz," *Excélsior* (Mexico), May 14, 1975.
27. Interview with Othón Salazar, Mexico, May 1993.
28. *Diario de los debates,* Dec. 1, 1952, p. 2.
29. *El Nacional* (Mexico), Feb. 7, 1953.
30. Jorge Alberto Lozoya, *El ejército mexicano* (Mexico, 1976), p. 63.

31. Rodríguez, *Adolfo Ruiz Cortines,* p. 144.
32. Aguilar interview.
33. Santos, *Memorias,* pp. 895–97.
34. Pablo Gonzalez Casanova, *La democracia en México* (Mexico, 1967), p. 185.
35. Interview with David Vargas Bravo, Mexico, Sept. 1995.
36. Mena Brito interview, June 7, 1995.
37. See Moisés González Navarro, *La Confederación Nacional Campesina* (Mexico, 1968), pp. 212–13; Francisco Martínez de la Vega, "Los motivos de Jaramillo," *Siempre!* (Mexico), Apr. 3, 1954.
38. González Casanova, *Democracia,* p. 184.
39. *Siempre!* vol. 1, no. 1, June 27, 1953.
40. Jacinto B. Treviño, *Memorias* (Mexico, 1961), pp. 263–64.
41. *El Nacional,* July 5, 1958.
42. Carlos Monsiváis, *Carlos Monsiváis* (Mexico, 1966), pp. 40–41.
43. Alonso Lujambio, "El dilema de Christlieb Ibarrola: Cuatro cartas a Gustavo Díaz Ordaz," *Estudios* (Mexico), no. 38, fall 1994, p. 57.
44. Rodríguez, *Adolfo Ruiz Cortines,* p. 141.
45. Lujambio, "Dilema de Christlieb Ibarrola," p. 55.
46. Salazar interview.
47. Ibid.
48. Farías, *Así lo recuerdo,* p. 76.
49. Mena Brito interview, Feb. 9, 1994.
50. Ortiz Mena interview.
51. See Olga Pellicer de Brody and José Luis Reyna, "El afianzamiento de la estabilidad política," *Historia de la Revolución Mexicana* (Mexico, 1978), vol. 22, pp. 173–83.
52. Luis Gómez Z., *Sucesos y remembranzas,* 2 vols. (Mexico, 1979), vol. 1, pp. 455–83; see also Antonio Alonso, *El movimiento ferrocarrilero en México, 1958–1959* (Mexico, 1972).
53. Interview with Adolfo Orive Alba, Mexico, Mar. 1992.
54. Lázaro Cárdenas, "Apuntes 1941–1956," in *Obras,* 4 vols. (Mexico, 1973), vol. 2, p. 348.
55. Ibid., p. 561.
56. Rodríguez, *Adolfo Ruiz Cortines,* p. 29.
57. Cárdenas, "Apuntes," vol. 2, p. 647.

Chapter 20

1. Armando de María y Campos, *Un ciudadano: Cómo es y cómo piensa Adolfo López Mateos* (Mexico, 1958), pp. 67–70.
2. Interview with Antonio Mena Brito, Mexico, Feb. 9, 1994.
3. Ibid.
4. Ibid.
5. Interview with Celestina Vargas Bervera, Mexico, Sept. 1992.
6. Clemente Díaz de la Vega, *Adolfo López Mateos: Vida y obra* (Toluca, 1986), p. 46.
7. Interview with Gustavo G. Velázquez, Mexico, Sept. 1992.
8. Quoted in Díaz de la Vega, *Adolfo López Mateos,* p. 56.
9. Luis M. Farías, *Así lo recuerdo* (Mexico, 1992), p. 176.

10. Interview with Roberto Barrios, Mexico, Sept. 1992.
11. Interview with Víctor Manuel Villegas, Mexico, Nov. 1992.
12. Interview with Gabriel Figueroa, Mexico, Mar. 1992.
13. Mena Brito interview.
14. "Discurso 'La Constitución Mexicana de 1917,' pronunciado por Adolfo López Mateos en el acto de la Asociación Nacional de Abogados y la Facultad de Derecho, el día 9 de febrero de 1955," in De Maria y Campos, *Un ciudadano,* p. 330.
15. Farías, *Así lo recuerdo,* p. 194.
16. Interview with Luis Gómez Z., Mexico, Sept. 17, 1995.
17. Interview with David Vargas Bravo, Mexico, Oct. 1995; see Juan Sánchez Borreguí, *Memorias personales y sindicales: La administración y el sindicato de los Ferrocarriles nacionales de México* (Mexico, 1982), pp. 172–200.
18. Ibid.
19. Interview with Antonio Ortiz Mena, Mexico, Oct. 1995.
20. Vicente Lombardo Toledano, "El Derecho a la Huelga: La tranquilidad social," *Siempre!* (Mexico), no. 298, Mar. 10, 1959; Luis Gómez Z., *Sucesos y remembranzas,* 2 vols. (Mexico, 1979), vol. 1, pp. 455–83.
21. Carlos Monsiváis, *Carlos Monsiváis* (Mexico, 1966), p. 42.
22. See Olga Pellicer de Brody and José Luis Reyna, "El afianzamiento de la estabilidad política," *Historia de la Revolución Mexicana,* 23 vols. (Mexico, 1978), vol. 22, pp. 173–83; see also Antonio Alonso, *El movimiento ferrocarrilero en México, 1958–1959* (Mexico, 1972).
23. José Alvarado, "El crepúsculo, Vallejo y la Poesía," *Siempre!* no. 303, Apr. 15, 1959.
24. Quoted in Gómez Z., *Sucesos y remembranzas,* vol. 1, p. 483.
25. "La Cámara fija su posición en el conflicto rielero," *Siempre!* no. 303, Apr. 15, 1959.
26. Gómez Z. interview.
27. *Política* (Mexico), Sept. 15, 1961; see also Pellicer de Brody and Reyna, "Afianzamiento," p. 61.
28. Interview with Carlos Monsiváis by the author, Mexico, Oct. 1995; see also Carlos Fuentes, *Tiempo Mexicano* (Mexico, 1971).
29. Fuentes, *Tiempo*, p. 118.
30. Monsiváis, *Carlos Monsiváis,* p. 42.
31. Ibid., pp. 44–45.
32. Quoted in Gabriel Zaid, *El progreso improductivo* (Mexico, 1979), p. 232.
33. Tomás Calvillo, *El Navismo o los motivos de la dignidad* (Mexico, 1986), p. 27; interview with Salvador Nava, San Luis Potosí, July 1991.
34. Calvillo, *Navismo*, pp. 17–78.
35. Lázaro Cárdenas, "Apuntes 1957–66" *Obras*, 4 vols. (Mexico, 1973), vol. 3, p. 57.
36. Elena Poniatowska, *Palabras cruzadas* (Mexico, 1961), pp. 26–40.
37. Cárdenas, "Apuntes," vol. 3, p. 123.
38. Fuentes, "Lázaro Cárdenas," *Tiempo,* p. 107.
39. Mena Brito interview, Feb. 9, 1994.
40. Cárdenas, "Apuntes," pp. 125, 144.

41. Fuentes, "Radiografía de una época," *Tiempo*, pp. 61–85.
42. Fuentes, "Nueve años: 1953–1962," *Siempre!* Aug. 8, 1962.
43. *Tiempo* (Mexico), June 30, 1960, pp. 10–12.
44. Ibid., July 7, 1960, p. 12.
45. Pellicer de Brody, *México y la Revolución Cubana* (Mexico, 1973), pp. 13–45.
46. Ortiz Mena interview, Oct. 1995.
47. Olga Pellicer de Brody and Esteban L. Mancilla, "El entendimiento con los Estados Unidos y la Gestación del Desarrollo Estabilizador," *Historia de la Revolución Mexicana,* vol. 23, pp. 237–42.
48. Daniel Cosío Villegas, "Mi general en México," *Ensayos y notas* (Mexico, 1966), vol. 2, pp. 411–19.
49. Interview with Noberto Aguirre Palancares, 1978.
50. "¿Por cuál camino Sr. Presidente?" *El Universal* (Mexico), Nov. 24, 1960.
51. Interview with Juan Sánchez Navarro, Mexico, Sept. 1995.
52. Víctor Manuel Villegas interview.

Chapter 21

1. Interview with Guadalupe and María Díaz-Ordaz, Mexico, Oct. 7, 1995; interview with Luis Castañeda Guzmán, Oaxaca, April 3, 1995.
2. Interview with Mariela Morales, Oaxaca, April 4, 1995; interview with José Melgar Castillejos, Mexico, April 7, 1993.
3. Interview with Francisco Herrera Muzgo, Oaxaca, April 3, 1995.
4. Castañeda Guzmán interview; interview with Susana Franco, Puebla, Feb. 13, 1995.
5. Interview with Urbano de Loya, Oaxaca, Oct. 10, 1994.
6. María and Guadalupe Díaz Ordaz interview; interview with María del Carmen Cervantes, Oaxaca, Mar. 1995.
7. Interview with Gastón García Cantú, Mexico, June 5, 1993.
8. De Loya interview; interview with Trinidad Torres Flores, Puebla, April 1993.
9. Salvador Novo, "Cartas a un amigo," Mexico, Mar. 5, 1966.
10. De Loya interview.
11. Interview with Gonzalo Bautista O'Farrill, Puebla, Dec. 1992.
12. Interview with Rómulo O'Farrill, Jr., Mexico, 1995.
13. José Cabrera Parra, *Díaz Ordaz y el 68* (Mexico, 1982), p. 143.
14. Interview with Susana Franco, Puebla, Feb. 1995.
15. Interview with Gonzalo Bautista O'Farrill, Mexico, May 1995.
16. Interview with Herminio Vázquez Caballero, Mexico, Dec. 1993.
17. Interview with Gustavo Díaz Ordaz, Jr., Mexico, Aug. 29, 1995.
18. Interview with Arturo Escamilla, Puebla, Apr. 1993.
19. De Loya interview.
20. *Diario de los debates de la Cámara de Senadores,* Dec. 20, 1951.
21. Cabrera Parra, *Díaz Ordaz y el 68*, p. 27.
22. Interview with Agustín Arriaga Rivera, Mexico, May 1994.
23. Interview with Emilio Bolaños Cacho, Mexico, May 24, 1994.
24. Interview with Antonio Mena Brito, Mexico, Feb. 9, 1994.
25. Vázquez Caballero interview.

26. Farías, *Así lo recuerdo,* pp. 235–36.
27. Ibid., pp. 56–57.
28. Ibid., p. 55.
29. Julio Scherer, *Los presidentes* (Mexico, 1986), p. 15.
30. De Loya interview.
31. See Farías, *Así lo recuerdo.*
32. Vázquez Caballero interview.
33. Cabrera Parra, *Díaz Ordaz y el 68*, p. 32.
34. Interview with Gustavo and Alfredo Díaz Ordaz, Mexico, May 1992.
35. Vázquez Caballero interview.
36. Ibid.
37. Vázquez Caballero interview.
38. Vázquez Caballero interview.
39. Ibid.
40. Cabrera, Parra, *Díaz Ordaz y el 68*, p. 50.
41. Díaz Ordaz, Jr., interview.
42. Cabrera, Parra, *Díaz Ordaz y el 68*, p. 92.
43. Ibid., p. 147.
44. Ibid., p. 39.
45. Interview with Antonio Ortiz Mena, Mexico, Oct. 20, 1987.
46. Ibid.
47. "Discurso pronunciado por el Presidente de México, Lic. Gustavo Díaz Ordaz, ante el Congreso de los Estados Unidos de América, reunido en sesión conjunta en la Cámara de representantes en la ciudad de Washington, el 27 de octubre de 1967," in *Gustavo Díaz Ordaz: Su pensamiento, su palabra* (Mexico, 1988), pp. 63–69.
48. Evelyn P. Stevens, *Protesta y respuesta en México* (Mexico, 1979), p. 118; Ricardo Pozas Horcasitas, *La democracia en blanco: El movimiento médico en México, 1964–1965* (Mexico, 1993), pp. 93–94; Norberto Treviño Zapata, *El movimiento médico en México: 1964–1965: Crónica documental y reflexiones* (Mexico, 1989), p. 24; interview with Ismael Cosío Villegas, Mexico, 1978.
49. Ibid.
50. Raúl Cruz Zapata, *Carlos A. Madrazo: Biografía política* (Mexico, 1988), pp. 65–71; Jorge Bernardo Soto, "Por qué son importantes las reformas en la estructura del PRI," *Política* (Mexico), Nov.–Dec. 1965.
51. Alonso Lujambio, "El dilema de Cristlieb Ibarrola: Cuatro cartas a Gustavo Díaz Ordaz," *Estudios* (Mexico), no. 38, fall 1994, pp. 49–75.
52. Ortiz Mena interview.
53. *Diario de México* (Mexico), June 23, 1966.
54. Ibid., July 3, 1966.
55. Arnaldo Orfila Reynal to the author, Oct. 10, 1995.
56. Mena Brito interview.
57. "Informe presidencial del 1st de septiembre de 1967," in *Gustavo Díaz Ordaz: Discurso de toma de posesión y capítulo político de todos y cada uno de sus informes al Congreso* (Mexico, n.d.), p. 89.
58. Arriaga Rivera interview.
59. Jaime Labastida to the author, Sept. 26, 1995.
60. Arriaga Rivera interview.

61. "Informe presidencial del 1. de septiembre de 1966," in *Gustavo Díaz Ordaz: Discurso,* p. 68.
62. *Revista de la Universidad de México* (Mexico), vol. 23, no. 1, Sept. 1968, pp. 4–5; see also Elena Poniatowska, *La noche de Tlatelolco* (Mexico, 1994), pp. 275–76; Raúl Alvarez Garín, Gilberto Guevara Niebla, et al., *Pensar el 68* (Mexico, 1993), pp. 257–59.
63. *Revista de la Universidad,* p. 6; see also Poniatowska, *Noche de Tlatelolco,* p. 277; Aurora Cano Andaluz, comp., *1968 Antología Periodística* (Mexico, 1993).
64. Carlos Monsiváis, "Javier Barros Sierra: ¡Viva la discrepancia!" in Álvarez Garín and Guevara Niebla, *Pensar,* p. 101.
65. *Díaz Ordaz: Su pensamiento,* pp. 75–76.
66. Poniatowska, *Noche de Tlatelolco,* pp. 31, 33, 40, 41, 44, 61; *Revista de la Universidad,* p. 13.
67. Poniatowska, *Noche de Tlatelolco,* p. 35; interview with Luis González de Alba, Mexico, Oct. 3, 1995.
68. González de Alba interview.
69. Poniatowska, *Noche de Tlatelolco,* p. 33.
70. Ibid., p. 90.
71. *Revista de la Universidad,* p. 15.
72. Ibid., pp. 16–17.
73. Poniatowska, *Noche de Tlatelolco,* p. 63.
74. González de Alba interview.
75. Poniatowska, *Noche de Tlatelolco,* pp. 64–73.
76. Ricardo Garibay, *Cómo se gana la vida* (Mexico, 1992), pp. 273–75.
77. Interview with Juan Sánchez Navarro, Mexico, Oct. 1992.
78. Unpublished memoirs of Gustavo Díaz Ordaz.
79. Daniel Cosío Villegas, *Memorias* (Mexico, 1977), pp. 260–61.
80. Díaz Ordaz memoirs.
81. Garibay, *Cómo se gana,* p. 274.
82. Díaz Ordaz memoirs.
83. Ibid.
84. Ibid.
85. Ibid.
86. Poniatowska, *Noche de Tlatelolco,* p. 34.
87. Díaz Ordaz memoirs.
88. *Revista de la Universidad,* p. 21.
89. Díaz Ordaz memoirs.
90. Cano, *Antología Periodística,* p. 118.
91. *Díaz Ordaz: Su pensamiento,* pp. 115–30.
92. "El C.N.H. insiste en continuar el movimiento," *Novedades* (Mexico), Sept. 3, 1968.
93. Daniel Cosío Villegas, *Labor periodística* (Mexico, 1972), pp. 211–13.
94. Poniatowska, *Noche de Tlatelolco,* pp. 60–61.
95. Díaz Ordaz memoirs.
96. Ibid.
97. *Revista de la Universidad,* p. 27.
98. See Cruz, *Carlos A. Madrazo.*
99. Monsiváis, "Javier Barros Sierra," p. 102.

100. Garibay, *Cómo se gana,* p. 279.
101. Díaz Ordaz memoirs.
102. *Excélsior* (Mexico), Oct. 4, 1968.
103. Poniatowska, *Noche Tlatelolco,* pp. 167, 173–76, 181–83, 216–18, 237–38, 243, 273; see also Daniel Cazés, coord., *Memorial del 68: Relato a muchas voces* (Mexico, 1994); Cano, *Antología Periodística,* pp. 237, 243.
104. Poniatowska, *Noche de Tlatelolco,* p. 209.
105. Ibid., p. 172.
106. Gilberto Guevara Niebla, "El 2 de octubre," in Álvarez Garín, Guevara Niebla, *Pensar el 68,* p. 118.
107. Interview with Luis González de Alba, Mexico, Oct. 3, 1995.
108. Poniatowska, *Noche de Tlatelolco,* p. 240.
109. Raúl Alvarez Garín and Gilberto Guevara Niebla, *Pensar el 68* (Mexico, 1988), p. 123.
110. Poniatowska, *Noche de Tlatelolco,* pp. 177, 242.
111. Ibid., p. 184.
112. Ibid., p. 229.
113. Octavio Paz, *Posdata* (Mexico, 1970), pp. 105–55.
114. Poniatowska, *Noche de Tlatelolco,* p. 199.
115. Díaz Ordaz memoirs.
116. Cazés, *Memorial,* p. 126.
117. Scherer, *Los Presidentes,* p. 61.
118. Marcelino García Barragán's testimony, *Proceso* (Mexico), no. 104, Oct. 30, 1978, pp. 6–8.
119. Scherer, *Los Presidentes,* p. 20.
120. *Excélsior*, Sept. 2, 1969.
121. Cosío Villegas, *Labor periodística,* pp. 211–13.
122. Interview with Irma Serrano, Mexico, Mar. 19, 1993.
123. Díaz Ordaz, Jr., interview.
124. *Proceso*, no. 24, April 16, 1977, pp. 6–7.
125. Luis Tomás Cervantes Cabeza de Vaca, "Ya vienen por mí," in Álvarez Garín and Guevara Niebla, *Pensar el 68,* pp. 193–96; see also Poniatowska, *Noche de Tlatelolco*, pp. 106–8, 114–15, 117–18.
126. Ibid.

Part V

1. Unpublished memoirs of Gustavo Díaz Ordaz.
2. Ibid.
3. Daniel Cosío Villegas, *Labor periodística e imaginaria* (Mexico, 1971), p. 220.
4. Cited in Elena Poniatowska, *La noche de Tlatelolco* (Mexico, 1994), p. 91.

Chapter 22

1. Interview with Ernesto P. Uruchurtu, Mexico, Nov. 1993.
2. Interview with Antonio Ortiz Mena, Mexico, Oct. 20, 1987.
3. A phrase coined by Carlos Fuentes. Another editor and writer, Fernando Benítez, said that the country had only two paths open to it: "either Echeverría or fascism."

4. Interview with Carlos Monsiváis, Mexico, Sep. 1995.
5. Gabriel Zaid, "Carta a Carlos Fuentes," *Cómo leer en bicicleta* (Mexico, 1975), pp. 106–11.
6. The two friends—the author and Héctor Aguilar Camín—published the account, "La saña y el terror," *Siempre!* (Mexico), June 30, 1971.
7. Daniel Cosío Villegas, "L. E. Divagar entre las naciones," *Excélsior* (Mexico), Mar. 24, 1973.
8. Gonzalo N. Santos, *Memorias* (Mexico, 1977), pp. 919–20.
9. Witnessed by Fausto Zerón-Medina.
10. Adrian Lajous in *Mexico, From Boom to Bust, 1940–1982,* second program of a three-part series broadcast by WGBH (Boston), Nov. 23, 1988.
11. Hugo Margáin in ibid.
12. Gabriel Zaid, *La economía presidencial* (Mexico, 1987), pp. 11–30.
13. Ibid., pp. 33–40.
14. Enrique Krauze, *Daniel Cosío Villegas: Una biografía intelectual* (Mexico, 1980), p. 258.
15. *Proceso* (Mexico), Oct. 2, 1995.
16. Krauze, *Daniel Cosío Villegas,* p. 259.
17. Daniel Cosío Villegas, *El estilo personal de gobernar* (Mexico, 1974), p. 125.
18. Interview with Daniel Cosío Villegas, Mexico, Feb. 1976.
19. Lázaro Cárdenas, "Apuntes 1967–1970," in *Obras*, 4 vols. (Mexico, 1979), vol. 4, pp. 210–27.
20. Julio Scherer García, *Los presidentes* (Mexico, 1986), pp. 52–53.
21. Interview with Manuel Gómez Morín, Mexico, July 1971.
22. Gabriel Zaid, "El 18 Brumario de Luis Echeverría," *Vuelta* (Mexico), no. 2, Dec. 1976; also *El progreso improductivo* (Mexico, 1979), pp. 272–75.
23. Vicente Leñero, *Los periodistas* (Mexico, 1978).

Chapter 23

1. José López Portillo, *Mis Tiempos*, 2 vols. (Mexico, 1988), vol. 1, 20–23.
2. Ibid., pp. 7, 9.
3. Ibid., p. 262.
4. Ibid., p. 64.
5. Ibid., pp. 108–9.
6. Personal statement by López Portillo, Mexico, Mar. 8, 1978.
7. Enrique Krauze, "Jesús Reyes Heroles, cambiar para conservar," in *Por una democracia sin adjetivos* (Mexico, 1986), pp. 168–97.
8. Gabriel Zaid, "Un presidente apostador," *La economía presidencial* (Mexico, 1987), pp. 66–83.
9. Enrique Krauze, "El timón y la tormenta," *Democracia sin adjetivos,* pp. 17–43.
10. Cited in Gabriel Zaid, *De los libros al poder* (Mexico, 1988), pp. 83–84.
11. Jesús Silva Herzog, José Ángel Gurría, José Andrés de Oteyza, Jorge Díaz Serrano, *Mexico, From Boom to Bust, 1940–1982,* second program of a three-part series broadcast by WGBH (Boston), Nov. 23, 1988.

Chapter 24

1. Enrique Krauze, *Por una democracia sin adjetivos* (Mexico, 1986), pp. 44–75.

2. Ibid., pp. 96–98, 99–111.
3. Gabriel Zaid, *La economía presidencial,* pp. 125–42, 173–78.
4. Krauze, *Democracia sin adjetivos,* pp. 112–42.
5. Personal testimony, witnesses Héctor Aguilar Camín and Carlos Monsiváis.
6. Interview with Miguel de la Madrid, Mexico, April 1987.
7. Ibid.

Chapter 25

1. Enrique Krauze, "Los obreros y el poder," *Textos heréticos* (Mexico, 1992), pp. 75–82.
2. Interview with Pedro Aspe Armella, Mexico, Oct. 1994.
3. Enrique Krauze, "Fines de siglo en México," *Textos,* pp. 119–24.
4. Interview with Luis Donaldo Colosio, Mexico, Jan. 1994.
5. Gabriel Zaid, *La nueva economía presidencial* (Mexico, 1994), p. 157.
6. Krauze, *Textos,* pp. 83–93.
7. Gabriel Zaid, "Sobregiros de confianza," published in *Contenido* (Mexico), Nov. 1991, and later in *Nueva economía,* p. 105.
8. Interview with José María Córdoba, Mexico, 1992.

Chapter 26

1. Andrés Oppenheimer, *Bordering on Chaos* (Boston, New York, 1996), p. 36.
2. Enrique Krauze, "José Pérez Méndez," *Reforma,* (Mexico), Jan. 9, 1994.
3. Luis González y González, "El subsuelo indígena," *La república restaurada: Vida social,* vol. 3 (Mexico, 1974), in Daniel Cosío Villegas, ed., *Historia moderna de México,* pp. 278–81; see also Juan Pedro Viqueira Albán, *María Candelaria, india natural de Cancuc* (Mexico, 1993).
4. Interview with Samuel Ruiz, San Cristóbal de las Casas, July 27, 1994.
5. Interview with Gonzalo Ituarte, July 26, 1994.
6. *Proceso* (Mexico), August 7, 21, 1995.
7. *Ibid.*, July 24, 1995.
8. Carlos Tello Díaz, *La rebelión de las cañadas* (Mexico, 1995), pp. 97–98.
9. *Proceso* (Mexico), August 7, 21, 1995.
10. Tello Díaz, *Rebelión de las cañadas,* pp. 123, 136.
11. Ibid., pp. 127, 170.
12. Ibid., pp. 127–39.
13. Oppenheimer, *Chaos,* p. 25.
14. "Declaración de la Selva Lacandona," *La Jornada* (Mexico), Jan. 2, 1994.
15. Enrique Krauze, "Redención o democracia," *Reforma,* Feb. 1994.
16. Gabriel Zaid, "La Guerrilla Postmoderna," "Enfoque," *Reforma,* May 15, 1994.
17. Interview with Luis Donaldo Colosio by the author, Mexico, Dec. 1993.
18. Gabriel Zaid, "Muerte y resurrección de la cultura católica," *Vuelta* (Mexico), no. 156, Nov. 1989, pp. 9–23.
19. Gabriel Zaid, *La economía presidencial* (Mexico, 1987), p. 60.
20. Ejército Zapatista de Liberación Nacional, "Chiapas: El Sureste en dos vientos, una tormenta y una profecia. 27 enero 1994," in *EZLN: documentos y comunicados, 1 de enero–8 de agosto de 1994* (Mexico, 1994), sections 2 and 3, pp. 60, 62.

21. "Declaración de la Selva."
22. Tello Díaz, *Rebelión de las cañadas*, p. 146.
23. Ejército Zapatista de Liberación Nacional, "Al Consejo 500 años de Resistencia Indígena, 6 de Febrero," in *EZLN: Documentos y comunicados*, p. 120.
24. Zaid, *Economía presidencial*, p. 149–52.

INDEX